ADVANCED
CALCULUS

Revised Edition

ADVANCED
CALCULUS

Revised Edition

Lynn Harold Loomis

Shlomo Sternberg

Harvard University, USA

World Scientific

NEW JERSEY • LONDON • SINGAPORE • BEIJING • SHANGHAI • HONG KONG • TAIPEI • CHENNAI

Published by

World Scientific Publishing Co. Pte. Ltd.
5 Toh Tuck Link, Singapore 596224
USA office: 27 Warren Street, Suite 401-402, Hackensack, NJ 07601
UK office: 57 Shelton Street, Covent Garden, London WC2H 9HE

British Library Cataloguing-in-Publication Data
A catalogue record for this book is available from the British Library.

The original edition was published in 1968 by Addison-Wesley Publishing Company, Inc.
In 1990, Jones and Bartlett Publishers, Inc. published the revised edition.

ADVANCED CALCULUS
Revised Edition

ISBN 978-981-4583-92-3
ISBN 978-981-4583-93-0 (pbk)

Printed in Singapore

PREFACE

This book is based on an honors course in advanced calculus that we gave in the 1960's. The foundational material, presented in the unstarred sections of Chapters 1 through 11, was normally covered, but different applications of this basic material were stressed from year to year, and the book therefore contains more material than was covered in any one year. It can accordingly be used (with omissions) as a text for a year's course in advanced calculus, or as a text for a three-semester introduction to analysis.

These prerequisites are a good grounding in the calculus of one variable from a mathematically rigorous point of view, together with some acquaintance with linear algebra. The reader should be familiar with limit and continuity type arguments and have a certain amount of mathematical sophistication. As possible introductory texts, we mention *Differential and Integral Calculus* by R. Courant, *Calculus* by T. Apostol, *Calculus* by M. Spivak, and *Pure Mathematics* by G. Hardy. The reader should also have some experience with partial derivatives.

In overall plan the book divides roughly into a first half which develops the calculus (principally the differential calculus) in the setting of normed vector spaces, and a second half which deals with the calculus of differentiable manifolds.

Vector space calculus is treated in two chapters, the differential calculus in Chapter 3, and the basic theory of ordinary differential equations in Chapter 6. The other early chapters are auxiliary. The first two chapters develop the necessary purely algebraic theory of vector spaces, Chapter 4 presents the material on compactness and completeness needed for the more substantive results of the calculus, and Chapter 5 contains a brief account of the extra structure encountered in scalar product spaces. Chapter 7 is devoted to multilinear (tensor) algebra and is, in the main, a reference chapter for later use. Chapter 8 deals with the theory of (Riemann) integration on Euclidean spaces and includes (in exercise form) the fundamental facts about the Fourier transform. Chapters 9 and 10 develop the differential and integral calculus on manifolds, while Chapter 11 treats the exterior calculus of E. Cartan.

The first eleven chapters form a logical unit, each chapter depending on the results of the preceding chapters. (Of course, many chapters contain material that can be omitted on first reading; this is generally found in starred sections.)

On the other hand, Chapters 12, 13, and the latter parts of Chapters 6 and 11 are independent of each other, and are to be regarded as illustrative applications of the methods developed in the earlier chapters. Presented here are elementary Sturm-Liouville theory and Fourier series, elementary differential geometry, potential theory, and classical mechanics. We usually covered only one or two of these topics in our one-year course.

We have not hesitated to present the same material more than once from different points of view. For example, although we have selected the contraction mapping fixed-point theorem as our basic approach to the implicit-function theorem, we have also outlined a "Newton's method" proof in the text and have sketched still a third proof in the exercises. Similarly, the calculus of variations is encountered twice—once in the context of the differential calculus of an infinite-dimensional vector space and later in the context of classical mechanics. The notion of a submanifold of a vector space is introduced in the early chapters, while the invariant definition of a manifold is given later on.

In the introductory treatment of vector space theory, we are more careful and precise than is customary. In fact, this level of precision of language is not maintained in the later chapters. Our feeling is that in linear algebra, where the concepts are so clear and the axioms so familiar, it is pedagogically sound to illustrate various subtle points, such as distinguishing between spaces that are normally identified, discussing the naturality of various maps, and so on. Later on, when overly precise language would be more cumbersome, the reader should be able to produce for himself a more precise version of any assertions that he finds to be formulated too loosely. Similarly, the proofs in the first few chapters are presented in more formal detail. Again, the philosophy is that once the student has mastered the notion of what constitutes a formal mathematical proof, it is safe and more convenient to present arguments in the usual mathematical colloquialisms.

While the level of formality decreases, the level of mathematical sophistication does not. Thus increasingly abstract and sophisticated mathematical objects are introduced. It has been our experience that Chapter 9 contains the concepts most difficult for students to absorb, especially the notions of the tangent space to a manifold and the Lie derivative of various objects with respect to a vector field.

There are exercises of many different kinds spread throughout the book. Some are in the nature of routine applications. Others ask the reader to fill in or extend various proofs of results presented in the text. Sometimes whole topics, such as the Fourier transform or the residue calculus, are presented in exercise form. Due to the rather abstract nature of the textual material, the student is strongly advised to work out as many of the exercises as he possibly can.

Any enterprise of this nature owes much to many people besides the authors, but we particularly wish to acknowledge the help of L. Ahlfors, A. Gleason, R. Kulkami, R. Rasala, and G. Mackey and the general influence of the book by Dieudonné. We also wish to thank the staff of Jones and Bartlett for their invaluable help in preparing this revised edition.

Cambridge, Massachusetts L.H.L.
1968, 1989 S.S.

CONTENTS

CHAPTER 0

INTRODUCTION

This preliminary chapter contains a short exposition of the set theory that forms the substratum of mathematical thinking today. It begins with a brief discussion of logic, so that set theory can be discussed with some precision, and continues with a review of the way in which mathematical objects can be defined as sets. The chapter ends with four sections which treat specific set-theoretic topics.

It is intended that this material be used mainly for reference. Some of it will be familiar to the reader and some of it will probably be new. We suggest that he read the chapter through "lightly" at first, and then refer back to it for details as needed.

1. LOGIC: QUANTIFIERS

A *statement* is a sentence which is true or false as it stands. Thus '$1 < 2$' and '$4 + 3 = 5$' are, respectively, true and false mathematical statements. Many sentences occurring in mathematics contain variables and are therefore not true or false as they stand, but become statements when the variables are given values. Simple examples are '$x < 4$', '$x < y$', 'x is an integer', '$3x^2 + y^2 = 10$'. Such sentences will be called *statement frames.* If $P(x)$ is a frame containing the one variable 'x', then $P(5)$ is the statement obtained by replacing 'x' in $P(x)$ by the numeral '5'. For example, if $P(x)$ is '$x < 4$', then $P(5)$ is '$5 < 4$', $P(\sqrt{2})$ is '$\sqrt{2} < 4$', and so on.

Another way to obtain a statement from the frame $P(x)$ is to assert that $P(x)$ is always true. We do this by prefixing the phrase 'for every x'. Thus, 'for every x, $x < 4$' is a false statement, and 'for every x, $x^2 - 1 = (x - 1)(x + 1)$' is a true statement. This prefixing phrase is called a *universal quantifier.* Synonymous phrases are 'for each x' and 'for all x', and the symbol customarily used is '$(\forall x)$', which can be read in any of these ways. One frequently presents sentences containing variables as being always true without explicitly writing the universal quantifiers. For instance, the associative law for the addition of numbers is often written

$$x + (y + z) = (x + y) + z,$$

where it is understood that the equation is true for all x, y and z. Thus the actual statement being made is

$$(\forall x)(\forall y)(\forall z)[x + (y + z) = (x + y) + z].$$

Finally, we can convert the frame $P(x)$ into a statement by asserting that it is sometimes true, which we do by writing 'there exists an x such that $P(x)$'. This process is called *existential quantification*. Synonymous prefixing phrases here are 'there is an x such that', 'for some x', and, symbolically, '$(\exists x)$'.

The statement '$(\forall x)(x < 4)$' still contains the variable 'x', of course, but 'x' is no longer *free* to be given values, and is now called a *bound* variable. Roughly speaking, quantified variables are bound and unquantified variables are free. The notation '$P(x)$' is used only when 'x' is free in the sentence being discussed.

Now suppose that we have a sentence $P(x, y)$ containing *two* free variables. Clearly, we need two quantifiers to obtain a statement from this sentence. This brings us to a *very* important observation. *If quantifiers of both types are used, then the order in which they are written affects the meaning of the statement;* $(\exists y)(\forall x)P(x, y)$ and $(\forall x)(\exists y)P(x, y)$ *say different things.* The first says that one y can be found that works for all x: "there exists a y such that for all x . . .". The second says that for each x a y can be found that works: "for each x there exists a y such that . . .". But in the second case, it may very well happen that when x is changed, the y that can be found will also have to be changed. The existence of a single y that serves for all x is thus the stronger statement. For example, it is true that $(\forall x)(\exists y)(x < y)$ and false that $(\exists y)(\forall x)(x < y)$. The reader must be absolutely clear on this point; his whole mathematical future is at stake. The second statement says that there exists a y, call it y_0, such that $(\forall x)(x < y_0)$, that is, such that every number is less than y_0. This is false; $y_0 + 1$, in particular, is not less than y_0. The first statement says that for each x we can find a *corresponding* y. And we can: take $y = x + 1$.

On the other hand, among a group of quantifiers of the same type the order does not affect the meaning. Thus '$(\forall x)(\forall y)$' and '$(\forall y)(\forall x)$' have the same meaning. We often abbreviate such clumps of similar quantifiers by using the quantification symbol only once, as in '$(\forall x, y)$', which can be read 'for every x and y'. Thus the strictly correct '$(\forall x)(\forall y)(\forall z)[x + (y + z) = (x + y) + z]$' receives the slightly more idiomatic rendition '$(\forall x, y, z)[x + (y + z) = (x + y) + z]$'. The situation is clearly the same for a group of existential quantifiers.

The beginning student generally feels that the prefixing phrases 'for every x there exists a y such that' and 'there exists a y such that for every x' sound artificial and are unidiomatic. This is indeed the case, but this awkwardness is the price that has to be paid for the order of the quantifiers to be fixed, so that the meaning of the quantified statement is clear and unambiguous. Quantifiers do occur in ordinary idiomatic discourse, but their idiomatic occurrences often house ambiguity. The following two sentences are good examples of such ambiguous idiomatic usage: "Every x is less than some y" and "Some y is greater than every x". If a poll were taken, it would be found that most men on the street feel that these two sentences say the same thing, but half will feel that the common assertion is false and half will think it true! The trouble here is that the matrix is preceded by one quantifier and followed by another, and the poor reader doesn't know which to take as the inside, or first applied, quantifier. The two possible symbolic renditions of our first sentence, '$[(\forall x)(x < y)](\exists y)$' and '$(\forall x)[(x < y)(\exists y)]$', are respectively false and true. Mathematicians do use hanging quantifiers in the interests of more idiomatic writing, but only if they are sure the reader will understand their order of application, either from the context or by comparison with standard usage. In general, a hanging quantifier would probably be read as the inside, or first applied, quantifier, and with this understanding our two ambiguous sentences become true and false in that order.

After this apology the reader should be able to tolerate the definition of sequential convergence. It involves three quantifiers and runs as follows: The sequence $\{x_n\}$ converges to x if $(\forall \epsilon)(\exists N)(\forall n)(\text{if } n > N \text{ then } |x_n - x| < \epsilon)$. In exactly the same format, we define a function f to be continuous at a if $(\forall \epsilon)(\exists \delta)(\forall x)(\text{if } |x - a| < \delta \text{ then } |f(x) - f(a)| < \epsilon)$. We often omit an inside universal quantifier by displaying the final frame, so that the universal quantification is understood. Thus we define f to be continuous at a if for every ϵ there is a δ such that

$$\text{if} \quad |x - a| < \delta, \qquad \text{then} \quad |f(x) - f(a)| < \epsilon.$$

We shall study these definitions later. We remark only that it is perfectly possible to build up an intuitive understanding of what these and similar quantified statements actually say.

2. THE LOGICAL CONNECTIVES

When the word 'and' is inserted between two sentences, the resulting sentence is true if both constituent sentences are true and is false otherwise. That is, the "truth value", T or F, of the compound sentence depends only on the truth values of the constituent sentences. We can thus describe the way 'and' acts in compounding sentences in the simple "truth table"

P	Q	P and Q
T	T	T
T	F	F
F	T	F
F	F	F

where 'P' and 'Q' stand for arbitrary statement frames. Words like 'and' are called logical connectives. It is often convenient to use symbols for connectives, and a standard symbol for 'and' is the ampersand '&'. Thus 'P & Q' is read 'P and Q'.

Another logical connective is the word 'or'. Unfortunately, this word is used ambiguously in ordinary discourse. Sometimes it is used in the *exclusive* sense, where 'P or Q' means that one of P and Q is true, but not both, and sometimes it is used in the *inclusive* sense that at least one is true, and possibly both are true. Mathematics cannot tolerate any fundamental ambiguity, and in mathematics 'or' is *always* used in the latter way. We thus have the truth table

P	Q	P or Q
T	T	T
T	F	T
F	T	T
F	F	F

The above two connectives are *binary*, in the sense that they combine *two* sentences to form one new sentence. The word 'not' applies to one sentence and really shouldn't be considered a connective at all; nevertheless, it is called a *unary* connective. A standard symbol for 'not' is '$\sim$'. Its truth table is obviously

P	$\sim P$
T	F
F	T

In idiomatic usage the word 'not' is generally buried in the interior of a sentence. We write 'x is not equal to y' rather than 'not (x is equal to y)'. However, for the purpose of logical manipulation, the negation sign (the word 'not' or a symbol like '$\sim$') precedes the sentence being negated. We shall, of course, continue to write '$x \neq y$', but keep in mind that this is idiomatic for 'not $(x = y)$' or '$\sim(x = y)$'.

We come now to the troublesome 'if . . . , then . . .' connective, which we write as either 'if P, then Q' or '$P \Rightarrow Q$'. This is almost always applied in the universally quantified context $(\forall x)(P(x) \Rightarrow Q(x))$, and its meaning is best unraveled by a study of this usage. We consider 'if $x < 3$, then $x < 5$' to be a true sentence. More exactly, it is true for all x, so that the universal quantification $(\forall x)(x < 3 \Rightarrow x < 5)$ is a true statement. This conclusion forces us to agree that, in particular, '$2 < 3 \Rightarrow 2 < 5$', '$4 < 3 \Rightarrow 4 < 5$', and '$6 < 3 \Rightarrow 6 < 5$' are all true statements. The truth table for '$\Rightarrow$' thus contains the values entered below.

P	Q	$P \Rightarrow Q$
T	T	T
T	F	-
F	T	T
F	F	T

On the other hand, we consider '$x < 7 \Rightarrow x < 5$' to be a false sentence, and therefore have to agree that '$6 < 7 \Rightarrow 6 < 5$' is false. Thus the remaining row in the table above gives the value 'F' for $P \Rightarrow Q$.

Combinations of frame variables and logical connectives such as we have been considering are called *truth-functional forms.* We can further combine the elementary forms such as '$P \Rightarrow Q$' and '$\sim P$' by connectives to construct *composite* forms such as '$\sim(P \Rightarrow Q)$' and '$(P \Rightarrow Q)$ & $(Q \Rightarrow P)$'. A sentence has a given (truth-functional) form if it can be obtained from that form by substitution. Thus '$x < y$ or $\sim(x < y)$' has the form 'P or $\sim P$', since it is obtained from this form by substituting the sentence '$x < y$' for the sentence variable 'P'. Composite truth-functional forms have truth tables that can be worked out by combining the elementary tables. For example, '$\sim(P \Rightarrow Q)$' has the table below, the truth value for the whole form being in the column under the connective which is applied last ('$\sim$' in this example).

P	Q	$\sim$	$(P \Rightarrow Q)$
T	T	F	T
T	F	T	F
F	T	F	T
F	F	F	T

Thus $\sim(P \Rightarrow Q)$ is true only when P is true and Q is false.

A truth-functional form such as 'P or $(\sim P)$' which is always true (i.e., has only 'T' in the final column of its truth table) is called a *tautology* or a *tautologous form.* The reader can check that

$$(P \,\&\, (P \Rightarrow Q)) \Rightarrow Q \qquad \text{and} \qquad ((P \Rightarrow Q) \,\&\, (Q \Rightarrow R)) \Rightarrow (P \Rightarrow R)$$

are also tautologous. Indeed, any valid principle of reasoning that does not involve quantifiers must be expressed by a tautologous form.

The 'if and only if' form '$P \Leftrightarrow Q$', or 'P if and only if Q', or 'P iff Q', is an abbreviation for '$(P \Rightarrow Q)$ & $(Q \Rightarrow P)$'. Its truth table works out to be

P	Q	$P \Leftrightarrow Q$
T	T	T
T	F	F
F	T	F
F	F	T

That is, $P \Leftrightarrow Q$ is true if P and Q have the same truth values, and is false otherwise.

Two truth-functional forms A and B are said to be *equivalent* if (the final columns of) their truth tables are the same, and, in view of the table for '$\Leftrightarrow$', we see that A and B *are equivalent if* $A \Leftrightarrow B$ *is tautologous,* and conversely.

Replacing a sentence obtained by substitution in a form A by the equivalent sentence obtained by the same substitutions in an equivalent form B is a device much used in logical reasoning. Thus to prove a statement P true, it suffices to prove the statement $\sim P$ false, since 'P' and '$\sim(\sim P)$' are equivalent forms. Other important equivalences are

$$\begin{aligned} \sim(P \text{ or } Q) &\Leftrightarrow (\sim P) \;\&\; (\sim Q), \\ (P \Rightarrow Q) &\Leftrightarrow Q \text{ or } (\sim P), \\ \sim(P \Rightarrow Q) &\Leftrightarrow P \;\&\; (\sim Q). \end{aligned}$$

A bit of conventional sloppiness which we shall indulge in for smoother idiom is the use of 'if' instead of the correct 'if and only if' in definitions. We define f to be continuous at x *if* so-and-so, meaning, of course, that f is continuous at x *if and only if* so-and-so. This causes no difficulty, since it is clear that 'if and only if' is meant when a definition is being given.

3. NEGATIONS OF QUANTIFIERS

The combinations '$\sim(\forall x)$' and '$(\exists x)\sim$' have the same meanings: something is *not always true* if and only if it is *sometimes false*. Similarly, '$\sim(\exists y)$' and '$(\forall y)\sim$' have the same meanings. These equivalences can be applied to move a negation sign past each quantifier in a string of quantifiers, giving the following important practical rule:

> *In taking the negation of a statement beginning with a string of quantifiers, we simply change each quantifier to the opposite kind and move the negation sign to the end of the string.*

Thus

$$\sim(\forall x)(\exists y)(\forall z)P(x, y, z) \;\Leftrightarrow\; (\exists x)(\forall y)(\exists z)\sim P(x, y, z).$$

There are other principles of quantificational reasoning that can be isolated and which we shall occasionally mention, but none seem worth formalizing here.

4. SETS

It is present-day practice to define every mathematical object as a set of some kind or other, and we must examine this fundamental notion, however briefly.

A *set* is a collection of objects that is itself considered an entity. The objects in the collection are called the *elements* or *members* of the set. The symbol for 'is a member of' is '$\in$' (a sort of capital epsilon), so that '$x \in A$' is read "x is a member of A", "x is an element of A", "x belongs to A", or "x is in A".

We use the equals sign '$=$' in mathematics to mean logical identity; $A = B$ means that A *is* B. Now a set A is considered to be the same object as a set B if and only if A and B have exactly the same members. That is, '$A = B$' means that $(\forall x)(x \in A \Leftrightarrow x \in B)$.

We say that a set A is a *subset* of a set B, or that A is *included* in B (or that B is a *superset* of A) if every element of A is an element of B. The symbol for inclusion is '$\subset$'. Thus '$A \subset B$' means that $(\forall x)(x \in A \Rightarrow x \in B)$. Clearly,

$$(A = B) \Leftrightarrow (A \subset B) \text{ and } (B \subset A).$$

This is a frequently used way of establishing set identity: we prove that $A = B$ by proving that $A \subset B$ and that $B \subset A$. If the reader thinks about the above equivalence, he will see that it depends first on the equivalence of the truth-functional forms '$P \Leftrightarrow Q$' and '$(P \Rightarrow Q)$ & $(Q \Rightarrow P)$', and then on the obvious quantificational equivalence between '$(\forall x)(R$ & $S)$' and '$(\forall x)R$ & $(\forall x)S$'.

We define a set by specifying its members. If the set is finite, the members can actually be listed, and the notation used is braces surrounding a membership list. For example $\{1, 4, 7\}$ is the set containing the three numbers 1, 4, 7, $\{x\}$ is the *unit set* of x (the set having only the one object x as a member), and $\{x, y\}$ is the *pair set* of x and y. We can abuse this notation to name some infinite sets. Thus $\{2, 4, 6, 8, \ldots\}$ would certainly be considered the set of all even positive integers. But infinite sets are generally defined by statement frames. If $P(x)$ is a frame containing the free variable 'x', then $\{x : P(x)\}$ is the set of all x such that $P(x)$ is true. In other words, $\{x : P(x)\}$ is that set A such that

$$y \in A \Leftrightarrow P(y).$$

For example, $\{x : x^2 < 9\}$ is the set of all real numbers x such that $x^2 < 9$, that is, the open interval $(-3, 3)$, and $y \in \{x : x^2 < 9\} \Leftrightarrow y^2 < 9$. A statement frame $P(x)$ can be thought of as stating a *property* that an object x may or may not have, and $\{x : P(x)\}$ is the set of all objects having that property.

We need the *empty set* $\emptyset$, in much the same way that we need zero in arithmetic. If $P(x)$ is *never* true, then $\{x : P(x)\} = \emptyset$. For example,

$$\{x : x \neq x\} = \emptyset.$$

When we said earlier that all mathematical objects are customarily considered sets, it was taken for granted that the reader understands the distinction between an object and a name of that object. To be on the safe side, we add a few words. A chair is not the same thing as the word 'chair', and the number 4 is a mathematical object that is not the same thing as the numeral '4'. The numeral '4' is a name of the number 4, as also are 'four', '2 + 2', and 'IV'. According to our present viewpoint, 4 itself is taken to be some specific set. There is no need in this course to carry logical analysis this far, but some readers may be interested to know that we usually define 4 as $\{0, 1, 2, 3\}$. Similarly, $2 = \{0, 1\}$, $1 = \{0\}$, and 0 is the empty set $\emptyset$.

It should be clear from the above discussion and our exposition thus far that we are using a symbol surrounded by single quotation marks as a name of that symbol (the symbol itself being a name of something else). Thus ' '4' ' is a name of '4' (which is itself a name of the number 4). This is strictly correct usage, but mathematicians almost universally mishandle it. It is accurate to write: let x be the number; call this number 'x'. However, the latter is almost always written: call this number x. This imprecision causes no difficulty to the reading mathematician, and it often saves the printed page from a shower of quotation marks. There is, however, a potential victim of such ambiguous treatment of symbols. This is the person who has never realized that mathematics is not about symbols but about objects to which the symbols refer. Since by now the present reader has safely avoided this pitfall, we can relax and occasionally omit the strictly necessary quotation marks.

In order to avoid overworking the word 'set', we use many synonyms, such as 'class', 'collection', 'family' and 'aggregate'. Thus we might say, "Let $\mathcal{A}$ be a family of classes of sets". If a shoe store is a collection of pairs of shoes, then a chain of shoe stores is such a three-level object.

5. RESTRICTED VARIABLES

A variable used in mathematics is not allowed to take all objects as values; it can only take as values the members of a certain set, called the *domain* of the variable. The domain is sometimes explicitly indicated, but is often only implied. For example, the letter 'n' is customarily used to specify an integer, so that '$(\forall n)P(n)$' would automatically be read "for every integer n, $P(n)$". However, sometimes n is taken to be a positive integer. In case of possible ambiguity or doubt, we would indicate the restriction explicitly and write '$(\forall n \in \mathbb{Z})P(n)$', where '$\mathbb{Z}$' is the standard symbol for the set of all integers. The quantifier is read, literally, "for all n in $\mathbb{Z}$", and more freely, "for every integer n". Similarly, '$(\exists n \in \mathbb{Z})P(n)$' is read "there exists an n in $\mathbb{Z}$ such that $P(n)$" or "there exists an integer n such that $P(n)$". Note that the symbol '$\in$' is here read as the preposition 'in'. The above quantifiers are called *restricted* quantifiers.

In the same way, we have restricted set formation, both implicit and explicit, as in '$\{n : P(n)\}$' and '$\{n \in \mathbb{Z} : P(n)\}$', both of which are read "the set of all integers n such that $P(n)$".

Restricted variables can be defined as abbreviations of unrestricted variables by

$$\begin{aligned}
(\forall x \in A)P(x) \quad &\Leftrightarrow \quad (\forall x)(x \in A \Rightarrow P(x)), \\
(\exists x \in A)P(x) \quad &\Leftrightarrow \quad (\exists x)(x \in A \ \& \ P(x)), \\
\{x \in A : P(x)\} \quad &= \quad \{x : x \in A \ \& \ P(x)\}.
\end{aligned}$$

Although there is never any ambiguity in sentences containing explicitly restricted variables, it sometimes helps the eye to see the structure of the sentence if the restricting phrases are written in superscript position, as in $(\forall \epsilon^{>0})(\exists n^{\in \mathbb{Z}})$. Some restriction was implicit on page 1. If the reader agreed that $(\forall x)(x^2 - 1 = (x - 1)(x + 1))$ was true, he probably took x to be a real number.

6. ORDERED PAIRS AND RELATIONS

Ordered pairs are basic tools, as the reader knows from analytic geometry. According to our general principle, the ordered pair $\langle a, b\rangle$ is taken to be a certain set, but here again we don't care which particular set it is so long as it guarantees the crucial characterizing property:

$$\langle x, y\rangle = \langle a, b\rangle \Leftrightarrow x = a \text{ and } y = b.$$

Thus $\langle 1, 3\rangle \neq \langle 3, 1\rangle$.

The notion of a correspondence, or relation, and the special case of a mapping, or function, is fundamental to mathematics. A correspondence is a pairing of objects such that given any two objects x and y, the pair $\langle x, y\rangle$ either does or does not correspond. A particular correspondence (relation) is generally presented by a statement frame $P(x, y)$ having two free variables, with x and y corresponding if any only if $P(x, y)$ is true. Given any relation (correspondence), the set of all ordered pairs $\langle x, y\rangle$ of corresponding elements is called its *graph*.

Now a relation is a mathematical object, and, as we have said several times, it is current practice to regard every mathematical object as a set of some sort or other. Since the graph of a relation is a set (of ordered pairs), it is efficient and customary to take the graph *to be* the relation. *Thus a relation (correspondence) is simply a set of ordered pairs.* If R is a relation, then we say that x has the relation R to y, and we write 'xRy', if and only if $\langle x, y\rangle \in R$. We also say that x corresponds to y under R. The set of all *first* elements occurring in the ordered pairs of a relation R is called the *domain* of R and is designated dom R or $\mathfrak{D}(R)$. Thus

$$\text{dom } R = \{x : (\exists y)\langle x, y\rangle \in R\}.$$

The set of second elements is called the *range* of R:

$$\text{range } R = \{y : (\exists x)\langle x, y\rangle \in R\}.$$

The *inverse*, R^{-1}, of a relation R is the set of ordered pairs obtained by reversing those of R:

$$R^{-1} = \{\langle x, y\rangle : \langle y, x\rangle \in R\}.$$

A statement frame $P(x, y)$ having two free variables actually determines a *pair* of mutually inverse relations R & S, called the *graphs* of P, as follows:

$$R = \{\langle x, y\rangle : P(x, y)\}, \qquad S = \{\langle y, x\rangle : P(x, y)\}.$$

A two-variable frame together with a choice of which variable is considered to be first might be called a *directed* frame. Then a directed frame would have a uniquely determined relation for its graph. The relation of strict inequality on the real number system $\mathbb{R}$ would be considered the set $\{\langle x, y\rangle : x < y\}$, since the variables in '$x < y$' have a natural order.

The set $A \times B = \{\langle x, y\rangle : x \in A \;\&\; y \in B\}$ of *all* ordered pairs with first element in A and second element in B is called the *Cartesian product* of the sets A and B. A relation R is always a subset of dom $R \times$ range R. If the two "factor spaces" are the same, we can use exponential notation: $A^2 = A \times A$.

The Cartesian product $\mathbb{R}^2 = \mathbb{R} \times \mathbb{R}$ is the "analytic plane". Analytic geometry rests upon the one-to-one coordinate correspondence between $\mathbb{R}^2$ and the Euclidean plane $\mathbb{E}^2$ (determined by an axis system in the latter), which enables us to treat geometric questions algebraically and algebraic questions geometrically. In particular, since a relation between sets of real numbers is a subset of $\mathbb{R}^2$, we can "picture" it by the corresponding subset of the Euclidean plane, or of any model of the Euclidean plane, such as this page. A simple Cartesian product is shown in Fig. 0.1 ($A \cup B$ is the *union* of the sets A and B).

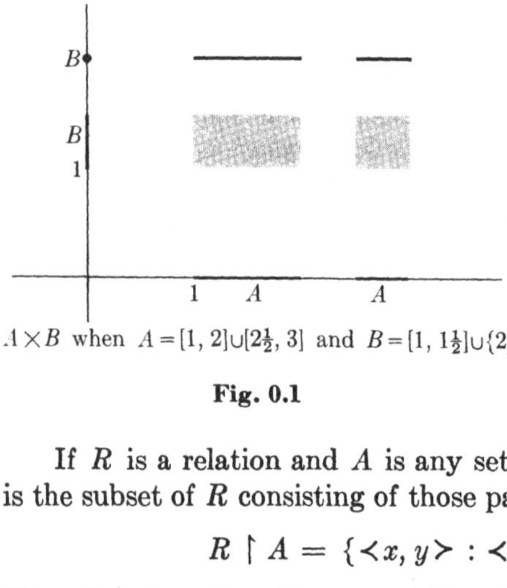

$A \times B$ when $A = [1, 2] \cup [2\frac{1}{2}, 3]$ and $B = [1, 1\frac{1}{2}] \cup \{2\}$

Fig. 0.1

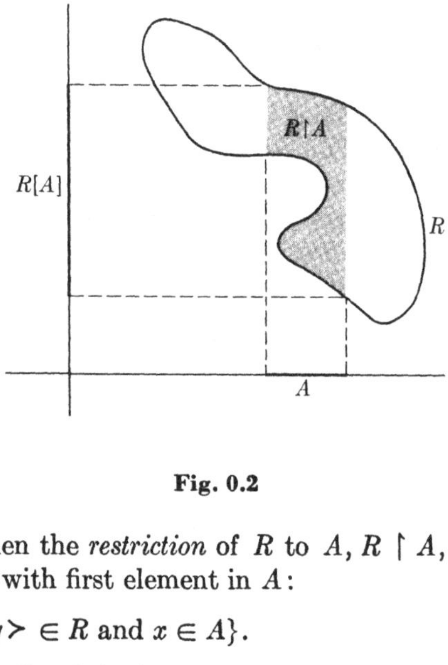

Fig. 0.2

If R is a relation and A is any set, then the *restriction* of R to A, $R \restriction A$, is the subset of R consisting of those pairs with first element in A:

$$R \restriction A = \{\langle x, y\rangle : \langle x, y\rangle \in R \text{ and } x \in A\}.$$

Thus $R \restriction A = R \cap (A \times \text{range } R)$, where $C \cap D$ is the *intersection* of the sets C and D.

If R is a relation and A is any set, then the *image of A under R*, $R[A]$, is the set of second elements of ordered pairs in R whose first elements are in A:

$$R[A] = \{y : (\exists x)(x \in A \;\&\; \langle x, y\rangle \in R)\}.$$

Thus $R[A] = \text{range } (R \restriction A)$, as shown in Fig. 0.2.

7. FUNCTIONS AND MAPPINGS

A *function* is a relation f such that each domain element x is paired with exactly one range element y. This property can be expressed as follows:

$$\langle x, y\rangle \in f \text{ and } \langle x, z\rangle \in f \Rightarrow y = z.$$

The y which is thus uniquely determined by f and x is designated $f(x)$:

$$y = f(x) \Leftrightarrow \langle x, y \rangle \in f.$$

One tends to think of a function as being active and a relation which is not a function as being passive. A function f *acts on* an element x in its domain to give $f(x)$. We take x and *apply* f to it; indeed we often call a function an *operator*. On the other hand, if R is a relation but not a function, then there is in general no particular y related to an element x in its domain, and the pairing of x and y is viewed more passively.

We often define a function f by specifying its value $f(x)$ for each x in its domain, and in this connection a stopped arrow notation is used to indicate the pairing. Thus $x \mapsto x^2$ is the function assigning to each number x its square x^2.

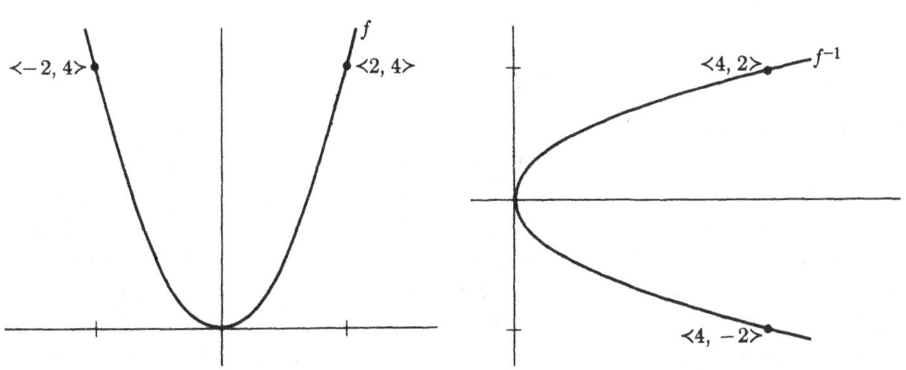

Fig. 0.3

If we want it to be understood that f is this function, we can write "Consider the function $f: x \mapsto x^2$". The domain of f must be understood for this notation to be meaningful.

If f is a function, then f^{-1} is of course a relation, but in general it is not a function. For example, if f is the function $x \mapsto x^2$, then f^{-1} contains the pairs $\langle 4, 2 \rangle$ and $\langle 4, -2 \rangle$ and so is not a function (see Fig. 0.3). If f^{-1} *is* a function, we say that f is *one-to-one* and that f is a one-to-one correspondence between its domain and its range. Each $x \in \text{dom} f$ corresponds to only one $y \in \text{range} f$ (f is a function), *and* each $y \in \text{range} f$ corresponds to only one $x \in \text{dom} f$ (f^{-1} is a function).

The notation

$$f: A \to B$$

is read "a (the) function f on A into B" or "the function f from A to B". The notation implies that f is a function, that $\text{dom} f = A$, and that $\text{range} f \subset B$. Many people feel that the very notion of function should include all these ingredients; that is, a function should be considered an ordered triple $\langle f, A, B \rangle$, where f is a function according to our more limited definition, A is the domain of f, and B is a superset of the range of f, which we shall call the *codomain* of f in this context. We shall use the terms 'map', 'mapping', and 'transformation' for such a triple, so that the notation $f: A \to B$ in its totality presents a mapping. Moreover, when there is no question about which set is the codomain, we shall often call the function f itself a mapping, since the triple $\langle f, A, B\rangle$ is then determined by f. The two arrow notations can be combined, as in: "Define $f: \mathbb{R} \to \mathbb{R}$ by $x \mapsto x^2$".

A mapping $f: A \to B$ is said to be *injective* if f is one-to-one, *surjective* if range $f = B$, and *bijective* if it is both injective and surjective. A bijective mapping $f: A \to B$ is thus a one-to-one correspondence between its domain A and its codomain B. Of course, a function is always surjective onto its range R, and the statement that f is surjective means that $R = B$, where B is the understood codomain.

8. PRODUCT SETS; INDEX NOTATION

One of the characteristic habits of the modern mathematician is that as soon as a new kind of object has been defined and discussed a little, he immediately looks at the set of all such objects. With the notion of a function from A to S well in hand, we naturally consider the set of all functions from A to S, which we designate S^A. Thus $\mathbb{R}^{\mathbb{R}}$ is the set of all real-valued functions of one real variable, and $S^{\mathbb{Z}^+}$ is the set of all infinite sequences in S. (It is understood that an infinite sequence is nothing but a function whose domain is the set $\mathbb{Z}^+$ of all positive integers.) Similarly, if we set $\bar{n} = \{1, \ldots, n\}$, then $S^{\bar{n}}$ is the set of all finite sequences of length n in S.

If B is a subset of S, then its *characteristic function* (relative to S) is the function on S, usually designated χ_B, which has the constant value 1 on B and the constant value 0 off B. The set of all characteristic functions of subsets of S is thus 2^S (since $2 = \{0, 1\}$). But because this collection of functions is in a natural one-to-one correspondence with the collection of all subsets of S, χ_B corresponding to B, we tend to identify the two collections. Thus 2^S is also interpreted as the set of all subsets of S. We shall spend most of the remainder of this section discussing further similar definitional ambiguities which mathematicians tolerate.

The ordered triple $\langle x, y, z\rangle$ is usually defined to be the ordered pair $\langle\langle x, y\rangle, z\rangle$. The reason for this definition is probably that a function of two variables x and y is ordinarily considered a function of the single ordered pair variable $\langle x, y\rangle$, so that, for example, a real-valued function of two real variables is a subset of $(\mathbb{R} \times \mathbb{R}) \times \mathbb{R}$. But we also consider such a function a subset of Cartesian 3-space $\mathbb{R}^3$. Therefore, we define $\mathbb{R}^3$ as $(\mathbb{R} \times \mathbb{R}) \times \mathbb{R}$; that is, we define the ordered triple $\langle x, y, z\rangle$ as $\langle\langle x, y\rangle, z\rangle$.

On the other hand, the ordered triple $\langle x, y, z\rangle$ could also be regarded as the finite sequence $\{\langle 1, x\rangle, \langle 2, y\rangle, \langle 3, z\rangle\}$, which, of course, is a different object. These two models for an ordered triple serve equally well, and, again, mathematicians tend to slur over the distinction. We shall have more to say on this point later when we discuss natural isomorphisms (Section 1.6). For the moment we shall simply regard $\mathbb{R}^3$ and $\mathbb{R}^{\overline{3}}$ as being the same; an ordered triple is something which can be "viewed" as being either an ordered pair of which the first element is an ordered pair or as a sequence of length 3 (or, for that matter, as an ordered pair of which the *second* element is an ordered pair).

Similarly, we pretend that Cartesian 4-space $\mathbb{R}^4$ is $\mathbb{R}^{\overline{4}}$, $\mathbb{R}^2 \times \mathbb{R}^2$, or $\mathbb{R}^1 \times \mathbb{R}^3 = \mathbb{R} \times ((\mathbb{R} \times \mathbb{R}) \times \mathbb{R})$, etc. Clearly, we are in effect assuming an associative law for ordered pair formation that we don't really have.

This kind of ambiguity, where we tend to identify two objects that really are distinct, is a necessary corollary of deciding exactly what things are. It is one of the prices we pay for the precision of set theory; in days when mathematics was vaguer, there would have been a single fuzzy notion.

The device of indices, which is used frequently in mathematics, also has ambiguous implications which we should examine. An indexed collection, as a set, is nothing but the range set of a function, the indexing function, and a particular indexed object, say x_i, is simply the value of that function at the domain element i. If the set of indices is I, the indexed set is designated $\{x_i : i \in I\}$ or $\{x_i\}_{i \in I}$ (or $\{x_i\}_{i=1}^{\infty}$ in case $I = \mathbb{Z}^+$). However, this notation suggests that we view the indexed set as being obtained by letting the index run through the index set I and collecting the indexed objects. That is, an indexed set is viewed as being the set *together with* the indexing function. This ambivalence is reflected in the fact that the same notation frequently designates the mapping. Thus we refer to the *sequence* $\{x_n\}_{n=1}^{\infty}$, where, of course, the sequence is the mapping $n \mapsto x_n$. We believe that if the reader examines his idea of a sequence he will find this ambiguity present. He means neither just the set nor just the mapping, but the mapping with emphasis on its range, or the range "together with" the mapping. But since set theory cannot reflect these nuances in any simple and graceful way, we shall take an indexed set *to be* the indexing function. Of course, the same range object may be repeated with different indices; there is no implication that an indexing is one-to-one. Note also that indexing imposes no restriction on the set being indexed; any set can at least be self-indexed (by the identity function).

Except for the ambiguous '$\{x_i : i \in I\}$', there is no universally used notation for the indexing function. Since x_i is the value of the function at i, we might think of 'x_i' as another way of writing '$x(i)$', in which case we designate the function 'x' or '$\mathbf{x}$'. We certainly do this in the case of ordered n-tuplets when we say, "Consider the n-tuplet $\mathbf{x} = \langle x_1, \ldots, x_n \rangle$". On the other hand, there is no compelling reason to use this notation. We can call the indexing function anything we want; if it is f, then of course $f(i) = x_i$ for all i.

We come now to the general definition of Cartesian product. Earlier we argued (in a special case) that the Cartesian product $A \times B \times C$ is the set of all ordered triples $\mathbf{x} = \langle x_1, x_2, x_3 \rangle$ such that $x_1 \in A$, $x_2 \in B$, and $x_3 \in C$. More generally, $A_1 \times A_2 \times \cdots \times A_n$, or $\prod_{i=1}^{n} A_i$, is the set of ordered n-tuples $\mathbf{x} = \langle x_1, \ldots, x_n \rangle$ such that $x_i \in A_i$ for $i = 1, \ldots, n$. If we interpret an ordered n-tuplet as a function on $\bar{n} = \{1, \ldots, n\}$, we have

> $\prod_{i=1}^{n} A_i$ is the set of all functions $\mathbf{x}$ with domain $\bar{n}$ such that $x_i \in A_i$ for all $i \in \bar{n}$.

This rephrasal generalizes almost verbatim to give us the notion of the Cartesian product of an arbitrary indexed collection of sets.

> **Definition.** The Cartesian product $\prod_{i \in I} S_i$ of the indexed collection of sets $\{S_i : i \in I\}$ is the set of all functions f with domain the index set I such that $f(i) \in S_i$ for all $i \in I$.

We can also use the notation $\prod \{S_i : i \in I\}$ for the product and f_i for the value $f(i)$.

9. COMPOSITION

If we are given maps $f: A \to B$ and $g: B \to C$, then the *composition* of g with f, $g \circ f$, is the map of A into C defined by

$$(g \circ f)(x) = g(f(x)) \qquad \text{for all} \quad x \in A.$$

This is the function of a function operation of elementary calculus. If f and g are the maps from $\mathbb{R}$ to $\mathbb{R}$ defined by $f(x) = x^{1/3} + 1$ and $g(x) = x^2$, then $f \circ g(x) = (x^2)^{1/3} + 1 = x^{2/3} + 1$, and $g \circ f(x) = (x^{1/3} + 1)^2 = x^{2/3} + 2x^{1/3} + 1$. Note that the codomain of f must be the domain of g in order for $g \circ f$ to be defined. This operation is perhaps the basic binary operation of mathematics.

> **Lemma.** Composition satisfies the associative law:
>
> $$f \circ (g \circ h) = (f \circ g) \circ h.$$

Proof. $(f \circ (g \circ h))(x) = f((g \circ h)(x)) = f(g(h(x))) = (f \circ g)(h(x)) = ((f \circ g) \circ h)(x)$ for all $x \in \operatorname{dom} h$. □

If A is a set, the identity map $I_A: A \to A$ is the mapping taking every $x \in A$ to itself. Thus $I_A = \{\langle x, x \rangle : x \in A\}$. If f maps A into B, then clearly

$$f \circ I_A = f = I_B \circ f.$$

If $g: B \to A$ is such that $g \circ f = I_A$, then we say that g is a *left inverse* of f and that f is a *right inverse* of g.

> **Lemma.** If the mapping $f: A \to B$ has both a right inverse h and a left inverse g, they must necessarily be equal.

Proof. This is just algebraic juggling and works for any associative operation. We have

$$h = I_A \circ h = (g \circ f) \circ h = g \circ (f \circ h) = g \circ I_B = g. \ \square$$

In this case we call the uniquely determined map $g: B \to A$ such that $f \circ g = I_B$ and $g \circ f = I_A$ the *inverse* of f. We then have:

Theorem. A mapping $f: A \to B$ has an inverse if and only if it is bijective, in which case its inverse is its relational inverse f^{-1}.

Proof. If f is bijective, then the relational inverse f^{-1} is a function from B to A, and the equations $f \circ f^{-1} = I_B$ and $f^{-1} \circ f = I_A$ are obvious. On the other hand, if $f \circ g = I_B$, then f is surjective, since then every y in B can be written $y = f(g(y))$. And if $g \circ f = I_A$, then f is injective, for then the equation $f(x) = f(y)$ implies that $x = g(f(x)) = g(f(y)) = y$. Thus f is bijective if it has an inverse. □

Now let $\mathfrak{S}(A)$ be the set of all bijections $f: A \to A$. Then $\mathfrak{S}(A)$ is closed under the binary operation of composition and

1) $f \circ (g \circ h) = (f \circ g) \circ h$ for all $f, g, h \in \mathfrak{S}$;
2) there exists a unique $I \in \mathfrak{S}(A)$ such that $f \circ I = I \circ f = f$ for all $f \in \mathfrak{S}$;
3) for each $f \in \mathfrak{S}$ there exists a unique $g \in \mathfrak{S}$ such that $f \circ g = g \circ f = I$.

Any set G closed under a binary operation having these properties is called a *group* with respect to that operation. Thus $\mathfrak{S}(A)$ is a group with respect to composition.

Composition can also be defined for relations as follows. If $R \subset A \times B$ and $S \subset B \times C$, then $S \circ R \subset A \times C$ is defined by

$$\langle x, z \rangle \in S \circ R \Leftrightarrow (\exists y^{\in B})(\langle x, y \rangle \in R \;\&\; \langle y, z \rangle \in S).$$

If R and S are mappings, this definition agrees with our earlier one.

10. DUALITY

There is another elementary but important phenomenon called *duality* which occurs in practically all branches of mathematics. Let $F: A \times B \to C$ be any function of two variables. It is obvious that if x is held fixed, then $F(x, y)$ is a function of the one variable y. That is, for each fixed x there is a function $h^x: B \to C$ defined by $h^x(y) = F(x, y)$. Then $x \mapsto h^x$ is a mapping φ of A into C^B. Similarly, each $y \in B$ yields a function $g_y \in C^A$, where $g_y(x) = F(x, y)$, and $y \mapsto g_y$ is a mapping θ from B to C^A.

Now suppose conversely that we are given a mapping $\varphi: A \to C^B$. For each $x \in A$ we designate the corresponding value of φ in index notation as h^x, so that h^x is a function from B to C, and we define $F: A \times B \to C$ by $F(x, y) = h^x(y)$. We are now back where we started. Thus the mappings $\varphi: A \to C^B$, $F: A \times B \to C$, and $\theta: B \to C^A$ are equivalent, and can be thought of as three different ways of viewing the same phenomenon. The extreme mappings φ and θ will be said to be *dual* to each other.

The mapping φ is the indexed family of functions $\{h^x : x \in A\} \subset C^B$. Now suppose that $\mathfrak{F} \subset C^B$ is an *un*indexed collection of functions on B into C, and define $F: \mathfrak{F} \times B \to C$ by $F(f, y) = f(y)$. Then $\theta: B \to C^{\mathfrak{F}}$ is defined by $g_y(f) = f(y)$. What is happening here is simply that in the expression $f(y)$ we regard *both* symbols as variables, so that $f(y)$ is a function on $\mathfrak{F} \times B$. Then when we hold y fixed, we have a function on $\mathfrak{F}$ mapping $\mathfrak{F}$ into C.

We shall see some important applications of this duality principle as our subject develops. For example, an $m \times n$ matrix is a function $\mathbf{t} = \{t_{ij}\}$ in $\mathbb{R}^{\overline{m} \times \overline{n}}$. We picture the matrix as a rectangular array of numbers, where 'i' is the row index and 'j' is the column index, so that t_{ij} is the number at the intersection of the ith row and the jth column. If we hold i fixed, we get the n-tuple forming the ith row, and the matrix can therefore be interpreted as an m-tuple of row n-tuples. Similarly (dually), it can be viewed as an n-tuple of column m-tuples.

In the same vein, an n-tuple $\langle f_1, \ldots, f_n \rangle$ of functions from A to B can be regarded as a single n-tuple-valued function from A to $B^{\overline{n}}$,

$$a \mapsto \langle f_1(a), \ldots, f_n(a) \rangle .$$

In a somewhat different application, duality will allow us to regard a finite-dimensional vector space V as being its own second conjugate space $(V^*)^*$.

It is instructive to look at elementary Euclidean geometry from this point of view. Today we regard a straight line as being a set of geometric points. An older and more neutral view is to take points and lines as being two different kinds of primitive objects. Accordingly, let A be the set of all points (so that A is the Euclidean plane as we now view it), and let B be the set of all straight lines. Let F be the *incidence* function: $F(p, l) = 1$ if p and l are incident (p is "on" l, l is "on" p) and $F(p, l) = 0$ otherwise. Thus F maps $A \times B$ into $\{0, 1\}$. Then for each $l \in B$ the function $g_l(p) = F(p, l)$ is the characteristic function of the set of points that we think of as being the line l ($g_l(p)$ has the value 1 if p is on l and 0 if p is not on l.) Thus each line determines the set of points that are on it. But, dually, each point p determines the set of lines l "on" *it*, through *its* characteristic function $h^p(l)$. Thus, in complete duality we can regard a line as being a set of points and a point as being a set of lines. This duality aspect of geometry is basic in projective geometry.

It is sometimes awkward to invent new notation for the "partial" function obtained by holding a variable fixed in a function of several variables, as we did above when we set $g_y(x) = F(x, y)$, and there is another device that is frequently useful in this situation. This is to put a dot in the position of the "varying variable". Thus $F(a, \cdot)$ is the function of one variable obtained from $F(x, y)$ by holding x fixed at the value a, so that in our beginning discussion of duality we have

$$h^x = F(x, \cdot), \qquad g_y = F(\cdot\,, y).$$

If f is a function of one variable, we can then write $f = f(\cdot)$, and so express the above equations also as $h^x(\cdot) = F(x, \cdot)$, $g_y(\cdot) = F(\cdot\,, y)$. The flaw in this notation is that we can't indicate substitution without losing meaning. Thus the value of the function $F(x, \cdot)$ at b is $F(x, b)$, but from this evaluation we cannot read backward and tell what function was evaluated. We are therefore forced to some such cumbersome notation as $F(x, \cdot)|_b$, which can get out of hand. Nevertheless, the dot device is often helpful when it can be used without evaluation difficulties. In addition to eliminating the need for temporary notation, as mentioned above, it can also be used, in situations where it is strictly speaking superfluous, to direct the eye at once to the position of the variable.

For example, later on $D_\xi F$ will designate the directional derivative of the function F in the (fixed) direction ξ. This is a function whose value at α is $D_\xi F(\alpha)$, and the notation $D_\xi F(\cdot)$ makes this implicitly understood fact explicit.

11. THE BOOLEAN OPERATIONS

Let S be a fixed domain, and let $\mathfrak{F}$ be a family of subsets of S. The *union* of $\mathfrak{F}$, or the *union of all the sets in* $\mathfrak{F}$, is the set of all elements belonging to at least one set in $\mathfrak{F}$. We designate the union $\bigcup\mathfrak{F}$ or $\bigcup_{A\in\mathfrak{F}} A$, and thus we have

$$\bigcup\mathfrak{F} = \{x : (\exists A^{\in\mathfrak{F}})(x \in A)\}, \qquad y \in \bigcup\mathfrak{F} \Leftrightarrow (\exists A^{\in\mathfrak{F}})(y \in A).$$

We often consider the family $\mathfrak{F}$ to be indexed. That is, we assume given a set I (the set of indices) and a surjective mapping $i \mapsto A_i$ from I to $\mathfrak{F}$, so that $\mathfrak{F} = \{A_i : i \in I\}$. Then the union of the indexed collection is designated $\bigcup_{i\in I} A_i$ or $\bigcup\{A_i : i \in I\}$. The device of indices has both technical and psychological advantages, and we shall generally use it.

If $\mathfrak{F}$ is finite, and either it or the index set is listed, then a different notation is used for its union. If $\mathfrak{F} = \{A, B\}$, we designate the union $A \cup B$, a notation that displays the listed names. Note that here we have $x \in A \cup B \Leftrightarrow x \in A$ or $x \in B$. If $\mathfrak{F} = \{A_i : i = 1, \ldots, n\}$, we generally write '$A_1 \cup A_2 \cup \cdots \cup A_n$' or '$\bigcup_{i=1}^n A_i$' for $\bigcup\mathfrak{F}$.

The *intersection* of the indexed family $\{A_i\}_{i\in I}$, designated $\bigcap_{i\in I} A_i$, is the set of all points that lie in every A_i. Thus

$$x \in \bigcap_{i\in I} A_i \;\Leftrightarrow\; (\forall_i^{\in I})(x \in A_i).$$

For an unindexed family $\mathfrak{F}$ we use the notation $\bigcap\mathfrak{F}$ or $\bigcap_{A\in\mathfrak{F}} A$, and if $\mathfrak{F} = \{A, B\}$, then $\bigcap\mathfrak{F} = A \cap B$.

The complement, A', of a subset of S is the set of elements $x \in S$ not in A: $A' = \{x^{\in S} : x \notin A\}$. The law of De Morgan states that *the complement of an intersection is the union of the complements:*

$$\Big(\bigcap_{i\in I} A_i\Big)' = \bigcup_{i\in I} (A_i').$$

This an immediate consequence of the rule for negating quantifiers. It is the equivalence between 'not always in' and 'sometimes not in': $[\sim(\forall i)(x \in A_i) \Leftrightarrow (\exists i)(x \notin A_i)]$ says exactly that

$$x \in \left(\bigcap_i A_i\right)' \Leftrightarrow x \in \bigcup_i (A_i').$$

If we set $B_i = A_i'$ and take complements again, we obtain the dual form: $(\bigcup_{i \in I} B_i)' = \bigcap_{i \in I}(B_i')$.

Other principles of quantification yield the laws

$$B \cap \left(\bigcup_{i \in I} A_i\right) = \bigcup_{i \in I} (B \cap A_i)$$

from $P \;\&\; (\exists x)Q(x) \Leftrightarrow (\exists x)(P \;\&\; Q(x))$,

$$B \cup \left(\bigcap_{i \in I} A_i\right) = \bigcap_{i \in I} (B \cup A_i),$$
$$B \cap \left(\bigcap_{i \in I} A_i\right) = \bigcap_{i \in I} (B \cap A_i),$$
$$B \cup \left(\bigcup_{i \in I} A_i\right) = \bigcup_{i \in I} (B \cup A_i).$$

In the case of two sets, these laws imply the following familiar laws of set algebra:

$$(A \cup B)' = A' \cap B', \qquad (A \cap B)' = A' \cup B' \qquad \text{(De Morgan)},$$
$$A \cap (B \cup C) = (A \cap B) \cup (A \cap C),$$
$$A \cup (B \cap C) = (A \cup B) \cap (A \cup C).$$

Even here, thinking in terms of indices makes the laws more intuitive. Thus

$$(A_1 \cap A_2)' = A_1' \cup A_2'$$

is obvious when thought of as the equivalence between 'not always in' and 'sometimes not in'.

The family $\mathfrak{F}$ is *disjoint* if distinct sets in $\mathfrak{F}$ have no elements in common, i.e., if $(\forall X, Y^{\in \mathfrak{F}})(X \neq Y \Rightarrow X \cap Y = \emptyset)$. For an indexed family $\{A_i\}_{i \in I}$ the condition becomes $i \neq j \Rightarrow A_i \cap A_j = \emptyset$. If $\mathfrak{F} = \{A, B\}$, we simply say that A and B are disjoint.

Given $f\colon U \to V$ and an indexed family $\{B_i\}$ of subsets of V, we have the following important identities:

$$f^{-1}\left[\bigcup_i B_i\right] = \bigcup_i f^{-1}[B_i], \qquad f^{-1}\left[\bigcap_i B_i\right] = \bigcap_i f^{-1}[B_i],$$

and, for a single set $B \subset V$,

$$f^{-1}[B'] = (f^{-1}[B])'.$$

For example,

$$x \in f^{-1}\left[\bigcap_i B_i\right] \Leftrightarrow f(x) \in \bigcap_i B_i \Leftrightarrow (\forall i)(f(x) \in B_i)$$
$$\Leftrightarrow (\forall i)(x \in f^{-1}[B_i]) \Leftrightarrow x \in \bigcap_i f^{-1}[B_i].$$

The first, but not the other two, of the three identities above remains valid when f is replaced by any relation R. It follows from the commutative law, $(\exists x)(\exists y)A \Leftrightarrow (\exists y)(\exists x)A$. The second identity fails for a general R because '$(\exists x)(\forall y)$' and '$(\forall y)(\exists x)$' have different meanings.

12. PARTITIONS AND EQUIVALENCE RELATIONS

A *partition* of a set A is a disjoint family $\mathfrak{F}$ of sets whose union is A. We call the elements of $\mathfrak{F}$ 'fibers', and we say that $\mathfrak{F}$ *fibers* A or is a *fibering* of A. For example, the set of straight lines parallel to a given line in the Euclidean plane is a fibering of the plane. If '$\bar{x}$' designates the unique fiber containing the point x, then $x \mapsto \bar{x}$ is a surjective mapping $\pi: A \to \mathfrak{F}$ which we call the *projection* of A on $\mathfrak{F}$. Passing from a set A to a fibering $\mathfrak{F}$ of A is one of the principal ways of forming new mathematical objects.

Any function f automatically fibers its domain into sets on which f is constant. If A is the Euclidean plane and $f(p)$ is the x-coordinate of the point p in some coordinate system, then f is constant on each vertical line; more exactly, $f^{-1}(x)$ is a vertical line for every x in $\mathbb{R}$. Moreover, $x \mapsto f^{-1}(x)$ is a bijection from $\mathbb{R}$ to the set of all fibers (vertical lines). In general, if $f: A \to B$ is any surjective mapping, and if for each value y in B we set

$$A_y = f^{-1}(y) = \{x \in A : f(x) = y\},$$

then $\mathfrak{F} = \{A_y : y \in B\}$ is a fibering of A and $\varphi: y \mapsto A_y$ is a bijection from B to $\mathfrak{F}$. Also $\varphi \circ f$ is the projection $\pi: A \to \mathfrak{F}$, since $\varphi \circ f(x) = \varphi(f(x))$ is the set $\bar{x}$ of all z in A such that $f(z) = f(x)$.

The above process of generating a fibering of A from a function on A is relatively trivial. A more important way of obtaining a fibering of A is from an equality-like relation on A called an equivalence relation. An *equivalence relation* $\sim$ on A is a binary relation which is *reflexive* ($x \sim x$ for every $x \in A$), *symmetric* ($x \sim y \Rightarrow y \sim x$), and *transitive* ($x \sim y$ and $y \sim z \Rightarrow x \sim z$). Every fibering $\mathfrak{F}$ of A generates a relation $\sim$ by the stipulation that $x \sim y$ if and only if x and y are in the same fiber, and obviously $\sim$ is an equivalence relation. The most important fact to be established in this section is the converse.

Theorem. Every equivalence relation $\sim$ on A is the equivalence relation of a fibering.

Proof. We obviously have to define $\bar{x}$ as the set of elements y equivalent to x, $\bar{x} = \{y : y \sim x\}$, and our problem is to show that the family $\mathfrak{F}$ of all subsets of A obtained this way is a fibering.

The reflexive, symmetric, and transitive laws become

$$x \in \bar{x}, \qquad x \in \bar{y} \Rightarrow y \in \bar{x}, \qquad \text{and} \qquad x \in \bar{y} \text{ and } y \in \bar{z} \Rightarrow x \in \bar{z}.$$

Reflexivity thus implies that $\mathfrak{F}$ covers A. Transitivity says that if $y \in \bar{z}$, then $x \in \bar{y} \Rightarrow x \in \bar{z}$; that is, if $y \in \bar{z}$, then $\bar{y} \subset \bar{z}$. But also, if $y \in \bar{z}$, then $z \in \bar{y}$ by symmetry, and so $\bar{z} \subset \bar{y}$. Thus $y \in \bar{z}$ implies $\bar{y} = \bar{z}$. Therefore, if two of our sets $\bar{a}$ and $\bar{b}$ have a point x in common, then $\bar{a} = \bar{x} = \bar{b}$. In other words, if $\bar{a}$ is not the set $\bar{b}$, then $\bar{a}$ and $\bar{b}$ are disjoint, and we have a fibering. □

The fundamental role this argument plays in mathematics is due to the fact that in many important situations equivalence relations occur as the primary object, and then are used to define partitions and functions. We give two examples.

Let $\mathbb{Z}$ be the integers (positive, negative, and zero). A fraction 'm/n' can be considered an ordered pair $\langle m, n \rangle$ of integers with $n \neq 0$. The set of all fractions is thus $\mathbb{Z} \times (\mathbb{Z} - \{0\})$. Two fractions $\langle m, n \rangle$ and $\langle p, q \rangle$ are "equal" if and only if $mq = np$, and equality is checked to be an equivalence relation. The equivalence class $\overline{\langle m, n \rangle}$ is the object taken to be the rational number m/n. Thus the rational number system $\mathbb{Q}$ is the set of fibers in a partition of $\mathbb{Z} \times (\mathbb{Z} - \{0\})$.

Next, we choose a fixed integer $p \in \mathbb{Z}$ and define a relation E on $\mathbb{Z}$ by $mEn \Leftrightarrow p$ divides $m - n$. Then E is an equivalence relation, and the set $\mathbb{Z}_p$ of its equivalence classes is called the integers modulo p. It is easy to see that mEn if and only if m and n have the same remainder when divided by p, so that in this case there is an easily calculated function f, where $f(m)$ is the remainder after dividing m by p, which defines the fibering. The set of possible remainders is $\{0, 1, \ldots, p - 1\}$, so that $\mathbb{Z}_p$ contains p elements.

A function on a set A can be "factored" through a fibering of A by the following theorem.

Theorem. Let g be a function on A, and let $\mathfrak{F}$ be a fibering of A. Then g is constant on each fiber of $\mathfrak{F}$ if and only if there exists a function $\bar{g}$ on $\mathfrak{F}$ such that $g = \bar{g} \circ \pi$.

Proof. If g is constant on each fiber of $\mathfrak{F}$, then the association of this unique value with the fiber defines the function $\bar{g}$, and clearly $g = \bar{g} \circ \pi$. The converse is obvious. □

CHAPTER 1

VECTOR SPACES

The calculus of functions of more than one variable unites the calculus of one variable, which the reader presumably knows, with the theory of vector spaces, and the adequacy of its treatment depends directly on the extent to which vector space theory really is used. The theories of differential equations and differential geometry are similarly based on a mixture of calculus and vector space theory. Such "vector calculus" and its applications constitute the subject matter of this book, and in order for our treatment to be completely satisfactory, we shall have to spend considerable time at the beginning studying vector spaces themselves. This we do principally in the first two chapters. The present chapter is devoted to general vector spaces and the next chapter to finite-dimensional spaces.

We begin this chapter by introducing the basic concepts of the subject—vector spaces, vector subspaces, linear combinations, and linear transformations—and then relate these notions to the lines and planes of geometry. Next we establish the most elementary formal properties of linear transformations and Cartesian product vector spaces, and take a brief look at quotient vector spaces. This brings us to our first major objective, the study of direct sum decompositions, which we undertake in the fifth section. The chapter concludes with a preliminary examination of bilinearity.

1. FUNDAMENTAL NOTIONS

Vector spaces and subspaces. The reader probably has already had some contact with the notion of a vector space. Most beginning calculus texts discuss geometric vectors, which are represented by "arrows" drawn from a chosen origin O. These vectors are added geometrically by the parallelogram rule: The sum of the vector $\overrightarrow{OA}$ (represented by the arrow from O to A) and the vector $\overrightarrow{OB}$ is the vector $\overrightarrow{OP}$, where P is the vertex opposite O in the parallelogram having OA and OB as two sides (Fig. 1.1). Vectors can also be multiplied by numbers: $x(\overrightarrow{OA})$ is that vector $\overrightarrow{OB}$ such that B is on the line through O and A, the distance from O to B is $|x|$ times the distance from O to A, and B and A are on the same side of O if x is positive, and on opposite sides if x is negative

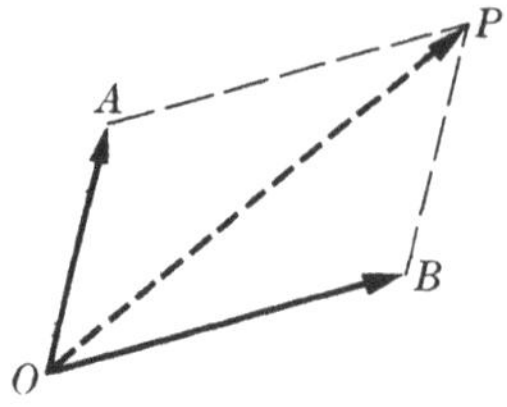

Fig. 1.1

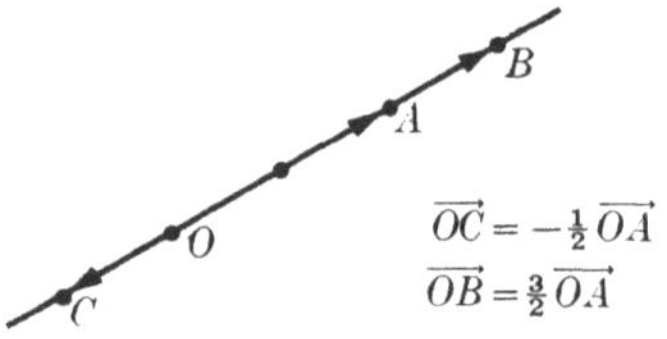

Fig. 1.2

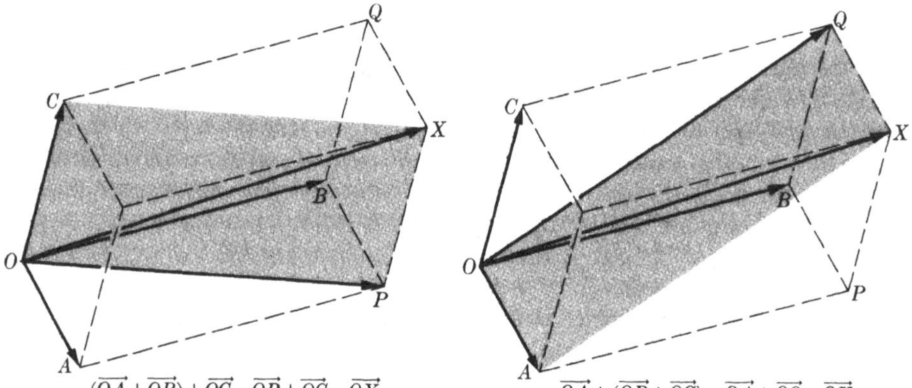

$(\overrightarrow{OA}+\overrightarrow{OB})+\overrightarrow{OC}=\overrightarrow{OP}+\overrightarrow{OC}=\overrightarrow{OX}$ $\qquad$ $\overrightarrow{OA}+(\overrightarrow{OB}+\overrightarrow{OC})=\overrightarrow{OA}+\overrightarrow{OQ}=\overrightarrow{OX}$

Fig. 1.3

(Fig. 1.2). These two vector operations satisfy certain laws of algebra, which we shall soon state in the definition. The geometric proofs of these laws are generally sketchy, consisting more of plausibility arguments than of airtight logic. For example, the geometric figure in Fig. 1.3 is the essence of the usual proof that vector addition is associative. In each case the final vector $\overrightarrow{OX}$ is represented by the diagonal starting from O in the parallelepiped constructed from the three edges OA, OB, and OC. The set of all geometric vectors, together with these two operations and the laws of algebra that they satisfy, constitutes one example of a vector space. We shall return to this situation in Section 2.

The reader may also have seen coordinate triples treated as vectors. In this system a three-dimensional vector is an ordered triple of numbers $\langle x_1, x_2, x_3 \rangle$ which we think of geometrically as the coordinates of a point in space. Addition is now *algebraically* defined,

$$\langle x_1, x_2, x_3 \rangle + \langle y_1, y_2, y_3 \rangle = \langle x_1 + y_1, x_2 + y_2, x_3 + y_3 \rangle,$$

as is multiplication by numbers, $t\langle x_1, x_2, x_3 \rangle = \langle tx_1, tx_2, tx_3 \rangle$. The vector laws are much easier to prove for these objects, since they are almost algebraic formalities. The set $\mathbb{R}^3$ of all ordered triples of numbers, together with these two operations, is a second example of a vector space.

If we think of an ordered triple $<x_1, x_2, x_3>$ as a function $\mathbf{x}$ with domain the set of integers from 1 to 3, where x_i is the value of the function $\mathbf{x}$ at i (see Section 0.8), then this vector space suggests a general type called a *function space,* which we shall examine after the definition. For the moment we remark only that we defined the sum of the triple $\mathbf{x}$ and the triple $\mathbf{y}$ as that triple $\mathbf{z}$ such that $z_i = x_i + y_i$ for every i.

A vector space, then, is a collection of objects that can be added to each other and multiplied by numbers, subject to certain laws of algebra. In this context a number is often called a *scalar.*

Definition. Let V be a set, and let there be given a mapping $<\alpha, \beta> \mapsto \alpha + \beta$ from $V \times V$ to V, called *addition,* and a mapping $<x, \alpha> \mapsto x\alpha$ from $\mathbb{R} \times V$ to V, called *multiplication by scalars.* Then V is a vector space with respect to these two operations if:

A1. $\alpha + (\beta + \gamma) = (\alpha + \beta) + \gamma$ for all $\alpha, \beta, \gamma \in V$.

A2. $\alpha + \beta = \beta + \alpha$ for all $\alpha, \beta \in V$.

A3. There exists an element $0 \in V$ such that $\alpha + 0 = \alpha$ for all $\alpha \in V$.

A4. For every $\alpha \in V$ there exists a $\beta \in V$ such that $\alpha + \beta = 0$.

S1. $(xy)\alpha = x(y\alpha)$ for all $x, y \in \mathbb{R}$, $\alpha \in V$.

S2. $(x + y)\alpha = x\alpha + y\alpha$ for all $x, y \in \mathbb{R}$, $\alpha \in V$.

S3. $x(\alpha + \beta) = x\alpha + x\beta$ for all $x \in \mathbb{R}$, $\alpha, \beta \in V$.

S4. $1\alpha = \alpha$ for all $\alpha \in V$.

In contexts where it is clear (as it generally is) which operations are intended, we refer simply to the vector space V.

Certain further properties of a vector space follow directly from the axioms. Thus the zero element postulated in A3 is unique, and for each α the β of A4 is unique, and is called $-\alpha$. Also $0\alpha = 0$, $x0 = 0$, and $(-1)\alpha = -\alpha$. These elementary consequences are considered in the exercises.

Our standard example of a vector space will be the set $V = \mathbb{R}^A$ of all real-valued functions on a set A under the natural operations of addition of two functions and multiplication of a function by a number. This generalizes the example $\mathbb{R}^{\{1,2,3\}} = \mathbb{R}^3$ that we looked at above. Remember that a function f in $\mathbb{R}^A$ is simply a mathematical object of a certain kind. We are saying that two of these objects can be added together in a natural way to form a third such object, and that the set of all such objects then satisfies the above laws for addition. Of course, $f + g$ is defined as the function whose value at a is $f(a) + g(a)$, so that $(f + g)(a) = f(a) + g(a)$ for all a in A. For example, in $\mathbb{R}^3$ we defined the sum $\mathbf{x} + \mathbf{y}$ as that triple whose value at i is $x_i + y_i$ for all i. Similarly, cf is the function defined by $(cf)(a) = c(f(a))$ for all a. Laws A1 through S4 follow at once from these definitions and the corresponding laws of algebra for the real number system. For example, the equation $(s + t)f = sf + tf$ means that $((s+t)f)(a) = (sf+tf)(a)$ for all $a \in A$. But

$$((s+t)f)(a) = (s+t)(f(a)) = s(f(a)) + t(f(a)) = (sf)(a) + (tf)(a) = (sf+tf)(a),$$

where we have used the definition of scalar multiplication in $\mathbb{R}^A$, the distributive law in $\mathbb{R}$, the definition of scalar multiplication in $\mathbb{R}^A$, and the definition of addition in $\mathbb{R}^A$, in that order. Thus we have S2, and the other laws follow similarly.

The set A can be anything at all. If $A = \mathbb{R}$, then $V = \mathbb{R}^{\mathbb{R}}$ is the vector space of all real-valued functions of one real variable. If $A = \mathbb{R} \times \mathbb{R}$, then $V = \mathbb{R}^{\mathbb{R}\times\mathbb{R}}$ is the space of all real-valued functions of two real variables. If $A = \{1, 2\} = \bar{2}$, then $V = \mathbb{R}^{\bar{2}} = \mathbb{R}^2$ is the Cartesian plane, and if $A = \{1, \ldots, n\} = \bar{n}$, then $V = \mathbb{R}^{\bar{n}}$ is Cartesian n-space. If A contains a single point, then $\mathbb{R}^A$ is a natural bijective image of $\mathbb{R}$ itself, and of course $\mathbb{R}$ is trivially a vector space with respect to its own operations.

Now let V be any vector space, and suppose that W is a nonempty subset of V that is closed under the operations of V. That is, if α and β are in W, then so is $\alpha + \beta$, and if α is in W, then so is $x\alpha$ for every scalar x. For example, let V be the vector space $\mathbb{R}^{[a,b]}$ of all real-valued functions on the closed interval $[a, b] \subset \mathbb{R}$, and let W be the set $\mathcal{C}([a, b])$ of all *continuous* real-valued functions on $[a, b]$. Then W is a subset of V that is closed under the operations of V, since $f+g$ and cf are continuous whenever f and g are. Or let V be Cartesian 2-space $\mathbb{R}^2$, and let W be the set of ordered pairs $\mathbf{x} = \langle x_1, x_2 \rangle$ such that $x_1 + x_2 = 0$. Clearly, W is closed under the operations of V.

Such a subset W is always a vector space in its own right. The universally quantified laws A1, A2, and S1 through S4 hold in W because they hold in the larger set V. And since there is *some* β in W, it follows that $0 = 0\beta$ is in W because W is closed under multiplication by scalars. For the same reason, if α is in W, then so is $-\alpha = (-1)\alpha$. Therefore, A3 and A4 also hold, and we see that W is a vector space. We have proved the following lemma.

Lemma. If W is a nonempty subset of a vector space V which is closed under the operations of V, then W is itself a vector space.

We call W a *subspace* of V. Thus $\mathcal{C}([a, b])$ is a subspace of $\mathbb{R}^{[a,b]}$, and the pairs $\langle x_1, x_2 \rangle$ such that $x_1 + x_2 = 0$ form a subspace of $\mathbb{R}^2$. Subspaces will be with us from now to the end.

A subspace of a vector space $\mathbb{R}^A$ is called a *function space*. In other words, a function space is a collection of real-valued functions on a common domain which is closed under addition and multiplication by scalars.

What we have defined so far ought to be called the notion of a *real* vector space or a vector space over $\mathbb{R}$. There is an analogous notion of a *complex* vector space, for which the scalars are the complex numbers. Then laws S1 through S4 refer to multiplication by *complex* numbers, and the space $\mathbb{C}^A$ of all complex-valued functions on A is the standard example. In fact, if the reader knew what is meant by a field F, we could give a single general definition of a vector space over F, where scalar multiplication is by the elements of F, and the standard example is the space $V = F^A$ of all functions from A to F. Throughout this book it will be understood that a vector space is a real vector space unless explicitly stated otherwise. However, much of the analysis holds as well for complex vector spaces, and most of the pure algebra is valid for any scalar field F.

EXERCISES

1.1 Sketch the geometric figure representing law S3,

$$x(\overrightarrow{OA} + \overrightarrow{OB}) = x(\overrightarrow{OA}) + x(\overrightarrow{OB}),$$

for geometric vectors. Assume that $x > 1$.

1.2 Prove S3 for $\mathbb{R}^3$ using the explicit displayed form $\{x_1, x_2, x_3\}$ for ordered triples.

1.3 The vector 0 postulated in A3 is unique, as elementary algebraic fiddling will show. For suppose that $0'$ also satisfies A3. Then

$$\begin{aligned} 0' &= 0' + 0 && \text{(A3 for 0)} \\ &= 0 + 0' && \text{(A2)} \\ &= 0 && \text{(A3 for } 0'\text{).} \end{aligned}$$

Show by similar algebraic juggling that, given α, the β postulated in A4 is unique. This unique β is designated $-\alpha$.

1.4 Prove similarly that $0\alpha = 0$, $x0 = 0$, and $(-1)\alpha = -\alpha$.

1.5 Prove that if $x\alpha = 0$, then either $x = 0$ or $\alpha = 0$.

1.6 Prove S1 for a function space $\mathbb{R}^A$. Prove S3.

1.7 Given that α is any vector in a vector space V, show that the set $\{x\alpha : x \in \mathbb{R}\}$ of all scalar multiples of α is a subspace of V.

1.8 Given that α and β are any two vectors in V, show that the set of all vectors $x\alpha + y\beta$, where x and y are any real numbers, is a subspace of V.

1.9 Show that the set of triples $\mathbf{x}$ in $\mathbb{R}^3$ such that $x_1 - x_2 + 2x_3 = 0$ is a subspace M. If N is the similar subspace $\{\mathbf{x} : x_1 + x_2 + x_3 = 0\}$, find a nonzero vector $\mathbf{a}$ in $M \cap N$. Show that $M \cap N$ is the set $\{x\mathbf{a} : x \in \mathbb{R}\}$ of all scalar multiples of $\mathbf{a}$.

1.10 Let A be the open interval $(0, 1)$, and let V be $\mathbb{R}^A$. Given a point x in $(0, 1)$, let V_x be the set of functions in V that have a derivative at x. Show that V_x is a subspace of V.

1.11 For any subsets A and B of a vector space V we define the set sum $A + B$ by $A + B = \{\alpha + \beta : \alpha \in A \text{ and } \beta \in B\}$. Show that $(A + B) + C = A + (B + C)$.

1.12 If $A \subset V$ and $X \subset \mathbb{R}$, we similarly define $XA = \{x\alpha : x \in X \text{ and } \alpha \in A\}$. Show that a nonvoid set A is a subspace if and only if $A + A = A$ and $\mathbb{R}A = A$.

1.13 Let V be $\mathbb{R}^2$, and let M be the line through the origin with slope k. Let $\mathbf{x}$ be any nonzero vector in M. Show that M is the subspace $\mathbb{R}\mathbf{x} = \{t\mathbf{x} : t \in \mathbb{R}\}$.

1.14 Show that any other line L with the same slope k is of the form $M + \mathbf{a}$ for some $\mathbf{a}$.

1.15 Let M be a subspace of a vector space V, and let α and β be any two vectors in V. Given $A = \alpha + M$ and $B = \beta + M$, show that either $A = B$ or $A \cap B = \emptyset$. Show also that $A + B = (\alpha + \beta) + M$.

1.16 State more carefully and prove what is meant by "a subspace of a subspace is a subspace".

1.17 Prove that the intersection of two subspaces of a vector space is always itself a subspace.

1.18 Prove more generally that the intersection $W = \bigcap_{i \in I} W_i$ of any family $\{W_i : i \in I\}$ of subspaces of V is a subspace of V.

1.19 Let V again be $\mathbb{R}^{(0,1)}$, and let W be the set of all functions f in V such that $f'(x)$ exists for every x in $(0, 1)$. Show that W is the intersection of the collection of subspaces of the form V_x that were considered in Exercise 1.10.

1.20 Let V be a function space $\mathbb{R}^A$, and for a point a in A let W_a be the set of functions such that $f(a) = 0$. W_a is clearly a subspace. For a subset $B \subset A$ let W_B be the set of functions f in V such that $f = 0$ on B. Show that W_B is the intersection $\bigcap_{a \in B} W_a$.

1.21 Supposing again that X and Y are subspaces of V, show that if $X + Y = V$ and $X \cap Y = \{0\}$, then for every vector ζ in V there is a unique pair of vectors $\xi \in X$ and $\eta \in Y$ such that $\zeta = \xi + \eta$.

1.22 Show that if X and Y are subspaces of a vector space V, then the union $X \cup Y$ can only be a subspace if either $X \subset Y$ or $Y \subset X$.

Linear combinations and linear span. Because of the commutative and associative laws for vector addition, the sum of a finite set of vectors is the same for all possible ways of adding them. For example, the sum of the three vectors $\alpha_a, \alpha_b, \alpha_c$ can be calculated in 12 ways, all of which give the same result:

$$(\alpha_a + \alpha_b) + \alpha_c = \alpha_c + (\alpha_a + \alpha_b) = (\alpha_c + \alpha_a) + \alpha_b = \alpha_b + (\alpha_c + \alpha_a), \quad \text{etc.}$$

Therefore, if $I = \{a, b, c\}$ is the set of indices used, the notation $\sum_{i \in I} \alpha_i$, which indicates the sum without telling us how we got it, is unambiguous. In general, for any finite indexed set of vectors $\{\alpha_i : i \in I\}$ there is a uniquely determined sum vector $\sum_{i \in I} \alpha_i$ which we can compute by ordering and grouping the α_i's in any way.

The index set I is often a block of integers $\overline{n} = \{1, \ldots, n\}$. In this case the vectors α_i form an n-tuple $\{\alpha_i\}_1^n$, and unless directed to do otherwise we would add them in their natural order and write the sum as $\sum_{i=1}^n \alpha_i$. Note that the way they are *grouped* is still left arbitrary.

Frequently, however, we have to use indexed sets that are not ordered. For example, the general polynomial of degree at most 5 in the two variables 's' and 't' is

$$\sum_{0 \leq i+j \leq 5} c_{ij} s^i t^j,$$

and the finite set of monomials $\{s^i t^j\}_{i+j \leq 5}$ has no natural order.

* The formal proof that the sum of a finite collection of vectors is independent of how we add them is by induction. We give it only for the interested reader.

In order to avoid looking silly, we begin the induction with two vectors, in which case the commutative law $\alpha_a + \alpha_b = \alpha_b + \alpha_a$ displays the identity of all possible sums. Suppose then that the assertion is true for index sets having fewer than n elements, and consider a collection $\{\alpha_i : i \in I\}$ having n members. Let β and γ be the sum of these vectors computed in two ways. In the computation of β there was a last addition performed, so that $\beta = (\sum_{i \in J_1} \alpha_i) + (\sum_{i \in J_2} \alpha_i)$, where $\{J_1, J_2\}$ partitions I and where we can write these two partial sums without showing how they were formed, since by our inductive hypothesis all possible ways of adding them give the same result.

Similarly, $\gamma = (\sum_{i \in K_1} \alpha_i) + (\sum_{i \in K_2} \alpha_i)$. Now set

$$L_{jk} = J_j \cap K_k \qquad \text{and} \qquad \xi_{jk} = \sum_{i \in L_{jk}} \alpha_i,$$

where it is understood that $\xi_{jk} = 0$ if L_{jk} is empty (see Exercise 1.37). Then $\sum_{i \in J_1} = \xi_{11} + \xi_{12}$ by the inductive hypothesis, and similarly for the other three sums. Thus

$$\beta = (\xi_{11} + \xi_{12}) + (\xi_{21} + \xi_{22}) = (\xi_{11} + \xi_{21}) + (\xi_{12} + \xi_{22}) = \gamma,$$

which completes our proof. *

A vector β is called a *linear combination* of a subset A of the vector space V if β is a finite sum $\sum x_i\alpha_i$, where the vectors α_i are all in A and the scalars x_i are arbitrary. Thus, if A is the subset $\{t^n\}_0^\infty \subset \mathbb{R}^{\mathbb{R}}$ of all "monomials", then a function f is a linear combination of the functions in A if and only if f is a polynomial function $f(t) = \sum_1^n c_i t^i$. If A is finite, it is often useful to take the indexed set $\{\alpha_i\}$ to be the whole of A, and to simply use a 0-coefficient for any vector missing from the sum. Thus, if A is the subset $\{\sin t, \cos t, e^t\}$ of $\mathbb{R}^{\mathbb{R}}$, then we can consider A an ordered triple in the listed ordering, and the function $3 \sin t - e^t = 3 \cdot \sin t + 0 \cdot \cos t + (-1)e^t$ is the linear combination of the triple A having the coefficient triple $\langle 3, 0, -1 \rangle$.

Consider now the set L of all linear combinations of the two vectors $\langle 1, 1, 1 \rangle$ and $\langle 0, 1, -1 \rangle$ in $\mathbb{R}^3$. It is the set of all vectors $s\langle 1, 1, 1 \rangle + t\langle 0, 1, -1 \rangle = \langle s, s+t, s-t \rangle$, where s and t are any real numbers. Thus $L = \{\langle s, s+t, s-t \rangle : \langle s, t \rangle \in \mathbb{R}^2\}$. It will be clear on inspection that L is closed under addition and scalar multiplication, and therefore is a subspace of $\mathbb{R}^3$. Also, L contains each of the two given vectors, with coefficient pairs $\langle 1, 0 \rangle$ and $\langle 0, 1 \rangle$, respectively. Finally, any subspace M of $\mathbb{R}^3$ which contains each of the two given vectors will also contain all of their linear combinations, and so will include L. That is, *L is the smallest subspace* of $\mathbb{R}^3$ containing $\langle 1, 1, 1 \rangle$ and $\langle 0, 1, -1 \rangle$. It is called the *linear span* of the two vectors, or the subspace *generated by* the two vectors. In general, we have the following theorem.

Theorem 1.1. If A is a nonempty subset of a vector space V, then the set $L(A)$ of all linear combinations of the vectors in A is a subspace, and it is the smallest subspace of V which includes the set A.

Proof. Suppose first that A is finite. We can assume that we have indexed A in some way, so that $A = \{\alpha_i : i \in I\}$ for some finite index set I, and every element of $L(A)$ is of the form $\sum_{i \in I} x_i \alpha_i$. Then we have

$$\left(\sum x_i \alpha_i\right) + \left(\sum y_i \alpha_i\right) = \sum (x_i + y_i) \alpha_i$$

because the left-hand side becomes $\sum_i (x_i \alpha_i + y_i \alpha_i)$ when it is regrouped by pairs, and then S2 gives the right-hand side. We also have

$$c\left(\sum x_i \alpha_i\right) = \sum (c x_i) \alpha_i$$

by S3 and mathematical induction. Thus $L(A)$ is closed under addition and multiplication by scalars and hence is a subspace. Moreover, $L(A)$ contains each α_i (why?) and so includes A. Finally, if a subspace W includes A, then it contains each linear combination $\sum x_i \alpha_i$, so it includes $L(A)$. Therefore, $L(A)$ can be directly characterized as the uniquely determined smallest subspace which includes the set A.

If A is infinite, we obviously can't use a single finite listing. However, the sum $(\sum_1^n x_i \alpha_i) + (\sum_1^m y_j \beta_j)$ of two linear combinations of elements of A is clearly a finite sum of scalars times elements of A. If we wish, we can rewrite it as $\sum_1^{n+m} x_i \alpha_i$, where we have set $\beta_j = \alpha_{n+j}$ and $y_j = x_{n+j}$ for $j = 1, \ldots, m$. In any case, $L(A)$ is again closed under addition and multiplication by scalars and so is a subspace. □

We call $L(A)$ the *linear span* of A. If $L(A) = V$, we say that A *spans* V; V is *finite-dimensional* if it has a finite spanning set.

If $V = \mathbb{R}^3$, and if δ^1, δ^2, and δ^3 are the "unit points on the axes", $\delta^1 = \langle 1, 0, 0 \rangle$, $\delta^2 = \langle 0, 1, 0 \rangle$, and $\delta^3 = \langle 0, 0, 1 \rangle$, then $\{\delta^i\}_1^3$ spans V, since $\mathbf{x} = \langle x_1, x_2, x_3 \rangle = \langle x_1, 0, 0 \rangle + \langle 0, x_2, 0 \rangle + \langle 0, 0, x_3 \rangle = x_1 \delta^1 + x_2 \delta^2 + x_3 \delta^3 = \sum_1^3 x_i \delta^i$ for every $\mathbf{x}$ in $\mathbb{R}^3$. More generally, if $V = \mathbb{R}^n$ and δ^j is the n-tuple having 1 in the jth place and 0 elsewhere, then we have similarly that $\mathbf{x} = \langle x_1, \ldots, x_n \rangle = \sum_{i=1}^n x_i \delta^i$, so that $\{\delta^i\}_1^n$ spans $\mathbb{R}^n$. Thus $\mathbb{R}^n$ is finite-dimensional. In general, a function space on an infinite set A will not be finite-dimensional. For example, it is true but not obvious that $\mathcal{C}([a, b])$ has no finite spanning set.

EXERCISES

1.23 Given $\alpha = \langle 1, 1, 1 \rangle$, $\beta = \langle 0, 1, -1 \rangle$, $\gamma = \langle 2, 0, 1 \rangle$, compute the linear combinations $\alpha + \beta + \gamma$, $3\alpha - 2\beta + \gamma$, $x\alpha + y\beta + z\gamma$. Find x, y, and z such that $x\alpha + y\beta + z\gamma = \langle 0, 0, 1 \rangle = \delta^3$. Do the same for δ^1 and δ^2.

1.24 Given $\alpha = \langle 1, 1, 1 \rangle$, $\beta = \langle 0, 1, -1 \rangle$, $\gamma = \langle 1, 0, 2 \rangle$, show that each of α, β, γ is a linear combination of the other two. Show that it is impossible to find coefficients x, y, and z such that $x\alpha + y\beta + z\gamma = \delta^1$.

1.25 a) Find the linear combination of the set $A = <t, t^2 - 1, t^2 + 1>$ with coefficient triple $<2, -1, 1>$. Do the same for $<0, 1, 1>$.

b) Find the coefficient triple for which the linear combination of the triple A is $(t + 1)^2$. Do the same for 1.

c) Show in fact that any polynomial of degree ≤ 2 is a linear combination of A.

1.26 Find the linear combination f of $\{e^t, e^{-t}\} \subset \mathbb{R}^{\mathbb{R}}$ such that $f(0) = 1$ and $f'(0) = 2$.

1.27 Find a linear combination f of $\sin x$, $\cos x$, and e^x such that $f(0) = 0, f'(0) = 1$, and $f''(0) = 1$.

1.28 Suppose that $a \sin x + b \cos x + ce^x$ is the zero function. Prove that $a = b = c = 0$.

1.29 Prove that $<1, 1>$ and $<1, 2>$ span $\mathbb{R}^2$.

1.30 Show that the subspace $M = \{\mathbf{x} : x_1 + x_2 = 0\} \subset \mathbb{R}^2$ is spanned by one vector.

1.31 Let M be the subspace $\{\mathbf{x} : x_1 - x_2 + 2x_3 = 0\}$ in $\mathbb{R}^3$. Find two vectors $\mathbf{a}$ and $\mathbf{b}$ in M neither of which is a scalar multiple of the other. Then show that M is the linear span of $\mathbf{a}$ and $\mathbf{b}$.

1.32 Find the intersection of the linear span of $<1, 1, 1>$ and $<0, 1, -1>$ in $\mathbb{R}^3$ with the coordinate subspace $x_2 = 0$. Exhibit this intersection as a linear span.

1.33 Do the above exercise with the coordinate space replaced by

$$M = \{\mathbf{x} : x_1 + x_2 = 0\}.$$

1.34 By Theorem 1.1 the linear span $L(A)$ of an arbitrary subset A of a vector space V has the following two properties:

i) $L(A)$ is a subspace of V which includes A;

ii) If M is any subspace which includes A, then $L(A) \subset M$.

Using only (i) and (ii), show that

a) $A \subset B \Rightarrow L(A) \subset L(B)$;

b) $L(L(A)) = L(A)$.

1.35 Show that

a) if M and N are subspaces of V, then so is $M + N$;

b) for any subsets $A, B \subset V$, $L(A \cup B) = L(A) + L(B)$.

1.36 Remembering (Exercise 1.18) that the intersection of any family of subspaces is a subspace, show that the linear span $L(A)$ of a subset A of a vector space V is the intersection of all the subspaces of V that include A. This alternative characterization is sometimes taken as the definition of linear span.

1.37 By convention, the sum of an empty set of vectors is taken to be the zero vector. This is necessary if Theorem 1.1 is to be strictly correct. Why? What about the preceding problem?

Linear transformations. The general function space $\mathbb{R}^A$ and the subspace $\mathcal{C}([a, b])$ of $\mathbb{R}^{[a,b]}$ both have the property that in addition to being closed under the vector operations, they are also closed under the operation of multiplication of two functions. That is, the pointwise product of two functions is again a function $[(fg)(a) = f(a)g(a)]$, and the product of two continuous functions is continuous. With respect to these three operations, addition, multiplication, and scalar multiplication, $\mathbb{R}^A$ and $\mathcal{C}([a, b])$ are examples of *algebras*. If the reader noticed this extra operation, he may have wondered why, at least in the context of function spaces, we bother with the notion of vector space. Why not study all three operations? The answer is that the vector operations are exactly the operations that are "preserved" by many of the most important mappings of sets of functions. For example, define $T: \mathcal{C}([a, b]) \to \mathbb{R}$ by $T(f) = \int_a^b f(t)\, dt$. Then the laws of the integral calculus say that $T(f + g) = T(f) + T(g)$ and $T(cf) = cT(f)$. Thus T "preserves" the vector operations. Or we can say that T "commutes" with the vector operations, since plus followed by T equals T followed by plus. However, T does not preserve multiplication: it is *not* true in general that $T(fg) = T(f)T(g)$.

Another example is the mapping $T: \mathbf{x} \mapsto \mathbf{y}$ from $\mathbb{R}^3$ to $\mathbb{R}^2$ defined by $y_1 = 2x_1 - x_2 + x_3$, $y_2 = x_1 + 3x_2 - 5x_3$, for which we can again verify that $T(\mathbf{x} + \mathbf{y}) = T(\mathbf{x}) + T(\mathbf{y})$ and $T(c\mathbf{x}) = cT(\mathbf{x})$. The theory of the solvability of systems of linear equations is essentially the theory of such mappings T; thus we have another important type of mapping that preserves the vector operations (but not products).

These remarks suggest that we study vector spaces in part so that we can study mappings which preserve the vector operations. Such mappings are called linear transformations.

Definition. If V and W are vector spaces, then a mapping $T: V \to W$ is a *linear transformation* or a *linear map* if $T(\alpha + \beta) = T(\alpha) + T(\beta)$ for all $\alpha, \beta \in V$, and $T(x\alpha) = xT(\alpha)$ for all $\alpha \in V$, $x \in \mathbb{R}$.

These two conditions on T can be combined into the single equation

$$T(x\alpha + y\beta) = xT(\alpha) + yT(\beta) \qquad \text{for all} \quad \alpha, \beta \in V \quad \text{and all} \quad x, y \in \mathbb{R}.$$

Moreover, this equation can be extended to any finite sum by induction, so that if T is linear, then

$$T\left(\sum_{i \in I} x_i \alpha_i\right) = \sum_{i \in I} x_i T(\alpha_i)$$

for any linear combination $\sum x_i \alpha_i$. For example, $\int_a^b \left(\sum_1^n c_i f_i\right) = \sum_1^n c_i \int_a^b f_i$.

EXERCISES

1.38 Show that the most general linear map from $\mathbb{R}$ to $\mathbb{R}$ is multiplication by a constant.

1.39 For a fixed α in V the mapping $x \mapsto x\alpha$ from $\mathbb{R}$ to V is linear. Why?

1.40 Why is this true for $\alpha \mapsto x\alpha$ when x is fixed?

1.41 Show that every linear mapping from $\mathbb{R}$ to V is of the form $x \mapsto x\alpha$ for a fixed vector α in V.

1.42 Show that every linear mapping from $\mathbb{R}^2$ to V is of the form $\langle x_1, x_2 \rangle \mapsto x_1\alpha_1 + x_2\alpha_2$ for a fixed pair of vectors α_1 and α_2 in V. What is the range of this mapping?

1.43 Show that the map $f \mapsto \int_a^b f(t)\,dt$ from $\mathcal{C}([a, b])$ to $\mathbb{R}$ does not preserve products.

1.44 Let g be any fixed function in $\mathbb{R}^A$. Prove that the mapping $T: \mathbb{R}^A \to \mathbb{R}^A$ defined by $T(f) = gf$ is linear.

1.45 Let φ be any mapping from a set A to a set B. Show that composition by φ is a linear mapping from $\mathbb{R}^B$ to $\mathbb{R}^A$. That is, show that $T: \mathbb{R}^B \to \mathbb{R}^A$ defined by $T(f) = f \circ \varphi$ is linear.

In order to acquire a supply of examples, we shall find all linear transformations having $\mathbb{R}^n$ as domain space. It may be well to start by looking at one such transformation. Suppose we choose some fixed triple of functions $\{f_i\}_1^3$ in the space $\mathbb{R}^{\mathbb{R}}$ of all real-valued functions on $\mathbb{R}$, say $f_1(t) = \sin t$, $f_2(t) = \cos t$, and $f_3(t) = e^t = \exp(t)$. Then for each triple of numbers $\mathbf{x} = \{x_i\}_1^3$ in $\mathbb{R}^3$ we have the linear combination $\sum_{i=1}^3 x_i f_i$ with $\{x_i\}$ as coefficients. This is the element of $\mathbb{R}^{\mathbb{R}}$ whose value at t is $\sum_1^3 x_i f_i(t) = x_1 \sin t + x_2 \cos t + x_3 e^t$. Different coefficient triples give different functions, and the mapping $\mathbf{x} \mapsto \sum_{i=1}^3 x_i f_i = x_1 \sin + x_2 \cos + x_3 \exp$ is thus a mapping from $\mathbb{R}^3$ to $\mathbb{R}^{\mathbb{R}}$. It is clearly linear. If we call this mapping T, then we can recover the determining triple of functions from T as the images of the "unit points" δ^i in $\mathbb{R}^3$; $T(\delta^j) = \sum \delta_i^j f_i = f_j$, and so $T(\delta^1) = \sin$, $T(\delta^2) = \cos$, and $T(\delta^3) = \exp$. We are going to see that *every* linear mapping from $\mathbb{R}^3$ to $\mathbb{R}^{\mathbb{R}}$ is of this form.

In the following theorem $\{\delta^i\}_1^n$ is the spanning set for $\mathbb{R}^n$ that we defined earlier, so that $\mathbf{x} = \sum_1^n x_i \delta^i$ for every n-tuple $\mathbf{x} = \langle x_1, \ldots, x_n \rangle$ in $\mathbb{R}^n$.

Theorem 1.2. If $\{\beta_j\}_1^n$ is any fixed n-tuple of vectors in a vector space W, then the "linear combination mapping" $\mathbf{x} \mapsto \sum_1^n x_i\beta_i$ is a linear transformation T from $\mathbb{R}^n$ to W, and $T(\delta^j) = \beta_j$ for $j = 1, \ldots, n$. Conversely, if T is any linear mapping from $\mathbb{R}^n$ to W, and if we set $\beta_j = T(\delta^j)$ for $j = 1, \ldots, n$, then T is the linear combination mapping $\mathbf{x} \mapsto \sum_1^n x_i\beta_i$.

Proof. The linearity of the linear combination map T follows by exactly the same argument that we used in Theorem 1.1 to show that $L(A)$ is a subspace. Thus

$$T(\mathbf{x} + \mathbf{y}) = \sum_1^n (x_i + y_i)\beta_i = \sum_1^n (x_i\beta_i + y_i\beta_i)$$
$$= \sum_1^n x_i\beta_i + \sum_1^n y_i\beta_i = T(\mathbf{x}) + T(\mathbf{y}),$$

and

$$T(s\mathbf{x}) = \sum_1^n (sx_i)\beta_i = \sum_1^n s(x_i\beta_i) = s\sum_1^n x_i\beta_i = sT(\mathbf{x}).$$

Also $T(\delta^j) = \sum_{i=1}^n \delta_i^j\beta_i = \beta_j$, since $\delta_j^j = 1$ and $\delta_i^j = 0$ for $i \neq j$.

Conversely, if $T: \mathbb{R}^n \to W$ is linear, and if we set $\beta_j = T(\delta^j)$ for all j, then for any $\mathbf{x} = \langle x_1, \ldots, x_n \rangle$ in $\mathbb{R}^n$ we have $T(\mathbf{x}) = T(\sum_1^n x_i \delta^i) = \sum_1^n x_i T(\delta^i) = \sum_1^n x_i\beta_i$. Thus T is the mapping $\mathbf{x} \mapsto \sum_1^n x_i\beta_i$. □

This is a tremendously important theorem, simple though it may seem, and the reader is urged to fix it in his mind. To this end we shall invent some terminology that we shall stay with for the first three chapters. If $\boldsymbol{\alpha} = \{\alpha_1, \ldots, \alpha_n\}$ is an n-tuple of vectors in a vector space W, let $L_{\boldsymbol{\alpha}}$ be the corresponding linear combination mapping $\mathbf{x} \mapsto \sum_1^n x_i\alpha_i$ from $\mathbb{R}^n$ to W. Note that the n-tuple $\boldsymbol{\alpha}$ itself is an element of W^n. If T is any linear mapping from $\mathbb{R}^n$ to W, we shall call the n-tuple $\{T(\delta^i)\}_1^n$ the *skeleton* of T. In these terms the theorem can be restated as follows.

Theorem 1.2′. For each n-tuple $\boldsymbol{\alpha}$ in W^n, the map $L_{\boldsymbol{\alpha}}: \mathbb{R}^n \to W$ is linear and its skeleton is $\boldsymbol{\alpha}$. Conversely, if T is any linear map from $\mathbb{R}^n$ to W, then $T = L_{\boldsymbol{\beta}}$ where $\boldsymbol{\beta}$ is the skeleton of T.

Or again:

Theorem 1.2″. The map $\boldsymbol{\alpha} \mapsto L_{\boldsymbol{\alpha}}$ is a bijection from W^n to the set of all linear maps T from $\mathbb{R}^n$ to W, and $T \mapsto$ skeleton (T) is its inverse.

A linear transformation from a vector space V to the scalar field $\mathbb{R}$ is called a *linear functional* on V. Thus $f \mapsto \int_a^b f(t)\,dt$ is a linear functional on $V = \mathcal{C}([a, b])$. The above theorem is particularly simple for a linear functional F: since $W = \mathbb{R}$, each vector $\beta_i = F(\delta^i)$ in the skeleton of F is simply a number b_i, and the skeleton $\{b_i\}_1^n$ is thus an element of $\mathbb{R}^n$. In this case we would write $F(\mathbf{x}) = \sum_1^n b_i x_i$, putting the numerical coefficient 'b_i' before the variable 'x_i'. Thus $F(\mathbf{x}) = 3x_1 - x_2 + 4x_3$ is the linear functional on $\mathbb{R}^3$ with skeleton $\langle 3, -1, 4 \rangle$. The set of all linear functionals on $\mathbb{R}^n$ is in a natural one-to-one correspondence with $\mathbb{R}^n$ itself; we get $\mathbf{b}$ from F by $b_i = F(\delta^i)$ for all i, and we get F from $\mathbf{b}$ by $F(\mathbf{x}) = \sum b_i x_i$ for all $\mathbf{x}$ in $\mathbb{R}^n$.

We next consider the case where the codomain space of T is a Cartesian space $\mathbb{R}^m$, and in order to keep the two spaces clear in our minds, we shall, for the moment, take the domain space to be $\mathbb{R}^3$. Each vector $\beta_i = T(\delta^i)$ in the skeleton of T is now an m-tuple of numbers. If we picture this m-tuple as a column of numbers, then the three m-tuples β_i can be pictured as a *rectangular array* of numbers, consisting of three columns each of m numbers. Let t_{ij} be the ith number in the jth column. Then the doubly indexed set of numbers $\{t_{ij}\}$ is called the *matrix* of the transformation T. We call it an m-by-3 (an $m \times 3$) matrix because the pictured rectangular array has m rows and three columns. The matrix determines T uniquely, since its columns form the skeleton of T. The identity $T(\mathbf{x}) = \sum_1^3 x_j T(\delta^j) = \sum_1^3 x_j\beta_j$ allows the m-tuple $T(\mathbf{x})$ to be calculated explicitly from $\mathbf{x}$ and the matrix $\{t_{ij}\}$. Picture multiplying the column m-tuple β_j by the scalar x_j and then adding across the three columns at the ith row, as below:

$$\underset{\mathbf{y}}{\begin{pmatrix} \vdots \\ y_i \\ \vdots \end{pmatrix}} = x_1 \underset{\beta_1}{\begin{pmatrix} \vdots \\ t_{i1} \\ \vdots \end{pmatrix}} + x_2 \underset{\beta_2}{\begin{pmatrix} \vdots \\ t_{i2} \\ \vdots \end{pmatrix}} + x_3 \underset{\beta_3}{\begin{pmatrix} \vdots \\ t_{i3} \\ \vdots \end{pmatrix}}.$$

Since t_{ij} is the ith number in the m-tuple β_j, the ith number in the m-tuple $\sum_{j=1}^3 x_j\beta_j$ is $\sum_{j=1}^3 x_j t_{ij}$. That is, if we let $\mathbf{y}$ be the m-tuple $T(\mathbf{x})$, then

$$y_i = \sum_{j=1}^{3} t_{ij}x_j \qquad \text{for} \quad i = 1, \ldots, m,$$

and this set of m scalar equations is equivalent to the one-vector equation $\mathbf{y} = T(\mathbf{x})$.

We can now replace three by n in the above discussion without changing anything except the diagram, and thus obtain the following specialization of Theorem 1.2.

Theorem 1.3. Every linear mapping T from $\mathbb{R}^n$ to $\mathbb{R}^m$ determines the $m \times n$ matrix $\mathbf{t} = \{t_{ij}\}$ having the skeleton of T as its columns, and the expression of the equation $\mathbf{y} = T(\mathbf{x})$ in linear combination form is equivalent to the m scalar equations

$$y_i = \sum_{j=1}^{n} t_{ij}x_j \qquad \text{for} \quad i = 1, \ldots, m.$$

Conversely, each $m \times n$ matrix $\mathbf{t}$ determines the linear combination mapping having the columns of $\mathbf{t}$ as its skeleton, and the mapping $\mathbf{t} \mapsto T$ is therefore a bijection from the set of all $m \times n$ matrices to the set of all linear maps from $\mathbb{R}^n$ to $\mathbb{R}^m$.

A linear functional F on $\mathbb{R}^n$ is a linear mapping from $\mathbb{R}^n$ to $\mathbb{R}^1$, so it must be expressed by a $1 \times n$ matrix. That is, the n-tuple $\mathbf{b}$ in $\mathbb{R}^n$ which is the skeleton of F is viewed as a matrix of one row and n columns.

As a final example of linear maps, we look at an important class of special linear functionals defined on any function space, the so-called *coordinate functionals*. If $V = \mathbb{R}^I$ and $i \in I$, then the ith coordinate functional π_i is simply *evaluation at* i, so that $\pi_i(f) = f(i)$. These functionals are obviously linear. In fact, *the vector operations on functions were defined to make them linear;* since $sf + tg$ is defined to be that function whose value at i is $sf(i) + tg(i)$ for all i, we see that $sf + tg$ is by definition that function such that $\pi_i(sf + tg) = s\pi_i(f) + t\pi_i(g)$ for all i!

If V is $\mathbb{R}^n$, then π_j is the mapping $\mathbf{x} = \langle x_1, \ldots, x_n \rangle \mapsto x_j$. In this case we know from the theorem that π_j must be of the form $\pi_j(\mathbf{x}) = \sum_1^n b_i x_i$ for some n-tuple $\mathbf{b}$. What is $\mathbf{b}$?

The general form of the linearity property, $T(\sum x_i\alpha_i) = \sum x_i T(\alpha_i)$, shows that T and T^{-1} both carry subspaces into subspaces.

Theorem 1.4. If $T: V \to W$ is linear, then the T-image of the linear span of any subset $A \subset V$ is the linear span of the T-image of A: $T[L(A)] = L(T[A])$. In particular, if A is a subspace, then so is $T[A]$. Furthermore, if Y is a subspace of W, then $T^{-1}[Y]$ is a subspace of V.

Proof. According to the formula $T(\sum x_i\alpha_i) = \sum x_i T(\alpha_i)$, a vector in W is the T-image of a linear combination on A if and only if it is a linear combination on $T[A]$. That is, $T[L(A)] = L(T[A])$. If A is a subspace, then $A = L(A)$ and $T[A] = L(T[A])$, a subspace of W. Finally, if Y is a subspace of W and $\{\alpha_i\} \subset T^{-1}[Y]$, then $T(\sum x_i\alpha_i) = \sum x_i T(\alpha_i) \in L(Y) = Y$. Thus $\sum x_i\alpha_i \in T^{-1}[Y]$ and $T^{-1}[Y]$ is its own linear span. □

The subspace $T^{-1}(0) = \{\alpha \in V : T(\alpha) = 0\}$ is called the *null space, or kernel*, of T, and is designated $N(T)$ or $\mathfrak{N}(T)$. The range of T is the subspace $T[V]$ of W. It is designated $R(T)$ or $\mathfrak{R}(T)$.

Lemma 1.1. A linear mapping T is injective if and only if its null space is $\{0\}$.

Proof. If T is injective, and if $\alpha \neq 0$, then $T(\alpha) \neq T(0) = 0$ and the null space accordingly contains only 0. On the other hand, if $N(T) = \{0\}$, then whenever $\alpha \neq \beta$, we have $\alpha - \beta \neq 0$, $T(\alpha) - T(\beta) = T(\alpha - \beta) \neq 0$, and $T(\alpha) \neq T(\beta)$; this shows that T is injective. □

A linear map $T: V \to W$ which is bijective is called an *isomorphism*. Two vector spaces V and W are *isomorphic* if and only if there exists an isomorphism between them.

For example, the map $\langle c_1, \ldots, c_n \rangle \to \sum_0^{n-1} c_{i+1}x^i$ is an isomorphism of $\mathbb{R}^n$ with the vector space of all polynomials of degree $< n$.

Isomorphic spaces "have the same form", and are identical as abstract vector spaces. That is, they cannot be distinguished from each other solely on the basis of vector properties which they do or do not have.

When a linear transformation is from V to itself, special things can happen. One possibility is that T can map a vector α essentially to itself, $T(\alpha) = x\alpha$ for some x in $\mathbb{R}$. In this case α is called an *eigenvector* (proper vector, characteristic vector), and x is the corresponding *eigenvalue*.

EXERCISES

1.46 In the situation of Exercise 1.45, show that T is an isomorphism if φ is bijective by showing that

a) φ injective $\Rightarrow T$ surjective,
b) φ surjective $\Rightarrow T$ injective.

1.47 Find the linear functional l on $\mathbb{R}^2$ such that $l(\langle 1, 1\rangle) = 0$ and $l(\langle 1, 2\rangle) = 1$. That is, find $\mathbf{b} = \langle b_1, b_2\rangle$ in $\mathbb{R}^2$ such that l is the linear combination map

$$\mathbf{x} \to b_1x_1 + b_2x_2.$$

1.48 Do the same for $l(\langle 2, 1\rangle) = -3$ and $l(\langle 1, 2\rangle) = 4$.

1.49 Find the linear $T: \mathbb{R}^2 \to \mathbb{R}^{\mathbb{R}}$ such that $T(\langle 1, 1\rangle) = t^2$ and $T(\langle 1, 2\rangle) = t^3$. That is, find the functions $f_1(t)$ and $f_2(t)$ such that T is the linear combination map $\mathbf{x} \to x_1f_1 + x_2f_2$.

1.50 Let T be the linear map from $\mathbb{R}^2$ to $\mathbb{R}^3$ such that $T(\delta^1) = \langle 2, -1, 1\rangle$, $T(\delta^2) = \langle 1, 0, 3\rangle$. Write down the matrix of T in standard rectangular form. Determine whether or not δ^1 is in the range of T.

1.51 Let T be the linear map from $\mathbb{R}^3$ to $\mathbb{R}^3$ whose matrix is

$$\begin{bmatrix} 1 & 2 & 3 \\ 2 & 0 & -1 \\ 3 & -1 & 1 \end{bmatrix}.$$

Find $T(\mathbf{x})$ when $\mathbf{x} = \langle 1, 1, 0\rangle$; do the same for $\mathbf{x} = \langle 3, -2, 1\rangle$.

1.52 Let M be the linear span of $\langle 1, -1, 0\rangle$ and $\langle 0, 1, 1\rangle$. Find the subspace $T[M]$ by finding two vectors spanning it, where T is as in the above exercise.

1.53 Let T be the map $\langle x, y\rangle \to \langle x + 2y, y\rangle$ from $\mathbb{R}^2$ to itself. Show that T is a linear combination mapping, and write down its matrix in standard form.

1.54 Do the same for $T: \langle x, y, z\rangle \to \langle x - z, x + z, y\rangle$ from $\mathbb{R}^3$ to itself.

1.55 Find a linear transformation T from $\mathbb{R}^3$ to itself whose range space is the span of $\langle 1, -1, 0\rangle$ and $\langle -1, 0, 2\rangle$.

1.56 Find two linear functionals on $\mathbb{R}^4$ the intersection of whose null spaces is the linear span of $\langle 1, 1, 1, 1\rangle$ and $\langle 1, 0, -1, 0\rangle$. You now have in hand a linear *transformation* whose null space is the above span. What is it?

1.57 Let $V = \mathcal{C}([a, b])$ be the space of continuous real-valued functions on $[a, b]$, also designated $\mathcal{C}^0([a, b])$, and let $W = \mathcal{C}^1([a, b])$ be those having continuous first derivatives. Let $D: W \to V$ be differentiation ($Df \doteq f'$), and define T on V by $T(f) = F$, where $F(x) = \int_a^x f(t)\, dt$. By stating appropriate theorems of the calculus, show that D and T are linear, T maps into W, and D is a left inverse of T ($D \circ T$ is the identity on V).

1.58 In the above exercise, identify the range of T and the null space of D. We know that D is surjective and that T is injective. Why?

1.59 Let V be the linear span of the functions $\sin x$ and $\cos x$. Then the operation of differentiation D is a linear transformation from V to V. Prove that D is an isomorphism from V to V. Show that $D^2 = -I$ on V.

1.60 a) As the reader would guess, $\mathcal{C}^3(\mathbb{R})$ is the set of real-valued functions on $\mathbb{R}$ having continuous derivatives up to and including the third. Show that $f \to f'''$ is a surjective linear map T from $\mathcal{C}^3(\mathbb{R})$ to $\mathcal{C}(\mathbb{R})$.

b) For any fixed a in $\mathbb{R}$ show that $f \to \langle f(a), f'(a), f''(a)\rangle$ is an isomorphism from the null space $N(T)$ to $\mathbb{R}^3$. [*Hint:* Apply Taylor's formula with remainder.]

1.61 An integral analogue of the matrix equations $y_i = \sum_j t_{ij}x_j$, $i = 1, \ldots, m$, is the equation

$$g(s) = \int_0^1 K(s, t)f(t)\,dt, \qquad s \in [0, 1].$$

Assuming that $K(s, t)$ is defined on the square $[0, 1] \times [0, 1]$ and is continuous as a function of t for each s, check that $f \to g$ is a linear mapping from $\mathcal{C}([0, 1])$ to $\mathbb{R}^{[0,1]}$.

1.62 For a finite set $A = \{\alpha_i\}$, Theorem 1.1 is a corollary of Theorem 1.4. Why?

1.63 Show that the inverse of an isomorphism is linear (and hence is an isomorphism).

1.64 Find the eigenvectors and eigenvalues of $T: \mathbb{R}^2 \to \mathbb{R}^2$ if the matrix of T is

$$\begin{bmatrix} 1 & -1 \\ -2 & 0 \end{bmatrix}.$$

Since every scalar multiple $x\alpha$ of an eigenvector α is clearly also an eigenvector, it will suffice to find one vector in each "eigendirection". This is a problem in elementary algebra.

1.65 Find the eigenvectors and eigenvalues of the transformations T whose matrices are

$$\begin{bmatrix} 1 & 1 \\ -2 & 0 \end{bmatrix}, \quad \begin{bmatrix} -1 & -1 \\ -1 & -1 \end{bmatrix}, \quad \begin{bmatrix} 1 & -1 \\ -2 & 2 \end{bmatrix}, \quad \begin{bmatrix} 2 & 1 \\ -4 & -2 \end{bmatrix}.$$

1.66 The five transformations in the above two exercises exhibit four different kinds of behavior according to the number of distinct eigendirections they have. What are the possibilities?

1.67 Let V be the vector space of polynomials of degree ≤ 3 and define $T: V \to V$ by $f \to tf'(t)$. Find the eigenvectors and eigenvalues of T.

2. VECTOR SPACES AND GEOMETRY

The familiar coordinate systems of analytic geometry allow us to consider geometric entities such as lines and planes in vector settings, and these geometric notions give us valuable intuitions about vector spaces. Before looking at the vector forms of these geometric ideas, we shall briefly review the construction of the coordinate correspondence for three-dimensional Euclidean space. As usual, the confident reader can skip it.

We start with the line. A coordinate correspondence between a line L and the real number system $\mathbb{R}$ is determined by choosing arbitrarily on L a zero point O and a unit point Q distinct from O. Then to each point X on L is assigned the number x such that $|x|$ is the distance from O to X, measured in terms of the segment OQ as unit, and x is positive or negative according as X and Q are on the same side of O or on opposite sides. The mapping $X \mapsto x$ is the coordinate correspondence. Now consider three-dimensional Euclidean space $\mathbb{E}^3$. We want to set up a coordinate correspondence between $\mathbb{E}^3$ and the Cartesian vector space $\mathbb{R}^3$. We first choose arbitrarily a zero point O and three unit points Q_1, Q_2, and Q_3 in such a way that the four points do not lie in a plane. Each of the unit points Q_i determines a line L_i through O and a coordinate correspondence on this line, as defined above. The three lines L_1, L_2, and L_3 are called the coordinate axes. Consider now any point X in $\mathbb{E}^3$. The plane through X parallel to L_2 and L_3 intersects L_1 at a point X_1, and therefore determines a number x_1, the coordinate of X_1 on L_1. In a similar way, X determines points X_2 on L_2 and X_3 on L_3 which have coordinates x_2 and x_3, respectively. Altogether X determines a triple

$$\mathbf{x} = \langle x_1, x_2, x_3 \rangle$$

in $\mathbb{R}^3$, and we have thus defined a mapping $\theta: X \mapsto \mathbf{x}$ from $\mathbb{E}^3$ to $\mathbb{R}^3$ (see Fig. 1.4). We call θ the coordinate correspondence defined by the axis system. The convention implicit in our notation above is that $\theta(Y)$ is $\mathbf{y}$, $\theta(A)$ is $\mathbf{a}$, etc. Note that the unit point Q_1 on L_1 has the coordinate triple $\delta^1 = \langle 1, 0, 0 \rangle$, and similarly, that

$$\theta(Q_2) = \delta^2 = \langle 0, 1, 0 \rangle$$

and

$$\theta(Q_3) = \delta^3 = \langle 0, 0, 1 \rangle.$$

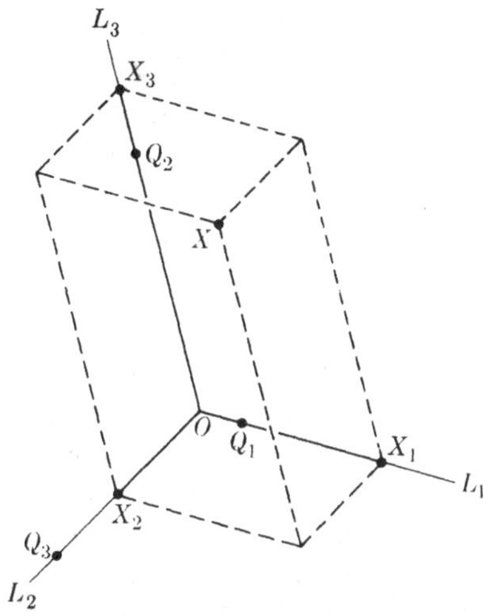

Fig. 1.4

There are certain basic facts about the coordinate correspondence that have to be proved as theorems of geometry before the correspondence can be used to treat geometric questions algebraically. These geometric theorems are quite tricky, and are almost impossible to discuss adequately on the basis of the usual secondary school treatment of geometry. We shall therefore simply assume them. They are:

1) θ is a bijection from $\mathbb{E}^3$ to $\mathbb{R}^3$.

2) Two line segments AB and XY are equal in length and parallel, and the direction from A to B is the same as that from X to Y if and only if $\mathbf{b} - \mathbf{a} = \mathbf{y} - \mathbf{x}$ (in the vector space $\mathbb{R}^3$). This relationship between line segments is important enough to formalize. A *directed* line segment is a geometric line segment, together with a choice of one of the two directions along it. If we interpret AB as the directed line segment from A to B, and if we define the directed line segments AB and XY to be *equivalent* (and write $AB \sim XY$) if they are equal in length, parallel, and similarly directed, then (2) can be restated:

$$AB \sim XY \Leftrightarrow \mathbf{b} - \mathbf{a} = \mathbf{y} - \mathbf{x}.$$

3) If $X \neq O$, then Y is on the line through O and X in $\mathbb{E}^3$ if and only if $\mathbf{y} = t\mathbf{x}$ for some t in $\mathbb{R}$. Moreover, this t is the coordinate of Y with respect to X as unit point on the line through O and X.

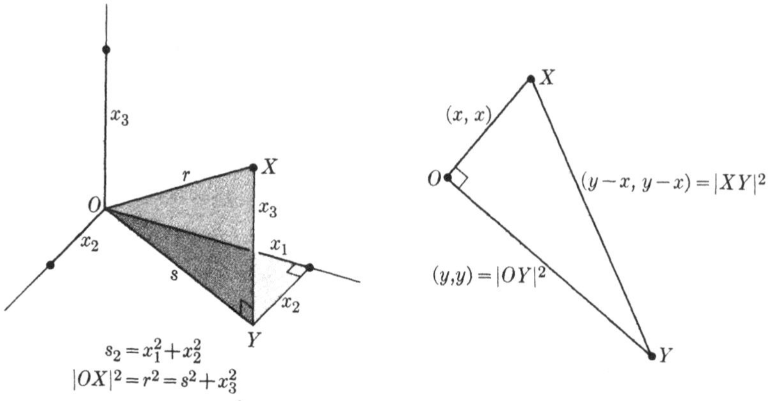

Fig. 1.5

4) If the axis system in $\mathbb{E}^3$ is *Cartesian*, that is, if the axes are mutually perpendicular and a common unit of distance is used, then the length $|OX|$ of the segment OX is given by the so-called *Euclidean norm* on $\mathbb{R}^3$, $|OX| = (\sum_1^3 x_i{}^2)^{1/2}$. This follows directly from the Pythagorean theorem. Then this formula and a second application of the Pythagorean theorem to the triangle OXY imply that the segments OX and OY are perpendicular if and only if the *scalar product* $(\mathbf{x}, \mathbf{y}) = \sum_{i=1}^3 x_i y_i$ has the value 0 (see Fig. 1.5).

In applying this result, it is useful to note that the scalar product $(\mathbf{x}, \mathbf{y})$ is linear as a function of either vector variable when the other is held fixed. Thus

$$(c\mathbf{x} + d\mathbf{y}, \mathbf{z}) = \sum_1^3 (cx_i + dy_i)z_i = c\sum_1^3 x_i z_i + d\sum_1^3 y_i z_i = c(\mathbf{x}, \mathbf{z}) + d(\mathbf{y}, \mathbf{z}).$$

Exactly the same theorems hold for the coordinate correspondence between the Euclidean plane $\mathbb{E}^2$ and the Cartesian 2-space $\mathbb{R}^2$, except that now, of course, $(\mathbf{x}, \mathbf{y}) = \sum_1^2 x_i y_i = x_1 y_1 + x_2 y_2$.

We can easily obtain the equations for lines and planes in $\mathbb{E}^3$ from these basic theorems. First, we see from (2) and (3) that if fixed points A and B are given, with $A \neq O$, then the line through B parallel to the segment OA contains the point X if and only if there exists a scalar t such that $\mathbf{x} - \mathbf{b} = t\mathbf{a}$ (see Fig. 1.6). Therefore, the equation of this line is

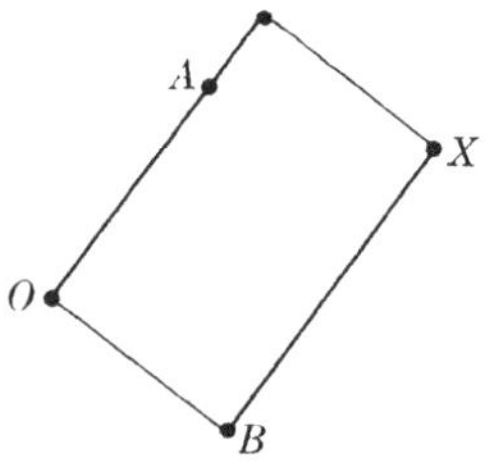

Fig. 1.6

$$\mathbf{x} = t\mathbf{a} + \mathbf{b}.$$

This vector equation is equivalent to the three numerical equations $x_i = a_i t + b_i$, $i = 1, 2, 3$. These are customarily called the *parametric equations* of the line, since they present the coordinate triple $\mathbf{x}$ of the varying point X on the line as functions of the "parameter" t.

Next, we know that the plane through B perpendicular to the direction of the segment OA contains the point X if and only if $BX \perp OA$, and it therefore follows from (2) and (4) that the plane contains X if and only if $(\mathbf{x} - \mathbf{b}, \mathbf{a}) = 0$ (see Fig. 1.7). But $(\mathbf{x} - \mathbf{b}, \mathbf{a}) = (\mathbf{x}, \mathbf{a}) - (\mathbf{b}, \mathbf{a})$ by the linearity of the scalar product in its first variable, and if we set $l = (\mathbf{b}, \mathbf{a})$, we see that the equation of the plane is

$$(\mathbf{x}, \mathbf{a}) = l \qquad \text{or} \qquad \sum_1^3 a_i x_i = l.$$

That is, a point X is on the plane through B perpendicular to the direction of OA if and only if this equation holds for its coordinate triple $\mathbf{x}$. Conversely, if $\mathbf{a} \neq 0$, then we can retrace the steps taken above to show that the set of points X in $\mathbb{E}^3$ whose coordinate triples $\mathbf{x}$ satisfy $(\mathbf{x}, \mathbf{a}) = l$ is a plane.

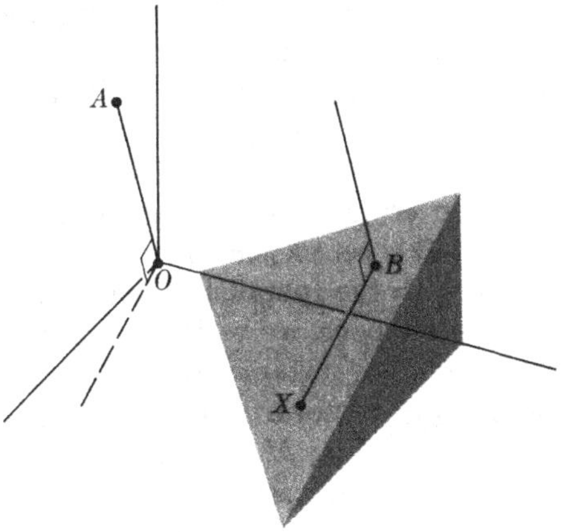

Fig. 1.7

The fact that $\mathbb{R}^3$ has the natural scalar product $(\mathbf{x}, \mathbf{y})$ is of course extremely important, both algebraically and geometrically. However, most vector spaces do not have natural scalar products, and we shall deliberately neglect scalar products in our early vector theory (but shall return to them in Chapter 5). This leads us to seek a different interpretation of the equation $\sum_1^3 a_i x_i = l$. We saw in Section 1 that $\mathbf{x} \mapsto \sum_1^3 a_i x_i$ is the most general linear functional f on $\mathbb{R}^3$. Therefore, given any plane M in $\mathbb{E}^3$, there is a nonzero linear functional f on $\mathbb{R}^3$ and a number l such that the equation of M is $f(\mathbf{x}) = l$. And conversely, given any nonzero linear functional $f\colon \mathbb{R}^3 \to \mathbb{R}$ and any $l \in \mathbb{R}$, the locus of $f(\mathbf{x}) = l$ is a plane M in $\mathbb{E}^3$. The reader will remember that we obtain the coefficient triple $\mathbf{a}$ from f by $a_i = f(\delta^i)$, since then $f(\mathbf{x}) = f(\sum_1^3 x_i \delta^i) = \sum_1^3 x_i f(\delta^i) = \sum_1^3 x_i a_i$.

Finally, we seek the vector form of the notion of parallel translation. In plane geometry when we are considering two congruent figures that are parallel and similarly oriented, we often think of obtaining one from the other by "sliding the plane along itself" in such a way that all lines remain parallel to their original positions. This description of a *parallel translation* of the plane can be more elegantly stated as the condition that every directed line segment slides to an equivalent one. If X slides to Y and O slides to B, then OX slides to BY, so that $OX \sim BY$ and $\mathbf{x} = \mathbf{y} - \mathbf{b}$ by (2). Therefore, the coordinate form of such a parallel sliding is the mapping $\mathbf{x} \mapsto \mathbf{y} = \mathbf{x} + \mathbf{b}$.

Conversely, for any $\mathbf{b}$ in $\mathbb{R}^2$ the plane mapping defined by $\mathbf{x} \mapsto \mathbf{y} = \mathbf{x} + \mathbf{b}$ is easily seen to be a parallel translation. These considerations hold equally well for parallel translations of the Euclidean space $\mathbb{E}^3$.

It is geometrically clear that under a parallel translation planes map to parallel planes and lines map to parallel lines, and now we can expect an easy algebraic proof. Consider, for example, the plane M with equation $f(\mathbf{x}) = l$; let us ask what happens to M under the translation $\mathbf{x} \mapsto \mathbf{y} = \mathbf{x} + \mathbf{b}$. Since $\mathbf{x} = \mathbf{y} - \mathbf{b}$, we see that a point $\mathbf{x}$ is on M if and only if its translate $\mathbf{y}$ satisfies the equation $f(\mathbf{y} - \mathbf{b}) = l$ or, since f is linear, the equation $f(\mathbf{y}) = l'$, where $l' = l + f(\mathbf{b})$. But this is the equation of a plane N. Thus the translate of M is the plane N.

It is natural to transfer all this geometric terminology from sets in $\mathbb{E}^3$ to the corresponding sets in $\mathbb{R}^3$ and therefore to speak of the set of ordered triples $\mathbf{x}$ satisfying $f(\mathbf{x}) = l$ as a set of *points* in $\mathbb{R}^3$ forming a *plane* in $\mathbb{R}^3$, and to call the mapping $\mathbf{x} \mapsto \mathbf{x} + \mathbf{b}$ the (parallel) translation of $\mathbb{R}^3$ through $\mathbf{b}$, etc. Moreover, since $\mathbb{R}^3$ is a vector space, we would expect these geometric ideas to interplay with vector notions. For instance, translation through $\mathbf{b}$ is simply the operation of adding the constant vector $\mathbf{b}\colon \mathbf{x} \mapsto \mathbf{x} + \mathbf{b}$. Thus if M is a plane, then the plane N obtained by translating M through $\mathbf{b}$ is just the vector set sum $M + \mathbf{b}$. If the equation of M is $f(\mathbf{x}) = l$, then the plane M goes through 0 if and only if $l = 0$, in which case M is a vector *subspace* of $\mathbb{R}^3$ (the null space of f). It is easy to see that any plane M is a translate of a plane through 0. Similarly, the line $\{t\mathbf{a} + \mathbf{b} : t \in \mathbb{R}\}$ is the translate through $\mathbf{b}$ of the line $\{t\mathbf{a} : t \in \mathbb{R}\}$, and this second line is a subspace, the linear span of the one vector $\mathbf{a}$. *Thus planes and lines in* $\mathbb{R}^3$ *are translates of subspaces.*

These notions all carry over to an arbitrary real vector space in a perfectly satisfactory way and with additional dimensional variety. A plane in $\mathbb{R}^3$ through 0 is a vector space which is *two-dimensional* in a strictly algebraic sense which we shall discuss in the next chapter, and a line is similarly one-dimensional. In $\mathbb{R}^3$ there are no proper subspaces other than planes and lines through 0, but in a vector space V with dimension $n > 3$ proper subspaces occur with all dimensions from 1 to $n - 1$. We shall therefore use the term "plane" loosely to refer to any translate of a subspace, whatever its dimension. More properly, translates of vector subspaces are called *affine subspaces.*

We shall see that if V is a finite-dimensional space with dimension n, then the null space of a nonzero linear functional f is always $(n - 1)$-dimensional, and therefore it cannot be a Euclidean-like two-dimensional plane except when

$n = 3$. We use the term *hyperplane* for such a null space or one of its translates. Thus, in general, a hyperplane is a set with the equation $f(\mathbf{x}) = l$, where f is a nonzero linear functional. It is a proper affine subspace (plane) which is maximal in the sense that the only affine subspace properly including it is the whole of V. In $\mathbb{R}^3$ hyperplanes are ordinary geometric planes, and in $\mathbb{R}^2$ hyperplanes are lines!

EXERCISES

2.1 Assuming the theorem $AB \sim XY \Leftrightarrow \mathbf{b} - \mathbf{a} = \mathbf{y} - \mathbf{x}$, show that $\overrightarrow{OC}$ is the sum of $\overrightarrow{OA}$ and $\overrightarrow{OB}$, as defined in the preliminary discussion of Section 1, if and only if $\mathbf{c} = \mathbf{b} + \mathbf{a}$. Considering also our assumed geometric theorem (3), show that the mapping $\mathbf{x} \mapsto \overrightarrow{OX}$ from $\mathbb{R}^3$ to the vector space of geometric vectors is linear and hence an isomorphism.

2.2 Let L be the line in the Cartesian plane $\mathbb{R}^2$ with equation $x_2 = 3x_1$. Express L in parametric form as $\mathbf{x} = t\mathbf{a}$ for a suitable ordered pair $\mathbf{a}$.

2.3 Let V be any vector space, and let α and β be distinct vectors. Show that the line through α and β has the parametric equation

$$\xi = t\beta + (1 - t)\alpha, \qquad t \in \mathbb{R}.$$

Show also that the *segment* from α to β is the image of $[0, 1]$ in the above mapping.

2.4 According to the Pythagorean theorem, a triangle with side lengths a, b, and c has a right angle at the vertex "opposite c" if and only if $c^2 = a^2 + b^2$.

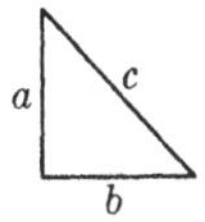

Prove from this that in a Cartesian coordinate system in $\mathbb{E}^3$ the length $|OX|$ of a segment OX is given by

$$|OX|^2 = \sum_1^3 x_i^2,$$

where $\mathbf{x} = \langle x_1, x_2, x_3 \rangle$ is the coordinate triple of the point X. Next use our geometric theorem (2) to conclude that

$$OX \perp OY \quad \text{if and only if} \quad (\mathbf{x}, \mathbf{y}) = 0, \qquad \text{where} \qquad (\mathbf{x}, \mathbf{y}) = \sum_1^3 x_i y_i.$$

(Use the bilinearity of $(\mathbf{x}, \mathbf{y})$ to expand $|X - Y|^2$.)

2.5 More generally, the law of cosine says that in any triangle labeled as indicated, $c^2 = a^2 + b^2 - 2ab \cos \theta$.

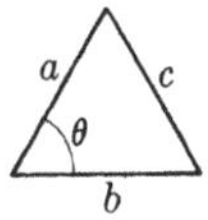

Apply this law to the diagram

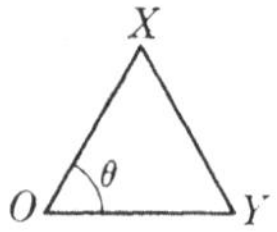

to prove that

$$(\mathbf{x}, \mathbf{y}) = 2|\mathbf{x}|\,|\mathbf{y}| \cos \theta,$$

where $(\mathbf{x}, \mathbf{y})$ is the scalar product $\sum_1^3 x_i y_i$, $|\mathbf{x}| = (\mathbf{x}, \mathbf{x})^{1/2} = |OX|$, etc.

2.6 Given a nonzero linear functional $f\colon \mathbb{R}^3 \to \mathbb{R}$, and given $k \in \mathbb{R}$, show that the set of points X in $\mathbb{E}^3$ such that $f(\mathbf{x}) = k$ is a plane. [*Hint:* Find a $\mathbf{b}$ in $\mathbb{R}^3$ such that $f(\mathbf{b}) = k$, and throw the equation $f(\mathbf{x}) = k$ into the form $(\mathbf{x} - \mathbf{b}, \mathbf{a}) = 0$, etc.]

2.7 Show that for any $\mathbf{b}$ in $\mathbb{R}^3$ the mapping $X \mapsto Y$ from $\mathbb{E}^3$ to itself defined by $\mathbf{y} = \mathbf{x} + \mathbf{b}$ is a parallel translation. That is, show that if $X \mapsto Y$ and $Z \mapsto W$, then $XZ \sim YW$.

2.8 Let M be the set in $\mathbb{R}^3$ with equation $3x_1 - x_2 + x_3 = 2$. Find triplets $\mathbf{a}$ and $\mathbf{b}$ such that M is the plane through $\mathbf{b}$ perpendicular to the direction of $\mathbf{a}$. What is the equation of the plane $P = M + \langle 1, 2, 1 \rangle$?

2.9 Continuing the above exercise, what is the condition on the triplet $\mathbf{b}$ in order for $N = M + \mathbf{b}$ to pass through the origin? What is the equation of N?

2.10 Show that if the plane M in $\mathbb{R}^3$ has the equation $f(\mathbf{x}) = l$, then M is a translate of the null space N of the linear functional f. Show that any two translates M and P of N are either identical or disjoint. What is the condition on the ordered triple $\mathbf{b}$ in order that $M + \mathbf{b} = M$?

2.11 Generalize the above exercise to hyperplanes in $\mathbb{R}^n$.

2.12 Let N be the subspace (plane through the origin) in $\mathbb{R}^3$ with equation $f(\mathbf{x}) = 0$. Let M and P be any two planes obtained from N by parallel translation. Show that $Q = M + P$ is a third such plane. If M and P have the equations $f(\mathbf{x}) = l_1$ and $f(\mathbf{x}) = l_2$, find the equation for Q.

2.13 If M is the plane in $\mathbb{R}^3$ with equation $f(\mathbf{x}) = l$, and if r is any nonzero number, show that the set product rM is a plane parallel to M.

2.14 In view of the above two exercises, discuss how we might consider the set of all parallel translates of the plane N with equation $f(\mathbf{x}) = 0$ as forming a new vector space.

2.15 Let L be the subspace (line through the origin) in $\mathbb{R}^3$ with parametric equation $\mathbf{x} = t\mathbf{a}$. Discuss the set of all parallel translates of L in the spirit of the above three exercises.

2.16 The best object to take as "being" the geometric vector $\overrightarrow{AB}$ is the equivalence class of all directed line segments XY such that $XY \sim AB$. Assuming whatever you need from properties (1) through (4), show that this *is* an equivalence relation on the set of all directed line segments (Section 0.12).

2.17 Assuming that the geometric vector $\overrightarrow{AB}$ is defined as in the above exercise, show that, strictly speaking, it is actually the mapping of the plane (or space) into itself that we have called the parallel translation through $\overrightarrow{AB}$. Show also that $\overrightarrow{AB} + \overrightarrow{CD}$ is the composition of the two translations.

3. PRODUCT SPACES AND HOM(V, W)

Product spaces. If W is a vector space and A is an arbitrary set, then the set $V = W^A$ of all W-valued functions on A is a vector space in exactly the same way that $\mathbb{R}^A$ is. Addition is the natural addition of functions, $(f + g)(a) = f(a) + g(a)$, and, similarly, $(xf)(a) = x(f(a))$ for every function f and scalar x. Laws A1 through S4 follow just as before and for exactly the same reasons. For variety, let us check the associative law for addition. The equation $f + (g + h) = (f + g) + h$ means that $(f + (g + h))(a) = ((f + g) + h)(a)$ for all $a \in A$. But

$$\begin{aligned}(f + (g + h))(a) &= f(a) + (g + h)(a)\\ &= f(a) + (g(a) + h(a)) = (f(a) + g(a)) + h(a)\\ &= (f + g)(a) + h(a) = ((f + g) + h)(a),\end{aligned}$$

where the middle equality in this chain of five holds by the associative law for W and the other four are applications of the definition of addition. Thus the associative law for addition holds in W^A because it holds in W, and the other laws follow in exactly the same way. As before, we let π_i be evaluation at i, so that $\pi_i(f) = f(i)$. Now, however, π_i is *vector valued* rather than scalar valued, because it is a mapping from V to W, and we call it the ith *coordinate projection* rather than the ith coordinate functional. Again these maps are all linear. In fact, as before, the natural vector operations on W^A are uniquely defined by the requirement that the projections π_i all be linear. We call the value $f(j) = \pi_j(f)$ the jth *coordinate* of the vector f. Here the analogue of Cartesian n-space is the set $W^{\bar{n}}$ of all n-tuples $\boldsymbol{\alpha} = \langle \alpha_1, \ldots, \alpha_n \rangle$ of vectors in W; it is also designated W^n. Clearly, α_j is the jth coordinate of the n-tuple $\boldsymbol{\alpha}$.

There is no reason why we must use the same space W at each index, as we did above. In fact, if $W_1, \ldots, W_n$ are *any* n vector spaces, then the set of all n-tuples $\boldsymbol{\alpha} = \langle \alpha_1, \ldots, \alpha_n \rangle$ such that $\alpha_j \in W_j$ for $j = 1, \ldots, n$ is a vector space under the same definitions of the operations and for the same reasons. That is, the Cartesian product $W = W_1 \times W_2 \times \cdots \times W_n$ is also a vector space of vector-valued functions. Such finite products will be very important to us. Of course, $\mathbb{R}^n$ is the product $\prod_1^n W_i$ with each $W_i = \mathbb{R}$; but $\mathbb{R}^n$ can also be considered $\mathbb{R}^m \times \mathbb{R}^{n-m}$, or more generally, $\prod_1^p W_i$, where $W_i = \mathbb{R}^{m_i}$ and $\sum_1^p m_i = n$. However, the most important use of finite product spaces arises from the fact that the study of certain phenomena on a vector space V may lead in a natural way to a collection $\{V_i\}_1^n$ of *subspaces* of V such that V is isomorphic to the product $\prod_1^n V_i$. Then the extra structure that V acquires when we regard it as the product space $\prod_1^n V_i$ is used to study the phenomena in question. This is the theory of direct sums, and we shall investigate it in Section 5.

Later in the course we shall need to consider a general Cartesian product of vector spaces. We remind the reader that if $\{W_i : i \in I\}$ is any indexed collection of vector spaces, then the Cartesian product $\prod_{i \in I} W_i$ of these vector spaces is defined as the set of all functions f with domain I such that $f(i) \in W_i$ for all $i \in I$ (see Section 0.8).

The following is a simple concrete example to keep in mind. Let S be the ordinary unit sphere in $\mathbb{R}^3$, $S = \{\mathbf{x} : \sum_1^3 x_i^2 = 1\}$, and for each point $\mathbf{x}$ on S let $W_\mathbf{x}$ be the *subspace* of $\mathbb{R}^3$ *tangent* to S at $\mathbf{x}$. By this we mean the subspace (plane through O) parallel to the tangent plane to S at $\mathbf{x}$, so that the translate $W_\mathbf{x} + \mathbf{x}$ is the tangent plane (see Fig. 1.8). A function f in the product space $W = \prod_{\mathbf{x} \in S} W_\mathbf{x}$ is a function which assigns to each point $\mathbf{x}$ on S a vector in $W_\mathbf{x}$, that is, a vector parallel to the tangent plane to S at $\mathbf{x}$. Such a function is called a *vector field* on S. Thus the product set W is the set of all vector fields on S, and W itself is a vector space, as the next theorem states.

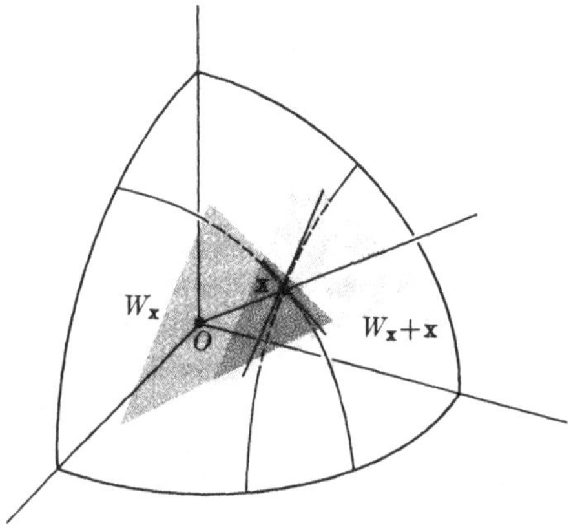

Fig. 1.8

Of course, the jth coordinate projection on $W = \prod_{i \in S} W_i$ is evaluation at j, $\pi_j(f) = f(j)$, and the natural vector operations on W are uniquely defined by the requirement that the coordinate projections all be linear. Thus $f + g$ must be that element of W whose value at j, $\pi_j(f + g)$, is $\pi_j(f) + \pi_j(g) = f(j) + g(j)$ for all $j \in I$, and similarly for multiplication by scalars.

Theorem 3.1. The Cartesian product of a collection of vector spaces can be made into a vector space in exactly one way so that the coordinate projections are all linear.

Proof. With the vector operations determined uniquely as above, the proofs of A1 through S4 that we sampled earlier hold verbatim. They did not require that the functions being added have all their values in the same space, but only that the values at a given domain element i all lie in the same space. □

Hom(V, W). Linear transformations have the simple but important properties that the sum of two linear transformations is linear and the composition of two linear transformations is linear. These imprecise statements are in essence the theme of this section, although they need bolstering by conditions on domains and codomains. Their proofs are simple formal algebraic arguments, but the objects being discussed will increase in conceptual complexity.

If W is a vector space and A is any set, we know that the space W^A of all mappings $f: A \to W$ is a vector space of functions (now vector valued) in the same way that $\mathbb{R}^A$ is. If A is itself a vector space V, we naturally single out for special study the subset of W^V consisting of all linear mappings. We designate this subset $\mathrm{Hom}(V, W)$. The following elementary theorems summarize its basic algebraic properties.

Theorem 3.2. $\mathrm{Hom}(V, W)$ is a vector subspace of W^V.

Proof. The theorem is an easy formality. If S and T are in $\mathrm{Hom}(V, W)$, then

$$\begin{aligned}(S + T)(x\alpha + y\beta) &= S(x\alpha + y\beta) + T(x\alpha + y\beta)\\ &= xS(\alpha) + yS(\beta) + xT(\alpha) + yT(\beta) = x(S + T)(\alpha) + y(S + T)(\beta),\end{aligned}$$

so $S + T$ is linear and $\mathrm{Hom}(V, W)$ is closed under addition. The reader should be sure he knows the justification for each step in the above continued equality. The closure of $\mathrm{Hom}(V, W)$ under multiplication by scalars follows similarly, and since $\mathrm{Hom}(V, W)$ contains the zero transformation, and so is nonempty, it is a subspace. □

Theorem 3.3. The composition of linear maps is linear: if $T \in \mathrm{Hom}(V, W)$ and $S \in \mathrm{Hom}(W, X)$, then $S \circ T \in \mathrm{Hom}(V, X)$. Moreover, composition is distributive over addition, under the obvious hypotheses on domains and codomains:

$$(S_1 + S_2) \circ T = S_1 \circ T + S_2 \circ T \quad \text{and} \quad S \circ (T_1 + T_2) = S \circ T_1 + S \circ T_2.$$

Finally, composition commutes with scalar multiplication:

$$c(S \circ T) = (cS) \circ T = S \circ (cT).$$

Proof. We have

$$\begin{aligned}S \circ T(x\alpha + y\beta) &= S(T(x\alpha + y\beta)) = S(xT(\alpha) + yT(\beta))\\ &= xS(T(\alpha)) + yS(T(\beta)) = x(S \circ T)(\alpha) + y(S \circ T)(\beta),\end{aligned}$$

so $S \circ T$ is linear. The two distributive laws will be left to the reader. □

Corollary. If $T \in \mathrm{Hom}(V, W)$ is fixed, then composition on the right by T is a linear transformation from the vector space $\mathrm{Hom}(W, X)$ to the vector space $\mathrm{Hom}(V, X)$. It is an isomorphism if T is an isomorphism.

Proof. The algebraic properties of composition stated in the theorem can be combined as follows:

$$\begin{aligned}(c_1S_1 + c_2S_2) \circ T &= c_1(S_1 \circ T) + c_2(S_2 \circ T),\\ S \circ (c_1T_1 + c_2T_2) &= c_1(S \circ T_1) + c_2(S \circ T_2).\end{aligned}$$

The first equation says exactly that composition on the right by a fixed T is a linear transformation. (Write $S \circ T$ as $\mathfrak{T}(S)$ if the equations still don't look right.) If T is an isomorphism, then composition by T^{-1} "undoes" composition by T, and so is its inverse. □

The second equation implies a similar corollary about composition on the left by a fixed S.

Theorem 3.4. If W is a product vector space, $W = \prod_i W_i$, then a mapping T from a vector space V to W is linear if and only if $\pi_i \circ T$ is linear for each coordinate projection π_i.

Proof. If T is linear, then $\pi_i \circ T$ is linear by the above theorem. Now suppose, conversely, that all the maps $\pi_i \circ T$ are linear. Then

$$\begin{aligned}\pi_i(T(x\alpha + y\beta)) = \pi_i \circ T(x\alpha + y\beta) &= x(\pi_i \circ T)(\alpha) + y(\pi_i \circ T)(\beta)\\ &= x\pi_i(T(\alpha)) + y\pi_i(T(\beta)) = \pi_i(xT(\alpha) + yT(\beta)).\end{aligned}$$

But if $\pi_i(f) = \pi_i(g)$ for all i, then $f = g$. Therefore, $T(x\alpha + y\beta) = xT(\alpha) + yT(\beta)$, and T is linear. □

If T is a linear mapping from $\mathbb{R}^n$ to W whose skeleton is $\{\beta_j\}^n$, then $\pi_i \circ T$ has skeleton $\{\pi_i(\beta_j)\}_{j=1}^n$. If W is $\mathbb{R}^m$, then π_i is the ith coordinate functional $\mathbf{y} \mapsto y_i$, and β_j is the jth column in the matrix $\mathbf{t} = \{t_{ij}\}$ of T. Thus $\pi_i(\beta_j) = t_{ij}$, and $\pi_i \circ T$ is the linear functional whose skeleton is the ith row of the matrix of T.

In the discussion centering around Theorem 1.3, we replaced the vector equation $\mathbf{y} = T(\mathbf{x})$ by the equivalent set of m scalar equations $y_i = \sum_{j=1}^n t_{ij}x_j$, which we obtained by reading off the ith coordinate in the vector equation. But in "reading off" the ith coordinate we were applying the coordinate mapping π_i, or in more algebraic terms, we were replacing the linear map T by the set of linear maps $\{\pi_i \circ T\}$, which is equivalent to it by the above theorem.

Now consider in particular the space Hom(V, V), which we may as well designate 'Hom(V)'. In addition to being a vector space, it is also closed under composition, which we consider a multiplication operation. Since composition of functions is *always* associative (see Section 0.9), we thus have for multiplication the laws

$$\begin{aligned}A \circ (B \circ C) &= (A \circ B) \circ C,\\ A \circ (B + C) &= (A \circ B) + (A \circ C),\\ (A + B) \circ C &= (A \circ C) + (B \circ C),\\ k(A \circ B) &= (kA) \circ B = A \circ (kB).\end{aligned}$$

Any vector space which has in addition to the vector operations an operation of multiplication related to the vector operations in the above ways is called an *algebra*. Thus,

Theorem 3.5. Hom(V) is an algebra.

We noticed earlier that certain real-valued function spaces are also algebras. Examples were $\mathbb{R}^A$ and $\mathcal{C}([0, 1])$. In these cases multiplication is commutative, but in the case of Hom(V) multiplication is not commutative unless V is a trivial space ($V = \{0\}$) or V is isomorphic to $\mathbb{R}$. We shall check this later when we examine the finite-dimensional theory in greater detail.

Product projections and injections. In addition to the coordinate projections, there is a second class of simple linear mappings that is of basic importance in the handling of a Cartesian product space $W = \prod_{k \in K} W_k$. These are, for each j, the mapping θ_j taking a vector $\alpha \in W_j$ to the function in the product space having the value α at the index j and 0 elsewhere. For example, θ_2 for $W_1 \times W_2 \times W_3$ is the mapping $\alpha \mapsto \langle 0, \alpha, 0 \rangle$ from W_2 to W. Or if we view $\mathbb{R}^3$ as $\mathbb{R} \times \mathbb{R}^2$, then θ_2 is the mapping $\langle x_2, x_3 \rangle \mapsto \langle 0, \langle x_2, x_3 \rangle \rangle = \langle 0, x_2, x_3 \rangle$. We call θ_j the *injection* of W_j into $\prod_k W_k$. The linearity of θ_j is probably obvious. The mappings π_j and θ_j are clearly connected, and the following *projection-injection* identities state their exact relationship. If I_j is the identity transformation on W_j, then

$$\pi_j \circ \theta_j = I_j \qquad \text{and} \qquad \pi_j \circ \theta_i = 0 \qquad \text{if} \quad i \neq j.$$

If K is finite and I is the identity on the product space W, then

$$\sum_{k \in K} \theta_k \circ \pi_k = I.$$

In the case $\prod_{i=1}^3 W_i$, we have $\theta_2 \circ \pi_2(\langle \alpha_1, \alpha_2, \alpha_3 \rangle) = \langle 0, \alpha_2, 0 \rangle$, and the identity simply says that $\langle \alpha_1, 0, 0 \rangle + \langle 0, \alpha_2, 0 \rangle + \langle 0, 0, \alpha_3 \rangle = \langle \alpha_1, \alpha_2, \alpha_3 \rangle$ for all $\alpha_1, \alpha_2, \alpha_3$. These identities will probably be clear to the reader, and we leave the formal proofs as an exercise.

The coordinate projections π_j are useful in the study of any product space, but because of the limitation in the above identity, the injections θ_j are of interest principally in the case of finite products. Together they enable us to decompose and reassemble linear maps whose domains or codomains are finite product spaces.

For a simple example, consider the T in $\mathrm{Hom}(\mathbb{R}^3, \mathbb{R}^2)$ whose matrix is

$$\begin{bmatrix} 2 & -1 & 1 \\ 1 & 1 & 4 \end{bmatrix}.$$

Then $\pi_1 \circ T$ is the linear functional whose skeleton $\langle 2, -1, 1 \rangle$ is the first row in the matrix of T, and we know that we can visualize its expression in equation form, $y_1 = 2x_1 - x_2 + x_3$, as being obtained from the vector equation $\mathbf{y} = T(\mathbf{x})$ by "reading off the first row". Thus we "decompose" T into the two linear functionals $l_i = \pi_i \circ T$. Then, speaking loosely, we have the reassembly $T = \langle l_1, l_2 \rangle$; more exactly, $T(\mathbf{x}) = \langle 2x_1 - x_2 + x_3, x_1 + x_2 + 4x_3 \rangle = \langle l_1(\mathbf{x}), l_2(\mathbf{x}) \rangle$ for all $\mathbf{x}$. However, we want to present this reassembly as the action of the linear maps θ_1 and θ_2. We have

$$\langle l_1(\mathbf{x}), l_2(\mathbf{x}) \rangle = \theta_1(l_1(\mathbf{x})) + \theta_2(l_2(\mathbf{x})) = (\theta_1 \circ \pi_1 + \theta_2 \circ \pi_2)(T(\mathbf{x})) = T(\mathbf{x}),$$

which shows that the decomposition and reassembly of T is an expression of the identity $\sum \theta_i \circ \pi_i = I$. In general, if $T \in \mathrm{Hom}(V, W)$ and $W = \prod_i W_i$, then $T_i = \pi_i \circ T$ is in $\mathrm{Hom}(V, W_i)$ for each i, and T_i can be considered "the part of T going into W_i", since $T_i(\alpha)$ is the ith coordinate of $T(\alpha)$ for each α. Then we can reassemble the T_i's to form T again by $T = \sum \theta_i \circ T_i$, for $\sum \theta_i \circ T_i = (\sum \theta_i \circ \pi_i) \circ T = I \circ T = T$. Moreover, any finite collection of T_i's on a common domain can be put together in this way to make a T. For example, we can assemble an m-tuple $\{T_i\}_1^m$ of linear maps on a common domain V to form a single m-tuple-valued linear map T. Given α in V, we simply define $T(\alpha)$ as that m-tuple whose ith coordinate is $T_i(\alpha)$ for $i = 1, \ldots, m$, and then check that T is linear. Thus without having to calculate, we see from this assembly principle that $T: \mathbf{x} \mapsto \langle 2x_1 - x_2 + x_3, x_1 + x_2 + 4x_3 \rangle$ is a linear mapping from $\mathbb{R}^3$ to $\mathbb{R}^2$, since we have formed T by assembling the two linear functionals $l_1(\mathbf{x}) = 2x_1 - x_2 + x_3$ and $l_2(\mathbf{x}) = x_1 + x_2 + 4x_3$ to form a single ordered-pair-valued map. This very intuitive process has an equally simple formal justification. We rigorize our discussion in the following theorem.

Theorem 3.6. If T_i is in $\mathrm{Hom}(V, W_i)$ for each i in a finite index set I, and if W is the product space $\prod_{i \in I} W_i$, then there is a uniquely determined T in $\mathrm{Hom}(V, W)$ such that $T_i = \pi_i \circ T$ for all i in I.

Proof. If T exists such that $T_i = \pi_i \circ T$ for each i, then $T = I_W \circ T = (\sum \theta_i \circ \pi_i) \circ T = \sum \theta_i \circ (\pi_i \circ T) = \sum \theta_i \circ T_i$. Thus T is uniquely determined as $\sum \theta_i \circ T_i$. Moreover, this T does have the required property, since then

$$\pi_j \circ T = \pi_j \circ \left(\sum \theta_i \circ T_i\right) = \sum_i (\pi_j \circ \theta_i) \circ T_i = I_j \circ T_j = T_j. \ \square$$

In the same way, we can decompose a linear T whose *domain* is a product space $V = \prod_{j=1}^n V_j$ into the maps $T_j = T \circ \theta_j$ with domains V_j, and then reassemble these maps to form T by the identity $T = \sum_{j=1}^n T_j \circ \pi_j$ (check it mentally!). Moreover, a finite collection of maps into a common *codomain* space can be put together to form a single map on the product of the domain spaces. Thus an n-tuple of maps $\{T_i\}_1^n$ into W defines a single map T into W, where the domain of T is the product of the domains of the T_i's, by the equation $T(\langle \alpha_1, \ldots, \alpha_n \rangle) = \sum_1^n T_i(\alpha_i)$ or $T = \sum_1^n T_i \circ \pi_i$. For example, if $T_1 : \mathbb{R} \to \mathbb{R}^2$ is the map $t \mapsto t\langle 2, 1 \rangle = \langle 2t, t \rangle$, and T_2 and T_3 are similarly the maps $t \mapsto t\langle -1, 1 \rangle$ and $t \mapsto t\langle 1, 4 \rangle$, then $T = \sum_1^3 T_i \circ \pi_i$ is the mapping from $\mathbb{R}^3$ to $\mathbb{R}^2$ whose matrix is

$$\begin{bmatrix} 2 & -1 & 1 \\ 1 & 1 & 4 \end{bmatrix}.$$

Again there is a simple formal argument, and we shall ask the reader to write out the proof of the following theorem.

Theorem 3.7. If T_j is in $\mathrm{Hom}(V_j, W)$ for each j in a finite index set J, and if $V = \prod_{j \in J} V_j$, then there exists a unique T in $\mathrm{Hom}(V, W)$ such that $T \circ \theta_j = T_j$ for each j in J.

Finally we should mention that Theorem 3.6 holds for *all* product spaces, finite or not, and states a property that *characterizes* product spaces. We shall investigate this situation in the exercises. The proof of the general case of Theorem 3.6 has to get along without the injections θ_j; instead, it is an application of Theorem 3.4.

The reader may feel that we are being overly formal in using the projections π_i and the injections θ_i to give algebraic formulations of processes that are easily visualized directly, such as reading off the scalar "components" of a vector equation. However, the mappings

$$\mathbf{x} \mapsto x_i \qquad \text{and} \qquad x_i \mapsto \langle 0, \ldots, 0, x_i, 0, \ldots, 0 \rangle$$

are clearly fundamental devices, and making their relationships explicit now will be helpful to us later on when we have to handle their occurrences in more complicated situations.

EXERCISES

3.1 Show that $\mathbb{R}^m \times \mathbb{R}^n$ is isomorphic to $\mathbb{R}^{n+m}$.

3.2 Show more generally that if $\sum_1^m n_i = n$, then $\prod_{i=1}^m \mathbb{R}^{n_i}$ is isomorphic to $\mathbb{R}^n$.

3.3 Show that if $\{B, C\}$ is a partitioning of A, then $\mathbb{R}^A$ and $\mathbb{R}^B \times \mathbb{R}^C$ are isomorphic.

3.4 Generalize the above to the case where $\{A_i\}_1^n$ partitions A.

3.5 Show that a mapping T from a vector space V to a vector space W is linear if and only if (the graph of) T is a subspace of $V \times W$.

3.6 Let S and T be nonzero linear maps from V to W. The definition of the map $S + T$ is *not* the same as the set sum of (the graphs of) S and T as subspaces of $V \times W$. Show that the set sum of (the graphs of) S and T *cannot be* a graph unless $S = T$.

3.7 Give the justification for each step of the calculation in Theorem 3.2.

3.8 Prove the distributive laws given in Theorem 3.3.

3.9 Let $D: \mathcal{C}^1([a, b]) \to \mathcal{C}([a, b])$ be differentiation, and let $S: \mathcal{C}([a, b]) \to \mathbb{R}$ be the definite integral map $f \mapsto \int_a^b f$. Compute the composition $S \circ D$.

3.10 We know that the general linear functional F on $\mathbb{R}^2$ is the map $\mathbf{x} \mapsto a_1x_1 + a_2x_2$ determined by the pair $\mathbf{a}$ in $\mathbb{R}^2$, and that the general linear map T in $\text{Hom}(\mathbb{R}^2)$ is determined by a matrix

$$\mathbf{t} = \begin{bmatrix} t_{11} & t_{12} \\ t_{21} & t_{22} \end{bmatrix}.$$

Then $F \circ T$ is another linear functional, and hence is of the form $\mathbf{x} \mapsto b_1x_1 + b_2x_2$ for some $\mathbf{b}$ in $\mathbb{R}^2$. Compute $\mathbf{b}$ from $\mathbf{t}$ and $\mathbf{a}$. Your computation should show you that $\mathbf{a} \mapsto \mathbf{b}$ is linear. What is its matrix?

3.11 Given S and T in $\text{Hom}(\mathbb{R}^2)$ whose matrices are

$$\begin{bmatrix} 1 & 2 \\ 3 & 4 \end{bmatrix} \quad \text{and} \quad \begin{bmatrix} 2 & 1 \\ 0 & 1 \end{bmatrix},$$

respectively, find the matrix of $S \circ T$ in $\text{Hom}(\mathbb{R}^2)$.

3.12 Given S and T in Hom($\mathbb{R}^2$) whose matrices are

$$\mathbf{s} = \begin{bmatrix} s_{11} & s_{12} \\ s_{21} & s_{22} \end{bmatrix} \quad \text{and} \quad \mathbf{t} = \begin{bmatrix} t_{11} & t_{12} \\ t_{21} & t_{22} \end{bmatrix},$$

find the matrix of $S \circ T$.

3.13 With the above answer in mind, what would you guess the matrix of $S \circ T$ is if S and T are in Hom($\mathbb{R}^3$)? Verify your guess.

3.14 We know that if $T \in \text{Hom}(V, W)$ is an isomorphism, then T^{-1} is an isomorphism in Hom(W, V). Prove that

$$S \circ T \text{ surjective} \Rightarrow S \text{ surjective}, \qquad S \circ T \text{ injective} \Rightarrow T \text{ injective},$$

and, therefore, that if $T \in \text{Hom}(V, W)$, $S \in \text{Hom}(W, V)$, and

$$S \circ T = I_V, \qquad T \circ S = I_{\ },$$

then T is an isomorphism.

3.15 Show that if S^{-1} and T^{-1} exist, then $(S \circ T)^{-1}$ exists and equals $T^{-1} \circ S^{-1}$. Give a more careful statement of this result.

3.16 Show that if S and T in Hom V commute with each other, then the null space of T, $N = N(T)$, and its range $R = R(T)$ are invariant under S ($S[N] \subset N$ and $S[R] \subset R$).

3.17 Show that if α is an eigenvector of T and S commutes with T, then $S(\alpha)$ is an eigenvector of T and has the same eigenvalue.

3.18 Show that if S commutes with T and T^{-1} exists, then S commutes with T^{-1}.

3.19 Given that α is an eigenvector of T with eigenvalue x, show that α is also an eigenvector of $T^2 = T \circ T$, of T^n, and of T^{-1} (if T is invertible) and that the corresponding eigenvalues are x^2, x^n, and $1/x$.

Given that $p(t)$ is a polynomial in t, define the operator $p(T)$, and under the above hypotheses, show that α is an eigenvector of $p(T)$ with eigenvalue $p(x)$.

3.20 If S and T are in Hom V, we say that S *doubly commutes* with T (and write S cc T) if S commutes with every A in Hom V which commutes with T. Fix T, and set $\{T\}'' = \{S : S \text{ cc } T\}$. Show that $\{T\}''$ is a commutative subalgebra of Hom V.

3.21 Given T in Hom V and α in V, let N be the linear span of the "trajectory of α under T" (the set $\{T^n\alpha : n \in \mathbb{Z}^+\}$). Show that N is invariant under T.

3.22 A transformation T in Hom V such that $T^n = 0$ for some n is said to be *nilpotent*. Show that if T is nilpotent, then $I - T$ is invertible. [*Hint:* The power series

$$\frac{1}{1 - x} = \sum_0^\infty x^n$$

is a finite sum if x is replaced by T.]

3.23 Suppose that T is nilpotent, that S commutes with T, and that S^{-1} exists, where $S, T \in \text{Hom } V$. Show that $(S - T)^{-1}$ exists.

3.24 Let φ be an isomorphism from a vector space V to a vector space W. Show that $T \to \varphi \circ T \circ \varphi^{-1}$ is an *algebra* isomorphism from the algebra Hom V to the algebra Hom W.

3.25 Show the π_j's and θ_j's explicitly for $\mathbb{R}^3 = \mathbb{R} \times \mathbb{R} \times \mathbb{R}$ using the stopped arrow notation. Also write out the identity $\sum \theta_j \circ \pi_j = I$ in explicit form.

3.26 Do the same for $\mathbb{R}^5 = \mathbb{R}^2 \times \mathbb{R}^3$.

3.27 Show that the first two projection-injection identities ($\pi_i \circ \theta_i = I_i$ and $\pi_j \circ \theta_i = 0$ if $j \neq i$) are simply a restatement of the definition of θ_i. Show that the linearity of θ_i follows formally from these identities and Theorem 3.4.

3.28 Prove the identity $\sum \theta_i \circ \pi_i = I$ by applying π_j to the equation and remembering that $f = g$ if $\pi_j(f) = \pi_j(g)$ for all j (this being just the equation $f(j) = g(j)$ for all j).

3.29 Prove the general case of Theorem 3.6. We are given an indexed collection of linear maps $\{T_i : i \in I\}$ with common domain V and codomains $\{W_i : i \in I\}$. The first question is how to define $T: V \to W = \prod_i W_i$. Do this by defining $T(\xi)$ suitably for each $\xi \in V$ and then applying Theorem 3.4 to conclude that T is linear.

3.30 Prove Theorem 3.7.

3.31 We know without calculation that the map

$$T: \mathbf{x} \mapsto \langle 3x_1 - x_2 + x_3, x_2 + x_3, x_1 - 5x_3, 2x_1 \rangle$$

from $\mathbb{R}^3$ to $\mathbb{R}^4$ is linear. Why? (Cite relevant theorems from the text.)

3.32 Write down the matrix for the transformation T in the above example, and then write down the mappings $T \circ \theta_i$ from $\mathbb{R}$ to $\mathbb{R}^4$ (for $i = 1, 2, 3$) in explicit ordered quadruplet form.

3.33 Let $W = \prod_1^n W_i$ be a finite product vector space and set $p_i = \theta_i \circ \pi_i$, so that p_i is in Hom W for all i. Prove from the projection-injection identities that $\sum_1^n p_i = I$ (the identity map on W), $p_i \circ p_j = 0$ if $i \neq j$, and $p_i \circ p_i = p_i$. Identify the range $R_i = R(p_i)$.

3.34 In the context of the above exercise, define T in Hom W as

$$\sum_{m=1}^{n} m p_m.$$

Show that α is an eigenvector of T if and only if α is in one of the subspaces R_i and that then the eigenvalue of α is i.

3.35 In the same situation show that the polynomial

$$\prod_{j=1}^{n} (T - jI) = (T - I) \circ \cdots \circ (T - nI)$$

is the zero transformation.

3.36 Theorems 3.6 and 3.7 can be combined if $T \in \text{Hom}(V, W)$, where *both* V and W are product spaces:

$$V = \prod_1^n V_j \quad \text{and} \quad W = \prod_1^m W_i.$$

State and prove a theorem which says that such a T can be decomposed into a doubly indexed family $\{T_{ij}\}$ when $T_{ij} \in \text{Hom}(V_i, W_j)$ and conversely that any such doubly indexed family can be assembled to form a single T form V to W.

3.37 Apply your theorem to the special case where $V = \mathbb{R}^n$ and $W = \mathbb{R}^m$ (that is, $V_i = W_j = \mathbb{R}$ for all i and j). Now T_{ij} is from $\mathbb{R}$ to $\mathbb{R}$ and hence is simply multiplication by a number t_{ij}. Show that the indexed collection $\{t_{ij}\}$ of these numbers is the matrix of T.

3.38 Given an m-tuple of vector spaces $\{W_i\}_1^m$, suppose that there are a vector space X and maps p_i in $\mathrm{Hom}(X, W_i)$, $i = 1, \ldots, m$, with the following property:

> **P.** For any m-tuple of linear maps $\{T_i\}$ from a common domain space V to the above spaces W_i (so that $T_i \in \mathrm{Hom}(V, W_i)$, $i = 1, \ldots, m$), there is a unique T in $\mathrm{Hom}(V, X)$ such that $T_i = p_i \circ T$, $i = 1, \ldots, m$.

Prove that there is a "canonical" isomorphism from

$$W = \prod_1^m W_i \quad \text{to} \quad X$$

under which the given maps p_i become the projections π_i. [*Remark:* The product space W itself has property P by Theorem 3.6, and this exercise therefore shows that P is an abstract characterization of the product space.]

4. AFFINE SUBSPACES AND QUOTIENT SPACES

In this section we shall look at the "planes" in a vector space V and see what happens to them when we translate them, intersect them with each other, take their images under linear maps, and so on. Then we shall confine ourselves to the set of all planes that are translates of a fixed subspace and discover that this set itself is a vector space in the most obvious way. Some of this material has been anticipated in Section 2.

Affine subspaces. If N is a subspace of a vector space V and α is any vector of V, then the set $N + \alpha = \{\xi + \alpha : \xi \in N\}$ is called either the *coset* of N containing α or the *affine subspace* of V *through* α and *parallel to* N. The set $N + \alpha$ is also called the *translate* of N through α. We saw in Section 2 that affine subspaces are the general objects that we want to call planes. If N is given and fixed in a discussion, we shall use the notation $\bar{\alpha} = N + \alpha$ (see Section 0.12).

We begin with a list of some simple properties of affine subspaces. Some of these will generalize observations already made in Section 2, and the proofs of some will be left as exercises.

1) With a fixed subspace N assumed, if $\gamma \in \bar{\alpha}$, then $\bar{\gamma} = \bar{\alpha}$. For if $\gamma = \alpha + \eta_0$, then $\gamma + \eta = \alpha + (\eta_0 + \eta) \in \bar{\alpha}$, so $\bar{\gamma} \subset \bar{\alpha}$. Also $\alpha + \eta = \gamma + (\eta - \eta_0) \in \bar{\gamma}$, so $\bar{\alpha} \subset \bar{\gamma}$. Thus $\bar{\alpha} = \bar{\gamma}$.

2) With N fixed, for any α and β, either $\bar{\alpha} = \bar{\beta}$ or $\bar{\alpha}$ and $\bar{\beta}$ are disjoint. For if $\bar{\alpha}$ and $\bar{\beta}$ are not disjoint, then there exists a γ in each, and $\bar{\alpha} = \bar{\gamma} = \bar{\beta}$ by (1). The reader may find it illuminating to compare these calculations with the more general ones of Section 0.12. Here $\alpha \sim \beta$ if and only if $\alpha - \beta \in N$.

3) Now let $\mathcal{A}$ be the collection of *all* affine subspaces of V; $\mathcal{A}$ is thus the set of *all* cosets of *all* vector subspaces of V. Then the intersection of any sub-

family of $\mathcal{A}$ is either empty or itself an affine subspace. In fact, if $\{A_i\}_{i\in I}$ is an indexed collection of affine subspaces and A_i is a coset of the vector subspace W_i for each $i \in I$, then $\bigcap_{i\in I} A_i$ is either empty or a coset of the vector subspace $\bigcap_{i\in I} W_i$. For if $\beta \in \bigcap_{i\in I} A_i$, then (1) implies that $A_i = \beta + W_i$ for all i, and then $\bigcap A_i = \beta + \bigcap W_i$.

4) If $A, B \in \mathcal{A}$, then $A + B \in \mathcal{A}$. That is, the set sum of any two affine subspaces is itself an affine subspace.

5) If $A \in \mathcal{A}$ and $T \in \text{Hom}(V, W)$, then $T[A]$ is an affine subspace of W. In particular, if $t \in \mathbb{R}$, then $tA \in \mathcal{A}$.

6) If B is an affine subspace of W and $T \in \text{Hom}(V, W)$, then $T^{-1}[B]$ is either empty or an affine subspace of V.

7) For a fixed $\alpha \in V$ the *translation* of V *through* α is the mapping $S_\alpha: V \to V$ defined by $S_\alpha(\xi) = \xi + \alpha$ for all $\xi \in V$. Translation is *not* linear; for example, $S_\alpha(0) = \alpha$. It is clear, however, that translation carries affine subspaces into affine subspaces. Thus $S_\alpha(A) = A + \alpha$ and $S_\alpha(\beta + W) = (\alpha + \beta) + W$.

8) *An affine transformation* from a vector space V to a vector space W is a linear mapping from V to W followed by a translation in W. Thus an affine transformation is of the form $\xi \mapsto T(\xi) + \beta$, where $T \in \text{Hom}(V, W)$ and $\beta \in W$. Note that $\xi \mapsto T(\xi + \alpha)$ is affine, since

$$T(\xi + \alpha) = T(\xi) + \beta, \qquad \text{where} \qquad \beta = T(\alpha).$$

It follows from (5) and (7) that an affine transformation carries affine subspaces of V into affine subspaces of W.

Quotient space. Now fix a subspace N of V, and consider the set W of all translates (cosets) of N. We are going to see that W itself is a vector space in the most natural way possible. Addition will be set addition, and scalar multiplication will be set multiplication (except in one special case). For example, if N is a line through the origin in $\mathbb{R}^3$, then W consists of all lines in $\mathbb{R}^3$ parallel to N. We are saying that this set of parallel lines will automatically turn out to be a vector space: the set sums of any two of the lines in W turn out to be a line in W! And if $L \in W$ and $t \neq 0$, then the set product tL is a line in W. The translates of L fiber $\mathbb{R}^3$, and the set of fibers is a natural vector space.

During this discussion it will be helpful temporarily to indicate set sums by '$+_s$' and set products by '$\cdot_s$'. With N fixed, it follows from (2) above that two cosets are disjoint or identical, so that the set W of all cosets is a fibering of V in the general case, just as it was in our example of the parallel lines. From (4) or by a direct calculation we know that $\bar{\alpha} +_s \bar{\beta} = \overline{\alpha + \beta}$. Thus W is closed under set addition, and, naturally, we take this to be our operation of addition on W. That is, we define $+$ on W by $\bar{\alpha} + \bar{\beta} = \bar{\alpha} +_s \bar{\beta}$. Then the natural map $\pi: \alpha \mapsto \bar{\alpha}$ from V to W preserves addition, $\pi(\alpha + \beta) = \pi(\alpha) + \pi(\beta)$, since this is just our equation $\overline{\alpha + \beta} = \bar{\alpha} + \bar{\beta}$ above. Similarly, if $t \in \mathbb{R}$, then the set product $t \cdot_s \bar{\alpha}$ is either $\overline{t\alpha}$ or $\{0\}$. Hence if we define $t\bar{\alpha}$ as the set product when $t \neq 0$ and as $\bar{0} = N$ when $t = 0$, then π also preserves scalar multiplication,

$$\pi(t\alpha) = t\pi(\alpha).$$

We thus have two vectorlike operations on the set W of all cosets of N, and we naturally expect W to turn out to be a vector space. We could prove this by verifying all the laws, but it is more elegant to notice the general setting for such a verification proof.

Theorem 4.1. Let V be a vector space, and let W be a set having two vectorlike operations, which we designate in the usual way. Suppose that there exists a surjective mapping $T: V \to W$ which preserves the operations: $T(s\alpha + t\beta) = sT(\alpha) + tT(\beta)$. Then W is a vector space.

Proof. We have to check laws A1 through S4. However, one example should make it clear to the reader how to proceed. We show that $T(0)$ satisfies A3 and hence is the zero vector of W. Since every $\beta \in W$ is of the form $T(\alpha)$, we have

$$T(0) + \beta = T(0) + T(\alpha) = T(0 + \alpha) = T(\alpha) = \beta,$$

which is A3. We shall ask the reader to check more of the laws in the exercises. □

Theorem 4.2. The set of cosets of a fixed subspace N of a vector space V themselves form a vector space, called the quotient space V/N, under the above natural operations, and the projection π is a surjective linear map from V to V/N.

Theorem 4.3. If T is in $\text{Hom}(V, W)$, and if the null space of T includes the subspace $M \subset V$, then T has a unique factorization through V/M. That is, there exists a unique transformation S in $\text{Hom}(V/M, W)$ such that $T = S \circ \pi$.

Proof. Since T is zero on M, it follows that T is constant on each coset A of M, so that $T[A]$ contains only one vector. If we define $S(A)$ to be the unique vector in $T[A]$, then $S(\bar{\alpha}) = T(\alpha)$, so $S \circ \pi = T$ by definition. Conversely, if $T = R \circ \pi$, then $R(\bar{\alpha}) = R \circ \pi(\alpha) = T(\alpha)$, and R is our above S. The linearity of S is practically obvious. Thus

$$S(\bar{\alpha} + \bar{\beta}) = S(\overline{\alpha + \beta}) = T(\alpha + \beta) = T(\alpha) + T(\beta) = S(\bar{\alpha}) + S(\bar{\beta}),$$

and homogeneity follows similarly. This completes the proof. □

One more remark is of interest here. If N is invariant under a linear map T in $\text{Hom}\, V$ (that is, $T[N] \subset N$), then for each α in V, $T[\bar{\alpha}]$ is a subset of the coset $\overline{T(\alpha)}$, for

$$T[\bar{\alpha}] = T[\alpha + N] = T(\alpha) +_s T[N] \subset T(\alpha) +_s N = \overline{T(\alpha)}.$$

There is therefore a map $S: V/N \to V/N$ defined by the requirement that $S(\bar{\alpha}) = \overline{T(\alpha)}$ (or $S \circ \pi = \pi \circ T$), and it is easy to check that S is linear. Therefore,

Theorem 4.4. If a subspace N of a vector space V is carried into itself by a transformation T in Hom V, then there is a unique transformation S in $\text{Hom}(V/N)$ such that $S \circ \pi = \pi \circ T$.

EXERCISES

4.1 Prove properties (4), (5), and (6) of affine subspaces.

4.2 Choose an origin O in the Euclidean plane $\mathbb{E}^2$ (your sheet of paper), and let L_1 and L_2 be two parallel lines not containing O. Let X and Y be distinct points on L_1 and Z any point on L_2. Draw the figure giving the geometric sums

$$\overrightarrow{OX} + \overrightarrow{OZ} \qquad \text{and} \qquad \overrightarrow{OY} + \overrightarrow{OZ}$$

(parallelogram rule), and state the theorem from plane geometry that says that these two sum points are on a third line L_3 parallel to L_1 and L_2.

4.3 a) Prove the associative law for addition for Theorem 4.1.
b) Prove also laws A4 and S2.

4.4 Return now to Exercise 2.1 and reexamine the situation in the light of Theorem 4.1. Show, finally, how we really know that the geometric vectors form a vector space.

4.5 Prove that the mapping S of Theorem 4.3 is injective if and only if N is the null space of T.

4.6 We know from Exercise 4.5 that if T is a surjective element of $\text{Hom}(V, W)$ and N is the null space of T, then the S of Theorem 4.3 is an *isomorphism* from V/N to W. Its inverse S^{-1} assigns a coset of N to each η in W. Show that the process of "indefinite integration" is an example of such a map S^{-1}. This is the process of calculating an integral and adding an arbitrary constant, as in

$$\int \sin x \, dx = -\cos x + c.$$

4.7 Suppose that N and M are subspaces of a vector space V and that $N \subset M$. Show that then M/N is a subspace of V/N and that V/M is naturally isomorphic to the quotient space $(V/N)/(M/N)$. [*Hint:* Every coset of N is a subset of some coset of M.]

4.8 Suppose that N and M are any subspaces of a vector space V. Prove that $(M + N)/N$ is naturally isomorphic to $M/(M \cap N)$. (Start with the fact that each coset of $M \cap N$ is included in a unique coset of N.)

4.9 Prove that the map S of Theorem 4.4 is linear.

4.10 Given $T \in \text{Hom}\, V$, show that $T^2 = 0$ ($T^2 = T \circ T$) if and only if $R(T) \subset N(T)$.

4.11 Suppose that $T \in \text{Hom}\, V$ and the subspace N are such that T is the identity on N and also on V/N. The latter assumption is that the S of Theorem 4.4 is the identity on V/N. Set $R = T - I$, and use the above exercise to show that $R^2 = 0$. Show that if $T = I + R$ and $R^2 = 0$, then there is a subspace N such that T is the identity on N and also on V/N.

4.12 We now view the above situation a little differently. Supposing that T is the identity on N and on V/N, and setting $R = I - T$, show that there exists a $K \in \mathrm{Hom}(V/N, V)$ such that $R = K \circ \pi$. Show that for any coset A of N the action of T on A can be viewed as translation through $K(A)$. That is, if $\xi \in A$ and $\eta = K(A)$, then $T(\xi) = \xi + \eta$.

4.13 Consider the map $T: \langle x_1, x_2 \rangle \mapsto \langle x_1 + 2x_2, x_2 \rangle$ in $\mathrm{Hom}\, \mathbb{R}^2$, and let N be the null space of $R = T - I$. Identify N and show that T is the identity on N and on $\mathbb{R}^2/N$. Find the map K of the above exercise. Such a mapping T is called a *shear transformation* of V parallel to N. Draw the unit square and its image under T.

4.14 If we remember that the linear span $L(A)$ of a subset A of a vector space V can be defined as the intersection of all the subspaces of V that include A, then the fact that the intersection of any collection of *affine* subspaces of a vector space V is either an affine subspace or empty suggests that we define the *affine span* $M(A)$ of a nonempty subset $A \subset V$ as the intersection of all *affine* subspaces including A. Then we know from (3) in our list of affine properties that $M(A)$ is an affine subspace, and by its definition above that it is the smallest affine subspace including A. We now naturally wonder whether $M(A)$ can be directly described in terms of linear combinations. Show first that if $\alpha \in A$, then $M(A) = L(A - \alpha) + \alpha$; then prove that $M(A)$ is the set of all linear combinations $\sum x_i\alpha_i$ on A such that $\sum x_i = 1$.

4.15 Show that the linear span of a set B is the affine span of $B \cup \{0\}$.

4.16 Show that $M(A + \gamma) = M(A) + \gamma$ for any γ in V and that $M(xA) = xM(A)$ for any x in $\mathbb{R}$.

5. DIRECT SUMS

We come now to the heart of the chapter. It frequently happens that the study of some phenomenon on a vector space V leads to a finite collection of subspaces $\{V_i\}$ such that V is naturally isomorphic to the product space $\prod_i V_i$. Under this isomorphism the maps $\theta_i \circ \pi_i$ on the product space become certain maps P_i in $\mathrm{Hom}\, V$, and the projection-injection identities are reflected in the identities $\sum P_i = I$, $P_j \circ P_j = P_j$ for all j, and $P_i \circ P_j = 0$ if $i \neq j$. Also, $V_i = \text{range}\, P_i$. The product structure that V thus acquires is then used to study the phenomenon that gave rise to it. For example, this is the way that we unravel the structure of a linear transformation in $\mathrm{Hom}\, V$, the study of which is one of the central problems in linear algebra.

Direct sums. If $V_1, \ldots, V_n$ are subspaces of the vector space V, then the mapping $\pi: \langle \alpha_1, \ldots, \alpha_n \rangle \mapsto \sum_1^n \alpha_i$ is a linear transformation from $\prod_1^n V_i$ to V, since it is the sum $\pi = \sum_1^n \pi_i$ of the coordinate projections.

> **Definition.** We shall say that the V_i's are *independent* if π is injective and that V is the *direct sum* of the V_i's if π is an isomorphism. We express the latter relationship by writing $V = V_1 \oplus \cdots \oplus V_n = \bigoplus_1^n V_i$.

Thus $V = \bigoplus_{i=1}^n V_i$ if and only if π is injective and surjective, i.e., if and only if the subspaces $\{V_i\}_1^n$ are both independent and span V. A useful restatement of the direct sum condition is that each $\alpha \in V$ is *uniquely* expressible as a sum $\sum_1^n \alpha_i$, with $\alpha_i \in V_i$ for all i; α has *some* such expression because the V_i's span V, and the expression is *unique* by their independence.

For example, let $V = \mathcal{C}(\mathbb{R})$ be the space of real-valued continuous functions on $\mathbb{R}$, let V_e be the subset of *even* functions (functions f such that $f(-x) = f(x)$ for all x), and let V_o be the subset of *odd* functions (functions such that $f(-x) = -f(x)$ for all x). It is clear that V_e and V_o are subspaces of V, and we claim that $V = V_e \oplus V_o$. To see this, note that for any f in V, $g(x) = (f(x) + f(-x))/2$ is even, $h(x) = (f(x) - f(-x))/2$ is odd, and $f = g + h$. Thus $V = V_e + V_o$. Moreover, this decomposition of f is unique, for if $f = g_1 + h_1$ also, where g_1 is even and h_1 is odd, then $g - g_1 = h_1 - h$, and therefore $g - g_1 = 0 = h_1 - h$, since the only function that is both even and odd is zero. The even-odd components of e^x are the hyperbolic cosine and sine functions:

$$e^x = \frac{(e^x + e^{-x})}{2} + \frac{(e^x - e^{-x})}{2} = \cosh x + \sinh x.$$

Since π is injective if and only if its null space is $\{0\}$ (Lemma 1.1), we have:

Lemma 5.1. The independence of the subspaces $\{V_i\}_1^n$ is equivalent to the property that if $\alpha_i \in V_i$ for all i and $\sum_1^n \alpha_i = 0$, then $\alpha_i = 0$ for all i.

Corollary. If the subspaces $\{V_i\}_1^n$ are independent, $\alpha_i \in V_i$ for all i, and $\sum_1^n \alpha_i$ is an element of V_j, then $\alpha_i = 0$ for $i \neq j$.

We leave the proof to the reader.

The case of two subspaces is particularly simple.

Lemma 5.2. The subspaces M and N of V are independent if and only if $M \cap N = \{0\}$.

Proof. If $\alpha \in M$, $\beta \in N$, and $\alpha + \beta = 0$, then $\alpha = -\beta \in M \cap N$. If $M \cap N = \{0\}$, this will further imply that $\alpha = \beta = 0$, so M and N are independent. On the other hand, if $0 \neq \beta \in M \cap N$, and if we set $\alpha = -\beta$, then $\alpha \in M$, $\beta \in N$, and $\alpha + \beta = 0$, so M and N are not independent. □

Note that the first argument above is simply the general form of the uniqueness argument we gave earlier for the even-odd decomposition of a function on $\mathbb{R}$.

Corollary. $V = M \oplus N$ if and only if $V = M + N$ and $M \cap N = \{0\}$.

Definition. If $V = M \oplus N$, then M and N are called *complementary* subspaces, and each is a *complement* of the other.

Warning: A subspace M of V does not have a unique complementary subspace unless M is trivial (that is, $M = \{0\}$ or $M = V$). If we view $\mathbb{R}^3$ as coordinatized Euclidean 3-space, then M is a proper subspace if and only if M is a plane containing the origin or M is a line through the origin (see Fig. 1.9). If M and N are

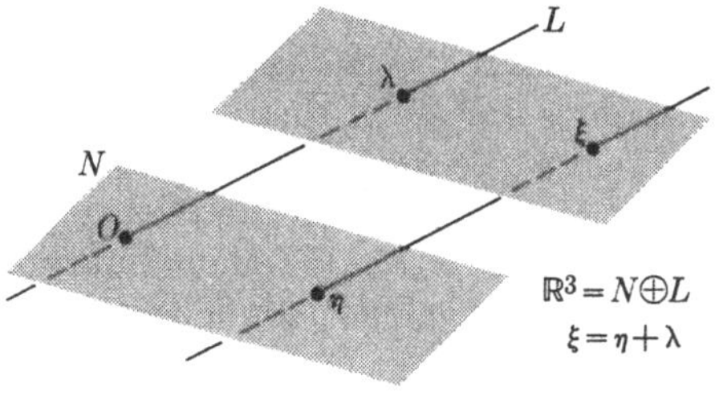

Fig. 1.9

proper subspaces one of which is a plane and the other a line not lying in that plane, then M and N are complementary subspaces. Moreover, these are the only nontrivial complementary pairs in $\mathbb{R}^3$. The reader will be asked to prove some of these facts in the exercises and they all will be clear by the middle of the next chapter.

The following lemma is technically useful.

Lemma 5.3. If V_1 and V_0 are independent subspaces of V and $\{V_i\}_2^n$ are independent subspaces of V_0, then $\{V_i\}_1^n$ are independent subspaces of V.

Proof. If $\alpha_i \in V_i$ for all i and $\sum_1^n \alpha_i = 0$, then, setting $\alpha_0 = \sum_2^n \alpha_i$, we have $\alpha_1 + \alpha_0 = 0$, with $\alpha_0 \in V_0$. Therefore, $\alpha_1 = \alpha_0 = 0$ by the independence of V_1 and V_0. But then $\alpha_2 = \alpha_3 = \cdots = \alpha_n = 0$ by the independence of $\{V_i\}_2^n$, and we are done (Lemma 5.1). □

Corollary. $V = V_1 \oplus V_0$ and $V_0 = \bigoplus_{i=2}^n V_i$ together imply that $V = \bigoplus_{i=1}^n V_i$.

Projections. If $V = \bigoplus_{i=1}^n V_i$, if π is the isomorphism $\langle \alpha_1, \ldots, \alpha_n \rangle \mapsto \alpha = \sum_1^n \alpha_i$, and if π_j is the jth projection map $\langle \alpha_1, \ldots, \alpha_n \rangle \mapsto \alpha_j$ from $\prod_{i=1}^n V_i$ to V_j, then $(\pi_j \circ \pi^{-1})(\alpha) = \alpha_j$.

Definition. We call α_j the jth *component* of α, and we call the linear map $P_j = \pi_j \circ \pi^{-1}$ the *projection of V onto V_j* (with respect to the given direct sum decomposition of V). Since each α in V is uniquely expressible as a sum $\alpha = \sum_1^n \alpha_i$, with α_i in V_i for all i, we can view $P_j(\alpha) = \alpha_j$ as "the part of α in V_j".

This use of the word "projection" is different from its use in the Cartesian product situation, and each is different from its use in the quotient space context (Section 0.12). It is apparent that these three uses are related, and the ambiguity causes little confusion since the proper meaning is always clear from the context.

Theorem 5.1. If the maps P_i are the above projections, then range $P_i = V_i$, $P_i \circ P_j = 0$ for $i \neq j$, and $\sum_1^n P_i = I$.

Proof. Since π is an isomorphism and $P_j = \pi_j \circ \pi^{-1}$, we have range $P_j =$ range $\pi_j = V_j$. Next, it follows directly from the corollary to Lemma 5.1 that if $\alpha \in V_j$, then $P_i(\alpha) = 0$ for $i \neq j$, and so $P_i \circ P_j = 0$ for $i \neq j$. Finally, $\sum_1^n P_i = \sum_1^n \pi_i \circ \pi^{-1} = (\sum_1^n \pi_i) \circ \pi^{-1} = \pi \circ \pi^{-1} = I$, and we are done. □

The above projection properties are clearly the reflection in V of the projection-injection identities for the isomorphic space $\prod_1^n V_i$.

A converse theorem is also true.

Theorem 5.2. If $\{P_i\}_1^n \subset \text{Hom } V$ satisfy $\sum_1^n P_i = I$ and $P_i \circ P_j = 0$ for $i \neq j$, and if we set $V_i = \text{range } P_i$, then $V = \bigoplus_{i=1}^n V_i$, and P_i is the corresponding projection on V_i.

Proof. The equation $\alpha = I(\alpha) = \sum_1^n P_i(\alpha)$ shows that the subspaces $\{V_i\}_1^n$ span V. Next, if $\beta \in V_j$, then $P_i(\beta) = 0$ for $i \neq j$, since $\beta \in \text{range } P_j$ and $P_i \circ P_j = 0$ if $i \neq j$. Then also $P_j(\beta) = (I - \sum_{i \neq j} P_i)(\beta) = I(\beta) = \beta$. Now consider $\alpha = \sum_1^n \alpha_i$ for any choice of $\alpha_i \in V_i$. Using the above two facts, we have $P_j(\alpha) = P_j(\sum_{i=1}^n \alpha_i) = \sum_{i=1}^n P_j(\alpha_i) = \alpha_j$. Therefore, $\alpha = 0$ implies that $\alpha_j = P_j(0) = 0$ for all j, and the subspaces V_i are independent. Consequently, $V = \bigoplus_1^n V_i$. Finally, the fact that $\alpha = \sum P_i(\alpha)$ and $P_i(\alpha) \in V_i$ for all i shows that $P_j(\alpha)$ is the jth component of α for every α and therefore that P_j is the projection of V onto V_j. □

There is an intrinsic characterization of the kind of map that is a projection.

Lemma 5.4. The projections P_i are idempotent ($P_i^2 = P_i$), or, equivalently, each is the identity on its range. The null space of P_i is the sum of the spaces V_j for $j \neq i$.

Proof. $P_j^2 = P_j \circ (I - \sum_{i \neq j} P_i) = P_j \circ I = P_j$. Since this can be rewritten as $P_j(P_j(\alpha)) = P_j(\alpha)$ for every α in V, it says exactly that P_j is the identity on its range.

Now set $W_i = \sum_{j \neq i} V_j$, and note that if $\beta \in W_i$, then $P_i(\beta) = 0$ since $P_i[V_j] = 0$ for $j \neq i$. Thus $W_i \subset N(P_i)$. Conversely, if $P_i(\alpha) = 0$, then $\alpha = I(\alpha) = \sum_1^n P_j(\alpha) = \sum_{j \neq i} P_j(\alpha) \in W_i$. Thus $N(P_i) \subset W_i$, and the two spaces are equal. □

Conversely:

Lemma 5.5. If $P \in \text{Hom}(V)$ is idempotent, then V is the direct sum of its range and null space, and P is the corresponding projection on its range.

Proof. Setting $Q = I - P$, we have $PQ = P - P^2 = 0$. Therefore, V is the direct sum of the ranges of P and Q, and P is the corresponding projection on its range, by the above theorem. Moreover, the range of Q is the null space of P, by the corollary. □

If $V = M \oplus N$ and P is the corresponding projection on M, we call P *the projection on M along N*. The projection P is not determined by M alone, since M does not determine N. A pair P and Q in Hom V such that $P + Q = I$ and $PQ = QP = 0$ is called a pair of complementary projections.

In the above discussion we have neglected another fine point. Strictly speaking, when we form the sum $\pi = \sum_1^n \pi_i$, we are treating each π_j as though it were from $\prod_1^n V_i$ to V, whereas actually the codomain of π_j is V_j. And we want P_j to be from V to V, whereas $\pi_j \circ \pi^{-1}$ has codomain V_j, so the equation $P_j = \pi_j \circ \pi^{-1}$ can't quite be true either. To repair these flaws we have to introduce the injection $\iota_j \colon V_j \to V$, which is the identity map on V_j, but which views V_j as a subspace of V and so takes V as its codomain. If our concept of a mapping includes a codomain possibly larger than the range, then we have to admit such identity injections. Then, setting $\bar{\pi}_j = \iota_j \circ \pi_j$, we have the correct equations $\pi = \sum_1^n \bar{\pi}_i$ and $P_j = \bar{\pi}_j \circ \pi^{-1}$.

EXERCISES

5.1 Prove the corollary to Lemma 5.1.

5.2 Let α be the vector $\langle 1, 1, 1 \rangle$ in $\mathbb{R}^3$, and let $M = \mathbb{R}\alpha$ be its one-dimensional span. Show that each of the three coordinate planes is a complement of M.

5.3 Show that a finite product space $V = \prod_1^n V_i$ has subspaces $\{W_i\}_1^n$ such that W_i is isomorphic to V_i and $V = \bigoplus_1^n W_i$. Show how the corresponding projections $\{P_i\}$ are related to the π_i's and θ_i's.

5.4 If $T \in \mathrm{Hom}(V, W)$, show that (the graph of) T is a complement of $W' = \{0\} \times W$ in $V \times W$.

5.5 If l is a linear functional on V ($l \in \mathrm{Hom}(V, \mathbb{R}) = V^*$), and if α is a vector in V such that $l(\alpha) \neq 0$, show that $V = N \oplus M$, where N is the null space of l and $M = \mathbb{R}\alpha$ is the linear span of α. What does this result say about complements in $\mathbb{R}^3$?

5.6 Show that any complement M of a subspace N of a vector space V is isomorphic to the quotient space V/N.

5.7 We suppose again that every subspace has a complement. Show that if $T \in \mathrm{Hom}\, V$ is *not* injective, then there is a nonzero S in $\mathrm{Hom}\, V$ such that $T \circ S = 0$. Show that if $T \in \mathrm{Hom}\, V$ is *not* surjective, then there is a nonzero S in $\mathrm{Hom}\, V$ such that $S \circ T = 0$.

5.8 Using the above exercise for half the arguments, show that $T \in \mathrm{Hom}\, V$ is *injective* if and only if $T \circ S = 0 \Rightarrow S = 0$ and that T is *surjective* if and only if $S \circ T = 0 \Rightarrow S = 0$. We thus have characterizations of injectivity and surjectivity that are *formal*, in the sense that they do not refer to the fact that S and T are transformations, but refer only to the algebraic properties of S and T as elements of an algebra.

5.9 Let M and N be complementary subspaces of a vector space V, and let X be a subspace such that $X \cap N = \{0\}$. Show that there is a linear injection from X to M. [*Hint:* Consider the projection P of V onto M along N.] Show that any two complements of a subspace N are isomorphic by showing that the above injection is surjective if and only if X is a complement of N.

5.10 Going back to the first point of the preceding exercise, let Y be a complement of $P[X]$ in M. Show that $X \cap Y = \{0\}$ and that $X \oplus Y$ is a complement of N.

5.11 Let M be a proper subspace of V, and let $\{\alpha_i : i \in I\}$ be a finite set in V. Set $L = L(\{\alpha_i\})$, and suppose that $M + L = V$. Show that there is a subset $J \subset I$ such that $\{\alpha_i : i \in J\}$ spans a complement of M. [*Hint:* Consider a largest possible subset J such that $M \cap L(\{\alpha_i\}_J) = \{0\}$.]

5.12 Given $T \in \text{Hom}(V, W)$ and $S \in \text{Hom}(W, X)$, show that

a) $S \circ T$ is surjective $\Leftrightarrow$ S is surjective and $R(T) + N(S) = W$;

b) $S \circ T$ is injective $\Leftrightarrow$ T is injective and $R(T) \cap N(S) = \{0\}$;

c) $S \circ T$ is an isomorphism $\Leftrightarrow$ S is surjective, T is injective, and $W = R(T) \oplus N(S)$.

5.13 Assuming that every subspace of V has a complement, show that $T \in \text{Hom}\, V$ satisfies $T^2 = 0$ if and only if V has a direct sum decomposition $V = M \oplus N$ such that $T = 0$ on N and $T[M] \subset N$.

5.14 Suppose next that $T^3 = 0$ but $T^2 \neq 0$. Show that V can be written as $V = V_1 \oplus V_2 \oplus V_3$, where $T[V_1] \subset V_2$, $T[V_2] \subset V_3$, and $T = 0$ on V_3. (Assume again that any subspace of a vector space has a complement.)

5.15 We now suppose that $T^n = 0$ but $T^{n-1} \neq 0$. Set $N_i =$ null space (T^i) for $i = 1, \ldots, n-1$, and let V_1 be a complement of N_{n-1} in V. Show first that

$$T[V_1] \cap N_{n-2} = \{0\}$$

and that $T[V_1] \subset N_{n-1}$. Extend $T[V_1]$ to a complement V_2 of N_{n-2} in N_{n-1}, and show that in this way we can construct subspaces $V_1, \ldots, V_n$ such that

$$V = \bigoplus_1^n V_i, \qquad T[V_i] \subset V_{i+1} \quad \text{for} \quad i < n,$$

and

$$T[V_n] = \{0\}.$$

On solving a linear equation. Many important problems in mathematics are in the following general form. A linear operator $T: V \to W$ is given, and for a given $\eta \in W$ the equation $T(\xi) = \eta$ is to be solved for $\xi \in V$. In our terms, the condition that there exist a solution is exactly the condition that η be in the range space of T. In special circumstances this condition can be given more or less useful equivalent alternative formulations. Let us suppose that we know how to recognize $R(T)$, in which case we may as well make it the new codomain, and so assume that T is surjective. There still remains the problem of determining what we mean by solving the equation. The universal principle running through all the important instances of the problem is that a solution process calculates a right inverse to T, that is, a linear operator $S: W \to V$ such that $T \circ S = I_W$, the identity on W. Thus a solution process picks one solution vector $\xi \in V$ for each $\eta \in W$ in such a way that the solving ξ varies linearly with η. Taking this as our meaning of solving, we have the following fundamental reformulation.

Theorem 5.3. Let T be a surjective linear map from the vector space V to the vector space W, and let N be its null space. Then a subspace M is a complement of N if and only if the restriction of T to M is an isomorphism from M to W. The mapping $M \mapsto (T \upharpoonright M)^{-1}$ is a bijection from the set of all such complementary subspaces M to the set of all linear right inverses of T.

Proof. It should be clear that a subspace M is the range of a linear right inverse of T (a map S such that $T \circ S = I_W$) if and only if $T \upharpoonright M$ is an isomorphism to W, in which case $S = (T \upharpoonright M)^{-1}$. Strictly speaking, the right inverse must be from W to V and therefore must be $R = \iota_M \circ S$, where ι_M is the identity injection from M to V. Then $(R \circ T)^2 = R \circ (T \circ R) \circ T = R \circ I_W \circ T = R \circ T$, and $R \circ T$ is a projection whose range is M and whose null space is N (since R is injective). Thus $V = M \oplus N$. Conversely, if $V = M \oplus N$, then $T \upharpoonright M$ is injective because $M \cap N = \{0\}$ and surjective because $M + N = V$ implies that $W = T[V] = T[M + N] = T[M] + T[N] = T[M] + \{0\} = T[M]$. □

Polynomials in *T*. The material in this subsection will be used in our study of differential equations with constant coefficients and in the proof of the diagonalizability of a symmetric matrix. In linear algebra it is basic in almost any approach to the canonical forms of matrices.

If $p_1(t) = \sum_0^m a_i t^i$ and $p_2(t) = \sum_0^n b_j t^j$ are any two polynomials, then their product is the polynomial

$$p(t) = p_1(t)p_2(t) = \sum_0^{m+n} c_k t^k,$$

where $c_k = \sum_{i+j=k} a_i b_j = \sum_{i=0}^k a_i b_{k-i}$. Now let T be any fixed element of $\mathrm{Hom}(V)$, and for any polynomial $q(t)$ let $q(T)$ be the transformation obtained by replacing t by T. That is, if $q(t) = \sum_0^l c_k t^k$, then $q(T) = \sum_0^l c_k T^k$, where, of course, T^l is the composition product $T \circ T \circ \cdots \circ T$ with l factors. Then the bilinearity of composition (Theorem 3.3) shows that if $p(t) = p_1(t)p_2(t)$, then $p(T) = p_1(T) \circ p_2(T)$. In particular, any two polynomials in T commute with each other under composition. More simply, the commutative law for addition implies that

$$\text{if} \quad p(t) = p_1(t) + p_2(t), \qquad \text{then} \quad p(T) = p_1(T) + p_2(T).$$

The mapping $p(t) \mapsto p(T)$ from the algebra of polynomials to the algebra $\mathrm{Hom}(V)$ thus preserves addition, multiplication, and (obviously) scalar multiplication. That is, it preserves all the operations of an algebra and is therefore what is called an (algebra) homomorphism.

The word "homomorphism" is a general term describing a mapping θ between two algebraic systems of the same kind such that θ preserves the operations of the system. Thus a homomorphism between vector spaces is simply a linear transformation, and a homomorphism between groups is a mapping preserving the one group operation. An accessible, but not really typical, example of the latter is the logarithm function, which is a homomorphism from the multiplicative group of positive real numbers to the additive group of $\mathbb{R}$. The logarithm function is actually a bijective homomorphism and is therefore a group *isomorphism.*

If this were a course in algebra, we would show that the division algorithm and the properties of the degree of a polynomial imply the following theorem. (However, see Exercises 5.16 through 5.20.)

Theorem 5.4. If $p_1(t)$ and $p_2(t)$ are relatively prime polynomials, then there exist polynomials $a_1(t)$ and $a_2(t)$ such that

$$a_1(t)p_1(t) + a_2(t)p_2(t) = 1.$$

By relatively prime we mean having no common factors except constants. We shall assume this theorem and the results of the discussion preceding it in proving our next theorem.

We say that a subspace $M \subset V$ is *invariant* under $T \in \mathrm{Hom}(V)$ if $T[M] \subset M$ [that is, $T \restriction M \in \mathrm{Hom}(M)$].

Theorem 5.5. Let T be any transformation in Hom V, and let q be any polynomial. Then the null space N of $q(T)$ is invariant under T, and if $q = q_1q_2$ is any factorization of q into relatively prime factors and N_1 and N_2 are the null space of $q_1(T)$ and $q_2(T)$, respectively, then $N = N_1 \oplus N_2$.

Proof. Since $T \circ q(T) = q(T) \circ T$, we see that if $q(T)(\alpha) = 0$, then $q(T)(T\alpha) = T(q(T)(\alpha)) = 0$, so $T[N] \subset N$. Note also that since $q(T) = q_1(T) \circ q_2(T)$, it follows that any α in N_2 is also in N, so $N_2 \subset N$. Similarly, $N_1 \subset N$. We can therefore replace V by N and T by $T \restriction N$; hence we can assume that $T \in \mathrm{Hom}\, N$ and $q(T) = q_1(T) \circ q_2(T) = 0$.

Now choose polynomials a_1 and a_2 so that $a_1q_1 + a_2q_2 = 1$. Since $p \mapsto p(T)$ is an algebraic homomorphism, we then have

$$a_1(T) \circ q_1(T) + a_2(T) \circ q_2(T) = I.$$

Set $A_1 = a_1(T)$, etc., so that $A_1 \circ Q_1 + A_2 \circ Q_2 = I$, $Q_1 \circ Q_2 = 0$, and all the operators A_i, Q_i commute with each other. Finally, set $P_i = A_i \circ Q_i = Q_i \circ A_i$ for $i = 1, 2$. Then $P_1 + P_2 = I$ and $P_1P_2 = P_2P_1 = 0$. Thus P_1 and P_2 are projections, and N is the direct sum of their ranges: $N = V_1 \oplus V_2$. Since each range is the null space of the other projection, we can rewrite this as $N = N_1 \oplus N_2$, where $N_i = N(P_i)$. It remains for us to show that $N(P_i) = N(Q_i)$. Note first that since $Q_1 \circ P_2 = Q_1 \circ Q_2 \circ A_2 = 0$, we have $Q_1 = Q_1 \circ I = Q_1 \circ (P_1 + P_2) = Q_1 \circ P_1$. Then the two identities $P_i = A_i \circ Q_i$ and $Q_i = Q_i \circ P_i$ show that the null space of each of P_i and Q_i is included in the other, and so they are equal. This completes the proof of the theorem. □

Corollary. Let $p(t) = \prod_{i=1}^m p_i(t)$ be a factorization of the polynomial $p(t)$ into relatively prime factors, let T be an element of $\mathrm{Hom}(V)$, and set $N_i = N(p_i(T))$ for $i = 1, \ldots, m$ and $N = N(p(T))$. Then N and all the N_i are invariant under T, and $N = \bigoplus_{i=1}^m N_i$.

Proof. The proof is by induction on m. The theorem is the case $m = 2$, and if we set $q = \prod_2^m p_i(t)$ and $M = N(q(T))$, then the theorem implies that $N = N_1 \oplus M$ and that N_1 and M are invariant under T. Restricting T to M, we see that the inductive hypothesis implies that $M = \bigoplus_{i=2}^m N_i$ and that N_i is invariant under T for $i = 2, \ldots, m$. The corollary to Lemma 5.3 then yields our result. □

EXERCISES

5.16 Presumably the reader knows (or can see) that the degree $d(P)$ of a polynomial P satisfies the laws

$$d(P+Q) \le \max\{d(P), d(Q)\},$$
$$d(P \cdot Q) = d(P) + d(Q) \qquad \text{if both } P \text{ and } Q \text{ are nonzero.}$$

The degree of the zero polynomial is undefined. (It would have to be $-\infty$!) By induction on the degree of P, prove that for any two polynomials P and D, with $D \neq 0$, there are polynomials Q and R such that $P = DQ + R$ and $d(R) < d(D)$ *or* $R = 0$. [*Hint:* If $d(P) < d(D)$, we can take Q and R as what? If $d(P) \ge d(D)$, and if the leading terms of P and D are ax^n and bx^m, respectively, with $n \ge m$, then the polynomial

$$P' = P - \left(\frac{a}{b}\right) x^{n-m} D$$

has degree less than $d(P)$, so $P' = DQ' + R'$ by the inductive hypothesis. Now finish the proof.]

5.17 Assuming the above result, prove that R and Q are uniquely determined by P and D. (Assume also that $P = DQ' + R'$, and prove from the properties of degree that $R' = R$ and $Q' = Q$.) These two results together constitute the ***division algorithm*** for polynomials.

5.18 If P is any polynomial

$$P(x) = \sum_0^n a_n x^n,$$

and if t is any number, then of course $P(t)$ is the number

$$\sum_0^n a_n t^n.$$

Prove from the division algorithm that for any polynomial P and any number t there is a polynomial Q such that

$$P(x) = (x - t)Q(x) + P(t),$$

and therefore that $P(x)$ is divisible by $(x - t)$ if and only if $P(t) = 0$.

5.19 Let P and Q be nonzero polynomials, and choose polynomials A_0 and B_0 such that among all the polynomials of the form $AP + BQ$ the polynomial

$$D = A_0P + B_0Q$$

is nonzero and has minimum degree. Prove that D is a factor of both P and Q. (Suppose that D does not divide P and apply the division algorithm to get a contradiction with the choice of A_0 and B_0.)

5.20 Let P and Q be nonzero relatively prime polynomials. This means that if E is a common factor of P and Q ($P = EP', Q = EQ'$), then E is a constant. Prove that there are polynomials A and B such that $A(x)P(x) + B(x)Q(x) = 1$. (Apply the above exercise.)

5.21 In the context of Theorem 5.5, show that the restriction of $q_2(T) = Q_2$ to N_1 is an isomorphism (from N_1 to N_1).

5.22 An involution on V is a mapping $T \in \text{Hom } V$ such that $T^2 = I$. Show that if T is an involution, then V is a direct sum $V = V_1 \oplus V_2$, where $T(\xi) = \xi$ for every $\xi \in V_1$ ($T = I$ on V_1) and $T(\xi) = -\xi$ for every $\xi \in V_2$ ($T = -I$ on V_2). (Apply Theorem 5.5.)

5.23 We noticed earlier (in an exercise) that if φ is any mapping from a set A to a set B, then $f \mapsto f \circ \varphi$ is a linear map T_φ from $\mathbb{R}^B$ to $\mathbb{R}^A$. Show now that if $\psi: B \to C$, then

$$T_{\psi \circ \varphi} = T_\varphi \circ T_\psi.$$

(This should turn out to be a direct consequence of the associativity of composition.)

5.24 Let A be any set, and let $\varphi: A \to A$ be such that $\varphi \circ \varphi(a) = a$ for every a. Then $T_\varphi: f \mapsto f \circ \varphi$ is an involution on $V = \mathbb{R}^A$ (since $T_{\varphi \circ \psi} = T_\psi \circ T_\varphi$). Show that the decomposition of $\mathbb{R}^{\mathbb{R}}$ as the direct sum of the subspace of even functions and the subspace of odd functions arises from an involution on $\mathbb{R}^{\mathbb{R}}$ defined by such a map $\varphi: \mathbb{R} \to \mathbb{R}$.

5.25 Let V be a subspace of $\mathbb{R}^{\mathbb{R}}$ consisting of differentiable functions, and suppose that V is invariant under differentiation ($f \in V \Rightarrow Df \in V$). Suppose also that on V the linear operator $D \in \text{Hom } V$ satisfies $D^2 - 2D - 3I = 0$. Prove that V is the direct sum of two subspaces M and N such that $D = 3I$ on M and $D = -I$ on N. Actually, it follows that M is the linear span of a single vector, and similarly for N. Find these two functions, if you can. ($f' = 3f \Rightarrow f = $?)

***Block decompositions of linear maps.** Given T in Hom V and a direct sum decomposition $V = \bigoplus_1^n V_i$, with corresponding projections $\{P_i\}_1^n$, we can consider the maps $T_{ij} = P_i \circ T \circ P_j$. Although T_{ij} is from V to V, we may also want to consider it as being from V_j to V_i (in which case, strictly speaking, what is it?). We picture the T_{ij}'s arranged schematically in a rectangular array similar to a matrix, as indicated below for $n = 2$.

T_{11}	T_{12}
T_{21}	T_{22}

Furthermore, since $T = \sum_{i,j} T_{ij}$, we call the doubly indexed family the block decomposition of T associated with the given direct sum decomposition of V.

More generally, if $T \in \text{Hom}(V, W)$ and W also has a direct sum decomposition $W = \bigoplus_{i=1}^m W_i$, with corresponding projections $\{Q_i\}_1^m$, then the family $\{T_{ij}\}$ defined by $T_{ij} = Q_i \circ T \circ P_j$ and pictured as an $m \times n$ rectangular array is the block decomposition of T with respect to the two direct sum decompositions.

Whenever T in Hom V has a special relationship to a particular direct sum decomposition of V, the corresponding block diagram may have features that display these special properties in a vivid way; this then helps us to understand the nature of T better and to calculate with it more easily.

For example, if $V = V_1 \oplus V_2$, then V_1 is invariant under T (i.e., $T[V_1] \subset V_1$) if and only if the block diagram is *upper triangular*, as shown in the following diagram.

$$\begin{array}{c|c} T_{11} & T_{12} \\ \hline 0 & T_{22} \end{array}$$

Suppose, next, that $T^2 = 0$. Letting V_1 be the range of T, and supposing that V_1 has a complement V_2, the reader should clearly see that the corresponding block diagram is

$$\begin{array}{c|c} 0 & T_{12} \\ \hline 0 & 0 \end{array}$$

This form is called *strictly* upper triangular; it is upper triangular and also zero on the main diagonal. Conversely, if T has some strictly upper-triangular 2×2 block diagram, then $T^2 = 0$.

If R is a composition product, $R = ST$, then its block components can be computed in terms of those of S and T. Thus

$$R_{ik} = P_iRP_k = P_iSTP_k = P_iS\left(\sum_{j=1}^{n} P_j\right)TP_k = \sum_{j=1}^{n} S_{ij}T_{jk}.$$

We have used the identities $I = \sum_{j=1}^{n} P_j$ and $P_j = P_j^2$. The 2×2 case is pictured below.

$$\begin{array}{c|c} S_{11}T_{11} + S_{12}T_{21} & S_{11}T_{12} + S_{12}T_{22} \\ \hline S_{21}T_{11} + S_{22}T_{21} & S_{21}T_{12} + S_{22}T_{22} \end{array}$$

From this we can read off a fact that will be useful to us later: If T is 2×2 upper triangular ($T_{21} = 0$), and if T_{ii} is invertible as a map from V_i to V_i ($i = 1, 2$), then T is invertible and its inverse is

$$\begin{array}{c|c} T_{11}^{-1} & -T_{11}^{-1}T_{12}T_{22}^{-1} \\ \hline 0 & T_{22}^{-1} \end{array}$$

We find this solution by simply setting the product diagram equal to

$$\begin{array}{c|c} I_1 & 0 \\ \hline 0 & I_2 \end{array}$$

and solving; but of course with the diagram in hand it can simply be checked to be correct.

EXERCISES

5.26 Show that if $T \in \operatorname{Hom} V$, if $V = \bigoplus_1^n V_i$, and if $\{P_i\}_1^n$ are the corresponding projections, then the sum of the transformation $T_{ij} = P_i \circ T \circ P_j$ is T.

5.27 If S and T are in Hom V and $\{S_{ij}\}$, $\{T_{ij}\}$ are their block components with respect to some direct sum decomposition of V, show that $S_{ij} \circ T_{lk} = 0$ if $j \neq l$.

5.28 Verify that if T has an upper-triangular block diagram with respect to the direct sum decomposition $V = V_1 \oplus V_2$, then V_1 is invariant under T.

5.29 Verify that if the diagram is strictly upper triangular, then $T^2 = 0$.

5.30 Show that if $V = V_1 \oplus V_2 \oplus V_3$ and $T \in \text{Hom } V$, then the subspaces V_i are all invariant under T if and only if the block diagram for T is

$$\begin{bmatrix} T_{11} & 0 & 0 \\ 0 & T_{22} & 0 \\ 0 & 0 & T_{33} \end{bmatrix}.$$

Show that T is invertible if and only if T_{ii} is invertible (as an element of Hom V_i) for each i.

5.31 Supposing that T has an upper-triangular 2×2 block diagram and that T_{ii} is invertible as an element of Hom V_i for $i = 1, 2$, verify that T is invertible by forming the 2×2 block diagram that is the product of the diagram for T and the diagram given in the text as the inverse of T.

5.32 Supposing that T is as in the preceding exercise, show that $S = T^{-1}$ *must* have the given block diagram by considering the two equations $T \circ S = I$ and $S \circ T = I$ in their block form.

5.33 What would strictly upper triangular mean for a 3×3 block diagram? What is the corresponding property of T? Show that T has this property if and only if it has a strictly upper-triangular block diagram. (See Exercise 5.14.)

5.34 Suppose that T in Hom V satisfies $T^n = 0$ (but $T^{n-1} \neq 0$). Show that T has a strictly upper-triangular $n \times n$ block decomposition. (Apply Exercise 5.15.)

6. BILINEARITY

Bilinear mappings. The notion of a *bilinear* mapping is important to the understanding of linear algebra because it is the vector setting for the duality principle (Section 0.10).

Definition. If U, V, and W are vector spaces, then a mapping

$$\omega\colon \langle \xi, \eta \rangle \mapsto \omega(\xi, \eta)$$

from $U \times V$ to W is *bilinear* if it is linear in each variable *when the other variable is held fixed.*

That is, if we hold ξ fixed, then $\eta \mapsto \omega(\xi, \eta)$ is linear [and so belongs to $\text{Hom}(V, W)$]; if we hold η fixed, then similarly $\omega(\xi, \eta)$ is in $\text{Hom}(U, W)$ as a function of ξ. This is *not* the same notion as linearity on the product vector space $U \times V$. For example, $\langle x, y \rangle \to x + y$ is a linear mapping from $\mathbb{R}^2$ to $\mathbb{R}$, but it is not bilinear. If y is held fixed, then the mapping $x \mapsto x + y$ is affine (translation through y), but it is not linear unless y is 0. On the other hand, $\langle x, y \rangle \mapsto xy$ is a bilinear mapping from $\mathbb{R}^2$ to $\mathbb{R}$, but it is not linear. If y

is held fixed, then the mapping $x \mapsto yx$ is linear. But the sum of two ordered couples does not map to the sum of their images:

$$<x, y> + <u, v> = <x + u, y + v> \mapsto (x + u)(y + v),$$

which is not the sum of the images, $xy + uv$. Similarly, the scalar product $(\mathbf{x}, \mathbf{y}) = \sum_1^n x_i y_i$ is bilinear from $\mathbb{R}^n \times \mathbb{R}^n$ to $\mathbb{R}$, as we observed in Section 2.

The linear meaning of bilinearity is partially explained in the following theorem.

Theorem 6.1. If $\omega: U \times V \to W$ is bilinear, then, by duality, ω is equivalent to a linear mapping from U to $\text{Hom}(V, W)$ and also to a linear mapping from V to $\text{Hom}(U, W)$.

Proof. For each fixed $\eta \in V$ let ω_η be the mapping $\xi \mapsto \omega(\xi, \eta)$. That is, $\omega_\eta(\xi) = \omega(\xi, \eta)$. Then $\omega_\eta \in \text{Hom}(U, W)$ by the bilinear hypothesis. The mapping $\eta \mapsto \omega_\eta$ is thus from V to $\text{Hom}(U, W)$, and *its* linearity is due to the linearity of ω in η when ξ is held fixed:

$$\omega_{c\eta+d\zeta}(\xi) = \omega(\xi, c\eta + d\zeta) = c\omega(\xi, \eta) + d\omega(\xi, \zeta) = c\omega_\eta(\xi) + d\omega_\zeta(\xi),$$

so that

$$\omega_{c\eta+d\zeta} = c\omega_\eta + d\omega_\zeta.$$

Similarly, if we define ω^ξ by $\omega^\xi(\eta) = \omega(\xi, \eta)$, then $\xi \mapsto \omega^\xi$ is a linear mapping from U to $\text{Hom}(V, W)$. Conversely, if $\varphi: U \to \text{Hom}(V, W)$ is linear, then the function ω defined by $\omega(\xi, \eta) = \varphi(\xi)(\eta)$ is bilinear. Moreover, $\omega^\xi = \varphi(\xi)$, so that φ is the mapping $\xi \mapsto \omega^\xi$. □

We shall see that bilinearity occurs frequently. Sometimes the reinterpretation provided by the above theorem provides new insights; at other times it seems less helpful.

For example, the composition map $<S, T> \mapsto S \circ T$ is bilinear, and the corollary of Theorem 3.3, which in effect states that composition on the right by a fixed T is a linear map, is simply part of an explicit statement of the bilinearity. But the linear map $T \mapsto$ composition by T is a complicated object that we have no need for except in the case $W = \mathbb{R}$.

On the other hand, the linear combination formula $\sum_1^n x_i \alpha_i$ and Theorem 1.2 do receive new illumination.

Theorem 6.2. The mapping $\omega(\mathbf{x}, \boldsymbol{\alpha}) = \sum_1^n x_i \alpha_i$ is bilinear from $\mathbb{R}^n \times V^n$ to V. The mapping $\boldsymbol{\alpha} \mapsto \omega_{\boldsymbol{\alpha}}$ is therefore a linear mapping from V^n to $\text{Hom}(\mathbb{R}^n, V)$, and, in fact, is an isomorphism.

Proof. The linearity of ω in $\mathbf{x}$ for a fixed $\boldsymbol{\alpha}$ was proved in Theorem 1.2, and its linearity in $\boldsymbol{\alpha}$ for a fixed $\mathbf{x}$ is seen in the same way. Then $\boldsymbol{\alpha} \mapsto \omega_{\boldsymbol{\alpha}}$ is linear by Theorem 6.1. Its bijectivity was implicit in Theorem 1.2. □

It should be remarked that we can use any finite index set I just as well as the special set $\bar{n}$ and conclude that $\omega(\mathbf{x}, \boldsymbol{\alpha}) = \sum_{i \in I} x_i \alpha_i$ is bilinear from $\mathbb{R}^I \times V^I$

to V and that $\boldsymbol{\alpha} \mapsto \omega_{\boldsymbol{\alpha}}$ is an isomorphism from V^I to $\mathrm{Hom}(\mathbb{R}^I, V)$. Also note that $\omega_{\boldsymbol{\alpha}} = L_{\boldsymbol{\alpha}}$ in the terminology of Section 1.

Corollary. The scalar product $(\mathbf{x}, \mathbf{a}) = \sum_1^n x_i a_i$ is bilinear from $\mathbb{R}^n \times \mathbb{R}^n$ to $\mathbb{R}$; therefore, $\mathbf{a} \mapsto \omega_{\mathbf{a}} = L_{\mathbf{a}}$ is an isomorphism from $\mathbb{R}^n$ to $\mathrm{Hom}(\mathbb{R}^n, \mathbb{R})$.

Natural isomorphisms. We often find two vector spaces related to each other in such a way that a particular isomorphism between them is singled out. This phenomenon is hard to pin down in general terms but easy to describe by examples.

Duality is one source of such "natural" isomorphisms. For example, an $m \times n$ matrix $\{t_{ij}\}$ is a real-valued function of the two variables $\langle i, j \rangle$, and as such it is an element of the Cartesian space $\mathbb{R}^{\overline{m} \times \overline{n}}$. We can also view $\{t_{ij}\}$ as a sequence of n column vectors in $\mathbb{R}^m$. This is the dual point of view where we hold j fixed and obtain a function of i for each j. From this point of view $\{t_{ij}\}$ is an element of $(\mathbb{R}^{\overline{m}})^{\overline{n}}$. This correspondence between $\mathbb{R}^{\overline{m} \times \overline{n}}$ and $(\mathbb{R}^{\overline{m}})^{\overline{n}}$ is clearly an isomorphism, and is an example of a natural isomorphism.

We review next the various ways of looking at Cartesian n-space itself. One standard way of defining an ordered n-tuplet is by induction. The ordered triplet $\langle x, y, z \rangle$ is defined as the ordered pair $\langle \langle x, y \rangle, z \rangle$, and the ordered n-tuplet $\langle x_1, \ldots, x_n \rangle$ is defined as $\langle \langle x_1, \ldots, x_{n-1} \rangle, x_n \rangle$. Thus we define $\mathbb{R}^n$ inductively by setting $\mathbb{R}^1 = \mathbb{R}$ and $\mathbb{R}^n = \mathbb{R}^{n-1} \times \mathbb{R}$.

The ordered n-tuplet can also be defined as the function on $\overline{n} = \{1, \ldots, n\}$ which assigns x_i to i. Then

$$\langle x_1, \ldots, x_n \rangle = \{\langle 1, x_1 \rangle, \ldots, \langle n, x_n \rangle\},$$

and Cartesian n-space is $\mathbb{R}^{\overline{n}} = \mathbb{R}^{\{1, \ldots, n\}}$.

Finally, we often wish to view Cartesian $(n + m)$-space as the Cartesian product of Cartesian n-space with Cartesian m-space, so we now take

$$\langle x_1, \ldots, x_{n+m} \rangle \quad \text{as} \quad \langle \langle x_1, \ldots, x_n \rangle, \langle x_{n+1}, \ldots, x_{n+m} \rangle \rangle$$

and $\mathbb{R}^{n+m}$ as $\mathbb{R}^n \times \mathbb{R}^m$.

Here again if we pair two different models for the same n-tuplet, we have an obvious natural isomorphism between the corresponding models for Cartesian n-space.

Finally, the characteristic properties of Cartesian product spaces given in Theorems 3.6 and 3.7 yield natural isomorphisms. Theorem 3.6 says that an n-tuple of linear maps $\{T_i\}_1^n$ on a common domain V is equivalent to a single n-tuple-valued map T, where $T(\xi) = \langle T_1(\xi), \ldots, T_n(\xi) \rangle$ for all $\xi \in V$. (This is duality again! $T_i(\xi)$ is a function of the two variables i and ξ.) And it is not hard to see that this identification of T with $\{T_i\}_1^n$ is an isomorphism from $\prod_i \mathrm{Hom}(V, W_i)$ to $\mathrm{Hom}(V, \prod_i W_i)$.

Similarly, Theorem 3.7 identifies an n-tuple of linear maps $\{T_i\}_1^n$ into a common codomain V with a single linear map T of an n-tuple variable, and this identification is a natural isomorphism from $\prod_1^n \mathrm{Hom}(W_i, V)$ to $\mathrm{Hom}(\prod_1^n W_i, V)$.

An arbitrary isomorphism between two vector spaces identifies them in a transient way. For the moment we think of the vector spaces as representing the same abstract space, but only so long as the isomorphism is before us. If we shift to a different isomorphism between them, we obtain a new temporary identification. Natural isomorphisms, on the other hand, effect permanent identifications, and we think of paired objects as being two aspects of the same object in a deeper sense. Thus we think of a matrix as "being" either a sequence of row vectors, a sequence of column vectors, or a single function of two integer indices. We shall take a final look at this question at the end of Section 3 in the next chapter.

*We can now make the ultimate dissection of the theorems centering around the linear combination formula. Laws S1 through S3 state *exactly* that the scalar product $x\alpha$ is bilinear. More precisely, they state that the mapping $S: \langle x, \alpha \rangle \mapsto x\alpha$ from $\mathbb{R} \times W$ to W is bilinear. In the language of Theorem 6.1, $x\alpha = \omega_\alpha(x)$, and from that theorem we conclude that the mapping $\alpha \mapsto \omega_\alpha$ is an isomorphism from W to $\text{Hom}(\mathbb{R}, W)$.

This isomorphism between W and $\text{Hom}(\mathbb{R}, W)$ extends to an isomorphism from W^n to $(\text{Hom}(\mathbb{R}, W))^n$, which in turn is naturally isomorphic to $\text{Hom}(\mathbb{R}^n, W)$ by the second Cartesian product isomorphism. Thus W^n is naturally isomorphic to $\text{Hom}(\mathbb{R}^n, W)$; the mapping is $\boldsymbol{\alpha} \mapsto L_{\boldsymbol{\alpha}}$, where $L_{\boldsymbol{\alpha}}(\mathbf{x}) = \sum_1^n x_i\alpha_i$.

In particular, $\mathbb{R}^n$ is naturally isomorphic to the space $\text{Hom}(\mathbb{R}^n, \mathbb{R})$ of all linear functionals on $\mathbb{R}^n$, the n-tuple $\mathbf{a}$ corresponding to the functional $\omega_{\mathbf{a}}$ defined by $\omega_{\mathbf{a}}(\mathbf{x}) = \sum_1^n a_ix_i$.

Also, $(\mathbb{R}^m)^n$ is naturally isomorphic to $\text{Hom}(\mathbb{R}^n, \mathbb{R}^m)$. And since $\mathbb{R}^{\overline{m}\times\overline{n}}$ is naturally isomorphic to $(\mathbb{R}^m)^n$, it follows that the spaces $\mathbb{R}^{\overline{m}\times\overline{n}}$ and $\text{Hom}(\mathbb{R}^n, \mathbb{R}^m)$ are naturally isomorphic. This is simply our natural association of a transformation T in $\text{Hom}(\mathbb{R}^n, \mathbb{R}^m)$ to an $m \times n$ matrix $\{t_{ij}\}$.

CHAPTER 2

FINITE-DIMENSIONAL VECTOR SPACES

We have defined a vector space to be finite-dimensional if it has a finite spanning set. In this chapter we shall focus our attention on such spaces, although this restriction is unnecessary for some of our discussion. We shall see that we can assign to each finite-dimensional space V a unique integer, called the dimension of V, which satisfies our intuitive requirements about dimensionality and which becomes a principal tool in the deeper explorations into the nature of such spaces. A number of "dimensional identities" are crucial in these further investigations. We shall find that the dual space of all linear functionals on V, $V^* = \text{Hom}(V, \mathbb{R})$, plays a more satisfactory role in finite-dimensional theory than in the context of general vector spaces. (However, we shall see later in the book that when we add limit theory to our algebra, there are certain special infinite-dimensional vector spaces for which the dual space plays an equally important role.) A finite-dimensional space can be characterized as a vector space isomorphic to some Cartesian space $\mathbb{R}^n$, and such an isomorphism allows a transformation T in Hom V to be "transferred" to $\mathbb{R}^n$, whereupon it acquires a matrix. The theory of linear transformations on such spaces is therefore mirrored completely by the theory of matrices. In this chapter we shall push much deeper into the nature of this relationship than we did in Chapter 1. We also include a section on matrix computations, a brief section describing the trace and determinant functions, and a short discussion of the diagonalization of a quadratic form.

1. BASES

Consider again a fixed finite indexed set of vectors $\boldsymbol{\alpha} = \{\alpha_i : i \in I\}$ in V and the corresponding linear combination map $L_{\boldsymbol{\alpha}}: \mathbf{x} \mapsto \sum x_i\alpha_i$ from $\mathbb{R}^I$ to V having $\boldsymbol{\alpha}$ as skeleton.

Definition. The finite indexed set $\{\alpha_i : i \in I\}$ is *independent* if the above mapping $L_{\boldsymbol{\alpha}}$ is injective, and $\{\alpha_i\}$ is a *basis* for V if $L_{\boldsymbol{\alpha}}$ is an isomorphism (onto V). In this situation we call $\{\alpha_i : i \in I\}$ an *ordered* basis or *frame* if $I = \bar{n} = \{1, \ldots, n\}$ for some positive integer n.

Thus $\{\alpha_i : i \in I\}$ is a basis if and only if for each $\xi \in V$ there exists a unique indexed "coefficient" set $\mathbf{x} = \{x_i : i \in I\} \in \mathbb{R}^I$ such that $\xi = \sum x_i\alpha_i$. The numbers x_i always exist because $\{\alpha_i : i \in I\}$ spans V, and $\mathbf{x}$ is unique because $L_{\boldsymbol{\alpha}}$ is injective.

For example, we can check directly that $\mathbf{b}^1 = <2, 1>$ and $\mathbf{b}^2 = <1, -3>$ form a basis for $\mathbb{R}^2$. The problem is to show that for each $\mathbf{y} \in \mathbb{R}^2$ there is a unique $\mathbf{x}$ such that

$$\mathbf{y} = \sum_1^2 x_i \mathbf{b}^i = x_1<2, 1> + x_2<1, -3> = <2x_1 + x_2, x_1 - 3x_2>.$$

Since this vector equation is equivalent to the two scalar equations $y_1 = 2x_1 + x_2$ and $y_2 = x_1 - 3x_2$, we can find the unique solution $x_1 = (3y_1 + y_2)/7$, $x_2 = (y_1 - 2y_2)/7$ by the usual elimination method of secondary school algebra.

The form of these definitions is dictated by our interpretation of the linear combination formula as a linear mapping. The more usual definition of independence is a corollary.

Lemma 1.1. The independence of the finite indexed set $\{\alpha_i : i \in I\}$ is equivalent to the property that $\sum_I x_i\alpha_i = 0$ only if all the coefficients x_i are 0.

Proof. This is the property that the null space of $L_{\boldsymbol{\alpha}}$ consist only of 0, and is thus equivalent to the injectivity of $L_{\boldsymbol{\alpha}}$, that is, to the independence of $\{\alpha_i\}$, by Lemma 1.1 of Chapter 1. □

If $\{\alpha_i\}_1^n$ is an ordered basis (frame) for V, the unique n-tuple $\mathbf{x}$ such that $\xi = \sum_1^n x_i\alpha_i$ is called the *coordinate n-tuple* of ξ (with respect to the basis $\{\alpha_i\}$), and x_i is the ith *coordinate* of ξ. We call $x_i\alpha_i$ (and sometimes x_i) the ith *component* of ξ. The mapping $L_{\boldsymbol{\alpha}}$ will be called a *basis isomorphism*, and its inverse $L_{\boldsymbol{\alpha}}^{-1}$, which assigns to each vector $\xi \in V$ its unique coordinate n-tuple $\mathbf{x}$, is a *coordinate isomorphism*. The linear functional $\xi \mapsto x_j$ is the jth *coordinate functional*; it is the composition of the coordinate isomorphism $\xi \mapsto \mathbf{x}$ with the jth coordinate projection $\mathbf{x} \mapsto x_j$ on $\mathbb{R}^n$. We shall see in Section 3 that the n coordinate functionals form a basis for $V^* = \text{Hom}(V, \mathbb{R})$.

In the above paragraph we took the index set I to be $\overline{n} = \{1, \ldots, n\}$ and used the language of n-tuples. The only difference for an arbitrary finite index set is that we speak of a coordinate function $\mathbf{x} = \{x_i : i \in I\}$ instead of a coordinate n-tuple.

Our first concern will be to show that every finite-dimensional (finitely spanned) vector space has a basis. We start with some remarks about indices.

We note first that a finite indexed set $\{\alpha_i : i \in I\}$ can be independent only if the indexing is injective as a mapping into V, for if $\alpha_k = \alpha_l$, then $\sum x_i\alpha_i = 0$, where $x_k = 1$, $x_l = -1$, and $x_i = 0$ for the remaining indices. Also, if $\{\alpha_i : i \in I\}$ is independent and $J \subset I$, then $\{\alpha_i : i \in J\}$ is independent, since if $\sum_J x_i\alpha_i = 0$, and if we set $x_i = 0$ for $i \in I - J$, then $\sum_I x_i\alpha_i = 0$, and so each x_j is 0. A finite *unindexed* set is said to be independent if it is independent in some (necessarily bijective) indexing. It will of course then be independent with respect to *any* bijective indexing. An arbitrary set is independent if every finite subset is independent. It follows that a set A is *dependent* (not independent) if and only if there exist *distinct* elements $\alpha_1, \ldots, \alpha_n$ in A and scalars $x_1, \ldots, x_n$ not all zero such that $\sum_1^n x_i\alpha_i = 0$. An unindexed basis would be defined in the obvious way. However, a set can always be regarded as being indexed, by itself if necessary!

Lemma 1.2. If B is an independent subset of a vector space V and β is any vector not in the linear span $L(B)$, then $B \cup \{\beta\}$ is independent.

Proof. Otherwise there is a zero linear combination, $x\beta + \sum_1^n x_i\beta_i = 0$, where $\beta_1, \ldots, \beta_n$ are distinct elements of B and the coefficients are not all 0. But then x cannot be zero: if it were, the equation would contradict the independence of B. We can therefore divide by x and solve for β, so that $\beta \in L(B)$, a contradiction. □

The reader will remember that we call a vector space V finite-dimensional if it has a finite spanning set $\{\alpha_i\}_1^n$. We can use the above lemma to construct a basis for such a V by choosing some of the α_i's. We simply run through the sequence $\{\alpha_i\}_1^n$ and choose those members that increase the linear span of the preceding choices. We end up with a spanning set since $\{\alpha_i\}_1^n$ spans, and our subsequence is independent at each step, by the lemma. In the same way we can extend an independent set $\{\beta_i\}_1^n$ to a basis by choosing some members of a spanning set $\{\alpha_i\}_1^n$. This procedure is intuitive, but it is messy to set up rigorously. We shall therefore proceed differently.

Theorem 1.1. Any minimal finite spanning set is a basis, and therefore any finite-dimensional vector space V has a basis. More generally, if $\{\beta_j : j \in J\}$ is a finite independent set and $\{\alpha_i : i \in I\}$ is a finite spanning set, and if K is a smallest subset of I such that $\{\beta_j\}_J \cup \{\alpha_i\}_K$ spans, then this collection is independent and a basis. Therefore, any finite independent subset of a finite-dimensional space can be extended to a basis.

Proof. It is sufficient to prove the second assertion, since it includes the first as a special case. If $\{\beta_j\}_J \cup \{\alpha_i\}_K$ is not independent, then there is a nontrivial zero linear combination $\sum_J y_j\beta_j + \sum_K x_i\alpha_i = 0$. If every x_i were zero, this equation would contradict the independence of $\{\beta_j\}_J$. Therefore, some x_k is not zero, and we can solve the equation for α_k. That is, if we set $L = K - \{k\}$, then the linear span of $\{\beta_j\}_J \cup \{\alpha_i\}_L$ contains α_k. It therefore includes the whole original spanning set and hence is V. But this contradicts the minimal nature of K, since L is a proper subset of K. Consequently, $\{\beta_j\}_J \cup \{\alpha_i\}_K$ is independent. □

We next note that $\mathbb{R}^n$ itself has a very special basis. In the indexing map $i \mapsto \alpha_i$ the vector α_j corresponds to the index j, but under the linear combination map $\mathbf{x} \mapsto \sum x_i\alpha_i$ the vector α_j corresponds to the function δ^j which has the value 1 at j and the value 0 elsewhere, so that $\sum_i \delta_i^j\alpha_i = \alpha_j$. This function

δ^j is called a *Kronecker delta function.* It is clearly the characteristic function χ_B of the one-point set $B = \{j\}$, and the symbol 'δ^j' is ambiguous, just as 'χ_B' is ambiguous; in each case the meaning depends on what domain is implicit from the context. We have already used the delta functions on $\mathbb{R}^n$ in proving Theorem 1.2 of Chapter 1.

Theorem 1.2. The Kronecker functions $\{\delta^j\}_{j=1}^n$ form a basis for $\mathbb{R}^n$.

Proof. Since $\sum_i^n x_i\delta^i(j) = x_j$ by the definition of δ^i, we see that $\sum_1^n x_i\delta^i$ is the n-tuple $\mathbf{x}$ itself, so the linear combination mapping $L_{\boldsymbol{\delta}}: \mathbf{x} \mapsto \sum_1^n x_i\delta^i$ is the identity mapping $\mathbf{x} \mapsto \mathbf{x}$, a trivial isomorphism. □

Among all possible indexed bases for $\mathbb{R}^n$, the Kronecker basis is thus singled out by the fact that its basis isomorphism is the identity; for this reason it is called the standard basis or the *natural* basis for $\mathbb{R}^n$. The same holds for $\mathbb{R}^I$ for any finite set I.

Finally, we shall draw some elementary conclusions from the existence of a basis.

Theorem 1.3. If $T \in \text{Hom}(V, W)$ is an isomorphism and $\boldsymbol{\alpha} = \{\alpha_i : i \in I\}$ is a basis for V, then $\{T(\alpha_i) : i \in I\}$ is a basis for W.

Proof. By hypothesis $L_{\boldsymbol{\alpha}}$ is an isomorphism in $\text{Hom}(\mathbb{R}^n, V)$, and so $T \circ L_{\boldsymbol{\alpha}}$ is an isomorphism in $\text{Hom}(\mathbb{R}^n, W)$. Its skeleton $\{T(\alpha_i)\}$ is therefore a basis for W. □

We can view any basis $\{\alpha_i\}$ as the image of the standard basis $\{\delta^i\}$ under the basis isomorphism. Conversely, any isomorphism $\theta: \mathbb{R}^I \to B$ becomes a basis isomorphism for the basis $\alpha_j = \theta(\delta^j)$.

Theorem 1.4. If X and Y are complementary subspaces of a vector space V, then the union of a basis for X and a basis for Y is a basis for V. Conversely, if a basis for V is partitioned into two sets, with linear spans X and Y, respectively, then X and Y are complementary subspaces of V.

Proof. We prove only the first statement. If $\{\alpha_i : i \in J\}$ is a basis for X and $\{\alpha_i : i \in K\}$ is a basis for Y, then it is clear that $\{\alpha_i : i \in J \cup K\}$ spans V, since its span includes both X and Y, and so $X + Y = V$. Suppose then that $\sum_{J \cup K} x_i\alpha_i = 0$. Setting $\xi = \sum_J x_i\alpha_i$ and $\eta = \sum_K x_i\alpha_i$, we see that $\xi \in X$, $\eta \in Y$, and $\xi + \eta = 0$. But then $\xi = \eta = 0$, since X and Y are complementary. And then $x_i = 0$ for $i \in J$ because $\{\alpha_i\}_J$ is independent, and $x_i = 0$ for $i \in K$ because $\{\alpha_i\}_K$ is independent. Therefore, $\{\alpha_i\}_{J \cup K}$ is a basis for V. We leave the converse argument as an exercise. □

Corollary. If $V = \bigoplus_1^n V_i$ and B_i is a basis for V_i, then $B = \bigcup_1^n B_i$ is a basis for V.

Proof. We see from the theorem that $B_1 \cup B_2$ is a basis for $V_1 \oplus V_2$. Proceeding inductively we see that $\bigcup_{i=1}^j B_i$ is a basis for $\bigoplus_{i=1}^j V_i$ for $j = 2, \ldots, n$, and the corollary is the case $j = n$. □

If we follow a coordinate isomorphism by a linear combination map, we get the mapping of the following existence theorem, which we state only in n-tuple form.

Theorem 1.5. If $\boldsymbol{\beta} = \{\beta_i\}_1^n$ is an ordered basis for the vector space V, and if $\{\alpha_i\}_1^n$ is any n-tuple of vectors in a vector space W, then there exists a unique $S \in \mathrm{Hom}(V, W)$ such that $S(\beta_i) = \alpha_i$ for $i = 1, \ldots, n$.

Proof. By hypothesis $L_{\boldsymbol{\beta}}$ is an isomorphism in $\mathrm{Hom}(\mathbb{R}^n, V)$, and so $S = L_{\boldsymbol{\alpha}} \circ (L_{\boldsymbol{\beta}})^{-1}$ is an element of $\mathrm{Hom}(V, W)$ such that $S(\beta_i) = L_{\boldsymbol{\alpha}}(\delta^i) = \alpha_i$. Conversely, if $S \in \mathrm{Hom}(V, W)$ is such that $S(\beta_i) = \alpha_i$ for all i, then $S \circ L_{\boldsymbol{\beta}}(\delta^i) = \alpha_i$ for all i, so that $S \circ L_{\boldsymbol{\beta}} = L_{\boldsymbol{\alpha}}$. Thus S is uniquely determined as $L_{\boldsymbol{\alpha}} \circ (L_{\boldsymbol{\beta}})^{-1}$. □

It is natural to ask how the unique S above varies with the n-tuple $\{\alpha_i\}$. The answer is: linearly and "isomorphically".

Theorem 1.6. Let $\{\beta_i\}_1^n$ be a fixed ordered basis for the vector space V, and for each n-tuple $\boldsymbol{\alpha} = \{\alpha_i\}_1^n$ chosen from the vector space W let $S_{\boldsymbol{\alpha}} \in \mathrm{Hom}(V, W)$ be the unique transformation defined above. Then the map $\boldsymbol{\alpha} \mapsto S_{\boldsymbol{\alpha}}$ is an isomorphism from W^n to $\mathrm{Hom}(V, W)$.

Proof. As above, $S_{\boldsymbol{\alpha}} = L_{\boldsymbol{\alpha}} \circ \theta^{-1}$, where θ is the basis isomorphism $L_{\boldsymbol{\beta}}$. Now we know from Theorem 6.2 of Chapter 1 that $\boldsymbol{\alpha} \mapsto L_{\boldsymbol{\alpha}}$ is an isomorphism from W^n to $\mathrm{Hom}(\mathbb{R}^n, W)$, and composition on the right by the fixed coordinate isomorphism θ^{-1} is an isomorphism from $\mathrm{Hom}(\mathbb{R}^n, W)$ to $\mathrm{Hom}(V, W)$ by the corollary to Theorem 3.3 of Chapter 1. Composing these two isomorphisms gives us the theorem. □

***Infinite bases.** Most vector spaces do not have finite bases, and it is natural to try to extend the above discussion to index sets I that may be infinite. The Kronecker functions $\{\delta^i : i \in I\}$ have the same definitions, but they no longer span $\mathbb{R}^I$. By definition f is a linear combination of the functions δ^i if and only if f is of the form $\sum_{i \in I_1} c_i \delta^i$, where I_1 is a finite subset of I. But then $f = 0$ outside of I_1. Conversely, if $f \in \mathbb{R}^I$ is 0 except on a finite set I_1, then $f = \sum_{i \in I_1} f(i)\delta^i$. The linear span of $\{\delta^i : i \in I\}$ is thus exactly the set of all functions of $\mathbb{R}^I$ that are zero except on a finite set. We shall designate this subspace $\mathbb{R}_I$.

If $\{\alpha_i : i \in I\}$ is an indexed set of vectors in V and $f \in \mathbb{R}_I$, then the sum $\sum_{i \in I} f(i)\alpha_i$ becomes meaningful if we adopt the reasonable convention that the sum of an arbitrary number of 0's is 0. Then $\sum_{i \in I} = \sum_{i \in I_0}$, where I_0 is any finite subset of I outside of which f is zero.

With this convention, $L_{\boldsymbol{\alpha}}: f \mapsto \sum_i f(i)\alpha_i$ is a linear map from $\mathbb{R}_I$ to V, as in Theorem 1.2 of Chapter 1. And with the same convention, $\sum_{i \in I} f(i)\alpha_i$ is an elegant expression for the general linear combination of the vectors α_i. Instead of choosing a finite subset I_1 and numbers c_i for just those indices i in I_1, we define c_i for all $i \in I$, but with the stipulation that $c_i = 0$ for all but a finite number of indices. That is, we take $\mathbf{c} = \{c_i : i \in I\}$ as a function in $\mathbb{R}_I$.

We make the same definitions of *independence* and *basis* as before. Then $\{\alpha_i : i \in I\}$ is a basis for V if and only if $L_{\mathbf{a}}: \mathbb{R}_I \to V$ is an isomorphism, i.e., if and only if for each $\xi \in V$ there exists a unique $\mathbf{x} \in \mathbb{R}_I$ such that $\xi = \sum_i x_i\alpha_i$.

By using an axiom of set theory called the axiom of choice, it can be shown that *every* vector space has a basis in this sense and that any independent set can be extended to a basis. Then Theorems 1.4 and 1.5 hold with only minor changes in notation. In particular, if a basis for a subspace M of V is extended to a basis for V, then the linear span of the added part is a subspace N complementary to M. Thus, in a purely algebraic sense, every subspace has complementary subspaces. We assume this fact in some of our exercises.

The above sums are always finite (despite appearances), and the above notion of basis is purely algebraic. However, infinite bases in this sense are not very useful in analysis, and we shall therefore concentrate for the present on spaces that have finite bases (i.e., are finite-dimensional). Then in one important context later on we shall discuss infinite bases where the sums are genuinely infinite by virtue of limit theory.

EXERCISES

1.1 Show by a direct computation that $\{\langle 1, -1\rangle, \langle 0, 1\rangle\}$ is a basis for $\mathbb{R}^2$.

1.2 The student must realize that the ith coordinate of a vector depends on the whole basis and not just on the ith basis vector. Prove this for the second coordinate of vectors in $\mathbb{R}^2$ using the standard basis and the basis of the above exercise.

1.3 Show that $\{\langle 1, 1\rangle, \langle 1, 2\rangle\}$ is a basis for $V = \mathbb{R}^2$. The basis isomorphism from $\mathbb{R}^2$ to V is now from $\mathbb{R}^2$ to $\mathbb{R}^2$. Find its matrix. Find the matrix of the coordinate isomorphism. Compute the coordinates, *with respect to this basis*, of $\langle -1, 1\rangle$, $\langle 0, 1\rangle$, $\langle 2, 3\rangle$.

1.4 Show that $\{\mathbf{b}^i\}_1^3$, where $\mathbf{b}^1 = \langle 1, 0, 0\rangle$, $\mathbf{b}^2 = \langle 1, 1, 0\rangle$, and $\mathbf{b}^3 = \langle 1, 1, 1\rangle$, is a basis for $\mathbb{R}^3$.

1.5 In the above exercise find the three linear functionals l_i that are the coordinate functionals with respect to the given basis. Since

$$\mathbf{x} = \sum_1^3 l_i(\mathbf{x})\mathbf{b}^i,$$

finding the l_i is equivalent to solving $\mathbf{x} = \sum_1^3 y_i\mathbf{b}^i$ for the y_i's in terms of $\mathbf{x} = \langle x_1, x_2, x_3\rangle$.

1.6 Show that any set of polynomials no two of which have the same degree is independent.

1.7 Show that if $\{\alpha_i\}_1^n$ is an independent subset of V and T in $\operatorname{Hom}(V, W)$ is injective, then $\{T(\alpha_i)\}_1^n$ is an independent subset of W.

1.8 Show that if T is any element of $\operatorname{Hom}(V, W)$ and $\{T(\alpha_i)\}_1^n$ is independent in W, then $\{\alpha_i\}_1^n$ is independent in V.

1.9 Later on we are going to call a vector space V n-dimensional if every basis for V contains exactly n elements. If V is the span of a single vector α, so that $V = \mathbb{R}\alpha$, then V is clearly one-dimensional.

Let $\{V_i\}_1^n$ be a collection of one-dimensional subspaces of a vector space V, and choose a nonzero vector α_i in V_i for each i. Prove that $\{\alpha_i\}_1^n$ is independent if and only if the subspaces $\{V_i\}_1^n$ are independent and that $\{\alpha_i\}_1^n$ is a basis if and only if $V = \bigoplus_1^n V_i$.

1.10 Finish the proof of Theorem 1.4.

1.11 Give a proof of Theorem 1.4 based on the existence of isomorphisms.

1.12 The reader would guess, and we shall prove in the next section, that every subspace of a finite-dimensional space is finite-dimensional. Prove now that a subspace N of a finite-dimensional vector space V is finite-dimensional if and only if it has a complement M. (Work from a combination of Theorems 1.1 and 1.4 and direct sum projections.)

1.13 Since $\{\mathbf{b}^i\}_1^3 = \{\langle 1, 0, 0\rangle, \langle 1, 1, 0\rangle, \langle 1, 1, 1\rangle\}$ is a basis for $\mathbb{R}^3$, there is a unique T in $\text{Hom}(\mathbb{R}^3, \mathbb{R}^2)$ such that $T(\mathbf{b}^1) = \langle 1, 0\rangle$, $T(\mathbf{b}^2) = \langle 0, 1\rangle$, and $T(\mathbf{b}^3) = \langle 1, 1\rangle$. Find the matrix of T. (Find $T(\delta^i)$ for $i = 1, 2, 3$.)

1.14 Find, similarly, the S in Hom $\mathbb{R}^3$ such that $S(\mathbf{b}^i) = \delta^i$ for $i = 1, 2, 3$.

1.15 Show that the infinite sequence $\{t^n\}_0^\infty$ is a basis for the vector space of all polynomials.

2. DIMENSION

The concept of dimension rests on the fact that two different bases for the same space always contain the same number of elements. This number, which is then the number of elements in every basis for V, is called the *dimension* of V. It tells all there is to know about V to within isomorphism: There exists an isomorphism between two spaces if and only if they have the same dimension. We shall consider only finite dimensions. If V is not finite-dimensional, its dimension is an infinite cardinal number, a concept with which the reader is probably unfamiliar.

Lemma 2.1. If V is finite-dimensional and T in Hom V is surjective, then T is an isomorphism.

Proof. Let n be the smallest number of elements that can span V. That is, there is some spanning set $\{\alpha_i\}_1^n$ and none with fewer than n elements. Then $\{\alpha_i\}_1^n$ is a basis, by Theorem 1.1, and the linear combination map $\theta: \mathbf{x} \mapsto \sum_1^n x_i\alpha_i$ is accordingly a basis isomorphism. But $\{\beta_i\}_1^n = \{T(\alpha_i)\}_1^n$ also spans, since T is surjective, and so $T \circ \theta$ is also a basis isomorphism, for the same reason. Then $T = (T \circ \theta) \circ \theta^{-1}$ is an isomorphism. □

Theorem 2.1. If V is finite-dimensional, then all bases for V contain the same number of elements.

Proof. Two bases with n and m elements determine basis isomorphisms $\theta: \mathbb{R}^n \to V$ and $\varphi: \mathbb{R}^m \to V$. Suppose that $m < n$ and, viewing $\mathbb{R}^n$ as $\mathbb{R}^m \times \mathbb{R}^{n-m}$, let π be the projection of $\mathbb{R}^n$ onto $\mathbb{R}^m$,

$$\pi(\langle x_1, \ldots, x_m, \ldots, x_n \rangle) = \langle x_1, \ldots, x_m \rangle.$$

Since $T = \theta^{-1} \circ \varphi$ is an isomorphism from $\mathbb{R}^m$ to $\mathbb{R}^n$ and $T \circ \pi: \mathbb{R}^n \to \mathbb{R}^n$ is therefore surjective, it follows from the lemma that $T \circ \pi$ is an isomorphism. Then $\pi = T^{-1} \circ (T \circ \pi)$ is an isomorphism. But it isn't, because $\pi(\delta^n) = 0$, and we have a contradiction. Therefore no basis can be smaller than any other basis. □

The integer that is the number of elements in every basis for V is of course called the *dimension* of V, and we designate it $d(V)$. Since the standard basis $\{\delta^i\}_1^n$ for $\mathbb{R}^n$ has n elements, we see that $\mathbb{R}^n$ is n-dimensional in this precise sense.

Corollary. Two finite-dimensional vector spaces are isomorphic if and only if they have the same dimension.

Proof. If T is an isomorphism from V to W and B is a basis for V, then $T[B]$ is a basis for W by Theorem 1.3. Therefore $d(V) = \#B = \#T[B] = d(W)$, where $\#A$ is the number of elements in A. Conversely, if $d(V) = d(W) = n$, then V and W are each isomorphic to $\mathbb{R}^n$ and so to each other. □

Theorem 2.2. Every subspace M of a finite-dimensional vector space V is finite-dimensional.

Proof. Let $\mathcal{A}$ be the family of finite independent subsets of M. By Theorem 1.1, if $A \in \mathcal{A}$, then A can be extended to a basis for V, and so $\#A \leq d(V)$. Thus $\{\#A : A \in \mathcal{A}\}$ is a finite set of integers, and we can choose $B \in \mathcal{A}$ such that $n = \#B$ is the maximum of this finite set. But then $L(B) = M$, because otherwise for any $\alpha \in M - L(B)$ we have $B \cup \{\alpha\} \in \mathcal{A}$, by Lemma 1.2, and

$$\#(B \cup \{\alpha\}) = n + 1,$$

contradicting the maximal nature of n. Thus M is finitely spanned. □

Corollary. Every subspace M of a finite-dimensional space V has a complement.

Proof. Use Theorem 1.1 to extend a basis for M to a basis for V, and let N be the linear span of the added vectors. Then apply Theorem 1.4. □

Dimensional identities. We now prove two basic dimensional identities. We will always assume V finite-dimensional.

Lemma 2.2. If V_1 and V_2 are complementary subspaces of V, then $d(V) = d(V_1) + d(V_2)$. More generally, if $V = \bigoplus_1^n V_i$ then $d(V) = \sum_1^n d(V_i)$.

Proof. This follows at once from Theorem 1.4 and its corollary. □

Theorem 2.3. If U and W are subspaces of a finite-dimensional vector space, then $d(U + W) + d(U \cap W) = d(U) + d(W)$.

Proof. Let V be a complement of $U \cap W$ in U. We start by showing that then V is also a complement of W in $U + W$. First

$$V + W = V + ((U \cap W) + W) = (V + (U \cap W)) + W = U + W.$$

We have used the obvious fact that the sum of a vector space and a subspace is the vector space. Next,

$$V \cap W = (V \cap U) \cap W = V \cap (U \cap W) = \{0\},$$

because V is a complement of $U \cap W$ in U. We thus have both $V + W = U + W$ and $V \cap W = \{0\}$, and so V is a complement of W in $U + W$ by the corollary of Lemma 5.2 of Chapter 1.

The theorem is now a corollary of the above lemma. We have

$$\begin{aligned} d(U) + d(W) = (d(U \cap W) + d(V)) + d(W) &= d(U \cap W) + (d(V) + d(W)) \\ &= d(U \cap W) + d(U + W). \ \square \end{aligned}$$

Theorem 2.4. Let V be finite-dimensional, and let W be any vector space. Let $T \in \mathrm{Hom}(V, W)$ have null space N (in V) and range R (in W). Then R is finite-dimensional and $d(V) = d(N) + d(R)$.

Proof. Let U be a complement of N in V. Then we know that $T \restriction U$ is an isomorphism onto R. (See Theorem 5.3 of Chapter 1.) Therefore, R is finite-dimensional and $d(R) + d(N) = d(U) + d(N) = d(V)$ by our first identity. $\square$

Corollary. If W is finite-dimensional and $d(W) = d(V)$, then T is injective if and only if it is surjective, so that in this case injectivity, surjectivity, and bijectivity are all equivalent.

Proof. T is surjective if and only if $R = W$. But this is equivalent to $d(R) = d(W)$, and if $d(W) = d(V)$, then the theorem shows this is turn to be equivalent to $d(N) = 0$, that is, to $N = \{0\}$. $\square$

Theorem 2.5. If $d(V) = n$ and $d(W) = m$, then $\mathrm{Hom}(V, W)$ is finite-dimensional and its dimension is mn.

Proof. By Theorem 1.6, $\mathrm{Hom}(V, W)$ is isomorphic to W^n which is the direct sum of the n subspaces isomorphic to W under the injections θ_i for $i = 1, \ldots, n$. The dimension of W^n is therefore $\sum_1^n m = mn$ by Lemma 2.2. $\square$

Another proof of Theorem 2.5 will be available in Section 4.

EXERCISES

2.1 Prove that if $d(V) = n$, then any spanning subset of n elements is a basis.

2.2 Prove that if $d(V) = n$, then any independent subset of n elements is a basis.

2.3 Show that if $d(V) = n$ and W is a subspace of the same dimension, then $W = V$.

2.4 Prove by using dimensional identities that if f is a nonzero linear functional on an n-dimensional space V, then its null space has dimension $n - 1$.

2.5 Prove by using dimensional identities that if f is a linear functional on a finite-dimensional space V, and if α is a vector not in its null space N, then $V = N \oplus \mathbb{R}\alpha$.

2.6 Given that N is an $(n - 1)$-dimensional subspace of an n-dimensional vector space V, show that N is the null space of a linear functional.

2.7 Let X and Y be subspaces of a finite-dimensional vector space V, and suppose that T in $\text{Hom}(V, W)$ has null space $N = X \cap Y$. Show that $T[X + Y] = T[X] \oplus T(Y)$, and then deduce Theorem 2.3 from Lemma 2.2 and Theorem 2.4. This proof still depends on the existence of a T having $N = X \cap Y$ as its null space. Do we know of any such T?

2.8 Show that if V is finite-dimensional and $S, T \in \text{Hom}\, V$, then

$$S \circ T = I \Rightarrow T \text{ is invertible.}$$

Show also that $T \circ S = I \Rightarrow T$ is invertible.

2.9 A subspace N of a vector space V has *finite codimension n* if the quotient space V/N is finite-dimensional, with dimension n. Show that a subspace N has finite codimension n if and only if N has a complementary subspace M of dimension n. (Move a basis for V/N back into V.) *Do not* assume V to be finite-dimensional.

2.10 Show that if N_1 and N_2 are subspaces of a vector space V with finite codimensions, then $N = N_1 \cap N_2$ has finite codimension and

$$\text{cod}(N) \leq \text{cod}(N_1) + \text{cod}(N_2).$$

(Consider the mapping $\xi \mapsto \langle \bar{\xi}_1, \bar{\xi}_2 \rangle$ when $\bar{\xi}_i$ is the coset of N_i containing ξ.)

2.11 In the above exercise, suppose that $\text{cod}(N_1) = \text{cod}(N_2)$, that is, $d(V/N_1) = d(V/N_2)$. Prove that $d(N_1/N) = d(N_2/N)$.

2.12 Given nonzero vectors β in V and f in V^* such that $f(\beta) \neq 0$, show that some scalar multiple of the mapping $\xi \mapsto f(\xi)\beta$ is a projection. Prove that any projection having a one-dimensional range arises in this way.

2.13 We know that the choice of an origin O in Euclidean 3-space $\mathbb{E}^3$ induces a vector space structure in $\mathbb{E}^3$ (under the correspondence $X \mapsto \overrightarrow{OX}$) and that this vector space is three-dimensional. Show that a geometric plane through O becomes a two-dimensional subspace.

2.14 An m-dimensional plane M is a translate $N + \alpha_0$ of an m-dimensional subspace N. Let $\{\beta_i\}_1^m$ be any basis of N, and set $\alpha_i = \beta_i + \alpha_0$. Show that M is exactly the set of linear combinations

$$\sum_{i=0}^{m} x_i\alpha_i \qquad \text{such that} \qquad \sum_0^m x_i = 1.$$

2.15 Show that Exercise 2.14 is a corollary of Exercise 4.14 of Chapter 1.

2.16 Show, conversely, that if a plane M is the affine span of $m + 1$ elements, then its dimension is $\leq m$.

2.17 From the above two exercises concoct a direct definition of the dimension of an affine subspace.

2.18 Write a small essay suggested by the following definition. An $(m+1)$-tuple $\{\alpha_i\}_0^m$ is *affinely independent* if the conditions

$$\sum_0^m x_i\alpha_i = 0 \quad \text{and} \quad \sum_0^m x_i = 0$$

together imply that

$$x_i = 0 \quad \text{for all} \quad i.$$

2.19 A polynomial on a vector space V is a real-valued function on V which can be represented as a finite sum of finite products of linear functionals. Define the *degree* of a polynomial; define a *homogeneous polynomial of degree k*. Show that the set of homogeneous polynomials of degree k is a vector space X_k.

2.20 Continuing the above exercise, show that if $k_1 < k_2 < \cdots < k_N$, then the vector spaces $\{X_{k_i}\}_1^N$ are independent subspaces of the vector space of all polynomials. [Assume that a polynomial $p(t)$ of a real variable can be the zero function only if all its coefficients are 0. For any polynomial P on V consider the polynomials $p_\alpha(t) = P(t\alpha)$.]

2.21 Let $\langle\alpha, \beta\rangle$ be a basis for the two-dimensional space V, and let $\langle\lambda, \mu\rangle$ be the corresponding coordinate projections (dual basis in V^*). Show that every polynomial on V "is a polynomial in the two variables λ and μ".

2.22 Let $\langle\alpha, \beta\rangle$ be a basis for a two-dimensional vector space V, and let $\langle\lambda, \mu\rangle$ be the corresponding coordinate projections (dual basis for V^*). Show that

$$\langle\lambda^2, \lambda\mu, \mu^2\rangle$$

is a basis for the vector space of homogeneous polynomials on V of degree 2. Similarly, compute the dimension of the space of homogeneous polynomials of degree 3 on a two-dimensional vector space.

2.23 Let V and W be two-dimensional vector spaces, and let F be a mapping from V to W. Using coordinate systems, define the notion of F being quadratic and then show that it is independent of coordinate systems. Generalize the above exercise to higher dimensions and also to higher degrees.

2.24 Now let $F: V \to W$ be a mapping between two-dimensional spaces such that for any $\mathbf{u}, \mathbf{v} \in V$ and any $l \in W^*$, $l(F(t\mathbf{u} + \mathbf{v}))$ is a quadratic function of t, that is, of the form $at^2 + bt + c$. Show that F is quadratic according to your definition in the above exercises.

3. THE DUAL SPACE

Although throughout this section all spaces will be assumed finite-dimensional, many of the definitions and properties are valid for infinite-dimensional spaces as well. But for such spaces there is a difference between purely algebraic situations and situations in which algebra is mixed with hypotheses of continuity. One of the blessings of finite dimensionality is the absence of this complication. As the reader has probably surmised from the number of special linear functionals we have met, particularly the coordinate functionals, the space $\mathrm{Hom}(V, \mathbb{R})$ of *all* linear functionals on V plays a special role.

Definition. The *dual space* (or *conjugate space*) V^* of the vector space V is the vector space $\mathrm{Hom}(V, \mathbb{R})$ of all linear mappings from V to $\mathbb{R}$. Its elements are called linear *functionals*.

We are going to see that in a certain sense V is in turn the dual space of V^* (V and $(V^*)^*$ are naturally isomorphic), so that the two spaces are symmetrically related. We shall briefly study the notion of *annihilation* (orthogonality) which has its origins in this setting, and then see that there is a natural isomorphism between $\mathrm{Hom}(V, W)$ and $\mathrm{Hom}(W^*, V^*)$. This gives the mathematician a new tool to use in studying a linear transformation T in $\mathrm{Hom}(V, W)$; the relationship between T and its image T^* exposes new properties of T itself.

Dual bases. At the outset one naturally wonders how big a space V^* is, and we settle the question immediately.

Theorem 3.1. Let $\{\beta_i\}_1^n$ be an ordered basis for V, and let ε_j be the corresponding jth coordinate functional on V: $\varepsilon_j(\xi) = x_j$, where $\xi = \sum_1^n x_i\beta_i$. Then $\{\varepsilon_j\}_1^n$ is an ordered basis for V^*.

Proof. Let us first make the proof by a direct elementary calculation.

a) *Independence.* Suppose that $\sum_1^n c_j\varepsilon_j = 0$, that is, $\sum_1^n c_j\varepsilon_j(\xi) = 0$ for all $\xi \in V$. Taking $\xi = \beta_i$ and remembering that the coordinate n-tuple of β_i is δ^i, we see that the above equation reduces to $c_i = 0$, and this for all i. Therefore, $\{\varepsilon_j\}_1^n$ is independent.

b) *Spanning.* First note that the basis expansion $\xi = \sum x_i\beta_i$ can be rewritten $\xi = \sum \varepsilon_i(\xi)\beta_i$. Then for any $\lambda \in V^*$ we have $\lambda(\xi) = \sum_1^n l_i\varepsilon_i(\xi)$, where we have set $l_i = \lambda(\beta_i)$. That is, $\lambda = \sum l_i\varepsilon_i$. This shows that $\{\varepsilon_j\}_1^n$ spans V^*, and, together with (a), that it is a basis. □

Definition. The basis $\{\varepsilon_j\}$ for V^* is called the *dual* of the basis $\{\beta_i\}$ for V.

As usual, one of our fundamental isomorphisms is lurking behind all this, but we shall leave its exposure to an exercise.

Corollary. $d(V^*) = d(V)$.

The three equations

$$\xi = \sum \varepsilon_i(\xi)\beta_i, \qquad \lambda = \sum \lambda(\beta_i)\varepsilon_i, \qquad \lambda(\xi) = \sum \lambda(\beta_i)\cdot\varepsilon_i(\xi)$$

are worth looking at. The first two are symmetrically related, each presenting the basis expansion of a vector with its coefficients computed by applying the corresponding element of the dual basis to the vector. The third is symmetric itself between ξ and λ.

Since a finite-dimensional space V and its dual space V^* have the same dimension, they are of course isomorphic. In fact, each basis for V defines an isomorphism, for we have the associated coordinate isomorphism from V to $\mathbb{R}^n$, the dual basis isomorphism from $\mathbb{R}^n$ to V^*, and therefore the composite isomorphism from V to V^*. This isomorphism varies with the basis, however, and there is in general no natural isomorphism between V and V^*.

It is another matter with Cartesian space $\mathbb{R}^n$ because it has a standard basis, and therefore a standard isomorphism with its dual space $(\mathbb{R}^n)^*$. It is not hard to see that this is the isomorphism $\mathbf{a} \mapsto L_{\mathbf{a}}$, where $L_{\mathbf{a}}(\mathbf{x}) = \sum_1^n a_i x_i$, that we discussed in Section 1.6. We can therefore feel free to identify $\mathbb{R}^n$ with $(\mathbb{R}^n)^*$, only keeping in mind that when we think of an n-tuple $\mathbf{a}$ as a linear functional, we mean the functional $L_{\mathbf{a}}(\mathbf{x}) = \sum_1^n a_i x_i$.

The second conjugate space. Despite the fact that V and V^* are not naturally isomorphic in general, we shall now see that V *is* naturally isomorphic to $V^{**} = (V^*)^*$.

Theorem 3.2. The function $\omega: V \times V^* \to \mathbb{R}$ defined by $\omega(\xi, f) = f(\xi)$ is bilinear, and the mapping $\xi \mapsto \omega^\xi$ from V to V^{**} is a natural isomorphism.

Proof. In this context we generally set $\xi^{**} = \omega^\xi$, so that ξ^{**} is defined by $\xi^{**}(f) = f(\xi)$ for all $f \in V^*$. The bilinearity of ω should be clear, and Theorem 6.1 of Chapter 1 therefore applies. The reader might like to run through a direct check of the linearity of $\xi \mapsto \xi^{**}$ starting with $(c_1\xi_1 + c_2\xi_2)^{**}(f)$.

There still is the question of the injectivity of this mapping. If $\alpha \neq 0$, we can find $f \in V^*$ so that $f(\alpha) \neq 0$. One way is to make α the first vector of an ordered basis and to take f as the first functional in the dual basis; then $f(\alpha) = 1$. Since $\alpha^{**}(f) = f(\alpha) \neq 0$, we see in particular that $\alpha^{**} \neq 0$. The mapping $\xi \to \xi^{**}$ is thus injective, and it is then bijective by the corollary of Theorem 2.4. □

If we think of V^{**} as being naturally identified with V in this way, the two spaces V and V^* are symmetrically related to each other. Each is the dual of the other. In the expression '$f(\xi)$' we think of *both* symbols as variables and then hold one or the other fixed for the two interpretations. In such a situation we often use a more symmetric symbolism, such as $\langle \xi, f \rangle$, to indicate our intention to treat both symbols as variables.

Lemma 3.1. If $\{\lambda_i\}$ is the basis in V^* dual to the basis $\{\alpha_i\}$ in V, then $\{\alpha_i^{**}\}$ is the basis in V^{**} dual to the basis $\{\lambda_i\}$ in V^*.

Proof. We have $\alpha_i^{**}(\lambda_j) = \lambda_j(\alpha_i) = \delta_j^i$, which shows that α_i^{**} is the ith coordinate projection. In case the reader has forgotten, the basis expansion $f = \sum c_j\lambda_j$ implies that $\alpha_i^{**}(f) = f(\alpha_i) = (\sum c_j\lambda_j)(\alpha_i) = c_i$, so that α_i^{**} is the mapping $f \mapsto c_i$. □

Annihilator subspaces. It is in this dual situation that *orthogonality* first naturally appears. However, we shall save the term 'orthogonal' for the latter context in which V and V^* have been identified through a scalar product, and shall speak here of the *annihilator* of a set rather than its orthogonal complement.

Definition. If $A \subset V$, the *annihilator* of A, A°, is the set of all f in V^* such that $f(\alpha) = 0$ for all α in A. Similarly, if $A \subset V^*$, then

$$A^\circ = \{\alpha \in V : f(\alpha) = 0 \text{ for all } f \in A\}.$$

If we view V as $(V^*)^*$, the second definition is included in the first.

The following properties are easily established and will be left as exercises:

1) A° is always a subspace.
2) $A \subset B \Rightarrow B^\circ \subset A^\circ$.
3) $(L(A))^\circ = A^\circ$.
4) $(A \cup B)^\circ = A^\circ \cap B^\circ$.
5) $A \subset A^{\circ\circ}$.

We now add one more crucial dimensional identity to those of the last section.

Theorem 3.3. If W is a subspace of V, then $d(V) = d(W) + d(W^\circ)$.

Proof. Let $\{\beta_i\}_1^m$ be a basis for W, and extend it to a basis $\{\beta_i\}_1^n$ for V. Let $\{\lambda_i\}_1^n$ be the dual basis in V^*. We claim that then $\{\lambda_i\}_{m+1}^n$ is a basis for W°. First, if $j > m$, then $\lambda_j(\beta_i) = 0$ for $i = 1, \ldots, m$, and so λ_j is in W° by (3) above. Thus $\{\lambda_{m+1}, \ldots, \lambda_n\} \subset W^\circ$. Now suppose that $f \in W^\circ$, and let $f = \sum_{j=1}^n c_j\lambda_j$ be its (dual) basis expansion. Then for each $i \leq m$ we have $c_i = f(\beta_i) = 0$, since $\beta_i \in W$ and $f \in W^\circ$; therefore, $f = \sum_{m+1}^n c_j\lambda_j$. Thus every f in W° is in the span of $\{\lambda_i\}_{m+1}^n$. Altogether, we have shown that W° is the span of $\{\lambda_i\}_{m+1}^n$, as claimed. Then $d(W^\circ) + d(W) = (n - m) + m = n = d(V)$, and we are done. □

Corollary. $A^{\circ\circ} = L(A)$ for every subset $A \subset V$.

Proof. Since $(L(A))^\circ = A^\circ$, we have $d(L(A)) + d(A^\circ) = d(V)$, by the theorem. Also $d(A^\circ) + d(A^{\circ\circ}) = d(V^*) = d(V)$. Thus $d(A^{\circ\circ}) = d(L(A))$, and since $L(A) \subset A^{\circ\circ}$, by (5) above, we have $L(A) = A^{\circ\circ}$. □

The adjoint of T. We shall now see that with every T in $\mathrm{Hom}(V, W)$ there is naturally associated an element of $\mathrm{Hom}(W^*, V^*)$ which we call the *adjoint* of T and designate T^*. One consequence of the intimate relationship between T and T^* is that the range of T^* is exactly the annihilator of the null space of T. Combined with our dimensional identities, this implies that the ranges of T and T^* have the same dimension. And later on, after we have established the connection between matrix representations of T and T^*, this turns into the very mysterious fact that the dimension of the linear span of the row vectors of an m-by-n matrix is the same as the dimension of the linear span of its column vectors, which gives us our notion of the *rank* of a matrix. In Chapter 5 we shall study a situation (Hilbert space) in which we are given a fixed fundamental isomorphism between V and V^*. If T is in Hom V, then of course T^* is in Hom V^*, and we can use this isomorphism to "transfer" T^* into Hom V. But now T can be compared with its (transferred) adjoint T^*, and they may be equal. That is, T may be *self-adjoint*. It turns out that the self-adjoint transformations are "nice" ones, as we shall see for ourselves in simple cases, and also, fortunately, that many important linear maps arising from theoretical physics are self-adjoint.

If $T \in \text{Hom}(V, W)$ and $l \in W^*$, then of course $l \circ T \in V^*$. Moreover, the mapping $l \mapsto l \circ T$ (T fixed) is a linear mapping from W^* to V^* by the corollary to Theorem 3.3 of Chapter 1. This mapping is called the *adjoint* of T and is designated T^*. Thus $T^* \in \text{Hom}(W^*, V^*)$ and $T^*(l) = l \circ T$ for all $l \in W^*$.

Theorem 3.4. The mapping $T \mapsto T^*$ is an isomorphism from the vector space $\text{Hom}(V, W)$ to the vector space $\text{Hom}(W^*, V^*)$. Also $(T \circ S)^* = S^* \circ T^*$ under the relevant hypotheses on domains and codomains.

Proof. Everything we have said above through the linearity of $T \mapsto T^*$ is a consequence of the bilinearity of $\omega(l, T) = l \circ T$. The map we have called T^* is simply ω_T, and the linearity of $T \mapsto T^*$ thus follows from Theorem 6.1 of Chapter 1. Again the reader might benefit from a direct linearity check, beginning with $(c_1T_1 + c_2T_2)^*(l)$.

To see that $T \mapsto T^*$ is injective, we take any $T \neq 0$ and choose $\alpha \in V$ so that $T(\alpha) \neq 0$. We then choose $l \in W^*$ so that $l(T(\alpha)) \neq 0$. Since $l(T(\alpha)) = (T^*(l))(\alpha)$, we have verified that $T^* \neq 0$.

Next, if $d(V) = m$ and $d(W) = n$, then also $d(V^*) = m$ and $d(W^*) = n$ by the corollary of Theorem 3.1, and $d(\text{Hom}(V, W)) = mn = d(\text{Hom}(W^*, V^*))$ by Theorem 2.5. The injective map $T \mapsto T^*$ is thus an isomorphism (by the corollary of Theorem 2.4).

Finally, $(T \circ S)^*l = l \circ (T \circ S) = (l \circ T) \circ S = S^*(l \circ T) = S^*(T^*(l)) = (S^* \circ T^*)l$, so that $(T \circ S)^* = S^* \circ T^*$. □

The reader would probably guess that T^{**} becomes identified with T under the identification of V with V^{**}. This is so, and it is actually the reason for calling the isomorphism $\xi \mapsto \xi^{**}$ natural. We shall return to this question at the end of the section. Meanwhile, we record an important elementary identity.

Theorem 3.5. $(R(T^*))^\circ = N(T)$ and $N(T^*) = (R(T))^\circ$.

Proof. The following statements are definitionally equivalent in pairs as they occur: $l \in N(T^*)$, $T^*(l) = 0$, $l \circ T = 0$, $l(T(\xi)) = 0$ for all $\xi \in V$, $l \in (R(T))^\circ$. Therefore, $N(T^*) = (R(T))^\circ$. The other proof is similar and will be left to the reader. [Start with $\alpha \in N(T)$ and end with $\alpha \in (R(T^*))^\circ$.] □

The *rank* of a linear transformation is the dimension of its range space.

Corollary. The rank of T^* is equal to the rank of T.

Proof. The dimensions of $R(T)$ and $(N(T))^\circ$ are each $d(V) - d(N(T))$ by Theorems 2.4 and 3.3, and the second is $d(R(T^*))$ by the above theorem. Therefore, $d(R(T)) = d(R(T^*))$. □

Dyads. Consider any T in $\mathrm{Hom}(V, W)$ whose range M is one-dimensional. If β is a nonzero vector in M, then $x \mapsto x\beta$ is a basis isomorphism $\theta: \mathbb{R} \to M$ and $\theta^{-1} \circ T: V \to \mathbb{R}$ is a linear functional $\lambda \in V^*$. Then $T = \theta \circ \lambda$ and $T(\xi) = \lambda(\xi)\beta$ for all ξ. We write this as $T = \lambda(\cdot)\beta$, and call any such T a *dyad*.

Lemma 3.2. If T is the dyad $\lambda(\cdot)\beta$, then T^* is the dyad $\beta^{**}(\cdot)\lambda$.

Proof. $(T^*(l))(\xi) = (l \circ T)(\xi) = l(T(\xi)) = l(\lambda(\xi)\beta) = l(\beta)\lambda(\xi)$, so that $T^*(l) = l(\beta)\lambda = \beta^{**}(l)\lambda$, and $T^* = \beta^{**}(\cdot)\lambda$. □

***Natural isomorphisms again.** We are now in a position to illustrate more precisely the notion of a natural isomorphism. We saw above that among all the isomorphisms from a finite-dimensional vector space V to its second dual, we could single one out naturally, namely, the map $\xi \mapsto \xi^{**}$, where $\xi^{**}(f) = f(\xi)$ for all f in V^*. Let us call this isomorphism φ_V. The technical meaning of the word 'natural' pertains to the collection $\{\varphi_V\}$ of all these isomorphisms; we found a way to choose one isomorphism φ_V for each space V, and the proof that this is a "natural" choice lies in the smooth way the various φ_V's relate to each other. To see what we mean by this, consider two finite-dimensional spaces V and W and a map T in $\mathrm{Hom}(V, W)$. Then T^* is in $\mathrm{Hom}(W^*, V^*)$ and $T^{**} = (T^*)^*$ is in $\mathrm{Hom}(V^{**}, W^{**})$. The setting for the four maps T, T^{**}, φ_V, and φ_W can be displayed in a diagram as follows:

$$\begin{array}{ccc} V & \xrightarrow{\quad T \quad} & W \\ {\scriptstyle \varphi_V}\big\downarrow & & \big\downarrow{\scriptstyle \varphi_W} \\ V^{**} & \xrightarrow{\quad T^{**} \quad} & W^{**} \end{array}$$

The diagram indicates two maps, $\varphi_W \circ T$ and $T^{**} \circ \varphi_V$, from V to W^{**}, and we define the collection of isomorphisms $\{\varphi_V\}$ to be *natural* if these two maps are always equal for any V, W and T. This is the condition that the two ways of going around the diagram give the same result, i.e., that *the diagram be commutative.*

Put another way, it is the condition that T "become" T^{**} when V is identified with V^{**} (by φ_V) and W is identified with W^{**} (by φ_W). We leave its proof as an exercise.

EXERCISES

3.1 Let θ be an isomorphism from a vector space V to $\mathbb{R}^n$. Show that the functionals $\{\pi_i \circ \theta\}_1^n$ form a basis for V^*.

3.2 Show that the standard isomorphism from $\mathbb{R}^n$ to $(\mathbb{R}^n)^*$ that we get by composing the coordinate isomorphism for the standard basis for $\mathbb{R}^n$ (the *identity*) with the dual basis isomorphism for $(\mathbb{R}^n)^*$ is just our friend $\mathbf{a} \mapsto l_{\mathbf{a}}$, where $l_{\mathbf{a}}(\mathbf{x}) = \sum_1^n a_i x_i$. (Show that the dual basis isomorphism is $\mathbf{a} \mapsto \sum_1^n a_i \pi_i$.)

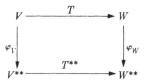

3.3 We know from Theorem 1.6 that a choice of a basis $\{\beta_i\}$ for V defines an isomorphism from W^n to $\mathrm{Hom}(V, W)$ for any vector space W. Apply this fact and Theorem 1.3 to obtain a basis in V^*, and show that this basis is the dual basis of $\{\beta_i\}$.

3.4 Prove the properties of A° that are listed in the text.

3.5 Find (a basis for) the annihilator of $\langle 1, 1, 1\rangle$ in $\mathbb{R}^3$. (Use the isomorphism of $(\mathbb{R}^3)^*$ with $\mathbb{R}^3$ to express the basis vectors as triples.)

3.6 Find (a basis for) the annihilator of $\{\langle 1, 1, 1\rangle, \langle 1, 2, 3\rangle\}$ in $\mathbb{R}^3$.

3.7 Find (a basis for) the annihilator of $\{\langle 1, 1, 1, 1\rangle, \langle 1, 2, 3, 4\rangle\}$ in $\mathbb{R}^4$.

3.8 Show that if $V = M \oplus N$, then $V^* = M^\circ \oplus N^\circ$.

3.9 Show that if M is any subspace of an n-dimensional vector space V and $d(M) = m$, then M can be viewed as being the linear span of an independent subset of m elements of V *or* as being the annihilator of (the intersection of the null spaces of) an independent subset of $n - m$ elements of V^*.

3.10 If $B = \{f_i\}_1^m$ is a finite collection of linear functionals on V ($B \subset V^*$), then its annihilator B° is simply the intersection $N = \bigcap_1^n N_i$ of the null spaces $N_i = N(f_i)$ of the functionals f_i. State the dual of Theorem 3.3 in this context. That is, take W as the linear span of the functionals f_i, so that $W \subset V^*$ and $W^\circ \subset V$. State the dual of the corollary.

3.11 Show that the following theorem is a consequence of the corollary of Theorem 3.3.

> **Theorem.** Let N be the intersection $\bigcap_1^n N_i$ of the null spaces of a set $\{f_i\}_1^n$ of linear functionals on V, and suppose that g in V^* is zero on N. Then g is a linear combination of the set $\{f_i\}_1^n$.

3.12 A corollary of Theorem 3.3 is that if W is a proper subspace of V, then there is at least one nonzero linear functional f in V^* such that $f = 0$ on W. Prove this fact directly by elementary means. (You are allowed to construct a suitable basis.)

3.13 An m-tuple of linear functionals $\{f_i\}_1^m$ on a vector space V defines a linear mapping $\alpha \mapsto \langle f_1(\alpha), \ldots, f_m(\alpha)\rangle$ from V to $\mathbb{R}^m$. What theorem is being applied here? Prove that the range of this linear mapping is the whole of $\mathbb{R}^m$ if and only if $\{f_i\}_1^m$ is an independent set of functionals. [*Hint:* If the range is a proper subspace W, there is a nonzero m-tuple $\mathbf{a}$ such that $\sum_1^n a_i x_i = 0$ for all $\mathbf{x} \in W$.]

3.14 Continuing the above exercise, what is the null space N of the linear mapping $\alpha \mapsto \langle f_1(\alpha), \ldots, f_m(\alpha)\rangle$? If g is a linear functional which is zero on N, show that g is a linear combination of the f_i, now as a corollary of the above exercise and Theorem 4.3 of Chapter 1. (Assume the set $\{f_i\}_1^m$ independent.)

3.15 Write out from scratch the proof that T^* is linear [for a given T in $\mathrm{Hom}(V, W)$]. Also prove directly that $T \mapsto T^*$ is linear.

3.16 Prove the other half of Theorem 3.5.

3.17 Let θ_i be the isomorphism $\alpha \mapsto \alpha^{**}$ from V_i to V_i^{**} for $i = 1, 2$, and suppose given T in $\mathrm{Hom}(V_1, V_2)$. The loose statement $T = T^{**}$ means exactly that

$$T^{**} = \theta_2 \circ T \circ \theta_1^{-1} \quad \text{or} \quad T^{**} \circ \theta_1 = \theta_2 \circ T.$$

Prove this identity. As usual, do this by proving that it holds for each α in V_1.

3.18 Let $\theta\colon \mathbb{R}^n \to V$ be a basis isomorphism. Prove that the adjoint θ^* is the coordinate isomorphism for the dual basis if $(\mathbb{R}^n)^*$ is identified with $\mathbb{R}^n$ in the natural way.

3.19 Let ω be any bilinear functional on $V \times W$. Then the two associated linear transformations are $T\colon V \to W^*$ defined by $(T(\xi))(\eta) = \omega(\xi, \eta)$ and $S\colon W \to V^*$ defined by $(S(\eta))(\xi) = \omega(\xi, \eta)$. Prove that $S = T^*$ if W is identified with W^{**}.

3.20 Suppose that f in $(\mathbb{R}^m)^*$ has coordinate m-tuple $\mathbf{a}$ $[f(\mathbf{y}) = \sum_1^m a_i y_i]$ and that T in $\mathrm{Hom}(\mathbb{R}^n, \mathbb{R}^m)$ has matrix $\mathbf{t} = \{t_{ij}\}$. Write out the explicit expression of the number $f(T(\mathbf{x}))$ in terms of all these coordinates. Rearrange the sum so that it appears in the form

$$g(\mathbf{x}) = \sum_1^n b_i x_i,$$

and then read off the formula for $\mathbf{b}$ in terms of $\mathbf{a}$.

4. MATRICES

Matrices and linear transformations. The reader has already learned something about matrices and their relationship to linear transformations from Chapter 1; we shall begin our more systematic discussion by reviewing this earlier material. By popular conception a matrix is a rectangular array of numbers such as

$$\begin{bmatrix} t_{11} & t_{12} & \cdots & t_{1n} \\ t_{21} & t_{22} & \cdots & t_{2n} \\ \vdots & & & \vdots \\ t_{m1} & t_{m2} & \cdots & t_{mn} \end{bmatrix}.$$

Note that the first index numbers the rows and the second index numbers the columns. If there are m rows and n columns in the array, it is called an m-by-n ($m \times n$) matrix. This notion is inexact. A rectangular array is a way of *picturing* a matrix, but a matrix is really a function, just as a sequence is a function. With the notation $\overline{m} = \{1, \ldots, m\}$, the above matrix is a function assigning a number to every pair of integers $\langle i, j\rangle$ in $\overline{m} \times \overline{n}$. It is thus an element of the set $\mathbb{R}^{\overline{m}\times\overline{n}}$. The addition of two $m \times n$ matrices is performed in the obvious place-by-place way, and is merely the addition of two functions in $\mathbb{R}^{\overline{m}\times\overline{n}}$; the same is true for scalar multiplication. The set of all $m \times n$ matrices is thus the vector space $\mathbb{R}^{\overline{m}\times\overline{n}}$, a Cartesian space with a rather fancy finite index set. We shall use the customary index notation t_{ij} for the value $\mathbf{t}(i, j)$ of the function $\mathbf{t}$ at $\langle i, j\rangle$, and we shall also write $\{t_{ij}\}$ for $\mathbf{t}$, just as we do for sequences and other indexed collections.

The additional properties of matrices stem from the correspondence between $m \times n$ matrices $\{t_{ij}\}$ and transformations $T \in \mathrm{Hom}(\mathbb{R}^n, \mathbb{R}^m)$.

The following theorem restates results from the first chapter. See Theorems 1.2, 1.3, and 6.2 of Chapter 1 and the discussion of the linear combination map at the end of Section 1.6.

Theorem 4.1. Let $\{t_{ij}\}$ be an m-by-n matrix, and let $\mathbf{t}^j$ be the m-tuple that is its jth column for $j = 1, \ldots, n$. Then there is a unique T in $\mathrm{Hom}(\mathbb{R}^n, \mathbb{R}^m)$ such that skeleton $T = \{\mathbf{t}^j\}$, i.e., such that $T(\delta^j) = \mathbf{t}^j$ for all j. T is defined as the linear combination mapping $\mathbf{x} \mapsto \mathbf{y} = \sum_{j=1}^n x_j \mathbf{t}^j$, and an equivalent presentation of T is the collection of scalar equations

$$y_i = \sum_{j=1}^{n} t_{ij} x_j \qquad \text{for} \quad i = 1, \ldots, m.$$

Each T in $\mathrm{Hom}(\mathbb{R}^n, \mathbb{R}^m)$ arises this way, and the bijection $\{t_{ij}\} \mapsto T$ from $\mathbb{R}^{\overline{m} \times \overline{n}}$ to $\mathrm{Hom}(\mathbb{R}^n, \mathbb{R}^m)$ is a natural isomorphism.

The only additional remark called for here is that in identifying an $m \times n$ matrix with an n-tuple of m-tuples, we are making use of one of the standard identifications of duality (Section 0.10). We are treating the natural isomorphism between the really distinct spaces $\mathbb{R}^{\overline{m} \times \overline{n}}$ and $(\mathbb{R}^{\overline{m}})^{\overline{n}}$ as though it were the identity.

We can also relate T to $\{t_{ij}\}$ by way of the *rows* of $\{t_{ij}\}$. As above, taking ith coordinates in the m-tuple equation $\mathbf{y} = \sum_{j=1}^n \mathbf{x}_j \mathbf{t}^j$, we get the equivalent and familiar system of numerical (scalar) equations $y_i = \sum_{j=1}^n t_{ij} x_j$ for $i = 1, \ldots, m$. Now the mapping $\mathbf{x} \mapsto \sum_{j=1}^n c_j x_j$ from $\mathbb{R}^n$ to $\mathbb{R}$ is the most general linear functional on $\mathbb{R}^n$. In the above numerical equations, therefore, we have simply used the m rows of the matrix $\{t_{ij}\}$ to present the m-tuple of linear functionals on $\mathbb{R}^n$ which is equivalent to the single m-tuple-valued linear mapping T in $\mathrm{Hom}(\mathbb{R}^n, \mathbb{R}^m)$ by Theorem 3.6 of Chapter 1.

The choice of ordered bases for arbitrary finite-dimensional spaces V and W allows us to transfer the above theorem to $\mathrm{Hom}(V, W)$. Since we are now going to correlate a matrix $\mathbf{t}$ in $\mathbb{R}^{\overline{m} \times \overline{n}}$ with a transformation T in $\mathrm{Hom}(V, W)$, we shall designate the transformation in $\mathrm{Hom}(\mathbb{R}^n, \mathbb{R}^m)$ discussed above by $\overline{T}$.

Theorem 4.2. Let $\{\alpha_j\}_1^n$ and $\{\beta_i\}_1^m$ be ordered bases for the vector spaces V and W, respectively. For each matrix $\{t_{ij}\}$ in $\mathbb{R}^{\overline{m} \times \overline{n}}$ let T be the unique element of $\mathrm{Hom}(V, W)$ such that $T(\alpha_j) = \sum_{i=1}^m t_{ij} \beta_i$ for $j = 1, \ldots, n$. Then the mapping $\{t_{ij}\} \mapsto T$ is an isomorphism from $\mathbb{R}^{\overline{m} \times \overline{n}}$ to $\mathrm{Hom}(V, W)$.

Proof. We simply combine the isomorphism $\{t_{ij}\} \mapsto \overline{T}$ of the above theorem with the isomorphism $\overline{T} \mapsto T = \psi \circ \overline{T} \circ \varphi^{-1}$ from $\mathrm{Hom}(\mathbb{R}^n, \mathbb{R}^m)$ to $\mathrm{Hom}(V, W)$, where φ and ψ are the two given basis isomorphisms. Then T is the transformation described in the theorem, for $T(\alpha_j) = \psi(\overline{T}(\varphi^{-1}(\alpha_j))) = \psi(\overline{T}(\delta^j)) = \psi(\mathbf{t}^j) = \sum_{i=1}^m t_{ij} \beta_i$. The map $\{t_{ij}\} \mapsto T$ is the composition of two isomorphisms and so is an isomorphism. □

It is instructive to look at what we have just done in a slightly different way. Given the matrix $\{t_{ij}\}$, let τ_j be the vector in W whose coordinate m-tuple is the jth column $\mathbf{t}^j$ of the matrix, so that $\tau_j = \sum_{i=1}^m t_{ij} \beta_i$. Then let T be the unique element of $\mathrm{Hom}(V, W)$ such that $T(\alpha_j) = \tau_j$ for $j = 1, \ldots, n$. Now we have obtained T from $\{t_{ij}\}$ in the following two steps: T corresponds to the n-tuple

$\{\tau_j\}_1^n$ under the isomorphism from $\text{Hom}(V, W)$ to W^n given by Theorem 1.6, and $\{\tau_j\}_1^n$ corresponds to the matrix $\{t_{ij}\}$ by extension of the coordinate isomorphism between W and $\mathbb{R}^m$ to its product isomorphism from W^n to $(\mathbb{R}^m)^n$.

Corollary. If $\mathbf{y}$ is the coordinate m-tuple of the vector η in W and $\mathbf{x}$ is the coordinate n-tuple of ξ in V (with respect to the given bases), then $\eta = T(\xi)$ if and only if $y_i = \sum_{j=1}^n t_{ij}x_j$ for $i = 1, \ldots, m$.

Proof. We know that the scalar equations are equivalent to $\mathbf{y} = \overline{T}(\mathbf{x})$, which is the equation $\mathbf{y} = \psi^{-1} \circ T \circ \varphi(\mathbf{x})$. The isomorphism ψ converts this to the equation $\eta = T(\xi)$. □

Our problem now is to discover the matrix analogues of relationship between linear transformations. For transformations between the Cartesian spaces $\mathbb{R}^n$ this is a fairly direct, uncomplicated business, because, as we know, the matrix here is a natural alter ego for the transformation. But when we leave the Cartesian spaces, a transformation T no longer has a matrix in any natural way, and only acquires one when bases are chosen and a corresponding $\overline{T}$ on Cartesian spaces is thereby obtained. All matrices now are determined *with respect to chosen bases*, and all calculations are complicated by the necessary presence of the basis and coordinate isomorphisms. There are two ways of handling this situation. The first, which we shall follow in general, is to describe things directly for the general space V and simply to accept the necessarily more complicated statements involving bases and dual bases and the corresponding loss in transparency. The other possibility is first to read off the answers for the Cartesian spaces and then to transcribe them via coordinate isomorphisms.

Lemma 4.1. The matrix element t_{kj} can be obtained from T by the formula

$$t_{kj} = \mu_k(T(\alpha_j)),$$

where μ_k is the kth element of the dual basis in W^*.

Proof. $\mu_k(T(\alpha_j)) = \mu_k(\sum_{i=1}^m t_{ij}\beta_i) = \sum_i t_{ij}\mu_k(\beta_i) = \sum_i t_{ij}\, \delta_i^k = t_{kj}$. □

In terms of Cartesian spaces, $\overline{T}(\delta^j)$ is the jth column m-tuple $\mathbf{t}^j$ in the matrix $\{t_{ij}\}$ of $\overline{T}$, and t_{kj} is the kth coordinate of $\mathbf{t}^j$. From the point of view of linear maps, the kth coordinate is obtained by applying the kth coordinate projection π_k, so that $t_{kj} = \pi_k(\overline{T}(\delta^j))$. Under the basis isomorphisms, π_k becomes μ_k, $\overline{T}$ becomes T, δ^j becomes α_j, and the Cartesian identity becomes the identity of the lemma.

The transpose. The *transpose* of the $m \times n$ matrix $\{t_{ij}\}$ is the $n \times m$ matrix $\{t_{ij}^*\}$ defined by $t_{ij}^* = t_{ji}$ for all i, j. The rows of $\mathbf{t}^*$ are of course the columns of $\mathbf{t}$, and conversely.

Theorem 4.3. The matrix of T^* with respect to the dual bases in W^* and V^* is the transpose of the matrix of T.

Proof. If $\mathbf{s}$ is the matrix of T^*, then Lemmas 3.1 and 4.1 imply that

$$\begin{aligned} s_{ji} &= \alpha_j^{**}(T^*(\mu_i)) = \alpha_j^{**}(\mu_i \circ T) \\ &= (\mu_i \circ T)(\alpha_j) = \mu_i(T(\alpha_j)) = t_{ij}. \ \square \end{aligned}$$

Definition. The *row space* of the matrix $\{t_{ij}\} \in \mathbb{R}^{\overline{m}\times\overline{n}}$ is the subspace of $\mathbb{R}^n$ spanned by the m row vectors. The *column space* is similarly the span of the n column vectors in $\mathbb{R}^m$.

Corollary. The row and column spaces of a matrix have the same dimension.

Proof. If T is the element of $\operatorname{Hom}(\mathbb{R}^n, \mathbb{R}^m)$ defined by $T(\delta^j) = \mathbf{t}^j$, then the set $\{\mathbf{t}^j\}_1^n$ of column vectors in the matrix $\{t_{ij}\}$ is the image under T of the standard basis of $\mathbb{R}^n$, and so its span, which we have called the column space of the matrix, is exactly the range of T. In particular, the dimension of the column space is $d(R(T)) = \operatorname{rank} T$.

Since the matrix of T^* is the transpose $\mathbf{t}^*$ of the matrix $\mathbf{t}$, we have, similarly, that rank T^* is the dimension of the column space of $\mathbf{t}^*$. But the column space of $\mathbf{t}^*$ is the row space of $\mathbf{t}$, and the assertion of the corollary is thus reduced to the identity rank $T^* =$ rank T, which is the corollary of Theorem 3.5. $\square$

This common dimension is called the *rank* of the matrix.

Matrix products. If $T \in \operatorname{Hom}(\mathbb{R}^n, \mathbb{R}^m)$ and $S \in \operatorname{Hom}(\mathbb{R}^m, \mathbb{R}^l)$, then of course $R = S \circ T \in \operatorname{Hom}(\mathbb{R}^n, \mathbb{R}^l)$, and it certainly should be possible to calculate the matrix $\mathbf{r}$ of R from the matrices $\mathbf{s}$ and $\mathbf{t}$ of S and T, respectively. To make this computation, we set $\mathbf{y} = T(\mathbf{x})$ and $\mathbf{z} = S(\mathbf{y})$, so that $\mathbf{z} = (S \circ T)(\mathbf{x}) = R(\mathbf{x})$. The equivalent scalar equations in terms of the matrices $\mathbf{t}$ and $\mathbf{s}$ are

$$y_i = \sum_{h=1}^{n} t_{ih}x_h \qquad \text{and} \qquad z_k = \sum_{i=1}^{m} s_{ki}y_i,$$

so that

$$z_k = \sum_{i=1}^{m} s_{ki} \sum_{h=1}^{n} t_{ih}x_h = \sum_{h=1}^{n} \left(\sum_{i=1}^{m} s_{ki}t_{ih} \right) x_h.$$

But $z_k = \sum_{h=1}^{n} r_{kh}x_h$ for $k = 1, \ldots, l$. Taking $\mathbf{x}$ as δ^j, we have

$$r_{kj} = \sum_{i=1}^{m} s_{ki}t_{ij} \qquad \text{for all} \quad k \text{ and } j.$$

We thus have found the formula for the matrix $\mathbf{r}$ of the map $R = S \circ T: \mathbf{x} \to \mathbf{z}$. Of course, $\mathbf{r}$ is defined to be the product of the matrices $\mathbf{s}$ and $\mathbf{t}$, and we write $\mathbf{r} = \mathbf{s} \cdot \mathbf{t}$ or $\mathbf{r} = \mathbf{st}$.

Note that in order for the product $\mathbf{st}$ to be defined, the number of columns in the left factor must equal the number of rows in the right factor. We get the element r_{kj} by going across the kth row of $\mathbf{s}$ and simultaneously down the jth column of $\mathbf{t}$, multiplying corresponding elements as we go, and adding the resulting products. This process is illustrated in Fig. 2.1. In terms of the scalar product $(\mathbf{x}, \mathbf{y}) = \sum_1^n x_i y_i$ on $\mathbb{R}^n$, we see that the element r_{kj} in $\mathbf{r} = \mathbf{st}$ is the scalar product of the kth row of $\mathbf{s}$ and the jth column of $\mathbf{t}$.

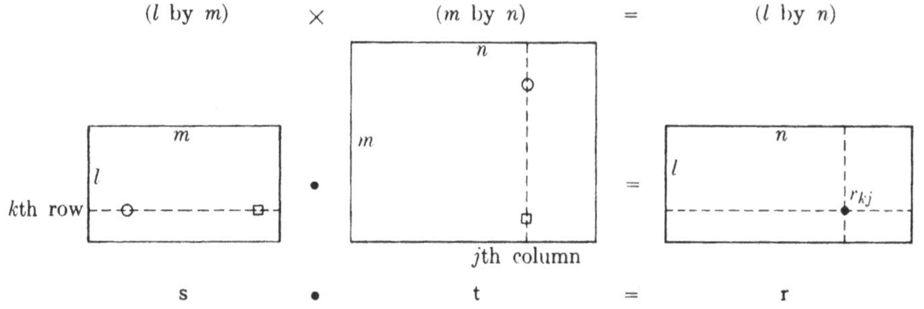

Fig. 2.1

Since we have defined the product of two matrices as the matrix of the product of the corresponding transformations, i.e., so that the mapping $T \mapsto \{t_{ij}\}$ preserves products ($S \circ T \mapsto \mathbf{st}$), it follows from the general principle of Theorem 4.1 of Chapter 1 that the algebraic laws satisfied by composition of transformations will automatically hold for the product of matrices. For example, we know without making an explicit computation that matrix multiplication is associative. Then for square matrices we have the following theorem.

Theorem 4.4. The set M_n of square $n \times n$ matrices is an algebra naturally isomorphic to the algebra $\mathrm{Hom}(\mathbb{R}^n)$.

Proof. We already know that $T \mapsto \{t_{ij}\}$ is a natural linear isomorphism from $\mathrm{Hom}(\mathbb{R}^n)$ to M_n (Theorem 4.1), and we have defined the product of matrices so that the mapping also preserves multiplication. The laws of algebra (for an algebra) therefore follow for M_n from our observation in Theorem 3.5 of Chapter 1 that they hold for $\mathrm{Hom}(\mathbb{R}^n)$. □

The identity I in $\mathrm{Hom}(\mathbb{R}^n)$ takes the basis vector δ^j into itself, and therefore its matrix $\mathbf{e}$ has δ^j for its jth column: $\mathbf{e}^j = \delta^j$. Thus $e_{ij} = \delta_i^j = 1$ if $i = j$ and $e_{ij} = \delta_i^j = 0$ if $i \neq j$. That is, the matrix $\mathbf{e}$ is 1 along the main diagonal (from upper left to lower right) and 0 elsewhere. Since $I \mapsto \mathbf{e}$ under the algebra isomorphism $T \mapsto \mathbf{t}$, we know that $\mathbf{e}$ is the identity for matrix multiplication. Of course, we can check this directly: $\sum_{j=1}^n t_{ij} e_{jk} = t_{ik}$, and similarly for multiplying by $\mathbf{e}$ on the left. The symbol '$\mathbf{e}$' is ambiguous in that we have used it to denote the identity in the space $\mathbb{R}^{\overline{n} \times \overline{n}}$ of square $n \times n$ matrices for any n.

Corollary. A square $n \times n$ matrix $\mathbf{t}$ has a multiplicative inverse if and only if its rank is n.

Proof. By the theorem there exists an $\mathbf{s} \in M_n$ such that $\mathbf{st} = \mathbf{ts} = \mathbf{e}$ if and only if there exists an $S \in \mathrm{Hom}(\mathbb{R}^n)$ such that $S \circ T = T \circ S = I$. But such an S exists if and only if T is an isomorphism, and by the corollary to Theorem 2.4 this is equivalent to the dimension of the range of T being n. But this dimension is the rank of $\mathbf{t}$, and the argument is complete. □

A square matrix (or a transformation in Hom V) is said to be *nonsingular* if it is invertible.

Theorem 4.5. If $\{\alpha_i\}_1^n$, $\{\beta_j\}_1^m$, and $\{\gamma_k\}_1^l$ are ordered bases for the vector spaces U, V, and W, respectively, and if $T \in \mathrm{Hom}(U, V)$ and $S \in \mathrm{Hom}(V, W)$, then the matrix of $S \circ T$ is the product of the matrices of S and T (with respect to the given bases).

Proof. By definition the matrix of $S \circ T$ is the matrix of $\overline{S \circ T} = \chi^{-1} \circ (S \circ T) \circ \varphi$ in $\mathrm{Hom}(\mathbb{R}^n, \mathbb{R}^l)$, where φ and χ are the given basis isomorphisms for U and W. But if ψ is the basis isomorphism for V, we have

$$\overline{S \circ T} = (\chi^{-1} \circ S \circ \psi) \circ (\psi^{-1} \circ T \circ \varphi) = \bar{S} \circ \bar{T},$$

and therefore its matrix is the product of the matrices of $\bar{S}$ and $\bar{T}$ by the definition of matrix multiplication. The latter are the matrices of S and T with respect to the given bases. Putting these observations together, we have the theorem. □

There is a simple relationship between matrix products and transposition.

Theorem 4.6. If the matrix product $\mathbf{st}$ is defined, then so is $\mathbf{t}^*\mathbf{s}^*$, and $\mathbf{t}^*\mathbf{s}^* = (\mathbf{st})^*$.

Proof. A direct calculation is easy. We have

$$(st)^*_{jk} = (st)_{kj} = \sum_{i=1}^m s_{ki}t_{ij} = \sum_{i=1}^m t^*_{ji}s^*_{ik} = (t^*s^*)_{jk}.$$

Thus $(\mathbf{st})^* = \mathbf{t}^*\mathbf{s}^*$, as asserted. □

This identity is clearly the matrix form of the transformation identity $(S \circ T)^* = T^* \circ S^*$, and it can be deduced from the latter identity if desired.

Cartesian vectors as matrices. We can view an n-tuple $\mathbf{x} = \langle x_1, \ldots, x_n \rangle$ as being alternatively either an $n \times 1$ matrix, in which case we call it a *column vector*, or a $1 \times n$ matrix, in which case we call it a *row vector*. Of course, these identifications are natural isomorphisms. The point of doing this is, in part, that then the equations $y_i = \sum_{j=1}^n t_{ij}x_j$ say exactly that the column vector $\mathbf{y}$ is the matrix product of $\mathbf{t}$ and the column vector $\mathbf{x}$, that is, $\mathbf{y} = \mathbf{t} \cdot \mathbf{x}$. *The linear map* $T: \mathbb{R}^n \to \mathbb{R}^m$ *becomes left multiplication by the fixed matrix* $\mathbf{t}$ *when* $\mathbb{R}^n$ *is viewed as the space of* $n \times 1$ *column vectors.* For this reason we shall take the column vector as the standard matrix interpretation of an n-tuple $\mathbf{x}$; then $\mathbf{x}^*$ is the corresponding row vector.

In particular, a linear functional $F \in (\mathbb{R}^n)^*$ becomes left multiplication by *its* matrix, which is of course $1 \times n$ (F being from $\mathbb{R}^n$ to $\mathbb{R}^1$), and therefore is simply the row matrix interpretation of an n-tuple in $\mathbb{R}^n$. That is, in the natural isomorphism $\mathbf{a} \mapsto L_\mathbf{a}$ from $\mathbb{R}^n$ to $(\mathbb{R}^n)^*$, where $L_\mathbf{a}(\mathbf{x}) = \sum_1^n a_i x_i$, the functional $L_\mathbf{a}$ can now be interpreted as left matrix multiplication by the n-tuple $\mathbf{a}$ viewed as the *row* vector $\mathbf{a}^*$. The matrix product of the row vector ($1 \times n$ matrix) $\mathbf{a}^*$ and the column vector ($n \times 1$ matrix) $\mathbf{x}$ is a 1×1 matrix $\mathbf{a}^* \cdot \mathbf{x}$, that is, a number.

Let us now see what these observations say about T^*. The number $L_\mathbf{a}(T(\mathbf{x}))$ is the 1×1 matrix $\mathbf{a}^*\mathbf{t}\mathbf{x}$. Since $L_\mathbf{a}(T(\mathbf{x})) = (T^*(L_\mathbf{a}))(\mathbf{x})$ by the definition of T^*, we see that the functional $T^*(L_\mathbf{a})$ is left multiplication by the row vector $\mathbf{a}^*\mathbf{t}$. Since the row vector form of $L_\mathbf{a}$ is $\mathbf{a}^*$ and the row vector form of $T^*(L_\mathbf{a})$ is $\mathbf{a}^*\mathbf{t}$, this shows that when the functionals on $\mathbb{R}^n$ are interpreted as *row* vectors, T^* becomes *right* multiplication by $\mathbf{t}$. This only repeats something we already know. If we take transposes to throw the row vectors into the standard column vector form for n-tuples, it shows that T^* is left multiplication by $\mathbf{t}^*$, and so gives another proof that the matrix of T^* is t^*.

Change of basis. If $\varphi: \mathbf{x} \mapsto \xi = \sum_1^n x_i \beta_i$ and $\theta: \mathbf{y} \mapsto \xi = \sum_1^n y_i \beta'_i$ are two basis isomorphisms for V, then $A = \theta^{-1} \circ \varphi$ is the isomorphism in $\mathrm{Hom}(\mathbb{R}^n)$ which takes the coordinate n-tuple $\mathbf{x}$ of a vector ξ with respect to the basis $\{\beta_i\}$ into the coordinate n-tuple $\mathbf{y}$ of the *same* vector with respect to the basis $\{\beta'_i\}$. The isomorphism A is called the "change of coordinates" isomorphism. In terms of the matrix $\mathbf{a}$ of A, we have $\mathbf{y} = \mathbf{a}\mathbf{x}$, as above.

The change of coordinate map $A = \theta^{-1} \circ \varphi$ should not be confused with the similar looking $T = \theta \circ \varphi^{-1}$. The latter is a mapping on V, and is the element of $\mathrm{Hom}(V)$ which takes each β_i to β'_i.

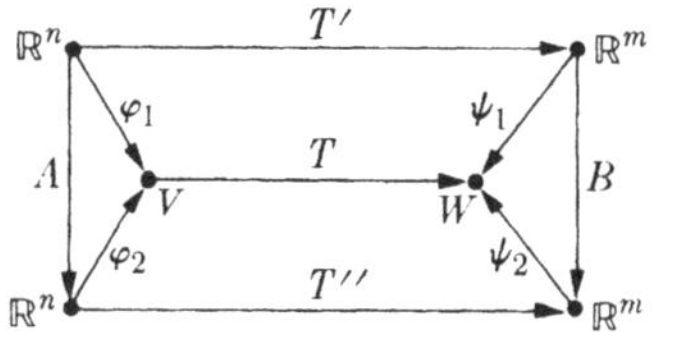

Fig. 2.2

We now want to see what happens to the matrix of a transformation $T \in \mathrm{Hom}(V, W)$ when we change bases in its domain and codomain spaces. Suppose then that φ_1 and φ_2 are basis isomorphisms from $\mathbb{R}^n$ to V, that ψ_1 and ψ_2 are basis isomorphisms from $\mathbb{R}^m$ to W, and that $\mathbf{t}'$ and $\mathbf{t}''$ are the matrices of T with respect to the first and second bases, respectively. That is, $\mathbf{t}'$ is the matrix of $T' = (\psi_1)^{-1} \circ T \circ \varphi_1 \in \mathrm{Hom}(\mathbb{R}^n, \mathbb{R}^m)$, and similarly for $\mathbf{t}''$. The mapping $A = \varphi_2^{-1} \circ \varphi_1 \in \mathrm{Hom}(\mathbb{R}^n)$ is the change of coordinates transformation for V: if $\mathbf{x}$ is the coordinate n-tuple of a vector ξ with respect to the first basis [that is, $\xi = \varphi_1(\mathbf{x})$], then $A(\mathbf{x})$ is its coordinate n-tuple with respect to the second basis. Similarly, let B be the change of coordinates map $\psi_2^{-1} \circ \psi_1$ for W. The diagram in Fig. 2.2 will help keep the various relationships of these spaces and mappings straight. We say that the diagram is commutative, which means that any two paths between two points represent the same map. By selecting various pairs of paths, we can read off all the identities which hold for the nine maps $T, T', T'', \varphi_1, \varphi_2, A, \psi_1, \psi_2, B$. For example, T'' can be obtained by going backward along A, forward along T', and then forward along B. That is, $T'' = B \circ T' \circ A^{-1}$. Since these "outside maps" are all maps of Cartesian spaces, we can then read off the corresponding matrix identity

$$\mathbf{t}'' = \mathbf{b}\mathbf{t}'\mathbf{a}^{-1},$$

showing how the matrix of T with respect to the second pair of bases is obtained from its matrix with respect to the first pair.

What we have actually done in reading off the above identity from the diagram is to eliminate certain retraced steps in the longer path which the definitions would give us. Thus from the definitions we get

$$B \circ T' \circ A^{-1} = (\psi_2^{-1} \circ \psi_1) \circ (\psi_1^{-1} \circ T \circ \varphi_1) \circ (\varphi_1^{-1} \circ \varphi_2) = \psi_2^{-1} \circ T \circ \varphi_2 = T''.$$

In the above situation the domain and codomain spaces were different, and the two basis changes were independent of each other. If $W = V$, so that $T \in \mathrm{Hom}(V)$, then of course we consider only one basis change and the formula becomes

$$\mathbf{t}'' = \mathbf{a} \cdot \mathbf{t}' \cdot \mathbf{a}^{-1}.$$

Now consider a linear functional $F \in V^*$. If $\mathbf{f}''$ and $\mathbf{f}'$ are its coordinate n-tuples considered as column vectors ($n \times 1$ matrices), then the matrices of F with respect to the two bases are the row vectors $(\mathbf{f}')^*$ and $(\mathbf{f}'')^*$, as we saw earlier. Also, there is no change of basis in the range space since here $W = \mathbb{R}$, with its permanent natural basis vector 1. Therefore, $\mathbf{b} = \mathbf{e}$ in the formula $\mathbf{t}'' = \mathbf{b}\mathbf{t}'\mathbf{a}^{-1}$, and we have $(\mathbf{f}'')^* = (\mathbf{f}')^*\mathbf{a}^{-1}$ or

$$\mathbf{f}'' = (\mathbf{a}^{-1})^*\mathbf{f}'.$$

We want to compare this with the change of coordinates of a vector $\xi \in V$, which, as we saw earlier, is given by

$$\mathbf{x}'' = \mathbf{a}\mathbf{x}'.$$

These changes go in the *oppositive* directions (with a transposition thrown in). For reasons largely historical, functionals F in V^* are called *covariant* vectors, and since the matrix for a change of coordinates in V is the transpose of the inverse of the matrix for the corresponding change of coordinates in V^*, the vectors ξ in V are called *contravariant* vectors. These terms are used in classical tensor analysis and differential geometry.

The isomorphism $\{t_{ij}\} \mapsto T$, being from a Cartesian space $\mathbb{R}^{\bar{m}\times\bar{n}}$, is automatically a basis isomorphism. Its basis in $\mathrm{Hom}(V, W)$ is the image under the isomorphism of the standard basis in $\mathbb{R}^{\bar{m}\times\bar{n}}$, where the latter is the set of Kronecker functions δ^{kl} defined by $\delta^{kl}(i, j) = 0$ if $\langle k, l\rangle \neq \langle i, j\rangle$ and $\delta^{kl}(k, l) = 1$. (Remember that in $\mathbb{R}^A$, δ^a is that function such that $\delta^a(b) = 0$

if $b \neq a$ and $\delta^a(a) = 1$. Here $A = \overline{m} \times \overline{n}$ and the elements a of A are ordered pairs $a = \langle k, l \rangle$.) The function δ^{kl} is that matrix whose columns are all 0 except for the lth, and the lth column is the m-tuple δ^k. The corresponding transformation D_{kl} thus takes every basis vector α_j to 0 except α_l and takes α_l to β_k. That is, $D_{kl}(\alpha_j) = 0$ if $j \neq l$, and $D_{kl}(\alpha_l) = \beta_k$. Again, D_{kl} takes the lth basis vector in V to the kth basis vector in W and takes the other basis vectors in V to 0.

If $\xi = \sum x_i\alpha_i$, it follows that $D_{kl}(\xi) = x_l\beta_k$.

Since $\{D_{kl}\}$ is the basis defined by the isomorphism $\{t_{ij}\} \mapsto T$, it follows that $\{t_{ij}\}$ is the coordinate set of T with respect to this basis; it is the image of T under the coordinate isomorphism. It is interesting to see how this basis expansion of T automatically appears. We have

$$T(\xi) = T\left(\sum_{j=1}^{n} x_j\alpha_j\right) = \sum_{j=1}^{n} x_jT(\alpha_j) = \sum_{i,j} t_{ij}x_j\beta_i = \sum_{i,j} t_{ij}D_{ij}(\xi),$$

so that

$$T = \sum_{i,j} t_{ij}D_{ij}.$$

Our original discussion of the dual basis in V^* was a special case of the present situation. There we had $\mathrm{Hom}(V, \mathbb{R}) = V^*$, with the permanent standard basis 1 for $\mathbb{R}$. The basis for V^* corresponding to the basis $\{\alpha_i\}$ for V therefore consists of those maps D_l taking α_l to 1 and α_j to 0 for $j \neq l$. Then $D_l(\xi) = D_l(\sum x_j\alpha_j) = x_l$, and D_l is the lth coordinate functional ε_l.

Finally, we note that the matrix expression of $T \in \mathrm{Hom}(\mathbb{R}^n, \mathbb{R}^m)$ is very suggestive of the block decompositions of T that we discussed earlier in Section 1.5. In the exercises we shall ask the reader to show that in fact $T_{kl} = t_{kl}D_{kl}$.

EXERCISES

4.1 Prove that if $\omega: V \times V \to \mathbb{R}$ is a bilinear functional on V and $T: V \to V^*$ is the corresponding linear transformation defined by $(T(\eta))(\xi) = \omega(\xi, \eta)$, then for any basis $\{\alpha_i\}$ for V the matrix $t_{ij} = \omega(\alpha_i, \alpha_j)$ is the matrix of T.

4.2 Verify that the row and column rank of the following matrix are both 1:

$$\begin{bmatrix} -5 & 2 & 3 \\ -10 & 4 & 6 \end{bmatrix}.$$

4.3 Show by a direct calculation that if the row rank of a 2×3 matrix is 1, then so is its column rank.

4.4 Let $\{f_i\}_1^3$ be a linearly dependent set of $\mathcal{C}^2$-functions (twice continuously differentiable real-valued functions) on $\mathbb{R}$. Show that the three triples $\langle f_i(x), f_i'(x), f_i''(x) \rangle$ are dependent for any x. Prove therefore that $\sin t$, $\cos t$, and e^t are linearly independent. (Compute the derivative triples for a well-chosen x.)

4.5 Compute

$$\begin{bmatrix} 5 & -2 & 3 & 1 \\ -4 & 1 & 2 & -3 \\ 1 & 6 & -1 & 4 \end{bmatrix} \times \begin{bmatrix} 1 & 3 \\ 2 & -1 \\ -3 & 0 \\ 4 & 2 \end{bmatrix}.$$

4.6 Compute

$$\begin{bmatrix} a & b \\ c & d \end{bmatrix} \times \begin{bmatrix} d & -b \\ -c & a \end{bmatrix}.$$

From your answer give a necessary and sufficient condition for

$$\begin{bmatrix} a & b \\ c & d \end{bmatrix}^{-1}$$

to exist.

4.7 A matrix $\mathbf{a}$ is *idempotent* if $\mathbf{a}^2 = \mathbf{a}$. Find a basis for the vector space $\mathbb{R}^{2\times 2}$ of all 2×2 matrices consisting entirely of idempotents.

4.8 By a direct calculation show that

$$\begin{bmatrix} 1 & 2 \\ -2 & 3 \end{bmatrix}$$

is invertible and find its inverse.

4.9 Show by explicitly solving the equation

$$\begin{bmatrix} a & b \\ c & d \end{bmatrix} \cdot \begin{bmatrix} x & y \\ z & w \end{bmatrix} = \begin{bmatrix} 1 & 0 \\ 0 & 1 \end{bmatrix}$$

that the matrix on the left is invertible if and only if (the determinant) $ad - bc$ is not zero.

4.10 Find a nonzero 2×2 matrix

$$\begin{bmatrix} a & b \\ c & d \end{bmatrix}$$

whose square is zero.

4.11 Find *all* 2×2 matrices whose squares are zero.

4.12 Prove by computing matrix products that matrix multiplication is associative.

4.13 Similarly, prove directly the distributive law, $(\mathbf{r} + \mathbf{s}) \cdot \mathbf{t} = \mathbf{r} \cdot \mathbf{t} + \mathbf{s} \cdot \mathbf{t}$.

4.14 Show that left matrix multiplication by a fixed $\mathbf{r}$ in $\mathbb{R}^{\bar{m}\times\bar{n}}$ is a linear transformation from $\mathbb{R}^{\bar{n}\times\bar{p}}$ to $\mathbb{R}^{\bar{m}\times\bar{p}}$. What theorem in Chapter 1 does this mirror?

4.15 Show that the rank of a product of two matrices is at most the minimum of their ranks. (Remember that the rank of a matrix is the dimension of the range space of its associated T.)

4.16 Let $\mathbf{a}$ be an $m \times n$ matrix, and let $\mathbf{b}$ be $n \times m$. If $m > n$, show that $\mathbf{a} \cdot \mathbf{b}$ cannot be the identity $\mathbf{e}$ $(m \times m)$.

4.17 Let Z be the subset of 2×2 matrices of the form

$$\begin{bmatrix} a & b \\ -b & a \end{bmatrix}.$$

Prove that Z is a *subalgebra* of $\mathbb{R}^{2\times 2}$ (that is, Z is closed under addition, scalar multiplication, and matrix multiplication). Show that in fact Z is isomorphic to the complex number system.

4.18 A matrix (necessarily square) which is equal to its transpose is said to be *symmetric*. As a square array it is symmetric about the main diagonal. Show that for any $m \times n$ matrix $\mathbf{t}$ the product $\mathbf{t} \cdot \mathbf{t}^*$ is meaningful and symmetric.

4.19 Show that if $\mathbf{s}$ and $\mathbf{t}$ are symmetric $n \times n$ matrices, and if they commute, then $\mathbf{s} \cdot \mathbf{t}$ is symmetric. (Do not try to answer this by writing out matrix products.) Show conversely that if $\mathbf{s}$, $\mathbf{t}$, and $\mathbf{s} \cdot \mathbf{t}$ are all symmetric, then $\mathbf{s}$ and $\mathbf{t}$ commute.

4.20 Suppose that T in Hom $\mathbb{R}^2$ has a symmetric matrix and that T is not of the form cI. Show that T has exactly two eigenvectors (up to scalar multiples). What does the matrix of T become with respect to the "eigenbasis" for $\mathbb{R}^2$ consisting of these two eigenvectors?

4.21 Show that the symmetric 2×2 matrix $\mathbf{t}$ has a symmetric square root $\mathbf{s}$ ($\mathbf{s}^2 = \mathbf{t}$) if and only if its eigenvalues are nonnegative. (Assume the above exercise.)

4.22 Suppose that $\mathbf{t}$ is a 2×2 matrix such that $\mathbf{t}^* = \mathbf{t}^{-1}$. Show that $\mathbf{t}$ has one of the forms

$$\begin{bmatrix} a & b \\ -b & a \end{bmatrix}, \quad \begin{bmatrix} a & b \\ b & -a \end{bmatrix},$$

where $a^2 + b^2 = 1$.

4.23 Prove that multiplication by the above $\mathbf{t}$ is a Euclidean isometry. That is, show that if $\mathbf{y} = \mathbf{t} \cdot \mathbf{x}$, where $\mathbf{x}$ and $\mathbf{y} \in \mathbb{R}^2$, then $\|x\| = \|y\|$, where $\|x\| = (x_1^2 + x_2^2)^{1/2}$.

4.24 Let $\{D_{kl}\}$ be the basis for $\operatorname{Hom}(V, W)$ defined in the text. Taking $W = V$, show that these operators satisfy the very important multiplication rules

$$D_{ij} \circ D_{kl} = 0 \quad \text{if } j \neq k,$$
$$D_{ik} \circ D_{kl} = D_{il}.$$

4.25 Keeping the above identities in mind, show that if $l \neq m$, then there are transformations S and T in Hom V such that

$$S \circ T - T \circ S = D_{lm}.$$

Also find S and T such that

$$S \circ T - T \circ S = D_{ll} - D_{mm}.$$

4.26 Given T in Hom $\mathbb{R}^n$, we know from Chapter 1 that $T = \sum_{i,j} T_{ij}$, where $T_{ij} = P_i T P_j$ and $P_i = \theta_i \pi_i$. Now we also have

$$T = \sum_{kl} t_{kl} D_{kl}.$$

Show from the definition of D_{ij} in the text that $P_i D_{ij} P_j = D_{ij}$ and that $P_i D_{kl} P_j = 0$ if either $i \neq k$ or $j \neq l$. Conclude that $T_{ij} = t_{ij} D_{ij}$.

5. TRACE AND DETERMINANT

Our aim in this short section is to acquaint the reader with two very special real-valued functions on Hom V and to describe some of their properties.

Theorem 5.1. If V is an n-dimensional vector space, there is exactly one linear functional λ on the vector space $\mathrm{Hom}(V)$ with the property that $\lambda(S \circ T) = \lambda(T \circ S)$ for all S, T in $\mathrm{Hom}(V)$ and normalized so that $\lambda(I) = n$. If a basis is chosen for V and the corresponding matrix of T is $\{t_{ij}\}$, then $\lambda(T) = \sum_{i=1}^{n} t_{ii}$, the sum of the elements on the main diagonal.

Proof. If we choose a basis and define $\lambda(T)$ as $\sum_1^n t_{ii}$, then it is clear that λ is a linear functional on $\mathrm{Hom}(V)$ and that $\lambda(I) = n$. Moreover,

$$\lambda(S \circ T) = \sum_{i=1}^{n} \left(\sum_{j=1}^{n} s_{ij} t_{ji} \right) = \sum_{i,j=1}^{n} s_{ij} t_{ji} = \sum_{i,j} t_{ji} s_{ij} = \lambda(T \circ S).$$

That is, each basis for V gives us a functional λ in $(\mathrm{Hom}\ V)^*$ such that $\lambda(S \circ T) = \lambda(T \circ S)$, $\lambda(I) = n$, and $\lambda(T) = \sum t_{ii}$ for the matrix representation of that basis.

Now suppose that μ is any element of $(\mathrm{Hom}(V))^*$ such that $\mu(S \circ T) = \mu(T \circ S)$ and $\mu(I) = n$. If we choose a basis for V and use the isomorphism $\theta\colon \{t_{ij}\} \mapsto T$ from $\mathbb{R}^{\overline{n} \times \overline{n}}$ to Hom V, we have a functional $\nu = \mu \circ \theta$ on $\mathbb{R}^{\overline{n} \times \overline{n}}$ ($\nu = \theta^* \mu$) such that $\nu(\mathbf{st}) = \nu(\mathbf{ts})$ and $\nu(\mathbf{e}) = n$. By Theorem 4.1 (or 3.1) ν is given by a matrix $\mathbf{c}$, $\nu(\mathbf{t}) = \sum_{i,j=1}^{n} c_{ij} t_{ij}$, and the equation $\nu(\mathbf{st} - \mathbf{ts}) = 0$ becomes $\sum_{i,j,k=1}^{n} c_{ij}(s_{ik} t_{kj} - s_{jk} t_{ki}) = 0$.

We are going to leave it as an exercise for the reader to show that if $l \neq m$, then very simple special matrices $\mathbf{s}$ and $\mathbf{t}$ can be chosen so that this sum reduces to $c_{lm} = 0$, and, by a different choice, to $c_{ll} - c_{mm} = 0$.

Together with the requirement that $\nu(\mathbf{e}) = n$, this implies that $c_{lm} = 0$ for $l \neq m$ and $c_{mm} = 1$ for $m = 1, \ldots, n$. That is, $\nu(\mathbf{t}) = \sum_1^n t_{mm}$, and ν is the λ of the basis being used. Altogether this shows that there is a unique λ in $(\mathrm{Hom}\ V)^*$ such that $\lambda(S \circ T) = \lambda(T \circ S)$ for all S and T and $\lambda(I) = n$, and that $\lambda(T)$ has the diagonal evaluation as $\sum t_{ii}$ in every basis. □

This unique λ is called the *trace functional*, and $\lambda(T)$ is the *trace* of T. It is usually designated $\mathrm{tr}(T)$.

The determinant function $\Delta(T)$ on Hom V is much more complicated, and we shall not prove that it exists until Chapter 7. Its geometric meaning is as follows. First, $|\Delta(T)|$ is the factor by which T multiplies volumes. More precisely, if we define a "volume" v for subsets of V by choosing a basis and using the coordinate correspondence to transfer to V the "natural" volume on $\mathbb{R}^n$, then, for any figure $A \subset V$, $v(T[A]) = |\Delta(T)| \cdot v(A)$. This will be spelled out in Chapter 8. Second, $\Delta(T)$ is positive or negative according as T preserves or reverses *orientation*, which again is a sophisticated notion to be explained later. For the moment we shall list properties of $\Delta(T)$ that are related to this geometric interpretation, and we give a sufficient number to show the uniqueness of Δ.

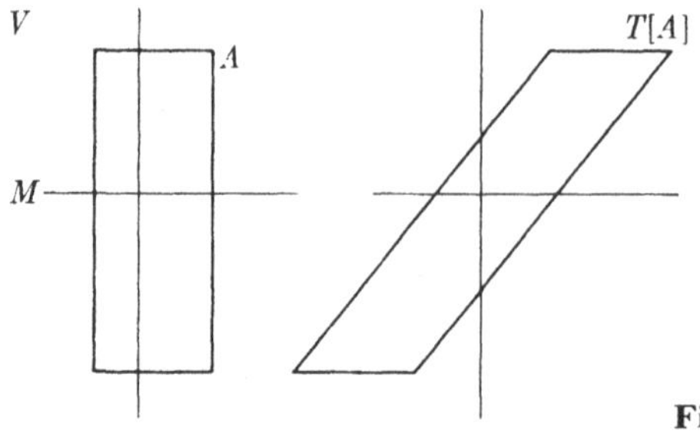

Fig. 2.3

We assume that for each finite-dimensional vector space V there is a function Δ (or Δ_V when there is any question about domain) from $\mathrm{Hom}(V)$ to $\mathbb{R}$ such that the following are true:

a) $\Delta(S \circ T) = \Delta(S)\,\Delta(T)$ for any S, T in $\mathrm{Hom}(V)$.

b) If a subspace N of V is invariant under T and T is the identity on N and on V/N (that is, $T[\bar{\alpha}] = \bar{\alpha}$ for each coset $\bar{\alpha} = \alpha + N$ of N), then $\Delta(T) = 1$. Such a T is a shearing of V along the planes parallel to N. In two dimensions it can be pictured as in Fig. 2.3.

c) If V is a direct sum $V = M + N$ of T-invariant subspaces M and N, and if $R = T \upharpoonright M$ and $S = T \upharpoonright N$, then $\Delta(T) = \Delta(R)\,\Delta(S)$. More exactly, $\Delta_V(T) = \Delta_M(R)\,\Delta_N(S)$.

d) If V is one-dimensional, so that any T in $\mathrm{Hom}(V)$ is simply multiplication by a constant c_T, then $\Delta(T)$ is that constant c_T.

e) If V is two-dimensional and T interchanges a pair of independent vectors, then $\Delta(T) = -1$. This is clearly a pure orientation-changing property.

The fact that Δ is uniquely determined by these properties will follow from our discussion in the next section, which will also give us a process for calculating Δ. This process is efficient for dimensions greater than two, but for T in $\mathrm{Hom}(\mathbb{R}^2)$ there is a simple formula for $\Delta(T)$ which every student should know by heart.

Theorem 5.2. If T is in $\mathrm{Hom}(\mathbb{R}^2)$ and $\{t_{ij}\}$ is its 2×2 matrix, then $\Delta(T) = t_{11}t_{22} - t_{12}t_{21}$.

This is a special case of a general formula, which we shall derive in Chapter 7, that expresses $\Delta(T)$ as a sum of $n!$ terms, each term being a product of n numbers from the matrix of T. This formula is too complicated to be useful in computations for large n, but for $n = 3$ it is about as easy to use as our row-reduction calculation in the next section, and for $n = 2$ it becomes the above simple expression. There are a few more properties of Δ with which every student should be familiar. They will all be proved in Chapter 7.

Theorem 5.3. If T is in $\mathrm{Hom}\ V$, then $\Delta(T^*) = \Delta(T)$. If θ is an isomorphism from V to W and $S = \theta \circ T \circ \theta^{-1}$, then $\Delta(S) = \Delta(T)$.

Theorem 5.4. The transformation T is nonsingular (invertible) if and only if $\Delta(T) \neq 0$.

In the next theorem we consider T in Hom $\mathbb{R}^n$, and we want to think of $\Delta(T)$ as a function of the matrix $\mathbf{t}$ of T. To emphasize this we shall use the notation $D(\mathbf{t}) = \Delta(T)$.

Theorem 5.5 (*Cramer's rule*). Given an $n \times n$ matrix $\mathbf{t}$ and an n-tuple $\mathbf{y}$, let $\mathbf{t} \,|_j\, \mathbf{y}$ be the matrix obtained by replacing the jth column of $\mathbf{t}$ by $\mathbf{y}$. Then

$$\mathbf{y} = \mathbf{t} \cdot \mathbf{x} \Rightarrow D(\mathbf{t})x_j = D(\mathbf{t} \,|_j\, \mathbf{y})$$

for all j.

If $\mathbf{t}$ is nonsingular $[D(\mathbf{t}) \neq 0]$, this becomes an explicit formula for the solution $\mathbf{x}$ of the equation $\mathbf{y} = \mathbf{t} \cdot \mathbf{x}$; it is theoretically important even in those cases when it is not useful in practice (large n).

EXERCISES

5.1 Finish Theorem 5.1 by applying Exercise 4.25.

5.2 It follows from our discussion of trace that $\mathrm{tr}(T) = \sum t_{ii}$ is independent of the basis. Show that this fact follows directly from

$$\mathrm{tr}(\mathbf{t} \cdot \mathbf{s}) = \mathrm{tr}(\mathbf{s} \cdot \mathbf{t})$$

and the change of basis formula in the preceding section.

5.3 Show by direct computation that the function $d(\mathbf{t}) = t_{11}t_{22} - t_{12}t_{21}$ satisfies $d(\mathbf{s} \cdot \mathbf{t}) = d(\mathbf{s})\, d(\mathbf{t})$ (where $\mathbf{s}$ and $\mathbf{t}$ are 2×2 matrices). Conclude that if V is two-dimensional and $d(T)$ is defined for T in Hom V by choosing a basis and setting $d(T) = d(\mathbf{t})$, then $d(T)$ is actually independent of the basis.

5.4 Continuing the above exercise, show that $d(T) = \Delta(T)$ in any of the following cases:

1) T interchanges two independent vectors.
2) T has two eigenvectors.
3) T has a matrix of the form

$$\begin{bmatrix} 1 & a \\ 0 & 1 \end{bmatrix}.$$

Show next that if T has none of the above forms, then $T = R \circ S$, where S is of type (1) and R is of type (2) or (3). [*Hint:* Suppose $T(\alpha) = \beta$, with α and β independent. Let S interchange α and β, and consider $R = T \circ S$.] Show finally that $d(T) = \Delta(T)$ for all T in Hom V. (V is two-dimensional.)

5.5 If $\mathbf{t}$ is symmetric and 2×2, show that there is a 2×2 matrix $\mathbf{s}$ such that $\mathbf{s}^* = \mathbf{s}^{-1}$, $\Delta(\mathbf{s}) = 1$, and $\mathbf{sts}^{-1}$ is diagonal.

5.6 Assuming Theorem 5.2, verify Theorem 5.4 for the 2×2 case.

5.7 Assuming Theorem 5.2, verify Theorem 5.5 for the 2×2 case.

5.8 In this exercise we suppose that the reader remembers what a continuous function of a real variable is. Suppose that the 2×2 matrix function

$$\mathbf{a}(t) = \begin{bmatrix} a_{11}(t) & a_{12}(t) \\ a_{21}(t) & a_{22}(t) \end{bmatrix}$$

has continuous components $a_{ij}(t)$ for $t \in (0, 1)$, and suppose that $\mathbf{a}(t)$ is nonsingular for every t. Show that the solution $\mathbf{y}(t)$ to the linear equation $\mathbf{a}(t) \cdot \mathbf{y}(t) = \mathbf{x}(t)$ has continuous components $y_1(t)$ and $y_2(t)$ if the functions $x_1(t)$ and $x_2(t)$ are continuous.

5.9 A homogeneous second-order linear differential equation is an equation of the form

$$y'' + a_1 y' + a_0 y = 0,$$

where $a_1 = a_1(t)$ and $a_0 = a_0(t)$ are continuous functions. A solution is a $\mathcal{C}^2$-function f (i.e., a twice continuously differentiable function) such that $f''(t) + a_1(t)f'(t) + a_0(t)f(t) = 0$. Suppose that f and g are $\mathcal{C}^2$-functions [on $(0, 1)$, say] such that the 2×2 matrix

$$\begin{bmatrix} f(t) & g(t) \\ f'(t) & g'(t) \end{bmatrix}$$

is always nonsingular. Show that there is a homogeneous second-order differential equation of which they are both solutions.

5.10 In the above exercise show that the space of all solutions is a two-dimensional vector space. That is, show that if $h(t)$ is any third solution, then h is a linear combination of f and g.

5.11 By a "linear motion" of the Cartesian plane $\mathbb{R}^2$ into itself we shall mean a continuous map $x \mapsto \mathbf{t}(x)$ from $[0, 1]$ to the set of 2×2 nonsingular matrices such that $\mathbf{t}(0) = \mathbf{e}$. Show that $\Delta(\mathbf{t}(1)) > 0$.

5.12 Show that if $\Delta(\mathbf{s}) = 1$, then there is a linear motion whose final matrix $\mathbf{t}(1)$ is $\mathbf{s}$.

6. MATRIX COMPUTATIONS

The computational process by which the reader learned to solve systems of linear equations in secondary school algebra was undoubtedly "elimination by successive substitutions". The first equation is solved for the first unknown, and the solution expression is substituted for the first unknown in the remaining equations, thereby eliminating the first unknown from the remaining equations. Next, the second equation is solved for the second unknown, and this unknown is then eliminated from the remaining equations. In this way the unknowns are eliminated one at a time, and a solution is obtained.

This same procedure also solves the following additional problems:

1) to obtain an explicit basis for the linear span of a set of m vectors in $\mathbb{R}^n$; therefore, in particular,
2) to find the dimension of such a subspace;
3) to compute the determinant of an $m \times m$ matrix;
4) to compute the inverse of an invertible $m \times m$ matrix.

In this section we shall briefly study this process and the solutions to these problems.

We start by noting that the kinds of changes we are going to make on a finite sequence of vectors do not alter its span.

Lemma 6.1. Let $\{\alpha_i\}_1^m$ be any m-tuple of vectors in a vector space, and let $\{\beta_i\}_1^m$ be obtained from $\{\alpha_i\}_1^m$ by any one of the following elementary operations:

1) interchanging two vectors;
2) multiplying some α_i by a nonzero scalar;
3) replacing α_i by $\alpha_i - x\alpha_j$ for some $j \neq i$ and some $x \in \mathbb{R}$.

Then

$$L(\{\beta_i\}_1^m) = L(\{\alpha_i\}_1^m).$$

Proof. If $\alpha_i' = \alpha_i - x\alpha_j$, then $\alpha_i = \alpha_i' + x\alpha_j$. Thus if $\{\beta_i\}_1^m$ is obtained from $\{\alpha_i\}_1^m$ by one operation of type (3), then $\{\alpha_i\}_1^m$ can be obtained from $\{\beta_i\}_1^m$ by one operation of type (3). In particular, each sequence is in the linear span of the other, and the two linear spans are therefore the same.

Similarly, each of the other operations can be undone by one of the same type, and the linear spans are unchanged. □

When we perform these operations on the sequence of *row* vectors in a matrix, we call them *elementary row operations*.

We define the *order* of an n-tuple $\mathbf{x} = \langle x_1, \ldots, x_n \rangle$ as the index of the first nonzero entry. Thus if $x_i = 0$ for $i < j$ and $x_j \neq 0$, then the order of $\mathbf{x}$ is j. The order of $\langle 0, 0, 0, 2, -1, 0 \rangle$ is 4.

Let $\{a_{ij}\}$ be an $m \times n$ matrix, let V be its row space, and let $n_1 < n_2 < \cdots < n_k$ be the integers that occur as orders of nonzero vectors in V. We are going to construct a basis for V consisting of k elements having exactly the above set of orders.

If every nonzero row in $\{a_{ij}\}$ has order $>p$, then every nonzero vector $\mathbf{x}$ in V has order $>p$, since $\mathbf{x}$ is a linear combination of these row vectors. Since some vector in V has the minimal order n_1, it follows that some row in $\{a_{ij}\}$ has order n_1. We move such a row to the top by interchanging two rows. We then multiply this row $\mathbf{x}$ by a constant, so that its first nonzero entry x_{n_1} is 1. Let $\mathbf{a}^1, \ldots, \mathbf{a}^n$ be the row vectors that we now have, so that $\mathbf{a}^1$ has order n_1 and $a_{n_1}^1 = 1$. We next subtract multiples of $\mathbf{a}^1$ from each of the other rows in such a way that the new ith row has 0 as its n_1-coordinate. Specifically, we replace $\mathbf{a}^i$ by $\mathbf{a}^i - a_{n_1}^i \cdot \mathbf{a}^1$ for $i > 1$. The matrix that we thus obtain has the property that its jth column is the zero m-tuple for each $j < n_1$ and its n_1th column is δ^1 in $\mathbb{R}^m$. Its first row has order n_1, and every other row has order $>n_1$. Its row space is still V. We again call it $\mathbf{a}$.

Now let $\mathbf{x} = \sum_1^m c_i \mathbf{a}^i$ be a vector in V with order n_2. Then $c_1 = 0$, for if $c_1 \neq 0$, then the order of $\mathbf{x}$ is n_1. Thus $\mathbf{x}$ is a linear combination of the second to the mth rows, and, just as in the first case, one of these rows must therefore have order n_2.

We now repeat the above process all over again, keying now on this vector. We bring it to the second row, make its n_2-coordinate 1, and subtract multiples of it from all the other rows (including the first), so that the resulting matrix has δ^2 for its n_2th column. Next we find a row with order n_3, bring it to the third row, and make the n_3th column δ^3, etc.

We exhibit this process below for one 3×4 matrix. This example is dishonest in that it has been chosen so that fractions will not occur through the application of (2). The reader will not be that lucky when he tries his hand. Our defense is that by keeping the matrices simple we make the process itself more apparent.

$$\begin{bmatrix} 0 & -1 & 2 & 3 \\ 2 & 2 & 4 & -2 \\ 2 & 4 & 0 & 3 \end{bmatrix} \xrightarrow[(1)]{} \begin{bmatrix} 2 & 2 & 4 & -2 \\ 0 & -1 & 2 & 3 \\ 2 & 4 & 0 & 3 \end{bmatrix} \xrightarrow[(2)]{} \begin{bmatrix} 1 & 1 & 2 & -1 \\ 0 & -1 & 2 & 3 \\ 2 & 4 & 0 & 3 \end{bmatrix}$$

$$\xrightarrow[(3)]{} \begin{bmatrix} 1 & 1 & 2 & -1 \\ 0 & -1 & 2 & 3 \\ 0 & 2 & -4 & 5 \end{bmatrix} \xrightarrow[(2)]{} \begin{bmatrix} 1 & 1 & 2 & -1 \\ 0 & 1 & -2 & -3 \\ 0 & 2 & -4 & 5 \end{bmatrix}$$

$$\xrightarrow[(3)]{} \begin{bmatrix} 1 & 0 & 4 & 2 \\ 0 & 1 & -2 & -3 \\ 0 & 0 & 0 & 11 \end{bmatrix} \xrightarrow[(1)]{} \begin{bmatrix} 1 & 0 & 4 & 2 \\ 0 & 1 & -2 & -3 \\ 0 & 0 & 0 & 1 \end{bmatrix}$$

$$\xrightarrow[(3)]{} \begin{bmatrix} 1 & 0 & 4 & 0 \\ 0 & 1 & -2 & 0 \\ 0 & 0 & 0 & 1 \end{bmatrix}$$

Note that from the final matrix we can tell that the orders in the row space are 1, 2, and 4, whereas the original matrix only displays the orders 1 and 2.

We end up with an $m \times n$ matrix having the same row space V and the following special structure:

1) For $1 \leq j \leq k$ the jth row has order n_j.
2) If $k < m$, the remaining $m - k$ rows are zero (since a nonzero row would have order $> n_k$, a contradiction).
3) The n_jth column is δ^j.

It follows that any linear combination of the first k rows with coefficients $c_1, \ldots, c_k$ has c_j in the n_jth place, and hence cannot be zero unless all the c_j's are zero. These k rows thus form a basis for V, solving problems (1) and (2).

Our final matrix is said to be in *row-reduced echelon* form. It can be shown to be uniquely determined by the space V and the above requirements relating its rows to the orders of the elements of V. Its rows form the *canonical* basis of V.

A typical row-reduced echelon matrix is shown in Fig. 2.4. This matrix is 8×11, its orders are 1, 4, 5, 7, 10, and its row space has dimension 5. It is entirely 0 below the broken line. The dashes in the first five lines represent arbitrary numbers, but any change in these remaining entries changes the spanned space V.

We shall now look for the significance of the row-reduction operations from the point of view of general linear theory. In this discussion it will be convenient to use the fact from Section 4 that if an n-tuplet in $\mathbb{R}^n$ is viewed as an $n \times 1$ matrix (i.e., as a column vector), then the system of linear equations $y_k = \sum_{j=1}^{n} a_{ij}x_j$, $i = 1, \ldots, m$, expresses exactly the single matrix equation $\mathbf{y} = \mathbf{a} \cdot \mathbf{x}$. Thus the associated linear transformation $A \in \mathrm{Hom}(\mathbb{R}^n, \mathbb{R}^m)$ is now viewed as being simply multiplication by the matrix $\mathbf{a}$; $\mathbf{y} = A(\mathbf{x})$ if and only if $\mathbf{y} = \mathbf{a} \cdot \mathbf{x}$.

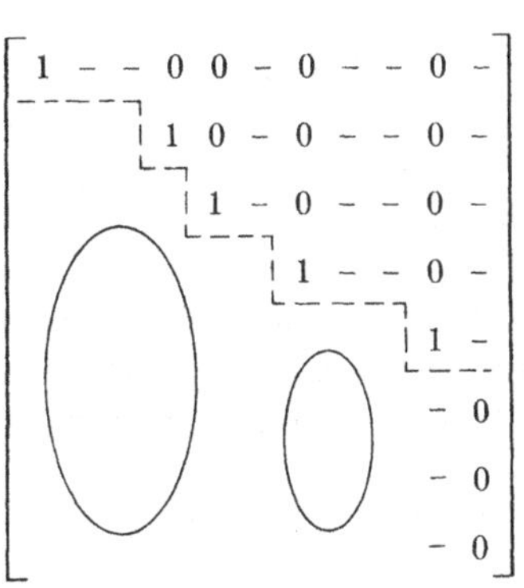

Fig. 2.4

We first note that each of our elementary row operations on an $m \times n$ matrix $\mathbf{a}$ is equivalent to premultiplication by a corresponding $m \times m$ elementary matrix $\mathbf{u}$. Supposing for the moment that this is so, we can find out what $\mathbf{u}$ is by using the $m \times m$ identity matrix $\mathbf{e}$. Since $\mathbf{u} \cdot \mathbf{a} = (\mathbf{u} \cdot \mathbf{e}) \cdot \mathbf{a}$, we see that the result of performing the operation on the matrix $\mathbf{a}$ can also be obtained by premultiplying $\mathbf{a}$ by the matrix $\mathbf{u} \cdot \mathbf{e}$. That is, *if* the elementary operation can be obtained as matrix multiplication by $\mathbf{u}$, then the multiplier is $\mathbf{u} \cdot \mathbf{e}$. This argument suggests that we should perform the operation on $\mathbf{e}$ and then see if premultiplying $\mathbf{a}$ by the resulting matrix performs the operation on $\mathbf{a}$.

If the elementary operation is interchanging the i_0th and j_0th rows, then performing it on $\mathbf{e}$ gives the matrix $\mathbf{a}$ with $u_{kk} = 1$ for $k \neq i_0$ and $k \neq j_0$, $u_{i_0j_0} = u_{j_0i_0} = 1$ and $u_{kl} = 0$ for all other indices. Moreover, examination of the sums defining the elements of the product matrix $\mathbf{u} \cdot \mathbf{a}$ will show that premultiplying by this $\mathbf{u}$ does just interchange the i_0th and j_0th rows of any $m \times n$ matrix $\mathbf{a}$.

In the same way, multiplying the i_0th row of $\mathbf{a}$ by c is equivalent to premultiplying by the matrix $\mathbf{u}$ which is the same as $\mathbf{e}$ except that $u_{i_0i_0} = c$. Finally, multiplying the j_0th row by x and adding it to the i_0th row is equivalent to premultiplying by the matrix $\mathbf{u}$ which is the identity $\mathbf{e}$ except that $u_{i_0j_0}$ is x instead of 0.

These three elementary matrices are indicated schematically in Fig. 2.5. Each has the value 1 on the main diagonal and 0 off the main diagonal except as indicated.

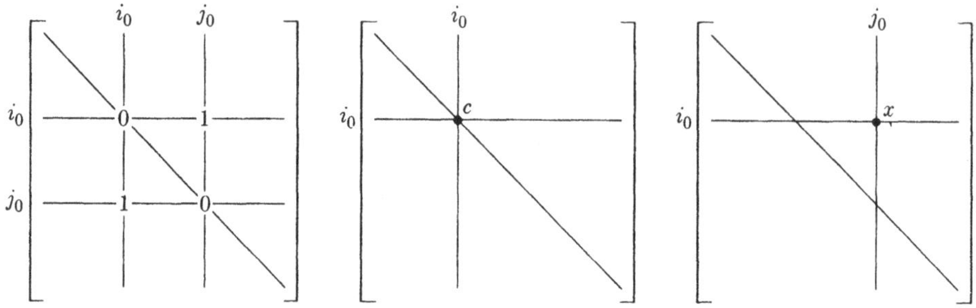

Fig. 2.5

These elementary matrices $\mathbf{u}$ are all nonsingular (invertible). The row interchange matrix is its own inverse. The inverse of multiplying the jth row by c is multiplying the same row by $1/c$. And the inverse of adding c times the jth row to the ith row is adding $-c$ times the jth row to the ith row.

If $\mathbf{u}^1, \mathbf{u}^2, \ldots, \mathbf{u}^p$ is a sequence of elementary matrices, and if

$$\mathbf{b} = \mathbf{u}^p \cdot \mathbf{u}^{p-1} \cdot \ldots \cdot \mathbf{u}^1,$$

then $\mathbf{b} \cdot \mathbf{a}$ is the matrix obtained from $\mathbf{a}$ by performing the corresponding sequence of elementary row operations on $\mathbf{a}$. If $\mathbf{u}^1, \ldots, \mathbf{u}^p$ is a sequence which row reduces $\mathbf{a}$, then $\mathbf{r} = \mathbf{b} \cdot \mathbf{a}$ is the resulting row-reduced echelon matrix.

Now suppose that $\mathbf{a}$ is a square $m \times m$ matrix and is nonsingular (invertible). Thus the dimension of the row space is m, and hence there are m different orders $n_1, \ldots, n_k$. That is, $k = m$, and since $1 \leq n_1 < n_2 < \cdots < n_m = m$, we must also have $n_i = i$, $i = 1, \ldots, m$. Remembering that the n_ith column in $\mathbf{r}$ is δ^i, we see that now the ith column in $\mathbf{r}$ is δ^i and therefore that $\mathbf{r}$ is simply the identity matrix $\mathbf{e}$. Thus $\mathbf{b} \cdot \mathbf{a} = \mathbf{e}$ and $\mathbf{b}$ is the inverse of $\mathbf{a}$.

Let us find the inverse of

$$\begin{bmatrix} 1 & 2 \\ 3 & 4 \end{bmatrix}$$

by this procedure. The row-reducing sequence is

$$\begin{bmatrix} 1 & 2 \\ 3 & 4 \end{bmatrix} \xrightarrow[(3)]{} \begin{bmatrix} 1 & 2 \\ 0 & -2 \end{bmatrix} \xrightarrow[(2)]{} \begin{bmatrix} 1 & 2 \\ 0 & 1 \end{bmatrix} \xrightarrow[(3)]{} \begin{bmatrix} 1 & 0 \\ 0 & 1 \end{bmatrix}.$$

The corresponding elementary matrices are

$$\begin{bmatrix} 1 & 0 \\ -3 & 1 \end{bmatrix}, \quad \begin{bmatrix} 1 & 0 \\ 0 & -\frac{1}{2} \end{bmatrix}, \quad \begin{bmatrix} 1 & -2 \\ 0 & 1 \end{bmatrix}.$$

The inverse is therefore the product

$$\begin{bmatrix} 1 & -2 \\ 0 & 1 \end{bmatrix}\begin{bmatrix} 1 & 0 \\ 0 & -\frac{1}{2} \end{bmatrix}\begin{bmatrix} 1 & 0 \\ -3 & 1 \end{bmatrix} = \begin{bmatrix} 1 & 1 \\ 0 & -\frac{1}{2} \end{bmatrix}\begin{bmatrix} 1 & 0 \\ -3 & 1 \end{bmatrix} = \begin{bmatrix} -2 & 1 \\ \frac{3}{2} & -\frac{1}{2} \end{bmatrix}.$$

Check it if you are in doubt.

Finally, since $\mathbf{b} \cdot \mathbf{e} = \mathbf{b}$, we see that we get $\mathbf{b}$ from $\mathbf{e}$ by applying the same row operations (gathered together as premultiplication by $\mathbf{b}$) that we used to reduce $\mathbf{a}$ to echelon form. This is probably the best way of computing the inverse of a matrix. To keep track of the operations, we can place $\mathbf{e}$ to the right of $\mathbf{a}$ to form a single $m \times 2m$ matrix $\mathbf{a} \mid \mathbf{e}$, and then row reduce it. In echelon form it will then be the $m \times 2m$ matrix $\mathbf{e} \mid \mathbf{b}$, and we can read off the inverse $\mathbf{b}$ of the original matrix $\mathbf{a}$.

Let us recompute the inverse of

$$\begin{bmatrix} 1 & 2 \\ 3 & 4 \end{bmatrix}$$

by this method. We row reduce

$$\left[\begin{array}{cc|cc} 1 & 2 & 1 & 0 \\ 3 & 4 & 0 & 1 \end{array}\right],$$

getting

$$\left[\begin{array}{cc|cc} 1 & 2 & 1 & 0 \\ 3 & 4 & 0 & 1 \end{array}\right] \xrightarrow[(3)]{} \left[\begin{array}{cc|cc} 1 & 2 & 1 & 0 \\ 0 & -2 & -3 & 1 \end{array}\right] \xrightarrow[(2)]{} \left[\begin{array}{cc|cc} 1 & 2 & 1 & 0 \\ 0 & 1 & \frac{3}{2} & -\frac{1}{2} \end{array}\right]$$

$$\xrightarrow[(3)]{} \left[\begin{array}{cc|cc} 1 & 0 & -2 & 1 \\ 0 & 1 & \frac{3}{2} & -\frac{1}{2} \end{array}\right],$$

from which we read off the inverse to be

$$\begin{bmatrix} -2 & 1 \\ \frac{3}{2} & -\frac{1}{2} \end{bmatrix}.$$

Finally we consider the problem of computing the *determinant* of a square $m \times m$ matrix. We use *two* elementary operations (one modified) as follows:

1′) interchanging two rows and simultaneously changing the sign of one of them;

2) as before, replacing some row α_i by $\alpha_i - x\alpha_j$ for some $j \neq i$.

When applied to the rows of a square matrix, these operations leave the determinant unchanged. This follows from the properties of determinants listed in Section 5, and its proof will be left as an exercise. Moreover, these properties will be trivial consequences of our definition of a determinant in Chapter 7.

Consider, then, a square $m \times m$ matrix $\{a_{ij}\}$. We interchange the first and pth rows to bring a row of minimal order n_1 to the top, and change the sign of the row being moved down (the first row here). We do *not* make the leading coefficient of the new first row 1; this elementary operation is not being used now. We *do* subtract multiples of the first row from the remaining rows, in order to make all the remaining entries in the n_1th column 0. The n_1th column is now $c_1\delta^1$, where c_1 is the leading coefficient in the first row. And the new matrix has the same determinant as the original matrix.

We continue as before, subject to the above modifications. We change the sign of a row moved downward in an interchange, we do *not* make leading coefficients 1, and we *do* clear out the n_j*th* column so that it becomes $c_j\delta^{n_j}$, where c_j is the leading coefficient of the jth row ($1 \le j \le k$). As before, the remaining $m - k$ rows are 0 (if $k < m$). Let us call this resulting matrix *semi*reduced. Note that we can find the corresponding reduced echelon matrix from it by k applications of (2); we simply multiply the jth row by $1/c_j$ for $j = 1, \ldots, k$. If $\mathbf{s}$ is the semireduced matrix which we obtained from $\mathbf{a}$ using (1′) and (3), then we shall show below that its determinant, and therefore the determinant of $\mathbf{a}$ also, is the *product of the entries on the main diagonal:* $\prod_{i=1}^m s_{ii}$. Recapitulating, we can compute the determinant of a square matrix $\mathbf{a}$ by using the operations (1′) and (3) to change $\mathbf{a}$ to a semireduced matrix $\mathbf{s}$, and then taking the product of the numbers on the main diagonal of $\mathbf{s}$.

If we apply this process to

$$\begin{bmatrix} 1 & 2 \\ 3 & 4 \end{bmatrix},$$

we get

$$\begin{bmatrix} 1 & 2 \\ 3 & 4 \end{bmatrix} \xrightarrow[(3)]{} \begin{bmatrix} 1 & 2 \\ 0 & -2 \end{bmatrix} \xrightarrow[(3)]{} \begin{bmatrix} 1 & 0 \\ 0 & -2 \end{bmatrix},$$

and the determinant is $1 \cdot (-2) = -2$. Our 2×2 determinant formula, applied to

$$\begin{bmatrix} 1 & 2 \\ 3 & 4 \end{bmatrix},$$

gives $1 \cdot 4 - 2 \cdot 3 = 4 - 6 = -2$.

If the original matrix $\{a_{ij}\}$ is nonsingular, so that $k = m$ and $n_i = i$ for $i = 1, \ldots, m$, then the jth column in the semireduced matrix is $c_j\delta^j$, so that $s_{jj} = c_j$, and we are claiming that the determinant is the product $\prod_{i=1}^m c_i$ of the leading coefficients.

To see this, note that if T is the transformation in $\mathrm{Hom}(\mathbb{R}^n)$ corresponding to our semireduced matrix, then $T(\delta^j) = c_j\delta^j$, so that $\mathbb{R}^n$ is the direct sum of n T-invariant, one-dimensional subspaces, on the jth of which T is multiplication by c_j. It follows from (c) and (d) of our list of determinant properties that $\Delta(T) = \prod_1^n c_j = \prod_1^n s_{jj}$. This is nonzero.

On the other hand, if $\{a_{ij}\}$ is singular, so that $k = d(V) < m$, then the mth row in the semireduced matrix is 0 and, in particular, $s_{mm} = 0$. The product $\prod_i s_{ii}$ is thus zero. Now, without altering the main diagonal, we can subtract multiples of the *columns* containing the leading row entries (the columns with indices n_j) to make the mth column a zero column. This process is equivalent to *post*multiplying by elementary matrices of type (2) and, therefore, again leaves the determinant unchanged. But now the transformation S of *this* matrix leaves $\mathbb{R}^{m-1}$ invariant (as the span of $\delta^1, \ldots, \delta^{m-1}$ in $\mathbb{R}^m$) and takes δ^m to 0, so that $\Delta(S) = 0$ by (c) in the list of determinant properties. So again the determinant is the product of the entries on the main diagonal of the semi-reduced matrix, zero in this case.

We have also found that a *matrix is nonsingular (invertible) if and only if its determinant is nonzero.*

EXERCISES

6.1 Compute the canonical basis of the row space of

$$\begin{bmatrix} 1 & 2 & 1 & 2 \\ 3 & 2 & 3 & 2 \\ -1 & -3 & 0 & 4 \\ 0 & 4 & -1 & -3 \end{bmatrix}.$$

6.2 Do the same for

$$\begin{bmatrix} 1 & 2 & 4 & 5 \\ 1 & 2 & 3 & 4 \\ -1 & -2 & 0 & 2 \end{bmatrix}.$$

6.3 Do the same for the above matrix but with a different first choice.

6.4 Calculate the inverse of

$$\begin{bmatrix} 1 & 2 & 3 \\ 2 & 3 & 4 \\ 3 & 4 & 7 \end{bmatrix}$$

by row reduction. Check your answer by multiplication.

6.5 Row reduce

$$\begin{bmatrix} 1 & 2 & 3 & y_1 \\ 2 & 3 & 4 & y_2 \\ 3 & 4 & 7 & y_3 \end{bmatrix}.$$

How does the fourth column in the row-reduced matrix compare with the inverse of

$$\begin{bmatrix} 1 & 2 & 3 \\ 2 & 3 & 4 \\ 3 & 4 & 7 \end{bmatrix}$$

computed in the above exercise? Explain.

6.6 Check whether or not $\langle 1, 1, 1, 1 \rangle$, $\langle 1, 2, 3, 4 \rangle$, $\langle 0, 1, 0, 1 \rangle$, and $\langle 4, 3, 2, 1 \rangle$ are linearly independent by row reducing. Part of one of the row-reducing operations is unnecessary for this check. What is it?

6.7 Let us call a k-tuple of vectors $\{\alpha_i\}_1^k$ in $\mathbb{R}^n$ *canonical* if the $k \times n$ matrix $\mathbf{a}$ with α_i as its ith row for all i is in row-reduced echelon form. Supposing that an n-tuple ξ is in the row space of $\mathbf{a}$, we can read off what its coordinates are with respect to the above canonical basis. What are they? How then can we check whether or not an arbitrary n-tuple ξ is in the row space?

6.8 Use the device of row reducing, as suggested in the above exercise, to determine whether or not $\delta^1 = \langle 1, 0, 0, 0 \rangle$ is in the span of $\langle 1, 1, 1, 1 \rangle$, $\langle 1, 2, 3, 4 \rangle$, and $\langle 2, 0, 1, -1 \rangle$. Do the same for $\langle 1, 2, 1, 2 \rangle$, and also for $\langle 1, 1, 0, 4 \rangle$.

6.9 Supposing that $a \neq 0$, show that

$$\begin{bmatrix} a & b \\ c & d \end{bmatrix}$$

is invertible if and only if $ad - bc \neq 0$ by reducing the matrix to echelon form.

6.10 Let $\mathbf{a}$ be an $m \times n$ matrix, and let $\mathbf{u}$ be the nonsingular matrix that row reduces $\mathbf{a}$, so that $\mathbf{r} = \mathbf{u} \cdot \mathbf{a}$ is the row-reduced echelon matrix obtained from $\mathbf{a}$. Suppose that $\mathbf{r}$ has $m - k > 0$ zero rows at the bottom (the kth row being nonzero). Show that the bottom $m - k$ rows of $\mathbf{u}$ span the annihilator (range $A)^\circ$ of the range of A. That is, $\mathbf{y} = \mathbf{ax}$ for some $\mathbf{x}$ if and only if

$$\sum_1^m c_i y_i = 0$$

for each m-tuple $\mathbf{c}$ in the bottom $m - k$ rows of $\mathbf{u}$. [*Hint:* The bottom row of $\mathbf{r}$ is obtained by applying the bottom row of $\mathbf{u}$ to the *columns* of $\mathbf{a}$.]

6.11 Remember that we find the row-reducing matrix $\mathbf{u}$ by applying to the $m \times m$ identity matrix $\mathbf{e}$ the row operations that reduce $\mathbf{a}$ to $\mathbf{r}$. That is, we row reduce the $m \times (n + m)$ juxtaposition matrix $\mathbf{a} \mid \mathbf{e}$ to $\mathbf{r} \mid \mathbf{u}$. Assuming the result stated in the above exercise, find the range of $A \in \text{Hom}(\mathbb{R}^3)$ as the null space of a functional if the matrix of A is

$$\begin{bmatrix} 1 & 2 & 3 \\ 2 & 3 & 4 \\ 3 & 5 & 7 \end{bmatrix}.$$

6.12 Similarly, find the range of A if the matrix of A is

$$\begin{bmatrix} 1 & 1 & 1 \\ 2 & 0 & 1 \\ 0 & 2 & 1 \\ 4 & 2 & 3 \end{bmatrix}.$$

6.13 Let $\mathbf{a}$ be an $m \times n$ matrix, and let $\mathbf{a}$ be row reduced to $\mathbf{r}$. Let A and R be the corresponding operators in $\text{Hom}(\mathbb{R}^n, \mathbb{R}^m)$ [so that $A(\mathbf{x}) = \mathbf{a} \cdot \mathbf{x}$]. Show that A and R have the same null space and that A^* and R^* have the same range space.

6.14 Show that solving a system of m linear equations in n unknowns is equivalent to solving a matrix equation

$$\mathbf{k} = \mathbf{tx}$$

for the n-tuple $\mathbf{x}$, given the $m \times n$ matrix $\mathbf{t}$ and the m-tuple $\mathbf{k}$. Let $T \in \text{Hom}(\mathbb{R}^n, \mathbb{R}^m)$ be multiplication by $\mathbf{t}$. Review the possibilities for a solution from our general linear theory for T (range, null space, affine subspace).

6.15 Let $\mathbf{b} = \mathbf{c} \mid \mathbf{d}$ be the $m \times (n+p)$ matrix obtained by juxtaposing the $m \times n$ matrix $\mathbf{c}$ and the $m \times p$ matrix $\mathbf{d}$. If $\mathbf{a}$ is an $l \times m$ matrix, show that

$$\mathbf{a} \cdot \mathbf{b} = \mathbf{ac} \mid \mathbf{ad}.$$

State the similar result concerning the expression of $\mathbf{b}$ as the juxtaposition of n column m-tuples. State the corresponding theorem for the "distributivity" of right multiplication over juxtaposition.

6.16 Let $\mathbf{a}$ be an $m \times n$ matrix and $\mathbf{k}$ a column m-tuple. Let $\mathbf{b} \mid \mathbf{l}$ be the $m \times (n+1)$ matrix obtained from the $m \times (n+1)$ juxtaposition matrix $\mathbf{a} \mid \mathbf{k}$ by row reduction. Show that $\mathbf{a} \cdot \mathbf{x} = \mathbf{k}$ if and only if $\mathbf{b} \cdot \mathbf{x} = \mathbf{l}$. Show that there is a solution $\mathbf{x}$ if and only if every row that is zero in $\mathbf{b}$ is zero in $\mathbf{l}$. Restate this condition in terms of the notion of *row rank*.

6.17 Let $\mathbf{b}$ be the row-reduced echelon matrix obtained from an $m \times n$ matrix $\mathbf{a}$. Thus $\mathbf{b} = \mathbf{u} \cdot \mathbf{a}$, where $\mathbf{u}$ is nonsingular, and B and A have the same null space (where $B \in \mathrm{Hom}(\mathbb{R}^n, \mathbb{R}^m)$ is multiplication by $\mathbf{b}$). We can read off from $\mathbf{b}$ a basis for a subspace $W \subset \mathbb{R}^n$ such that $B \restriction W$ is an isomorphism onto range B. What is this basis? We then know that the null space N of B is a complement of W. One complement of W, call it M, can be read off from W. What is M?

6.18 Continuing the above exercise, show that for each standard basis vector δ_i in M we can read off from the matrix $\mathbf{b}$ a vector α_i in W such that $\delta^i - \alpha_i \in N$. Show that these vectors $\{\delta^i - \alpha_i\}$ form a basis for N.

6.19 We still have to show that the modified elementary row operations leave the determinant of a square matrix unchanged, assuming the properties (a) through (e) from Section 5. First, show from (a), (c), (d), and (e) that if T in Hom $\mathbb{R}^2$ is defined by $T(\delta^1) = \delta^2$ and $T(\delta^2) = -\delta^1$, then $\Delta(T) = 1$. Do this by a very simple factorization, $T = R \circ S$, where (e) can be applied to S. Conclude that a type (1′) elementary matrix has determinant 1.

6.20 Show from the determinant property (b) that an elementary matrix of type (2) has determinant 1. Show, therefore, that the modified elementary row operations on a square matrix leave its determinant unchanged.

*7. THE DIAGONALIZATION OF A QUADRATIC FORM

As we mentioned earlier, one of the crucial problems of linear algebra is the analysis of the "structure" of a linear transformation T in Hom V. From the point of view of bases, every theorem in this area asserts that with the choice of a special basis for V the matrix of T can be given the such-and-such simple form. This is a very difficult part of the subject, and we are only making contact with it in this book, although Theorem 5.5 of Chapter 1 and its corollary form a cornerstone of the structural results.

In this section we are going to solve a simpler problem. In the above language it is the problem of choosing a basis for V making simple the matrix of a transformation T in $\mathrm{Hom}(V, V^*)$. Such a transformation is equivalent to a bilinear functional on V (by Theorem 6.1 of Chapter 1 and Theorem 3.2 of this chapter); we shall tackle the problem in this setting.

Let V be a finite-dimensional *real* vector space, and let $\omega: V \times V \to \mathbb{R}$ be a bilinear functional. If $\{\alpha_i\}_1^n$ is a basis for V, then ω determines a matrix $t_{ij} = \omega(\alpha_i, \alpha_j)$. We know that if $\omega_\eta(\xi) = \omega(\xi, \eta)$, then $\omega_\eta \in V^*$ and $\eta \mapsto \omega_\eta$ is a linear mapping T from V to V^*. We leave it as an exercise for the reader to show that $\{t_{ij}\}$ is the matrix of T with respect to the basis $\{\alpha_i\}$ for V and its dual basis for V^* (Exercise 4.1).

If $\xi = \sum_1^n x_i\alpha_i$ and $\eta = \sum_1^n y_j\alpha_j$, then

$$\omega(\xi, \eta) = \sum_{i,j} x_iy_j\omega(\alpha_i, \alpha_j) = \sum_{i,j} t_{ij}x_iy_j.$$

In particular, if we set $q(\xi) = \omega(\xi, \xi)$, then $q(\xi) = \sum_{i,j} t_{ij}x_ix_j$ is a homogeneous quadratic polynomial in the coordinates x_i.

For the rest of this section we assume that ω is symmetric: $\omega(\xi, \eta) = \omega(\eta, \xi)$. Then we can recover ω from the quadratic form q by

$$\omega(\xi, \eta) = \frac{q(\xi + \eta) - q(\xi - \eta)}{4},$$

as the reader can easily check. In particular, if the bilinear form ω is not identically zero, then there are vectors ξ such that $q(\xi) = \omega(\xi, \xi) \neq 0$.

What we want to do is to show that we can find a basis $\{\alpha_i\}_1^n$ for V such that $\omega(\alpha_i, \alpha_j) = 0$ if $i \neq j$ and $\omega(\alpha_i, \alpha_i)$ has one of the three values $0, \pm 1$. Borrowing from the standard usage of scalar product theory (see Chapter 5), we say that such a basis is *orthonormal*. Our proof that an orthonormal basis exists will be an induction on $n = \dim V$. If $n = 1$, then any nonzero vector β is a basis, and if $\omega(\beta, \beta) \neq 0$, then we can choose $\alpha = x\beta$ so that $x^2\omega(\beta, \beta) = \omega(\alpha, \alpha) = \pm 1$, the required value of x obviously being $x = |\omega(\beta, \beta)|^{-1/2}$. In the general case, if ω is the zero functional, then any basis will trivially be orthonormal, and we can therefore suppose that ω is not identically 0. Then there exists a β such that $\omega(\beta, \beta) \neq 0$, as we noted earlier. We set $\alpha_n = x\beta$, where x is chosen to make $q(\alpha_n) = \omega(\alpha_n, \alpha_n) = \pm 1$. The nonzero linear functional $f(\xi) = \omega(\xi, \alpha_n)$ has an $(n - 1)$-dimensional null space N, and if we let ω' be the restriction of ω to $N \times N$, then ω' has an orthonormal basis $\{\alpha_i\}_1^{n-1}$ by the inductive hypothesis. Also $\omega(\alpha_i, \alpha_n) = \omega(\alpha_n, \alpha_i) = 0$ if $i < n$, because α_i is in the null space of f. Therefore, $\{\alpha_i\}_1^n$ is an orthonormal basis for ω, and we have reached our goal:

Theorem 7.1. If ω is a symmetric bilinear functional on a finite-dimensional real vector space V, then V has an ω-orthonormal basis.

For an ω-orthonormal basis the expansion $\omega(\xi, \eta) = \sum x_iy_j\omega(\alpha_i, \alpha_j)$ reduces to

$$\omega(\xi, \eta) = \sum_{i=1}^{n} x_iy_iq(\alpha_i),$$

where $q(\alpha_i) = \pm 1$ or 0. If we let V_1 be the span of those basis vectors α_i for which $q(\alpha_i) = 1$, and similarly for V_{-1} and V_0, then we see that $q(\xi) > 0$ for every nonzero ξ in V_1, $q(\xi) < 0$ for every nonzero vector ξ in V_{-1}, and $q = 0$

on V_0. Furthermore, $V = V_1 \oplus V_{-1} \oplus V_0$, and the three subspaces are ω-orthonormal to each other (which means that $\omega(\xi, \eta) = 0$ if $\xi \in V_1$ and $\eta \in V_{-1}$, etc.). Finally, $q(\xi) \leq 0$ for every ξ in $V_{-1} \oplus V_0$.

If we choose another orthonormal basis $\{\beta_i\}$ and let W_1, W_{-1}, and W_0 be its corresponding subspaces, then W_1 may be different from V_1, but *their dimensions must be the same.* For $W_1 \cap (V_{-1} \oplus V_0) = \{0\}$, since any nonzero ξ in this intersection would yield the contradictory inequalities $q(\xi) > 0$ and $q(\xi) \leq 0$. Thus W_1 can be extended to a complement of $V_{-1} \oplus V_0$, and since V_1 *is* a complement, we have $d(W_1) \leq d(V_1)$. Similarly, $d(V_1) \leq d(W_1)$, and the dimensions therefore are equal. Incidentally, this shows that W_1 *is* a complement of $V_{-1} \oplus V_0$. In exactly the same way, we find that $d(W_{-1}) = d(V_{-1})$ and finally, by subtraction, that $d(W_0) = d(V_0)$. It is conventional to reorder an ω-orthonormal basis $\{\alpha_i\}_1^n$ so that all the α_i's with $q(\alpha_i) = 1$ come first, then those with $q(\alpha_i) = -1$, and finally those with $q(\alpha_i) = 0$. Our results above can then be stated as follows:

Theorem 7.2. If ω is a symmetric bilinear functional on a finite-dimensional space V, then there are integers n and p such that if $\{\alpha_i\}_1^m$ is any ω-orthonormal basis in conventional order, and if $\xi = \sum_1^m x_i\alpha_i$, then

$$q(\xi) = x_1^2 + \cdots + x_p^2 - x_{p+1}^2 - \cdots - x_{p+n}^2 = \sum_1^p x_i^2 - \sum_{p+1}^{p+n} x_i^2.$$

The integer $p - n$ is called the *signature* of the form q (or its associated symmetric bilinear functional ω), and $p + n$ is its *rank*. Note that $p + n$ is the dimension of the column space of the above matrix of q, and hence equals the dimension of the range of the related linear map T. Therefore, $p + n$ is the rank of every matrix of q.

An inductive proof that an orthonormal basis exists doesn't show us how to find one in practice. Let us suppose that we have the matrix $\{t_{ij}\}$ of ω with respect to *some* basis $\{\alpha_i\}_1^n$ before us, so that $\omega(\xi, \eta) = \sum x_iy_jt_{ij}$, where $\xi = \sum_1^n x_i\alpha_i$, $\eta = \sum_1^n y_i\alpha_i$, and $t_{ij} = \omega(\alpha_i, \alpha_j)$, and we want to know how to go about actually finding an orthonormal basis $\{\beta_i\}_1^n$. The main problem is to find an *orthogonal* basis; normalization is then trivial. The first objective is to find a vector β such that $\omega(\beta, \beta) \neq 0$. If some $t_{ii} = \omega(\alpha_i, \alpha_i)$ is not zero, we can take $\beta = \alpha_i$. If all $t_{ii} = 0$ and the form ω is not the zero form, there must be some $t_{ij} \neq 0$, say $t_{12} \neq 0$. If we set $\gamma_1 = \alpha_1 + \alpha_2$ and $\gamma_i = \alpha_i$ for $i > 1$, then $\{\gamma_i\}_1^n$ is a basis, and the matrix $s = \{s_{ij}\}$ of ω with respect to the basis $\{\gamma_i\}$ has

$$s_{11} = \omega(\gamma_1, \gamma_1) = \omega(\alpha_1 + \alpha_2, \alpha_1 + \alpha_2) = t_{11} + 2t_{12} + t_{22} = 2t_{12} \neq 0.$$

Similarly, $s_{ij} = t_{ij}$ if either i or j is greater than 1.

For example, if ω is the bilinear form on $\mathbb{R}^2$ defined by $\omega(\mathbf{x}, \mathbf{y}) = x_1y_2 + x_2y_1$, then its matrix $t_{ij} = \omega(\delta^i, \delta^j)$ is

$$\begin{bmatrix} 0 & 1 \\ 1 & 0 \end{bmatrix},$$

and we must change the basis to get $t_{11} \neq 0$. According to the above scheme, we set $\gamma_1 = \delta^1 + \delta^2$ and $\gamma_2 = \delta^2$ and get the new matrix $s_{ij} = \omega(\gamma_i, \gamma_j)$, which works out to

$$\begin{bmatrix} 2 & 1 \\ 1 & 0 \end{bmatrix}.$$

The next step is to find a basis for the null space of the functional $\omega(\xi, \gamma_1) = \sum x_i s_{1i}$. We do this by modifying $\gamma_2, \ldots, \gamma_n$; we replace γ_j by $\gamma_j + c\gamma_1$ and calculate c so that this vector is in the null space. Therefore, we want $0 = \omega(\gamma_j + c\gamma_1, \gamma_1) = s_{1j} + cs_{11}$, and so $c = -s_{1j}/s_{11}$. Note that we cannot take this orthogonalizing step until we have made $s_{11} \neq 0$. The new set still spans and thus is a basis, and the new matrix $\{r_{ij}\}$ has $r_{11} \neq 0$ and $r_{1j} = r_{j1} = 0$ for $j > 1$. We now simply repeat the whole procedure for the restriction of ω to this $(n-1)$-dimensional null space, with matrix $\{r_{ij} : 2 \leq i, j \leq n\}$, and so on. This is a long process, but until we normalize, it consists only of rational operations on the original matrix. We add, subtract, multiply, and divide, but we do not have to find roots of polynomial equations.

Continuing our above example, we set $\beta_1 = \gamma_1$, but we have to replace γ_2 by $\beta_2 = \gamma_2 - (s_{12}/s_{11})\gamma_1 = \gamma_2 - \frac{1}{2}\gamma_1$. The final matrix $r_{ij} = \omega(\beta_i, \beta_j)$ has

$$r_{11} = s_{11} = 2, \qquad r_{12} = \omega(\beta_1, \beta_2) = \omega(\gamma_1, \gamma_2 - \tfrac{1}{2}\gamma_1) = s_{12} - \tfrac{1}{2}s_{11} = 0,$$

and $r_{22} = \omega(\gamma_2 - \frac{1}{2}\gamma_1, \gamma_2 - \frac{1}{2}\gamma_1) = s_{22} - s_{12} + s_{11}/4 = -\frac{1}{2}$, so that

$$\{r_{ij}\} = \begin{bmatrix} 2 & 0 \\ 0 & -\frac{1}{2} \end{bmatrix}.$$

The final basis is $\beta_1 = \gamma_1 = \delta^1 + \delta^2$ and $\beta_2 = \gamma_2 - \frac{1}{2}\gamma_1 = \delta^2 - \frac{1}{2}(\delta^1 + \delta^2) = (\delta^2 - \delta^1)/2$.

The steps we had to take above are reminiscent of row reduction, but since we are changing bases simultaneously in the domain and range spaces of the transformation $T: V \to V^*$ associated with ω, each step involves simultaneously premultiplying and postmultiplying by an elementary matrix. That is, we are simultaneously row *and* column reducing. It should be intuitively clear that this has to be the case if we are to operate on a symmetric matrix in such a way as to keep it symmetric.

For additional information about quadratic forms, we go back to the change of basis formula for the matrix of a transformation: $\mathbf{t}'' = \mathbf{b} \cdot \mathbf{t}' \cdot \mathbf{a}^{-1}$. Here the transformation T associated with the form ω is from V to V^*, and so $b = (a^*)^{-1}$, according to our calculations in Section 4. Now one of the properties of the determinant function is that $\Delta(T^*) = \Delta(T)$, and so $\Delta(a^*) = \Delta(a)$. Therefore, if $\mathbf{t}$ and $\mathbf{s}$ are the matrices of a quadratic form with respect to a first and second basis in V, and if $\mathbf{a}$ is the change of basis matrix, then $\mathbf{s} = (\mathbf{a}^*)^{-1} \cdot \mathbf{t} \cdot \mathbf{a}^{-1}$ and $\Delta(s) = (\Delta(a^{-1}))^2\,\Delta(t)$. Therefore, a quadratic form has *parity*. If it is nonsingular, then its determinant is either always positive or always negative, and we can call it even or odd. In our continuing example, the beginning and final matrices

$$\begin{bmatrix} 0 & 1 \\ 1 & 0 \end{bmatrix} \quad \text{and} \quad \begin{bmatrix} 2 & 0 \\ 0 & -\frac{1}{2} \end{bmatrix}$$

both have determinant -1.

In the two-dimensional case, the determinant of a form with respect to an orthonormalized basis is $+1$ if the diagonal elements are both $+1$ or both -1, and -1 if they are of opposite sign. We can therefore read off the signature of a nonsingular form over a two-dimensional space *without orthonormalizing*. If the determinant $t_{11}t_{22} - (t_{12})^2$ is positive, the signature is ± 2, and we can determine which by looking at t_{11} (since t_{11} is then unchanged by our orthogonalizing procedure). Thus the signature is $+2$ or -2 depending on whether $t_{11} > 0$ or $t_{11} < 0$. If the determinant is negative, then the signature is 0. Thus the signature of the form $\omega(\mathbf{x}, \mathbf{y}) = x_1y_2 + x_2y_1$, with matrix

$$\begin{bmatrix} 0 & 1 \\ 1 & 0 \end{bmatrix},$$

is known to be 0, without any calculation.

Theorems 7.1 and 7.2 are important for the classification of critical points of real-valued functions on vector spaces. We shall see in Section 3.16 that the second differential of such a function F is a symmetric bilinear functional, and that the signature of its form has the same significance in determining the behavior of F near a point at which its first differential is zero that the sign of the second derivative has in the elementary calculus.

A quadratic form q is said to be *definite* if $q(\xi)$ is never zero except for $\xi = 0$. Then $q(\xi)$ must always have the same sign, and q is accordingly called *positive definite* or *negative definite*. Looking back to Theorem 7.2, it should be obvious that q is positive definite only if $p = d(V)$ and $n = 0$, and negative definite only if $n = d(V)$ and $p = 0$. A symmetric bilinear functional whose associated quadratic form is positive definite is called a *scalar product*. This is a very important notion on general vector spaces, and the whole of Chapter 5 is devoted to developing some of its implications.

CHAPTER 3

THE DIFFERENTIAL CALCULUS

Our algebraic background is now adequate for the differential calculus, but we still need some multidimensional limit theory. Roughly speaking, the differential calculus is the theory of linear approximations to nonlinear mappings, and we have to know what we mean by approximation in general vector settings. We shall therefore start this chapter by studying the notion of a measure of length, called a *norm*, for the vectors in a vector space V. We can then study the phenomenon suggested by the way in which a tangent plane to a surface approximates the surface near the point of tangency. This is the general theory of unique local linear approximations of mappings, called *differentials*. The collection of rules for computing differentials includes all the familiar laws of the differential calculus, and achieves the same goal of allowing complicated calculations to be performed in a routine way. However, the theory is richer in the multidimensional setting, and one new aspect which we must master is the interplay between the linear transformations which are differentials and their evaluations at given vectors, which are directional derivatives in general and partial derivatives when the vectors belong to a basis. In particular, when the spaces in question are finite-dimensional and are replaced by Cartesian spaces through a choice of bases, then the differential is entirely equivalent to its matrix, which is a certain matrix of partial derivatives called the Jacobian matrix of the mapping. Then the rules of the differential calculus are expressed in terms of matrix operations.

Maximum and minimum points of real-valued functions are found exactly as before, by computing the differential and setting it equal to zero. However, we shall neglect this subject, except in starred sections. It also is much richer than its one-variable counterpart, and in certain infinite-dimensional situations it becomes the subject called the calculus of variations.

Finally, we shall begin our study of the inverse-mapping theorem and the implicit-function theorem. The inverse-mapping theorem states that if a mapping between vector spaces is continuously differentiable, and if its differential at a point α is invertible (as a linear transformation), then the mapping itself is invertible in the neighborhood of α. The implicit-function theorem states that if a continuously differentiable vector-valued function G of two vector variables is set equal to zero, and if the second partial differential of G is invertible (as a linear mapping) at a point $\langle\alpha, \beta\rangle$ where $G(\alpha, \beta) = 0$, then the equation

$G(\xi\eta) = 0$ can be solved for η in terms of ξ near this point. That is, there is a uniquely determined mapping $\eta = F(\xi)$ defined near α such that $\beta = F(\alpha)$ and such that $G(\xi, F(\xi)) = 0$ in the neighborhood of α. These two theorems are fundamental to the further development of analysis. They are deeper results than our work up to this point in that they depend on a special property of vector spaces called *completeness;* we shall have to put off part of their proofs to the next chapter, where we shall study completeness in a fairly systematic way.

In a number of starred sections at the end of the chapter we present some harder material that we do not expect the reader to master. However, he should try to get a rough idea of what is going on.

1. REVIEW IN $\mathbb{R}$

Every student of the calculus is presumed to be familiar with the properties of the real number system and the theory of limits. But we shall need more than familiarity at this point. It will be absolutely essential that the student understand the ϵ-definitions and be able to work with them.

To be on the safe side, we shall review some of this material in the setting of limits of functions; the confident reader can skip it. We suppose that all the functions we consider are defined at least on an open interval containing a, except possibly at a itself. The need for this exception is shown by the difference quotients of the calculus, which are not defined at the point near which their behavior is crucial.

Definition. $f(x)$ approaches l as x approaches a (in symbols, $f(x) \to l$ as $x \to a$) if for every positive ϵ there exists a positive δ such that

$$0 < |x - a| < \delta \Rightarrow |f(x) - l| < \epsilon.$$

We also say that l is the limit of $f(x)$ as x approaches a and write $\lim_{x\to a} f(x) = l$. The displayed statement in the definition is understood to be universally quantified in x, so that the definition really begins with the three quantifiers $(\forall\epsilon^{>0})(\exists\delta^{>0})(\forall x)$. These prefixing quantifiers make the definition sound artificial and unidiomatic when read as ordinary prose, but the reader will remember from our introductory discussion of quantification that this artificiality is absolutely necessary in order for the meaning of the sentence to be clear and unambiguous. *Any change in the order of the quantifiers* $(\forall\epsilon)(\exists\delta)(\forall x)$ *changes the meaning of the statement.*

The meaning of the inner universal quantification

$$(\forall x)(0 < |x - a| < \delta \Rightarrow |f(x) - l| < \epsilon)$$

is intuitive and easily pictured (see Fig. 3.1).

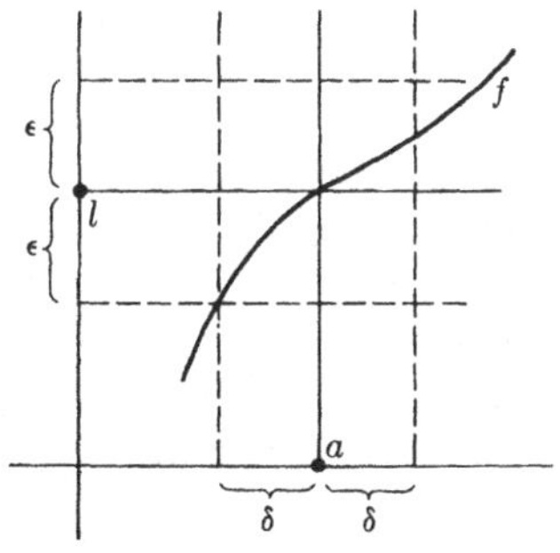

Fig. 3.1

For all x closer to a than δ the value of f at x is closer to l than ϵ. The definition begins by stating that such a positive δ can be found for each positive ϵ. Of course, δ will vary with ϵ; if ϵ is made smaller, we will generally have to go closer to a, that is, we will have to take δ smaller, before all the values of f on $(a - \delta, a + \delta) - \{a\}$ become ϵ close to l.

The variables 'ϵ' and 'δ' are almost always restricted to positive real numbers, and from now on we shall let this restriction be implicit unless there seems to be some special call for explicitness. Thus we shall write simply $(\forall\epsilon)(\exists\delta)\ldots$

The definition of convergence is used in various ways. In the simplest situations we are given one or more functions having limits at a, say, $f(x) \to u$ and $g(x) \to v$, and we want to prove that some other function h has a limit w at a. In such cases we always try to find an inequality expressing the quantity we wish to make small, $|h(x) - w|$, in terms of the quantities which we know can be made small, $|f(x) - u|$ and $|g(x) - v|$.

For example, suppose that $h = f + g$. Since $f(x)$ is close to u and $g(x)$ is close to v, clearly $h(x)$ is close to $w = u + v$. But how close? Since $h(x) - w = (f(x) - u) + (g(x) - v)$, we have

$$|h(x) - w| \leq |f(x) - u| + |g(x) - v|.$$

From this it is clear that in order to make $|h(x) - w|$ less than ϵ it is sufficient to make each of $|f(x) - u|$ and $|g(x) - v|$ less than $\epsilon/2$. Therefore, given any ϵ, we can take δ_1 so that $0 < |x - a| < \delta_1 \Rightarrow |f(x) - u| < \epsilon/2$, and δ_2 so that $0 < |x - a| < \delta_2 \Rightarrow |g(x) - v| < \epsilon/2$, and we can then take δ as the *smaller* of these two numbers, so that if $0 < |x - a| < \delta$, then *both* inequalities hold. Thus

$$0 < |x - a| < \delta \;\Rightarrow\; |h(x) - w| \leq |f(x) - u| + |g(x) - v| < \frac{\epsilon}{2} + \frac{\epsilon}{2} = \epsilon,$$

and we have found the desired δ for the function h.

Suppose next that $u \neq 0$ and that $h = 1/f$. Clearly, $h(x)$ is close to $w = 1/u$ when $f(x)$ is close to u, and so we try to express $h(x) - w$ in terms of $f(x) - u$. Thus

$$h(x) - w = \frac{1}{f(x)} - \frac{1}{u} = \frac{u - f(x)}{f(x)u},$$

and so $|h(x) - w| \leq |f(x) - u|/|f(x)u|$. The trouble here is that the denominator is variable, and if it should happen to be very small, it might cancel the smallness of $|f(x) - u|$ and not force a small quotient. But the answer to this problem is easy. Since $f(x)$ is close to u and u is not zero, $f(x)$ cannot be close to zero. For instance, if $f(x)$ is closer to u than $|u|/2$, then $f(x)$ must be farther from 0 than $|u|/2$. We therefore choose δ_1 so that $0 < |x - a| < \delta_1 \Rightarrow |f(x) - u| < |u|/2$, from which it follows that $|f(x)| > |u|/2$. Then

$$|h(x) - w| < 2|f(x) - u|/|u|^2,$$

and now, given any ϵ, we take δ_2 so that

$$0 < |x - a| < \delta_2 \Rightarrow |f(x) - u| < \epsilon|u|^2/2.$$

Again taking δ as the smaller of δ_1 and δ_2, so that *both* inequalities will hold simultaneously when $0 < |x - a| < \delta$, we have

$$0 < |x - a| < \delta \Rightarrow |h(x) - w| < 2|f(x) - u|/|u|^2 < 2\epsilon|u|^2/2|u|^2 = \epsilon,$$

and again we have found our δ for the function h.

We have tried to show how one would think about these situations. The actual proof that would be written down would only show the choice of δ. Thus,

Lemma 1.1. If $f(x) \to u$ and $g(x) \to v$ as $x \to a$, then $f(x) + g(x) \to u + v$ as $x \to a$.

Proof. Given ϵ, choose δ_1 so that $0 < |x - a| < \delta_1 \Rightarrow |f(x) - a| < \epsilon/2$ (by the assumed convergence of f to u at a), and, similarly, choose δ_2 so that $0 < |x - a| < \delta_2 \Rightarrow |g(x) - v| < \epsilon/2$. Take δ as the smaller of δ_1 and δ_2. Then

$$\begin{aligned} 0 < |x - a| < \delta \Rightarrow &|(f(x) + g(x)) - (u + v)| \\ &\leq |f(x) - u| + |g(x) - v| < \epsilon/2 + \epsilon/2 = \epsilon. \end{aligned}$$

Thus we have proved that for every ϵ there is a δ such that

$$0 < |x - a| < \delta \Rightarrow |(f(x) + g(x)) - (u + v)| < \epsilon,$$

and we are done. □

In addition to understanding ϵ-techniques in limit theory, it is necessary to understand and to be able to use the fundamental property of the real number system called the *least upper bound* property. In the following statement of the property the semi-infinite interval $(-\infty, a]$ is of course the subset $\{x \in \mathbb{R} : x \leq a\}$.

If A is a nonempty subset of $\mathbb{R}$ such that $A \subset (-\infty, a]$ for some a, then there exists a uniquely determined smallest number b such that $A \subset (-\infty, b]$.

A number a such that $A \subset (-\infty, a]$ is called an *upper bound* of A; clearly, a is an upper bound of A if and only if every x in A is less than or equal to a. A set having an upper bound is said to be bounded above. The property says that a nonempty set A which is bounded above has a *least* upper bound (lub). If we reverse the order relation by multiplying everything by -1, then we have the alternative formulation which asserts that a nonempty subset of $\mathbb{R}$ that is bounded below has a *greatest lower bound* (glb). The least upper bound of the interval $(0, 1)$ is 1. The least upper bound of $[0, 1]$ is also 1. The greatest lower bound of $\{1/n : n \text{ a positive integer}\}$ is 0. Furthermore, lub $\{x : x \text{ is a positive rational number and } x^2 < 2\} = \sqrt{2}$, glb $\{e^x : x \in \mathbb{R}\} = 0$, and lub $\{e^x : x \text{ is rational and } x < \sqrt{2}\} = e^{\sqrt{2}}$.

EXERCISES

1.1 Prove that if $f(x) \to l$ and $f(x) \to m$ as $x \to a$, then $l = m$. We can therefore talk about *the* limit of f as $x \to a$.

1.2 Prove that if $f(x) \to l$ and $g(x) \to m$ (as $x \to a$), then $f(x)\,g(x) \to lm$ as $x \to a$.

1.3 Prove that $|x - a| \leq |a|/2 \Rightarrow |x| \geq |a|/2$.

1.4 Prove (in detail) the greatest lower bound property from the least upper bound property.

1.5 Show that lub $A = x$ if and only if x is an upper bound of A and, for every positive ϵ, $x - \epsilon$ is not an upper bound of A.

1.6 Let A and B be subsets of $\mathbb{R}$ that are nonempty and bounded above. Show that $A + B$ is nonempty and bounded above and that lub $(A + B) =$ lub $A +$ lub B.

1.7 Formulate and prove a correct theorem about the least upper bound of the product of two sets.

1.8 Define the notion of a *one-sided* limit for a function whose domain is a subset of $\mathbb{R}$. For example, we want to be able to discuss the limit of $f(x)$ as x approaches *a from below*, which we might designate

$$\lim_{x \uparrow a} f(x).$$

1.9 If the domain of a real-valued function f is an interval, say $[a, b]$, we say that f is an *increasing* function if

$$x < y \Rightarrow f(x) \leq f(y).$$

Prove that an increasing function has one-sided limits everywhere.

1.10 Let $[a, b]$ be a closed interval in $\mathbb{R}$, and let $f\colon [a, b] \to \mathbb{R}$ be increasing. Show that $\lim_{x \to y} f(x) = f(y)$ for all y in $[a, b]$ (f is continuous on $[a, b]$) if and only if the range of f does not omit any subinterval $(c, d) \subset [f(a), f(b)]$. [*Hint:* Suppose the range omits (c, d), and set $y =$ lub $\{x : f(x) \leq c\}$. Then $f(x) \not\to f(y)$ as $x \to y$.]

1.11 A set that intersects every open subinterval of an interval $[s, t]$ is said to be *dense* in $[s, t]$. Show that if $f\colon [a, b] \to \mathbb{R}$ is increasing and range f is dense in $[f(a), f(b)]$, then range $f = [f(a), f(b)]$. (For any z between $f(a)$ and $f(b)$ set $y =$ lub $\{x : f(x) \leq z\}$, etc.)

1.12 Assuming the results of the above two exercises, show that if f is a continuous strictly increasing function from $[a, b]$ to $\mathbb{R}$, and if $r = f(a)$ and $s = f(b)$, then f^{-1} is a continuous strictly increasing function from $[r, s]$ to $\mathbb{R}$. [A function f is continuous if $f(x) \to f(y)$ as $x \to y$ for every y in its domain; it is strictly increasing if $x < y \Rightarrow f(x) < f(y)$.]

1.13 Argue somewhat as in Exercise 1.11 above to prove that if $f\colon [a, b] \to \mathbb{R}$ is continuous on $[a, b]$, then the range of f includes $[f(a), f(b)]$. This is the *intermediate-value theorem.*

1.14 Suppose the function $q\colon \mathbb{R} \to \mathbb{R}$ satisfies $q(xy) = q(x)\,q(y)$ for all $x, y \in \mathbb{R}$. Note that $q(x) = x^n$ (n a positive integer) and $q(x) = |x|^r$ (r any real number) satisfy this "functional equation". So does $q(x) \equiv 0$ ($r = -\infty$?). Show that if q satisfies the functional equation and $q(x) > 1$ for $x > 1$, then there is a real number $r > 1$ such that $q(x) = x^r$ for all positive x.

1.15 Show that if q is continuous and satisfies the functional equation $q(xy) = q(x)\,q(y)$ for all $x, y \in \mathbb{R}$, and if there is at least one point a where $q(a) \neq 0, 1$, then $q(x) \equiv x^r$ for all positive x. Conclude that if also q is nonnegative, then $q(x) \equiv |x|^r$ on $\mathbb{R}$.

1.16 Show that if $q(x) \equiv |x|^r$, and if $q(x+y) \leq q(x) + q(y)$, then $r \leq 1$. (Try $y = 1$ and x large; what is $q'(x)$ like if $r > 1$?)

2. NORMS

In the limit theory of $\mathbb{R}$, as reviewed briefly above, the absolute-value function is used prominently in expressions like '$|x - y|$' to designate the distance between two numbers, here between x and y. The definition of the convergence of $f(x)$ to u is simply a careful statement of what it means to say that the distance $|f(x) - u|$ tends to zero as the distance $|x - a|$ tends to zero. The properties of $|x|$ which we have used in our proofs are

1) $|x| > 0$ if $x \neq 0$, and $|0| = 0$;
2) $|xy| = |x|\,|y|$;
3) $|x + y| \leq |x| + |y|$.

The limit theory of vector spaces is studied in terms of functions called *norms*, which serve as multidimensional analogues of the absolute-value function on $\mathbb{R}$. Thus, if $p: V \to \mathbb{R}$ is a norm, then we want to interpret $p(\alpha)$ as the "size" of α and $p(\alpha - \beta)$ as the "distance" between α and β. However, if V is not one-dimensional, there is no one notion of size that is most natural. For example, if f is a positive continuous function on $[a, b]$, and if we ask the reader for a number which could be used as a measure of how "large" f is, there are two possibilities that will probably occur to him: the maximum value of f and the area under the graph of f. Certainly, f must be considered small if $\max f$ is small. But also, we would have to agree that f is small in a different sense if its area is small. These are two examples of norms on the vector space $V = \mathcal{C}([a, b])$ of all continuous functions on $[a, b]$:

$$p(f) = \max\{|f(t)| : t \in [a, b]\} \qquad \text{and} \qquad q(f) = \int_a^b |f(t)|\,dt.$$

Note that f can be small in the second sense and not in the first.

In order to be useful, a notion of size for a vector must have properties analogous to those of the absolute-value function on $\mathbb{R}$.

Definition. A *norm* is a real-valued function p on a vector space V such that

n1. $p(\alpha) > 0$ if $\alpha \neq 0$ (positivity);

n2. $p(x\alpha) = |x|p(\alpha)$ for all $\alpha \in V$, $x \in \mathbb{R}$ (homogeneity);

n3. $p(\alpha + \beta) \leq p(\alpha) + p(\beta)$ for all $\alpha, \beta \in V$ (triangle inequality).

A *normed linear space* (nls), or *normed vector space*, is a vector space V together with a norm p on V. A normed linear space is thus really a *pair*

$\prec V, p \succ$, but generally we speak simply of the normed linear space V, a definite norm on V then being understood.

It has been customary to designate the norm of α by $\|\alpha\|$, presumably to suggest the analogy with absolute value. The triangle inequality n3 then becomes $\|\alpha + \beta\| \leq \|\alpha\| + \|\beta\|$, which is almost identical in form with the basic absolute-value inequality $|x + y| \leq |x| + |y|$. Similarly, n2 becomes $\|x\alpha\| = |x|\,\|\alpha\|$, analogous to $|xy| = |x|\,|y|$ in $\mathbb{R}$. Furthermore, $\|\alpha - \beta\|$ is similarly interpreted as the distance between α and β. This is reasonable since if we set $\alpha = \xi - \eta$ and $\beta = \eta - \zeta$, then n3 becomes the usual triangle inequality of geometry:

$$\|\xi - \zeta\| \leq \|\xi - \eta\| + \|\eta - \zeta\|.$$

We shall use both the double bar notation and the "p"-notation for norms; each is on occasion superior to the other.

The most commonly used norms on $\mathbb{R}^n$ are $\|\mathbf{x}\|_1 = \sum_1^n |x_i|$, the Euclidean norm $\|\mathbf{x}\|_2 = (\sum_1^n x_i^2)^{1/2}$, and $\|\mathbf{x}\|_\infty = \max\,\{|x_i|\}_i$. Similar norms on the infinite-dimensional vector space $\mathcal{C}([a, b])$ of all continuous real-valued functions on $[a, b]$ are

$$\|f\|_1 = \int_a^b |f(t)|\,dt,$$

$$\|f\|_2 = \left(\int_a^b |f(t)|^2\,dt\right)^{1/2},$$

$$\|f\|_\infty = \max\,\{|f(t)| : a \leq t \leq b\}.$$

It should be easy for the reader to check that $\|\ \|_1$ is a norm in both cases above, and we shall take up the so-called uniform norms $\|\ \|_\infty$ in the next paragraph. The Euclidean norms $\|\ \|_2$ are trickier; their properties depend on scalar product considerations. These will be discussed in Chapter 5. Meanwhile, so that the reader can use the Euclidean norm $\|\ \|_2$ on $\mathbb{R}^n$, we shall ask him to prove the triangle inequality for it (the other axioms being obvious) by brute force in an exercise. On $\mathbb{R}$ itself the absolute value is a norm, and it is the only norm to within a constant multiple.

We can transfer the above norms on $\mathbb{R}^n$ to arbitrary finite-dimensional spaces by the following general remark.

Lemma 2.1. If p is a norm on a vector space W and T is an injective linear map from a vector space V to W, then $p \circ T$ is a norm on V.

Proof. The proof is left to the reader.

Uniform norms. The two norms $\|\ \|_\infty$ considered above are special cases of a very general situation. Let A be an arbitrary nonempty set, and let $\mathcal{B}(A, \mathbb{R})$ be the set of all bounded functions $f\colon A \to \mathbb{R}$. That is, $f \in \mathcal{B}(A, \mathbb{R})$ if and only if $f \in \mathbb{R}^A$ and range $f \subset [-b, b]$ for some $b \in \mathbb{R}$. This is the same as saying that range $|f| \subset [0, b]$, and we call any such b a *bound* of $|f|$. The set $\mathcal{B}(A, \mathbb{R})$ is a vector space V, since if $|f|$ and $|g|$ are bounded by b and c, respectively, then $|xf + yg|$ is bounded by $|x|b + |y|c$. The *uniform norm* $\|f\|_\infty$ is defined as the *smallest* bound of $|f|$. That is,

$$\|f\|_\infty = \text{lub } \{|f(p)| : p \in A\}.$$

Of course, it has to be checked that $\| \ \|_\infty$ is a norm. For any p in A,

$$|f(p) + g(p)| \leq |f(p)| + |g(p)| \leq \|f\|_\infty + \|g\|_\infty.$$

Thus $\|f\|_\infty + \|g\|_\infty$ is a bound of $|f + g|$ and is therefore greater than or equal to the smallest such bound, which is $\|f + g\|_\infty$. This gives the triangle inequality. Next we note that if $x \neq 0$, then b bounds $|f|$ if and only if $|x|b$ bounds $|xf|$, and it follows that $\|xf\|_\infty = |x| \, \|f\|_\infty$. Finally, $\|f\|_\infty \geq 0$, and $\|f\|_\infty = 0$ only if f is the zero function.

We can replace $\mathbb{R}$ by any normed linear space W in the above discussion. A function $f: A \to W$ is bounded by b if and only if $\|f(p)\| \leq b$ for all p in A, and we define the corresponding uniform norm on $\mathcal{B}(A, W)$ by

$$\|f\|_\infty = \text{lub } \{\|f(p)\| : p \in A\}.$$

If $f \in \mathcal{C}([0, 1])$, then we know that the continuous function f assumes the least upper bound of its range as a value (that is, f "assumes its maximum value"), so that then $\|f\|_\infty$ is the maximum value of $|f|$. In general, however, the definition must be given in terms of lub.

Balls. Remembering that $\|\alpha - \xi\|$ is interpreted as the distance from α to ξ, it is natural to define the *open ball of radius r about the center* α as $\{\xi : \|\alpha - \xi\| < r\}$. We designate this ball $B_r(\alpha)$. Translation through β preserves distance,

$$\|T_\beta(\xi) - T_\beta(\eta)\| = \|(\xi + \beta) - (\eta + \beta)\| = \|\xi - \eta\|,$$

and therefore $\xi \in B_r(\alpha)$ if and only if $\xi + \beta \in B_r(\alpha + \beta)$. That is, translation through β carries $B_r(\alpha)$ into $B_r(\alpha + \beta)$: $T_\beta[B_r(\alpha)] = B_r(\alpha + \beta)$. Also, scalar multiplication by c multiplies all distances by c, and it follows in a similar way that $cB_r(\alpha) = B_{cr}(c\alpha)$.

Although $B_r(\alpha)$ behaves like a ball, the actual set being defined is different for different norms, and some of them "look unspherelike". The unit balls about the origin in $\mathbb{R}^2$ for the three norms $\| \ \|_1$, $\| \ \|_2$, and $\| \ \|_\infty$ are shown in Fig. 3.2.

A subset A of a nls V is *bounded* if it lies in some ball, say $B_r(\alpha)$. Then it also lies in a ball about the origin, namely $B_{r+\|\alpha\|}(0)$. This is simply the fact that if $\|\xi - \alpha\| < r$, then $\|\xi\| < r + \|\alpha\|$, which we get from the triangle inequality upon rewriting $\|\xi\|$ as $\|(\xi - \alpha) + \alpha\|$.

The radius of the largest ball about a vector β which does not touch a set A is naturally called the *distance from* β *to* A. It is clearly glb $\{\|\xi - \beta\| : \xi \in A\}$ (see Fig. 3.3).

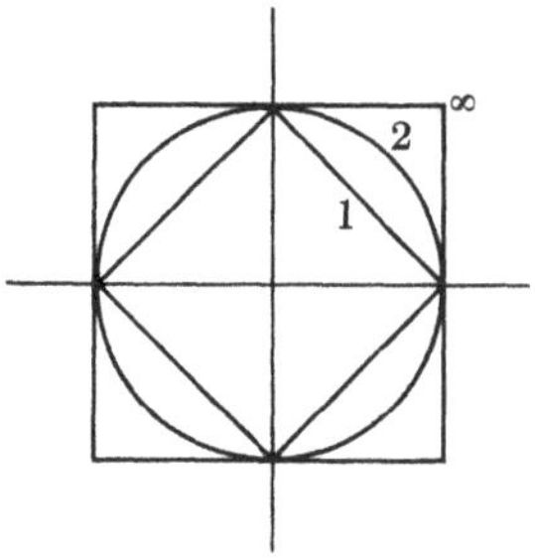

Fig. 3.2

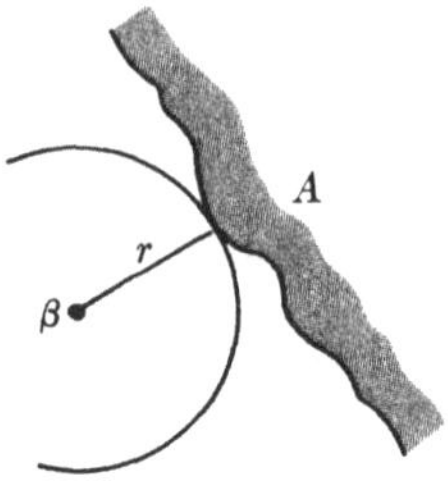

$\rho(\beta, A) = r$

Fig. 3.3

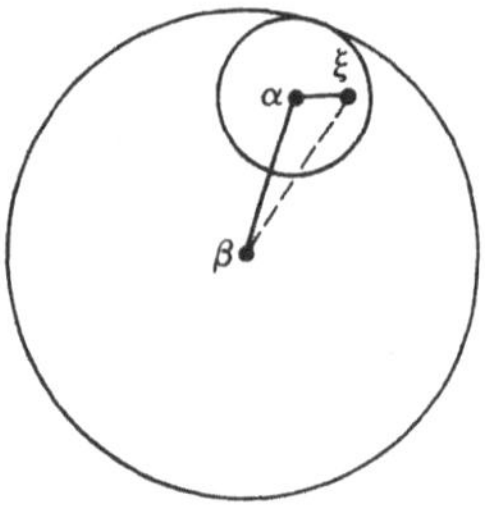

Fig. 3.4

A point α is an *interior* point of a set A if some ball about α is included in A. This is equivalent to saying that the distance from α to the complement of A is positive (supposing that A is not the whole of V), and should coincide with the reader's intuitive notion of what an "inside" point should be. A subset A of a normed linear space is said to be *open* if every point of A is an interior point.

If our language is to be consistent, an open ball should be an open set. It is: if $\alpha \in B_r(\beta)$, then $\|\alpha - \beta\| < r$, and then $B_\delta(\alpha) \subset B_r(\beta)$, provided that $\delta \leq r - \|\alpha - \beta\|$, by virtue of the triangle inequality (see Fig. 3.4). The reader should write down the detailed proof. He has to show that if $\xi \in B_\delta(\alpha)$, then $\xi \in B_r(\beta)$. Our intuitions about distances are quite trustworthy, but they should always be checked by a computation. The reader probably can see by a mental argument that the union of any collection of open sets is open. In particular, the union of any collection of open balls is open (Fig. 3.5), and this is probably the most intuitive way of visualizing an open set. (See Exercise 2.9.)

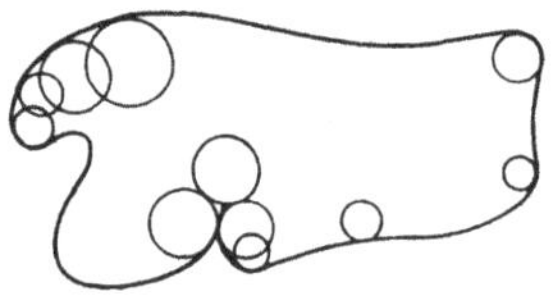

Fig. 3.5

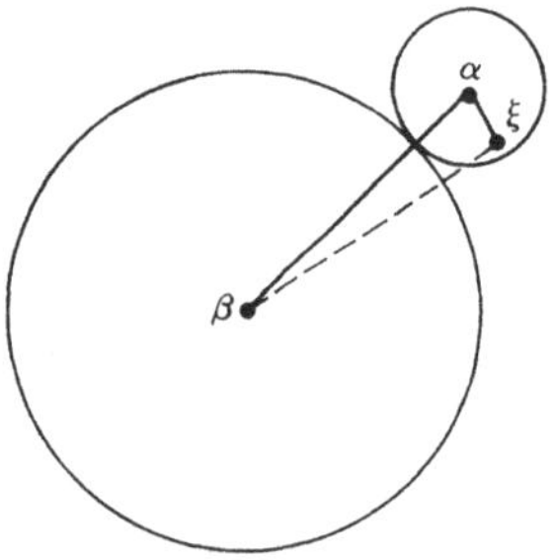

Fig. 3.6

A subset C is said to be *closed* if its complement C' is open.

Our discussion above shows that a nonempty set C is closed if and only if every point not in it is at a positive distance from it: $\alpha \notin C \Rightarrow \rho(\alpha, C) > 0$. The so-called *closed ball of radius r about β*, $B = \{\xi : \|\xi - \beta\| \leq r\}$, is a closed set. As Fig. 3.6 suggests, the proof is another application of the triangle inequality.

EXERCISES

2.1 Show that if $\|\xi - \alpha\| \leq \|\alpha\|/2$, then $\|\xi\| \geq \|\alpha\|/2$.

2.2 Prove in detail that

$$\|\mathbf{x}\|_1 = \sum_1^n |x_i|$$

is a norm on $\mathbb{R}^n$. Also prove that

$$\|f\|_1 = \int_a^b |f(t)|\, dt$$

is a norm on $\mathcal{C}([a, b])$.

2.3 For $\mathbf{x}$ in $\mathbb{R}^n$ let $|x|$ be the Euclidean length

$$|\mathbf{x}| = \left[\sum_1^n x_i^2\right]^{1/2},$$

and let $(\mathbf{x}, \mathbf{y})$ be the scalar product

$$(\mathbf{x}, \mathbf{y}) = \sum_1^n x_i y_i.$$

The *Schwarz inequality* says that

$$|(\mathbf{x}, \mathbf{y})| \leq |\mathbf{x}|\,|\mathbf{y}|$$

and that the inequality is strict if $\mathbf{x}$ and $\mathbf{y}$ are independent.

a) Prove the Schwarz inequality for the case $n = 2$ by squaring and canceling.
b) Now prove it for the general n in the same way.

2.4 Continuing the above exercise, prove that the Euclidean length $|\mathbf{x}|$ is a norm. The crucial step is the triangle inequality, $|\mathbf{x} + \mathbf{y}| \leq |\mathbf{x}| + |\mathbf{y}|$. Reduce it to the Schwarz inequality by squaring and canceling. This is of course our two-norm $\|\mathbf{x}\|_2$.

2.5 Prove that the unit balls for the norms $\|\ \|_1$ and $\|\ \|_\infty$ on $\mathbb{R}^2$ are as shown in Fig. 3.2.

2.6 Prove that an open ball is an open set.

2.7 Prove that a closed ball is a closed set.

2.8 Give an example of a subset of $\mathbb{R}^2$ that is neither open nor closed.

2.9 Show from the definition of an open set that any open set is the union of a family (perhaps very large!) of open balls. Show that any union of open sets is open. Conclude, therefore, that a set is open if and only if it is a union of open balls.

2.10 A subset A of a normed linear space V is said to be *convex* if A includes the line segment joining any two of its points. We know that the line segment from α to β is the image of $[0, 1]$ under the mapping $t \to t\beta + (1 - t)\alpha$. Thus A is convex if and only if $\alpha, \beta \in A$ and $t \in [0, 1] \Rightarrow t\beta + (1 - t)\alpha \in A$. Prove that every ball $B_r(\gamma)$ in a normed linear space V is convex.

2.11 A *seminorm* is the same as a norm except that the positivity condition n1 is relaxed to nonnegativity:

n1′. $p(\alpha) \geq 0$ for all α.

Thus $p(\alpha)$ may be 0 for some nonzero α. Every norm is in particular a seminorm. Prove:

a) If p is a seminorm on a vector space W and T is a linear mapping from V to W, then $p \circ T$ is a seminorm on V.

b) $p \circ T$ is a norm if and only if T is injective and p is a norm on range T.

2.12 Show that the sum of two seminorms is a seminorm.

2.13 Prove from the above two exercises (and *not* by a direct calculation) that

$$q(f) = \|f'\|_\infty + |f(t_0)|$$

is a seminorm on the space $\mathcal{C}^1([a, b])$ of all continuously differentiable real-valued functions on $[a, b]$, where t_0 is a fixed point in $[a, b]$. Prove that q is a norm.

2.14 Show that the sum of two bounded sets is bounded.

2.15 Prove that the sum $B_r(\alpha) + B_s(\beta)$ is *exactly* the ball $B_{r+s}(\alpha + \beta)$.

3. CONTINUITY

Let V and W be any two normed linear spaces. We shall designate both norms by $\| \ \|$. This ambiguous usage does not cause confusion. It is like the ambiguous use of "0" for the zero elements of all the vector spaces under consideration. If we replace the absolute value sign $| \ |$ by the general norm symbol $\| \ \|$ in the definition we gave earlier for the limit of a real-valued function of a real variable, it becomes verbatim the corresponding definition of convergence in the general setting. However, we shall repeat the definition and take the occasion to relax the hypothesis on the domain of f. Accordingly, let A by *any* subset of V, and let f be any mapping from A to W.

Definition. We say that $f(\xi)$ approaches β as ξ approaches α, and write $f(\xi) \to \beta$ as $\xi \to \alpha$, if for every ϵ there is a δ such that

$$\xi \in A \text{ and } 0 < \|\xi - \alpha\| < \delta \Rightarrow \|f(\xi) - \beta\| < \epsilon.$$

If $\alpha \in A$ and $f(\xi) \to f(\alpha)$ as $\xi \to \alpha$, then we say that f is *continuous at* α. We can then drop the requirement that $\xi \neq \alpha$ and have the direct ϵ,δ-characterization of continuity: f is continuous at α if for every ϵ there exists a δ such that $\|\xi - \alpha\| < \delta \Rightarrow \|f(\xi) - f(\alpha)\| < \epsilon$. It is understood here that ξ is universally quantified over the domain A of f. We say that f is *continuous* if f is continuous at every point α in its domain. If the absolute value of a number is replaced by the norm of a vector, the limit theorems that we sampled in Section 1 hold verbatim for normed linear spaces. We shall ask the reader to write out a few of these transcriptions in the exercises.

There is a property stronger than continuity at α which is much simpler to use when it is available. We say that f is *Lipschitz continuous* at α if there is a constant c such that $\|f(\xi) - f(\alpha)\| \leq c\|\xi - \alpha\|$ for all ξ sufficiently close to α.

That is, there are constants c and r such that

$$\|\xi - \alpha\| < r \Rightarrow \|f(\xi) - f(\alpha)\| \leq c\|\xi - \alpha\|.$$

The point is that now we can take δ simply as ϵ/c (provided ϵ is small enough so that this makes $\delta \leq r$; otherwise we have to set $\delta = \min\{\epsilon/c, r\}$). We say that f is a *Lipschitz function* (on its domain A) if there is a constant c such that $\|f(\xi) - f(\eta)\| \leq c\|\xi - \eta\|$ for all ξ, η in A. For a *linear* map $T: V \to W$ the Lipschitz inequality is more simply written as

$$\|T(\zeta)\| \leq c\|\zeta\|$$

for all $\zeta \in V$; we just use the fact that now $T(\xi) - T(\eta) = T(\xi - \eta)$ and set $\zeta = \xi - \eta$. In this context it is conventional to call T a *bounded* linear mapping rather than a Lipschitz linear mapping, and any such c is called a *bound* of T.

We know from the beginning calculus that if f is a continuous real-valued function on $[a, b]$ (that is, if $f \in \mathcal{C}([a, b])$), then $|\int_a^b f(x)\,dx| \leq m(b - a)$, where m is the maximum value of $|f(x)|$. But this is just the uniform norm of f, so that the inequality can be rewritten as $|\int_a^b f| \leq (b - a)\|f\|_\infty$. This shows that if the uniform norm is used on $\mathcal{C}([a, b])$, then $f \mapsto \int_a^b f$ is a bounded linear functional, with bound $b - a$.

It should immediately be pointed out that this is not the same notion of boundedness we discussed earlier. There we called a real-valued function bounded if its range was a bounded subset of $\mathbb{R}$. The analogue here would be to call a vector-valued function bounded if its range is norm bounded. But a nonzero linear transformation *cannot* be bounded in this sense, because

$$\|T(x\alpha)\| = |x|\,\|T(\alpha)\|.$$

The present definition amounts to the boundedness in the earlier sense of the *quotient* $T(\alpha)/\|\alpha\|$ (on $V - \{0\}$). It turns out that for a linear map T, being continuous and being Lipschitz are the same thing.

Theorem 3.1. Let T be a linear mapping from a normed linear space V to a normed linear space W. Then the following conditions are equivalent:

1) T is continuous at one point;
2) T is continuous;
3) T is bounded.

Proof. (1) $\Rightarrow$ (3). Suppose T is continuous at α_0. Then, taking $\epsilon = 1$, there exists δ such that $\|\alpha - \alpha_0\| < \delta \Rightarrow \|T(\alpha) - T(\alpha_0)\| < 1$. Setting $\xi = \alpha - \alpha_0$ and using the additivity of T, we have $\|\xi\| < \delta \Rightarrow \|T(\xi)\| < 1$. Now for any nonzero η, $\xi = \delta\eta/2\|\eta\|$ has norm $\delta/2$. Therefore, $\|T(\xi)\| < 1$. But $\|T(\xi)\| = \delta\|T(\eta)\|/2\|\eta\|$, giving $\|T(\eta)\| < 2\|\eta\|/\delta$. Thus T is bounded by $C = 2/\delta$.

(3) $\Rightarrow$ (2). Suppose $\|T(\xi)\| \leq C\|\xi\|$ for all ξ. Then for any α_0 and any ϵ we can take $\delta = \epsilon/C$ and have

$$\|\alpha - \alpha_0\| < \delta \Rightarrow \|T(\alpha) - T(\alpha_0)\| = \|T(\alpha - \alpha_0)\| \leq C\|\alpha - \alpha_0\| < C\delta = \epsilon.$$

(2) $\Rightarrow$ (1). Trivial. □

In the lemma below we prove that the norm function is a Lipschitz function from V to $\mathbb{R}$.

Lemma 3.1. For all $\alpha, \beta \in V$, $\big|\, \|\alpha\| - \|\beta\| \,\big| \leq \|\alpha - \beta\|$.

Proof. We have $\|\alpha\| = \|(\alpha - \beta) + \beta\| \leq \|\alpha - \beta\| + \|\beta\|$, so that $\|\alpha\| - \|\beta\| \leq \|\alpha - \beta\|$. Similarly, $\|\beta\| - \|\alpha\| \leq \|\beta - \alpha\| = \|\alpha - \beta\|$. This pair of inequalities is equivalent to the lemma. □

Other Lipschitz mappings will appear when we study mappings with continuous differentials. Roughly speaking, the Lipschitz property lies between continuity and continuous differentiability, and it is frequently the condition that we actually apply under the hypothesis of continuous differentiability.

The smallest bound of a bounded linear transformation T is called its norm. That is,

$$\|T\| = \text{lub } \{\|T(\alpha)\|/\|\alpha\| : \alpha \neq 0\}.$$

For example, let $T: \mathcal{C}([a, b]) \to \mathbb{R}$ be the Riemann integral, $T(f) = \int_a^b f(x)\, dx$. We saw earlier that if we use the uniform norm $\|f\|_\infty$ on $\mathcal{C}([a, b])$, then T is bounded by $b - a : |T(f)| \leq (b - a)\|f\|_\infty$. On the other hand, there is no smaller bound, because $\int_a^b 1 = b - a = (b - a)\|1\|_\infty$. Thus $\|T\| = b - a$. Other formulations of the above definition are useful. Since

$$\|T(\alpha)\|/\|\alpha\| = \|T(\alpha/\|\alpha\|)\|$$

by homogeneity, and since $\beta = \alpha/\|\alpha\|$ has norm 1, we have

$$\|T\| = \text{lub } \{\|T(\beta)\| : \|\beta\| = 1\}.$$

Finally, if $\|\gamma\| \leq 1$, then $\gamma = x\beta$, where $\|\beta\| = 1$ and $|x| \leq 1$, and

$$\|F(\gamma)\| = |x|\, \|F(\beta)\| \leq \|F(\beta)\|.$$

We therefore have an inefficient but still useful characterization:

$$\|T\| = \text{lub } \{\|T(\gamma)\| : \|\gamma\| \leq 1\}.$$

These last two formulations are *uniform norms*. Thus, if B_1 is the closed unit ball $\{\xi : \|\xi\| \leq 1\}$, we see that a linear T is bounded if and only if $T \upharpoonright B_1$ is bounded in the old sense, and then

$$\|T\| = \|T \upharpoonright B_1\|_\infty.$$

A linear map $T: V \to W$ is *bounded below* by b if $\|T(\xi)\| \geq b\|\xi\|$ for all ξ in V. If T has a bounded inverse and $m = \|T^{-1}\|$, then T is bounded below by $1/m$, for $\|T^{-1}(\eta)\| \leq m\|\eta\|$ for all $\eta \in W$ if and only if $\|\xi\| \leq m\|T(\xi)\|$ for all $\xi \in V$.

If V is finite-dimensional, then it is true, conversely, that if T is bounded below, then it is invertible (why?), but in general this does not follow.

If V and W are normed linear spaces, then $\mathrm{Hom}(V, W)$ is defined to be the set of all *bounded* linear maps $T: V \to W$. The results of Section 2.3 all remain true, but require some additional arguments.

Theorem 3.2. $\mathrm{Hom}(V, W)$ is itself a normed linear space if $\|T\|$ is defined as above, as the smallest bound for T.

Proof. This follows from the uniform norm discussion of Section 2 by virtue of the identity $\|T\| = \|T \restriction B_1\|_\infty$. □

Theorem 3.3. If U, V, and W are normed linear spaces, and if $T \in \mathrm{Hom}(U, V)$ and $S \in \mathrm{Hom}(V, W)$, then $S \circ T \in \mathrm{Hom}(U, W)$ and $\|S \circ T\| \leq \|S\| \, \|T\|$. It follows that composition on the right by a fixed T is a bounded linear transformation from $\mathrm{Hom}(V, W)$ to $\mathrm{Hom}(U, W)$, and similarly for composition on the left by a fixed S.

Proof

$$\|(S \circ T)(\alpha)\| = \|S(T(\alpha))\| \leq \|S\| \, \|T(\alpha)\| \leq \|S\|(\|T\| \, \|\alpha\|) = (\|S\| \cdot \|T\|)(\|\alpha\|).$$

Thus $S \circ T$ is bounded by $\|S\| \cdot \|T\|$ and everything else follows at once. □

As before, the conjugate space V^* is $\mathrm{Hom}(V, \mathbb{R})$, now the space of all *bounded* linear functionals.

EXERCISES

3.1 Write out the ϵ,δ-proofs of the following limit theorems.

1) Let V and W be normed linear spaces, and let F and G be mappings from V to W. If $\lim_{\xi \to \alpha} F(\xi) = \mu$ and $\lim_{\xi \to \alpha} G(\xi) = \nu$, then $\lim_{\xi \to \alpha} (F + G)(\xi) = \mu + \nu$.
2) Given $F: V \to W$ and $g: V \to \mathbb{R}$, if $F(\xi) \to \mu$ and $g(\xi) \to b$ as $\xi \to \alpha$, then $(gF)(\xi) \to b\mu$.

3.2 Prove that if $F(\xi) \to \mu$ as $\xi \to \alpha$ and $G(\eta) \to \lambda$ as $\eta \to \mu$, then $G \circ F(\xi) \to \lambda$ as $\xi \to \alpha$. Give a careful, complete statement of the theorem you have proved.

3.3 Suppose that A is an open subset of a nls V and that $\alpha_0 \in A$. Suppose that $F: A \to \mathbb{R}$ is such that $\lim_{\alpha \to \alpha_0} F(\alpha) = b \neq 0$. Prove that $1/F(\alpha) \to 1/b$ as $\alpha \to \alpha_0$ (ϵ,δ-proof).

3.4 The function $f(x) = |x|^r$ is continuous at $x = 0$ for any positive r. Prove that f is *not* Lipschitz continuous at $x = 0$ if $r < 1$. Prove, however, that f *is* Lipschitz continuous at $x = a$ if $a > 0$. (Use the mean-value theorem.)

3.5 Use the mean-value theorem of the calculus and the definition of the derivative to show that if f is a real-valued function on an interval I, and if f' exists everywhere, then f is a Lipschitz mapping if and only if f' is a bounded function. Show also that then $\|f'\|_\infty$ is the smallest Lipschitz constant C.

3.6 The "working rules" for $\|T\|$ are

1) $\|T(\xi)\| \leq \|T\| \, \|\xi\|$ for all ξ;

2) $\|T(\xi)\| \leq b\|\xi\|$, all $\xi \implies \|T\| \leq b$.

Prove these rules.

3.7 Prove that if we use the one-norm $\|\mathbf{x}\|_1 = \sum_1^n |x_i|$ on $\mathbb{R}^n$, then the norm of the linear functional

$$L_{\mathbf{a}} : \mathbf{x} \to \sum_1^n a_i x_i$$

is $\|\mathbf{a}\|_\infty$.

3.8 Prove similarly that if $\|\mathbf{x}\| = \|\mathbf{x}\|_\infty$, then $\|L_{\mathbf{a}}\| = \|\mathbf{a}\|_1$.

3.9 Use the above exercises to show that if $\|\mathbf{x}\|$ on $\mathbb{R}^n$ is the one-norm, then

$$\|\mathbf{x}\| = \text{lub } \{|f(\mathbf{x})| : f \in (\mathbb{R}^n)^* \text{ and } \|f\| \leq 1\}.$$

3.10 Show that if T in $\text{Hom}(\mathbb{R}^n, \mathbb{R}^m)$ has matrix $\mathbf{t} = \{t_{ij}\}$, and if we use the one-norm $\|\mathbf{x}\|_1$ on $\mathbb{R}^n$ and the uniform norm $\|\mathbf{y}\|_\infty$ on $\mathbb{R}^m$, then $\|T\| = \|\mathbf{t}\|_\infty$.

3.11 Show that the meaning of '$\text{Hom}(V, W)$' has changed by giving an example of a linear mapping that fails to be bounded. There is one in the text.

3.12 For a fixed ξ in V define the mapping $\text{ev}_\xi : \text{Hom}(V, W) \to W$ by $\text{ev}_\xi(T) = T(\xi)$. Prove that ev_ξ is a bounded linear mapping.

3.13 In the above exercise it is in fact true that $\|\text{ev}_\xi\| = \|\xi\|$, but to prove this we need a new theorem.

> **Theorem.** Given ξ in the normed linear space V, there exists a functional f in V^* such that $\|f\| = 1$ and $|f(\xi)| = \|\xi\|$.

Assuming this theorem, prove that $\|\text{ev}_\xi\| = \|\xi\|$. [*Hint:* Presumably you have already shown that $\|\text{ev}_\xi\| \leq \|\xi\|$. You now need a T in $\text{Hom}(V, W)$ such that $\|T\| = 1$ and $\|T(\xi)\| = \|\xi\|$. Consider a suitable dyad.]

3.14 Let $\mathbf{t} = \{t_{ij}\}$ be a square matrix, and define $\|\mathbf{t}\|$ as $\max_i (\sum_j |t_{ij}|)$. Prove that this is a norm on the space $\mathbb{R}^{\overline{n} \times \overline{n}}$ of all $n \times n$ matrices. Prove that $\|\mathbf{st}\| \leq \|\mathbf{s}\| \cdot \|\mathbf{t}\|$. Compute the norm of the identity matrix.

3.15 Let V be the normed linear space $\mathbb{R}^n$ under the uniform norm $\|\mathbf{x}\|_\infty = \max \{|x_i|\}$. If $T \in \text{Hom } V$, prove that $\|T\|$ is the norm of its matrix $\|\mathbf{t}\|$ as defined in the above exercise. That is, show that

$$\|T\| = \max_i \left[\sum_{j=1}^n |t_{ij}| \right].$$

(Show first that $\|\mathbf{t}\|$ is an upper bound of T, and then show that $\|T(\mathbf{x})\| = \|\mathbf{t}\| \, \|\mathbf{x}\|$ for a specially chosen $\mathbf{x}$.) Does part of the previous exercise now become superfluous?

3.16 Assume the following fact: If $f \in \mathcal{C}([0, 1])$ and $\|f\|_1 = a$, then given ϵ, there is a function $g \in \mathcal{C}([0, 1])$ such that

$$\|g\|_\infty = 1 \quad \text{and} \quad \int_0^1 fg > a - \epsilon.$$

Let $K(s, t)$ be continuous on $[0, 1] \times [0, 1]$ and bounded by b. Define T: $\mathcal{C}([0, 1]) \to \mathcal{B}([0, 1])$ by $Th = k$, where

$$k(s) = \int_0^1 K(s, t)\, h(t)\, dt.$$

If V and W are the normal linear spaces $\mathcal{C}$ and $\mathcal{B}$ under the uniform norms, prove that

$$\|T\| = \operatorname*{lub}_s \int |K(s, t)|\, dt.$$

[*Hint:* Proceed as in the above exercise.]

3.17 Let V and W be normed linear spaces, and let A be any subset of V containing more than one point. Let $\mathcal{L}(A, W)$ be the set of all Lipschitz mappings from A to W. For f in $\mathcal{L}(A, W)$, let $p(f)$ be the smallest Lipschitz constant for f. That is,

$$p(f) = \operatorname*{lub}_{\xi \neq \eta} \frac{\|f(\xi) - f(\eta)\|}{\|\xi - \eta\|}.$$

Prove that $\mathcal{L}(A, W)$ is a vector space V and that p is a seminorm on V.

3.18 Continuing the above exercise, show that if α is any fixed point of A, then $p(f) + \|f(\alpha)\|$ is a norm on V.

3.19 Let K be a mapping from a subset A of a normed linear space V to V which differs from the identity by a Lipschitz mapping with constant c less than 1. We may as well take $c = \frac{1}{2}$, and then our hypothesis is that

$$\|K(\xi) - K(\eta) - (\xi - \eta)\| \leq \tfrac{1}{2}\|\xi - \eta\|.$$

Prove that K is injective and that its inverse is a Lipschitz mapping with constant 2.

3.20 Continuing the above exercise, suppose in addition that the domain A of K is an open subset of V and that $K[C]$ is a closed set whenever C is a closed ball lying in A. Prove that if $C = C_r(\alpha)$, the closed ball of radius r about α, is a subset of A, then $K[C]$ includes the ball $B = B_{r/7}(\gamma)$, where $\gamma = K(\alpha)$. This proof is elementary but tricky. If there is a point v of B not in $K[C]$, then since $K[C]$ is closed, there is a largest ball B' about v disjoint from $K[C]$ and a point $\eta = K(\xi)$ in $K[C]$ as close to B' as we wish. Now if we change ξ by adding $v - \eta$, the change in the value of K will approximate $v - \eta$ closely enough to force the new value of K to be in B'. If we can also show that the new value $\xi + (v - \eta)$ is in C, then this new value of K is in $K[C]$, and we have our contradiction.

Draw a picture. Obviously, the radius ρ of B' is at most $r/7$. Show that if $\eta = K(\xi)$ is chosen so that $\|v - \eta\| \leq 3/2\rho$, then the above assertions follow from the triangle inequality, and the Lipschitz inequality displayed in Exercise 3.19. You have to prove that

$$\|K(\xi + (v - \eta)) - v\| < \rho$$

and

$$\|(\xi + (v - \eta)) - \alpha\| \leq r.$$

3.21 Assume the result of the above exercise and show that

$$B_{r/8}(\gamma) \subset K[B_r(\alpha)].$$

Show, therefore, that $K[A]$ is an open subset of V. State a theorem about the Lipschitz invertibility of K, including all the hypotheses on K that were used in the above exercises.

3.22 We shall see in the next chapter that if V and W are finite-dimensional spaces, then any continuous map from V to W takes bounded closed sets into bounded closed sets. Assuming this and the results of the above exercises, prove the following theorem.

Theorem. Let F be a mapping from an open subset A of a finite-dimensional normed linear space V to a finite-dimensional normed linear space W. Suppose that there is a T in $\mathrm{Hom}(V, W)$ such that T^{-1} exists and such that $F - T$ is Lipschitz on A, with constant $1/2m$, where $m = \|T^{-1}\|$. Then F is injective, its range $R = F[A]$ is an open subset of W, and its inverse F^{-1} is Lipschitz continuous, with constant $2m$.

4. EQUIVALENT NORMS

Two normed linear spaces V and W are *norm isomorphic* if there is a bijection T from V to W such that $T \in \mathrm{Hom}(V, W)$ and $T^{-1} \in \mathrm{Hom}(W, V)$. That is, an isomorphism is a linear isomorphism T such that both T and T^{-1} are continuous (bounded). As usual, we regard isomorphic spaces as being essentially the same. For two different norms on the same space we are led to the following definition.

Definition. Two norms p and q on the same vector space V are *equivalent* if there exist constants a and b such that $p \leq aq$ and $q \leq bp$.

Then $(1/b)q \leq p \leq aq$ and $(1/a)p \leq q \leq bp$, so that two norms are equivalent if and only if either can be bracketed by two multiples of the other. The above definition simply says that the *identity* map $\xi \mapsto \xi$ from V to V, considered as a map from the normed linear space $\langle V, p \rangle$ to the normed linear space $\langle V, q \rangle$, is bounded in both directions, and hence that these two normed linear spaces are isomorphic.

If V is infinite-dimensional, two norms will in general not be equivalent. For example, if $V = \mathcal{C}([0, 1])$ and $f_n(t) = t^n$, then $\|f_n\|_1 = 1/(n+1)$ and $\|f_n\|_\infty = 1$. Therefore, there is no constant a such that $\|f\|_\infty \leq a\|f\|_1$ for all $f \in \mathcal{C}[0, 1]$, and the norms $\| \; \|_\infty$ and $\| \; \|_1$ are not equivalent on $V = \mathcal{C}[a, b]$. This is why the very notion of a normed linear space depends on the assumption of a given norm.

However, we have the following theorem, which we shall prove in the next chapter by more sophisticated methods than we are using at present.

Theorem 4.1. On a finite-dimensional vector space V all norms are equivalent.

We shall need this theorem and also the following consequence of it occasionally in the present chapter.

Theorem 4.2. If V and W are finite-dimensional normed linear spaces, then every linear mapping T from V to W is necessarily bounded.

Proof. Because of the above theorem, it is sufficient to prove T bounded with respect to *some* pair of norms. Let $\theta\colon \mathbb{R}^n \to V$ and $\varphi\colon \mathbb{R}^m \to W$ be any basis isomorphisms, and let $\{t_{ij}\}$ be the matrix of $\overline{T} = \varphi^{-1} \circ T \circ \theta$ in $\mathrm{Hom}(\mathbb{R}^n, \mathbb{R}^m)$. Then

$$\|\overline{T}\mathbf{x}\|_\infty = \max_i \left|\sum_j t_{ij}x_j\right| \leq (\max_{i,j}|t_{ij}|)\left(\sum_1^n |x_j|\right) = b\|\mathbf{x}\|_1,$$

where $b = \max |t_{ij}|$. Now $q(\eta) = \|\varphi^{-1}(\eta)\|_\infty$ and $p(\xi) = \|\theta^{-1}(\xi)\|_1$ are norms on W and V respectively, by Lemma 2.1, and since

$$q(T(\xi)) = \|\overline{T}(\theta^{-1}\xi)\|_\infty \leq b\|\theta^{-1}\xi\|_1 = bp(\xi),$$

we see that T is bounded by b with respect to the norms p and q on V and W. □

If we change to an equivalent norm, we are merely passing through an isomorphism, and all continuous linear properties remain unchanged. For example:

Theorem 4.3. The vector space $\mathrm{Hom}(V, W)$ remains the same if either the domain norm or the range norm is replaced by an equivalent norm, and the two induced norms on $\mathrm{Hom}(V, W)$ are equivalent.

Proof. The proof is left to the reader.

We now ask what kind of a norm we might want on the Cartesian product $V \times W$ of two normed linear spaces. It is natural to try to choose the product norm so that the fundamental mappings relating the product space to the two factor spaces, the two projections π_i and the two injections θ_i, should be continuous. It turns out that these requirements determine the product norm uniquely to within equivalence. For if $\|\langle\alpha, \xi\rangle\|$ has these properties, then

$$\begin{aligned}\|\langle\alpha, \xi\rangle\| &= \|\langle\alpha, 0\rangle + \langle 0, \xi\rangle\| \leq \|\langle\alpha, 0\rangle\| + \|\langle 0, \xi\rangle\| \\ &\leq k_1\|\alpha\| + k_2\|\xi\| \leq k(\|\alpha\| + \|\xi\|),\end{aligned}$$

where k_i is a bound of the injection θ_i and k is the larger of k_1 and k_2. Also, $\|\alpha\| \leq c_1\|\langle\alpha, \xi\rangle\|$ and $\|\xi\| \leq c_2\|\langle\alpha, \xi\rangle\|$, by the boundedness of the projections π_i, and so $\|\alpha\| + \|\xi\| \leq c\|\langle\alpha, \xi\rangle\|$, where $c = c_1 + c_2$. Now $\|\alpha\| + \|\xi\|$ is clearly a norm $\|\ \|_1$ on $V \times W$, and our argument above shows that $\|\langle\alpha, \xi\rangle\|$ will satisfy our requirements if and only if it is equivalent to $\|\ \|_1$. Any such norm will be called a *product norm* for $V \times W$. The product norms most frequently used are the uniform (product) norm

$$\|\langle\alpha, \xi\rangle\|_\infty = \max\{\|\alpha\|, \|\xi\|\},$$

the Euclidean (product) norm $\|\langle\alpha, \xi\rangle\|_2 = (\|\alpha\|^2 + \|\xi\|^2)^{1/2}$, and the above sum (product) norm $\|\langle\alpha, \xi\rangle\|_1$. We shall leave the verification that the uniform and Euclidean norms actually are norms as exercises.

Each of these three product norms can be defined as well for n factor spaces as for two, and we gather the facts for this general case into a theorem.

Theorem 4.4. If $\{<V_i, p_i>\}_1^n$ is a finite set of normed linear spaces, then $\| \ \|_1$, $\| \ \|_2$, and $\| \ \|_\infty$, defined on $V = \prod_{i=1}^n V_i$ by $\|\boldsymbol{\alpha}\|_1 = \sum_1^n p_i(\alpha_i)$, $\|\boldsymbol{\alpha}\|_2 = (\sum_1^n p_i(\alpha_i)^2)^{1/2}$, and $\|\boldsymbol{\alpha}\|_\infty = \max\{p_i(\alpha_i) : i = 1, \ldots, n\}$, are equivalent norms on V, and each is a product norm in the sense that the projections π_i and the injections θ_i are all continuous.

*It looks above as though all we are doing is taking *any* norm $\| \ \|$ on $\mathbb{R}^n$ and then defining a norm $||| \ |||$ on the product space V by

$$|||\boldsymbol{\alpha}||| = \| <p_1(\alpha_1), \ldots, p_n(\alpha_n)> \|.$$

This is almost correct. The interested reader will discover, however, that $\| \ \|$ on $\mathbb{R}^n$ must have the property that if $|x_i| \leq |y_i|$ for $i = 1, \ldots, n$, then $\|\mathbf{x}\| \leq \|\mathbf{y}\|$ for the triangle inequality to follow for $||| \ |||$ in V. If we call such a norm on $\mathbb{R}^n$ an *increasing* norm, then the following is true.

If $\| \ \|$ is any increasing norm on $\mathbb{R}^n$, then $|||\boldsymbol{\alpha}||| = \| <p_1(\alpha_1), \ldots, p_n(\alpha_n)> \|$ is a product norm on $V = \prod_1^n V_i$.

However, we shall use only the 1-, 2-, ∞-product norms in this book.*

The triangle inequality, the continuity of addition, and our requirements on a product norm form a set of nearly equivalent conditions. In particular, we make the following observation.

Lemma 4.1. If V is a normed linear space, then the operation of addition is a bounded linear map from $V \times V$ to V.

Proof. The triangle inequality for the norm on V says exactly that addition is bounded by 1 when the sum norm is used on $V \times V$. □

A normed linear space V is a (norm) direct sum $\bigoplus_1^n V_i$ if the mapping $<x_1, \ldots, x_n> \mapsto \sum_1^n x_i$ is a norm isomorphism from $\prod_1^n V_i$ to V. That is, the given norm on V must be equivalent to the product norm it acquires when it is viewed as $\prod_1^n V_i$. If V is algebraically the direct sum $\bigoplus_1^n V_i$, we always have

$$\|x\| = \left\|\sum_1^n x_i\right\| \leq \sum_1^n \|x_i\|$$

by the triangle inequality for the norm on V, and the sum on the right is the one-norm for $\prod_1^n V_i$. Therefore, V will be the norm direct sum $\bigoplus_1^n V_i$ if, conversely, there is an n-tuple of constants $\{k_i\}$ such that $\|x_i\| \leq k_i\|x\|$ for all x. This is the same as saying that the projections $P_i : x \mapsto x_i$ are all bounded. Thus,

Theorem 4.5. If V is a normed linear space and V is algebraically the direct sum $V = \bigoplus_1^n V_i$, then $V = \bigoplus_1^n V_i$ as normed linear spaces if and only if the associated projections $\{P_i\}$ are all bounded.

EXERCISES

4.1 The fact that Hom(V, W) is unchanged when norms are replaced by equivalent norms can be viewed as a corollary of Theorem 3.3. Show that this is so.

4.2 Write down a string of quite obvious inequalities showing that the norms $\| \|_1$, $\| \|_2$, and $\| \|_\infty$ on $\mathbb{R}^n$ are equivalent. Discuss what happens as $n \to \infty$.

4.3 Let V be an n-dimensional vector space, and consider the collection of all norms on V of the form $p \circ \theta$, where $\theta: V \to \mathbb{R}^n$ is a coordinate isomorphism and p is one of the norms $\| \|_1$, $\| \|_2$, $\| \|_\infty$ on $\mathbb{R}^n$. Show that all of these norms are equivalent. (Use the above exercise and the reasoning in Theorem 4.2.)

4.4 Prove that $\|\langle \alpha, \xi \rangle\| = \max \{\|\alpha\|, \|\xi\|\}$ is a norm on $V \times W$.

4.5 Prove that $\|\langle \alpha, \xi \rangle\| = \|\alpha\| + \|\xi\|$ is a norm on $V \times W$.

4.6 Prove that $\|\langle \alpha, \xi \rangle\| = (\|\alpha\|^2 + \|\beta\|^2)^{1/2}$ is a norm on $V \times W$.

4.7 Assuming Exercises 4.4 through 4.6, prove by induction the corresponding part of Theorem 4.4.

4.8 Prove that if A is an open subset of $V \times W$, then $\pi_1[A]$ is an open subset of V.

4.9 Prove (ϵ, δ) that $\langle T, S \rangle \to S \circ T$ is a continuous map from

$$\text{Hom}(V_1, V_2) \times \text{Hom}(V_2, V_3) \quad \text{to} \quad \text{Hom}(V_1, V_3),$$

where the V_i are all normed linear spaces.

4.10 Let $\| \|$ be any increasing norm on $\mathbb{R}^n$; that is, $\|\mathbf{x}\| \leq \|\mathbf{y}\|$ if $x_i \leq y_i$ for all i. Let p_i be a norm on the vector space V_i for $i = 1, \ldots, n$. Show that

$$|||\boldsymbol{\alpha}||| = \|\langle p_1(\alpha_1), \ldots, p_n(\alpha_n) \rangle\|$$

is a norm on $V = \prod_1^n V_i$.

4.11 Suppose that $p: V \to \mathbb{R}$ is a nonnegative function such that $p(x\alpha) = |x|p(\alpha)$ for all x, α. This is surely a minimum requirement for any function purporting to be a measure of length of a vector.

a) Define continuity with respect to p and show that Theorem 3.1 is valid.
b) Our next requirement is that addition be continuous as a map from $V \times V$ to V, and we decide that continuity at 0 means that for every ϵ there is a δ such that

$$p(\alpha) < \delta \text{ and } p(\beta) < \delta \Rightarrow p(\alpha + \beta) < \epsilon.$$

Argue again as in Theorem 3.1 to show that there is a constant c such that

$$p(\alpha + \beta) \leq c(p(\alpha) + p(\beta)) \qquad \text{for all} \quad \alpha, \beta \in V.$$

4.12 Let V and W be normed linear spaces, and let $f: V \times W \to \mathbb{R}$ be bounded and bilinear. Let T be the corresponding linear map from V to W^*. Prove that T is bounded and that $\|T\|$ is the smallest bound to f, that is, the smallest b such that

$$|f(\alpha, \beta)| \leq b\|\alpha\| \, \|\beta\| \qquad \text{for all} \quad \alpha, \beta.$$

4.13 Let the normed linear space V be a norm direct sum $M \oplus N$. Prove that the subspaces M and N are closed sets in V. (The converse theorem is false.)

4.14 Let N be a closed subspace of the normed linear space V. If A is a coset $N + \alpha$, define $|||A|||$ as glb $\{\|\xi\| : \xi \in A\}$. Prove that $|||A|||$ is a norm on the quotient space V/N. Prove also that if $\bar{\xi}$ is the coset containing ξ, then the mapping $\xi \mapsto \bar{\xi}$ (the natural projection π of V onto V/N) is bounded by 1.

4.15 Let V and W be normed linear spaces, and let T in Hom(V, W) have a null space which includes the closed subspace N. Prove that the unique linear S from V/N to W defined by $T = S \circ \pi$ (Theorem 4.3 of Chapter 1) is bounded and that $\|S\| = \|T\|$.

4.16 Let N be a closed subspace of a normed linear space, and suppose that N has a finite-dimensional complement in the purely algebraic sense. Prove that then V is the norm direct sum $M \oplus N$. (Use the above exercise and Theorem 4.2 to prove that if P is the projection of V onto N along M, then P is bounded.)

4.17 Let N_1 and N_2 be closed subspaces of the normed linear space V, and suppose that they have the same finite codimension. Prove that N_1 and N_2 are norm isomorphic. (Assume the results of the above exercise and Exercise 2.11 of Chapter 2.)

4.18 Prove that if p is a seminorm on a vector space V, then its null set is a subspace N, p is constant on the cosets of N, and p factors: $p = q \circ \pi$, where q is a norm on V/N and π is the natural projection $\xi \mapsto \bar{\xi}$ of V onto V/N. Note that $\xi \mapsto \bar{\xi}$ is thus an isometric surjection from the seminormed space V to the normed space V/N. An isometry is a distance-preserving map.

5. INFINITESIMALS

The notion of an *infinitesimal* was abused in the early literature of the calculus, its treatment generally amounting to logical nonsense, and the term fell into such disrepute that many modern books avoid it completely. Nevertheless, it is a very useful idea, and we shall base our development of the differential upon the properties of two special classes of infinitesimals which we shall call "big oh" and "little oh" (and designate '$\mathcal{O}$' and 'o', respectively).

Originally an infinitesimal was considered to be a number that "is infinitely small but not zero". Of course, there is no such number. Later, an infinitesimal was considered to be a variable that approaches zero as its limit. However, we know that it is *functions* that have limits, and a variable can be considered to have a limit only if it is somehow considered to be a function. We end up looking at functions φ such that $\varphi(t) \to 0$ as $t \to 0$. The definition of derivative involves several such infinitesimals. If $f'(x)$ exists and has the value a, then the fundamental difference quotient $(f(x + t) - f(x))/t$ is the quotient of two infinitesimals, and, furthermore, $((f(x + t) - f(x))/t) - a$ also approaches 0 as $t \to 0$. This last function is not defined at 0, but we can get around this if we wish by multiplying through by t, obtaining

$$(f(x + t) - f(x)) - at = \varphi(t),$$

where $f(x + t) - f(x)$ is the "change in f" infinitesimal, at is a linear infinitesimal, and $\varphi(t)$ is an infinitesimal that approaches 0 *faster than* t (i.e., $\varphi(t)/t \to 0$ as $t \to 0$). If we divide the last equation by t again, we see that this property of the infinitesimal φ, that it converges to 0 faster than t as $t \to 0$, is exactly equivalent to the fact that the difference quotient of f converges to a. This makes it clear that the study of derivatives is included in the study of the rate at which infinitesimals get small, and the usefulness of this paraphrase will shortly become clear.

Definition. A subset A of a normed linear space V is a *neighborhood* of a point α if A includes some open ball about α. A *deleted neighborhood* of α is a neighborhood of α minus the point α itself.

We define special sets of functions $\mathfrak{I}$, $\mathfrak{O}$, and $\mathfrak{o}$ as follows. It will be assumed in these definitions that each function is from a neighborhood of 0 in a normed linear space V to a normed linear space W.

$f \in \mathfrak{I}$ if $f(0) = 0$ and f is continuous at 0. These functions are the *infinitesimals.*

$f \in \mathfrak{O}$ if $f(0) = 0$ and f is Lipschitz continuous at 0. That is, there exist positive constants r and c such that $\|f(\xi)\| \leq c\|\xi\|$ on $B_r(0)$.

$f \in \mathfrak{o}$ if $f(0) = 0$ and $\|f(\xi)\|/\|\xi\| \to 0$ as $\xi \to 0$.

When the spaces V and W are not understood, we specify them by writing $\mathfrak{O}(V, W)$, etc.

A simple set of functions from $\mathbb{R}$ to $\mathbb{R}$ makes the qualitative difference between these classes apparent. The function $f(x) = |x|^{1/2}$ is in $\mathfrak{I}(\mathbb{R}, \mathbb{R})$ but not in $\mathfrak{O}$, $g(x) = x$ is in $\mathfrak{O}$ and therefore in $\mathfrak{I}$ but not in $\mathfrak{o}$, and $h(x) = x^2$ is in all three classes (Fig. 3.7).

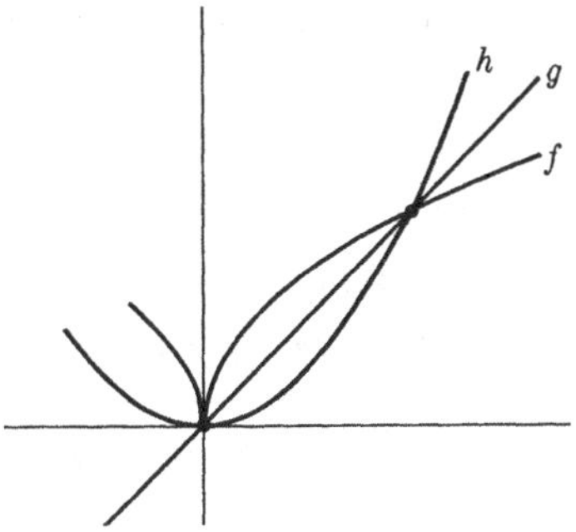

Fig. 3.7

It is clear that $\mathfrak{I}$, $\mathfrak{O}$, and $\mathfrak{o}$ are unchanged when the norms on V and W are replaced by equivalent norms.

Our previous notion of the sum of two functions does not apply to a pair of functions $f, g \in \mathfrak{I}(V, W)$ because their domains may be different. However, $f + g$ is defined on the intersection $\operatorname{dom} f \cap \operatorname{dom} g$, which is still a neighborhood of 0. Moreover, addition remains commutative and associative when extended in this way. It is clear that then $\mathfrak{I}(V, W)$ is almost a vector space. The only trouble occurs in connection with the equation $f + (-f) = 0$; the domain of the function on the left is $\operatorname{dom} f$, whereas we naturally take 0 to be the zero function on the whole of V.

*The way out of this difficulty is to identify two functions f and g in $\mathcal{I}$ if they are the same on some ball about 0. We define f and g to be *equivalent* ($f \sim g$) if and only if there exists a neighborhood of 0 on which $f = g$. We then check (in our minds) that this *is* an equivalence relation and that we now do have a vector space. Its elements are called *germs of functions at* 0. Strictly speaking, a germ is thus an equivalence class of functions, but in practice one tends to think of germs in terms of their representing functions, only keeping in mind that two functions are the same as germs when they agree on a neighborhood of 0.*

As one might guess from our introductory discussion, the algebraic properties of the three classes $\mathcal{I}$, $\mathcal{O}$, and $\mathfrak{o}$ are crucial for the differential calculus. We gather them together in the following theorem.

Theorem 5.1

1) $\mathfrak{o}(V, W) \subset \mathcal{O}(V, W) \subset \mathcal{I}(V, W)$, and each of the three classes is closed under addition and multiplication by scalars.
2) If $f \in \mathcal{O}(V, W)$, and if $g \in \mathcal{O}(W, X)$, then $g \circ f \in \mathcal{O}(V, X)$, where $\operatorname{dom} g \circ f = f^{-1}[\operatorname{dom} g]$.
3) If either f or g above is in $\mathfrak{o}$, then so is $g \circ f$.
4) If $f \in \mathcal{O}(V, W)$ and $g \in \mathcal{I}(V, \mathbb{R})$, then $fg \in \mathfrak{o}(V, W)$, and similarly if $f \in \mathcal{I}$ and $g \in \mathcal{O}$.
5) In (4) if either f or g is in $\mathfrak{o}$ and the other is merely bounded on a neighborhood of 0, then $fg \in \mathfrak{o}(V, W)$.
6) $\operatorname{Hom}(V, W) \subset \mathcal{O}(V, W)$.
7) $\operatorname{Hom}(V, W) \cap \mathfrak{o}(V, W) = \{0\}$.

Proof. Let $\mathfrak{L}_\epsilon(V, W)$ be the set of infinitesimals f such that $\|f(\xi)\| \leq \epsilon\|\xi\|$ on some ball about 0. Then $f \in \mathcal{O}$ if and only if f is in *some* $\mathfrak{L}_\epsilon$, and $f \in \mathfrak{o}$ if and only if f is in *every* $\mathfrak{L}_\epsilon$. Obviously, $\mathfrak{o} \subset \mathcal{O} \subset \mathcal{I}$.

1) If $\|f(\xi)\| \leq a\|\xi\|$ on $B_t(0)$ and $\|g(\xi)\| \leq b\|\xi\|$ on $B_u(0)$, then

$$\|f(\xi) + g(\xi)\| \leq (a + b)\|\xi\|$$

on $B_r(0)$, where $r = \min\{t, u\}$. Thus $\mathcal{O}$ is closed under addition. The closure of $\mathfrak{o}$ under addition follows similarly, or simply from the limit of a sum being the sum of the limits.

2) If $\|f(\xi)\| \leq a\|\xi\|$ when $\|\xi\| \leq t$ and $\|g(\eta)\| \leq b\|\eta\|$ when $\|\eta\| \leq u$, then

$$\|g(f(\xi))\| \leq b\|f(\xi)\| \leq ab\|\xi\|$$

when $\|\xi\| \leq t$ and $\|f(\xi)\| \leq u$, and so when $\|\xi\| \leq r = \min\{t, u/a\}$.

3) Now suppose that $f \in \mathfrak{o}$ in (2). Then, given ϵ, we can take $a = \epsilon/b$ and have

$$\|g(f(\xi))\| \leq \epsilon\|\xi\|$$

when $\|\xi\| \leq r$. Thus $g \circ f \in \mathfrak{o}$. The argument when $g \in \mathfrak{o}$ and $f \in \mathcal{O}$ is essentially the same.

4) Given $\|f(\xi)\| \leq c\|\xi\|$ on $B_r(0)$ and given ϵ, we choose δ such that $|g(\xi)| \leq \epsilon/c$ on $B_\delta(0)$ and have

$$\|f(\xi)g(\xi)\| \leq \epsilon\|\xi\|$$

when $\|\xi\| \leq \min(\delta, r)$. The other result follows similarly, as also does (5).

6) A bounded linear transformation is in $\mathcal{O}$ by definition.

7) Suppose that $f \in \mathrm{Hom}(V, W) \cap \mathfrak{o}(V, W)$. Take any $\alpha \neq 0$. Given ϵ, choose r so that $\|f(\xi)\| \leq \epsilon\|\xi\|$ on $B_r(0)$. Then write α as $\alpha = x\xi$, where $\|\xi\| < r$. (Find ξ and x.) Then

$$\|f(\alpha)\| = \|f(x\xi)\| = |x| \cdot \|f(\xi)\| \leq |x| \cdot \epsilon \cdot \|\xi\| = \epsilon\|\alpha\|.$$

Thus $\|f(\alpha)\| \leq \epsilon\|\alpha\|$ for every positive ϵ, and so $f(\alpha) = 0$. Thus $f = 0$, proving (7). □

Remark. The additivity of f was not used in this argument, only its homogeneity. It follows therefore that there is no homogeneous function (of degree 1) in $\mathfrak{o}$ except 0.

Sometimes when more than one variable is present it is necessary to indicate with respect to which variable a function is in $\mathcal{O}$ or $\mathfrak{o}$. We then write "$f(\xi) = \mathcal{O}(\xi)$" for "$f \in \mathcal{O}$", where "$\mathcal{O}(\xi)$" is used to designate an arbitrary element of $\mathcal{O}$.

The following rather curious lemma will be useful later in our proof of the differentiability of an implicitly defined function. It is understood that $\eta = f(\xi)$, where f is the function we are studying.

Lemma 5.1. If $\eta = \mathcal{O}(\xi) + \mathfrak{o}(\langle \xi, \eta \rangle)$ and also $\eta = \mathfrak{I}(\xi)$, then $\eta = \mathcal{O}(\xi)$.

Proof. The hypotheses imply that there are numbers b, r_1 and ρ such that $\|\eta\| \leq b\|\xi\| + \frac{1}{2}(\|\xi\| + \|\eta\|)$ if $\|\xi\| \leq r_1$ and $\|\xi\| + \|\eta\| \leq \rho$, and then that $\|\eta\| \leq \rho/2$ if $\|\xi\|$ is smaller than some r_2. If $\|\xi\| \leq r = \min\{r_1, r_2, \rho/2\}$, then all the conditions are met and $\|\eta\| \leq b\|\xi\| + \frac{1}{2}(\|\xi\| + \|\eta\|)$. But this is the inequality $\|\eta\| \leq (2b + 1)\|\xi\|$, and so $\eta = \mathcal{O}(\xi)$. □

We shall also need the following straightforward result.

Lemma 5.2. If $f \in \mathcal{O}(V, X)$ and $g \in \mathcal{O}(V, Y)$, then $\langle f, g \rangle \in \mathcal{O}(V, X \times Y)$. That is, $\langle \mathcal{O}(\xi), \mathcal{O}(\xi) \rangle = \mathcal{O}(\xi)$.

Proof. The proof is left to the reader.

EXERCISES

5.1 Prove in detail that the class $\mathfrak{I}(V, W)$ is unchanged if the norms on V and W are replaced by equivalent norms.

5.2 Do the same for $\mathcal{O}$ and $\mathfrak{o}$.

5.3 Prove (5) of the $\mathcal{O}\mathfrak{o}$-theorem (Theorem 5.1).

5.4 Prove also that if in (4) either f or g is in $\mathcal{O}$ and the other is merely bounded on a neighborhood of 0, then $fg \in \mathfrak{o}(V, W)$.

5.5 Prove Lemma 5.2. (Remember that $F = \langle F_1, F_2 \rangle$ is loose language for $F = \theta_1 \circ F_1 + \theta_2 \circ F_2$.) State the generalization to n functions. State the $\mathfrak{o}$-form of the theorem.

5.6 Given $F_1 \in \mathcal{O}(V_1, W)$ and $F_2 \in \mathcal{O}(V_2, W)$, define F from (a subset of) $V = V_1 \times V_2$ to W by $F(\alpha_1, \alpha_2) = F_1(\alpha_1) + F_2(\alpha_2)$. Prove that $F \in \mathcal{O}(V, W)$. (First state the defining equation as an identity involving the projections π_1 and π_2 and not involving explicit mention of the domain vectors α_1 and α_2.)

5.7 Given $F_1 \in \mathcal{O}(V_1, W)$ and $F_2 \in \mathcal{O}(V_2, \mathbb{R})$, define precisely what you mean by F_1F_2 and show that it is in $\mathfrak{o}(V_1 \times V_2, W)$.

5.8 Define the class $\mathcal{O}^n$ as follows: $f \in \mathcal{O}^n$ if $f \in \mathfrak{I}$ and $\|f(\xi)\|/\|\xi\|^n$ is bounded in some deleted ball about 0. (A deleted neighborhood of α is a neighborhood minus α.) State and prove a theorem about $f + g$ when $f \in \mathcal{O}^n$ and $g \in \mathcal{O}^m$.

5.9 State and prove a theorem about $f \circ g$ when $f \in \mathcal{O}^n$ and $g \in \mathcal{O}^m$.

5.10 State and prove a theorem about fg when $f \in \mathcal{O}^n$ and $g \in \mathcal{O}^m$.

5.11 Define a similar class $\mathfrak{o}^n$. State and prove a theorem about $f \circ g$ when $f \in \mathcal{O}^n$ and $g \in \mathfrak{o}^m$.

6. THE DIFFERENTIAL

Before considering the notion of the differential, we shall review some geometric material from the elementary calculus. We do this for motivation only; our subsequent theory is independent of the preliminary discussion.

In the elementary one-variable calculus the derivative $f'(a)$ of a function f at the point a has geometric meaning as the slope of the tangent line to the graph of f at the point a. (Of course, according to our notion of a function, the graph of f *is* f.) The tangent line thus has the (point-slope) equation $y - f(a) = f'(a)(x - a)$, and is the graph of the affine map $x \mapsto f'(a)(x - a) + f(a)$.

We ordinarily examine the nature of the curve f near the point $\langle a, f(a) \rangle$ by using new variables which are zero at this point. That is, we express everything in terms of $s = y - f(a)$ and $t = x - a$. This change of variables is simply the translation $\langle x, y \rangle \mapsto \langle t, s \rangle = \langle x - a, y - f(a) \rangle$ in the Cartesian plane $\mathbb{R}^2$ which brings the point of interest $\langle a, f(a) \rangle$ to the origin. If we picture the situation in a Euclidean plane, of which the next page is a satisfactory local model, then this translation in $\mathbb{R}^2$ is represented by a choice of new axes, the t- and s-axes, with origin at the point of tangency. Since $y = f(x)$

if and only if $s = f(a+t) - f(a)$, we see that the image of f under this translation is the function Δf_a defined by $\Delta f_a(t) = f(a+t) - f(a)$. (See Fig. 3.8.) Of course, Δf_a is simply our old friend the change in f brought about by changing x from a to $a+t$.

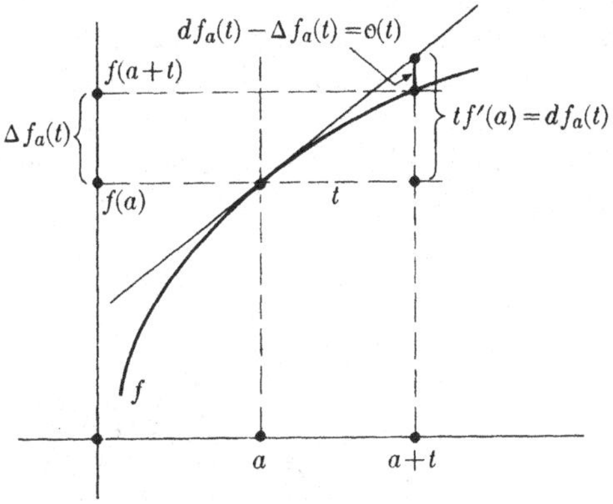

Fig. 3.8

Similarly, the equation $y - f(a) = f'(a)(x - a)$ becomes $s = f'(a)t$, and the tangent line accordingly translates to the line that is (the graph of) the linear functional $l\colon t \mapsto f'(a)t$ having the number $f'(a)$ as its skeleton (matrix). Remember that from the point of view of the geometric configuration (curve and tangent line) in the Euclidean plane, all that we are doing is choosing the natural axis system, with origin at the point of tangency. Then the curve is (the graph of) the function Δf_a, and the tangent line is (the graph of) the linear map l.

Now it follows from the definition of $f'(a)$ that l can also be characterized as *the* linear function that approximates Δf_a most closely. For, by definition,

$$\frac{\Delta f_a(t)}{t} \to f'(a) \qquad \text{as} \qquad t \to 0,$$

and this is exactly the same as saying that

$$\frac{\Delta f_a(t) - l(t)}{t} \to 0 \qquad \text{or} \qquad \Delta f_a - l \in \mathcal{o}.$$

But we know from the $\mathcal{O}\mathcal{o}$-theorem that the expression of the function Δf_a as the sum $l + \mathcal{o}$ is unique. This unique linear approximation l is called the *differential* of f at a and is designated df_a. Again, the differential of f at a is *the* linear function $l\colon \mathbb{R} \mapsto \mathbb{R}$ that approximates the actual change in f, Δf_a, in the sense that $\Delta f_a - l \in \mathcal{o}$; we saw above that if the derivative $f'(a)$ exists, then the differential of f at a exists and has $f'(a)$ as its skeleton (1×1 matrix).

Similarly, if f is a function of two variables, then (the graph of) f is a surface in Cartesian 3-space $\mathbb{R}^3 = \mathbb{R}^2 \times \mathbb{R}$, and the tangent plane to this surface at $\langle a, b, f(a, b)\rangle$ has the equation $z - f(a, b) = f_1(a, b)(x - a) + f_2(a, b)(x - b)$, where $f_1 = \partial f/\partial x$ and $f_2 = \partial f/\partial y$. If, as above, we set

$$\Delta f_{\langle a,b \rangle}(s, t) = f(a + s, b + t) - f(a, b)$$

and $l(s, t) = sf_1(a, b) + tf_2(a, b)$, then $\Delta f_{\langle a,b \rangle}$ is the change in f around a, b and l is the linear functional on $\mathbb{R}^2$ with matrix (skeleton) $\langle f_1(a, b), f_2(a, b) \rangle$. Moreover, it is a theorem of the standard calculus that if the partial derivatives of f are continuous, then again l approximates $\Delta f_{\langle a,b \rangle}$, with error in $\mathfrak{o}$. Here also l is called the differential of f at $\langle a, b \rangle$ and is designated $df_{\langle a,b \rangle}$ (Fig. 3.9). The notation in the figure has been changed to show the value at $\mathbf{t} = \langle t_1, t_2 \rangle$ of the differential $df_{\mathbf{a}}$ of f at $\mathbf{a} = \langle a_1, a_2 \rangle$.

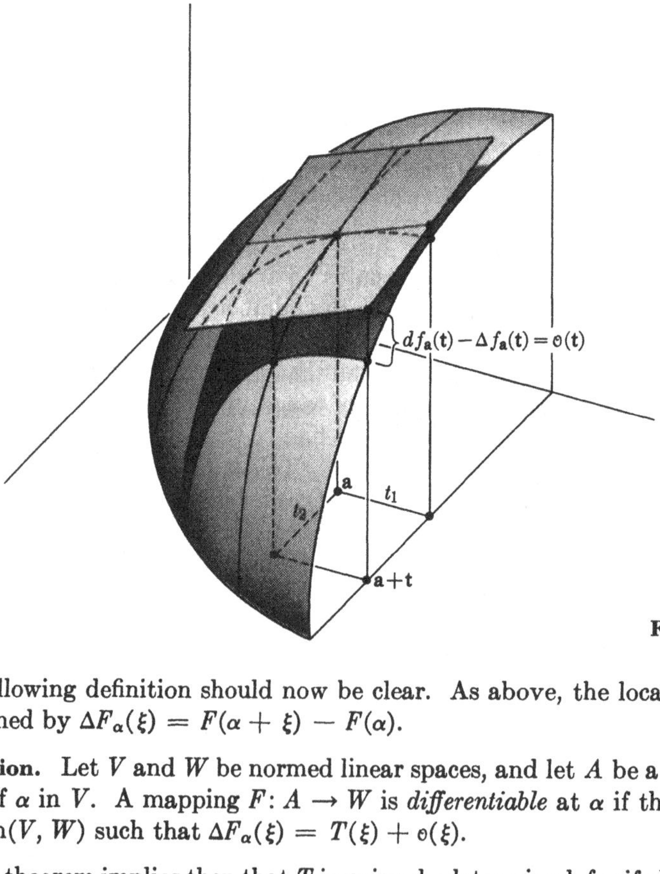

Fig. 3.9

The following definition should now be clear. As above, the local function ΔF_α is defined by $\Delta F_\alpha(\xi) = F(\alpha + \xi) - F(\alpha)$.

Definition. Let V and W be normed linear spaces, and let A be a neighborhood of α in V. A mapping $F: A \to W$ is *differentiable* at α if there is a T in $\text{Hom}(V, W)$ such that $\Delta F_\alpha(\xi) = T(\xi) + \mathfrak{o}(\xi)$.

The $\mathfrak{O}\mathfrak{o}$-theorem implies then that T is uniquely determined, for if also $\Delta F_\alpha = S + \mathfrak{o}$, then $T - S \in \mathfrak{o}$, and so $T - S = 0$ by (7) of the theorem. This uniquely determined T is called the *differential of F at α* and is designated dF_α. Thus

$$\Delta F_\alpha = dF_\alpha + \mathfrak{o},$$

where dF_α is the unique (bounded) linear approximation to ΔF_α.

*Our preliminary discussion should make it clear that this definition of the differential agrees with standard usage when the domain space is $\mathbb{R}^n$. However, in certain cases when the domain space is an infinite-dimensional function space, dF_α is called the *first variation* of F at α. This is due to the fact that although the early writers on the calculus of variations saw its analogy with the differential calculus, they did not realize that it was the same subject.*

We gather together in the next two theorems the familiar rules for differentiation. They follow immediately from the definition and the $\mathcal{O}o$-theorem.

It will be convenient to use the notation $\mathfrak{D}_\alpha(V, W)$ for the set of all mappings from neighborhoods of α in V to W that are differentiable at α.

Theorem 6.1

1) If $F \in \mathfrak{D}_\alpha(V, W)$, then $\Delta F_\alpha \in \mathcal{O}(V, W)$.
2) If $F, G \in \mathfrak{D}_\alpha(V, W)$, then $F + G \in \mathfrak{D}_\alpha(V, W)$ and $d(F+G)_\alpha = dF_\alpha + dG_\alpha$.
3) If $F \in \mathfrak{D}_\alpha(V, \mathbb{R})$ and $G \in \mathfrak{D}_\alpha(V, W)$, then $FG \in \mathfrak{D}_\alpha(V, W)$ and $d(FG)_\alpha = F(\alpha)\, dG_\alpha + dF_\alpha G(\alpha)$, the second term being a dyad.
4) If F is a constant function on V, then F is differentiable and $dF_\alpha = 0$.
5) If $F \in \text{Hom}(V, W)$, then F is differentiable at every $\alpha \in V$ and $dF_\alpha = F$.

Proof

1) $\Delta F_\alpha = dF_\alpha + o = \mathcal{O} + o = \mathcal{O}$ by (1) and (6) of the $\mathcal{O}o$-theorem.
2) It is clear that $\Delta(F+G)_\alpha = \Delta F_\alpha + \Delta G_\alpha$. Therefore, $\Delta(F+G)_\alpha = (dF_\alpha + o) + (dG_\alpha + o) = (dF_\alpha + dG_\alpha) + o$ by (1) of the $\mathcal{O}o$-theorem. Since $dF_\alpha + dG_\alpha \in \text{Hom}(V, W)$, we have (2).
3) $$\begin{aligned}\Delta(FG)_\alpha(\xi) &= F(\alpha+\xi)G(\alpha+\xi) - F(\alpha)G(\alpha)\\ &= \Delta F_\alpha(\xi)G(\alpha) + F(\alpha)\,\Delta G_\alpha(\xi) + \Delta F_\alpha(\xi)\,\Delta G_\alpha(\xi),\end{aligned}$$

   as the reader will see upon expanding and canceling. This is just the usual device of adding and subtracting middle terms in order to arrive at the form involving the Δ's. Thus

   $$\Delta(FG)_\alpha = (dF_\alpha + o)G(\alpha) + F(\alpha)(dG_\alpha + o) + \mathcal{O}\mathcal{O} = dF_\alpha G(\alpha) + F(\alpha)\, dG_\alpha + o$$

   by the $\mathcal{O}o$-theorem.
4) If $\Delta F_\alpha = 0$, then $dF_\alpha = 0$ by (7) of the $\mathcal{O}o$-theorem.
5) $\Delta F_\alpha(\xi) = F(\alpha+\xi) - F(\alpha) = F(\xi)$. Thus $\Delta F_\alpha = F \in \text{Hom}(V, W)$. □

The composite-function rule is somewhat more complicated.

Theorem 6.2. If $F \in \mathfrak{D}_\alpha(V, W)$ and $G \in \mathfrak{D}_{F(\alpha)}(W, X)$, then $G \circ F \in \mathfrak{D}_\alpha(V, X)$ and

$$d(G \circ F)_\alpha = dG_{F(\alpha)} \circ dF_\alpha.$$

Proof. We have

$$\begin{aligned}\Delta(G \circ F)_\alpha(\xi) &= G(F(\alpha + \xi)) - G(F(\alpha))\\ &= G(F(\alpha) + \Delta F_\alpha(\xi)) - G(F(\alpha))\\ &= \Delta G_{F(\alpha)}(\Delta F_\alpha(\xi))\\ &= dG_{F(\alpha)}(\Delta F_\alpha(\xi)) + \mathfrak{o}(\Delta F_\alpha(\xi))\\ &= dG_{F(\alpha)}(dF_\alpha(\xi)) + dG_{F(\alpha)}(\mathfrak{o}(\xi)) + \mathfrak{o} \circ \mathcal{O}\\ &= (dG_{F(\alpha)} \circ dF_\alpha)(\xi) + \mathcal{O} \circ \mathfrak{o} + \mathfrak{o} \circ \mathcal{O}.\end{aligned}$$

Thus $\Delta(G \circ F)_\alpha = dG_{F(\alpha)} \circ dF_\alpha + \mathfrak{o}$, and since $dG_{F(\alpha)} \circ dF_\alpha \in \text{Hom}(V, W)$, this proves the theorem. The reader should be able to justify each step taken in this chain of equalities. □

EXERCISES

6.1 The coordinate mapping $\langle x, y \rangle \mapsto x$ from $\mathbb{R}^2$ to $\mathbb{R}$ is differentiable. Why? What is its differential?

6.2 Prove that differentiation commutes with the application of bounded linear maps. That is, show that if $F: V \to W$ is differentiable at α and if $T \in \text{Hom}(W, X)$, then $T \circ F$ is differentiable at α and $d(T \circ F)_\alpha = T \circ dF_\alpha$.

6.3 Prove that $F \in \mathfrak{D}_\alpha(V, \mathbb{R})$ and $F(\alpha) \neq 0 \Rightarrow G = 1/F \in \mathfrak{D}_\alpha(V, \mathbb{R})$ and

$$dG_\alpha = \frac{-dF_\alpha}{(F(\alpha))^2}.$$

6.4 Let $F: V \to \mathbb{R}$ be differentiable at α, and let $f: \mathbb{R} \to \mathbb{R}$ be a function whose derivative exists at $a = F(\alpha)$. Prove that $f \circ F$ is differentiable at α and that

$$d(f \circ F)_\alpha = f'(a)\, dF_\alpha.$$

[Remember that the differential of f at a is simply multiplication by its derivative: $df_a(h) = hf'(a)$.] Show that the preceding problem is a special case.

6.5 Let V and W be normed linear spaces, and let $F: V \to W$ and $G: W \to V$ be continuous maps such that $G \circ F = I_V$ and $F \circ G = I_W$. Suppose that F is differentiable at α and that G is differentiable at $\beta = F(\alpha)$. Prove that

$$dG_\beta = (dF_\alpha)^{-1}.$$

6.6 Let $f: V \to \mathbb{R}$ be differentiable at α. Show that $g = f^n$ is differentiable at α and that

$$dg_\alpha = nf^{n-1}(\alpha)\, df_\alpha.$$

(Prove this both by an induction on the product rule and by the composite-function rule, assuming in the second case that $D_x x^n = nx^{n-1}$.)

6.7 Prove from the product rule by induction that if the n functions $f_i: V \to \mathbb{R}$, $i = 1, \ldots, n$, are all differentiable at α, then so is $f = \prod_1^n f_i$, and that

$$df_\alpha = \sum_{i=1}^{n} \left[\prod_{j \neq i} f_j(\alpha) \right] d(f_i)_\alpha.$$

6.8 A *monomial of degree n* on the normed linear space V is a product $\prod_1^n l_i$ of linear functionals ($l_i \in V^*$). A *homogeneous polynomial of degree n* is a finite sum of monomials of degree n. A polynomial of degree n is a sum of homogeneous polynomials P_i, $i = 0, \ldots, n$, where P_0 is a constant. Show from the above exercise and other known facts that a polynomial is differentiable everywhere.

6.9 Show that if $F_1: V \to W_1$ and $F_2: V \to W_2$ are both differentiable at α, then so is $F = \langle F_1, F_2 \rangle$ from V to $W = W_1 \times W_2$ (use the injections θ_1 and θ_2).

6.10 Show without using explicit computations, but using the results of earlier exercises instead, that the mapping $F = \mathbb{R}^2 \to \mathbb{R}^2$ defined by

$$\langle x, y \rangle \mapsto \langle (x - y)^2, (x + y)^3 \rangle$$

is everywhere differentiable. Now compute its differential at $\langle a, b \rangle$.

6.11 Let $F: V \to X$ and $G: W \to X$ be differentiable at α and β respectively, and define $K: V \times W \to X$ by

$$K(\xi, \eta) = F(\xi) + G(\eta).$$

Show that K is differentiable at $\langle \alpha, \beta \rangle$

a) by a direct Δ-calculation;
b) by using the projections π_1 and π_2 to express K in terms of F and G without explicit reference to the variable, and then applying the differentiation rules.

6.12 Now suppose given $F: V \to \mathbb{R}$ and $G: W \to X$, and define K by

$$K(\xi, \eta) = F(\xi)G(\eta).$$

Show that if F and G are differentiable at α and β respectively, then K is differentiable at $\langle \alpha, \beta \rangle$ in the manner of (b) in the above exercise.

6.13 Let V and W be normed linear spaces. Prove that the map $\langle \alpha, \beta \rangle \mapsto \|\alpha\| \, \|\beta\|$ from $V \times W$ to $\mathbb{R}$ is in $\mathfrak{o}(V \times W, \mathbb{R})$. Use the maximum norm on the product space.

Let $f: V \times W \to \mathbb{R}$ be bounded and bilinear. Here boundedness means that there is some b such that $|f(\alpha, \beta)| \leq b\|\alpha\| \, \|\beta\|$ for all α, β. Prove that f is differentiable everywhere and find its differential.

6.14 Let f and g be differentiable functions from $\mathbb{R}$ to $\mathbb{R}$. We know from the composite-function rule of the ordinary calculus that

$$(f \circ g)'(a) = f'(g(a))g'(a).$$

Our composite-function rule says that

$$d(f \circ g)_a = df_{g(a)} \circ dg_a,$$

where df_x is the linear mapping $t \to f'(x)t$. Show that these two statements are equivalent.

6.15 Prove that $f(x, y) = \|\langle x, y\rangle\|_1 = |x| + |y|$ is differentiable except on the coordinate axes (that is, $df_{\langle a,b\rangle}$ exists if a and b are *both* nonzero).

6.16 Comparing the shapes of the unit balls for $\| \|_1$ and $\| \|_\infty$ on $\mathbb{R}^2$, guess from the above the theorem about the differentiability of $\| \|_\infty$. Prove it.

6.17 Let V and W be fixed normed linear spaces, let X_d be the set of all maps from V to W that are differentiable at 0, let X_0 be the set of all maps from V to W that belong to $\mathfrak{o}(V, W)$, and let X_l be $\text{Hom}(V, W)$. Prove that X_d and X_0 are vector spaces and that $X_d = X_0 \oplus X_l$.

6.18 Let F be a Lipschitz function with constant C which is differentiable at a point α. Prove that $\|dF_\alpha\| \leq C$.

7. DIRECTIONAL DERIVATIVES; THE MEAN-VALUE THEOREM

Directional derivatives form the connecting link between differentials and the derivatives of the elementary calculus, and, although they add one more concept that has to be fitted into the scheme of things, the reader should find them intuitively satisfying and technically useful.

A continuous function f from an interval $I \subset \mathbb{R}$ to a normed linear space W can have a derivative $f'(x)$ at a point $x \in I$ in exactly the sense of the elementary calculus:

$$f'(x) = \lim_{t \to 0} \frac{f(x + t) - f(x)}{t}.$$

The range of such a function f is a curve or arc in W, and it is conventional to call f itself a *parametrized arc* when we want to keep this geometric notion in mind. We shall also call $f'(x)$, if it exists, the *tangent vector* to the arc f at x. This terminology fits our geometric intuition, as Fig. 3.10 suggests. For simplicity we have set $x = 0$ and $f(x) = 0$. If $f'(x)$ exists, we say that the parametrized arc f is *smooth* at x. We also say that f is smooth at $\alpha = f(x)$, but this terminology is ambiguous if f is not injective (i.e., if the arc crosses itself). An arc is smooth if it is smooth at every value of the parameter.

We naturally wonder about the relationship between the existence of the tangent vector $f'(x)$ and the differentiability of f at x. If df_x exists, then, being a linear map on $\mathbb{R}$, it is simply multiplication "by" the fixed vector α that is its skeleton, $df_x(h) = h\, df_x(1) = h\alpha$, and we expect α to be the tangent vector.

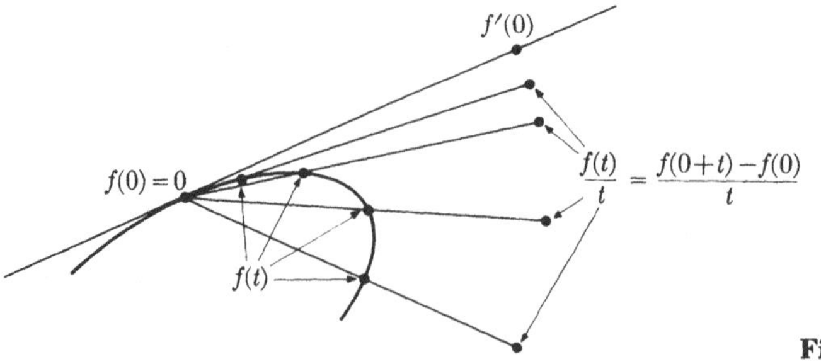

Fig. 3.10

$f'(x)$. We showed this and also the converse result for the ordinary calculus in our preliminary discussion in Section 6. Actually, our argument was valid for vector-valued functions, but we shall repeat it anyway.

When we think of a vector-valued function of a real variable as being an arc, we often use Greek letters like 'λ' and 'γ' for the function, as we do below. This of course does not in any way change what is being proved, but is slightly suggestive of a geometric interpretation.

Theorem 7.1. A parametrized arc $\gamma: [a, b] \to V$ is differentiable at $x \in (a, b)$ if and only if the tangent vector (derivative) $\alpha = \gamma'(x)$ exists, in which case the tangent vector is the skeleton of the differential, $d\gamma_x(h) = h\gamma'(x) = h\alpha$.

Proof. If the parametrized arc $\gamma: [a, b] \to V$ is differentiable at $x \in (a, b)$, then $d\gamma_x(h) = hd\gamma_x(1) = h\alpha$, where $\alpha = d\gamma_x(1)$. Since $\Delta\gamma_x - d\gamma_x \in \mathfrak{o}$, this gives $\|\Delta\gamma_x(h) - h\alpha\|/|h| \to 0$, and so $\Delta\gamma_x(h)/h \to \alpha$ as $h \to 0$. Thus α is the derivative $\gamma'(x)$ in the ordinary sense. By reversing the above steps we see that the existence of $\gamma'(x)$ implies the differentiability of γ at x. □

Now let F be a function from an open set A in a normed linear space V to a normed linear space W. One way to study the behavior of F in the neighborhood of a point α in A is to consider how it behaves on each straight line through α. That is, we study F by temporarily restricting it to a one-dimensional domain. The advantage gained in doing this is that the restricted F is then simply a parametrized arc, and its differential is simply multiplication by its ordinary derivative.

For any nonzero $\xi \in V$ the straight line through α in the direction ξ has the parametric representation $t \mapsto \alpha + t\xi$. The restriction of F to this line is the parametrized arc $\gamma: \gamma(t) = F(\alpha + t\xi)$. Its tangent vector (derivative) at the origin $t = 0$, if it exists, is called the *derivative of F in the direction ξ at α*, or the derivative of F *with respect to ξ at α*, and is designated $D_\xi F(\alpha)$. Clearly,

$$D_\xi F(\alpha) = \lim_{t \to 0} \frac{F(\alpha + t\xi) - F(\alpha)}{t}.$$

Comparing this with our original definition of f', we see that the tangent vector $\gamma'(x)$ to a parametrized arc γ is the directional derivative $D_1\gamma(x)$ with respect to the standard basis vector 1 in $\mathbb{R}$.

Strictly speaking, we are misusing the word "direction", because different vectors can have the same direction. Thus, if $\eta = c\xi$ with $c > 0$, then η and ξ point in the same direction, but, because $D_\xi F(\alpha)$ is linear in ξ (as we shall see in a moment), their associated derivatives are different: $D_\eta F(\alpha) = cD_\xi F(\alpha)$.

We now want to establish the relationship between directional derivatives, which are vectors, and differentials, which are linear maps. We saw above that for an arc γ differentiability is equivalent to the existence of $\gamma'(x) = D_1\gamma(x)$. In the general case the relationship is not as simple as it is for arcs, but in one direction everything goes smoothly.

Theorem 7.2. *If F is differentiable at α, and if λ is any smooth arc through α, with $\alpha = \lambda(x)$, then $\gamma = F \circ \lambda$ is smooth at x, and $\gamma'(x) = dF_\alpha(\lambda'(x))$. In particular, if F is differentiable at α, then every directional derivative $D_\xi F(\alpha)$ exists, and $D_\xi F(\alpha) = dF_\alpha(\xi)$.*

Proof. The smoothness of γ is equivalent to its differentiability at x and therefore follows from the composite-function theorem. Moreover, $\gamma'(x) = d\gamma_x(1) = d(F \circ \lambda)_x(1) = dF_\alpha(d\lambda_x(1)) = dF_\alpha(\lambda'(x))$. If λ is the parametrized line $\lambda(t) = \alpha + t\xi$, then it has the constant derivative ξ, and since $\alpha = \lambda(0)$ here, the above formula becomes $\gamma'(0) = dF_\alpha(\xi)$. That is, $D_\xi F(\alpha) = \gamma'(0) = dF_\alpha(\xi)$. □

It is not true, conversely, that the existence of all the directional derivatives $D_\xi F(\alpha)$ of a function F at a point α implies the differentiability of F at α. The easiest counterexample involves the notion of a homogenous function. We say that a function $F: V \to W$ is *homogeneous* if $F(x\xi) = xF(\xi)$ for all x and ξ. For such a function the directional derivative $D_\xi F(0)$ exists because the arc $\gamma(t) = F(0 + t\xi) = tF(\xi)$ is linear, and $\gamma'(0) = F(\xi)$. Thus, *all of the directional derivatives of a homogeneous function F exist at 0 and $D_\xi F(0) = F(\xi)$.* If F is also differentiable at 0, then $dF_0(\xi) = D_\xi F(0) = F(\xi)$ and $F = dF_0$. *Thus a differentiable homogeneous function must be linear.* Therefore, any nonlinear homogeneous function F will be a function such that $D_\xi F(0)$ exists for all ξ but dF_0 does not exist. Taking the simplest possible situation, define $F: \mathbb{R}^2 \to \mathbb{R}$ by $F(x, y) = x^3/(x^2 + y^2)$ if $\langle x, y\rangle \neq \langle 0, 0\rangle$ and $F(0, 0) = 0$. Then

$$F(tx, ty) = tF(x, y),$$

so that F is homogeneous, but F is not linear.

However, if V is finite-dimensional, and if for each ξ in a spanning set of vectors the directional derivative $D_\xi F(\alpha)$ exists and is a continuous function of α on an open set A, then F is continuously differentiable on A. The proof of this fact depends on the mean-value theorem, which we take up next, but we shall not complete it until Section 9 (Theorem 9.3).

The reader will remember the mean-value theorem as a cornerstone of the calculus, and this is just as true in our general theory. We shall apply it in the next section to give the proof of the general form of the above-mentioned theorem, and practically all of our more advanced work will depend on it. The ordinary mean-value theorem does not have an exact analogue here. Instead we shall prove a theorem that in the one-variable calculus is an easy consequence of the mean-value theorem.

Theorem 7.3. *Let f be a continuous function (parametrized arc) from a closed interval $[a, b]$ to a normed linear space, and suppose that $f'(t)$ exists and that $\|f'(t)\| \leq m$ for all $t \in (a, b)$. Then $\|f(b) - f(a)\| \leq m(b - a)$.*

Proof. Fix $\epsilon > 0$, and let A be the set of points $x \in [a, b]$ such that

$$\|f(x) - f(a)\| \leq (m + \epsilon)(x - a) + \epsilon.$$

A includes at least a small interval $[a, c]$, because f is continuous at a. Set $l = \text{lub } A$. Then $\|f(l) - f(a)\| \leq (m + \epsilon)(l - a) + \epsilon$ by the continuity of f at l. Thus $l \in A$, and $a < l \leq b$. We claim that $l = b$. For if $l < b$, then $f'(l)$ exists and $\|f'(l)\| \leq m$. Therefore, there is a δ such that

$$\|[f(x) - f(l)]/(x - l)\| < m + \epsilon$$

when $|x - l| \leq \delta$. It follows that

$$\begin{aligned}\|f(l + \delta) - f(a)\| &\leq \|f(l + \delta) - f(l)\| + \|f(l) - f(a)\| \\ &\leq (m + \epsilon)\delta + (m + \epsilon)(l - a) + \epsilon \\ &= (m + \epsilon)(l + \delta - a) + \epsilon,\end{aligned}$$

so that $l + \delta \in A$, a contradiction. Therefore, $l = b$. We thus have

$$\|f(b) - f(a)\| \leq (m + \epsilon)(b - a) + \epsilon,$$

and, since ϵ is arbitrary, $\|f(b) - f(a)\| \leq m(b - a)$. □

The following more general version of the mean-value theorem is the form in which it is ordinarily applied. As usual, F and G are from a subset of V to W.

Theorem 7.4. If F is differentiable in the ball $B_r(\alpha)$, and if $\|dF_\beta\| \leq \epsilon$ for every β in this ball, then $\|\Delta F_\beta(\xi)\| \leq \epsilon\|\xi\|$ whenever β and $\beta + \xi$ are in the ball. More generally, the same result holds if the ball $B_r(\alpha)$ is replaced by any convex set C.

Proof. The segment from β to $\beta + \xi$ is the range of the parametrized arc $\lambda(t) = \beta + t\xi$ from $[0, 1]$ to V. If β and $\beta + \xi$ are in the ball $B_r(\alpha)$, then this segment is a subset of the ball. Setting $\gamma(t) = F(\beta + t\xi)$, we then have $\gamma'(x) = dF_{\beta+x\xi}(\lambda'(x)) = dF_{\beta+x\xi}(\xi)$, from Theorem 7.2. Therefore, $\|\gamma'(x)\| \leq \epsilon\|\xi\|$ on $[0, 1]$, and the mean-value theorem then implies that

$$\|\Delta F_\beta(\xi)\| = \|F(\beta + \xi) - F(\beta)\| = \|\gamma(1) - \gamma(0)\| \leq \epsilon\|\xi\|(1 - 0) = \epsilon\|\xi\|,$$

which is the desired inequality. The only property of $B_r(\alpha)$ that we have used is that it includes the line segment joining any two of its points. This is the definition of convexity, and the theorem is therefore true for any convex set. □

Corollary. If G is differentiable on the convex set C, if $T \in \text{Hom}(V, W)$, and if $\|dG_\beta - T\| \leq \epsilon$ for all β in C, then $\|\Delta G_\beta(\xi) - T(\xi)\| \leq \epsilon\|\xi\|$ whenever β and $\beta + \xi$ are in C.

Proof. Set $F = G - T$, and note that $dF_\beta = dG_\beta - T$ and $\Delta F_\beta = \Delta G_\beta - T$. □

We end this section with a few words about notation. Notice the reversal of the positions of the variables in the identity $(D_\xi F)(\alpha) = dF_\alpha(\xi)$. This difference has practical importance. We have a function of the two variables 'α' and 'ξ' which we can convert to a function of one variable by holding the other variable fixed; it is convenient technically to put the fixed variable in subscript position. Thus we think of $dF_\alpha(\xi)$ with α held fixed and have the function dF_α in $\mathrm{Hom}(V, W)$, whereas in $(D_\xi F)(\alpha)$ we hold ξ fixed and have the directional derivative $D_\xi F\colon A \to W$ in the fixed direction ξ as a function of α, generalizing the notation for any ordinary partial derivative $\partial F/\partial x_i(\mathbf{a})$ as a function of $\mathbf{a}$. We can also express this implication of the subscript position of a variable in the dot notation (Section 0.10): when we write $D_\xi F(\alpha)$, we are thinking of the value at α of the function $D_\xi F(\cdot)$.

Still a third notation that we shall use in later chapters puts the function symbol in subscript position. We write

$$J_F(\alpha) = dF_\alpha.$$

This notation implies that the mapping F is going to be fixed through a discussion and gets it "out of the way" by putting it in subscript position.

If F is differentiable at each point of the open set A, then we naturally consider dF to be the map $\alpha \mapsto dF_\alpha$ from A to $\mathrm{Hom}(V, W)$. In the "J"-notation, $dF = J_F$. Later in this chapter we are going to consider the differentiability of *this* map at α. This notion of the *second differential* $d^2F_\alpha = d(dF)_\alpha$ is probably confusing at first sight, and a preliminary look at it now may ease the later discussion. We simply have a new map $G = dF$ from an open set A in a normed linear space V to a normed linear space $X = \mathrm{Hom}(V, W)$, and we consider its differentiability at α. If $dG_\alpha = d(dF)_\alpha$ exists, it is a linear map from V to $\mathrm{Hom}(V, W)$, and there *is* something special now. Referring back to Theorem 6.1 of Chapter 1, we know that $dG_\alpha = d^2F_\alpha$ is equivalent by duality to a bilinear mapping ω from $V \times V$ to W: since $dG_\alpha(\xi)$ is itself a transformation in $\mathrm{Hom}(V, W)$, we can evaluate it at η, and we define ω by

$$\omega(\xi, \eta) = (dG_\alpha(\xi))(\eta).$$

The dot notation may be helpful here. The mapping $\alpha \mapsto dF_\alpha$ is simply $dF_{(\cdot)}$, and we have defined G by $G(\cdot) = dF_{(\cdot)}$. Later, the fact that $dG_\alpha(\xi)$ is a mapping can be emphasized by writing it as $dG_\alpha(\xi)(\cdot)$. In each case here we have a function of one variable, and the dot only reminds us of that fact and shows us where we shall put the variable when indicating an evaluation. In the case of ω we have the original use of the dot, as in $\omega(\xi, \cdot) = dG_\alpha(\xi)$.

EXERCISES

7.1 Given $f\colon \mathbb{R} \to \mathbb{R}$ such that $f'(a)$ exists, show that the "directional derivative" $D_b f(a)$ has the value $bf'(a)$, by a direct evaluation of the limit of the difference quotient.

7.2 Let f be a real-valued function on an n-dimensional space V, and suppose that f is differentiable at $\alpha \in V$. Show that the directions ξ in which the derivative $D_\xi F(\alpha)$ is zero make up an $(n - 1)$-dimensional subspace of V (or the whole of V). What similar conclusions can be drawn if f maps V to a two-dimensional space W?

7.3 a) Show by a direct argument on limits that if f and g are two functions from an interval $I \subset \mathbb{R}$ to a normed linear space V, and if $f'(x)$ and $g'(x)$ both exist, then $(f+g)'(x)$ exists and $(f+g)'(x) = f'(x) + g'(x)$.

b) Prove the same result as a corollary of Theorems 7.1 and 7.2 and the differentiation rules of Section 6.

7.4 a) Given $f: I \to V$ and $g: I \to W$, show by a direct limit argument that if $f'(x)$ and $g'(x)$ both exist, and if $F = \langle f, g \rangle : I \to V \times W$, then $F'(x)$ exists and $F'(x) = \langle f'(x), g'(x) \rangle$.

b) Prove the same result from Theorems 7.1 and 7.2 and the differentiation rules of Section 6, using the exact relation $F = \theta_1 \circ f + \theta_2 \circ g$.

7.5 In the spirit of the above two exercises, state a product law for derivatives of arcs and prove it as in the (b) proofs above.

7.6 Find the tangent vector to the arc $\langle e^t, \sin t \rangle$ at $t = 0$; at $t = \pi/2$. [Apply Exercise 7.4(a).] What is the differential of the above parametrized arc at these two points? That is, if $f(t) = \langle e^t, \sin t \rangle$, what are df_0 and $df_{\pi/2}$?

7.7 Let $F: \mathbb{R}^2 \to \mathbb{R}^2$ be the mapping $\langle x, y \rangle \mapsto \langle 3x^2y, x^2y^3 \rangle$. Compute the directional derivative $D_{\langle 1,2 \rangle}F(3, -1)$

a) as the tangent vector at $\langle 3, -1 \rangle$ to the arc $f \circ \lambda$, where λ is the straight line through $\langle 3, -1 \rangle$ in the direction $\langle 1, 2 \rangle$;

b) by first computing $dF_{\langle 3,-1 \rangle}$ and then evaluating at $\langle 1, 2 \rangle$.

7.8 Let λ and μ be any two linear functionals on a vector space V. Evaluate the product $f(\xi) = \lambda(\xi)\mu(\xi)$ along the line $\xi = t\alpha$, and hence compute $D_\alpha f(\alpha)$. Now evaluate f along the general line $\xi = t\alpha + \beta$, and from it compute $D_\alpha f(\beta)$.

7.9 Work the above exercise by computing differentials.

7.10 If $f: \mathbb{R}^n \to \mathbb{R}$ is differentiable at $\mathbf{a}$, we know that its differential $df_{\mathbf{a}}$, being a linear functional on $\mathbb{R}^n$, is given by its skeleton n-tuple $\mathbf{l}$ according to the formula

$$df_{\mathbf{a}}(\mathbf{x}) = (\mathbf{l}, \mathbf{x}) = \sum_1^n l_i x_i.$$

In this context we call the n-tuple $\mathbf{l}$ the *gradient* of f at $\mathbf{a}$. Show from the Schwarz inequality (Exercise 2.3) that if we use vectors $\mathbf{y}$ of Euclidean length 1, then the directional derivative $D_{\mathbf{y}}f(\mathbf{a})$ is maximum when $\mathbf{y}$ points in the direction of the *gradient* of f.

7.11 Let W be a normed linear space, and let V be the set of parametrized arcs $\lambda: [-1, 1] \to W$ such that $\lambda(0) = 0$ and $\lambda'(0)$ exists. Show that V is a vector space and that $\lambda \to \lambda'(0)$ is a surjective linear mapping from V to W. Describe in words the elements of the quotient space V/N, where N is the null space of the above map.

7.12 Find another homogeneous nonlinear function. Evaluate its directional derivatives $D_\xi F(0)$, and show again that they do not make up a linear map.

7.13 Prove that if F is a differentiable mapping from an open ball B of a normed linear space V to a normed linear space W such that $dF_\alpha = 0$ for every α in B, then F is a constant function.

7.14 Generalize the above exercise to the case where the domain of F is an open set A with the property that any two points of A can be joined by a smooth arc lying in A.

Show by a counterexample that the result does not generalize to arbitrary open sets A as the domain of F.

7.15 Prove the following generalization of the mean-value theorem. Let f be a continuous mapping from the closed interval $[a, b]$ to a normed linear space V, and let g be a continuous real-valued function on $[a, b]$. Suppose that $f'(t)$ and $g'(t)$ both exist at all points of the open interval (a, b) and that $\|f'(t)\| \leq g'(t)$ on (a, b). Then

$$\|f(b) - f(a)\| \leq g(b) - g(a).$$

[Consider the points x such that $\|f(x) - f(a)\| \leq g(x) - g(a) + \epsilon(x - a) + \epsilon$.]

8. THE DIFFERENTIAL AND PRODUCT SPACES

In this section we shall relate the differentiation rules to the special configurations resulting from the expression of a vector space as a finite Cartesian product. When dealing with the range, this is a trivial consideration, but when the domain is a product space, we become involved with a deeper theorem. These general product considerations will be specialized to the $\mathbb{R}^n$-spaces in the next section, but they also have a more general usefulness, as we shall see in the later sections of this chapter and in later chapters.

We know that an m-tuple of functions on a common domain, $F^i: A \to W_i$, $i = 1, \ldots, m$, is equivalent to a single m-tuple-valued function

$$F: A \to W = \prod_1^m W_i,$$

$F(\alpha)$ being the m-tuple $\{F^i(\alpha)\}_1^m$ for each $\alpha \in A$. We now check the obviously necessary fact that F is differentiable at α if and only if each F^i is differentiable at α.

Theorem 8.1. Given $F^i: A \to W_i, i = 1, \ldots, m$, and $F = \langle F^1, \ldots, F^m \rangle$, then F is differentiable at α if and only if all the functions F^i are, in which case $dF_\alpha = \langle dF^1_\alpha, \ldots, dF^m_\alpha \rangle$.

Proof. Strictly speaking, $F = \sum_1^m \theta_i \circ F^i$, where θ_j is the injection of W_j into the product space $W = \prod_1^m W_i$ (see Section 1.3). Since each θ_i is linear and hence differentiable, with $d(\theta_i)_\alpha = \theta_i$, we see that if each F^i is differentiable at α, then so is F, and $dF_\alpha = \sum_1^m \theta_i \circ dF^i_\alpha$. Less exactly, this is the statement $dF_\alpha = \langle dF^1_\alpha, \ldots, dF^m_\alpha \rangle$. The converse follows similarly from $F^i = \pi_i \circ F$, where π_j is the projection of $\prod_1^m W_i$ onto W_j. □

Theorems 7.1 and 8.1 have the following obvious corollary (which can also be proved as easily by a direct inspection of the limits involved).

Lemma 8.1. If f_i is an arc from $[a, b]$ to W_i, for $i = 1, \ldots, n$, and if f is the n-tuple-valued arc $f = \langle f_1, \ldots, f_n \rangle$, then $f'(x)$ exists if and only if $f'_i(x)$ exists for each i, in which case $f'(x) = \langle f'_1(x), \ldots, f'_n(x) \rangle$.

When the domain space V is a product space $\prod_1^n V_j$ the situation is more complicated. A function $F(\xi_1, \ldots, \xi_n)$ of n vector variables does not decompose into an equivalent n-tuple of functions. Moreover, although its differential $dF_{\boldsymbol{\alpha}}$ does decompose into an equivalent n-tuple of partial differentials $\{dF^j_{\boldsymbol{\alpha}}\}$, we do not have the simple theorem that $dF_{\boldsymbol{\alpha}}$ exists if and only if the partial differentials $dF^j_{\boldsymbol{\alpha}}$ all exist.

Of course, we regard a function $F(\xi_1, \ldots, \xi_n)$ of n vector variables as being a function of the single n-tuple variable $\boldsymbol{\xi} = \langle \xi_1, \ldots, \xi_n \rangle$, so that in principle there is nothing new when we consider the differentiability of F. However, when we consider a composition $F \circ G$, the inner function G must now be an n-tuple-valued function $G = \langle g^1, \ldots, g^n \rangle$, where g^i is from an open subset A of some normed linear space X to V_i, and we naturally try to express the differential of $F \circ G$ in terms of the differentials dg^i. To accomplish this we need the *partial differentials* $dF^i_{\boldsymbol{\alpha}}$ of F. For the moment we shall define the jth partial differential of F at $\boldsymbol{\alpha} = \langle \alpha_1, \ldots, \alpha_n \rangle$ as the restriction of the differential $dF_{\boldsymbol{\alpha}}$ to V_j, considered as a subspace of $V = \prod_1^n V_i$. As usual, this really involves the injection θ_j of V_j into $\prod_1^n V_i$, and our formal (temporary) definition, accordingly, is

$$dF^j_{\boldsymbol{\alpha}} = dF_{\boldsymbol{\alpha}} \circ \theta_j.$$

Then, since $\boldsymbol{\xi} = \langle \xi_1, \ldots, \xi_n \rangle = \sum_1^n \theta_i(\xi_i)$, we have

$$dF_{\boldsymbol{\alpha}}(\boldsymbol{\xi}) = \sum_1^n dF^i_{\boldsymbol{\alpha}}(\xi_i).$$

Similarly, since $G = \langle g^1, \ldots, g^n \rangle = \sum_1^n \theta_i \circ g^i$, we have

$$d(F \circ G)_\gamma = \sum_1^n dF^i_{G(\gamma)} \circ dg^i_\gamma,$$

which we shall call the *general chain rule*. There is ambiguity in the "i"-superscripts in this formula: to be more proper we should write $(dF)^i_{\boldsymbol{\alpha}}$ and $d(g^i)_\gamma$.

We shall now work around to the real definition of a partial differential. Since

$$\Delta F_{\boldsymbol{\alpha}} \circ \theta_j = (dF_{\boldsymbol{\alpha}} + \mathfrak{o}) \circ \theta_j = dF_{\boldsymbol{\alpha}} \circ \theta_j + \mathfrak{o} = dF^j_{\boldsymbol{\alpha}} + \mathfrak{o},$$

we see that $dF^i_{\boldsymbol{\alpha}}$ can be directly characterized, independently of $dF_{\boldsymbol{\alpha}}$, as follows:

> $dF^i_{\boldsymbol{\alpha}}$ is the unique element T_i of $\mathrm{Hom}(V_i, W)$ such that $\Delta F_{\boldsymbol{\alpha}} \circ \theta_i = T_i + \mathfrak{o}$.

That is, $dF^i_{\boldsymbol{\alpha}}$ is the differential at α_i of the function of the one variable ξ_i obtained by holding the other variables in $F(\xi_1, \ldots, \xi_n)$ fixed at the values $\xi_j = \alpha_j$. This is important because in practice it is often such partial differentiability that we come upon as the primary phenomenon. We shall therefore take this direct characterization as our definition of $dF^i_{\boldsymbol{\alpha}}$, after which our motivating calculation above is the proof of the following lemma.

Lemma 8.2. If A is an open subset of a product space $V = \prod_1^n V_i$, and if $F: A \to W$ is differentiable at $\boldsymbol{\alpha}$, then all the partial differentials $dF^i_{\boldsymbol{\alpha}}$ exist and $dF^i_{\boldsymbol{\alpha}} = dF_{\boldsymbol{\alpha}} \circ \theta_i$.

The question then occurs as to whether the existence of all the partial differentials $dF^i_{\boldsymbol{\alpha}}$ implies the existence of $dF_{\boldsymbol{\alpha}}$. The answer in general is negative, as we shall see in the next section, but if all the partial differentials $dF^i_{\boldsymbol{\alpha}}$ exist for each $\boldsymbol{\alpha}$ in an open set A and are continuous functions of $\boldsymbol{\alpha}$, then F is continuously differentiable on A. Note that Lemma 8.2 and the projection-injection identities show us what $dF_{\boldsymbol{\alpha}}$ must be if it exists: $dF^i_{\boldsymbol{\alpha}} = dF_{\boldsymbol{\alpha}} \circ \theta_i$ and $\sum \theta_i \circ \pi_i = I$ together imply that $dF_{\boldsymbol{\alpha}} = \sum dF^i_{\boldsymbol{\alpha}} \circ \pi_i$.

Theorem 8.2. Let A be an open subset of the normed linear space $V = V_1 \times V_2$, and suppose that $F: A \to W$ has continuous partial differentials $dF^1_{<\alpha,\beta>}$ and $dF^2_{<\alpha,\beta>}$ on A. Then $dF_{<\alpha,\beta>}$ exists and is continuous on A, and $dF_{<\alpha,\beta>}(\xi, \eta) = dF^1_{<\alpha,\beta>}(\xi) + dF^2_{<\alpha,\beta>}(\eta)$.

Proof. We shall use the sum norm on $V = V_1 \times V_2$. Given ϵ, we choose δ so that $\|dF^i_{<\mu,\nu>} - dF^i_{<\alpha,\beta>}\| < \epsilon$ for every $<\mu, \nu>$ in the δ-ball about $<\alpha, \beta>$ and for $i = 1, 2$. Setting

$$G(\xi) = F(\alpha + \xi, \beta + \eta) - dF^1_{<\alpha,\beta>}(\xi),$$

we have

$$\|dG_\zeta\| = \|dF^1_{<\alpha+\zeta,\beta+\eta>} - dF^1_{<\alpha,\beta>}\| < \epsilon \quad \text{if} \quad \|<\zeta, \eta>\| < \delta,$$

and the corollary of Theorem 7.4 implies that

$$\|F(\alpha + \xi, \beta + \eta) - F(\alpha, \beta + \eta) - dF^1_{<\alpha,\beta>}(\xi)\| \le \epsilon\|\xi\|$$

when $\|<\xi, \eta>\| < \delta$. Arguing similarly with

$$H(\eta) = F(\alpha, \beta + \eta) - dF^2_{<\alpha,\beta>}(\eta),$$

we find that

$$\|F(\alpha, \beta + \eta) - F(\alpha, \beta) - dF^2_{<\alpha,\beta>}(\eta)\| \le \epsilon\|\eta\|$$

when $\|<0, \eta>\| < \delta$. Combining the two inequalities, we have

$$\|\Delta F_{<\alpha,\beta>}(\xi, \eta) - T(<\xi, \eta>)\| \le \epsilon\|<\xi, \eta>\|$$

when $\|<\xi, \eta>\| < \delta$, where $T = dF^1_{<\alpha,\beta>} \circ \pi_1 + dF^2_{<\alpha,\beta>} \circ \pi_2$. That is, $\Delta F_{<\alpha,\beta>} - T = \mathfrak{o}$, and so $dF_{<\alpha,\beta>}$ exists and equals T. □

The theorem for more than two factor spaces is a corollary.

Theorem 8.3. If A is an open subset of $\prod_1^n V_i$ and $F \to W$ is such that for each $i = 1, \ldots, n$ the partial differential $dF^i_{\boldsymbol{\alpha}}$ exists for all $\boldsymbol{\alpha} \in A$ and is continuous as a function of $\boldsymbol{\alpha} = <\alpha_1, \ldots, \alpha_n>$, then $dF_{\boldsymbol{\alpha}}$ exists and is continuous on A. If $\boldsymbol{\xi} = <\xi_1, \ldots, \xi_n>$, then $dF_{\boldsymbol{\alpha}}(\boldsymbol{\xi}) = \sum_1^n dF^i_{\boldsymbol{\alpha}}(\xi_i)$.

Proof. The existence and continuity of $dF^1_{\boldsymbol{\alpha}}$ and $dF^2_{\boldsymbol{\alpha}}$ imply by the theorem that $dF^1_{\boldsymbol{\alpha}} \circ \pi_1 + dF^2_{\boldsymbol{\alpha}} \circ \pi_2$ is the differential of F considered as a function of the first two variables when the others are held fixed. Since it is the sum of continuous functions, it is itself continuous in $\boldsymbol{\alpha}$, and we can now apply the theorem again to add $dF^3_{\mathbf{a}}$ to this sum partial differential, concluding that $\sum_1^3 dF^i_{\mathbf{a}} \circ \pi_i$ is the partial differential of F on the factor space $V_1 \times V_2 \times V_3$, and so on (which is colloquial for induction). □

As an illustration of the use of these two theorems, we shall deduce the general product rule (although a direct proof based on Δ-estimates is perfectly feasible). A general product is simply a bounded bilinear mapping $\omega: X \times Y \to W$, where X, Y, and W are all normed linear spaces. The boundedness inequality here is $\|\omega(\xi, \eta)\| \leq b\|\xi\| \, \|\eta\|$.

We first show that ω is differentiable.

Lemma 8.3. A bounded bilinear mapping $\omega: X \times Y \to W$ is everywhere differentiable and $d\omega_{<\alpha,\beta>}(\xi, \eta) = \omega(\alpha, \eta) + \omega(\xi, \beta)$.

Proof. With β held fixed, $g_\beta(\xi) = \omega(\xi, \beta)$ is in $\text{Hom}(X, W)$ and therefore is everywhere differentiable and equal to its own differential. That is, $d\omega^1$ exists and $d\omega^1_{<\alpha,\beta>}(\xi) = \omega(\xi, \beta)$. Since $\beta \mapsto g_\beta$ is a bounded linear mapping, $d\omega^1_{<\alpha,\beta>} = g_\beta$ is a continuous function of $<\alpha, \beta>$. Similarly, $d\omega^2_{<\alpha,\beta>}(\eta) = \omega(\alpha, \eta)$, and $d\omega^2$ is continuous. The lemma is now a direct corollary of Theorem 8.2. □

If $\omega(\xi, \eta)$ is thought of as a product of ξ and η, then the product of two functions $g(\zeta)$ and $h(\zeta)$ is $\omega(g(\zeta), h(\zeta))$, where g is from an open subset A of a normed linear space V to X and h is from A to Y. The product rule is now just what would be expected: the differential of the product is the first times the differential of the second plus the second times the differential of the first.

Theorem 8.4. If $g: A \to X$ and $h: A \to Y$ are differentiable at β, then so is the product $F(\zeta) = \omega(g(\zeta), h(\zeta))$ and

$$dF_\beta(\zeta) = \omega(g(\beta), dh_\beta(\zeta)) + \omega(dg_\beta(\zeta), h(\beta)).$$

Proof. This is a direct corollary of Theorem 8.1, Lemma 8.3, and the chain rule. □

EXERCISES

8.1 Find the tangent vector to the arc $<\sin t, \cos t, t^2>$ at $t = 0$; at $t = \pi/2$. What is the differential of the above parametrized arc at the two given points? That is, if $f(t) = <\sin t, \cos t, t^2>$, what are df_0 and $df_{\pi/2}$?

8.2 Give the detailed proof of Lemma 8.1.

8.3 The formula

$$dF_{\mathbf{a}}(\xi) = \sum_1^n dF^i_{\mathbf{a}}(\xi_i)$$

is probably obvious in view of the identity

$$\xi = \sum_1^n \theta_i(\xi_i)$$

and the definition of partial differentials, but write out an explicit, detailed proof anyway.

8.4 Let F be a differentiable mapping from an n-dimensional vector space V to a finite-dimensional vector space W, and define $G: V \times W \to W$ by $G(\xi, \eta) = \eta - F(\xi)$. Thus the graph of F in $V \times W$ is the null set of G. Show that the null space of $dG_{<\alpha,\beta>}$ has dimension n for every $<\alpha, \beta> \in V \times W$.

8.5 Let $F(\xi, \eta)$ be a continuously differentiable function defined on a product $A \times B$, where B is a ball and A is an open set. Suppose that $dF^2_{<\alpha,\beta>} = 0$ for all $<\alpha, \beta>$ in $A \times B$. Prove that F is independent of η. That is, show that there is a continuously differentiable function $G(\xi)$ defined on A such that $F(\xi, \eta) = G(\xi)$ on $A \times B$.

8.6 By considering a domain in $\mathbb{R}^2$ as indicated at the right, show that there exists a function $f(x, y)$ on an open set A in $\mathbb{R}^2$ such that

$$\frac{\partial f}{\partial y} = 0$$

everywhere and such that $f(x, y)$ is *not* a function of x alone.

8.7 Let $F(\xi, \eta, \zeta)$ be any function of three vector variables, and for fixed γ set $G(\xi, \eta) = F(\xi, \eta, \gamma)$. Prove that the partial differential $dF^1_{<\alpha,\beta,\gamma>}$ exists if and only if $dG^1_{<\alpha,\beta>}$ exists, in which case they are equal.

8.8 Give a more careful proof of Theorem 8.3. That is, state the inductive hypothesis and show that the theorem follows from it and Theorem 8.2. If you are meticulous in your argument, you will need a form of the above exercise.

8.9 Let f be a differentiable mapping from $\mathbb{R}^2$ to $\mathbb{R}$. Regarding $\mathbb{R}^2$ as $\mathbb{R} \times \mathbb{R}$, show that the two partial differentials of f are simply multiplication by its partial derivatives. Generalize to n dimensions. Show that the above is still true for a map F from $\mathbb{R}^2$ to a general vector space V, the partial derivatives now being vectors.

8.10 Give the details of the proof of Theorem 8.4.

9. THE DIFFERENTIAL AND $\mathbb{R}^n$

We shall now apply the results of the last two sections to mappings involving the Cartesian spaces $\mathbb{R}^n$, the bread and butter spaces of finite-dimensional theory. We start with the domain.

Theorem 9.1. If F is a mapping from (an open subset of) $\mathbb{R}^n$ to a normed linear space W, then the directional derivative of F in the direction of the jth standard basis vector δ^j is just the partial derivative $\partial F/\partial x_j$, and the jth partial differential is multiplication by $\partial F/\partial x_j$: $dF^j_{\mathbf{a}}(h) = h(\partial F/\partial x_j)(\mathbf{a})$. More exactly, if any one of the above three objects exists at $\mathbf{a}$, then they all do, with the above relationships.

Proof. We have

$$\begin{aligned}\frac{\partial F}{\partial x_j}(\mathbf{a}) &= \lim_{t\to 0}\frac{F(a_1,\ldots,a_j+t,\ldots,a_n)-F(a_1,\ldots,a_j,\ldots,a_n)}{t}\\ &= \lim_{t\to 0}\frac{F(\mathbf{a}+t\delta^j)-F(\mathbf{a})}{t} = D_{\delta^j}F(\mathbf{a}).\end{aligned}$$

Moreover, since the restriction of F to $\mathbf{a}+\mathbb{R}\delta^j$ is a parametrized arc whose differential at 0 is by definition the jth partial differential of F at $\mathbf{a}$ and whose tangent vector at 0 we have just computed to be $(\partial F/\partial x_j)(\mathbf{a})$, the remainder of the theorem follows from Theorem 7.1. □

Combining this theorem and Theorem 7.2, we obtain the following result.

Theorem 9.2. If $V = \mathbb{R}^n$ and F is differentiable at $\mathbf{a}$, then the partial derivatives $(\partial F/\partial x_j)(\mathbf{a})$ all exist and the n-tuple of partial derivatives at $\mathbf{a}$, $\{(\partial F/\partial x_j)(\mathbf{a})\}_1^n$, is the skeleton of $dF_{\mathbf{a}}$. In particular,

$$D_{\mathbf{y}}F(\mathbf{a}) = \sum_{j=1}^{n} y_j\frac{\partial F}{\partial x_j}(\mathbf{a}).$$

Proof. Since $dF_{\mathbf{a}}(\delta^i) = D_{\delta^i}F(\mathbf{a}) = (\partial F/\partial x_i)(\mathbf{a})$, as we noted above, we have

$$D_{\mathbf{y}}F(\mathbf{a}) = dF_{\mathbf{a}}(\mathbf{y}) = dF_{\mathbf{a}}\left(\sum_1^n y_i\delta^i\right) = \sum_1^n y_i\,dF_{\mathbf{a}}(\delta^i) = \sum_1^n y_i\frac{\partial F}{\partial x_i}(\mathbf{a}).$$

All that we have done here is to display $dF_{\mathbf{a}}$ as the linear combination mapping defined by its skeleton $\{dF_{\mathbf{a}}(\delta^i)\}$ (see Theorem 1.2 of Chapter 1), where $T(\delta^i) = dF_{\mathbf{a}}(\delta^i)$ is now recognized as the partial derivative $(\partial F/\partial x_i)(\mathbf{a})$. □

The above formula shows the barbarism of the classical notation for partial derivatives: note how it comes out if we try to evaluate $dF_{\mathbf{a}}(\mathbf{x})$. The notation $D_{\delta^j}F$ is precise but cumbersome. Other notations are F_j and D_jF. Each has its problems, but the second probably minimizes the difficulties. Using it, our formula reads $dF_{\mathbf{a}}(\mathbf{y}) = \sum_{j=1}^n y_jD_jF(\mathbf{a})$.

In the opposite direction we have the corresponding specialization of Theorem 8.3.

Theorem 9.3. If A is an open subset of $\mathbb{R}^n$, and if F is a mapping from A to a normed linear space W such that all of the partial derivatives $(\partial F/\partial x_j)(\mathbf{a})$ exist and are continuous on A, then F is continuously differentiable on A.

Proof. Since the jth partial differential of F is simply multiplication by $\partial F/\partial x_j$, we are (by Theorem 9.1) assuming the existence and continuity of all the partial differentials $dF_{\mathbf{a}}^j$ on A. Theorem 9.3 thus becomes a special case of Theorem 8.3. □

Now suppose that the range space of F is also a Cartesian space, so that F is a mapping from an open subset A of $\mathbb{R}^n$ to $\mathbb{R}^m$. Then $dF_{\mathbf{a}}$ is in $\mathrm{Hom}(\mathbb{R}^n, \mathbb{R}^m)$.

For computational purposes we want to represent linear maps from $\mathbb{R}^n$ to $\mathbb{R}^m$ by their matrices, and it is therefore of the utmost importance to find the matrix $\mathbf{t}$ of the differential $T = dF_{\mathbf{a}}$. This matrix is called the *Jacobian* matrix of F at $\mathbf{a}$.

The columns of $\mathbf{t}$ form the skeleton of $dF_{\mathbf{a}}$, and we saw above that this skeleton is the n-tuple of partial derivatives $(\partial F/\partial x_j)(\mathbf{a})$. If we write the m-tuple-valued F loosely as an m-tuple of functions, $F = \langle f_1, \ldots, f_m \rangle$, then according to Lemma 8.1, the jth column of $\mathbf{t}$ is the m-tuple

$$\frac{\partial F}{\partial x_j}(\mathbf{a}) = \left\langle \frac{\partial f_1}{\partial x_j}(\mathbf{a}), \ldots, \frac{\partial f_m}{\partial x_j}(\mathbf{a}) \right\rangle.$$

Thus,

Theorem 9.4. Let F be a mapping from an open subset of $\mathbb{R}^n$ to $\mathbb{R}^m$, and suppose that F is differentiable at $\mathbf{a}$. Then the matrix of $dF_{\mathbf{a}}$ (the Jacobian matrix of F at $\mathbf{a}$) is given by

$$t_{ij} = \frac{\partial f_i}{\partial x_j}(\mathbf{a}).$$

If we use the notation $y_i = f_i(\mathbf{x})$, we have

$$t_{ij} = \frac{\partial y_i}{\partial x_j}(\mathbf{a}).$$

If we also have a differentiable map $\mathbf{z} = G(\mathbf{y}) = \langle g_1(\mathbf{y}), \ldots, g_l(\mathbf{y}) \rangle$ from an open set $B \subset \mathbb{R}^m$ into $\mathbb{R}^l$, then $dG_{\mathbf{b}}$ has, similarly, the matrix

$$\frac{\partial g_k}{\partial y_i}(\mathbf{b}) = \frac{\partial z_k}{\partial y_i}(\mathbf{b}).$$

Also, if B contains $\mathbf{b} = F(\mathbf{a})$, then the composite-function rule

$$d(G \circ F)_{\mathbf{a}} = dG_{\mathbf{b}} \circ dF_{\mathbf{a}}$$

has the matrix form

$$\frac{\partial z_k}{\partial x_j}(\mathbf{a}) = \sum_{i=1}^{m} \frac{\partial z_k}{\partial y_i}(\mathbf{b}) \frac{\partial y_i}{\partial x_j}(\mathbf{a}),$$

or simply

$$\frac{\partial z_k}{\partial x_j} = \sum_{i=1}^{m} \frac{\partial z_k}{\partial y_i} \frac{\partial y_i}{\partial x_j}.$$

This is the usual form of the chain rule in the calculus. We see that it is merely the expression of the composition of linear maps as matrix multiplication.

We saw in Section 8 that the ordinary derivative $f'(a)$ of a function f of one real variable is the skeleton of the differential df_a, and it is perfectly reasonable to generalize this relationship and define the derivative $F'(\mathbf{a})$ of a function F of n real variables to be the skeleton of $dF_{\mathbf{a}}$, so that $F'(\mathbf{a})$ is the n-tuple of partial derivatives $\{(\partial F/\partial x_i)(\mathbf{a})\}_1^n$, as we saw above. In particular, if F is from an open subset of $\mathbb{R}^n$ to $\mathbb{R}^m$, then $F'(\mathbf{a})$ is the Jacobian matrix of F at $\mathbf{a}$. This gives the matrix chain rule the standard form

$$(G \circ F)'(\mathbf{a}) = G'(F(\mathbf{a}))F'(\mathbf{a}).$$

Some authors use the word 'derivative' for what we have called the differential, but this is a change from the traditional meaning in the one-variable case, and we prefer to maintain the distinction as discussed above: the differential dF_α is the linear map approximating ΔF_α and the derivative $F'(\mathbf{a})$ must be the matrix of this linear map when the domain and range spaces are Cartesian. However, we shall stay with the language of Jacobians.

Suppose now that A is an open subset of a finite-dimensional vector space V and that $H: A \to W$ is differentiable at $\alpha \in A$. Suppose that W is also finite-dimensional and that $\varphi: V \to \mathbb{R}^n$ and $\psi: W \to \mathbb{R}^m$ are any coordinate isomorphisms. If $\overline{A} = \varphi[A]$, then $\overline{A}$ is an open subset of $\mathbb{R}^n$ and $\overline{H} = \psi \circ H \circ \varphi^{-1}$ is a mapping from $\overline{A}$ to $\mathbb{R}^m$ which is differentiable at $\mathbf{a} = \varphi(\alpha)$, with $d\overline{H}_\mathbf{a} = \psi \circ dH_\alpha \circ \varphi^{-1}$. Then $d\overline{H}_\mathbf{a}$ is given by its Jacobian matrix $\{(\partial h^i/\partial x_j)(\mathbf{a})\}$, which we now call the *Jacobian matrix of H with respect to the chosen bases in V and W*. Change of bases in V and W changes the Jacobian matrix according to the rule given in Section 2.4.

If F is a mapping from $\mathbb{R}^n$ to itself, then the determinant of the Jacobian matrix $(\partial f^i/\partial x_j)(\mathbf{a})$ is called the *Jacobian* of F at $\mathbf{a}$. It is designated

$$\frac{\partial(f^1, \ldots, f^n)}{\partial(x_1, \ldots, x_n)}(\mathbf{a}), \qquad \text{or} \qquad \frac{\partial(y_1, \ldots, y_n)}{\partial(x_1, \ldots, x_n)}(\mathbf{a})$$

if it is understood that $y_i = f^i(\mathbf{x})$. Another notation is $J_F(\mathbf{a})$ (or simply $J(\mathbf{a})$ if F is understood). However, this is sometimes used to indicate the differential $dF_\mathbf{a}$, and we shall write $\det J_F(\mathbf{a})$ instead.

If $F(\mathbf{x}) = \langle x_1^2 - x_2^2, 2x_1x_2 \rangle$, then its Jacobian matrix is

$$\begin{bmatrix} 2x_1 & -2x_2 \\ 2x_2 & 2x_1 \end{bmatrix},$$

and $\det J(\mathbf{x}) = 4(x_1^2 + x_2^2) = 4(\|\mathbf{x}\|_2)^2$.

EXERCISES

9.1 By analogy with the notion of a parametrized arc, we define a smooth parametrized two-dimensional surface in a normed linear space W to be a continuously differentiable map Γ from a rectangle $I \times J$ in $\mathbb{R}^2$ to W. Suppose that $I \times J = [-1, 1] \times [-1, 1]$, and invent a definition of the tangent space to the range of Γ in W at the point $\Gamma(0, 0)$. Show that the two vectors

$$\frac{\partial \Gamma}{\partial x}(0, 0) \qquad \text{and} \qquad \frac{\partial \Gamma}{\partial y}(0, 0)$$

are a basis for this tangent space. (This should not have been your definition.)

9.2 Generalize the above exercise to a smooth parametrized n-dimensional surface in a normed linear space W.

9.3 Compute the Jacobian matrix of the mapping $\langle x, y\rangle \mapsto \langle x^2, y^2, (x+y)^2\rangle$. Show that its rank is two except at the origin.

9.4 Let $F = \langle f^1, f^2, f^3\rangle$ from $\mathbb{R}^3$ to $\mathbb{R}^3$ be defined by

$$f_1(x, y, z) = x + y + z, \qquad f_2(x, y, z) = x^2 + y^2 + z^2,$$

and

$$f_3(x, y, z) = x^3 + y^3 + z^3.$$

Compute the Jacobian of F at $\langle a, b, c\rangle$. Show that it is nonsingular unless two of the three coordinates are equal. Describe the locus of its singularities.

9.5 Compute the Jacobian of the mapping $F: \langle x, y\rangle \mapsto \langle (x+y)^2, y^3\rangle$ from $\mathbb{R}^2$ to $\mathbb{R}^2$ at $\langle 1, -1\rangle$; at $\langle 1, 0\rangle$; at $\langle a, b\rangle$. Compute the Jacobian of $G: \langle s, t\rangle \mapsto \langle s - t, s + t\rangle$ at $\langle u, v\rangle$.

9.6 In the above exercise compute the compositions $F \circ G$ and $G \circ F$. Compute the Jacobian of $F \circ G$ at $\langle y, v\rangle$. Compute the corresponding product of the Jacobians of F and G.

9.7 Compute the Jacobian matrix and determinant of the mapping T defined by $x = r\cos\theta$, $y = r\sin\theta$, $z = z$. Composing a function $f(x, y, z)$ with this mapping gives a new function:

$$g(r, \theta, z) = f(r\cos\theta, r\sin\theta, z).$$

That is, $g = f \circ T$. This composition (substitution) is called the change to cylindrical coordinates in $\mathbb{R}^3$.

9.8 Compute the Jacobian determinant of the polar coordinate transformation $\langle r, \theta\rangle \mapsto \langle x, y\rangle$, where $x = r\cos\theta$, $y = r\sin\theta$.

9.9 The transformation to spherical coordinates is given by $x = r\sin\varphi\cos\theta$, $y = r\sin\varphi\sin\theta$, $z = r\cos\theta$. Compute the Jacobian

$$\frac{\partial(x, y, z)}{\partial(r, \varphi, \theta)}.$$

9.10 Write out the chain rule for the following special cases:

$$dw/dt = ?, \qquad \text{where} \qquad w = F(x, y), \quad x = g(t), \quad y = h(t).$$

Find dw/dt when $w = F(x_1, \ldots, x_n)$ and $x_i = g_i(t)$, $i = 1, \ldots, n$. Find $\partial w/\partial u$ when $w = F(x, y)$, $x = g(u, v)$, $y = h(u, v)$. The special case where $g(u, v) = u$ can be rewritten

$$\frac{\partial}{\partial x} F(x, h(x, v)).$$

Compute it.

9.11 If $w = f(x, y)$, $x = r\cos\theta$, and $y = r\sin\theta$, show that

$$\left[r\frac{\partial w}{\partial r}\right]^2 + \left[\frac{\partial w}{\partial \theta}\right]^2 = \left[\frac{\partial w}{\partial x}\right]^2 + \left[\frac{\partial w}{\partial y}\right]^2.$$

10. ELEMENTARY APPLICATIONS

The elementary max-min theory from the standard calculus generalizes with little change, and we include a brief discussion of it at this point.

Theorem 10.1. Let F be a real-valued function defined on an open subset A of a normed linear space V, and suppose that F assumes a relative maximum value at a point α in A where dF_α exists. Then $dF_\alpha = 0$.

Proof. By definition $D_\xi F(\alpha)$ is the derivative $\gamma'(0)$ of the function $\gamma(t) = F(\alpha + t\xi)$, and the domain of γ is a neighborhood of 0 in $\mathbb{R}$. Since γ has a relative maximum value at 0, we have $\gamma'(0) = 0$ by the elementary calculus. Thus $dF_\alpha(\xi) = D_\xi F(\alpha) = 0$ for all ξ, and so $dF_\alpha = 0$. □

A point α such that $dF_\alpha = 0$ is called a *critical point*. The theorem states that a differentiable real-valued function can have an interior extremal value only at a critical point.

If V is $\mathbb{R}^n$, then the above argument shows that a real-valued function F can have a relative maximum (or minimum) at $\mathbf{a}$ only if the partial derivatives $(\partial F/\partial x_i)(\mathbf{a})$ are all zero, and, as in the elementary calculus, this often provides a way of calculating maximum (or minimum) values. Suppose, for example, that we want to show that the cube is the most efficient rectangular parallelepiped from the point of view of minimizing surface area for a given volume V. If the edges are x, y and z, we have $V = xyz$ and $A = 2(xy + xz + yz) = 2(xy + V/y + V/x)$. Then from $0 = \partial A/\partial x = 2(y - V/x^2)$, we see that $V = yx^2$, and, similarly, $\partial A/\partial y = 0$ implies that $V = xy^2$. Therefore, $yx^2 = xy^2$, and since neither x nor y can be 0, it follows that $x = y$. Then $V = yx^2 = x^3$, and $x = V^{1/3} = y$. Finally, substituting in $V = xyz$ shows that $z = V^{1/3}$. Our critical configuration is thus a cube, with minimum area $A = 6V^{2/3}$.

It was assumed above that A *has* an absolute minimum at some point $\langle x, y, z \rangle$. The reader might enjoy showing that $A \to \infty$ if any of x, y, z tends to 0 or ∞, which implies that the minimum does indeed exist.

We shall return to the problem of determining critical points in Sections 12, 15, and 16.

The condition $dF_\alpha = 0$ is necessary but not sufficient for an interior maximum or minimum. The reader will remember a sufficient condition from beginning calculus: If $f'(x) = 0$ and $f''(x) < 0$ (>0), then x is a relative maximum (minimum) point for f. We shall prove the corresponding general theorem in Section 16. There are more possibilities now; among them we have the analogous sufficient condition that if $dF_\alpha = 0$ and d^2F_α is negative (positive) definite as a quadratic form on V, then α is a relative maximum (minimum) point of F.

We consider next the notion of a tangent plane to a graph. The calculation of tangent lines to curves and tangent planes to surfaces is ordinarily considered a geometric application of the derivative, and we take this as sufficient justification for considering the general question here.

Let F be a mapping from an open subset A of a normed linear space V to a normed linear space W. When we view F as a graph in $V \times W$, we think of it as a "surface" S lying "over" the domain A, generalizing the geometric interpretation of the graph of a real-valued function of two real variables in $\mathbb{R}^3 = \mathbb{R}^2 \times \mathbb{R}$. The projection $\pi_1 \colon V \times W \to V$ projects S "down" onto A,

$$\langle \xi, F(\xi) \rangle \underset{\pi_1}{\mapsto} \xi,$$

and the mapping $\xi \mapsto \langle \xi, F(\xi) \rangle$ gives the point of S lying "over" ξ. Our geometric imagery views V as the plane (subspace) $V \times \{0\}$ in $V \times W$, just as we customarily visualize $\mathbb{R}$ as the real axis $\mathbb{R} \times \{0\}$ in $\mathbb{R}^2$.

We now assume that F is differentiable at α. Our preliminary discussion in Section 6 suggested that (the graph of) the linear function dF_α is the tangent plane to (the graph of) the function ΔF_α in $V \times W$, and that its translate M through $\langle \alpha, F(\alpha) \rangle$ is the tangent plane at $\langle \alpha, F(\alpha) \rangle$ to the surface S that is (the graph of) F. The equation of this plane is $\eta - F(\alpha) = dF_\alpha(\xi - \alpha)$, and it is accordingly (the graph of) the affine function $G(\xi) = dF_\alpha(\xi - \alpha) + F(\alpha)$. Now we know that dF_α is the unique T in $\mathrm{Hom}(V, W)$ such that $\Delta F_\alpha(\zeta) = T(\zeta) + \mathfrak{o}(\zeta)$, and if we set $\zeta = \xi - \alpha$, it is easy to see that this is the same as saying that G is the unique affine map from V to W such that

$$F(\xi) - G(\xi) = \mathfrak{o}(\xi - \alpha).$$

That is, M is the unique plane over V that "fits" the surface S around $\langle \alpha, F(\alpha) \rangle$ in the sense of $\mathfrak{o}$-approximation.

However, there is one further geometric fact that greatly strengthens our feeling that this *really* is the tangent plane.

Theorem 10.2. The plane with equation $\eta - F(\alpha) = dF_\alpha(\xi - \alpha)$ is exactly the union of all the straight lines through $\langle \alpha, F(\alpha) \rangle$ in $V \times W$ that are tangent to smooth curves on the surface $S = \text{graph } F$ passing through this point. In other words, the vectors in the subspace dF_α of $V \times W$ are exactly the tangent vectors to curves lying in S and passing through $\langle \alpha, F(\alpha) \rangle$.

Proof. This is nearly trivial. If $\langle \xi, \eta \rangle \in dF_\alpha$, then the arc

$$\gamma(t) = \langle \alpha + t\xi, F(\alpha + t\xi) \rangle$$

in S lying over the line $t \mapsto \alpha + t\xi$ in V has $\langle \xi, dF_\alpha(\xi) \rangle = \langle \xi, \eta \rangle$ as its tangent vector at $\langle \alpha, F(\alpha) \rangle$, by Lemma 8.1 and Theorem 8.2.

Conversely, if $t \mapsto \langle \lambda(t), F(\lambda(t)) \rangle$ is any smooth arc in S passing through α, with $\alpha = \lambda(t_0)$, then its tangent vector at $\langle \alpha, F(\alpha) \rangle$ is

$$\langle \lambda'(t_0), dF_\alpha(\lambda'(t_0)) \rangle,$$

a vector in (the graph of) dF_α. $\square$

As an example of the general tangent plane discussed above, let $F = \langle f_1, f_2 \rangle$ be the map from $\mathbb{R}^2$ to $\mathbb{R}^2$ defined by $f_1(\mathbf{x}) = (x_1^2 - x_2^2)/2$, $f_2(\mathbf{x}) = x_1x_2$. The graph of F is a surface over $\mathbb{R}^2$ in $\mathbb{R}^4 = \mathbb{R}^2 \times \mathbb{R}^2$. According to our above discussion, the tangent plane at $\langle \mathbf{a}, F(\mathbf{a}) \rangle$ has the equation $\mathbf{y} = dF_{\mathbf{a}}(\mathbf{x} - \mathbf{a}) + F(\mathbf{a})$. At $\mathbf{a} = \langle 1, 2 \rangle$ the Jacobian matrix of $dF_{\mathbf{a}}$ is

$$\begin{bmatrix} x_1 & -x_2 \\ x_2 & x_1 \end{bmatrix}_{\langle 1,2 \rangle} = \begin{bmatrix} 1 & -2 \\ 2 & 1 \end{bmatrix},$$

and $F(\mathbf{a}) = \langle -\frac{3}{2}, 2 \rangle$. The equation of the tangent plane M at $\langle 1, 2 \rangle$ is thus

$$\langle y_1, y_2 \rangle = \begin{bmatrix} 1 & -2 \\ 2 & 1 \end{bmatrix} \langle x_1 - 1, x_2 - 2 \rangle + \langle -\tfrac{3}{2}, 2 \rangle.$$

Computing the matrix product, we have the scalar equations

$$\begin{aligned} y_1 &= x_1 - 2x_2 + (-1 \quad +4 \quad -\tfrac{3}{2}) = x_1 - 2x_2 + \tfrac{3}{2}, \\ y_2 &= 2x_1 + x_2 + (-2 \quad -2 \quad +2) = 2x_1 + x_2 - 2. \end{aligned}$$

Note that these two equations present the affine space M as the intersection of the hyperplane in $\mathbb{R}^4$ consisting of all $\langle x_1, x_2, y_1, y_2 \rangle$ such that

$$x_1 - 2x_2 - y_1 = -\tfrac{3}{2},$$

with the hyperplane having the equation

$$2x_1 + x_2 - y_2 = 2.$$

EXERCISES

10.1 Find the maximum value of $f(x, y, z) = x + y + z$ on the ellipsoid

$$x^2 + 2y^2 + 3z^2 = 1.$$

10.2 Find the maximum value of the linear functional $f(\mathbf{x}) = \sum_1^n c_i x_i$ on the unit sphere $\sum_1^n x_i^2 = 1$.

10.3 Find the (minimum) distance between the two lines

$$\mathbf{x} = t\delta^1 \qquad \text{and} \qquad \mathbf{y} = s\langle 1, 1, 1 \rangle + \langle 1, 0, -1 \rangle$$

in $\mathbb{R}^3$.

10.4 Show that there is a uniquely determined pair of closest points on the two lines $\mathbf{x} = t\mathbf{a} + \mathbf{l}$ and $\mathbf{y} = s\mathbf{b} + \mathbf{m}$ in $\mathbb{R}^n$ unless $\mathbf{b} = k\mathbf{a}$ for some k. We assume that $\mathbf{a} \neq 0 \neq \mathbf{b}$. Remember that if $\mathbf{b}$ is not of the form $k\mathbf{a}$, then $|(\mathbf{a}, \mathbf{b})| < \|\mathbf{a}\|_2 \, \|\mathbf{b}\|_2$, according to the Schwarz inequality.

10.5 Show that the origin is the only critical point of $f(x, y, z) = xy + yz + zx$. Find a line through the origin along which 0 is a maximum point for f, and find another line along which 0 is a minimum point.

10.6 In the problem of minimizing the area of a rectangular parallelepiped of given volume V worked out in the text, it was assumed that

$$A = 2\left(xy + \frac{V}{x} + \frac{V}{y}\right)$$

has an absolute minimum at an interior point of the first quadrant. Prove this. Show first that $A \to \infty$ if $\langle x, y \rangle$ approaches the boundary in any way:

$$x \to 0, \quad x \to \infty, \quad y \to 0, \quad \text{or} \quad y \to \infty.$$

10.7 Let $F: \mathbb{R}^2 \to \mathbb{R}^2$ be the mapping defined by

$$y_1 = \sin(x_1 + x_2), \qquad y_2 = \cos(x_1 - x_2).$$

Find the equation of the tangent plane in $\mathbb{R}^4$ to the graph of F over the point $\mathbf{a} = \langle \pi/4, \pi/4 \rangle$.

10.8 Define $F: \mathbb{R}^3 \to \mathbb{R}^2$ by

$$y_1 = \sum_1^3 x_i^2, \qquad y_2 = \sum_1^3 x_i^3.$$

Find the equation of the tangent plane to the graph of F in $\mathbb{R}^5$ over $\mathbf{a} = \langle 1, 2, -1 \rangle$.

10.9 Let $\omega(\xi, \eta)$ be a bounded bilinear mapping from a product normed linear space $V \times W$ to a normed linear space X. Show that the equation of the tangent plane to the graph S of ω in $V \times W \times X$ at the point $\langle \alpha, \beta, \gamma \rangle \in S$ is

$$\zeta = \omega(\xi, \beta) + \omega(\alpha, \eta) + \omega(\alpha, \beta).$$

10.10 Let F be a bounded linear functional on the normed linear space V. Show that the equation of the tangent plane to the graph of F^3 in $V \times \mathbb{R}$ over the point α can be written in the form $y = F^2(\alpha)(3F(\xi) - 2F(\alpha))$.

10.11 Show that if the general equation for a tangent plane given in the text is applied to a mapping F in $\mathrm{Hom}(V, W)$, then it reduces to the equation for F itself $[\eta = F(\xi)]$, no matter where the point of tangency. (Naturally!)

10.12 Continuing Exercise 9.1, show that the tangent space to the range of Γ in W at $\Gamma(\mathbf{0})$ is the projection on W of the tangent space to the graph of Γ in $\mathbb{R}^2 \times W$ at the point $\langle \mathbf{0}, \Gamma(\mathbf{0}) \rangle$. Now define the tangent *plane* to range Γ in W at $\Gamma(\mathbf{0})$, and show that it is similarly the projection of the tangent plane to the graph of Γ.

10.13 Let $F: V \to W$ be differentiable at α. Show that the range of dF_α is the projection on W of the tangent space to the graph of F in $V \times W$ at the point $\langle \alpha, F(\alpha) \rangle$.

11. THE IMPLICIT-FUNCTION THEOREM

The formula for the Jacobian of a composite map that we obtained in Section 9 is reminiscent of the chain rule for the differential of a composite map that we derived earlier (Section 8). The Jacobian formula involves numbers (partial derivatives) that we multiply and add; the differential chain rule involves linear maps (partial differentials) that we *compose* and add. (The similarity becomes a full formal analogy if we use block decompositions.) Roughly speaking, the whole differential calculus goes this way. In the one-variable calculus a differential is a linear map from the one-dimensional space $\mathbb{R}$ to itself, and is therefore multiplication by a number, the derivative. In the many-variable calculus when we decompose with respect to one-dimensional subspaces, we get blocks of such numbers, i.e., Jacobian matrices. When we generalize the whole theory to vector spaces that are not one-dimensional, we get essentially the *same formulas* but with numbers replaced by linear maps (differentials) and multiplication by composition.

Thus the derivative of an inverse function is *the reciprocal* of the derivative of the function: if $g = f^{-1}$ and $b = f(a)$, then $g'(b) = 1/f'(a)$. The differential of an inverse map is the *composition inverse* of the differential of the map: if $G = F^{-1}$ and $F(\alpha) = \beta$, then $dG_\beta = (dF_\alpha)^{-1}$.

If the equation $g(x, y) = 0$ defines y implicitly as a function of x, $y = f(x)$, we learn to compute $f'(a)$ in the elementary calculus by differentiating

$$g(x, f(x)) \equiv 0,$$

and we get

$$\frac{\partial g}{\partial x}(a, b) + \frac{\partial g}{\partial y}(a, b)\, f'(a) = 0,$$

where $b = f(a)$. Hence

$$f'(a) = -\frac{\partial g/\partial x}{\partial g/\partial y}\cdot$$

We shall see below that if $G(\xi, \eta) = 0$ defines η as a function of ξ, $\eta = F(\xi)$, and if $\beta = F(\alpha)$, then we calculate the *differential* dF_α by differentiating the identity $G(\xi, F(\xi)) = 0$, and we get a formula formally identical to the above.

Finally, in exactly the same way, the so-called auxiliary variable method of solving max-min problems in the elementary calculus has the same formal structure as our later solution of a "constrained" maximum problem by Lagrange multipliers.

In this section we shall consider the existence and differentiability of functions implicitly defined. Suppose that we are given a (vector-valued) function $G(\xi, \eta)$ of two vector variables, and we want to know whether setting G equal to 0 defines η as a function of ξ, that is, whether there exists a unique function F such that $G(\xi, F(\xi))$ is identically zero. Supposing that such an "implicitly defined" function F exists and that everything is differentiable, we can try to compute the differential of F at α by differentiating the equation $G(\xi, F(\xi)) = 0$, or $G \circ \langle I, F\rangle = 0$. We get $dG^1_{\langle\alpha,\beta\rangle} \circ dI_\alpha + dG^2_{\langle\alpha,\beta\rangle} \circ dF_\alpha = 0$, where we have set $\beta = F(\alpha)$. If dG^2 is invertible, we can solve for dF_α, getting

$$dF_\alpha = -(dG^2_{\langle\alpha,\beta\rangle})^{-1} \circ dG^1_{\langle\alpha,\beta\rangle}.$$

Note that this has the same form as the corresponding expression from the elementary calculus that we reviewed above. If F is uniquely determined, then so is dF_α, and the above calculation therefore strongly suggests that we are going to need the existence of $(dG^2_{<\alpha,\beta>})^{-1}$ as a necessary condition for the existence of a uniquely defined implicit function around the point $<\alpha, \beta>$. Since β is $F(\alpha)$, we also need $G(\alpha, \beta) = 0$. These considerations will lead us to the right theorem, but we shall have to postpone part of its proof to the next chapter. What we can prove here is that if there is an implicitly defined function, then it must be differentiable.

Theorem 11.1. Let V, W, and X be normed linear spaces, and let G be a mapping from an open subset $A \times B$ of $V \times W$ to X. Suppose that F is a continuous mapping from A to B implicitly defined by the equation $G(\xi, \eta) = 0$, that is, satisfying $G(\xi, F(\xi)) = 0$ on A. Finally, suppose that G is differentiable at $<\alpha, \beta>$, where $\beta = F(\alpha)$, and that $dG^2_{<\alpha,\beta>}$ is invertible. Then F is differentiable at α and $dF_\alpha = -(dG^2_{<\alpha,\beta>})^{-1} \circ dG^1_{<\alpha,\beta>}$.

Proof. Set $\eta = \Delta F_\alpha(\xi)$, so that $G(\alpha + \xi, \beta + \eta) = G(\alpha + \xi, F(\alpha + \xi)) = 0$. Then

$$\begin{aligned} 0 &= G(\alpha + \xi, \beta + \eta) - G(\alpha, \beta) = \Delta G_{<\alpha,\beta>}(\xi, \eta) = dG_{<\alpha,\beta>}(\xi, \eta) + \mathfrak{o}(\xi, \eta) \\ &= dG^1_{<\alpha,\beta>}(\xi) + dG^2_{<\alpha,\beta>}(\eta) + \mathfrak{o}(\xi, \eta). \end{aligned}$$

Applying T^{-1} to this equation, where $T = dG^2_{<\alpha,\beta>}$, and solving for η, we get

$$\eta = -T^{-1}(dG^1_{<\alpha,\beta>}(\xi)) + \mathcal{O}(\mathfrak{o}(<\xi, \eta>)).$$

This equation is of the form $\eta = \mathcal{O}(\xi) + \mathfrak{o}(<\xi, \eta>)$, and since $\eta = \Delta F_\alpha(\xi)$ is an infinitesimal $\mathfrak{g}(\xi)$, by the continuity of F at α, Lemmas 5.1 and 5.2 imply first that $\eta = \mathcal{O}(\xi)$ and then that $<\xi, \eta> = \mathcal{O}(\xi)$. Thus $\mathcal{O}(\mathfrak{o}(<\xi, \eta>)) = \mathcal{O}((\mathfrak{o}\mathcal{O}(\xi))) = \mathfrak{o}(\xi)$, and we have

$$\Delta F_\alpha(\xi) = \eta = S(\xi) + \mathfrak{o}(\xi),$$

where $S = -(dG^2_{<\alpha,\beta>})^{-1} \circ dG^1_{<\alpha,\beta>}$, an element of $\mathrm{Hom}(V, W)$. Therefore, F is differentiable at α and dF_α has the asserted value. □

We shall show in the next chapter, as an application of the fixed-point theorem, that if V, W, and X are *finite-dimensional*, and if G is a *continuously differentiable* mapping from an open subset $A \times B$ of $V \times W$ to X such that at the point $<\alpha, \beta>$ we have both $G(\alpha, \beta) = 0$ and $dG^2_{<\alpha,\beta>}$ invertible, then there is a uniquely determined continuous mapping F from a neighborhood M of α to B such that $F(\alpha) = \beta$ and $G(\xi, F(\xi)) = 0$ on M. The same theorem is true for the more general class of *complete* normed linear spaces which we shall study in the next chapter. For these spaces it is also true that if T^{-1} exists, then so does S^{-1} for all S sufficiently close to T, and the mapping $S \mapsto S^{-1}$ is continuous. Therefore $dG^2_{<\mu,\nu>}$ is invertible for all $<\mu, \nu>$ sufficiently close to $<\alpha, \beta>$, and the above theorem then implies that F is differentiable on a neighborhood of α. Moreover, only continuous mappings are involved in the formula given by the theorem for $dF: \mu \mapsto dF_\mu$, and it follows that F is in fact continuously differentiable near α. These conclusions constitute the implicit-function theorem, which we now restate.

Theorem 11.2. Let V, W, and X be finite-dimensional (or, more generally, complete) normed linear spaces, let $A \times B$ be an open subset of $V \times W$, and let $G: A \times B \to X$ be continuously differentiable. Suppose that at the point $\langle \alpha, \beta \rangle$ in $A \times B$ we have both $G(\alpha, \beta) = 0$ and $dG^2_{\langle \alpha,\beta \rangle}$ invertible. Then there is a ball M about α and a uniquely defined continuously differentiable mapping F from M to B such that $F(\alpha) = \beta$ and $G(\xi, F(\xi)) = 0$ on M.

The so-called inverse-mapping theorem is a special case of the implicit-function theorem.

Theorem 11.3. Let H be a continuously differentiable mapping from an open subset B of a finite-dimensional (or complete) normed linear space W to a normed linear space V, and suppose that its differential is invertible at a point β. Then H itself is invertible near β. That is, there is a ball M about $\alpha = H(\beta)$ and a uniquely determined continuously differentiable function F from M to B such that $F(\alpha) = \beta$ and $H(F(\xi)) = \xi$ on M.

Proof. Set $G(\xi, \eta) = \xi - H(\eta)$. Then G is continuously differentiable from $V \times B$ to V and $dG^2_{\langle \alpha,\beta \rangle} = -dH_\beta$ is invertible. The implicit-function theorem then gives us a ball M about α and a uniquely determined continuously differentiable mapping F from M to B such that $F(\alpha) = \beta$ and $0 = G(\xi, F(\xi)) = \xi - H(F(\xi))$ on M. □

The inverse-mapping theorem is often given a slightly different formulation which we state as a corollary.

Corollary. Under the hypotheses of the above theorem there exists an open neighborhood U of β such that H is injective on U, $N = H[U]$ is open in V, and H^{-1} is continuously differentiable on N. (See Fig. 3.11.)

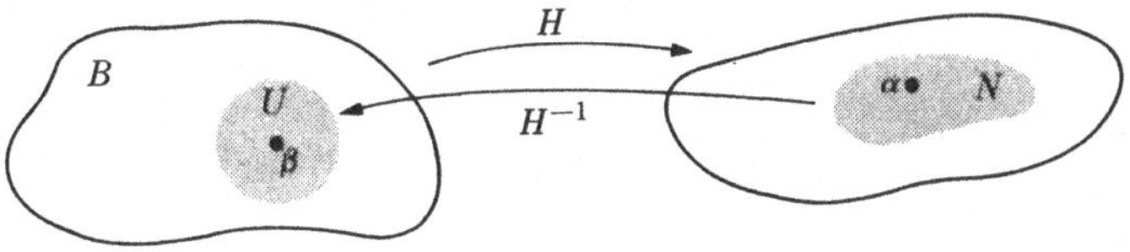

Fig. 3.11

Proof. The proof of the corollary is left as an exercise.

In practice we often have to apply the Cartesian formulations of these theorems. The student should certainly be able to write these down, but we shall state them anyway, starting with the simpler inverse-mapping theorem.

Theorem 11.4. Suppose that we are given n continuously differentiable real-valued functions $G_i(y_1, \ldots, y_n)$, $i = 1, \ldots, n$, of n real variables defined on a neighborhood B of a point $\mathbf{b}$ in $\mathbb{R}^n$ and suppose that the Jacobian determinant

$$\frac{\partial(G_1, \ldots, G_n)}{\partial(y_1, \ldots, y_n)}(\mathbf{b})$$

is not zero. Then there is a ball M about $\mathbf{a} = \mathbf{G}(\mathbf{b})$ in $\mathbb{R}^n$ and a uniquely determined n-tuple $\mathbf{F} = \langle F_1, \ldots, F_n \rangle$ of continuously differentiable real-valued functions defined on M such that $\mathbf{F}(\mathbf{a}) = \mathbf{b}$ and $\mathbf{G}(\mathbf{F}(\mathbf{x})) = \mathbf{x}$ on M for $i = 1, \ldots, n$. That is, $G_i(F_1(x_1, \ldots, x_n), \ldots, F_n(x_1, \ldots, x_n)) = x_i$ for all $\mathbf{x}$ in M and for $i = 1, \ldots, n$.

For example, if $\mathbf{x} = \langle y_1^3 + y_2^3, y_1^2 + y_2^2 \rangle$, then at the point $\mathbf{b} = \langle 1, 2 \rangle$ we have

$$\frac{\partial(x_1, x_2)}{\partial(y_1, y_2)} = \det \begin{bmatrix} 3y_1^2, & 3y_2^2 \\ 2y_1, & 2y_2 \end{bmatrix} \Bigg|_{\langle 1,2 \rangle}$$

$$= \det \begin{bmatrix} 3, & 12 \\ 2, & 4 \end{bmatrix} = -12 \neq 0,$$

and we therefore know without trying to solve explicitly that there is a unique solution for $\mathbf{y}$ in terms of $\mathbf{x}$ near $\mathbf{x} = \langle 1^3 + 2^3, 1^2 + 2^2 \rangle = \langle 9, 5 \rangle$. The reader would find it virtually impossible to solve for $\mathbf{y}$, since he would quickly discover that he had to solve a polynomial equation of degree 6. This clearly shows the power of the theorem: we are guaranteed the existence of a mapping which may be very difficult if not impossible to find explicitly. (However, in the next chapter we shall discover an iterative procedure for approximating the inverse mapping as closely as we want.)

Everything we have said here applies all the more to the implicit-function theorem, which we now state in Cartesian form.

Theorem 11.5. Suppose that we are given m continuously differentiable real-valued functions $G_i(\mathbf{x}, \mathbf{y}) = G_i(x_1, \ldots, x_n, y_1, \ldots, y_m)$ of $n + m$ real variables defined on an open subset $A \times B$ of $\mathbb{R}^{n+m}$ and an $(n + m)$-tuple $\langle \mathbf{a}, \mathbf{b} \rangle = \langle a_1, \ldots, a_n, b_1, \ldots, b_m \rangle$ such that $G_i(\mathbf{a}, \mathbf{b}) = 0$ for $i = 1, \ldots, m$, and such that the Jacobian determinant

$$\frac{\partial(G_1, \ldots, G_m)}{\partial(y_1, \ldots, y_m)} (\mathbf{a}, \mathbf{b})$$

is not zero. Then there is a ball M about $\mathbf{a}$ in $\mathbb{R}^n$ and a uniquely determined m-tuple $\mathbf{F} = \langle F_1, \ldots, F_m \rangle$ of continuously differentiable real-valued functions $F_j(\mathbf{x}) = F_j(x_1, \ldots, x_n)$ defined on M such that $\mathbf{b} = \mathbf{F}(\mathbf{a})$ and $G_i(\mathbf{x}, \mathbf{F}(\mathbf{x})) = 0$ on M for $i = 1, \ldots, m$. That is, $b_i = F_i(a_1, \ldots, a_n)$ for $i = 1, \ldots, m$, and $G_i(x_1, \ldots, x_n; F_1(x_1, \ldots, x_n), \ldots, F_m(x_1, \ldots, x_n))$ $= 0$ for all $\mathbf{x}$ in M and for $i = 1, \ldots, m$.

For example, the equations

$$x_1^3 + x_2^3 - y_1^2 - y_2^2 = 0,$$

$$x_1^2 - x_2^2 - y_1^3 - y_2^3 = 0$$

can be solved uniquely for $\mathbf{y}$ in terms of $\mathbf{x}$ near $\langle \mathbf{x}, \mathbf{y} \rangle = \langle 1, 1, 1, -1 \rangle$, because they hold *at* that point and because

$$\frac{\partial(G_1, G_2)}{\partial(y_1, y_2)} = \det\begin{bmatrix} -2y_1, & -2y_2 \\ -3y_1^2, & -3y_2^2 \end{bmatrix} = 6(y_1y_2^2 - y_2y_1^2)$$

has the value 12 there. Of course, we mean only that the solution functions exist, not that we can explicitly produce them.

EXERCISES

11.1 Show that $\langle x, y\rangle \mapsto \langle e^x + e^y, e^x + e^{-y}\rangle$ is locally invertible about any point $\langle a, b\rangle$, and compute the Jacobian matrix of the inverse map.

11.2 Show that $\langle u, v\rangle \mapsto \langle e^u + e^v, e^u - e^v\rangle$ is locally invertible about any point $\langle a, b\rangle$ in $\mathbb{R}^2$, by computing the Jacobian matrix. In this case the whole mapping is invertible, with an easily computed inverse. Make this calculation, compute the Jacobian matrix of the inverse map, and verify that the two matrices are inverses at the appropriate points.

11.3 Show that the mapping $\langle x, y, z\rangle \mapsto \langle \sin x, \cos y, e^z\rangle$ from $\mathbb{R}^3$ to $\mathbb{R}^3$ is locally invertible about $\langle 0, \pi/2, 0\rangle$. Show that

$$\langle x, y, z\rangle \mapsto \langle \sin(x + y + z), \cos(x - y + z), e^{(x+y-z)}\rangle$$

is locally invertible about $\langle \pi/4, -\pi/4, 0\rangle$.

11.4 Express the second map of the above exercise as the composition of two maps, and obtain your answer a second way.

11.5 Let $F: \langle x, y\rangle \mapsto \langle u, v\rangle$ be the mapping from $\mathbb{R}^2$ to $\mathbb{R}^2$ defined by $u = x^2 + y^2$, $v = 2xy$. Compute an inverse G of F, being careful to give the domain and range of G. How many inverse mappings are there? Compute the Jacobian matrices of F at $\langle 1, 2\rangle$ and of G at $\langle 5, 4\rangle$, and show by multiplying them that they are inverse.

11.6 Consider now the mapping $F: \langle x, y\rangle \mapsto \langle x^3, y^3\rangle$. Show that $dF_{\langle 0,0\rangle}$ is singular and yet that the mapping has an inverse G. What conclusion do we draw about the differentiability of G at the origin?

11.7 Define $F: \mathbb{R}^2 \to \mathbb{R}^2$ by $\langle x, y\rangle \mapsto \langle e^x \cos y, e^x \sin y\rangle$. Prove that F is locally invertible about every point.

11.8 Define $F: \mathbb{R}^3 \to \mathbb{R}^3$ by $\mathbf{x} \mapsto \mathbf{y}$ where

$$y_1 = x_1 + x_2^2 + (x_3 - 1)^4, \qquad y_2 = x_1^2 + x_2 + (x_3^3 - 3x_3), \qquad y_3 = x_1^3 + x_2^2 + x_3.$$

Prove that $\mathbf{x} \mapsto \mathbf{y} = F(\mathbf{x})$ is locally invertible about $\mathbf{x} = \langle 0, 0, 1\rangle$.

11.9 For a function $f: \mathbb{R} \to \mathbb{R}$ the proof of local invertibility around a point a where df_a is nonsingular is much simpler than the general case. Show first that the Jacobian matrix of f at a is the number $f'(a)$. We are therefore assuming that $f'(x)$ is continuous in a neighborhood of a and that $f'(a) \neq 0$. Prove that then f is strictly increasing (or decreasing) in an interval about a. Now finish the theorem. (See Exercise 1.12.)

11.10 Show that the equations

$$t^2 + x^3 + y^3 + z^3 = 0, \qquad t + x^2 + y^2 + z^2 = 4, \qquad 1 + x + y + z = 0$$

have differentiable solutions $x(t)$, $y(t)$, $z(t)$ around $\langle t, x, y, z \rangle = \langle 0, -1, 1, 0 \rangle$.

11.11 Show that the equations

$$e^x + e^{2y} + e^{3u} + e^{4v} = 4, \qquad e^x + e^y + e^u + e^v = 4$$

can be uniquely solved for u and v in terms of x and y around the point $\langle 0, 0, 0, 0 \rangle$.

11.12 Let S be the graph of the equation

$$xz + \sin(xy) + \cos(xz) = 1$$

in $\mathbb{R}^3$. Determine whether in the neighborhood of $(0, 1, 1)$ S is the graph of a differentiable function in any of the following forms:

$$z = f(x, y), \qquad x = g(y, z), \qquad y = h(x, z).$$

11.13 Given functions f and g from $\mathbb{R}^3$ to $\mathbb{R}$ such that $f(a, b, c) = 0$ and $g(a, b, c) = 0$, write down the condition on the partial derivatives of f and g that guarantees the existence of a unique pair of differentiable functions $y = h(x)$ and $z = k(x)$ satisfying

$$h(a) = b, \qquad k(a) = c,$$

and

$$f(x, y, z) = f(x, h(x), k(x)) = 0,$$
$$g(x, y, z) = g(x, h(x), k(x)) = 0 \qquad \text{around} \qquad \langle a, b, c \rangle.$$

11.14 Let $G(\xi, \eta, \zeta)$ be a continuously differentiable mapping from $V = \prod_1^3 V_i$ to W such that $dG^3_{\boldsymbol{\alpha}}: V_3 \to W$ is invertible and $G(\boldsymbol{\alpha}) = G(\alpha_1, \alpha_2, \alpha_3) = 0$. Prove that there exists a uniquely determined function $\zeta = F(\xi, \eta)$ defined around $\langle \alpha_1, \alpha_2 \rangle$ in $V_1 \times V_2$ such that $G(\xi, \eta, F(\xi, \eta)) = 0$ and $F(\alpha_1, \alpha_2) = \alpha_3$. Also show that

$$dF^1_{\langle \xi, \eta \rangle} = [-dG^3_{\langle \xi, \eta, \zeta \rangle}]^{-1}[dG^1_{\langle \xi, \eta, \zeta \rangle}],$$

where $\zeta = F(\xi, \eta)$.

11.15 Let $F(\xi, \eta)$ be a continuously differentiable function from $V \times W$ to X, and suppose that $dF^2_{\langle \alpha, \beta \rangle}$ is invertible. Setting $\gamma = F(\alpha, \beta)$, show that there is a product neighborhood $L \times M \times N$ of $\langle \gamma, \alpha, \beta \rangle$ in $X \times V \times W$ and a unique continuously differentiable mapping $G: L \times M \to N$ such that on $L \times M$, $F(\xi, G(\zeta, \xi)) = \zeta$.

11.16 Suppose that the equation $g(x, y, z) = 0$ can be solved for z in terms of x and y. This means that there is a function $f(x, y)$ such that $g(x, y, f(x, y)) = 0$. Suppose also that everything is differentiable and compute $\partial z/\partial x$.

11.17 Suppose that the equations

$$g(x, y, z) = 0 \qquad \text{and} \qquad h(x, y, z) = 0$$

can be solved for y and z as functions of x. Compute dy/dx.

11.18 Suppose that $g(x, y, u, v) = 0$ and $h(x, y, u, v) = 0$ can be solved for u and v as functions of x and y. Compute $\partial u/\partial x$.

11.19 Compute dz/dx where $x^3 + y^3 + z^3 = 0$ and $x^2 + y^2 + z^2 = 1$.

11.20 If $t^3 + x^3 + y^3 + z^3 = 0$ and $t^2 + x^2 + y^2 + z^2 = 1$, then $\partial z/\partial x$ is ambiguous. We are obviously going to think of two of the variables as functions of the other two.

Also z is going to be dependent and x independent. But is t or y going to be the other independent variable? Compute $\partial z/\partial x$ under each of these assumptions.

11.21 We are given four "physical variables" p, v, t, and φ such that each of them is a function of any two of the other three. Show that $\partial t/\partial p$ has two quite different meanings, and make explicit what the relationship between them is by labeling the various functions that are relevant and applying the implicit differentiation process.

11.22 Again the "one-dimensional" case is substantially simpler. Let G be a continuously differentiable mapping from $\mathbb{R}^2$ to $\mathbb{R}$ such that $G(a, b) = 0$ and

$$(\partial G/\partial y)(a, b) = G_2(a, b) > 0.$$

Show that there are positive numbers ϵ and δ such that for each c in $(a - \delta, a + \delta)$ the function $g(y) = G(c, y)$ is strictly increasing on $[b - \epsilon, b + \epsilon]$ and $G(c, b - \epsilon) < 0 < G(c, b + \epsilon)$. Conclude from the intermediate-value theorem (Exercise 1.13) that there exists a unique function $F\colon (a - \delta, a + \delta) \to (b - \epsilon, b + \epsilon)$ such that

$$G(x, F(x)) = 0.$$

11.23 By applying the same argument used in the above exercise a second time, prove that F is continuous.

11.24 In the inverse-function theorem show that $dF_\alpha = (dH_\beta)^{-1}$. That is, the differential of the inverse of H is the inverse of the differential of H. Show this

a) by applying the implicit-function theorem;
b) by a direct calculation from the identity $H(F(\xi)) = \xi$.

11.25 Again in the context of the inverse-mapping theorem, show that there is a neighborhood M of β in A such that $F(H(\eta)) = \eta$ on M. (Don't work at this. Just apply the theorem again.)

11.26 We continue in the context of the inverse-mapping theorem. Assume the result (from the next chapter) that if dH_β^{-1} exists, then so does dH_ξ^{-1}, for ξ sufficiently close to β. Show that there is an open neighborhood U of β in B such that H is injective on U, $H[U]$ is an open set N in V, and H^{-1} is continuously differentiable on N.

11.27 Use Exercise 3.21 to give a direct proof of the existence of a Lipschitz continuous local inverse in the context of the inverse-mapping theorem. [*Hint:* Apply Theorem 7.4.]

11.28 A direct proof of the differentiability of an inverse function is simpler than the implicit-function theorem proof. Work out such a proof, modeling your arguments in a general way upon those in Theorem 11.1.

11.29 Prove that the implicit-function theorem can be deduced from the inverse-function theorem as follows. Set

$$H(\xi, \eta) = \langle \xi, G(\xi, \eta) \rangle,$$

and show that $dH_{\langle\alpha,\beta\rangle}$ has the block diagram

$$\begin{array}{c|c} I & 0 \\ \hline dG^1 & dG^2 \end{array}$$

Then show that $dH_{\langle\alpha,\beta\rangle}{}^{-1}$ exists from the block diagram results of Chapter 1. Apply the inverse-mapping theorem.

12. SUBMANIFOLDS AND LAGRANGE MULTIPLIERS

If V and W are finite-dimensional spaces, with dimensions n and m, respectively, and if F is a continuous mapping from an open subset A of V to W, then (the graph of) F is a subset of $V \times W$ which we visualize as a kind of "n-dimensional surface" S spread out over A. (See Section 10.) We shall call F an n-dimensional *patch* in $V \times W$. More generally, if X is any $(n + m)$-dimensional vector space, we shall call a subset S an *n-dimensional patch* if there is an isomorphism φ from X to a product space $V \times W$ such that V is n-dimensional and $\varphi[S]$ is a patch in $V \times W$. That is, S becomes a patch in the above sense when X is considered to be $V \times W$. This means that if π_1 is the projection of $X = V \times W$ onto V, then $\pi_1[S]$ is an open subset A of V, and the restriction $\pi_1 \restriction S$ is one-to-one and has a continuous inverse. If π_2 is the projection on W, then $F = \pi_2 \circ (\pi_1 \restriction S)^{-1}$ is the map from A to W whose graph in $V \times W$ is S (when $V \times W$ is identified with X).

Now there are important surfaces that aren't such "patch" surfaces. Consider, for instance, the surface of the unit ball in $\mathbb{R}^3$, $S = \{\mathbf{x} : \sum_1^3 x^2 = 1\}$. S is obviously a two-dimensional surface in $\mathbb{R}^3$ which cannot be expressed as a graph, no matter how we try to express $\mathbb{R}^3$ as a direct sum. However, it should be equally clear that S *is* the union of overlapping surface patches. If α is any point on S, then any sufficiently small neighborhood N of α in $\mathbb{R}^3$ will intersect S in a patch; we take V as the subspace parallel to the tangent plane at α and W as the perpendicular line through 0. Moreover, this property of S is a completely adequate definition of what we mean by a submanifold.

A subset S of an $(n + m)$-dimensional vector space X is an *n-dimensional submanifold* of X if each α on S has a neighborhood N in X whose intersection with S is an n-dimensional patch.

We say that S is smooth if all these patches S_α are smooth, that is, if the function $F: A \to W$ whose graph in $V \times W$ is the patch S_α (when X is viewed as $V \times W$) is continuously differentiable for every such patch S_α.

The sphere we considered above is a two-dimensional smooth submanifold of $\mathbb{R}^3$.

Submanifolds are frequently presented as zero sets of mappings. For example, our sphere above is the zero set of the mapping G from $\mathbb{R}^3$ to $\mathbb{R}$ defined by $G(\mathbf{x}) = \sum_1^3 x_i^2 - 1$. It is obviously important to have a condition guaranteeing that such a null set is a submanifold.

Theorem 12.1. Let G be a continuously differentiable mapping from an open subset U of an $(n + m)$-dimensional vector space X to an m-dimensional vector space Y such that dG_α is surjective for every α in the zero set S of G. Then S is an n-dimensional submanifold of X.

Proof. Choose any point γ of S. Since dG_γ is surjective from the $(n + m)$-dimensional vector space X to the m-dimensional vector space Y, we know that the null space V of dG_γ has dimension n (Theorem 2.4, Chapter 2). Let W be any complement of V, and think of X as $V \times W$, so that G now becomes a function of two vector variables and γ is a point $\langle \alpha, \beta \rangle$ such that $G(\alpha, \beta) = 0$. The restriction of $dG_{\langle \alpha, \beta \rangle}$ to W is an isomorphism from W to Y; that is, $(dG^2_{\langle \alpha, \beta \rangle})^{-1}$ exists. Therefore, by the implicit-function theorem, there is a product neighborhood $S_\delta(\alpha) \times S_r(\beta)$ of $\langle \alpha, \beta \rangle$ in X whose intersection with S is the graph of a function on $S_\delta(\alpha)$. This proves our theorem. □

If S is a smooth submanifold, then the function F whose graph is the patch of S around γ (when X is viewed suitably as $V \times W$) is continuously differentiable, and therefore S has a uniquely determined n-dimensional tangent plane M at γ that fits S most closely around γ in the sense of our $\mathfrak{o}$-approximations. If $\gamma = 0$, this tangent plane is an n-dimensional subspace, and in general it is the translate through γ of a subspace N. We call N the tangent space of S at γ; its elements are exactly the vectors in X tangent to parametrized arcs drawn in S through γ. What we are going to do later is to describe an n-dimensional manifold S independently of any imbedding of S in a vector space. The tangent space to S at a point γ will still be an invaluable notion, but we are not going to be able to visualize it by an actual tangent plane in a space X carrying S. Instead, we will have to construct the vector space tangent to S at γ somehow.

The clue is provided by Theorem 10.2, which tells us that if S is imbedded as a submanifold in a vector space X, then each vector tangent to S at γ can be presented as the unique tangent vector at γ to some smooth curve lying in S. This mapping from the set of smooth curves in S through γ to the tangent space at γ is not injective; clearly, different curves can be tangent to each other at γ and so have the same tangent vector there. Therefore, the object in S that corresponds to a tangent vector at γ is an equivalence class of smooth curves through γ, and this will in fact be our *definition* of a tangent vector for a general manifold.

The notion of a submanifold allows us to consider in an elegant way a classical "constrained" maximum problem. We are given an open subset U of a finite-dimensional vector space X, a differentiable real-valued function F defined on U, and a submanifold S lying in U. We shall suppose that the submanifold S is the zero set of a continuously differentiable mapping G from U to a vector space Y such that dG_γ is surjective for each γ on S. We wish to consider the problem of maximizing (or minimizing) $F(\gamma)$ when γ is "constrained" to lie on S. We cannot expect to find such a maximum point γ_0 by setting $dF_\gamma = 0$ and solving for γ, because γ_0 will not be a critical point for F. Consider, for example, the function $g(\mathbf{x}) = \sum_1^3 x_i^2 - 1$ from $\mathbb{R}^3$ to $\mathbb{R}$ and $F(\mathbf{x}) = x_2$. Here the "surface" defined by $g = 0$ is the unit sphere $\sum_1^3 x_i^2 = 1$, and on this sphere F has its maximum value 1 at $\langle 0, 1, 0 \rangle$. But F is linear, and so $dF_\gamma = F$ can never be the zero transformation. The device known as Lagrange multipliers shows that we can nevertheless find such constrained critical points by solving $dL_\gamma = 0$ for a suitable function L.

Theorem 12.2. Suppose that F has a maximum value on S at the point γ. Then there is a functional l in Y^* such that γ is a critical point of the function $F - (l \circ G)$.

Proof. By the implicit-function theorem we can express X as $V \times W$ in such a way that the neighborhood of S around γ is the graph of a mapping H from an open set A in V to W. Thus, expressing F and G as functions on $V \times W$, we have $G(\xi, \eta) = 0$ near $\gamma = \langle \alpha, \beta \rangle$ if and only if $\eta = H(\xi)$, and the restriction of $F(\xi, \eta)$ to this zero surface is thus the function $K\colon A \to \mathbb{R}$ defined by $K(\xi) = F(\xi, H(\xi))$. By assumption α is a critical point for this function. Thus

$$0 = dK_\alpha = dF^1_{\langle\alpha,\beta\rangle} + dF^2_{\langle\alpha,\beta\rangle} \circ dH_\alpha.$$

Also from the identity $G(\xi, H(\xi)) = 0$, we get

$$0 = dG^1_{\langle\alpha,\beta\rangle} + dG^2_{\langle\alpha,\beta\rangle} \circ dH_\alpha.$$

Since $dG^2_{\langle\alpha,\beta\rangle}$ is invertible, we can solve the second equation for dH_α and substitute in the first, thus getting, dropping the subscripts for simplicity,

$$dF^1 - dF^2 \circ (dG^2)^{-1} \circ dG^1 = 0.$$

Let $l \in Y^*$ be the functional $dF^2 \circ (dG^2)^{-1}$. Then we have $dF^1 = l \circ dG^1$ and, by definition, $dF^2 = l \circ dG^2$. Composing the first equation (on the right) with $\pi_1\colon V \times W \to V$ and the second with π_2, and adding, we get $dF_{\langle\alpha,\beta\rangle} = l \circ dG_{\langle\alpha,\beta\rangle}$. That is, $d(F - l \circ G)_\gamma = 0$. □

Nothing we have said so far explains the phrase "Lagrange multipliers". This comes out of the Cartesian expression of the theorem, where we have U an open subset of a Cartesian space $\mathbb{R}^n$, $Y = \mathbb{R}^m$, $G = \langle g^1, \ldots, g^m \rangle$, and l in Y^* of the form $l_{\mathbf{c}} : l(\mathbf{y}) = \sum_1^m c_i y_i$. Then $F - l \circ G = F - \sum_1^m c_i g^i$, and $d(F - l \circ G)_{\mathbf{a}} = 0$ becomes

$$\frac{\partial F}{\partial x_j} - \sum_1^m c_i \frac{\partial g^i}{\partial x_j} = 0, \qquad j = 1, \ldots, n.$$

These n equations together with the m equations $G = \langle g^1, \ldots, g^m \rangle = 0$ give $m + n$ equations in the $m + n$ unknowns $x_1, \ldots, x_n, c_1, \ldots, c_m$.

Our original trivial example will show how this works out in practice. We want to maximize $F(\mathbf{x}) = x_2$ from $\mathbb{R}^3$ to $\mathbb{R}$ subject to the constraint $\sum_1^3 x_i^2 = 1$. Here $g(\mathbf{x}) = \sum_1^3 x_i^2 - 1$ is also from $\mathbb{R}^3$ to $\mathbb{R}$, and our method tells us to look for a critical point of $F - cg$ subject to $g = 0$. Our system of equations is

$$\begin{aligned} 0 - 2cx_1 &= 0, \\ 1 - 2cx_2 &= 0, \\ 0 - 2cx_3 &= 0, \\ \sum_1^3 x_i^2 &= 1. \end{aligned}$$

The first says that $c = 0$ or $x_1 = 0$, and the second implies that c cannot be 0. Therefore, $x_1 = x_3 = 0$, and the fourth equation then shows that $x_2 = \pm 1$.

Another example is our problem of minimizing the surface area $A = 2(xy + yz + zx)$ of a rectangular parallelepiped, subject to the constraint of a constant volume, $xyz = V$. The theorem says that the minimum point will be a critical point of $A - \lambda V$ for some λ, and, setting the differential of this function equal to zero, we get the equations

$$\begin{aligned} 2(y + z) - \lambda yz &= 0, \\ 2(x + z) - \lambda xz &= 0, \\ 2(x + y) - \lambda xy &= 0, \end{aligned}$$

together with the constraint

$$xyz = V.$$

The first three equations imply that $x = y = z$; the last then gives $V^{1/3}$ at the common value.

*13. FUNCTIONAL DEPENDENCE

The question, roughly, is this: If we are given a collection of continuous functions, all defined on some open set A, how can we tell whether or not some of them are functions of the rest? For example, if we are given three real-valued continuous functions f_1, f_2, and f_3, how can we tell whether or not some one of them is a function of the other two, say f_3 is a function of f_1 and f_2, which means that there is a function of two variables $g(x, y)$ such that $f_3(t) = g(f_1(t), f_2(t))$ for all t in the common domain A? If this happens, we say that f_3 is *functionally dependent* on f_1 and f_2. This is very nearly the same as asking when it will be the case that the range S of the mapping $F: t \mapsto \langle f_1(t), f_2(t), f_3(t) \rangle$ is a two-dimensional submanifold of $\mathbb{R}^3$. However, there are differences in these questions that are worth noting. If f_3 is functionally dependent on f_1 and f_2, then the range of F certainly lies *on* a two-dimensional submanifold of $\mathbb{R}^3$, namely, the graph of g. But this is no guarantee that it itself *forms* a two-dimensional submanifold. For example, both f_2 and f_3 might be functionally dependent on f_1, $f_2 = g \circ f_1$, and $f_3 = h \circ f_1$, in which case the range of F lies on the curve $\langle s, g(s), h(s) \rangle$ in $\mathbb{R}^3$, which is a one-dimensional submanifold. In the opposite direction, the range of F can be a two-dimensional submanifold M without f_3 being functionally dependent on f_2 and f_1. All we can conclude in this case is that *locally* one of the functions $\{f_i\}_1^3$ is a function of the other two, since *locally* M is a surface patch, in the language of the last section. But if we move a little bit away on the curving surface M to the neighborhood of another point, we may have to solve for a different one of the functions. Nevertheless, if $M = \text{range } F$ is a subset of a two-dimensional manifold, it is reasonable to say that the functions $\{f_i\}_1^3$ are *functionally dependent*, and we are led to examine this more natural notion.

If we assume that $F = \langle f_1, f_2, f_3 \rangle$ is continuously differentiable and that the rank of dF_α is 3 at some point α in A, then the implicit-function theorem implies that $F[A]$ includes a whole ball in $\mathbb{R}^3$ about the point $F(\alpha)$. Thus a necessary condition for $M = \text{range } F$ to lie on a two-dimensional submanifold in $\mathbb{R}^3$ is that the rank of dF_α be everywhere less than 3. We shall see, in fact, that if the rank of dF_α is 2 for all α, then $M = \text{range } F$ is essentially a two-dimensional manifold. (There is still a tiny difficulty that we shall explain later.) Our tools are going to be the implicit-function theorem and the following theorem, which could well have come much earlier, that the rank of T is a "lower semicontinuous" function of T.

Theorem 13.1. Let V and W be finite-dimensional vector spaces, normed in some way. Then for any T in $\text{Hom}(V, W)$ there is an ϵ such that

$$\|S - T\| < \epsilon \Rightarrow \text{rank } S \geq \text{rank } T.$$

Proof. Let T have null space N and range R, and let X be any complement of N in V. Then the restriction of T to X is an isomorphism to R, and hence is bounded below by some positive m. (Its inverse from R to X is bounded by some b, by Theorem 4.2, and we set $m = 1/b$.) Then if $\|S - T\| < m/2$, it follows that S is bounded below on X by $m/2$, for the inequalities

$$\|T(\alpha)\| \geq m\|\alpha\| \qquad \text{and} \qquad \|(S - T)(\alpha)\| \leq (m/2)\|\alpha\|$$

together imply that $\|S(\alpha)\| \geq (m/2)\|\alpha\|$. In particular, S is injective on X, and so $\text{rank } S = d(\text{range } S) \geq d(X) = d(R) = \text{rank } T$. □

We can now prove the general local theorem.

Theorem 13.2. Let V and W be finite-dimensional spaces, let r be an integer less than the dimension of W, and let F be a continuously differentiable map from an open subset $A \subset V$ to W such that the rank of $dF_\gamma = r$ for all γ in A. Then each point γ in A has a neighborhood U such that $F[U]$ is an r-dimensional patch submanifold of W.

Proof. For a fixed γ in A let V_1 and Y be the null space and range of dF_γ, let V_2 be a complement of V_1 in V, and view V as $V_1 \times V_2$. Then F becomes a function $F(\xi, \eta)$ of two variables, and if $\gamma = \langle \alpha, \beta \rangle$, then $dF^2_{\langle \alpha,\beta \rangle}$ is an isomorphism from V_2 to Y. At this point we can already choose the decomposition $W = W_1 \oplus W_2$ with respect to which $F[A]$ is going to be a graph (locally). We simply choose any direct sum decomposition $W = W_1 \oplus W_2$ such that W_2 is a complement of $Y = \text{range } dF_{\langle \alpha,\beta \rangle}$. Thus W_1 might be Y, but it doesn't have to be. Let P be the projection of W onto W_1, along W_2. Since Y is a complement of the null space of P, we know that $P \restriction Y$ is an isomorphism from Y to W_1. In particular, W_1 is r-dimensional, and

$$\text{rank } P \circ dF_{\langle \alpha,\beta \rangle} = r.$$

Moreover, and this is crucial, P is an isomorphism from the range of $dF_{<\xi,\eta>}$ to W_1 for all $<\xi, \eta>$ sufficiently close to $<\alpha, \beta>$. For the above rank theorem implies that rank $P \circ dF_{<\xi,\eta>} \geq$ rank $P \circ dF_{<\alpha,\beta>} = r$ on some neighborhood of $<\alpha, \beta>$. On the other hand, the range of $P \circ dF_{<\xi,\eta>}$ is included in the range of P, which is W_1, and so rank $P \circ dF_{<\xi,\eta>} \leq r$. Thus rank $P \circ dF_{<\xi,\eta>} = r$ for $<\xi, \eta>$ near $<\alpha, \beta>$, and since rank $dF_{<\xi,\eta>} = r$ by hypothesis, we see that P is an isomorphism on the range of any such $dF_{<\xi,\eta>}$.

Now define $H: W_1 \times A \to W_1$ as the mapping

$$<\zeta, \xi, \eta> \mapsto P \circ F(\xi, \eta) - \zeta.$$

If $\mu = P \circ F(\alpha, \beta)$, then $dH^3_{<\mu,\alpha,\beta>} = P \circ dF^2_{<\alpha,\beta>}$, which is an isomorphism from V_2 to W_1. Therefore, by the implicit-function theorem there exists a neighborhood $L \times M \times N$ of $<\mu, \alpha, \beta>$ and a uniquely determined continuously differentiable mapping G from $L \times M$ to N such that

$$H(\zeta, \xi, G(\zeta, \xi)) = 0$$

on $L \times M$. That is,

$$\zeta = P \circ F(\xi, G(\zeta, \xi))$$

on $L \times M$.

The remainder of our argument consists in showing that $F(\xi, G(\zeta, \xi))$ is a function of ζ alone. We start by differentiating the above equation with respect to ξ, getting

$$0 = P \circ (dF^1 + dF^2 \circ dG^2) = P \circ dF \circ <I, dG^2>.$$

As noted above, P is an isomorphism on the range of $dF_{<\xi,\eta>}$ for all $<\xi, \eta>$ sufficiently close to $<\alpha, \beta>$, and if we suppose that $L \times M$ is also taken small enough so that this holds, then the above equation implies that

$$dF_{<\xi,\eta>} \circ <I, dG^2> = 0$$

for all $<\zeta, \xi> \in L \times M$. But this is just the statement that the partial differential with respect to ξ of $F(\xi, G(\zeta, \xi))$ is identically 0, and hence that

$$F(\xi, G(\zeta, \xi))$$

is a continuously differentiable function K of ζ alone:

$$F(\xi, G(\zeta, \xi)) = K(\zeta).$$

Since $\eta = G(\zeta, \xi)$ and $\zeta = P \circ F(\xi, \eta)$, we thus have $F(\xi, \eta) = K(P \circ F(\xi, \eta))$, or

$$F = K \circ P \circ F,$$

and this holds on the open set U consisting of those points $<\xi, \eta>$ in $M \times N$ such that $P \circ F(\xi, \eta) \in L$. If we think of W as $W_1 \times W_2$, then F and K are ordered pairs of functions, $F = <F^1, F^2>$ and $K = <l, k>$, P is the mapping $<\zeta, \nu> \mapsto \zeta$, and the second component of the above equation is

$$F^2 = k \circ F^1.$$

Since $F^1[U] = P \circ F[U] = L$, the above equation says that $F[U]$ is the graph of the mapping k from L to W_2. Moreover, L is an open subset of the r-dimensional vector space W_1, and therefore $F[U]$ is an r-dimensional patch manifold in $W = W_1 \times W_2$. □

The above theorem includes the answer to our original question about functional dependence.

Corollary. Let $F = \{f^i\}_1^m$ be an m-tuple of continuously differentiable real-valued functions defined on an open subset A of a normed linear space V, and suppose that the rank of dF_α has the constant value r on A, where r is less than m. Then any point γ in A has a neighborhood U over which $m - r$ of the functions are functionally dependent on the remaining r.

Proof. By hypothesis the range Y of $dF_\gamma = \langle df_\gamma^1, \ldots, df_\gamma^m \rangle$ is an r-dimensional subspace of $\mathbb{R}^m$. We can therefore find a basis for a complementary subspace W_2 by choosing $m - r$ of the standard basis elements $\{\delta^i\}$, and we may as well renumber the functions f^i so that these are $\delta^{r+1}, \ldots, \delta^m$. Then the projection P of $\mathbb{R}^m$ onto $\mathbb{R}^r = L(\delta^1, \ldots, \delta^r)$ is an isomorphism from Y to $\mathbb{R}^r$ (since Y is a complement of its null space), and by the theorem there is a neighborhood U of γ over which $(I - P) \circ F$ is a function k of $P \circ F$. But this says exactly that $\langle f^{r+1}, \ldots, f^m \rangle = k \circ \langle f^1, \ldots, f^r \rangle$. That is, k is an $(m - r)$-tuple-valued function, $k = \langle k^{r+1}, \ldots, k^m \rangle$, and $f^j = k^j \circ \langle f^1, \ldots, f^r \rangle$ for $j = r + 1, \ldots, m$. □

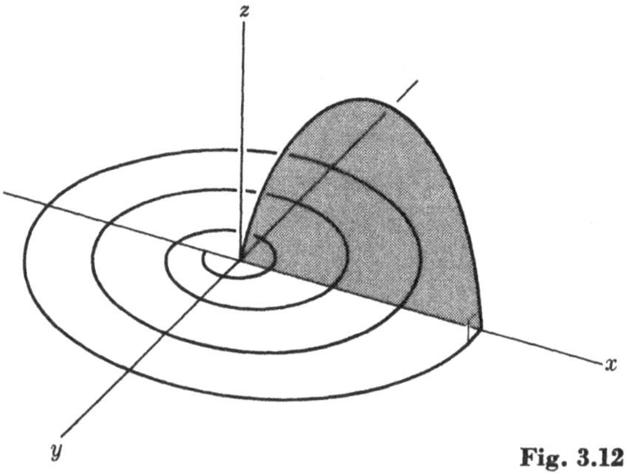

Fig. 3.12

We mentioned earlier in the section that there was a difficulty in concluding that if F is a continuously differentiable map from an open subset A of V to W whose differential has constant rank r less than $d(W)$, then $S = \text{range } F$ is an r-dimensional submanifold of S. The flaw can be described as follows. The definition of a submanifold S of X required that each point of S have a neighborhood in X whose intersection with S is a patch. In the case before us, what we can conclude is that if β is a point of S, then $\beta = F(\alpha)$ for some α in A, and α has a neighborhood U whose image under F is a patch. But this image may not be a full neighborhood of β in S, because S may curve back on itself in such a way as to intrude into every neighborhood of β. Consider, for example, the one-dimensional Γ imbedded in $\mathbb{R}^3$ suggested by following Fig. 3.12. The curve begins in the xz-plane along the z-axis, curves over, and when it comes to the xy-plane it starts spiraling in to the origin in the xy-plane (the point of change over from the xz-plane to the xy-plane is a singularity that we could smooth out). The origin is not a point having a neighborhood in $\mathbb{R}^3$ whose intersection with Γ is a one-patch, but the full curve is the image of $(-1, 1)$ under a continuously differentiable injection.

We would consider Γ to be a one-dimensional manifold without any difficulty, but something has gone wrong with its imbedding in $\mathbb{R}^3$, so it is not a one-dimensional *sub*manifold of $\mathbb{R}^3$.

*14. UNIFORM CONTINUITY AND FUNCTION-VALUED MAPPINGS

In the next chapter we shall see that a continuous function F whose domain is a bounded closed subset of a finite-dimensional vector space V is necessarily uniformly continuous. This means that given ϵ, there is a δ such that

$$\|\xi - \eta\| < \delta \Rightarrow \|F(\xi) - F(\eta)\| < \epsilon$$

for all vectors ξ and η in the domain of F.

The point is that δ depends only on ϵ and not, as in ordinary continuity, on the "anchor" point at which continuity is being asserted. This is a very important property. In this section we shall see that it underlies a class of theorems in which a point map is escalated to a function-valued map, and properties of the point map imply corresponding properties of the function-valued map. Such theorems have powerful applications, as we shall see in Section 15 and in Section 1 of Chapter 6. An application that we shall get immediately here is the theorem on differentiation under the integral sign. However, it is only Theorem 14.3 that will be used later in the book.

Suppose first that $F(\xi, \eta)$ is a bounded continuous function from a product open set $M \times N$ to a normed linear space X. Holding η fixed, we have a function $f_\eta(\xi) = F(\xi, \eta)$ which is a bounded continuous function on M, that is, an element of the normed linear space $Y = \mathcal{BC}(M, X)$ of all bounded continuous maps from M to X. This function is also indicated $F(\cdot, \eta)$, so that $f_\eta = F(\cdot, \eta)$. We are supposing that the uniform norm is being used on Y:

$$\|f_\eta\| = \text{lub}\,\{\|f_\eta(\xi)\| : \xi \in M\} = \text{lub}\,\{\|F(\xi, \eta)\| : \xi \in M\}.$$

Theorem 14.1. In the above context, if F is *uniformly* continuous, then the mapping $\eta \mapsto f_\eta$ (or $\eta \mapsto F(\cdot, \eta)$) is continuous, in fact, uniformly continuous, from N to Y.

Proof. Given ϵ, choose δ so that

$$\|\langle \xi, \eta \rangle - \langle \mu, \nu \rangle\| < \delta \Rightarrow \|F(\xi, \eta) - F(\mu, \nu)\| < \epsilon.$$

Taking $\mu = \xi$ and rewriting the right-hand side, we have

$$\|\eta - \nu\| < \delta \Rightarrow \|f_\eta(\xi) - f_\nu(\xi)\| < \epsilon$$

for all ξ. Thus

$$\|\eta - \nu\| < \delta \Rightarrow \|f_\eta - f_\nu\|_\infty \leq \epsilon. \ \square$$

We have proved that if a function of two variables is uniformly continuous, then the mappings obtained from it by the general duality principle are continuous. This phenomenon lies behind many well-known facts. For example:

Corollary. If $F(x, y)$ is a uniformly continuous real-valued function on the unit square $[0, 1] \times [0, 1]$ in $\mathbb{R}^2$, then $\int_0^1 F(x, y)\,dx$ is a continuous function of y.

Proof. The mapping $y \mapsto \int_0^1 F(x, y)\,dx$ is the composition of the bounded linear mapping $f \mapsto \int_0^1 f$ from $\mathcal{C}([0, 1]$ to $\mathbb{R}$ with the continuous mapping $y \to F(\cdot, y)$ from $[0, 1]$ to $\mathcal{C}[0, 1]$, and is continuous as the composition of continuous mappings. $\square$

We consider next the differentiability of the above duality-induced mapping.

Theorem 14.2. If F is a bounded continuous mapping from an open product set $M \times N$ of a normed linear space $V \times W$ to a normed linear space X, and if $dF^2_{\langle \alpha, \beta \rangle}$ exists and is a bounded uniformly continuous function of $\langle \alpha, \beta \rangle$ on $M \times N$, then $\varphi: \eta \to F(\cdot, \eta)$ is a differentiable mapping from N to $Y = \mathcal{BC}(M, X)$, and $[d\varphi_\beta(\eta)](\xi) = dF^2_{\langle \xi, \beta \rangle}(\eta)$.

Proof. Given ϵ, we choose δ by the uniform continuity of dF^2, so that

$$\|\mu - \nu\| < \delta \Rightarrow \|dF^2_{\langle \xi, \mu \rangle} - dF^2_{\langle \xi, \nu \rangle}\| < \epsilon$$

for all $\xi \in M$. The corollary to Theorem 7.4 then implies that

$$\|\eta\| < \delta \Rightarrow \|\Delta F^2_{\langle \xi, \beta \rangle}(\eta) - dF^2_{\langle \xi, \beta \rangle}(\eta)\| \leq \epsilon \|\eta\|$$

for all $\xi \in M$, all $\beta \in N$, and all η such that the line segment from β to $\beta + \eta$ is in N. We fix β and rewrite the right-hand side of the above inequality. This is the heart of the proof. First

$$\begin{aligned}\Delta F^2_{\langle \xi, \beta \rangle}(\eta) &= F(\xi, \beta + \eta) - F(\xi, \beta) \\ &= [f_{\beta+\eta} - f_\beta](\xi) = [\varphi(\beta + \eta) - \varphi(\beta)](\xi) = [\Delta\varphi_\beta(\eta)](\xi).\end{aligned}$$

Next we can check that if $\|dF^2_{\langle \mu, \nu \rangle}\| \leq b$ for $\langle \mu, \nu \rangle \in M \times N$, then the mapping T defined by the formula $[T(\eta)](\xi) = dF^2_{\langle \xi, \beta \rangle}(\eta)$ is an element of $\mathrm{Hom}(W, Y)$ of norm at most b. We leave the detailed verification of this as an exercise for the reader. The last displayed inequality now takes the form

$$\|\eta\| < \delta \Rightarrow \|[\Delta\varphi_\beta(\eta) - T(\eta)](\xi)\| \leq \epsilon\|\eta\|,$$

and hence

$$\|\eta\| < \delta \Rightarrow \|\Delta\varphi_\beta(\eta) - T(\eta)\|_\infty \leq \epsilon\|\eta\|.$$

This says exactly that the mapping φ is differentiable at β and $d\varphi_\beta = T$. □

The mapping φ is in fact continuously differentiable, as can be seen by arguing a little further in the above manner. The situation is very close to being an application of Theorem 14.1.

The classical theorem on differentiability under the integral sign is a corollary of the above theorem. We give a simple case. Note that if η is a real variable y, then the above formula for $d\varphi$ can be rewritten in terms of arc derivatives:

$$[\varphi'(b)](\xi) = \frac{\partial F}{\partial y}(\xi, b).$$

Corollary. If $F(x, y)$ is a continuous real-valued function on the unit square $[0, 1] \times [0, 1]$, and if $\partial F/\partial y$ exists and is a uniformly continuous function on the square, then $\int_0^1 F(x, y)\, dx$ is a differentiable function of y and its derivative is $\int_0^1 (\partial F/\partial y)(x, y)\, dx$.

Proof. The mapping $T\colon y \mapsto \int_0^1 F(x, y)\, dx$ is the composition of the bounded linear mapping $f \mapsto \int_0^1 f(x)\, dx$ from $\mathcal{C}([0, 1])$ to $\mathbb{R}$ with the differentiable mapping $\varphi\colon y \mapsto F(\cdot\,, y)$ from $[0, 1]$ to $\mathcal{C}([0, 1])$, and is therefore differentiable by the composite-function rule. Then Theorem 7.2 and the fact that the differential of a bounded linear map is itself give

$$T'(y) = \int_0^1 [\varphi'(y)](x)\, dx = \int_0^1 \frac{\partial F}{\partial y}(x, y)\, dx. \ \square$$

We come now to the situation of most importance to us, where a point to point map generates a function-to-function map by composition. Let A be an open set in a normed linear space V, let S be an arbitrary set, and let $\mathcal{A}$ be the set of bounded maps f from S to A. Then $\mathcal{A}$ is a subset of the normed linear space $\mathcal{B}(S, V)$ of all bounded functions from S to V under the uniform norm. A function $f \in \mathcal{A}$ will be an interior point of $\mathcal{A}$ if and only if the distance from the range of f to the boundary of $\mathcal{A}$ is a positive number δ, for this is clearly equivalent to saying that $\mathcal{A}$ includes a ball in $\mathcal{B}(S, V)$ about the point f. Now let g be any bounded mapping from A to a normed linear space W, and let $G\colon \mathcal{A} \to \mathcal{B}(S, W)$ be composition by g. That is, $h = G(f)$ if and only if $f \in \mathcal{A}$ and $h = g \circ f$. We can consider both the continuity and differentiability of G, but we shall only work out the differentiability theorem.

Theorem 14.3. Let the function $g\colon A \to W$ be differentiable at each point α in A, and let dg_α be a bounded uniformly continuous function of α. Then the mapping $G\colon \mathcal{A} \to \mathcal{B}(S, W)$ defined by $G(f) = g \circ f$ is differentiable at any interior point f in $\mathcal{A}$ and $dG_f\colon \mathcal{B}(S, V) \to \mathcal{B}(S, W)$ is defined by

$$[dG_f(h)](s) = dg_{f(s)}(h(s))$$

for all $s \in S$.

Proof. Given ϵ, choose δ by the uniform continuity of dg so that

$$\|\alpha - \beta\| < \delta \Rightarrow \|dg_\alpha - dg_\beta\| < \epsilon,$$

and then apply the corollary to Theorem 7.4 once more to conclude that

$$\|\xi\| < \delta \Rightarrow \|\Delta g_\alpha(\xi) - dg_\alpha(\xi)\| \le \epsilon\|\xi\|,$$

provided the line segment from α to $\alpha + \xi$ is in A. Now choose any fixed interior point f in $\mathcal{A}$, and choose $\delta' \le \delta$ so that $B_{\delta'}(f) \subset \mathcal{A}$. Then for any h in $\mathcal{B}(S, V)$,

$$\|h\|_\infty < \delta' \Rightarrow \|\Delta g_{f(s)}(h(s)) - dg_{f(s)}(h(s))\| \le \epsilon\|h(s)\|$$

for all $s \in S$. Define a map $T\colon \mathcal{B}(S, V) \mapsto \mathcal{B}(S, W)$ by $[T(h)](s) = dg_{f(s)}(h(s))$. Then the above displayed inequality can be rewritten as

$$\|h\|_\infty < \delta' \Rightarrow \|\Delta G_f(h) - T(h)\|_\infty \le \epsilon\|h\|_\infty.$$

That is, $\Delta G_f = T + \mathfrak{o}$. We will therefore be done when we have shown that $T \in \mathrm{Hom}(\mathcal{B}(S, V), \mathcal{B}(S, W))$.

First, we have

$$\begin{aligned}(T(h_1 + h_2))(s) &= dg_{f(s)}((h_1 + h_2)(s)) = dg_{f(s)}(h_1(s) + h_2(s)) \\ &= dg_{f(s)}(h_1(s)) + dg_{f(s)}(h_2(s)) = (T(h_1))(s) + (T(h_2))(s).\end{aligned}$$

Thus $T(h_1 + h_2) = T(h_1) + T(h_2)$, and homogeneity follows similarly. Second, if b is a bound to $\|dg_\alpha\|$ on A, then $\|T(h)\|_\infty = \mathrm{lub}\,\{\|(T(h))(s)\| : s \in S\} \le \mathrm{lub}\,\{\|dg_{f(s)}\| \cdot \|h(s)\| : s \in S\} \le b\|h\|_\infty$. Therefore, $\|T\| \le b$, and we are finished. □

In the above situation, if g is from $A \times U$ to W, so that $G(f)$ is the function h given by $h(t) = g(f(t), t)$, then nothing is changed except that the theorem is about dg^1 instead of dg. If, in addition, V is a product space $V_1 \times V_2$, so that f is of the form $\langle f_1, f_2 \rangle$ and $[G(f)](t) = g(f_1(t), f_2(t), t)$, then our rules about partial differentials give us the formula

$$[dG_f(h)](t) = dg^1_{f(t)}(h_1(t)) + dg^2_{f(t)}(h_2(t)).$$

*15. THE CALCULUS OF VARIATIONS

The problems of the calculus of variations are simply critical-point problems of a certain type with a characteristic twist in the way the condition $dF_\alpha = 0$ is used. We shall illustrate the subject by proving one of its standard theorems.

Since we want to solve a constrained maximum problem in which the domain is an infinite-dimensional vector space, a systematic discussion would start off with a more general form of the Lagrange multiplier theorem. However, for our purpose it is sufficient to note that if S is a closed plane $M + \alpha$, then the restriction of F to S is equivalent to a new function on the vector space M, and its differential at $\beta = \eta + \alpha$ in S is clearly just the restriction of dF_β to M. The requirement that β be a critical point for the constrained function is therefore simply the requirement that dF_β vanish on M.

Let F be a uniformly continuous differentiable real-valued function of three variables defined on (an open subset of) $W \times W \times \mathbb{R}$, where W is a normed linear space. Given a closed interval $[a, b] \subset \mathbb{R}$, let V be the normed linear space $\mathcal{C}^1([a, b], W)$ of smooth arcs $f\colon [a, b] \to W$, with $\|f\|$ taken as $\|f\|_\infty + \|f'\|_\infty$. The problem is to maximize the (nonlinear) functional $G(f) = \int_a^b F(f(t), f'(t), t)\, dt$, subject to the restraints $f(a) = \alpha$ and $f(b) = \beta$. That is, we consider only smooth arcs in W with fixed endpoints α and β, and we want to find that arc from α to β which maximizes (or minimizes) the integral. Now we can show that G is a continuously differentiable function from (an open subset of) V to $\mathbb{R}$. The easiest way to do this is to let X be the space $\mathcal{C}([a, b], W)$ of continuous arcs under the uniform norm, and to consider first the more general functional K from $X \times X$ to $\mathbb{R}$ defined by $K(f, g) = \int_a^b F(f(t), g(t), t)\, dt$. By Theorem 14.3 the integrand map $\langle f, g \rangle \mapsto F(f(\cdot), g(\cdot), \cdot)$ is differentiable from $X \times X$ to $\mathcal{C}([a, b])$ and its differential at $\langle f, g \rangle$ evaluated at $\langle h, k \rangle$ is the function

$$dF^1_{\langle f(t), g(t), t \rangle}(h(t)) + dF^2_{\langle f(t), g(t), t \rangle} k(t).$$

Since $f \mapsto \int_a^b f(t)$ is a bounded linear functional on $\mathcal{C}$, it is differentiable and equal to its differential. The composite-function rule therefore implies that K is differentiable and that

$$dK_{\langle f, g \rangle}(h, k) = \int_a^b [dF^1(h(t)) + dF^2(k(t))]\, dt,$$

where the partial differentials in the integrand are at the point $\langle f(t), g(t), t \rangle$. Now the pairs $\langle f, g \rangle$ such that f' exists and equals g form a closed subspace of $X \times X$ which is isomorphic to V. It is obvious that they form a subspace, but to see that it is closed requires the theory of the integral for parametrized arcs from Chapter 4, for it depends on the representation $f(t) = f(a) + \int_a^t f'(s)\, ds$ and the consequent norm inequality $\|f(t) - f(a)\| \le (t - a)\|f'\|_\infty$. Assuming this, we see that our original functional G is just the restriction of K to this subspace (isomorphic to) V, and hence is differentiable with

$$dG_f(h) = \int_a^b dF^1(h(t)) + dF^2(h'(t))\, dt.$$

This differential dG_f is called the *first variation* of G *about* f.

The fixed endpoints α and β for the arc f determine in turn a closed plane P in V, for the evaluation maps (coordinate projections) $\pi_x\colon f \mapsto f(x)$ are bounded and P is the intersection of the hyperplanes $\pi_a = \alpha$ and $\pi_b = \beta$. Since P is a translate of the subspace $M = \{f \in V\colon f(a) = f(b) = 0\}$, our constrained maximum equation is

$$dG_f(h) = \int_a^b [dF^1(h(t)) + dF^2(h'(t))]\,dt = 0$$

for all h in M.

We come now to the special trick of the calculus of variations, called the lemma of Du Bois-Reymond.

Suppose for simplicity that $W = \mathbb{R}$. Then F is a function $F(x, y, t)$ of three real variables, the partial differentials are equivalent to ordinary partial derivatives, and our critical-point equation is

$$dG_f(h) = \int_a^b \left(\frac{\partial F}{\partial x}\cdot h + \frac{\partial F}{\partial y}\cdot h'\right) = 0.$$

If we integrate the first term in the integral by parts and remember that $h(a) = h(b) = 0$, we see that the equation becomes

$$\int_a^b \left(\frac{\partial F}{\partial y} - \int \frac{\partial F}{\partial x}\right) g = 0,$$

where $g = h'$. Since h is an arbitrary continuously differentiable function except for the constraints $h(a) = h(b) = 0$, we see that g is an arbitrary continuous function except for the constraint $\int_a^b g(t)\,dt = 0$. That is, $\partial F/\partial y - \int \partial F/\partial x$ is orthogonal to the null space N of the linear functional $g \mapsto \int_a^b g(t)\,dt$. Since the one-dimensional space $N^\perp$ is clearly the set of constant functions,

$$\int_a^b g = 0 \Rightarrow \int_a^b Cg = 0\,,$$

our condition becomes

$$\frac{\partial F}{\partial y}\,(f(t), f'(t), t) = \int_0^t \frac{\partial F}{\partial x}\,(f(s), f'(s), s)\,ds + C.$$

This equation implies, in particular, that the left member is differentiable. This is not immediately apparent, since f' is only assumed to be continuous. Differentiating, we conclude finally that f is a critical point of the mapping G if and only if it is a solution of the differential equation

$$\frac{d}{dt}\frac{\partial F}{\partial y}\,(f(t), f'(t), t) = \frac{\partial F}{\partial x}\,(f(t), f'(t), t),$$

which is called the *Euler equation* of the variational problem. It is an ordinary differential equation for the unknown function f; when the indicated derivative is computed, it takes the form

$$\frac{\partial^2 F}{\partial y^2} f'' + \frac{\partial^2 F}{\partial y\,\partial x} f' + \frac{\partial^2 F}{\partial y\,\partial t} - \frac{\partial F}{\partial x} = 0.$$

If W is not $\mathbb{R}$, we get exactly the same result from the general form of the integration by parts formula (using Theorem 6.3) and a more sophisticated version of the above argument. (See Exercise 10.14 and 10.15 of Chapter 4.) That is, the smooth arc f with fixed endpoints α and β is a critical point of the mapping $g \mapsto \int_a^b F(g(t), g'(t), t)\,dt$ if and only if it satisfies the Euler differential equation

$$\frac{d}{dt}\, dF^2_{<f(t),f'(t),t>} = dF^1_{<f(t),f'(t),t>}.$$

This is now a vector-valued equation, with values in W^*. If W is finite-dimensional, with dimension n, then a choice of basis makes W^* into $\mathbb{R}^n$, and this vector equation is equivalent to n scalar equations

$$\frac{d}{dt}\frac{\partial F}{\partial y_i}(f(t), f'(t), t) = \frac{\partial F}{\partial x_i}(f(t), f'(t), t),$$

where F is now a function of $2n + 1$ real variables,

$$F(\mathbf{x}, \mathbf{y}, \mathbf{t}) = F(x_1, \ldots, x_n, y_1, \ldots, y_n, t).$$

Finally, let us see what happens to the simpler variational problem ($W = \mathbb{R}$) when the endpoints of f are not fixed. Now the critical-point equation is $dG_f(h) = 0$ for *all* h in V, and when we integrate by parts it becomes

$$\frac{\partial F}{\partial y}\cdot h\Big]_a^b + \int_a^b \left(\frac{\partial F}{\partial x} - \frac{d}{dt}\frac{\partial F}{\partial y}\right) h = 0$$

for all h in V. We can reason essentially as above, but a little more closely, to conclude that a function f is a critical point if and only if it satisfies the Euler equation

$$\frac{d}{dt}\left(\frac{\partial F}{\partial y}\right) - \frac{\partial F}{\partial x} = 0$$

and also the endpoint conditions

$$\left.\frac{\partial F}{\partial y}\right|_{t=a} = \left.\frac{\partial F}{\partial y}\right|_{t=b} = 0.$$

This has been only a quick look at the variational calculus, and the interested reader can pursue it further in treatises devoted to the subject. There are many more questions of the general type we have considered. For example, we may want neither fixed nor completely free endpoints but freedom subject to constraints. We shall take this up in Chapter 13 in the special case of the variational equations of mechanics. Or again, f may be a function of two or more variables and the integral may be a multiple integral. In this case the Euler equation may become a system of partial differential equations in the unknown f. Finally, there is the question of sufficient conditions for the critical function to give a maximum or minimum value to the integral. This will naturally involve a study of the second differential of the functional G, or its *second* variation, as it is known in this subject.

*16. THE SECOND DIFFERENTIAL AND THE CLASSIFICATION OF CRITICAL POINTS

Suppose that V and W are normed linear spaces, that A is an open subset of V, and that $F\colon A \to W$ is a continuously differentiable mapping. The first differential of F is the continuous mapping $dF\colon \gamma \mapsto dF_\gamma$ from A to $\text{Hom}(V, W)$. We now want to study the differentiability of *this* mapping at the point α. Presumably, we know what it means to say that dF is differentiable at α. By definition $d(dF)_\alpha$ is a bounded linear transformation T from V to $\text{Hom}(V, W)$ such that $\Delta(dF)_\alpha(\eta) - T(\eta) = \mathfrak{o}(\eta)$. That is, $dF_{\alpha+\eta} - dF_\alpha - T(\eta)$ is an element of $\text{Hom}(V, W)$ of norm less than $\epsilon\|\eta\|$ for η sufficiently small. We set $d^2F_\alpha = d(dF)_\alpha$ and repeat: $d^2F_\alpha = d^2F_\alpha(\cdot)$ is a linear map from V to $\text{Hom}(V, W)$, $d^2F_\alpha(\eta) = d^2F_\alpha(\eta)(\cdot)$ is an element of $\text{Hom}(V, W)$, and $d^2F_\alpha(\eta)(\xi)$ is a vector in W. Also, we know that d^2F_α is equivalent to a bounded bilinear map

$$\omega\colon V \times V \to W,$$

where $\omega(\eta, \xi) = d^2F_\alpha(\eta)(\xi)$.

The vector $d^2F_\alpha(\eta)(\xi)$ clearly ought to be some kind of second derivative of F at α, and the reader might even conjecture that it is the mixed derivative in the directions ξ and η.

Theorem 16.1. If $F\colon A \to W$ is continuously differentiable, and if the second differential d^2F_α exists, then for each fixed $\mu \in V$ the function $D_\mu F\colon \gamma \mapsto D_\mu F(\gamma)$ from A to W is differentiable at α and $D_\nu(D_\mu F)(\alpha) = (d^2F_\alpha(\nu))(\mu)$.

Proof. We use the evaluation-at-μ map $\text{ev}_\mu\colon \text{Hom}(V, W) \to W$ defined for a fixed μ in V by $\text{ev}_\mu(T) = T(\mu)$. It is a bounded linear mapping. Then

$$(D_\mu F)(\alpha) = dF_\alpha(\mu) = \text{ev}_\mu(dF_\alpha) = (\text{ev}_\mu \circ dF)(\alpha),$$

so that the function $D_\mu F$ is the composition $\text{ev}_\mu \circ dF$. It is differentiable at α because $d(dF)_\alpha$ exists and ev_μ is linear. Thus $\big(D_\nu(D_\mu F)\big)(\alpha) = d(D_\mu F)_\alpha(\nu) = d(\text{ev}_\mu \circ dF)_\alpha(\nu) = \big(\text{ev}_\mu \circ d(dF)_\alpha\big)(\nu) = \text{ev}_\mu[(d^2F_\alpha)(\nu)] = \big(d^2F_\alpha(\nu)\big)(\mu)$. □

The reader must remember in going through the above argument that $D_\mu F$ is the function $(D_\mu F)(\cdot)$, and he might prefer to use this notation, as follows:

$$\begin{aligned} D_\nu\big((D_\mu F)(\cdot)\big)\big|_\alpha &= d\big((D_\mu F)(\cdot)\big)_\alpha(\nu) = d(\text{ev}_\mu \circ dF_{(\cdot)})_\alpha(\nu) \\ &= [\text{ev}_\mu \circ d(dF_{(\cdot)})_\alpha](\nu) = \text{ev}_\mu\big(d^2F_\alpha(\nu)\big). \end{aligned}$$

If the domain space V is the Cartesian space $\mathbb{R}^n$, then the differentiability of $(D_{\delta^j}F)(\cdot) = (\partial F/\partial x_j)(\cdot)$ at $\mathbf{a}$ implies the existence of the second partial derivatives $(\partial^2 F/\partial x_i\, \partial x_j)(\mathbf{a})$ by Theorem 9.2, and with $\mathbf{b}$ and $\mathbf{c}$ fixed, we then have

$$\begin{aligned} D_{\mathbf{c}}(D_{\mathbf{b}}F) &= D_{\mathbf{c}}\left(\sum b_i \frac{\partial F}{\partial x_i}\right) = \sum b_i D_{\mathbf{c}} \frac{\partial F}{\partial x_i} \\ &= \sum b_i \left(\sum c_j \frac{\partial}{\partial x_j}\left(\frac{\partial F}{\partial x_i}\right)\right) = \sum_{i,j} b_i c_j \frac{\partial^2 F}{\partial x_j\, \partial x_i}. \end{aligned}$$

Thus,

Corollary 1. If $V = \mathbb{R}^n$ in the above theorem, then the existence of $d^2F_{\mathbf{a}}$ implies the existence of all the second partial derivatives $(\partial^2 F/\partial x_i\,\partial x_j)(\mathbf{a})$ and

$$d^2F_{\mathbf{a}}(\mathbf{b}, \mathbf{c}) = D_{\mathbf{b}}(D_{\mathbf{c}}F)(\mathbf{a}) = \sum_{i,j=1}^{n} b_i c_j \frac{\partial^2 F}{\partial x_j\,\partial x_i}(\mathbf{a}).$$

Moreover, from the above considerations and Theorem 9.3 we can also conclude that:

Theorem 16.2. If $V = \mathbb{R}^n$, and if all the second partial derivatives $(\partial^2 F/\partial x_i\,\partial x_j)(\mathbf{a})$ exist and are continuous on the open set A, then the second differential $d^2F_{\mathbf{a}}$ exists on A and is continuous.

Proof. We have directly from Theorem 9.3 that each first partial derivative $(\partial F/\partial x_j)(\cdot)$ is differentiable. But $\partial F/\partial x_j = \mathrm{ev}_{\delta^j} \circ dF$, and the corollary is then a consequence of the following general principle. □

Lemma. If $\{S_i\}_1^k$ is a finite collection of linear maps on a vector space W such that $S = \langle S_1, \ldots, S_k \rangle$ is invertible, then a mapping $F: A \to W$ is differentiable at α if and only if $S_i \circ F$ is differentiable at α for all i.

Proof. For then $S \circ F$ and $F = S^{-1} \circ S \circ F$ are differentiable, by Theorems 8.1 and 6.2. □

These considerations clearly extend to any number of differentiations. Thus, if $d^2F_{(\cdot)}: \mathbf{y} \to d^2F_{\mathbf{y}}$ is differentiable at $\mathbf{a}$, then for fixed $\mathbf{b}$ and $\mathbf{c}$ the evaluation $d^2F_{(\cdot)}(\mathbf{b}, \mathbf{c})$ is differentiable at $\mathbf{a}$ and the formula

$$(D_{\mathbf{c}}(D_{\mathbf{b}}F))(\cdot) = d^2F_{(\cdot)}(\mathbf{b}, \mathbf{a}) = \sum_{i,j} b_i c_j \frac{\partial^2 F}{\partial x_j\,\partial x_i}(\cdot)$$

shows (for special choices of $\mathbf{b}$ and $\mathbf{c}$) that all the second partials $(\partial^2 F/\partial x_j\,\partial x_i)(\cdot)$ are differentiable at $\mathbf{a}$, with

$$(D_{\mathbf{d}}D_{\mathbf{c}}D_{\mathbf{b}}F)(\mathbf{a}) = D_{\mathbf{d}}((D_{\mathbf{c}}D_{\mathbf{b}}F)(\cdot))|_{\mathbf{a}} = \sum_{i,j,k} b_i c_j d_k \frac{\partial^3 F}{\partial x_k\,\partial x_j\,\partial x_i}(\mathbf{a}).$$

Conversely, if all the third partials exist and are continuous on A, then the second partials are differentiable on A by Theorem 9.3, and then $d^2F_{(\cdot)}$ is differentiable by the lemma, since $(\partial^2 F/\partial x_i\,\partial x_j)(\cdot) = \mathrm{ev}_{\langle\delta^i,\delta^j\rangle} \circ d^2F_{(\cdot)}$.

As the reader will remember, it is crucially important in working with higher-order derivatives that $\partial^2 F/\partial x_i\,\partial x_j = \partial^2 F/\partial x_j\,\partial x_i$, and we very much need the same theorem here.

Theorem 16.3. The second differential is a symmetric function of its two arguments: $(d^2F_\alpha(\eta))(\xi) = (d^2F_\alpha(\xi))(\eta)$.

Proof. By the definition of $d(dF)_\alpha$, given ϵ, there is a δ such that

$$\|\Delta(dF)_\alpha(\eta) - d(dF)_\alpha(\eta)\| \leq \epsilon\|\eta\|$$

whenever $\|\eta\| \leq \delta$. Of course, $\Delta(dF)_\alpha(\eta) = dF_{\alpha+\eta} - dF_\alpha$. If we write down the same inequality with η replaced by $\eta + \zeta$, then the difference of the transformations in the left members of the two inequalities is

$$(dF_{\alpha+\eta} - dF_{\alpha+\eta+\zeta}) - d^2F_\alpha(-\zeta),$$

and the triangle inequality therefore implies that

$$\|(dF_{\alpha+\eta} - dF_{\alpha+\eta+\zeta}) - d^2F_\alpha(-\zeta)\| \leq \epsilon(\|\eta\| + \|\eta+\zeta\|) \leq 2\epsilon(\|\eta\| + \|\zeta\|),$$

provided that both η and $\eta + \zeta$ have norms at most δ. We shall take $\|\zeta\| \leq \delta/3$ and $\|\eta\| \leq 2\delta/3$. If we hold ζ fixed, and if we set $T = d^2F_\alpha(-\zeta)$ and $G(\xi) = F(\xi) - F(\xi + \zeta)$, then this inequality becomes $\|dG_{\alpha+\eta} - T\| \leq 2\epsilon(\|\eta\| + \|\zeta\|)$, and since it holds whenever $\|\eta\| \leq 2\delta/3$, we can apply the corollary to Theorem 7.4 and conclude that $\|\Delta G_{\alpha+\eta}(\xi) - T(\xi)\| \leq 2\epsilon(\|\eta\| + \|\zeta\|)\|\xi\|$, provided that η and $\eta + \xi$ have norms at most $2\delta/3$. This inequality therefore holds if η, ζ, and ξ all have norms at most $\delta/3$. If we now set $\zeta = -\eta$, we have

$$\Delta G_{\alpha+\eta} = \Delta F_{\alpha+\eta} - \Delta F_{\alpha+\eta+\zeta} = \Delta F_{\alpha+\eta} - \Delta F_\alpha,$$

and $\Delta G_{\alpha+\eta}(\xi) = F(\alpha + \eta + \xi) - F(\alpha + \eta) - F(\alpha + \xi) + F(\alpha)$. This function of η and ξ is called the *second difference* of F at α, and is designated $\Delta^2F_\alpha(\eta, \xi)$. *Note that it is symmetric in ξ and η.* Our final inequality can now be rewritten as

$$\|\Delta^2F_\alpha(\eta, \xi) - (d^2F_\alpha(\eta))(\xi)\| \leq 4\epsilon\|\eta\|\ \|\xi\|.$$

Reversing η and ξ, and using the symmetry of Δ^2F_α, we see that

$$\|(d^2F_\alpha(\eta))(\xi) - (d^2F_\alpha(\xi))(\eta)\| \leq 8\epsilon\|\eta\|\ \|\xi\|,$$

provided η and ξ have norms at most $\delta/3$. But now it follows by the usual homogeneity argument that this inequality holds for *all* η and ξ. Finally, since ϵ is arbitrary, the left-hand side is zero. □

The reader will remember from the elementary calculus that a critical point **a** for a function f [$f'(a) = 0$] is a relative extremum point if the second derivative $f''(a)$ exists and is not zero. In fact, if $f''(a) < 0$, then f has a relative maximum at a, because $f''(a) < 0$ implies that f' is decreasing in a neighborhood of a and the graph of f is therefore concave down in a neighborhood of a. Similarly, f has a relative minimum at a if $f'(a) = 0$ and $f''(a) > 0$. If $f''(a) = 0$, nothing can be concluded.

If f is a real-valued function defined on an open set A in a finite-dimensional vector space V, if $\alpha \in A$ is a critical point of f, and if d^2f_α exists and is a *non-*

singular element of $\text{Hom}(V, V^*)$, then we can draw similar conclusions about the behavior of f near α, only now there is a richer variety of possibilities. The reader is probably already familiar with what happens for a function f from $\mathbb{R}^2$ to $\mathbb{R}$. Then a may be a relative maximum point (a "cap" point on the graph of f), a relative minimum point, or a *saddle point* as shown in Fig. 3.13 for the graph of the translated function $\Delta f_{\mathbf{a}}$. However, it must be realized that new axes may have to be chosen for the orientation of the saddle to the axes to look as shown. Replacing f by $\Delta f_{\mathbf{a}}$ amounts to supposing that 0 is the critical point and that $f(0) = 0$. Note that if 0 is a saddle point, then there are two complementary subspaces, the coordinate axes in the Fig. 3.13, such that 0 is a relative maximum for f when f is restricted to one of them, and a relative minimum point for the restriction of f to the other.

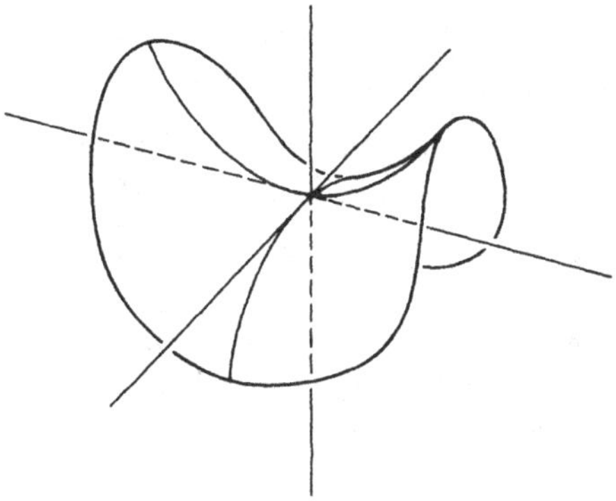

Fig. 3.13

We shall now investigate the general case and find that it is just like the two-dimensional case except that when there is a saddle point the subspace on which the critical point is a maximum point may have any dimension from 1 to $n - 1$ [where $d(V) = n$]. Moreover, this dimension is exactly the number of -1's in the standard orthonormal basis representation of the quadratic form $q(\xi) = \omega(\xi, \xi) = d^2 f_\alpha(\xi, \xi)$.

Our hypotheses, then, are that f is a continuously differentiable real-valued function on an open subset of a finite-dimensional normed linear space V, that $\alpha \in A$ is a critical point for f ($df_\alpha = 0$), and that the mapping $d^2 f_\alpha: V \to V^*$ exists and is nonsingular. This last hypothesis is equivalent to assuming that the bilinear form $\omega(\xi, \eta) = d^2 f_\alpha(\xi, \eta)$ has a nonsingular matrix with respect to any basis for V. We now use Theorem 7.1 of Chapter 2 to choose an ω-orthonormal basis $\{\alpha_i\}_1^n$. Remember that this means that $\omega(\alpha_i, \alpha_j) = 0$ if $i \neq j$, $\omega(\alpha_i, \alpha_i) = 1$ for $i = 1, \ldots, p$, and $\omega(\alpha_i, \alpha_i) = -1$ for $i = p + 1, \ldots, n$. There cannot be any 0 values for $\omega(\alpha_i, \alpha_i)$ because the matrix $t_{ij} = \omega(\alpha_i, \alpha_j)$ is nonsingular: if $\omega(\alpha_i, \alpha_i) = 0$, then the whole ith column is zero, the column space has dimension $\leq n - 1$, and the matrix is singular.

We can use the basis isomorphism φ to replace V by $\mathbb{R}^n$ (i.e., replace f by $f \circ \varphi$), and we can therefore suppose that $V = \mathbb{R}^n$ and that the standard basis is

ω-orthonormal, with $\omega(\mathbf{x}, \mathbf{y}) = \sum_1^p x_i y_i - \sum_{p+1}^n x_i y_i$. Since

$$\omega(\delta^i, \delta^j) = d^2 f_{\mathbf{a}}(\delta^i, \delta^j) = D_{\delta^i} D_{\delta^j} f(\mathbf{a}) = \frac{\partial^2 f}{\partial x_i\, \partial x_j}(\mathbf{a}),$$

our hypothesis of ω-orthogonality is that $(\partial^2 f/\partial x_i\, \partial x_j)(\mathbf{a}) = 0$ for $i \neq j$, $\partial^2 f/\partial x_i^2 = 1$ for $i = 1, \ldots, p$, and $\partial^2 f/\partial x_i^2 = -1$ for $i = p+1, \ldots, n$. Since p can have any value from 0 to n, there are $n + 1$ possibilities. We show first that if $p = n$, then $\mathbf{a}$ is a relative minimum of f. In this case the quadratic form q is said to be *positive definite*, since $q(\mathbf{x}) = \omega(\mathbf{x}, \mathbf{x})$ is positive for every nonzero $\mathbf{x}$. We also say that the bilinear form $\omega(\mathbf{x}, \mathbf{y}) = d^2 f_{\mathbf{a}}(\mathbf{x}, \mathbf{y})$ is *positive definite*, and, in the language of Chapter 5, that ω is a scalar product.

Theorem 16.4. Let f be a continuously differentiable real-valued function defined on an open subset A of $\mathbb{R}^n$, and let $\mathbf{a} \in A$ be a critical point of f at which $d^2 f$ exists and is positive definite. Then f has a relative minimum at $\mathbf{a}$.

Proof. We suppose, as above, that the standard basis $\{\delta^i\}_1^n$ is ω-orthonormal. By the definition of $d^2 f_{\mathbf{a}}$, given ϵ, there is a δ such that

$$|(df_{\mathbf{a}+\mathbf{y}}(\mathbf{x}) - df_{\mathbf{a}}(\mathbf{x})) - d^2 f_{\mathbf{a}}(\mathbf{x}, \mathbf{y})| \leq \epsilon \|\mathbf{x}\|\, \|\mathbf{y}\|$$

whenever $\|\mathbf{y}\| \leq \delta$. Now $df_{\mathbf{a}} = 0$, since $\mathbf{a}$ is a critical point of f, and $d^2 f_{\mathbf{a}}(\mathbf{x}, \mathbf{y}) = \sum_1^n x_i y_i$, by the assumption that $\{\delta^i\}_1^n$ is ω-orthonormal. Therefore, if we use the two-norm on $\mathbb{R}$ and set $\mathbf{y} = t\mathbf{x}$, we have

$$|df_{\mathbf{a}+t\mathbf{x}}(\mathbf{x}) - t\|\mathbf{x}\|^2| \leq \epsilon t \|\mathbf{x}\|^2.$$

Also, if $h(t) = f(\mathbf{a} + t\mathbf{x})$, then $h'(s) = df_{\mathbf{a}+s\mathbf{x}}(\mathbf{x})$, and this inequality therefore says that $(1 - \epsilon)t\|x\|^2 \leq h'(t) \leq (1 + \epsilon)t\|x\|^2$. Integrating, and remembering that $h(1) - h(0) = f(\mathbf{a} + \mathbf{x}) - f(\mathbf{a}) = \Delta f_{\mathbf{a}}(\mathbf{x})$, we have

$$\left(\frac{1-\epsilon}{2}\right) \|\mathbf{x}\|^2 \leq \Delta f_{\mathbf{a}}(\mathbf{x}) \leq \left(\frac{1+\epsilon}{2}\right) \|\mathbf{x}\|^2$$

whenever $\|x\| \leq \delta$. This shows not only that $\mathbf{a}$ is a relative minimum point but also that $\Delta f_{\mathbf{a}}$ lies between two very close paraboloids when $\mathbf{x}$ is sufficiently small. □

The above argument will work just as well in general. If

$$q(\mathbf{x}) = \sum_1^p x_i^2 - \sum_{p+1}^n x_i^2$$

is the quadratic form of the second differential and $\|\mathbf{x}\|_2^2 = \sum_1^n x_i^2$, then replacing $\|\mathbf{x}\|^2$ inside the absolute values in the above inequalities by $q(\mathbf{x})$, we conclude that

$$\frac{q(\mathbf{x}) - \epsilon\|\mathbf{x}\|^2}{2} \leq \Delta f_{\mathbf{a}}(\mathbf{x}) \leq \frac{q(\mathbf{x}) + \epsilon\|\mathbf{x}\|^2}{2},$$

or

$$\tfrac{1}{2}\left(\sum_1^p (1-\epsilon)x_i^2 - \sum_{p+1}^n (1+\epsilon)x_i^2\right) \le \Delta f_{\mathbf{a}}(\mathbf{x})$$
$$\le \tfrac{1}{2}\left(\sum_1^p (1+\epsilon)x_i^2 - \sum_{p+1}^n (1-\epsilon)x_i^2\right).$$

This shows that $\Delta f_{\mathbf{a}}$ lies between two very close quadratic surfaces of the same type when $\|\mathbf{x}\| \le \delta$. If $1 \le p \le n-1$ and $\mathbf{a} = 0$, then f has a relative minimum on the subspace $V_1 = L(\{\delta^i\}_1^p)$ and a relative maximum on the complementary space $V_2 = L(\{\delta^i\}_{p+1}^n)$.

According to our remarks at the end of Section 2.7, we can read off the type of a critical point for a function of *two* variables *without* orthonormalizing by looking at the determinant of the matrix of the (assumed nonsingular) form $d^2f_{\mathbf{a}}$. This determinant is

$$t_{11}t_{22} - (t_{12})^2 = \frac{\partial^2 f}{\partial x_1^2}\frac{\partial^2 f}{\partial x_2^2} - \left(\frac{\partial^2 f}{\partial x_1\,\partial x_2}\right)^2.$$

If it is positive, then $\mathbf{a}$ is either a relative minimum or a relative maximum. We can tell which by following f along a single line, say the x_1-axis. Thus, if $\partial^2 f/\partial x_1^2 < 0$, then $\mathbf{a}$ is a relative maximum point. On the other hand, if the above expression is negative, then $\mathbf{a}$ is a saddle point.

It is important for the calculus of variations that Theorem 16.4 remains true when the domain space is replaced by a space of the general type that we shall study in the next chapter, called a Banach space. The hypotheses now are that α is a critical point of f, that $q(\xi) = df_\alpha(\xi, \xi)$ is positive definite, and that the scalar product norm $q^{1/2}$ (see Chapter 5) is equivalent to the given norm on V. The proof remains virtually unchanged.

*17. HIGHER ORDER DIFFERENTIALS. THE TAYLOR FORMULA

We have seen that if V and W are normed linear spaces, and if F is a differentiable mapping from an open set A in V to W, then its differential $dF = dF_{(\cdot)}$ is a mapping from A to $\mathrm{Hom}(V, W)$. If *this* mapping is differentiable on A, then *its* differential $d(dF) = d(dF)_{(\cdot)}$ is a mapping from A to $\mathrm{Hom}(V, \mathrm{Hom}(V, W))$. We remember that an element of $\mathrm{Hom}(V, \mathrm{Hom}(V, W))$ is equivalent by duality to a bilinear mapping from $V \times V$ to W, and if we designate the space of all such bilinear mappings by $\mathrm{Hom}^2(V, W)$, then $d(dF)$ can be considered to be from A to $\mathrm{Hom}^2(V, W)$. We write $d(dF) = d^2F$, and call this mapping the second differential of F. In Section 16 we saw that $d^2F_\alpha(\xi, \eta) = D_\xi(D_\eta F)(\alpha)$, and that if $V = \mathbb{R}^n$, then

$$d^2F_{\mathbf{a}}(\mathbf{b}, \mathbf{c}) = D_{\mathbf{b}}D_{\mathbf{c}}F(\mathbf{a}) = \sum b_i c_j \frac{\partial^2 F}{\partial x_j\,\partial x_i}(\mathbf{a}).$$

The differentials of higher order are defined in the same way. If

$$d^2F\colon A \to \mathrm{Hom}^2(V, W)$$

is differentiable on A, then its differential, $d(d^2F) = d^3F$, is from A to $\mathrm{Hom}(V, \mathrm{Hom}^2(V, W)) = \mathrm{Hom}^3(V, W)$, the space of all trilinear mappings from $V^3 = V \times V \times V$ to W. Continuing inductively, we arrive at the notion of the nth differential of F on A as a mapping from A to $\mathrm{Hom}(V, \mathrm{Hom}^{n-1}(V, W)) = \mathrm{Hom}^n(V, W)$, the space of all n-linear mappings from V^n to W. The theorem that d^2F_α is a *symmetric* element of $\mathrm{Hom}^2(V, W)$ extends inductively to show that d^nF_α is a symmetric element of $\mathrm{Hom}^n(V, W)$. We shall omit this proof.

Our theorem on the evaluation of the second differential by mixed directional derivatives also generalizes by induction to give

$$D_{\xi_1}, \ldots, D_{\xi_n}F(\alpha) = d^nF_\alpha(\xi_1, \ldots, \xi_n),$$

for starting from the left-hand term, we have

$$\begin{aligned} D_{\xi_1}(D_{\xi_2}, \ldots, D_{\xi_n}F)(\cdot)|_\alpha &= d(D_{\xi_2}, \ldots, D_{\xi_n}F(\cdot))_\alpha(\xi_1) \\ &= d(d^{n-1}F_{(\cdot)}(\xi_2, \ldots, \xi_n))_\alpha(\xi_1) \\ &= d(\mathrm{ev}_{<\xi_2,\ldots,\xi_n>} \circ d^{n-1}F_{(\cdot)})_\alpha(\xi_1) \\ &= [\mathrm{ev}_{<\xi_2,\ldots,\xi_n>} \circ d(d^{n-1}F_{(\cdot)})_\alpha](\xi_1) \\ &= \mathrm{ev}_{<\xi_2,\ldots,\xi_n>}(d^nF_\alpha(\xi_1)) \\ &= (d^nF_\alpha(\xi_1))(\xi_2, \ldots, \xi_n) = d^nF_\alpha(\xi_1, \ldots, \xi_n). \end{aligned}$$

If $V = \mathbb{R}^n$, then our conclusions about partial derivatives extend inductively in the same way to show that F has continuous differentials on A up through order m if and only if all the mth-order partial derivatives $\partial^m F/\partial x_{i_1}, \ldots, \partial x_{i_m}$ exist and are continuous on A, with

$$d^mF_{\mathbf{a}}(c^1, \ldots, c^m) = \sum_{i_1,\ldots,i_m=1}^{n} c^1_{i_1}, \ldots, c^m_{i_m} \frac{\partial^m F}{\partial x_{i_1}, \ldots, \partial x_{i_m}}(\mathbf{a}).$$

We now consider the behavior of F along the line $t \mapsto \alpha + t\eta$, where, of course, α and η are fixed. If $\lambda(t) = F(\alpha + t\eta)$, then we can prove by induction that

$$d^j\lambda/dt^j = (D_\eta)^jF(\alpha + t\eta).$$

We know this to be true for $j = 1$ by Theorem 7.2, and assuming it for $j = m$, we have, by the same theorem,

$$\frac{d^{m+1}\lambda}{dt^{m+1}} = \left(\frac{d\lambda^m}{dt^m}\right)'(t) = d(D^m_\eta F)_{\alpha+t\eta}(\eta) = D_\eta(D^m_\eta F)(\alpha + t\eta) = D^{m+1}_\eta F(\alpha + t\eta).$$

Now suppose that F is real-valued ($W = \mathbb{R}$). We then have Taylor's formula:

$$\lambda(t) = \lambda(0) + t\lambda'(0) + \cdots + \frac{t^m}{m!}\lambda^{(m)}(0) + \frac{t^{m+1}}{(m+1)!}\lambda^{(m+1)}(kt)$$

for some k between 0 and 1. Taking $t = 1$ and substituting from above, we have

$$F(\alpha + \eta) = F(\alpha) + D_\eta F(\alpha) + \cdots + \frac{1}{m!}D^m_\eta F(\alpha) + \frac{1}{(m+1)!}D^{m+1}_\eta F(\alpha + k\eta),$$

which is the general Taylor formula in the normed linear space context. In terms of differentials, it is

$$F(\alpha+\eta) = F(\alpha) + dF_\alpha(\eta) + \cdots + \frac{1}{m!}\, d^m F_\alpha(\eta, \ldots, \eta)$$
$$+ \frac{1}{(m+1)!}\, d^{m+1} F_{\alpha+k\eta}(\eta, \ldots, \eta).$$

If $V = \mathbb{R}^n$, then $D_{\mathbf{y}}G = \sum_1^n y_i\, \partial G/\partial x_i$, and so the general term in the Taylor expansion is

$$\frac{1}{m!}\left(\sum_1^n y_i \frac{\partial}{\partial x_i}\right)^m F(\mathbf{a}) = \frac{1}{m!} \sum_{i_1, \ldots, i_m = 1}^{n} y_{i_1} \cdots y_{i_m} \frac{\partial^m F}{\partial x_{i_1} \cdots \partial x_{i_m}}(\mathbf{a}).$$

If $m = n = 2$, and if we use the notation $\mathbf{x} = \langle x, y \rangle$, $\mathbf{s} = \langle s, t \rangle$, then

$$\frac{1}{2!} D^2_{\mathbf{s}} F(\mathbf{a}) = \frac{1}{2}\left[s^2 \frac{\partial^2 F}{\partial x^2}(\mathbf{a}) + 2st \frac{\partial^2 F}{\partial x\, \partial y}(\mathbf{a}) + t^2 \frac{\partial^2 F}{\partial y^2}(\mathbf{a})\right].$$

The above description is logically simple, but it is inefficient in that it repeats identical terms such as $y_1 y_2(\partial^2 F/\partial x_1\, \partial x_2)$ and $y_2 y_1(\partial^2 F/\partial x_2\, \partial x_1)$. We conclude by describing for the interested reader the modern "multi-index" notation for this very complicated situation.

Remember that we are looking at the mth term of the Taylor formula for F, and that F has n variables.

For any n-tuple $\mathbf{k} = \langle k_1, \ldots, k_n \rangle$ of nonnegative integers, we define $|\mathbf{k}|$ as $\sum_1^n k_i$, and for $\mathbf{x} \in \mathbb{R}^n$, we set $\mathbf{x}^{\mathbf{k}} = x_1^{k_1} x_2^{k_2} \cdots x_n^{k_n}$. Also we set $F_{\mathbf{k}} = F_{k_1 k_2 \cdots k_n}$, or better, if $D_j F = \partial F/\partial x_j$, we set

$$D^{\mathbf{k}} F = D_1^{k_1} D_2^{k_2} \cdots D_n^{k_n} F = F_{\mathbf{k}}.$$

Finally, we set $\mathbf{k}! = k_1! k_2! \cdots k_n!$, and if $p \geq |\mathbf{k}|$, we set $\binom{p}{\mathbf{k}} = p!/\mathbf{k}!(p - |\mathbf{k}|)!$. Then the mth term of the Taylor expansion of F is

$$\frac{1}{m!} \sum_{|\mathbf{k}|=m} \binom{m}{\mathbf{k}} D^{\mathbf{k}} F(\mathbf{a}) \mathbf{x}^{\mathbf{k}},$$

which is surely a notational triumph.

The general Taylor formula is too cumbersome to be of much use in practice; it is principally of theoretical value. The Taylor expansions that we actually compute are generally found by other means, such as substitution of a polynomial (or power series) in a power series. For example,

$$\sin(x + y^2) = (x + y^2) - \frac{(x+y^2)^3}{3!} + \frac{(x+y^2)^5}{5!} \cdots$$
$$= x + y^2 - \frac{x^3}{3!} - \frac{x^2 y^2}{2} + \left(\frac{x^5}{5!} - \frac{xy^4}{2}\right) + \left(\frac{x^4 y^2}{4!} - \frac{y^6}{3!}\right) \cdots$$

A mapping from A to W which has continuous differentials of all orders through k is said to be of class C^k on A, and the collection of all such mappings is designated $C^k(A, W)$ or $\mathcal{C}^k(A, W)$. It is clear that $C^k(A, W)$ is a vector space (induction). Moreover, it can also be shown by induction that a composition of C^k-maps is itself of class C^k. This depends on recognizing the general form of the mth differential of a composition $F \circ G$ as being a finite sum, each term of which is a composition of functions chosen from $F, dF, \ldots, d^mF, G, dG, \ldots, d^mG$.

Functions of many variables are involved in these calculations, and it is simplest to treat each as a function of a single n-tuplet variable and to apply the obvious corollary of Theorem 8.1 that if $G^1, \ldots, G^n$ are of class C^k, then so is $G = \langle G^1, \ldots, G^n \rangle$, with $d^kG = \langle d^kG^1, \ldots, d^kG^n \rangle$. As a special case of composition, we can conclude that a product of C^k-maps is of class C^k.

We shall see in the next chapter that $\varphi: T \mapsto T^{-1}$ is a differentiable map on the open set of invertible elements in Hom V (if V is a Banach space) and that $d\varphi_T(H) = -T^{-1}HT^{-1}$. Since $\langle S, H, T \rangle \mapsto S^{-1}HT^{-1}$ then has continuous partial differentials, we can continue, and another induction shows that φ is of class C^k for every k and that $d^m\varphi_T(H_1, \ldots, H_m)$ is a finite sum of finite products of $T^{-1}, H_1, \ldots, H_m$. It then follows that a function F defined implicitly by a C^k-function G is also of class C^k, for its differential, as computed in the implicit-function theorem, is then a composition of maps of class C^{k-1}.

A mapping F which is of class C^k for all k is said to be of class C^∞, and it follows from our remarks above that the family of C^∞-maps is closed under all the operations that we have met in the calculus. If the domain of F is an open set in $\mathbb{R}^n$, then $F \in \mathcal{C}^\infty(A, W)$ if and only if all the partial derivatives of F exist and are continuous on A.

CHAPTER 4

COMPACTNESS AND COMPLETENESS

In this chapter we shall investigate two properties of subsets of a normed linear space V which are concerned with the fact that in a certain sense all the points which ought to be there really are there. These notions are largely independent of the algebraic structure of V, and we shall therefore study them in their own most natural setting, that of *metric spaces*. The stronger of these two properties, *compactness*, helps to explain why the theory of finite-dimensional spaces is so simple and satisfactory. The weaker property, *completeness*, is shared by important infinite-dimensional normed linear spaces, and allows us to treat these spaces in almost as satisfactory a way.

It is these properties that save the calculus from being largely a formal theory. They allow us to define crucial elements by limiting processes, and are responsible, for example, for an infinite series *having* a sum, a continuous real-valued function *assuming* a maximum value, and a definite integral *existing*. For the real number system itself, the compactness property is equivalent to the least upper bound property, which has already been an absolutely essential tool in our construction of the differential calculus in Chapter 3.

In Sections 8 through 10 we shall apply completeness to the calculus. The first of these sections is devoted to the existence and differentiability of functions defined by power series, and since we want to include power series in an operator T, we shall take the occasion to introduce and exploit the notion of a Banach algebra. Next we shall prove the contraction mapping fixed-point theorem, which is the missing ingredient in our unfinished proof of the implicit-function theorem in Chapter 3 and which will be the basis for the fundamental existence and uniqueness theorem for ordinary differential equations in Chapter 6. In Section 10 we shall prove a simple extension theorem for linear mappings into a complete normed linear space and apply it to construct the Riemann integral of a paramatrized arc.

1. METRIC SPACES; OPEN AND CLOSED SETS

In the preceding chapter we occasionally treated questions of convergence and continuity in situations where the domain was an arbitrary subset A of a normed linear space V. In such discussions the algebraic structure of V fades into the background, and the vector operations of V are used only to produce the combination $\|\alpha - \beta\|$, which is interpreted as the distance from α to β. If we distill out of these contexts what is essential to the convergence and continuity arguments, we find that we need a space A and a function $\rho: A \times A \to \mathbb{R}$, $\rho(x, y)$ being called the *distance* from x to y, such that

1) $\rho(x, y) > 0$ if $x \neq y$, and $\rho(x, x) = 0$;
2) $\rho(x, y) = \rho(y, x)$ for all $x, y \in A$;
3) $\rho(x, z) \leq \rho(x, y) + \rho(y, z)$ for all $x, y, z \in A$.

Any set A together with such a function ρ from $A \times A$ to $\mathbb{R}$ is called a *metric space*; the function ρ is the *metric*. It is obvious that a normed linear space is a metric space under the norm metric $\rho(\alpha, \beta) = \|\alpha - \beta\|$ and that any subset B of a metric space A is itself a metric space under $\rho \restriction B \times B$. If we start with a nice intuitive space, like $\mathbb{R}^n$ under one of its standard norms, and choose a weird subset B, it will be clear that a metric space can be a very odd object, and may fail to have almost any property one can think of.

Metric spaces very often arise in practice as subsets of normed linear spaces with the norm metric, but they come from other sources too. Even in the normed linear space context, metrics other than the norm metric are used. For example, S might be a two-dimensional spherical surface in $\mathbb{R}^3$, say $S = \{\mathbf{x} : \sum_1^3 x_i^2 = 1\}$, and $\rho(\mathbf{x}, \mathbf{y})$ might be the great circle distance from $\mathbf{x}$ to $\mathbf{y}$. Or, more generally, S might be any smooth two-dimensional surface in $\mathbb{R}^3$, and $\rho(\mathbf{x}, \mathbf{y})$ might be the length of the shortest curve connecting $\mathbf{x}$ to $\mathbf{y}$ in S.

In this chapter we shall adopt the metric space context for our arguments wherever it is appropriate, so that the student may become familiar with this more general but very intuitive notion. We begin by reproducing the basic definitions in the language of metrics. Because the scalar-vector dichotomy is not a factor in this context, we shall drop our convention that points be represented by Greek or boldface roman letters and shall use whatever letters we wish.

Definition. If X and Y are metric spaces, then $f: X \to Y$ is *continuous at* $a \in X$ if for every ϵ there is a δ such that

$$\rho(x, a) < \delta \Rightarrow \rho(f(x), f(a)) < \epsilon.$$

Here we have used the same symbol 'ρ' for metrics on different spaces, just as earlier we made ambiguous use of the norm symbol.

Definition. The *(open) ball of radius r about p*, $B_r(p)$, is simply the set of points whose distance from p is less that r:

$$B_r(p) = \{x : \rho(x, p) < r\}.$$

Definition. A subset $A \subset X$ is *open* if every point p in A is the center of some ball included in A, that is, if

$$(\forall p \in A)(\exists r^{>0})(B_r(p) \subset A).$$

Lemma 1.1. Every ball is open; in fact, if $q \in B_r(p)$ and $\delta = r - \rho(p, q)$, then $B_\delta(q) \subset B_r(p)$.

Proof. This amounts to the triangle inequality. For, if $x \in B_\delta(q)$, then $\rho(x, q) < \delta$ and $\rho(x, p) \leq \rho(x, q) + \rho(q, p) < \delta + \rho(p, q) = r$, so that $x \in B_r(p)$. Thus $B_\delta(q) \subset B_r(p)$. □

Lemma 1.2. If p is held fixed, then $\rho(p, x)$ is a continuous function of x.

Proof. A symbol-by-symbol paraphrase of Lemma 3.1 of Chapter 3 shows that $|\rho(p, x) - \rho(p, y)| \leq \rho(x, y)$, so that $\rho(p, x)$ is actually a Lipschitz function with constant 1. □

Theorem 1.1. The family $\mathfrak{I}$ of all open subsets of a metric space S has the following properties:

1) The union of any collection of open sets is open; that is, if $\{A_i : i \in I\} \subset \mathfrak{I}$, then $\bigcup_{i \in I} A_i \in \mathfrak{I}$.
2) The intersection of two open sets is open; that is, if $A, B \in \mathfrak{I}$, then $A \cap B \in \mathfrak{I}$.
3) $\emptyset, V \in \mathfrak{I}$.

Proof. These properties follow immediately from the definition. Thus any point p in $\bigcup_i A_i$ lies in some A_j, and therefore, since A_j is open, some ball about p is a subset of A_j and hence of the larger set $\bigcup_i A_i$. □

Corollary. A set is open if and only if it is a union of open balls.

Proof. This follows from the definition of open set, the lemma above, and property (1) of the theorem. □

The union of *all* the open subsets of an arbitrary set A is an open subset of A, by (1), and therefore is the *largest* open subset of A. It is called the interior of A and is designated A^{int}. Clearly, p is in A^{int} if and only if some ball about p is a subset of A, and it is helpful to visualize A^{int} as the union of all the balls that are in A.

Definition. A set A is *closed* if A' is open.

The theorem above and De Morgan's law (Section 0.11) then yield the following complementary set of properties for closed sets.

Theorem 1.2

1) The intersection of any family of closed sets is closed.
2) The union of two closed sets is closed.
3) $\emptyset$ and V are closed.

Proof. Suppose, for example, that $\{B_i : i \in I\}$ is a family of closed sets. Then the complement B_i' is open for each i, so that $\bigcup_i B_i'$ is open by the above theorem.

Also, $\bigcup_i B_i' = (\bigcap_i B_i)'$ by De Morgan's law (see Section 0.11). Thus $\bigcap_i B_i$ is the complement of an open set and is closed. □

Continuing our "complementary" development, we define the *closure*, $\overline{A}$, of an arbitrary set A as the intersection of all closed sets including A, and we have from (1) above that *$\overline{A}$ is the smallest closed set including A*. De Morgan's law implies the important identity

$$(\overline{A})' = (A')^{\text{int}}.$$

For F is a closed superset of A if and only if its complement $U = F'$ is an open subset of A'. By De Morgan's law the complement of the intersection of all such sets F is the union of all such sets U. That is, the complement of $\overline{A}$ is $(A')^{\text{int}}$.

This identity yields a direct characterization of closure:

Lemma 1.3. A point p is in $\overline{A}$ if and only if every ball about p intersects A.

Proof. A point p is *not* in $\overline{A}$ if and only if p is in the interior of A', that is, if and only if *some* ball about p does not intersect A. Negating the extreme members of this equivalence gives the lemma. □

Definition. The *boundary*, ∂A, of an arbitrary set A is the difference between its closure and its interior. Thus

$$\partial A = \overline{A} - A^{\text{int}}.$$

Since $A - B = A \cap B'$, we have the symmetric characterization $\partial A = \overline{A} \cap (\overline{A'})$. Therefore, $\partial A = \partial(A')$; also,

$$p \in \partial A \text{ if and only if every ball about } p \text{ intersects both } A \text{ and } A'.$$

Example. A ball $B_r(\alpha)$ is an open set. In a normed linear space the closure of $B_r(\alpha)$ is the closed ball about α of radius r, $\{\xi : \rho(\xi, \alpha) \leq r\}$. This is easily seen from Lemma 1.3. The boundary $\partial B_r(\alpha)$ is then the *spherical surface* of radius r about α, $\{\xi : \rho(\xi, \alpha) = r\}$. If some but not all of the points of this surface are added to the open ball, we obtain a set that is neither open nor closed. The student should expect that a random set he may encounter will be neither open nor closed.

Continuous functions furnish an important source of open and closed sets by the following lemma.

Lemma 1.4. If X and Y are metric spaces, and if f is a continuous mapping from X to Y, then $f^{-1}[A]$ is open in X whenever A is open in Y.

Proof. If $p \in f^{-1}[A]$, then $f(p) \in A$, and, since A is open, some ball $B_\epsilon(f(p))$ is a subset of A. But the continuity of f at p says exactly that there is a δ such that $f[B_\delta(p)] \subset B_\epsilon(f(p))$. In particular, $f[B_\delta(p)] \subset A$ and $B_\delta(p) \subset f^{-1}[A]$. Thus for each p in $f^{-1}[A]$ there is a ball about p included in $f^{-1}[A]$, and this set is therefore open. □

Since $f^{-1}[A'] = (f^{-1}[A])'$, we also have the following corollary.

Corollary. If $f: X \to Y$ is continuous, then $f^{-1}[C]$ is closed in X whenever C is closed in Y.

The converses of both of these results hold as well. As an example of the use of this lemma, consider for a fixed $\alpha \in X$ the continuous function $f: X \to \mathbb{R}$ defined by $f(\xi) = \rho(\xi, \alpha)$. The sets $(-r, r)$, $[0, r]$, and $\{r\}$ are respectively open, closed, and closed subsets of $\mathbb{R}$. Therefore, their inverse images under f—the ball $B_r(\alpha)$, the closed ball $\{\xi : \rho(\xi, \alpha) \leq r\}$, and the spherical surface $\{\xi : \rho(\xi, \alpha) = r\}$ —are respectively open, closed, and closed in X. In particular, the triangle inequality argument demonstrating directly that $B_r(\alpha)$ is open is now seen to be unnecessary by virtue of the triangle inequality argument that demonstrates the continuity of the distance function (Lemma 1.2).

It is *not* true that continuous functions take closed sets into closed sets in the forward direction. For example, if $f: \mathbb{R} \to \mathbb{R}$ is the arc tangent function, then $f[\mathbb{R}] = \text{range } f = (-\pi/2, \pi/2)$, which is not a closed subset of $\mathbb{R}$. The reader may feel that this example cheats and that we should only expect the f-image of a closed set to be a closed subset of the metric space that is the range of f. He might then consider $f(x) = 2x/(1 + x^2)$ from $\mathbb{R}$ to its range $[-1, 1]$. The set of positive integers $\mathbb{Z}^+$ is a closed subset of $\mathbb{R}$, but $f[\mathbb{Z}^+] = \{2n/(1 + n^2)\}_1^\infty$ is not closed in $[-1, 1]$, since 0 is clearly in its closure.

The *distance* between two nonempty sets A and B, $\rho(A, B)$, is defined as $\text{glb } \{\rho(a, b) : a \in A \text{ and } b \in B\}$. If A and B intersect, the distance is zero. If A and B are disjoint, the distance may still be zero. For example, the interior and exterior of a circle in the plane are disjoint *open* sets whose distance apart is zero. The x-axis and (the graph of) the function $f(x) = 1/x$ are disjoint *closed* sets whose distance apart is zero. As we have remarked earlier, a set A is closed if and only if every point not in A is a positive distance from A. More generally, for any set A a point p is in $\overline{A}$ if and only if $\rho(p, A) = 0$.

We list below some simple properties of the distance between subsets of a normed linear space.

1) Distance is unchanged by a translation: $\rho(A, B) = \rho(A + \gamma, B + \gamma)$ (because $\|(\alpha + \gamma) - (\beta + \gamma)\| = \|\alpha - \beta\|$).
2) $\rho(kA, kB) = |k|\rho(A, B)$ (because $\|k\alpha - k\beta\| = |k| \, \|\alpha - \beta\|$).
3) If N is a subspace, then the distance from B to N is unchanged when we translate B through a vector in N: $\rho(N, B) = \rho(N, B + \eta)$ if $\eta \in N$ (because $N - \eta = N$).
4) If $T \in \text{Hom}(V, W)$, then $\rho(T[A], T[B]) \leq \|T\|\rho(A, B)$ (because
$$\|T(\alpha) - T(\beta)\| \leq \|T\| \cdot \|\alpha - \beta\|).$$

Lemma 1.5. If N is a proper closed subspace and $0 < \epsilon < 1$, there exists an α such that $\|\alpha\| = 1$ and $\rho(\alpha, N) > 1 - \epsilon$.

Proof. Choose any $\beta \notin N$. Then $\rho(\beta, N) > 0$ (because N is closed), and there exists an $\eta \in N$ such that

$$\|\beta - \eta\| < \rho(\beta, N)/(1 - \epsilon)$$

[by the definition of $\rho(\beta, N)$]. Set $\alpha = (\beta - \eta)/\|\beta - \eta\|$. Then $\|\alpha\| = 1$ and

$$\begin{aligned}\rho(\alpha, N) &= \rho(\beta - \eta, N)/\|\beta - \eta\| \\ &= \rho(\beta, N)/\|\beta - \eta\| > \rho(\beta, N)(1 - \epsilon)/\rho(\beta, N) = 1 - \epsilon,\end{aligned}$$

by (2), (3), and the definition of η. □

The reader may feel that we ought to be able to improve this lemma. Surely, all we have to do is choose the point in N which is closest to β, and so obtain $\|\beta - \eta\| = \rho(\beta, N)$, giving finally a vector α such that $\|\alpha\| = 1$ and $\rho(\alpha, N) = 1$. However, this is a matter on which our intuition lets us down: if N is infinite-dimensional, there may not be a closest point η! For example, as we shall see later in the exercises of Chapter 5, if V is the space $\mathcal{C}([-1, 1])$ under the two-norm $\|f\| = (\int_{-1}^{1} f^2)^{1/2}$, and if N is the set of functions g in V such that $\int_0^1 g = 0$, then N is a closed subspace for which we cannot find such a "best" α. But if N is finite-dimensional, we can always find such a point, and if V is a Hilbert space, (see Chapter 5) we can also.

EXERCISES

1.1 Write out the proof of Lemma 1.2.

1.2 Prove (2) and (3) of Theorem 1.1.

1.3 Prove (2) of Theorem 1.2.

1.4 It is not true that the intersection of a sequence of open sets is necessarily open. Find a counterexample in $\mathbb{R}$.

1.5 Prove the corollary of Lemma 1.4.

1.6 Prove that $p \in \overline{A}$ if and only if $\rho(p, A) = 0$.

1.7 Let X and Y be metric spaces, and let $f: X \to Y$ have the property that $f^{-1}[B]$ is open in X whenever B is open in Y. Prove that f is continuous.

1.8 Show that $\rho(x, A) = \rho(x, \overline{A})$.

1.9 Show that $\rho(x, A)$ is a continuous function of x. (In fact, it is Lipschitz continuous.)

1.10 Invent metric spaces S (by choosing subsets of $\mathbb{R}^2$) having the following properties:

1) S has n points.
2) S is infinite and $\rho(x, y) \geq 1$ if $x \neq y$.
3) S has a ball $B_1(a)$ such that the closed ball $\{x: \rho(x, a) \leq 1\}$ is not the same as the closure of $B_1(a)$.

1.11 Prove that in a normed linear space a closed ball is the closure of the corresponding open ball.

1.12 Show that if $f: X \to Y$ and $g: Y \to Z$ are continuous (where X, Y, and Z are metric spaces), then so is $g \circ f$.

1.13 Let X and Y be metric spaces. Define the notion of a product metric on $Z = X \times Y$. Define a 1-metric ρ_1 and a uniform metric ρ_∞ on Z (showing that they *are* metrics) in analogy with the 1-norm and uniform norm on a product of normed linear spaces, and show that each is a product metric according to your definition above.

1.14 Do the same for a 2-metric ρ_2 on $Z = X \times Y$.

1.15 Let X and Y be metric spaces, and let V be a normed linear space. Let $f: X \to \mathbb{R}$ and $g: Y \to V$ be continuous maps. Prove that

$$\langle x, y \rangle \mapsto f(x)\, g(y)$$

is a continuous map from $X \times Y$ to V.

*2. TOPOLOGY

If X is an arbitrary set and $\mathfrak{I}$ is any family of subsets of X satisfying properties (1) through (3) in Theorem 1.1, then $\mathfrak{I}$ is called a *topology* on X. Theorem 1.1 thus asserts that the open subsets of a metric space X form a topology on X. The subsequent definitions of interior, closed set, and closure were purely topological in the sense that they depended only on the topology $\mathfrak{I}$, as were Theorem 1.2 and the identity $(\overline{A})' = (A')^{\text{int}}$. The study of the consequences of the existence of a topology is called *general topology*.

On the other hand, the definitions of balls and continuity given earlier were *metric* definitions, and therefore part of *metric space* theory. In metric spaces, then, we have not only the topology, but also our ϵ-definitions of continuity and balls and the spherical characterizations of closure and interior.

The reader may be surprised to be told now that although continuity and convergence were defined metrically, they also have purely topological characterizations and are therefore topological ideas. This is easy to see if one keeps in mind that in a metric space an open set is nothing but a union of balls. We have:

> f is continuous at p if and only if for every open set A containing $f(p)$ there exists an open set B containing p such that $f[B] \subset A$.

This *local* condition involving behavior around a single point p is more fluently rendered in terms of the notion of neighborhood. A set A is a *neighborhood* of a point p if $p \in A^{\text{int}}$. Then we have:

> f is continuous at p if and only if for every neighborhood N of $f(p)$, $f^{-1}[N]$ is a neighborhood of p.

Finally there is an elegant topological characterization of global continuity. Suppose that S_1 and S_2 are topological spaces. Then $f: S_1 \to S_2$ is continuous (everywhere) if and only if $f^{-1}[A]$ is open whenever A is open. Also, f is continuous if and only if $f^{-1}[B]$ is closed whenever B is closed. These conditions are not surprising in view of Lemma 1.4.

3. SEQUENTIAL CONVERGENCE

In addition to shifting to the more general point of view of metric space theory, we also want to add to our kit of tools the notion of sequential convergence, which the reader will probably remember from his previous encounter with the calculus. One of the principal reasons why metric space theory is simpler and more intuitive than general topology is that nearly all metric arguments can be presented in terms of sequential convergence, and in this chapter we shall partially make up for our previous neglect of this tool by using it constantly and in preference to other alternatives.

Definition. We say that the infinite sequence $\{x_n\}$ converges to the point a if for every ϵ there is an N such that

$$n > N \Rightarrow \rho(x_n, a) < \epsilon.$$

We also say that x_n approaches a as n approaches (or tends to) infinity, and we call a the limit of the sequence. In symbols we write $x_n \to a$ as $n \to \infty$, or $\lim_{n\to\infty} x_n = a$. Formally, this definition is practically identical with our earlier definition of function convergence, and where there are parallel theorems the arguments that we use in one situation will generally hold almost verbatim in the other. Thus the proof of Lemma 1.1 of Chapter 3 can be alternated slightly to give the following result.

Lemma 3.1. If $\{\xi_i\}$ and $\{\eta_i\}$ are two sequences in a normed linear space V, then

$$\xi_i \to \alpha \text{ and } \eta_i \to \beta \Rightarrow \xi_i + \eta_i \to \alpha + \beta.$$

The main difference is that we now choose N as max $\{N_1, N_2\}$ instead of choosing δ as min $\{\delta_1, \delta_2\}$. Similarly:

Lemma 3.2. If $\xi_i \to \alpha$ in V and $x_i \to a$ in $\mathbb{R}$, then $x_i\xi_i \to a\alpha$.

As before, the definition begins with three quantifiers, $(\forall\epsilon)(\exists N)(\forall n)$. A somewhat more idiomatic form can be obtained by rephrasing the definition in terms of balls and the notion of "almost all n". We say that $P(n)$ is true for *almost all* n if $P(n)$ is true for all but a finite number of integers n, or equivalently, if $(\exists N)(\forall n^{>N})P(n)$. Then we see that

$\lim x_n = a$ if and only if every ball about a contains almost all the x_n.

The following sequential characterization provides probably the most intuitive way of viewing the notion of closure and closed sets.

Theorem 3.1. A point x is in the closure $\overline{A}$ of a set A if and only if there is a sequence $\{x_n\}$ in A converging to x.

Therefore, a set A is closed if and only if every convergent sequence lying in A has its limit in A.

Proof. If $\{x_n\} \subset A$ and $x_n \to x$, then every ball about x contains almost every x_n, and so, in particular, intersects A. Thus $x \in \overline{A}$ by Lemma 1.3. Conversely, if $x \in \overline{A}$, then every ball about x intersects A, and we can construct a sequence in A that converges to x by choosing x_n as any point in $B_{1/n}(x) \cap A$. Since A is closed if and only if $A = \overline{A}$, the second statement of the theorem follows from the first. □

There is also a sequential characterization of continuity which helps greatly in using the notion of continuity in a flexible way. Let X and Y be metric spaces, and let f be any function from X to Y.

Theorem 3.2. The function f is continuous at a if and only if, for any sequence $\{x_n\}$ in X, if $x_n \to a$, then $f(x_n) \to f(a)$.

Proof. Suppose first that f is continuous at a, and let $\{x_n\}$ be any sequence converging to a. Then, given any ϵ, there is a δ such that

$$\rho(x\,,\, a) < \delta \Rightarrow \rho(f(x)\,,\, f(a)) < \epsilon,$$

by the continuity of f at a, and for this δ there is an N such that

$$n > N \Rightarrow \rho(x_n\,,\, a) < \delta,$$

because $x_n \to a$. Combining these implications, we see that given ϵ we have found N so that $n > N \Rightarrow \rho(f(x_n)\,,\, f(a)) < \epsilon$. That is, $f(x_n) \to f(a)$.

Now suppose that f is not continuous at a. In considering such a negation it is important that implicit universal quantifiers be made explicit. Thus, formally, we are assuming that $\sim(\forall\epsilon)(\exists\delta)(\forall x)(\rho(x\,,\, a) < \delta \Rightarrow \rho(f(x)\,,\, f(a)) < \epsilon)$, that is, that $(\exists\epsilon)(\forall\delta)(\exists x)(\rho(x\,,\, a) < \delta \;\&\; \rho(f(x)\,,\, f(a)) \geq \epsilon)$. Such symbolization will not be necessary after the reader has had some practice in computing logical negations; the experienced thinker will intuit the correct negation without a formal calculation. In any event, we now have a fixed ϵ, and for each δ of the form $\delta = 1/n$ we can let x_n be a corresponding x. We then have $\rho(x_n\,,\, a) < 1/n$ and $\rho(f(x_n)\,,\, f(a)) \geq \epsilon$ for all n. The first inequality shows that $x_n \to a$; the second shows that $f(x_n) \not\to f(a)$. Thus, if f is *not* continuous at a, then the sequential condition is *not* satisfied. □

The above type of argument is used very frequently and almost amounts to an automatic proof procedure in the relevant situations. We want to prove, say, that $(\forall x)(\exists y)(\forall z)P(x, y, z)$. Arguing by contradiction, we suppose this false, so that $(\exists x)(\forall y)(\exists z)\sim P(x, y, z)$. Then, instead of trying to use all numbers y, we let y run through some sequence converging to zero, such as $\{1/n\}$, and we choose one corresponding z, z_n, for each such y. We end up with $\sim P(x, 1/n, z_n)$ for the given x and all n, and we finish by arguing sequentially.

The reader will remember that two norms p and q on a vector space V are equivalent if and only if the identity map $\xi \mapsto \xi$ is continuous from $\langle V, p \rangle$ to $\langle V, q \rangle$ and also from $\langle V, q \rangle$ to $\langle V, p \rangle$. By virtue of the above theorem we now see that:

Theorem 3.3. The norms p and q are equivalent if and only if they yield exactly the same collection of convergent sequences.

Earlier we argued that a norm on a product $V \times W$ of two normed linear spaces should be equivalent to $\|\langle \alpha, \xi \rangle\|_1 = \|\alpha\| + \|\xi\|$. Now with respect to this sum norm it is clear that a sequence $\langle \alpha_n, \xi_n \rangle$ in $V \times W$ converges to $\langle \alpha, \xi \rangle$ if and only if $\alpha_n \to \alpha$ in V and $\xi_n \to \xi$ in W. We now see (again by Theorem 3.2) that:

Theorem 3.4. A product norm on $V \times W$ is any norm with the property that $\langle \alpha_n, \xi_n \rangle \to \langle \alpha, \xi \rangle$ in $V \times W$ if and only if $\alpha_n \to \alpha$ in V and $\xi_n \to \xi$ in W.

EXERCISES

3.1 Prove that a convergent sequence in a metric space has a unique limit. That is, show that if $x_n \to a$ and $x_n \to b$, then $a = b$.

3.2 Show that $x_n \to x$ in the metric space X if and only if $\rho(x_n, x) \to 0$ in $\mathbb{R}$.

3.3 Prove that if $x_n \to a$ in $\mathbb{R}$ and $x_n \geq 0$ for all n, then $a \geq 0$.

3.4 Prove that if $x_n \to 0$ in $\mathbb{R}$ and $|y_n| \leq x_n$ for all n, then $y_n \to 0$.

3.5 Give detailed ϵ, N-proofs of Lemmas 3.1 and 3.2.

3.6 By applying Theorem 3.2, prove that if X is a metric space, V is a normed linear space, and F and G are continuous maps from X to V, then $F + G$ is continuous. State and prove the similar theorem for a product FG.

3.7 Prove that continuity is preserved under composition by applying Theorem 3.2.

3.8 Show that (the range of) a sequence of points in a metric space is in general not a closed set. Show that it may be a closed set.

3.9 The fact that in a normed linear space the closure of an open ball includes the corresponding closed ball is practically trivial on the basis of Lemma 3.2 and Theorem 3.1. Show that this is so.

3.10 Show directly that if the maximum norm $\|\langle \alpha, \xi \rangle\| = \max \{\|\alpha\|, \|\xi\|\}$ is used on $V = V_1 \times V_2$, then it is true that

$$\langle \alpha_n, \xi_n \rangle \to \langle \alpha, \xi \rangle \quad \text{in} \quad V$$

if and only if

$$\alpha_n \to \alpha \quad \text{in} \quad V_1 \qquad \text{and} \qquad \xi_n \to \xi \quad \text{in} \quad V_2.$$

3.11 Show that if $\| \;\|$ is any increasing norm on $\mathbb{R}^2$ (see the remark after Theorem 4.3 of Chapter 3), then

$$\rho(\langle x_1, y_1\rangle, \langle x_2, y_2\rangle) = \|\langle \rho(x_1, x_2), \rho(y_1, y_2)\rangle\|$$

is a metric on the product $X \times Y$ of two metric spaces X and Y.

3.12 In the above exercise show that $\langle x_n, y_n\rangle \to \langle x, y\rangle$ in $X \times Y$ if and only if $x_n \to x$ in X and $y_n \to y$ in Y. This property would be our minimal requirement for a *product metric*.

3.13 Defining a product metric as above, use Theorem 3.2 to show that

$$\langle f, g\rangle : S \to X \times Y$$

is continuous if and only if $f: S \to X$ and $g: S \to Y$ are both continuous.

3.14 Let X, Y, and Z be metric spaces, and let $f: X \times Y \to Z$ be a mapping such that $f(x, y)$ is continuous in the variables separately. Suppose also that the continuity in x is *uniform over* y. That is, suppose that given ϵ and x_0, there is a δ such that

$$\rho(x, x_0) < \delta \Rightarrow \rho\big(f(x, y), f(x_0, y)\big) < \epsilon$$

for every value of y. Show that then f is continuous on $X \times Y$.

3.15 Define the function f on the closed unit square $[0, 1] \times [0, 1]$ by

$$f(0, 0) = 0, \qquad f(x, y) = \frac{xy}{(x+y)^2} \qquad \text{if} \quad \langle x, y\rangle \neq \langle 0, 0\rangle.$$

Then f is continuous as a function of x for each fixed value of y, and conversely. Show, however, that f is *not* continuous at the origin. That is, find a sequence $\langle x_n, y_n\rangle$ converging to $\langle 0, 0\rangle$ in the plane such that $f(x_n, y_n)$ does not converge to 0. This example shows that continuity of a function of two variables is a stronger property than continuity in each variable separately.

4. SEQUENTIAL COMPACTNESS

The reader is probably familiar with the idea of a subsequence. A subsequence of a sequence $\{x_n\}$ is a new sequence $\{y_m\}$ that is formed by selecting an infinite number, but generally not all, of the terms x_n, and counting them off in the order of the selected indices. Thus, if n_1 is the first selected n, n_2 the next, and so on, and if we set $y_m = x_{n_m}$, then we obtain the subsequence

$$\{y_1, y_2, \ldots, y_m, \ldots\} = \{x_{n_1}, x_{n_2}, \ldots, x_{n_m}, \ldots\} \qquad \text{or} \qquad \{y_m\}_m = \{x_{n_m}\}_m.$$

Strictly speaking, this counting off of the selected set of indices n is a sequence $m \mapsto n_m$ from $\mathbb{Z}^+$ to $\mathbb{Z}^+$ which preserves order: $n_{m+1} > n_m$ for all m. And the subsequence $m \mapsto x_{n_m}$ is the composition of the sequence $n \mapsto x_n$ and the selector sequence.

In order to avoid subscripts on subscripts, we may use the notation $n(m)$ instead of n_m. In either case we are being conventionally sloppy: we are using the same symbol 'n' as an integer-valued variable, when we write x_n, and as the selector function, when we write $n(m)$ or n_m. This is one of the standard notational ambiguities which we tolerate in elementary calculus, because the cure is considered worse than the disease. We *could* say: let f be a sequence, i.e., a function from $\mathbb{Z}^+$ to $\mathbb{R}$. Then a subsequence of f is a composition $f \circ g$, where g is a mapping from $\mathbb{Z}^+$ to $\mathbb{Z}^+$ such that $g(m+1) > g(m)$ for all m.

If you have grasped the idea of subsequence, you should be able to see that *any* infinite sequence of 0's and 1's, say $\{0, 1, 0, 0, 1, 0, 0, 0, 1, \ldots\}$, can be obtained as a subsequence of $\{0, 1, 0, 1, 0, 1, \ldots, [1 + (-1)^n]/2, \ldots\}$.

If $x_n \to a$, then it should be clear that every subsequence also converges to a. We leave the details as an exercise. On the other hand, if the sequence $\{x_n\}$ *does not* converge to a, then there is an ϵ such that for every N there is some larger n at which $\rho(x_n, a) \geq \epsilon$. Now we can choose such an n for every N, taking care that $n_{N+1} > n_N$, and thus choose a subsequence all of whose terms are at a distance at least ϵ from a. Then *this* sequence has *no* subsequence converging to a. Thus, if $\{x_n\}$ does *not* converge to a, then it has a subsequence *no* (sub)subsequence of which converges to a. Therefore,

Lemma 4.1. If the sequence $\{x_n\}$ and the point a are such that every subsequence of $\{x_n\}$ has itself a subsequence that converges to a, then $x_n \to a$.

This is a wild and unlikely sounding lemma, but we shall use it to prove a most important theorem (Theorem 4.2).

Definition. A subset A of a metric space is *sequentially compact* if every sequence in A has a subsequence that converges to a point of A.

Here, so to speak, we create convergence out of nothing. One would expect a compact set to have very powerful properties, and perhaps suspect that there aren't many such sets. We shall soon see, however, that every bounded closed subset of $\mathbb{R}^n$ is compact, and it is in the theory of finite-dimensional spaces that we most frequently use this notion. Sequential compactness in infinite-dimensional spaces is a much rarer phenomenon, but when it does occur it is very important, as we shall see in our brief look at Sturm-Liouville theory in Chapter 6.

We begin with a few simple but important general results.

Lemma 4.2. If A is a sequentially compact subset of a metric space S, then A is closed and bounded.

Proof. Suppose that $\{x_n\} \subset A$ and that $x_n \to b$. By the compactness of A there exists a subsequence $\{x_{n(i)}\}_i$ that converges to a point $a \in A$. But a subsequence of a convergent sequence converges to the same limit. Therefore, $a = b$ and $b \in A$. Thus A is closed.

Boundedness here will mean lying in some ball about a given point b. If A is not bounded, for each n there exists a point $x_n \in A$ such that $\rho(x_n, b) > n$. By compactness a subsequence $\{x_{n(i)}\}_i$ converges to a point $a \in A$, and

$$\rho(x_{n(i)}, b) \to \rho(a, b).$$

This clearly contradicts $\rho(x_{n(i)}, b) > n(i) \geq i$. □

Continuous functions carry compact sets into compact sets. The proof of the following result is left as an exercise.

Theorem 4.1. If f is continuous and A is a sequentially compact subset of its domain, then $f[A]$ is sequentially compact.

A nonempty compact set $A \subset \mathbb{R}$ contains maximum and minimum elements. This is because lub A is the limit of a sequence in A, and hence belongs to A itself, since A is closed. Combining this fact with the above theorem, we obtain the following well-known corollary.

Corollary. If f is a continuous real-valued function and dom (f) is nonempty and sequentially compact, then f is bounded and assumes maximum and minimum values.

The following very useful result is related to the above theorem.

Theorem 4.2. If f is continuous and bijective and dom (f) is sequentially compact, then f^{-1} is continuous.

Proof. We have to show that if $y_n \to y$ in the range of f, and if $x_n = f^{-1}(y_n)$ and $x = f^{-1}(y)$, then $x_n \to x$. It is sufficient to show that every subsequence $\{x_{n(i)}\}_i$ has itself a subsequence converging to x (by Lemma 4.1). But, since dom (f) is compact, there is a subsequence $\{x_{n(i(j))}\}_j$ converging to some z, and the continuity of f implies that $f(z) = \lim_{j\to\infty} f(x_{n(i(j))}) = \lim_{j\to\infty} y_{n(i(j))} = y$. Therefore, $z = f^{-1}(y) = x$, which is what we had to prove. Thus f^{-1} is continuous. □

We now take up the problem of showing that bounded closed sets in $\mathbb{R}^n$ are compact. We first prove it for $\mathbb{R}$ itself and then give an inductive argument for $\mathbb{R}^n$.

A sequence $\{x_n\} \subset \mathbb{R}$ is said to be *increasing* if $x_n \leq x_{n+1}$ for all n. It is *strictly increasing* if $x_n < x_{n+1}$ for all n. The notions of a *decreasing* sequence and a *strictly decreasing* sequence are obvious. A sequence which is either increasing or decreasing is said to be monotone. The relevance of these notions here lies in the following two lemmas.

Lemma 4.3. A bounded monotone sequence in $\mathbb{R}$ is convergent.

Proof. Suppose that $\{x_n\}$ is increasing and bounded above. Let l be the least upper bound of its range. That is, $x_n \leq l$ for all n, but for every ϵ, $l - \epsilon$ is not an upper bound, and so $l - \epsilon < x_N$ for some N. Then

$$n > N \Rightarrow l - \epsilon < x_N \leq x_n \leq l,$$

and so $|x_n - l| < \epsilon$. That is, $x_n \to l$ as $n \to \infty$. □

Lemma 4.4. Any sequence in $\mathbb{R}$ has a monotone subsequence.

Proof. Call x_n a *peak* term if it is greater than or equal to all later terms. If there are infinitely many peak terms, then they obviously form a decreasing subsequence. On the other hand, if there are only finitely many peak terms, then there is a last one x_{n_0} (or none at all), and then every later term is strictly less than some other still later term. We choose any n_1 greater than n_0, and then we can choose $n_2 > n_1$ so that $x_{n_1} < x_{n_2}$, etc. Therefore, in this case we can choose a strictly increasing subsequence. We have thus shown that any sequence $\{x_n\}$ in $\mathbb{R}$ has either a decreasing subsequence or a strictly increasing subsequence. □

Putting these two lemmas together, we have:

Theorem 4.3. Every bounded sequence in $\mathbb{R}$ has a convergent subsequence.

Now we can generalize to $\mathbb{R}^n$ by induction.

Theorem 4.4. Every bounded sequence in $\mathbb{R}^n$ has a convergent subsequence (using any product norm, say $\| \ \|_1$).

Proof. The above theorem is the case $n = 1$. Suppose then that the theorem is true for $n - 1$, and let $\{\mathbf{x}^m\}_m$ be a bounded sequence in $\mathbb{R}^n$. Thinking of $\mathbb{R}^n$ as $\mathbb{R}^{n-1} \times \mathbb{R}$, we have $\mathbf{x}^m = \langle \mathbf{y}^m, z_m \rangle$, and $\{\mathbf{y}^m\}_m$ is bounded in $\mathbb{R}^{n-1}$, because if $\mathbf{x} = \langle \mathbf{y}, z \rangle$, then $\|\mathbf{x}\|_1 = \|\mathbf{y}\|_1 + |z| \geq \|\mathbf{y}\|_1$. Therefore, there is a subsequence $\{\mathbf{y}^{n(i)}\}_i$ converging to some $\mathbf{y}$ in $\mathbb{R}^{n-1}$, by the inductive hypothesis. Since $\{z_{n(i)}\}$ is bounded in $\mathbb{R}$, *it* has a subsequence $\{z_{n(i(p))}\}_p$ converging to some x in $\mathbb{R}$. Of course, the corresponding subsubsequence $\{\mathbf{y}^{n(i(p))}\}_p$ still converges to $\mathbf{y}$ in $\mathbb{R}^{n-1}$, and then $\{\mathbf{x}^{n(i(p))}\}_p$ converges to $\mathbf{x} = \langle \mathbf{y}, z \rangle$ in $\mathbb{R}^n = \mathbb{R}^{n-1} \times \mathbb{R}$, since its two component sequences now converge to $\mathbf{y}$ and z, respectively. We have thus found a convergent subsequence of $\{\mathbf{x}^n\}$. □

Theorem 4.5. If A is a bounded closed subset of $\mathbb{R}^n$, then A is sequentially compact (in any product norm).

Proof. If $\{x_n\} \subset A$, then there is a subsequence $\{x_{n(i)}\}_i$ converging to some x in $\mathbb{R}^n$, by Theorem 4.4, and x is in A, since A is closed. Thus A is compact. □

We can now fill in one of the minor gaps in the last chapter.

Theorem 4.6. All norms on $\mathbb{R}^n$ are equivalent.

Proof. It is sufficient to prove that an arbitrary norm $\| \ \|$ is equivalent to $\| \ \|_1$. Setting $a = \max \{\|\delta^i\|\}_1^n$, we have

$$\|\mathbf{x}\| = \left\| \sum_1^n x_i \delta^i \right\| \leq \sum_1^n |x_i| \, \|\delta^i\| \leq a\|\mathbf{x}\|_1,$$

so one of our inequalities is trivial. We also have $\big| \|x\| - \|y\| \big| \leq \|x - y\| \leq a\|x - y\|_1$, so $\|x\|$ is a continuous function on $\mathbb{R}^n$ with respect to the one-norm. Now the unit one-sphere $S = \{x : \|\mathbf{x}\|_1 = 1\}$ is closed and bounded and so compact (in the one-norm). The restriction of the continuous function $\|x\|$ to this compact set S has a minimum value m, and m cannot be zero because S does not contain the zero vector. We thus have $\|x\| \geq m\|x\|_1$ on S, and so $\|x\| \geq m\|x\|_1$ on $\mathbb{R}^n$, by homogeneity. Altogether we have found positive constants a and m such that $m\| \ \|_1 \leq \| \ \| \leq a\| \ \|_1$. □

Composing with a coordinate isomorphism, we see that *all norms on any finite-dimensional vector space are equivalent.*

Corollary. If M is a finite-dimensional subspace of the normed linear space V, then M is a closed subspace of V.

Proof. Suppose that $\{\xi_n\} \subset M$ and $\xi_n \to \alpha \in V$. We have to show that α is in M. Now $\{\xi_n\}$ is a bounded subset of M, and its closure in M is therefore sequentially compact, by the theorem. Therefore, some subsequence converges to a point β in M as well as to α, and so $\alpha = \beta \in M$. □

EXERCISES

4.1 Prove by induction that if $f: \mathbb{Z}^+ \to \mathbb{Z}^+$ is such that $f(n+1) > f(n)$ for all n, then $f(n) \geq n$ for all n.

4.2 Prove carefully that if $x_n \to a$ as $n \to \infty$, then $x_{n(m)} \to a$ as $m \to \infty$ for any subsequence. The above exercise is useful in this proof.

4.3 Prove that if $\{x_n\}$ is an increasing sequence in $\mathbb{R}$ ($x_{n+1} \geq x_n$ for all n), and if $\{x_n\}$ has a convergent subsequence, then $\{x_n\}$ converges.

4.4 Give a more detailed version of the argument that if the sequence $\{x_n\}$ does not converge to a, then there is an ϵ and a subsequence $\{x_{n(m)}\}_m$ such that $\rho(x_{n(m)}, a) \geq \epsilon$ for all m.

4.5 Find a sequence in $\mathbb{R}$ having no convergent subsequence.

4.6 Find a nonconvergent sequence in $\mathbb{R}$ such that the set of limit points of convergent subsequence consists exactly of the number 1.

4.7 Show that there is a sequence $\{x_n\}$ in $[0, 1]$ such that for any $y \in [0, 1]$ there is a subsequence x_{n_m} converging to y.

4.8 Show that the set of limits of convergent subsequences of a sequence $\{x_n\}$ in a metric space X is a closed subset of X.

4.9 Prove Theorem 4.1.

4.10 Prove that the Cartesian product of two sequentially compact metric spaces is sequentially compact. (The proof is essentially in the text.)

4.11 A metric space is *boundedly compact* if every closed bounded set is sequentially compact. Prove that the Cartesian product of two boundedly compact metric spaces is boundedly compact (using, say, the maximum metric on the product space).

4.12 Prove that the sum $A + B$ of two sequentially compact subsets of a normed linear space is sequentially compact.

4.13 Prove that the sum $A + B$ of a closed set and a compact set is closed.

4.14 Show by an example in $\mathbb{R}$ that the sum of two closed sets need not be closed.

4.15 Let $\{C_n\}$ be a decreasing sequence ($C_{n+1} \subset C_n$ for all n) of nonempty closed subsets of a sequentially compact metric space S. Prove that $\bigcap_1^\infty C_n$ is nonempty.

4.16 Give an example of a decreasing sequence $\{C_n\}$ of nonempty closed subsets of a metric space such that $\bigcap_1^\infty C_n = \varnothing$.

4.17 Suppose the metric space S has the property that every decreasing sequence $\{C_n\}$ of nonempty closed subsets of S has nonempty intersection. Prove that then S must be sequentially compact. [*Hint:* Given any sequence $\{x_i\} \subset S$, let C_n be the closure of $\{x_i : i \geq n\}$.]

4.18 Let A be a sequentially compact subset of a nls V, and let B be obtained from A by drawing all line segments from points of A to the origin (that is,

$$B = \{t\alpha : \alpha \in A \text{ and } t \in [0, 1]\}).$$

Prove that B is compact.

4.19 Show by applying a compactness argument to Lemma 1.5 that if N is a proper closed subspace of a finite-dimensional vector space V, then there exists α in V such that $\|\alpha\| = \rho(\alpha, N) = 1$.

5. COMPACTNESS AND UNIFORMITY

The word 'uniform' is frequently used as a qualifying adjective in mathematics. Roughly speaking, it concerns a "point" property $P(y)$ which may or may not hold at each point y in a domain A and whose definition involves an existential quantifier. A typical form for $P(y)$ is $(\forall c)(\exists d)Q(y, c, d)$. Thus, if $P(y)$ is 'f is continuous at y', then $P(y)$ has the form $(\forall\epsilon)(\exists\delta)Q(y, \epsilon, \delta)$. The property holds on A if it holds for all y in A, that is, if

$$(\forall y^{\in A})[(\forall c)(\exists d)Q(y, c, d)].$$

Here d will, in general, depend both on y and c; if either y or c is changed, the corresponding d may have to be changed. Thus δ in the definition of continuity depends both on ϵ and on the point y at which continuity is being asserted. The property is said to hold *uniformly on A*, or *uniformly in y*, if a value d can be found that is independent of y (but still dependent on c). Thus the property holds uniformly in y if

$$(\forall c)(\exists d)(\forall y^{\in A})Q(y, c, d);$$

the uniformity of the property is expressed in the reversal of the order of the quantifiers $(\forall y^{\in A})$ and $(\exists d)$. Thus f is *uniformly continuous* on A if

$$(\forall\epsilon)(\exists\delta)(\forall y, z^{\in A})[\rho(y, z) < \delta \Rightarrow \rho(f(y), f(z)) < \epsilon].$$

Now δ is independent of the point at which continuity is being asserted, but still dependent on ϵ, of course.

We saw in Section 14 of the last chapter how much more powerful the point condition of continuity becomes when it holds uniformly. In the remainder of this section we shall discuss some other uniform notions, and shall see that the uniform property is often implied by the point property if the domain over which it holds is sequentially compact.

The formal statement forms we have examined above show clearly the distinction between uniformity and nonuniformity. However, in writing an argument, we would generally follow our more idiomatic practice of dropping out the inside universal quantifier. For example, a sequence of functions $\{f_n\} \subset W^A$ converges pointwise to $f: A \to W$ if it converges to f at every point p in A, that is, if for every point p in A and for every ϵ there is an N such that

$$n > N \Rightarrow \rho(f_n(p), f(p)) \leq \epsilon.$$

The sequence converges *uniformly* on A if an N exists that is independent of p, that is, if for every ϵ there is an N such that

$$n > N \Rightarrow \rho(f_n(p), f(p)) \leq \epsilon \qquad \text{for every} \quad p \text{ in } A.$$

When $\rho(\xi, \eta) = \|\xi - \eta\|$, saying that $\rho(f_n(p), f(p)) \leq \epsilon$ for all p is the same as saying that $\|f_n - f\|_\infty \leq \epsilon$. Thus $f_n \to f$ uniformly if and only if $\|f_n - f\|_\infty \to 0$; this is why the norm $\|f\|_\infty$ is called the uniform norm.

Pointwise convergence does not imply uniform convergence. Thus $f_n(x) = x^n$ on $A = (0, 1)$ converges pointwise to the zero function but does not converge uniformly.

Nor does continuity on A imply uniform continuity. The function $f(x) = 1/x$ is continuous on $(0, 1)$ but is not uniformly continuous. The function $\sin(1/x)$ is continuous and bounded on $(0, 1)$ but is not uniformly continuous. Compactness changes the latter situation, however.

Theorem 5.1. If f is continuous on A and A is compact, then f is uniformly continuous on A.

Proof. This is one of our "automatic" negation proofs. Uniform continuity (UC) is the property

$$(\forall \epsilon^{>0})(\exists \delta^{>0})(\forall x, y^{\in A})[\rho(x, y) < \delta \Rightarrow \rho(f(x), f(y)) < \epsilon].$$

Therefore, $\sim$UC $\Leftrightarrow (\exists \epsilon)(\forall \delta)(\exists x, y)[\rho(x, y) < \delta$ and $\rho(f(x), f(y)) \geq \epsilon]$. Take $\delta = 1/n$, with corresponding x_n and y_n. Thus, for all n, $\rho(x_n, y_n) < 1/n$ and $\rho(f(x_n), f(y_n)) \geq \epsilon$, where ϵ is a fixed positive number. Now $\{x_n\}$ has a convergent subsequence, say $x_{n(i)} \to x$, by the compactness of A. Since

$$\rho(y_{n(i)}, x_{n(i)}) < 1/i,$$

we also have $y_{n(i)} \to x$. By the continuity of f at x,

$$\rho(f(x_{n(i)}), f(y_{n(i)})) \leq \rho(f(x_{n(i)}), f(x)) + \rho(f(x), f(y_{n(i)})) \to 0,$$

which contradicts $\rho(f(x_{n(i)}), f(y_{n(i)})) \geq \epsilon$. This completes the proof by negation. □

The compactness of A does not, however, automatically convert the pointwise convergence of a sequence of functions on A into uniform convergence. The "piecewise linear" functions $f_n: [0, 1] \to [0, 1]$ defined by the graph shown in Fig. 4.1 converge pointwise to zero on the compact domain $[0, 1]$, but the convergence is not uniform. (However, see Exercise 5.4.)

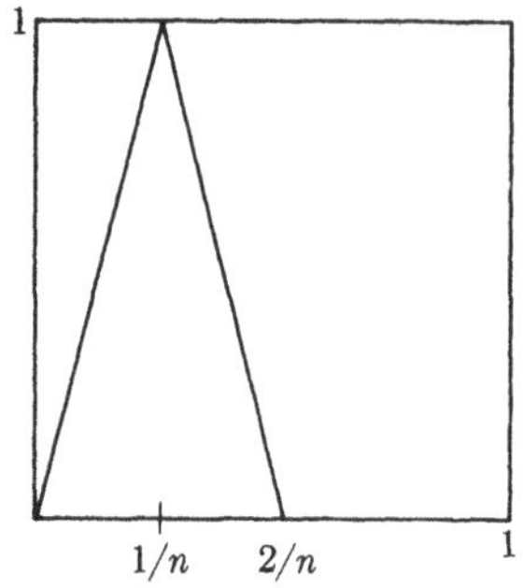

Fig. 4.1

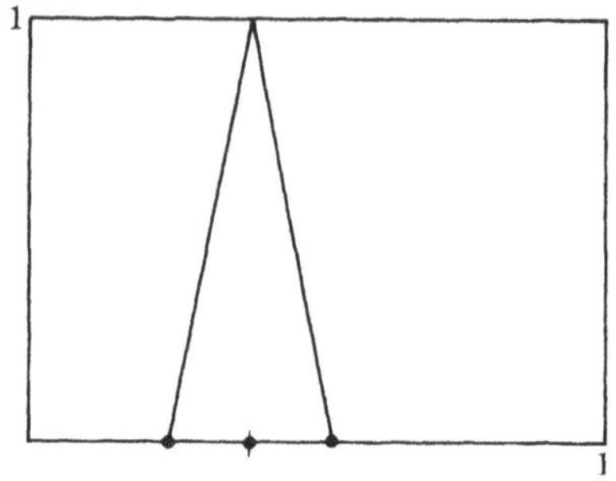

Fig. 4.2

We pointed out earlier that the distance between a pair of disjoint closed sets may be zero. However, if one of the closed sets is compact, then the distance must be positive.

Theorem 5.2. If A and C are disjoint nonempty closed sets, one of which is compact, then $\rho(A, C) > 0$.

Proof. The proof is by automatic contradiction, and is left to the reader.

This result is again a uniformity condition. Saying that a set A is disjoint from a closed set C is saying that $(\forall x^{\in A})(\exists r^{>0})(B_r(x) \cap C = \varnothing)$. Saying that $\rho(A, C) > 0$ is saying that $(\exists r^{>0})(\forall x^{\in A}) \ldots$

As a last consequence of sequential compactness, we shall establish a very powerful property which is taken as the definition of compactness in general topology. First, however, we need some preparatory work. If A is a subset of a metric space S, the *r-neighborhood of* A, $B_r[A]$, is simply the union of all the balls of radius r about points of A:

$$B_r[A] = \bigcup \{B_r(a) : a \in A\} = \{x : (\exists a^{\in A})(\rho(x, a) < r)\}.$$

A subset $A \subset S$ is *r-dense* in S if $S \subset B_r[A]$, that is, if each point of S is closer than r to some point of A.

A subset A of a metric space S is *dense* in S if $\overline{A} = S$. This is the same as saying that for every point p in S there are points of A arbitrarily close to p. The set $\mathbb{Q}$ of all rational numbers is a dense subset of the real number system $\mathbb{R}$, because any irrational real number x can be arbitrarily closely approximated by rational numbers. Since we do arithmetic in decimal notation, it is customary to use decimal approximations, and if $0 < x < 1$ and the decimal expansion of x is $x = \sum_1^\infty a_n/10^n$, where each a_n is an integer and $0 \leq a_n < 10$, then $\sum_1^N a_n/10^n$ is a rational number differing from x by less than 10^{-N}. Note that A is a dense subset of B if and only if A is r-dense in B for every positive r.

A set B is said to be *totally bounded* if for every positive r there is a finite set which is r-dense in B. Thus for every positive r the set B can be covered by a finite number of balls of radius r. For example, the $n - 1$ numbers $\{i/n\}_1^{n-1}$ are $(1/n)$-dense in the open interval $(0, 1)$ for each n, and so $(0, 1)$ is totally bounded.

Total boundedness is a much stronger property than boundedness, as the following lemma shows.

Lemma 5.1. If the normed linear space V is infinite-dimensional, then its closed unit ball $B_1 = \{\xi : \|\xi\| \leq 1\}$ cannot be covered by a finite number of balls of radius $\frac{1}{3}$.

Proof. Since V is not finite-dimensional, we can choose a sequence $\{\alpha_n\}$ such that α_{n+1} is not in the linear span M_n of $\{\alpha_1, \ldots, \alpha_n\}$, for each n. Since M_n is closed in V, by the corollary of Theorem 4.6, we can apply Lemma 1.5 to find a vector ξ_n in M_n such that $\|\xi_n\| = 1$ and $\rho(\xi_n, M_{n-1}) > \frac{2}{3}$ for all $n > 1$. We take $\xi_1 = \alpha_1/\|\alpha_1\|$, and we have a sequence $\{\xi_n\} \subset B_1$ such that

$$\|\xi_m - \xi_n\| > \tfrac{2}{3}$$

if $m \neq n$. Then no ball of radius $\frac{1}{3}$ can contain more than one ξ_1, proving the lemma. □

For a concrete example, let V be $\mathcal{C}([0, 1])$, and let f_n be the "peak" function sketched in Fig. 4.2, where the three points on the base are $1/(2n+2)$, $1/(2n+1)$, and $1/2n$. Then f_{n+1} is "disjoint" from f_n (that is, $f_{n+1}f_n = 0$), and we have $\|f_n\|_\infty = 1$ for all n and $\|f_n - f_m\|_\infty = 1$ if $n \neq m$. Thus no ball of radius $\frac{1}{2}$ can contain more than one of the functions f_n, and accordingly the closed unit ball in V cannot be covered by a finite number of balls of radius $\frac{1}{2}$.

Lemma 5.2. Every sequentially compact set A is totally bounded.

Proof. If A is not totally bounded, then there exists an r such that no finite subset F is r-dense in A. We can then define a sequence $\{p_n\}$ inductively by taking p_1 as any point of A, p_2 as any point of A not in $B_r(p_1)$, and p_n as any point of A not in $B_r[\bigcup_1^{n-1} p_i] = \bigcup_1^{n-1} B_r(p_i)$. Then $\{p_n\}$ is a sequence in A such that $\rho(p_i, p_j) \geq r$ for all $i \neq j$. But this sequence can have no convergent subsequence. Thus, if A is not totally bounded, then A is not sequentially compact, proving the lemma. □

Corollary. A normed linear space V is finite-dimensional if and only if its closed unit ball is sequentially compact.

Proof. This follows from Theorem 4.4 in one direction and from the above two lemmas in the other direction. □

Lemma 5.3. Suppose that A is sequentially compact and that $\{E_i : i \in I\}$ is an open covering of A (that is, $\{E_i\}$ is a family of open sets and $A \subset \bigcup_i E_i$). Then there exists an $r > 0$ with the property that for every point p in A the ball $B_r(p)$ is included in some E_j.

Proof. Otherwise, for every r there is a point p in A such that $B_r(p)$ is not a subset of any E_j. Take $r = 1/n$, with corresponding sequence $\{p_n\}$. Thus $B_{1/n}(p_n)$ is not a subset of any E_j. Since A is sequentially compact, $\{p_n\}$ has a convergent subsequence, $p_{n(m)} \to p$ as $m \to \infty$. Since $\{E_i\}$ covers A, some E_j contains p, and then $B_\epsilon(p) \subset E_j$ for some $\epsilon > 0$, since E_j is open. Taking m large enough so that $1/m < \epsilon/2$ and also $\rho(p_{n(m)}, p) < \epsilon/2$, we have

$$B_{1/n(m)}(p_{n(m)}) \subset B_\epsilon(p) \subset E_j,$$

contradicting the fact that $B_{1/n}(p_n)$ is not a subset of *any* E_i. The lemma has thus been proved. □

Theorem 5.3. If $\mathfrak{F}$ is an open covering of a sequentially compact set A, then some finite subfamily of $\mathfrak{F}$ covers A.

Proof. By the lemma immediately above there exists an $r > 0$ such that for every p in A the ball $B_r(p)$ lies entirely in some set of $\mathfrak{F}$, and by the first lemma there exist $p_1, \ldots, p_n$ in A such that $A \subset \bigcup_1^n B_r(p_i)$. Taking corresponding sets E_i in $\mathfrak{F}$ such that $B_r(p_i) \subset E_i$ for $i = 1, \ldots, n$, we clearly have $A \subset \bigcup_1^n E_i$. □

In general topology, a set A such that every open covering of A includes a finite covering is said to be compact or to have the Heine-Borel property. The above theorem says that in a metric space every sequentially compact set is compact. We shall see below that the reverse implication also holds, so that the two notions are in fact equivalent on a metric space.

Theorem 5.4. If A is a compact metric space, then A is sequentially compact.

Proof. Let $\{x_n\}$ be any sequence in A, and let $\mathfrak{F}$ be the collection of open balls B such that B contains only finitely many x_i. If $\mathfrak{F}$ were to cover A, then by compactness A would be the union of finitely many balls in $\mathfrak{F}$, and this would clearly imply that the whole of A contains only finitely many x_i, contradicting the fact that $\{x_i\}$ is an infinite sequence. Therefore, $\mathfrak{F}$ does not cover A, and so there is a point x in A such that every ball about x contains infinitely many of the x_i. More precisely, every ball about x contains x_i for infinitely many indices i. It can now be safely left to the reader to see that a subsequence of $\{x_n\}$ converges to x. □

EXERCISES

5.1 Show that $f_n(x) = x^n$ does not converge uniformly on $(0, 1)$.

5.2 Show that $f(x) = 1/x$ is not uniformly continuous on $(0, 1)$.

5.3 Define the notion of a function $K: X \times Y \to Y$ being uniformly Lipschitz in its second variable over its first variable.

5.4 Let S be a sequentially compact metric space, and let $\{f_n\}$ be a sequence of continuous real-valued functions on S that decreases pointwise to zero (that is, $\{f_n(p)\}$ is a decreasing sequence in $\mathbb{R}$ and $f_n(p) \to 0$ as $n \to \infty$ for each p in S). Prove that the convergence is uniform. (Try to apply Exercise 4.15.)

5.5 Restate the corollaries of Theorems 15.1 and 15.2 of Chapter 3, employing the weaker hypotheses that suffice by virtue of Theorem 5.1 of the present section.

5.6 Prove Theorem 5.2.

5.7 Prove that if A is an r-dense subset of a set X in a normed linear space V, and if B is an s-dense subset of a set $Y \subset V$, then $A + B$ is $(r + s)$-dense in $X + Y$. Conclude that the sum of two totally bounded subsets of V is totally bounded.

5.8 Suppose that the n points $\{p_i\}_1^n$ are r-dense in a metric space X. Let A be any subset of X. Show that A has a subset of at most n points that is $2r$-dense in A. Conclude that any subset of a totally bounded metric space is itself totally bounded.

5.9 Prove that the Cartesian product of two totally bounded metric spaces is totally bounded.

5.10 Show that if a metric space X has a dense subset A that is totally bounded, then X is totally bounded.

5.11 Show that if two continuous mappings f and g from a metric space X to a metric space Y are equal on a dense subset of X, then they are equal everywhere.

5.12 Write out in explicit quantified form involving the existence of balls the statement that the interiors of the sets $\{A_i\}$ cover the metric space A. Then show that the conclusion of Lemma 5.3 is another uniformity assertion.

5.13 Reprove the theorem that a continuous function on a compact domain is bounded on the basis of Theorem 5.3.

5.14 Reprove the theorem that a continuous function on a compact domain is uniformly continuous from Theorem 5.3.

6. EQUICONTINUITY

The application of sequential compactness that we shall make in an infinite-dimensional context revolves around the notion of an *equicontinuous* family of functions. If A and B are metric spaces, then a subset $\mathfrak{F} \subset B^A$ is said to be *equicontinuous* at p_0 in A if all the functions of $\mathfrak{F}$ are continuous at p_0 and if given ϵ, there is a δ which works for them all, i.e., such that

$$\rho(p, p_0) < \delta \Rightarrow \rho(f(p), f(p_0)) < \epsilon \qquad \text{for every } f \text{ in } \mathfrak{F}.$$

The family $\mathfrak{F}$ is *uniformly equicontinuous* if δ is also independent of p_0, and so is dependent *only* on ϵ. Our quantifier string is thus $(\forall\epsilon)(\exists\delta)(\forall p, q^{\in A})(\forall f^{\in \mathfrak{F}})$.

For example, given $m > 0$, let $\mathfrak{F}$ be a collection of functions f from $(0, 1)$ to $(0, 1)$ such that f' exists and $|f'| \leq m$ on $(0, 1)$. Then $|f(x) - f(y)| \leq m|x - y|$, by the ordinary mean-value theorem. Therefore, given any ϵ, we can take $\delta = \epsilon/m$ and have

$$|x - y| < \delta \Rightarrow |f(x) - f(y)| < \epsilon$$

for all $x, y \in (0, 1)$ and all $f \in \mathfrak{F}$. The collection $\mathfrak{F}$ is thus uniformly equicontinuous.

Theorem 6.1. If A and B are totally bounded metric spaces, and if $\mathfrak{F}$ is a *uniformly equicontinuous* subfamily of B^A, then $\mathfrak{F}$ is totally bounded in the uniform metric.

Proof. Given $\epsilon > 0$, choose δ so that for all f in $\mathfrak{F}$ and all p_1, p_2 in A, $\rho(p_1, p_2) < \delta \Rightarrow \rho(f(p_1), f(p_2)) < \epsilon/4$. Let D be a finite subset of A which is δ-dense in A, and let E be a finite subset of B which is $(\epsilon/4)$-dense in B. Let G be the set E^D of all functions on D into E. G is of course finite; in fact, $\#G = n^m$, where $m = \#D$ and $n = \#E$. Finally, for each $g \in G$ let $\mathfrak{F}_g$ be the set of all functions $f \in \mathfrak{F}$ such that

$$\rho(f(p), g(p)) < \epsilon/4 \qquad \text{for every} \quad p \in D.$$

We claim that the collections $\mathfrak{F}_g$ cover $\mathfrak{F}$ and that each $\mathfrak{F}_g$ has diameter at most ϵ. We will then obtain a finite ϵ-dense subset of $\mathfrak{F}$ by choosing one function from each nonempty $\mathfrak{F}_g$, and the theorem will be proved.

To show that every $f \in \mathfrak{F}$ is in some $\mathfrak{F}_g$, we simply construct a suitable g. For each p in D there exists a q in E whose distance from $f(p)$ is less than $\epsilon/4$. If we choose one such q in E for each p in D, we have a function g in G such that $f \in \mathfrak{F}_g$.

The final thing we have to show is that if $f, h \in \mathfrak{F}_g$, then $\rho(f, h) \leq \epsilon$. Since $\rho(h, g) < \epsilon/4$ on D and $\rho(f, g) < \epsilon/4$ on D, it follows that

$$\rho(f(p), h(p)) < \epsilon/2 \qquad \text{for every} \quad p \in D.$$

Then for any $p' \in A$ we have only to choose $p \in D$ such that $\rho(p', p) < \delta$, and we have

$$\begin{aligned}\rho(f(p'), h(p')) &\leq \rho(f(p'), f(p)) + \rho(f(p), h(p)) + \rho(h(p), h(p')) \\ &\leq \epsilon/4 + \epsilon/2 + \epsilon/4 = \epsilon. \ \square\end{aligned}$$

The above proof is a good example of a mathematical argument that is completely elementary but hard. When referring to mathematical reasoning, the words 'sophisticated' and 'difficult' are by no means equivalent.

7. COMPLETENESS

If $x_n \to a$ as $n \to \infty$, then the terms x_n obviously get close to each other as n gets large. On the other hand, if $\{x_n\}$ is a sequence whose terms get arbitrarily close to each other as $n \to \infty$, then $\{x_n\}$ clearly ought to converge to a limit. It may not, however; the desired limit point may be missing from the space. If a metric space S is such that every sequence which ought to converge actually does converge, then we say that S is *complete*. We now make this notion precise.

Definition. $\{x_n\}$ is a Cauchy sequence if for every ϵ there is an N such that

$$m > N \text{ and } n > N \ \Rightarrow\ \rho(x_m, x_n) < \epsilon.$$

Lemma 7.1. If $\{x_n\}$ is convergent, then $\{x_n\}$ is Cauchy.

Proof. Given ϵ, we choose N such that $n > N \Rightarrow \rho(x_n, a) < \epsilon/2$, where a is the limit of the sequence. Then if m and n are both greater than N, we have

$$\rho(x_m, x_n) \leq \rho(x_m, a) + \rho(a, x_n) < \epsilon/2 + \epsilon/2 = \epsilon. \ \square$$

Lemma 7.2. If $\{x_n\}$ is Cauchy, and if a subsequence is convergent, then $\{x_n\}$ itself converges.

Proof. Suppose that $x_{n(i)} \to a$ as $i \to \infty$. Given ϵ, we take N so that $m, n > N \Rightarrow \rho(x_n, x_m) < \epsilon$. Because $x_{n(i)} \to a$ as $i \to \infty$, we can choose an i such that $n(i) > N$ and $\rho(x_{n(i)}, a) < \epsilon$. Thus if $m > N$, we have

$$\rho(x_m, a) \leq \rho(x_m, x_{n(i)}) + \rho(x_{n(i)}, a) < 2\epsilon,$$

and so $x_m \to a$. □

Actually, of course, if $m, n > N \Rightarrow \rho(x_m, x_n) < \epsilon$, and if $x_n \to a$, then for any $m > N$ it is true that $\rho(x_m, a) \leq \epsilon$. Why?

Lemma 7.3. If A and B are metric spaces, and if T is a Lipschitz mapping from A to B, then T carries Cauchy sequences in A into Cauchy sequences in B. This is true in particular if A and B are normed linear spaces and T is an element of $\text{Hom}(A, B)$.

Proof. Let $\{x_n\}$ be a Cauchy sequence in A, and set $y_n = T(x_n)$. Given ϵ, choose N so that $m, n > N \Rightarrow \rho(x_m, x_n) < \epsilon/C$, when C is a Lipschitz constant for F. Then

$$m, n > N \;\Rightarrow\; \rho(y_m, y_n) = \rho\big(T(x_m), T(x_n)\big) \leq C\rho(x_m, x_n) < C\epsilon/C = \epsilon. \; \square$$

This lemma has a substantial generalization, as follows.

Theorem 7.1. If A and B are metric spaces, $\{x_n\}$ is Cauchy in A, and $F: A \to B$ is uniformly continuous, then $\{F(x_n)\}$ is Cauchy in B.

Proof. The proof is left as an exercise.

The student should try to acquire a good intuitive feel for the truth of these lemmas, after which the technical proofs become more or less obvious.

Definition. A metric space A is *complete* if every Cauchy sequence in A converges to a limit in A. A complete normed linear space is called a *Banach space.*

We are now going to list some important examples of Banach spaces. In each case a proof is necessary, so the list becomes a collection of theorems.

Theorem 7.2. $\mathbb{R}$ is complete.

Proof. Let $\{x_n\}$ be Cauchy in $\mathbb{R}$. Then $\{x_n\}$ is bounded (why?) and so, by Theorem 4.3, has a convergent subsequence. Lemma 7.2 then implies that $\{x_n\}$ is convergent. □

Theorem 7.3. If A is a complete metric space, and if f is a continuous bijective mapping from A to a metric space B such that f^{-1} is Lipschitz continuous, then B is complete. In particular, if V is a Banach space, and if T in $\text{Hom}(V, W)$ is invertible, then W is a Banach space.

Proof. Suppose that $\{y_n\}$ is a Cauchy sequence in B, and set $x_i = f^{-1}(y_i)$ for all i. Then $\{x_i\}$ is Cauchy in A, by Lemma 7.3, and so converges to some x in A, since A is complete. But then $y_n = f(x_n) \to f(x)$, because f is continuous. Thus every Cauchy sequence in B is convergent and B is complete. □

The Banach space assertion is a special case, because the invertibility of T means that T^{-1} exists in $\mathrm{Hom}(W, V)$ and hence is a Lipschitz mapping.

Corollary. If p and q are equivalent norms on V and $\langle V, p\rangle$ is complete, then so is $\langle V, q\rangle$.

Theorem 7.4. If V_1 and V_2 are Banach spaces, then so is $V_1 \times V_2$.

Proof. If $\{\langle \xi_n, \eta_n\rangle\}$ is Cauchy, then so are each of $\{\xi_n\}$ and $\{\eta_n\}$ (by Lemma 7.3, since the projections π_i are bounded). Then $\xi_n \to \alpha$ and $\eta_n \to \beta$ for some $\alpha \in V_1$ and $\beta \in V_2$. Thus $\langle \xi_n, \eta_n\rangle \to \langle \alpha, \beta\rangle$ in $V_1 \times V_2$. (See Theorem 3.4.) □

Corollary 1. If $\{V_i\}_1^n$ are Banach spaces, then so is $\prod_{i=1}^n V_i$.

Corollary 2. Every finite-dimensional vector space is a Banach space (in any norm).

Proof. $\mathbb{R}^n$ is complete (in the one-norm, say) by Theorem 7.2 and Corollary 1 above. We then impose a one-norm on V by choosing a basis, and apply the corollary of Theorem 7.3 to pass to any other norm. □

Theorem 7.5. Let W be a Banach space, let A be any set, and let $\mathcal{B}(A, W)$ be the vector space of all bounded functions from A to W with the uniform norm $\|f\|_\infty = \mathrm{lub}\,\{\|f(a)\| : a \in A\}$. Then $\mathcal{B}(A, W)$ is a Banach space.

Proof. Let $\{f_n\}$ be Cauchy, and choose any $a \in A$. Since $\|f_n(a) - f_m(a)\| \leq \|f_n - f_m\|_\infty$, it follows that $\{f_n(a)\}$ is Cauchy in W and so convergent. Define $g\colon A \to W$ by $g(a) = \lim f_n(a)$ for each $a \in A$. We have to show that g is bounded and that $f_n \to g$.

Given ϵ, we choose N so that $m, n > N \Rightarrow \|f_m - f_n\|_\infty < \epsilon$. Then

$$\|f_m(a) - g(a)\| = \lim_{n\to\infty} \|f_m(a) - f_n(a)\| \leq \epsilon.$$

Thus, if $m > N$, then $\|f_m(a) - g(a)\| \leq \epsilon$ for all $a \in A$, and hence $\|f_m - g\|_\infty \leq \epsilon$. This implies both that $f_m - g \in \mathcal{B}(A, W)$, and so

$$g = f_m - (f_m - g) \in \mathcal{B}(A, W),$$

and that $f_m \to g$ in the uniform norm. □

Theorem 7.6. If V is a normed linear space and W is a Banach space, then $\mathrm{Hom}(V, W)$ is a Banach space.

The method of proof is identical to that of the preceding theorem, and we leave it as an exercise. Boundedness here has a different meaning, but it is used in essentially the same way. One additional fact has to be established, namely, that the limit map (corresponding to g in the above theorem) is linear.

Theorem 7.7. A closed subset of a complete metric space is complete. A complete subset of any metric space is closed.

Proof. The proof is left to the reader.

It follows from Theorem 7.7 that a complete metric space A is *absolutely closed*, in the sense that no matter how we extend A to a larger metric space B, A is always a closed subset of B. Actually, this property is equivalent to completeness, for if A is not complete, then a very important construction of metric space theory shows that A can be completed. That is, we can construct a complete metric space B which includes A. Now, if A is not complete, then the closure of A in B, being complete, is different from A, and A is not absolutely closed.

See Exercise 7.21 through 7.23 for a construction of the completion of a metric space. The completion of a normed linear space is of course a Banach space.

Theorem 7.8. In the context of Theorem 7.5, let A be a metric space, let $\mathcal{C}(A, W)$ be the space of continuous functions from A to W, and set

$$\mathcal{BC}(A, W) = \mathcal{B}(A, W) \cap \mathcal{C}(A, W).$$

Then $\mathcal{BC}$ is a closed subspace of $\mathcal{B}$.

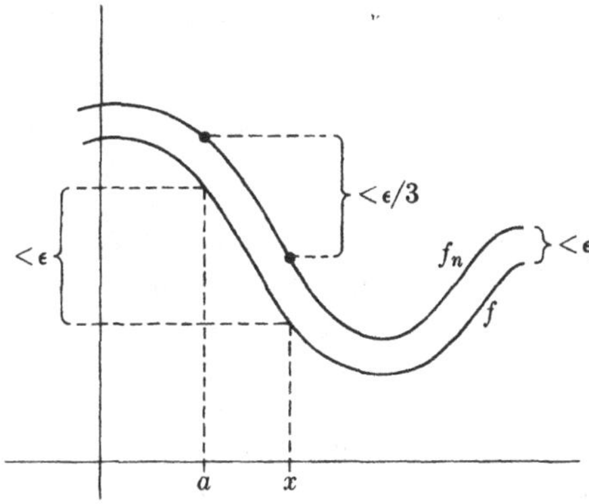

Fig. 4.3

Proof. We suppose that $\{f_n\} \subset \mathcal{BC}$ and that $\|f_n - g\|_\infty \to 0$, where $g \in \mathcal{B}$. We have to show that g is continuous. This is an application of a much used "up, over, and down" argument, which can be schematically indicated as in Fig. 4.3.

Given ϵ, we first choose any n such that $\|f_n - g\|_\infty < \epsilon/3$. Consider now any $a \in A$. Since f_n is continuous at a, there exists a δ such that

$$\rho(x, a) < \delta \Rightarrow \|f_n(x) - f_n(a)\| < \epsilon/3.$$

Then

$$\rho(x, a) < \delta \Rightarrow \|g(x) - g(a)\| \le \|g(x) - f_n(x)\| + \|f_n(x) - f_n(a)\| + \|f_n(a) - g(a)\| < \epsilon/3 + \epsilon/3 + \epsilon/3 = \epsilon.$$

Thus g is continuous at a for every $a \in A$, and so $g \in \mathcal{BC}$. □

This important classical result is traditionally stated as follows: *The limit of a uniformly convergent sequence of continuous functions is continuous.*

Remark. The proof was slightly more general. We actually showed that if $f_n \to f$ uniformly, and if each f_n is continuous at a, then f is continuous at a.

Corollary. $\mathcal{BC}(A, W)$ is a Banach space.

Theorem 7.9. If A is a sequentially compact metric space, then A is complete.

Proof. A Cauchy sequence in A has a subsequence converging to a limit in A, and therefore, by Lemma 7.2, itself converges to that limit. Thus A is complete. □

In Section 5 we proved that a compact set is also totally bounded. It can be shown, conversely, that a complete, totally bounded set A is sequentially compact, so that these two properties together are equivalent to compactness. The crucial fact is that if A is totally bounded, then every sequence in A has a Cauchy subsequence. If A is also complete, this Cauchy subsequence will converge to a point of A. Thus the fact that total boundedness and completeness together are equivalent to compactness follows directly from the next lemma.

Lemma 7.4. If A is totally bounded, then every sequence in A has a Cauchy subsequence.

Proof. Let $\{p_m\}$ be any sequence in A. Since A can be covered by a finite number of balls of radius 1, at least one ball in such a covering contains infinitely many of the points $\{p_m\}$. More precisely, there exists an infinite set $M_1 \subset \mathbb{Z}^+$ such that the set $\{p_m : m \in M_1\}$ lies in a single ball of radius 1. Suppose that $M_1, \ldots, M_n \subset \mathbb{Z}^+$ have been defined so that $M_{i+1} \subset M_i$ for $i = 1, \ldots, n-1$, M_n is infinite, and $\{p_m : m \in M_i\}$ is a subset of a ball of radius $1/i$ for $i = 1, \ldots, n$. Since A can be covered by a finite family of balls of radius $1/(n+1)$, at least one covering ball contains infinitely many points of the set $\{p_m : m \in M_n\}$. More precisely, there exists an infinite set $M_{n+1} \subset M_n$ such that $\{p_m : m \in M_{n+1}\}$ is a subset of a ball of radius $1/(n+1)$. We thus define an infinite sequence $\{M_n\}$ of subsets of $\mathbb{Z}^+$ having the above properties.

Now choose $m_1 \in M_1$, $m_2 \in M_2$ so that $m_2 > m_1$, and, in general, $m_{n+1} \in M_{n+1}$ so that $m_{n+1} > m_n$. Then the subsequence $\{p_{m_n}\}_n$ is Cauchy. For given ϵ, we can choose n so that $1/n < \epsilon/2$. Then $i, j > n \Rightarrow m_i, m_j \in M_n \Rightarrow \rho(p_{m_i}, p_{m_j}) < 2(1/n) < \epsilon$. This proves the lemma, and our theorem is a corollary. □

Theorem 7.10. A metric space S is sequentially compact if and only if S is totally bounded and complete.

The next three sections will be devoted to applications of completeness to the calculus, but before embarking on these vital matters we should say a few words about infinite series. As in the ordinary calculus, if $\{\xi_n\}$ is a sequence in a normed linear space V, we say that the series $\sum \xi_i$ converges and has the sum α, and write $\sum_1^\infty \xi_i = \alpha$, if the *sequence of partial sums converges to* α. This means that $\sigma_n \to \alpha$ as $n \to \infty$, where σ_n is the finite sum $\sum_1^n \xi_i$ for each n. We say that $\sum \xi_i$ *converges absolutely* if the series of norms $\sum \|\xi_i\|$ converges in $\mathbb{R}$. This is abuse of language unless it is true that every absolutely convergent series converges, and the importance of the notion stems from the following theorem.

Theorem 7.11. If V is a Banach space, then every absolutely convergent series in V is convergent.

Proof. Let $\sum \xi_i$ be absolutely convergent. This means that $\sum \|\xi_i\|$ converges in $\mathbb{R}$, i.e., that the sequence $\{s_n\}$ converges in $\mathbb{R}$, where $s_n = \sum_1^n \|\xi_i\|$. If $m < n$, then

$$\|\sigma_n - \sigma_m\| = \left\|\sum_{m+1}^{n} \xi_i\right\| \le \sum_{m+1}^{n} \|\xi_i\| = s_n - s_m.$$

Since $\{s_i\}$ is Cauchy in $\mathbb{R}$, this inequality shows that $\{\sigma_i\}$ is Cauchy in V and therefore, because V is complete, that $\{\sigma_n\}$ is convergent in V. That is, $\sum \xi_i$ is convergent in V. □

The reader will be asked to show in an exercise that, conversely, if a normed linear space V is such that every absolutely convergent series converges, then V is complete. This property therefore characterizes Banach spaces.

We shall make frequent use of the above theorem. For the moment we note just one corollary, the classical Weierstrass comparison test.

Corollary. If $\{f_n\}$ is a sequence of bounded real-valued (or W-valued, for some Banach space W) functions on a common domain A, and if there is a sequence $\{M_n\}$ of positive constants such that $\sum M_n$ is convergent and $\|f_n\|_\infty \le M_n$ for each n, then $\sum f_n$ is uniformly convergent.

Proof. The hypotheses imply that $\sum \|f_n\|_\infty$ converges, and so $\sum f_n$ converges in the Banach space $\mathcal{B}(A, W)$ by the theorem. But convergence in $\mathcal{B}(A, W)$ is uniform convergence. □

EXERCISES

7.1 Prove that a Cauchy sequence in a metric space is a bounded set.

7.2 Let V be a normed linear space. Prove that the sum of two Cauchy sequences in V is Cauchy.

7.3 Show also that if $\{\xi_n\}$ is Cauchy in V and $\{a_n\}$ is Cauchy in $\mathbb{R}$, then $\{a_n \xi_n\}$ is Cauchy in V.

7.4 Prove that if $\{\xi_n\}$ is a Cauchy sequence in a normed linear space V, then $\{\|\xi_n\|\}$ is a Cauchy sequence in $\mathbb{R}$.

7.5 Prove that if $\{x_n\}$ and $\{y_n\}$ are two Cauchy sequences in a metric space S, then $\{\rho(x_n, y_n)\}$ is a Cauchy sequence in $\mathbb{R}$.

7.6 Prove the statement made after the proof of Lemma 7.2.

7.7 The rational number system is an incomplete metric space. Prove this by exhibiting a Cauchy sequence of rational numbers that does not converge to a rational number.

7.8 Prove Theorem 7.1.

7.9 Deduce a strengthened form of Theorem 7.3 from Theorem 7.1.

7.10 Write out a careful proof of Theorem 7.6, modeled on the proof of Theorem 7.5.

7.11 Prove Theorem 7.7.

7.12 Let the metric space X have a dense subset Y such that every Cauchy sequence in Y is convergent in X. Prove that X is complete.

7.13 Show that the set W of all Cauchy sequences in a normed linear space V is itself a vector space and that a seminorm p can be defined on W by $p(\{\xi_n\}) = \lim \|\xi_n\|$. (Put this together from the material in the text and the preceding problems.)

7.14 Continuing the above exercise, for each $\xi \in V$, let ξ^c be the constant sequence all of whose terms are ξ. Show that $\theta: \xi \mapsto \xi^c$ is an isometric linear injection of V into W and that $\theta[V]$ is dense in W in terms of the seminorm from the above exercise.

7.15 Prove next that every Cauchy sequence in $\theta[V]$ is convergent in W. Put Exercises 4.18 of Chapter 3 and 7.12 through 7.14 of this chapter together to conclude that if N *is* the set of null Cauchy sequences in W, then W/N is a Banach space, and that $\xi \mapsto \xi^c$ is an isometric linear injection from V to a dense subspace of W/N. This constitutes the standard completion of the normed linear space V.

7.16 We shall now sketch a nonstandard way of forming the completion of a metric space S. Choose some point p_0 in S, and let V be the set of real-valued functions on S such that $f(p_0) = 0$ and f is a Lipschitz function. For f in V define $\|f\|$ as the smallest Lipschitz constant for f. That is,

$$\|f\| = \operatorname*{lub}_{p \neq q} \{|f(p) - f(q)|/\rho(p, q)\}.$$

Prove that V is a normed linear space under this norm. (V actually is complete, but we do not need this fact.)

7.17 Continuing the above exercise, we know that the dual space V^* of all bounded linear functionals on V is complete by Theorem 7.6. We now want to show that S can be isometrically imbedded in V^*; then the closure of S as a subset of V will be the desired completion of S. For each $p \in S$, let $\theta_p: V \to \mathbb{R}$ be "evaluation at p". That is, $\theta_p(f) = f(p)$. Show that $\theta_p \in V^*$ and that $\|\theta_p - \theta_q\| \leq \rho(p, q)$.

7.18 In order to conclude that the mapping $\theta: p \mapsto \theta_p$ is an isometry (i.e., is distance-preserving), we have to prove the opposite inequality $\|\theta_p - \theta_q\| \geq \rho(p, q)$. To do this, choose p and consider the special function $f(x) = \rho(p, x) - \rho(p, p_0)$. Show that f is in V and that $\|f\| = 1$ (from an early lemma in the chapter). Now apply the definition of $\|\theta_p - \theta_q\|$ and conclude that θ is an isometric injection of S into V^*. Then $\overline{\theta[S]}$ is our constructed completion.

7.19 Prove that if a normed linear space V has the property that every absolutely convergent series converges, then V is complete. (Let $\{\alpha_n\}$ be a Cauchy sequence. Show that there is a subsequence $\{\alpha_{n_i}\}_i$ such that if $\xi_i = \alpha_{n_{i+1}} - \alpha_{n_i}$, then $\|\xi_i\| < 2^{-i}$. Conclude that the subsequence converges and finish up.)

7.20 The above exercise gives a very useful criterion for V to be complete. Use it to prove that if V is a Banach space and N is a closed subspace, then V/N is a Banach space (see Exercise 4.14 of Chapter 3 for the norm on V/N).

7.21 Prove that the sum of a uniformly convergent series of infinitesimals (all on the same domain) is an infinitesimal.

8. A FIRST LOOK AT BANACH ALGEBRAS

When we were considering the implicit-function theorem and the inverse-function theorem in the last chapter, we saw how useful it is to know that if a transformation T has an inverse T^{-1}, then so does S whenever $\|S - T\|$ is small enough, and that the mapping $T \mapsto T^{-1}$ is continuous on the open set of all invertible elements. When the spaces in question are finite-dimensional, these facts can be made to follow from the continuity of the determinant function $T \mapsto \Delta(T)$ from Hom V to $\mathbb{R}$. It is also possible to produce them by arguing directly in terms of upper and lower bounds for T and its close approximations S. But the most natural, most elegant, and—in the case of Banach spaces—easiest way to prove these things is to show that if V is a Banach space and T in Hom V has norm less than one, then the sum of the geometric series $\sum_0^\infty T^n$ is the inverse of $I - T$, just as in the elementary calculus. But in making this argument, the fact that T is a linear transformation has little importance, and we shall digress for a moment to explore this situation.

Let us summarize the norm and algebraic properties of Hom V when V is a Banach space. First of all, we know that Hom V is also a Banach space. Second, it is an algebra. That is, it possesses an associative multiplication operation (composition) that relates to the linear operations according to the following laws:

$$S(T_1 + T_2) = ST_1 + ST_2,$$
$$(S_1 + S_2)T = S_1T + S_2T,$$
$$c(ST) = (cS)T = S(cT).$$

Finally, multiplication is related to the norm by

$$\|ST\| \le \|S\|\,\|T\| \qquad \text{and} \qquad \|I\| = 1.$$

This list of properties constitutes exactly the axioms for a *Banach algebra*.

Just as we can see certain properties of functions most clearly by forgetting that they are functions and considering them only as elements of a vector space, now it turns out that we can treat certain properties of transformations in Hom(V) most simply by forgetting the complicated nature of a linear transformation and considering it merely as an element of an abstract Banach algebra A.

The most important simple thing we can do in a Banach algebra that we couldn't do in a Banach space is to consider power series. The following theorem shows that the geometric series, in particular, plays the same central role here that it plays in elementary calculus. Since we are not thinking of the elements of A as transformations, we shall designate them by lower-case letters; e is the identity of A.

Theorem 8.1. If A is a Banach algebra, and if x in A has norm less than one, then $(e - x)$ is invertible and its inverse is the sum of the geometric series in x:

$$(e - x)^{-1} = \sum_0^\infty x^n.$$

Also, $\|e - (e - x)^{-1}\| \leq r/(1 - r)$, where $r = \|x\|$.

Proof. Since $\|x^n\| \leq \|x\|^n = r^n$, the series $\sum x^n$ is absolutely convergent when $\|x\| < 1$ by comparison with the ordinary geometric series $\sum r^n$. It is therefore convergent, and if $y = \sum_0^\infty x^n$, then

$$(e - x)y = \lim_{n\to\infty} (e - x) \sum_0^n x^i = \lim (e - x^{n+1}) = e,$$

since $\|x\|^{n+1} \leq r^{n+1} \to 0$. That is, $y = (e - x)^{-1}$. Finally,

$$\|e - (e - x)^{-1}\| = \left\|\sum_1^\infty x^n\right\| \leq \sum_1^\infty r^n = r/(1 - r). \square$$

Theorem 8.2. The set $\mathfrak{N}$ of invertible elements in a Banach algebra A is open and the mapping $x \mapsto x^{-1}$ is continuous from $\mathfrak{N}$ to $\mathfrak{N}$. In fact, if y^{-1} exists and $m = \|y^{-1}\|$, then $(y - h)^{-1}$ exists whenever $\|h\| < 1/m$ and $\|(y - h)^{-1} - y^{-1}\| \leq m^2\|h\|/(1 - m\|h\|)$.

Proof. Set $x = y^{-1}h$. Then $(y - h) = y(e - x)$, where $\|x\| = \|y^{-1}h\| \leq m\|h\|$, and so by the above theorem $y - h$ will be invertible, with $(y - h)^{-1} = (e - x)^{-1}y^{-1}$, provided $\|h\| < 1/m$. Then also

$$\|y^{-1} - (y - h)^{-1}\| \leq \|e - (e - x)^{-1}\| \cdot m,$$

and this is bounded above by

$$mr/(1 - r) \leq m^2\|h\|/(1 - m\|h\|),$$

by the last inequality in the above theorem. $\square$

Corollary. If V and W are Banach spaces, then the invertible elements in $\mathrm{Hom}(V, W)$ form an open set, and the map $T \mapsto T^{-1}$ is continuous on this domain.

Proof. Suppose that T^{-1} exists, and set $m = \|T^{-1}\|$. Then if $\|T - S\| < 1/m$, we have $\|I - T^{-1}S\| \leq \|T^{-1}\|\,\|T - S\| < 1$, and so $T^{-1}S = I - (I - T^{-1}S)$ is an invertible element of Hom V. Therefore, $S = T(T^{-1}S)$ is invertible and $S^{-1} = (T^{-1}S)^{-1}T^{-1}$. The continuity of $S \mapsto S^{-1}$ is left to the reader. $\square$

We saw above that the map $x \mapsto (e - x)^{-1}$ from the open unit ball $B_1(0)$ in a Banach algebra A to A is the sum of the geometric power series. We can define many other mappings by convergent power series, at hardly any greater effort.

Theorem 8.3. Let A be a Banach algebra. Let the sequence $\{a_n\} \subset A$ and the positive number δ be such that the sequence $\{\|a_n\|\delta^n\}$ is bounded. Then $\sum a_n x^n$ converges for x in the ball $B_\delta(0)$ in A, and if $0 < s < \delta$, then the series converges uniformly on $B_s(0)$.

Proof. Set $r = s/\delta$, and let b be a bound for the sequence $\{\|a_n\|\delta^n\}$. On the ball $B_s(0)$ we then have $\|a_n x^n\| \leq \|a_n\| s^n = \|a_n\|\delta^n r^n \leq br^n$, and the series therefore converges uniformly on this ball by comparison with the geometric series $b\sum r^n$, since $r < 1$. □

The series of most interest to us will have real coefficients. They are included in the above argument because the product of the vector x and the scalar t is the algebra product $(te)x$. In addition to dealing with the above geometric series, we shall be particularly interested in the exponential function $e^x = \sum_0^\infty x^n/n!$. The usual comparison arguments of the elementary calculus show just as easily here that this series converges for every x in A and uniformly on any ball.

It is natural to consider the differentiability of the maps from A to A defined by such convergent series, and we state the basic facts below, starting with a fundamental theorem on the differentiability of a limit of a sequence.

Theorem 8.4. Let $\{F^n\}$ be a sequence of maps from a ball B in a normed linear space V to a normed linear space W such that F^n converges pointwise to a map F on B and such that $\{dF^n_\alpha\}$ converges for each α and uniformly over α. Then F is differentiable on B and $dF_\beta = \lim dF^n_\beta$ for each β in B.

Proof. Fix β and set $T = \lim dF^n_\beta$. By the uniform convergence of $\{dF^n\}$, given ϵ, there is an N such that $\|dF^n_\alpha - dF^N_\alpha\| \leq \epsilon$ for all $n \geq N$ and for all α in B. It then follows from the mean-value theorem for differentials that

$$\|(\Delta F^n_\beta(\xi) - \Delta F^N_\beta(\xi)) - (dF^n_\beta(\xi) - dF^N_\beta(\xi))\| \leq 2\epsilon\|\xi\|$$

for all $n \geq N$ and all ξ such that $\beta + \xi \in B$. Letting $n \to \infty$ and regrouping, we have

$$\|(\Delta F_\beta(\xi) - T(\xi)) - (\Delta F^N_\beta(\xi) - dF^N_\beta(\xi))\| \leq 2\epsilon\|\xi\|$$

for all such ξ. But, by the definition of dF^N_β there is a δ such that

$$\|\Delta F^N_\beta(\xi) - dF^N_\beta(\xi)\| \leq \epsilon\|\xi\|$$

when $\|\xi\| < \delta$. Putting these last two inequalities together, we see that

$$\|\xi\| < \delta \Rightarrow \|\Delta F_\beta(\xi) - T(\xi)\| \leq 3\epsilon\|\xi\|.$$

Thus F is differentiable at β and $dF_\beta = T$. □

The remaining proofs are left as a set of exercises.

Lemma 8.1. Multiplication on a Banach algebra A is differentiable (from $A \times A$ to A). If we let p be the product function, so that $p(x, y) = xy$, then $dp_{<a,b>}(x, y) = ay + xb$.

Lemma 8.2. Let A be a commutative Banach algebra, and let p be the monomial function $p(x) = ax^n$. Then p is everywhere differentiable and $dp_y(x) = nay^{n-1}x$.

Lemma 8.3. If $\{\|a_n\|r^n\}$ is a bounded sequence in $\mathbb{R}$, then $\{n\|a_n\|s^n\}$ is bounded for any $0 < s < r$, and therefore $\sum na_nx^{n-1}$ converges uniformly on any ball in A smaller than $B_r(0)$.

Theorem 8.5. If A is a commutative Banach algebra and $\{a_n\} \subset A$ is such that $\{\|a_n\|r^n\}$ is bounded in $\mathbb{R}$, then $F(x) = \sum_0^\infty a_nx^n$ is defined and differentiable on the ball $B_r(0)$ in A, and

$$dF_y(x) = \left(\sum_1^\infty na_ny^{n-1}\right) \cdot x.$$

It is natural to call the element $\sum_1^\infty na_ny^{n-1}$ the derivative of F at y and to designate it $F'(y)$, although this departs from our rule that derivatives are vectors obtained as limits of difference quotients. The remarkable aspect of the above theorem is that for this kind of differentiable mapping from an open subset of A to A the linear transformation dF_y is multiplication by an element of A: $dF_y(x) = F'(y) \cdot x$.

In particular, the exponential function $\exp(x) = e^x = \sum_0^\infty x^n/n!$ is its own derivative, since $\sum_1^\infty nx^{n-1}/n! = \sum_0^\infty x^m/m!$, and from this fact (see the exercises) or from direct algebraic manipulation of the series in question, we can deduce the law of exponents $e^{x+y} = e^xe^y$. Remember, though, that this is on a *commutative* Banach algebra. The function $x \mapsto e^x = \sum_0^\infty x^n/n!$ can be defined just as easily on any Banach algebra A, but it is not nearly as pleasant when A is noncommutative. However, one thing that we can always do, and often thereby save the day, is to restrict the exponential mapping to a commutative subalgebra of A, say that generated by a single element x. For example, we can consider the parametrized arc $\gamma(t) = e^{tx}$ (x fixed) into *any* Banach algebra A, and, because its range lies in the commutative subalgebra X generated by x, we can apply Theorem 7.2 of Chapter 3 to conclude that γ is differentiable and that

$$\gamma'(t) = d\exp_{tx}(x) = xe^{tx}.$$

This can also easily be proved directly from the law of exponents:

$$\Delta\gamma_t(h) = e^{(t+h)x} - e^{tx} = e^{tx}(e^{hx} - 1),$$

and since it is clear from the series that $(e^{hx} - 1)/h \to x$ as $h \to 0$, we have that

$$\gamma'(t) = \lim_{h \to 0} \frac{\Delta\gamma_t(h)}{h} = xe^{tx}.$$

EXERCISES

8.1 Finish the proof of the corollary of Theorem 8.2.

8.2 Let A be a Banach algebra, and let $\{a_n\} \subset \mathbb{R}$ and $x \in A$ be such that $\sum a_i x^i$ converges. Suppose also that x satisfies a polynomial indentity $p(x) = \sum_0^n b_i x^i = 0$, where $\{b_i\} \subset \mathbb{R}$ and $b_n \neq 0$. Prove that the element $\sum_1^\infty a_i x^i$ is a polynomial in x of degree $\leq n - 1$. (Let M be the linear span of $\{x^i\}_0^{n-1}$, and show first that $x^i \in M$ for all i.)

8.3 Let A be any Banach algebra, let x be a fixed element in A, and let X be the smallest closed subalgebra of A containing x. Prove that X is a commutative Banach algebra. (The set of polynomials $p(x) = \sum_0^n a_i x^i$ is the smallest algebra containing x. Consider its closure in X.)

8.4 Prove Lemma 8.1. [*Hint:* $\langle x, y\rangle \mapsto xy$ is a bounded bilinear map.]

8.5 Prove Lemma 8.2 by making a direct Δ-estimate from the binomial expansion, as in the elementary calculus.

8.6 Prove Lemma 8.2 by induction from Lemma 8.1.

8.7 Let A be any Banach algebra. Prove that $p: x \mapsto x^3$ is differentiable and that $dp_a(x) = xa^2 + axa + a^2x$.

8.8 Prove by induction that if $q(x) = x^n$, then q is differentiable and

$$dq_a(x) = \sum_{i=0}^{n-1} a^i x a^{(n-1-i)}.$$

Deduce Lemma 8.2 as a corollary.

8.9 Let A be any Banach algebra. Prove that $r: x \mapsto x^{-1}$ is everywhere differentiable on the open set U of invertible elements and that

$$dr_a(x) = -a^{-1}xa^{-1}.$$

[*Hint:* Examine the proofs of Theorems 8.1 and 8.2.]

8.10 Let A be an open subset of a normed linear space V, and let F and G be mappings from A to a Banach *algebra* X that are differentiable at α. Prove that the product mapping FG is differentiable at α and that $d(FG_\alpha) = F(\alpha)\, dG_\alpha + dF_\alpha G(\alpha)$. Does it follow that $d(F^2)_\alpha = 2F(\alpha)\, dF_\alpha$?

8.11 Continuing the above exercise, show that if X is a commutative Banach algebra, then $d(F^n)_\alpha = nF^{n-1}(\alpha)\, dF_\alpha$.

8.12 Let $F: A \to X$ be a differentiable map from an open set A of a normed linear space to a Banach algebra X, and suppose that the element $F(\xi)$ is invertible in X for every ξ in A. Prove that the map $G: \xi \mapsto [F(\xi)]^{-1}$ is differentiable and that $dG_\alpha(\xi) = -F(\alpha)^{-1}\, dF_\alpha(\xi) F(\alpha)$. Show also that if F is a parametrized arc ($A = I \subset \mathbb{R}$), then $G'(\alpha) = -F(\alpha)^{-1} \cdot F'(\alpha) \cdot F(\alpha)^{-1}$.

8.13 Prove Lemma 8.3.

8.14 Prove Theorem 8.5 by showing that Lemma 8.3 makes Theorem 8.4 applicable.

8.15 Show that in Theorem 8.4 the convergence of F^n to F needs only to be assumed at one point, provided we know that the codomain space W is a Banach space.

8.16 We want to prove the law of exponents for the exponential function on a commutative Banach algebra. Show first that $(\exp(-x))(\exp x) \equiv e$ by applying Exercise 7.13 of Chapter 3, the above Exercise 8.10, and the fact that $d\exp_a(x) = (\exp a)x$.

8.17 Show that if X is a commutative Banach algebra and $F: X \to X$ is a differentiable map such that $dF_\alpha(\xi) = \xi F(\alpha)$, then $F(\xi) = \beta \exp \xi$ for some constant β. [Consider the differential of $F(\xi)\exp(-\xi)$.]

8.18 Now set $F(\xi) = \exp(\xi + \eta)$ and prove from the above exercise that

$$\exp(\xi + \eta) = \exp(\xi)\exp(\eta).$$

You will also need the fact that $\exp 0 = 1$.

8.19 Let z be a nilpotent element in a commutative Banach algebra X. That is, $z^p = 0$ for some positive integer p. Show by an elementary estimate based on the binomial expansion that if $\|x\| < 1$, then $\|x + z\|^n \leq kn^p\|x\|^{n-p}$ for $n > p$. The series of positive terms $\sum n^a r^n$ converges for $r < 1$ (by the ratio test). Show, therefore, that the series for $\log(1 - (x + z))$ and for $(1 - (x + z))^{-1}$ converge when $\|x\| < 1$.

8.20 Continuing the above exercise, show that $F(y) = \log(1 - y)$ is defined and differentiable on the ball $\|y - z\| < 1$ and that $dF_a(x) = -(1 - a)^{-1} \cdot x$. Show, therefore, that $\exp(\log(1 - y)) = 1 - y$ on this ball, either by applying the inverse mapping theorem or by applying the composite function rule for differentiating. Conclude that for every nilpotent element z in X there exists a u in X such that $\exp u = 1 - z$.

8.21 Let $X_1, \ldots, X_n$ be Banach algebras. Show that the product Banach space $X = \prod_1^n X_i$ becomes a Banach algebra if the product $\mathbf{xy} = \langle x, \ldots, x_n\rangle\langle y_1, \ldots, y_n\rangle$ is defined as $\langle x_1y_1, \ldots, x_ny_n\rangle$ and if the maximum norm is used on X.

8.22 In the above situation the projections π_i have now become bounded algebra homomorphisms. In fact, just as in our original vector definitions on a product space, our definition of multiplication on X was determined by the requirement that $\pi_i(\mathbf{xy}) = \pi_i(\mathbf{x})\pi_i(\mathbf{y})$ for all i. State and prove an algebra theorem analogous to Theorem 3.4 of Chapter 1.

8.23 Continuing the above discussion, suppose that the series $\sum \mathbf{a}_n\mathbf{x}^n$ converges in X, with sum $\mathbf{y}$. Show that then $\sum (a_n)_i(x_i)^n$ converges in X_i to y_i for each i, where, of course, $\mathbf{y} = \langle y_1, \ldots, y_n\rangle$. Conclude that $e^{\mathbf{x}} = \langle e^{x_1}, \ldots, e^{x_n}\rangle$ for any $\mathbf{x} = \langle x_1, \ldots, x_n\rangle$ in X.

8.24 Define the sine and cosine functions on a commutative Banach algebra, and show that $\sin' = \cos$, $\cos' = -\sin$, $\sin^2 + \cos^2 = e$.

9. THE CONTRACTION MAPPING FIXED-POINT THEOREM

In this section we shall prove the very simple and elegant *fixed-point* theorem for contraction mappings, and then shall use it to complete the proof of the implicit-function theorem. Later, in Chapter 6, it will be the basis of our proof of the fundamental existence and uniqueness theorem for ordinary differential equations. The section concludes with a comparison of the iterative procedure of the fixed-point theorem and that of Newton's method.

A mapping K from a metric space X to itself is a *contraction* if it is a Lipschitz mapping with constant less than 1; that is, if there is a constant C with $0 < C < 1$ such that $\rho(K(x), K(y)) \leq C\rho(x, y)$ for all $x, y \in X$. A fixed point of K is, of course, a point x such that $K(x) = x$.

A contraction K can have at most one fixed point, since if $K(x) = x$ and $K(y) = y$, then $\rho(x, y) = \rho(K(x), K(y)) \leq C\rho(x, y)$, and so $(1 - C)\rho(x, y) \leq 0$. Since $C < 1$, this implies that $\rho(x, y) = 0$ and $x = y$.

Theorem 9.1. Let X be a nonempty complete metric space, and let $K: X \to X$ be a contraction. Then K has a (unique) fixed point.

Proof. Choose any x_0 in X, and define the sequence $\{x_n\}_0^\infty$ inductively by setting $x_1 = K(x_0)$, $x_2 = K(x_1) = K^2(x_0)$, and $x_n = K(x_{n-1}) = K^n(x_0)$. Set $\delta = \rho(x_1, x_0)$. Then $\rho(x_2, x_1) = \rho(K(x_1), K(x_0)) \leq C\rho(x_1, x_0) = C\delta$, and, by induction,

$$\rho(x_{n+1}, x_n) = \rho(K(x_n), K(x_{n-1})) \leq C\rho(x_n, x_{n-1}) \leq C \cdot C^{n-1}\delta = C^n\delta.$$

It follows that $\{x_n\}$ is Cauchy, for if $m > n$, then

$$\rho(x_m, x_n) \leq \sum_n^{m-1} \rho(x_{i+1}, x_i) \leq \sum_n^{m-1} C^i\delta < C^n\delta/(1 - C),$$

and $C^n \to 0$ as $n \to \infty$, because $C < 1$. Since X is complete, $\{x_n\}$ converges to some a in X, and it then follows that $K(a) = \lim K(x_n) = \lim x_{n+1} = a$, so that a is a fixed point. □

In practice, we meet mappings K that are contractions only near some particular point p, and we have to establish that a suitable neighborhood of p is carried into itself by K. We show below that if K is a contraction on a ball about p, and if K doesn't move the center of p very far, then the theorem can be applied.

Corollary 1. Let D be a closed ball in a complete metric space X, and let $K: D \to X$ be a contraction which moves the center of D a distance at most $(1 - C)r$, where r is the radius of D and C is the contraction constant. Then K has a unique fixed point and it is in D.

Proof. We simply check that the range of K is actually in D. If p is the center of D and x is any point in D, then

$$\begin{aligned}\rho(K(x), p) &\leq \rho(K(x), K(p)) + \rho(K(p), p)\\ &\leq C\rho(x, p) + (1 - C)r \leq Cr + (1 - C)r = r. \ \square\end{aligned}$$

Corollary 2. Let B be an open ball in a complete metric space X, and let $K: B \to X$ be a contraction which moves the center of B a distance *less* than $(1 - C)r$, where r is the radius of B and C is the contraction constant. Then K has a unique fixed point.

Proof. Restrict K to any slightly smaller closed ball D concentric with B, and apply the above corollary. □

Corollary 3. Let K be a contraction on the complete metric space X, and suppose that K moves the point x a distance d. Then the distance from x to the fixed point is at most $d/(1 - C)$, where C is the contraction constant.

Proof. Let D be the closed ball about x of radius $r = d/(1 - C)$, and apply Corollary 1 to the restriction of K to D. It implies that the fixed point is in D. □

We now suppose that the contraction K contains a parameter s, so that K is now a function of two variables $K(s, x)$. We shall assume that K is a contraction in x uniformly over s, which means that $\rho(K(s, x), K(s, y)) \leq C\rho(x, y)$ for all x, y, and s, where $0 < C < 1$. We shall also assume that K is a continuous function of s for each fixed x.

Corollary 4. Let K be a mapping from $S \times X$ to X, where X is a complete metric space and S is any metric space, and suppose that $K(s, x)$ is a contraction in x uniformly over s and is continuous in s for each x. Then the fixed point p_s is a continuous function of s.

Proof. Given ϵ, we use the continuity of K in its first variable around the point $\langle t, p_t \rangle$ to choose δ, so that if $\rho(s, t) < \delta$, then the distance from $K(s, p_t)$ to $K(t, p_t)$ is at most ϵ. Since $K(t, p_t) = p_t$, this simply says that the contraction with parameter value s moves p_t a distance at most ϵ, and so the distance from p_t to the fixed point p_s is at most $\epsilon/(1 - C)$ by Corollary 3. That is, $\rho(s, t) < \delta \Rightarrow \rho(p_s, p_t) < \epsilon/(1 - C)$, where C is the uniform contraction constant, and the mapping $s \mapsto p_s$ is accordingly continuous at t. □

Combining Corollaries 2 and 4, we have the following theorem.

Theorem 9.2. Let B be a ball in a complete metric space X, let S be any metric space, and let K be a mapping from $S \times B$ to X which is a contraction in its second variable uniformly over its first variable and is continuous in its first variable for each value of its second variable. Suppose also that K moves the center of B a distance less than $(1 - C)r$ for every s in S, where r is the radius of B and C is the uniform contraction constant. Then for each s in S there is a unique p in B such that $K(s, p) = p$, and the mapping $s \mapsto p$ is continuous from S to B.

We can now complete the proof of the implicit-function theorem.

Theorem 9.3. Let V, W, and X be Banach spaces, let $A \times B$ be an open subset of $V \times W$, and let $G: A \times B \to X$ be continuous and have a continuous second partial differential. Suppose that the point $\langle \alpha, \beta \rangle$ in $A \times B$ is such that $G(\alpha, \beta) = 0$ and $dG^2_{\langle \alpha, \beta \rangle}$ is invertible. Then there are open balls M and N about α and β, respectively, such that for each ξ in M there is a unique η in N satisfying $G(\xi, \eta) = 0$. The function F thus uniquely defined near $\langle \alpha, \beta \rangle$ by the condition $G(\xi, F(\xi)) = 0$ is continuous.

Proof. Set $T = dG^2_{\langle \alpha, \beta \rangle}$ and $K(\xi, \eta) = \eta - T^{-1}(G(\xi, \eta))$. Then K is a continuous mapping from $A \times B$ to W such that $K(\alpha, \beta) = \beta$, and K has a continuous second partial differential such that $dK^2_{<\alpha,\beta>} = 0$. Because $dK^2_{<\mu,\nu>}$ is a continuous function of $<\mu, \nu>$, we can choose a product ball $M \times N$ about $<\alpha, \beta>$ on which $dK^2_{<\mu,\nu>}$ is bounded by $\frac{1}{2}$, and we can then decrease the ball M if necessary so that for μ in M we also have $\|K(\mu, \beta) - \beta\| < r/2$, where r is the radius of the ball N. The mean-value theorem for differentials implies that K is a contraction in its second variable with constant $\frac{1}{2}$. The preceding theorem therefore shows that for each ξ in M there is a unique η in N such that $K(\xi, \eta) = \eta$ and the mapping $F: \xi \mapsto \eta$ is continuous. Since $K(\xi, \eta) = \eta$ if and only if $G(\xi, \eta) = 0$, we are done. □

Theorems 8.2 and 9.3 complete the list of ingredients of the implicit-function theorem. (However, see Exercise 9.8.)

We next show, in the other direction, that if a contraction depending on a parameter is continuously differentiable, then the fixed point is a continuously differentiable function of the parameter.

Theorem 9.4. Let V and W be Banach spaces, and let K be a differentiable mapping from an open subset $A \times B$ of $V \times W$ to W which satisfies the hypotheses of Theorem 9.2. Then the function F from A to B uniquely defined by the equation $K(\xi, F(\xi)) = F(\xi)$ is differentiable.

Proof. The inequality $\|K(\xi, \eta') - K(\xi, \eta'')\| \le C\|\eta' - \eta''\|$ is equivalent to $\|dK^2_{<\alpha,\beta>}\| \le C$ for all $<\alpha, \beta>$ in $A \times B$. We now define G by $G(\xi, \eta) = \eta - K(\xi, \eta)$, and observe that $dG^2 = I - dK^2$ and that dG^2 is therefore invertible by Theorem 8.1. Since $G(\xi, F(\xi)) = 0$, it follows from Theorem 11.1 of Chapter 3 that F is differentiable and that its differential is obtained by differentiating the above equation. □

Corollary. If K is continuously differentiable, then so is F.

* We should emphasize that the fixed-point theorem not only has the implicit-function theorem as a consequence, but the proof of the fixed-point theorem gives an iterative procedure for actually finding the value of $F(\xi)$, once we know how to compute T^{-1} (where $T = dG^2_{<\alpha,\beta>}$). In fact, for a given value of ξ in a small enough ball about $<\alpha, \beta>$ consider the function $G(\xi, \cdot)$. If we set $K(\xi, \eta) = \eta - T^{-1}G(\xi, \eta)$, then the inductive procedure

$$\eta_{i+1} = K(\xi, \eta_i)$$

becomes

$$(\eta_{i+1} - \eta_i) = -T^{-1}G(\xi, \eta_i). \tag{9.1}$$

The meaning of this iterative procedure is easily seen by studying the graph of the situation where $V = W = R^1$. (See Fig. 4.4.) As was proved above, under suitable hypotheses, the series $\sum\|\eta_{i+1} - \eta_i\|$ converges geometrically.

It is instructive to compare this procedure with Newton's method of elementary calculus. There the iterative scheme (9.1) is replaced by

$$(\eta_{i+1} - \eta_i) = S_i^{-1}G(\xi, \eta_i), \tag{9.2}$$

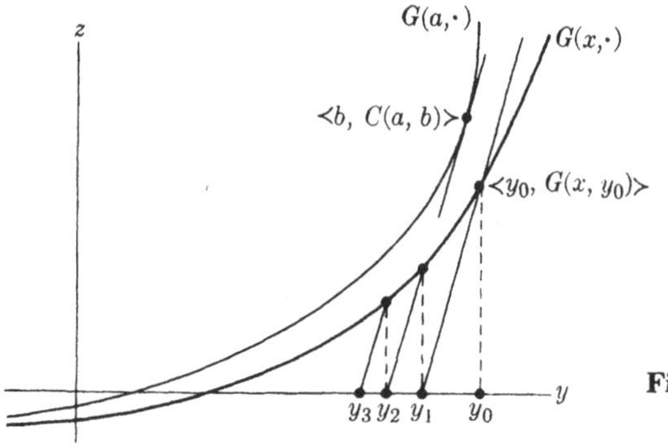

Fig. 4.4

where $S_i = dG^2_{\langle\xi,\eta_i\rangle}$. (See Fig. 4.5.) As we shall see, this procedure (when it works) converges much more rapidly than (9.1), but it suffers from the disadvantage that we must be able to compute the inverses of an infinite number of linear transformations S_i.

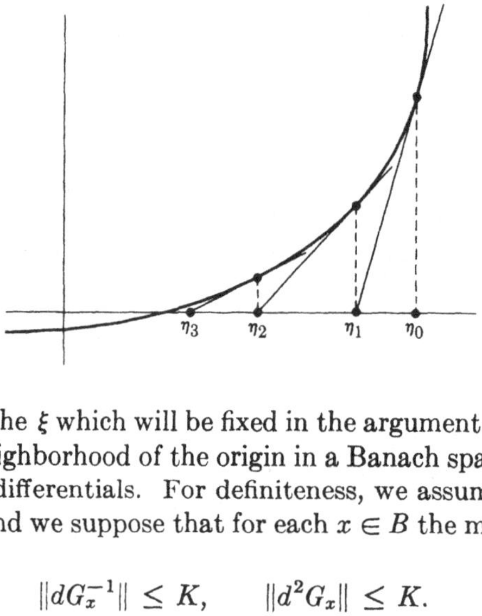

Fig. 4.5

Let us suppress the ξ which will be fixed in the argument and consider a map G defined in some neighborhood of the origin in a Banach space. Suppose that G has two continuous differentials. For definiteness, we assume that G is defined in the unit ball, B, and we suppose that for each $x \in B$ the map dG_x is invertible and, in fact,

$$\|dG_x^{-1}\| \leq K, \qquad \|d^2G_x\| \leq K.$$

Let $x_0 = 0$ and, assuming that x_n has been defined, we set

$$x_{n+1} = x_n - S_n^{-1}G(x_n),$$

where $S_n = dG_{x_n}$. We shall show that if $\|G(0)\|$ is sufficiently small (in terms of K), then the procedure is well defined (that is, $\|x_{n+1}\| < 1$) and converges rapidly. In fact, if τ is any real number between one and two (for instance $\tau = \frac{3}{2}$), we shall show that for some c (which can be made large if $\|G(0)\|$ is small)

$$\|x_n - x_{n-1}\| \leq e^{-c\tau^n}. \qquad (*)$$

Note that if we can establish (*) for large enough c, then $\|x_n\| \leq 1$ follows. In fact,

$$\|x_j\| \leq \sum_1^j e^{-c\tau^n} \leq \sum_1^\infty e^{-c\tau^n} \leq \sum_1^\infty e^{-cn(\tau-1)} = \frac{e^{-c(\tau-1)}}{1-e^{-c(\tau-1)}},$$

which is ≤ 1 if c is large. Let us try to prove (*) by induction. Assuming it true for n, we have

$$\begin{aligned}\|x_{n+1} - x_n\| &= \|S_n^{-1}G(x_n)\| \\ &\leq K\|G(x_{n-1} - S_{n-1}^{-1}G(x_{n-1}))\| \\ &\leq K\{\|G(x_{n-1}) - dG_{x_{n-1}}S_{n-1}^{-1}G(x_{n-1})\| + K\|x_n - x_{n-1}\|^2\}\end{aligned}$$

by Taylor's theorem. Now the first term on the right of the inequality vanishes, and we have

$$\|x_{n+1} - x_n\| \leq K^2\|x_n - x_{n-1}\|^2 \leq K^2e^{-2c\tau^n}.$$

For the induction to work we must have

$$K^2e^{-2c\tau^n} \leq e^{-c\tau^{n+1}}$$

or

$$K^2 \leq e^{c(2-\tau)\tau^n}. \tag{**}$$

Since $\tau < 2$, this last inequality can be arranged by choosing c sufficiently large. We must still verify (*) for $n = 1$. This says that

$$\|S_0^{-1}G(0)\| \leq e^{-c\tau}$$

or

$$\|G(0)\| \leq \frac{e^{-c\tau}}{K}. \tag{***}$$

In summary, for $1 < \tau < 2$ choose c so that $K^2 \leq e^{(2-\tau)c\tau}$ and

$$\frac{e^{-c(\tau-1)}}{1-e^{-c(\tau-1)}} \leq 1.$$

Then if (***) holds, the sequence x_n converges exponentially, that is, (*) holds. If $x = \lim x_i$, then $G(x) = \lim G(x_n) = \lim S_n(x_{n+1} - x_n) = 0$. This is Newton's method.

As a possible choice of c and τ, let $\tau = \frac{3}{2}$, and let c be given by $K^2 = e^{3c/4}$, so that (**) just holds. We may also assume that $K \geq 2^{3/4}$, so that $e^{3c/4} \geq 4^{3/4}$ or $e^c \geq 4$, which guarantees that $e^{-c/2} \leq \frac{1}{2}$, implying that $e^{-c/2}/(1 - e^{-c/2}) \leq 1$. Then (***) becomes the requirement $G(0) \leq K^{-5}$.

We end this section with an example of the fixed-point iterative procedure in its simplest context, that of the inverse-mapping theorem. We suppose that $H(0) = 0$ and that dH_0^{-1} exists, and we want to invert H near zero, i.e., solve the equation $H(\eta) - \xi = 0$ for η in terms of ξ. Our theory above tells us that the η corresponding to ξ will be the fixed point of the contraction $K(\xi, \eta) = \eta - T^{-1}H(\eta) + T^{-1}(\xi)$, where $T = dH_0$. In order to make our example as simple as possible, we shall take H from $\mathbb{R}^2$ to $\mathbb{R}^2$ and choose it so that $dH_0 = I$. Also, in order to avoid indices, we shall use the mongrel notation $\mathbf{x} = \langle x, y \rangle$, $\mathbf{u} = \langle u, v \rangle$.

Consider the mapping $\mathbf{x} = H(\mathbf{u})$ defined by $x = u + v^2$, $y = u^3 + v$. The Jacobian matrix

$$\begin{bmatrix} 1 & 2v \\ 3u^2 & 1 \end{bmatrix}$$

is clearly the identity at the origin. Moreover, in the expression $K(\mathbf{x}, \mathbf{u}) = \mathbf{x} + \mathbf{u} - H(\mathbf{u})$, the difference $H(\mathbf{u}) - \mathbf{u}$ is just the function $J(\mathbf{u}) = \langle v^2, u^3 \rangle$. This cancellation of the first-order terms is the practical expression of the fact that in forming $K(\xi, \eta) = \eta - T^{-1}G(\xi, \eta)$, we have acted to make $dK^2 = 0$ at the "center point" (the origin here). We naturally start the iteration with $\mathbf{u}_0 = 0$, and then our fixed-point sequence proceeds

$$\mathbf{u}_1 = K(\mathbf{x}, \mathbf{u}_0) = K(\mathbf{x}, 0), \ldots, \mathbf{u}_n = K(\mathbf{x}, \mathbf{u}_{n-1}).$$

Thus $\mathbf{u}_0 = 0$ and $\mathbf{u}_n = K(\mathbf{x}, \mathbf{u}_{n-1}) = \mathbf{x} - J(\mathbf{u}_{n-1})$, giving

$$\begin{aligned}
u_1 &= x, & v_1 &= y, \\
u_2 &= x - y^2, & v_2 &= y - x^3, \\
u_3 &= x - (y - x^3)^2, & v_3 &= y - (x - y^2)^3, \\
u_4 &= x - [y - (x - y^2)^3]^2, & v_4 &= y - [x - (y - x^3)^2]^3.
\end{aligned}$$

We are guaranteed that this sequence $\mathbf{u}_n$ will converge geometrically provided the starting point $\mathbf{x}$ is close enough to 0, and it seems clear that these two sequences of polynomials are computing the Taylor series expansions for the inverse functions $u(x, y)$ and $v(x, y)$. We shall ask the reader to prove this in an exercise. The two Taylor series start out

$$\begin{aligned}
u(x, y) &= x - y^2 - 2yx^3 + \cdots, \\
v(x, y) &= y - x^3 + 3x^2y^2 + \cdots
\end{aligned}$$

EXERCISES

9.1 Let B be a compact subset of a normed linear space such that $rB \subset B$ for all $r \in [0, 1]$. Suppose that $F: B \to B$ is a Lipschitz mapping with constant 1 (i.e., $\|F(\xi) - F(\eta)\| \leq \|\xi - \eta\|$ for all $\xi, \eta \in B$). Prove that F has a fixed point. [*Hint:* Consider first $G = rF$ for $0 < r < 1$.]

9.2 Give an example to show that the fixed point in the above exercise may not be unique.

9.3 Let X be a compact metric space, and let $K: X \to X$ "reduce each nonzero distance". That is, $\rho(K(x), K(y)) < \rho(x, y)$ if $x \neq y$. Prove that K has a unique fixed point. (Show that otherwise glb $\{\rho(K(x), x)\}$ is positive and achieved as a minimum. Then get a contradiction.)

9.4 Let K be a mapping from $S \times X$ to X, where X is a complete metric space and S is any metric space, and suppose that $K(s, x)$ is a contraction in s uniformly over x and is Lipschitz continuous in x uniformly over x. Show that the fixed point p_s is a Lipschitz continuous function of s. [*Hint:* Modify the ϵ,δ-beginning of the proof of Corollary 4 of Theorem 9.1.]

9.5 Let D be an open subset of a Banach space V, and let $K: D \to V$ be such that $I - K$ is Lipschitz with constant $\frac{1}{2}$.

a) Show that if $B_r(\alpha) \subset D$ and $\beta = K(\alpha)$, then $B_{r/2}(\beta) \subset K[D]$. (Apply a corollary of the fixed-point theorem to a certain simple contraction mapping.)

b) Conclude that K is injective and has an open range, and that K^{-1} is Lipschitz with constant 2.

9.6 Deduce an improved version of the result in Exercise 3.20, Chapter 3, from the result in the above exercise.

9.7 In the context of Theorem 9.3, show that $dG^2_{\langle\mu,\nu\rangle}$ is invertible if $\|dK^2_{\langle\mu,\nu\rangle}\| < 1$. (Do not be confused by the notation. We merely want to know that S is invertible if $\|I - T^{-1} \circ S\| < 1$.)

9.8 There is a slight discrepancy between the statements of Theorem 11.2 in Chapter 3 and Theorem 9.3. In the one case we assert the existence of a unique continuous mapping from a ball M, and in the other case, from the ball M to the ball N. Show that the requirement that the range be in N can be dropped by showing that two continuous solutions must agree on M. (Use the point-by-point uniqueness of Theorem 9.3.)

9.9 Compute the expression for dF_α from the identity $G(\xi, F(\xi)) = 0$ in Theorem 9.4, and show that if K is continuously differentiable, then all the maps involved in the solution expression are continuous and that $\alpha \mapsto dF_\alpha$ is therefore continuous.

9.10 Going back to the example worked out at the end of Section 9, show by induction that the polynomials $u_n - u_{n-1}$ and $v_n - v_{n-1}$ contain no terms of degree less than n.

9.11 Continuing the above exercise, show therefore that the power series defined by taking the terms of degree at most n from u_n is convergent in a ball about 0 and that its sum is the first component $u(x, y)$ of the mapping inverse to H.

9.12 The above conclusions hold generally. Let $J = \langle K, L\rangle$ be any mapping from a ball about 0 in $\mathbb{R}^2$ to $\mathbb{R}^2$ defined by the convergent power series

$$K(x, t) = \sum a_{ij}x^iy^j, \qquad L(x, y) = \sum b_{ij}x^iy^j$$

in which there are no terms of degree 0 or 1. With the conventions $\mathbf{x} = \langle x, y\rangle$ and $\mathbf{u} = \langle u, v\rangle$, consider the iterative sequence

$$\mathbf{u}_0 = 0, \qquad \mathbf{u}_n = \mathbf{x} - J(\mathbf{u}_{n-1}).$$

Make any necessary assumptions about what happens when one power series is substituted in another, and show by induction that $\mathbf{u}_n - \mathbf{u}_{n-1}$ contains no terms of degree less than n, and therefore that the $\mathbf{u}_n$ define a convergent power series whose sum is the function $\mathbf{u}(x, y) = \langle u(x, y), v(x, y)\rangle$ inverse to H in a neighborhood of 0. [Remember that $J(\eta) = H(\eta) - \eta$.]

9.13 Let A be a Banach algebra, and let x be an element of A of norm less than 1. Show that

$$(e - x)^{-1} = \prod_{i=1}^{\infty} (1 + x^{2^i}).$$

This means that if π_n is the partial product $\prod_1^n (1 + x^{2^i})$, then $\pi_n \to (e - x)^{-1}$. [*Hint:* Prove by induction that $(e - x)\pi_{n-1} = e - x^{2^n}$.]

This is another example of convergence at an exponential rate, like Newton's method in the text.

10. THE INTEGRAL OF A PARAMETRIZED ARC

In this section we shall make our final application of completeness. We first prove a very general extension theorem, and then apply it to the construction of the Riemann integral as an extension of an elementary integral defined for step functions.

Theorem 10.1. Let U be a subspace of a normed linear space V, and let T be a bounded linear mapping from U to a Banach space W. Then T has a uniquely determined extension to a bounded linear transformation S from the closure $\overline{U}$ to W. Moreover, $\|S\| = \|T\|$.

Proof. Fix $\alpha \in \overline{U}$ and choose $\{\xi_n\} \subset U$ so that $\xi_n \to \alpha$. Then $\{\xi_n\}$ is Cauchy and $\{T(\xi_n)\}$ is Cauchy (by the lemmas of Section 7), so that $\{T(\xi_n)\}$ converges to some $\beta \in W$. If $\{\eta_n\}$ is any other sequence in U converging to α, then $\xi_n - \eta_n \to 0$, $T(\xi_n) - T(\eta_n) = T(\xi_n - \eta_n) \to 0$, and so $T(\eta_n) \to \beta$ also. Thus β is independent of the sequence chosen, and, clearly, β must be the value $S(\alpha)$ at α of any continuous extension S of T. If $\alpha \in U$, then $\beta = \lim T(\alpha_n) = T(\alpha)$ by the continuity of T. We thus have S uniquely defined on $\overline{U}$ by the requirement that it be a continuous extension of T.

It remains to be shown that S is linear and bounded by $\|T\|$. For any $\alpha, \beta \in \overline{U}$ we choose $\{\xi_n\}, \{\eta_n\} \subset U$, so that $\xi_n \to \alpha$ and $\eta_n \to \beta$. Then $x\xi_n + y\eta_n \to x\xi + y\eta$, so that

$$S(x\alpha + y\beta) = \lim T(x\xi_n + y\eta_n) = x \lim T(\xi_n) + y \lim T(\eta_n) = xS(\alpha) + yS(\beta).$$

Thus S is linear. Finally,

$$\|S(\alpha)\| = \lim \|T(\xi_n)\| \leq \|T\| \lim \|\xi_n\| = \|T\| \cdot \|\alpha\|.$$

Thus $\|T\|$ is a bound for S, and, since S includes T, $\|S\| = \|T\|$. □

The above theorem has many applications, but we shall use it only once, to obtain the Riemann integral $\int_a^b f(t)\,dt$ of a continuous function f mapping a closed interval $[a, b]$ into a Banach space W as an extension of the trivial integral for step functions. If W is a normed linear space and $f\colon [a, b] \to W$ is a continuous function defined on a closed interval $[a, b] \subset \mathbb{R}$, we might expect to be able to define $\int_a^b f(t)\,dt$ as a suitable vector in W and to proceed with the integral calculus of vector-valued functions of one real variable. We haven't done this until now because we need the completeness of W to prove that the integral exists!

At first we shall integrate only certain elementary functions called step functions. A finite subset A of $[a, b]$ which contains the two endpoints a and b

will be called a *partition* of $[a, b]$. Thus A is (the range of) some finite sequence $\{t_i\}_0^n$, where $a = t_0 < t_1 < \cdots < t_n = b$, and A *subdivides* $[a, b]$ into a sequence of smaller intervals. To be definite, we shall take the *open* intervals (t_{i-1}, t_i), $i = 1, \ldots, n$, as the intervals of the subdivision. If A and B are partitions and $A \subset B$, we shall say that B is a *refinement* of A. Then each interval (s_{j-1}, s_j) of the B-subdivision is included in an interval (t_{i-1}, t_i) of the A-subdivision; t_{i-1} is the largest element of A which is less than or equal to s_{j-1}, and t_i is the smallest greater than or equal to s_j. A *step function* is simply a map $f\colon [a, b] \to W$ which is *constant* on the intervals of some subdivision $A = \{t_i\}_0^n$. That is, there exists a sequence of vectors $\{\alpha_i\}_1^n$ such that $f(\xi) = \alpha_i$ when $\xi \in (t_{i-1}, t_i)$. The values of f at the subdividing points may be among these values or they may be different.

For each step function f we define $\int_a^b f(t)\, dt$ as $\sum_{i=1}^n \alpha_i\, \Delta t_i$, where $f = \alpha_i$ on (t_{i-1}, t_i) and $\Delta t_i = t_i - t_{i-1}$. If f were real-valued, this would be simply the sum of the areas of the rectangles making up the region between the graph of f and the t-axis. Now f may be described as a step function in terms of many different subdivisions. For example, if f is constant on the intervals of A, and if we obtain B from A by adding one new point s, then f is constant on the (smaller) intervals of B. We have to be sure that the value of the integral of f doesn't change when we change the describing subdivision. In the case just mentioned this is easy to see. The one new point s lies in some interval (t_{i-1}, t_i), defined by the partition A. The contribution of this interval to the A-sum is $\alpha_i(t_i - t_{i-1})$, while in the B-sum it splits into $\alpha_i(t_i - s) + \alpha_i(s - t_{i-1})$. But this is the same vector. The remaining summands are the same in the two sums, and the integral is therefore unchanged. In general, suppose that f is a step function with respect to A and also with respect to C. Set $B = A \cup C$, the "common refinement" of A and C. We can pass from A to B in a sequence of steps at each of which we add one new point. As we have seen, the integral remains unchanged at each of these steps, and so it is the same for A as for B. It is similarly the same for C and B, and so for A and C. We have thus shown that $\int_a^b f$ is independent of the subdivision used to define f.

Now fix $[a, b]$ and W, and let $\mathcal{E}$ be the set of all step functions from $[a, b]$ to W. Then $\mathcal{E}$ is a vector space. For, if f and g in $\mathcal{E}$ are step functions relative to partitions A and B, then both functions are constant on the intervals of $C = A \cup B$, and therefore $xf + yg$ is also. Moreover, if $C = \{t_i\}_0^n$, and if on (t_{i-1}, t_i) we have $f = \alpha_i$ and $g = \beta_i$, so that $xf + yg = x\alpha_i + y\beta_i$ there, then the equation

$$\sum_{i=1}^n (x\alpha_i + y\beta_i)\, \Delta t_i = x\left(\sum_1^n \alpha_i\, \Delta t_i\right) + y\left(\sum_1^n \beta_i\, \Delta t_i\right)$$

is just $\int_a^b (xf + yg) = x \int_a^b f + y \int_a^b g$. The map $f \mapsto \int_a^b f$ is thus linear from $\mathcal{E}$ to W. Finally,

$$\left\| \int_a^b f \right\| = \left\| \sum_1^n \alpha_i\, \Delta t_i \right\| \le \sum_1^n \|\alpha_i\| \Delta t_i \le \|f\|_\infty (b - a),$$

where $\|f\|_\infty = \text{lub}\,\{\|f(t)\| : t \in [a, b]\} = \max\,\{\|\alpha_i\| : 1 \le i \le n\}$. That is, if we use on $\mathcal{E}$ the uniform norm defined from the norm of W, then the linear mapping $f \mapsto \int_a^b f$ is bounded by $(b - a)$. If W is *complete*, this transformation therefore has a unique bounded linear extension to the closure $\overline{\mathcal{E}}$ of $\mathcal{E}$ in $\mathcal{B}([a, b], W)$ by Theorem 10.1. But we can show that $\overline{\mathcal{E}}$ includes the space $\mathcal{C}([a, b], W)$ of all *continuous* functions from $[a, b]$ to W, and the integral of a continuous function is thus uniquely defined.

Lemma 10.1. $\mathcal{C}([a, b], W) \subset \overline{\mathcal{E}}$.

Proof. A continuous function f on $[a, b]$ is uniformly continuous (Theorem 5.1). That is, given $\epsilon^{>0}$, there exists $\delta^{>0}$ such that $|s - t| < \delta \Rightarrow \|f(s) - f(t)\| < \epsilon$. Now take any partition $A = \{t_i\}_0^n$ on $[a, b]$ such that $\Delta t_i = t_i - t_{i-1} < \delta$ for all i, and take α_i as any value of f on (t_{i-1}, t_i). Then $\|f(t) - \alpha_i\| < \epsilon$ on $[t_{i-1}, t_i]$. Thus, if g is the step function with value α_i on $(t_{i-1}, t_i]$ and $g(a) = \alpha_1$, then $\|f - g\|_\infty \le \epsilon$. Thus f is in $\overline{\mathcal{E}}$, as desired. □

Our main theorem is a recapitulation.

Theorem 10.2. If W is a Banach space and $V = \mathcal{C}([a, b], W)$ under the uniform norm, then there exists a $J \in \text{Hom}(V, W)$ uniquely determined by setting $J(f) = \lim \int_a^b f_n$, where $\{f_n\}$ is any sequence in $\mathcal{E}$ converging to f and $\int_a^b f_n$ is the integral on $\mathcal{E}$ defined above. Moreover, $\|J\| \le (b - a)$.

If f is elementary from $[a, b]$ to W and $c \in [a, b]$, then of course f is elementary on each of $[a, c]$ and $[c, b]$. If c is added to a subdivision A used in defining f, and if the sum defining $\int_a^b f$ with respect to $B = A \cup \{c\}$ is broken into two sums at c, we clearly have $\int_a^b f = \int_a^c f + \int_c^b f$. This same identity then follows for any continuous function f on $[a, b]$, since $\int_a^b f = \lim \int_a^b f_n = \lim\,(\int_a^c f_n + \int_c^b f_n) = \lim \int_a^c f_n + \lim \int_c^b f_n = \int_a^c f + \int_c^b f$.

The fundamental theorem of the calculus is still with us.

Theorem 10.3. If $f \in \mathcal{C}([a, b], W)$ and $F: [a, b] \to W$ is defined by $F(x) = \int_a^x f(t)\, dt$, then F' exists on (a, b) and $F'(x) = f(x)$.

Proof. By the continuity of f at x_0, for every ϵ there exists a δ such that

$$\|f(x_0) - f(x)\| < \epsilon$$

whenever $|x - x_0| < \delta$. But then

$$\left\|\int_{x_0}^x (f(x_0) - f(t))\, dt\right\| \le \epsilon |x - x_0|,$$

and since $\int_{x_0}^x f(x_0)\, dt = f(x_0)(x - x_0)$ by the definition of the integral for an elementary function, we see that

$$\left\|f(x_0) - \left(\int_{x_0}^x f(t)\, dt/(x - x_0)\right)\right\| < \epsilon.$$

Since $\int_{x_0}^x f(t)\, dt = F(x) - F(x_0)$, this is exactly the statement that the difference quotient for F converges to $f(x_0)$, as was to be proved. □

EXERCISES

10.1 Prove the following analogue of Theorem 10.1. Let A be a subset of a metric space B, let C be a complete metric space, and let $F: A \to C$ be uniformly continuous. Then F extends uniquely to a continuous map from $\bar{A}$ to C.

10.2 In Exercises 7.16 through 7.18 we have constructed a completion of S, namely, $\overline{\theta[S]}$ in V^*. Prove that this completion is unique to within isometry. That is, supposing that φ is some other isometric imbedding of S in a complete space X, show that the identification of the two images of S by $\varphi \circ \theta^{-1}$ (from $\theta[S]$ to $\varphi[S]$) extends to an isometric bijection from $\overline{\theta[S]}$ to $\overline{\varphi[S]}$. [*Hint:* Apply the above exercise.]

10.3 Suppose that S is a normed linear space X and that X is a dense subset of a complete metric space Y. This means, remember, that every point of Y is the limit of a sequence lying in the subset X. Prove that the vector space structure of X extends in a unique way to make Y a Banach space. Since we know from Exercise 7.18 that a metric space can be completed, this shows again that a normed linear space can always be completed to a Banach space.

10.4 In the elementary calculus, if f is continuous, then

$$\int_a^b f(t)\,dt = f(x)(b-a)$$

for some x in (a, b). Show that this is not true for vector-valued continuous functions f by considering the arc $f: [0, \pi] \to \mathbb{R}^2$ defined by

$$f(t) = \langle \sin t, \cos t \rangle.$$

10.5 Show that integration commutes with the application of linear transformations. That is, show that if f is a continuous function from $[a, b]$ to a Banach space W, and if $T \in \operatorname{Hom}(W, X)$, where X is a Banach space, then

$$\int_a^b T(f(t))\,dt = T\left[\int_a^b f(t)\,dt\right].$$

[*Hint:* Make the computation directly for step functions.]

10.6 State and prove the theorem suggested by the following identity:

$$\int_a^b \langle f(t), g(t) \rangle\,dt = \left\langle \int_a^b f(t)\,dt, \int_a^b g(t)\,dt \right\rangle.$$

(Apply the above exercise.)

10.7 Let W be *any* normed linear space, $\{\alpha_i\}_1^n$ a finite set of vectors in W, and $\{f_i\}_1^n$ a corresponding set of real-valued continuous functions on $[a, b]$. Define the arc γ by

$$\gamma(t) = \sum_1^n f_i(t)\alpha_i.$$

Prove that $\int_a^b \gamma(t)\,dt$ exists and equals

$$\sum_1^n \left[\int_a^b f_i(t)\,dt\right]\alpha_i.$$

10.8 Let f be a continuous function from $\mathbb{R}^2$ to a Banach space W. Describe how one might set up a theory of a double integral

$$\iint_{I\times J} f(s, t)\, ds\, dt,$$

where $I \times J$ is a closed rectangle.

10.9 Prove that if f_n converges uniformly to f, then

$$\int_a^b f_n(t)\, dt \to \int_a^b f(t)\, dt.$$

This is trivial if you have understood the definition and properties of the integral.

10.10 Suppose that $\{f_n\}$ is a sequence of smooth arcs from $[a, b]$ to a Banach space W such that $\sum_1^\infty f_n'(t)$ is uniformly convergent. Suppose also that $\sum_1^\infty f_n(a)$ is convergent. Prove that then $\sum f_n(t)$ is uniformly convergent, that $f = \sum_1^\infty f_n$ is smooth, and that $f' = \sum_1^\infty f_n'$. (Use the above exercise and the fundamental theorem of the calculus.)

10.11 Prove that even if W is not a Banach space, if the arc $f\colon [a, b] \to W$ has a continuous derivative, then $\int_a^b f'$ exists and equals $f(b) - f(a)$.

10.12 Let X be a normed linear space, and set $\langle l, \xi\rangle = l(\xi)$ for $\xi \in X$ and $l \in X^*$. Now let f and g be continuously differentiable functions (arcs) from the closed interval $[a, b]$ to X and X^*, respectively. Prove the integration by parts formula:

$$\langle g(b), f(b)\rangle - \langle g(a), f(a)\rangle = \int_a^b \langle f(t), g'(t)\rangle\, dt + \int_a^b \langle f'(t), g(t)\rangle\, dt.$$

[*Hint:* Apply Theorem 8.4 from Chapter 3.]

10.13 State the generalization of the above integration by parts formula that holds for any bounded bilinear mapping $\omega\colon V \times W \to X$, where X is a Banach space.

10.14 Let $t \mapsto l_t$ be a fixed continuous map from a closed interval $[a, b]$ to the dual W^* of a Banach space W. Suppose that for any continuous map g from $[a, b]$ to W

$$\int_a^b g(t)\, dt = 0 \Rightarrow \int_a^b l_t(g(t))\, dt = 0.$$

Show that there exists a fixed $L \in W^*$ such that

$$\int_a^b l_t(g(t))\, dt = L\left(\int_a^b g(t)\, dt\right)$$

for all continuous arcs $g\colon [a, b] \to W$. Show that it then follows that $l_t = L$ for all t.

10.15 Use the above exercise to deduce the general Euler equation of Section 3.15.

11. THE COMPLEX NUMBER SYSTEM

The complex number system $\mathbb{C}$ is the third basic number field that must be studied, after the rational numbers and the real numbers, and the reader surely has had some contact with it in the past.

Almost everybody views a complex number ξ as being equivalent to a pair of real numbers, the "real and imaginary parts" of ξ, and the complex number system $\mathbb{C}$ is thus viewed as being Cartesian 2-space $\mathbb{R}^2$ with some further structure. In particular, a complex-valued function is simply a certain kind of vector-valued function, and is equivalent to an ordered pair of real-valued functions, again its real and imaginary parts.

What distinguishes the complex number system $\mathbb{C}$ from its vector substratum $\mathbb{R}^2$ is the presence of an additional operation, complex multiplication. The vector operations of $\mathbb{R}^2$ together with this complex multiplication operation make $\mathbb{C}$ into a commutative algebra. Moreover, it turns out that $\langle 1, 0\rangle$ is the unique multiplicative identity in $\mathbb{C}$ and that every nonzero complex number ξ has a multiplicative inverse. These additional facts are summarized by saying that $\mathbb{C}$ is a *field*, and they allow us to use $\mathbb{C}$ as a new scalar field in vector space theory. In fact, the whole development of Chapters 1 and 2 remains valid when $\mathbb{R}$ is replaced everywhere by $\mathbb{C}$. Scalar multiplication is now multiplication by complex numbers. Thus $\mathbb{C}^n$ is the vector space of ordered n-tuples of complex numbers $\langle \xi_1, \ldots, \xi_n\rangle$, and the product of an n-tuple by a complex scalar α is defined by $\alpha\langle \xi_1, \ldots, \xi_n\rangle = \langle \alpha\xi_1, \ldots, \alpha\xi_n\rangle$, where $\alpha\xi_i$ is complex multiplication.

It is time to come to grips with complex multiplication. As the reader probably knows, it is given by an odd looking formula that is motivated by thinking of an element $\xi = \langle x_1, x_2\rangle$ as being in the form $x_1 + ix_2$, where $i^2 = -1$, and then using the ordinary laws of algebra. Then we have

$$\begin{aligned}\xi\eta &= (x_1 + ix_2)(y_1 + iy_2)\\ &= x_1y_1 + ix_1y_2 + ix_2y_1 + i^2x_2y_2 = (x_1y_1 - x_2y_2) + i(x_1y_2 + x_2y_1),\end{aligned}$$

and thus our definition is

$$\langle x_1, x_2\rangle\langle y_1, y_2\rangle = \langle x_1y_1 - x_2y_2, x_1y_2 + x_2y_1\rangle.$$

Of course, it has to be verified that this operation is commutative and satisfies the laws for an algebra. A straightforward check is possible but dull, and we shall indicate a neater way in the exercises.

The mapping $x \mapsto \langle x, 0\rangle$ is an isomorphic injection of the field $\mathbb{R}$ into the field $\mathbb{C}$. It clearly preserves sums, and the reader can check in his mind that it also preserves products. It is conventional to identify x with its image $\langle x, 0\rangle$, and so to view $\mathbb{R}$ as a subfield of $\mathbb{C}$.

The mysterious i can be identified in $\mathbb{C}$ as the pair $\langle 0, 1\rangle$, since then $i^2 = \langle 0, 1\rangle\langle 0, 1\rangle = \langle -1, 0\rangle$, which we have identified with -1. With these identifications we have $\langle x, y\rangle = \langle x, 0\rangle + \langle 0, y\rangle = \langle x, 0\rangle + \langle 0, 1\rangle\langle y, 0\rangle = x + iy$, and this is the way we shall write complex numbers from now on.

The mapping $x + iy \mapsto x - iy$ is a field isomorphism of $\mathbb{C}$ with itself. That is, it preserves both sums *and* products, as the reader can easily check. Such a self-isomorphism is called an *automorphism*. The above automorphism is called *complex conjugation*, and the image $x - iy$ of $\zeta = x + iy$ is called the *conjugate* of ζ, and is designated $\bar{\zeta}$. We shall ask the reader to show in an exercise that conjugation is the *only* automorphism of $\mathbb{C}$ (except the identity automorphism) which leaves the elements of the subfield $\mathbb{R}$ fixed.

The Euclidean norm of $\zeta = x + iy = \langle x, y \rangle$ is called the ***absolute value*** of ζ, and is designated $|\zeta|$, so that $|\zeta| = |x + iy| = (x^2 + y^2)^{1/2}$. This is reasonable because it then turns out that $|\zeta\gamma| = |\zeta|\,|\gamma|$. This can be verified by squaring and multiplying, but it is much more elegant first to notice the relationship between absolute value and the conjugation automorphism, namely,

$$\zeta\bar{\zeta} = |\zeta|^2$$

$[(x + iy)(x - iy) = x^2 - (iy)^2 = x^2 + y^2]$. Then $|\zeta\gamma|^2 = (\zeta\gamma)(\overline{\zeta\gamma}) = (\zeta\bar{\zeta})(\gamma\bar{\gamma}) = |\zeta|^2|\gamma|^2$, and taking square roots gives us our identity. The identity $\zeta\bar{\zeta} = |\zeta|^2$ also shows us that if $\zeta \neq 0$, then $\bar{\zeta}/|\zeta|^2$ is its multiplicative inverse.

Because the real number system $\mathbb{R}$ is a subfield of the complex number system $\mathbb{C}$, any vector space over $\mathbb{C}$ is automatically also a vector space over $\mathbb{R}$: multiplication by complex scalars includes multiplication by real scalars. And any complex linear transformation between complex vector spaces is automatically real linear. The converse, of course, does not hold. For example, a real linear mapping T from $\mathbb{R}^2$ to $\mathbb{R}^2$ is not in general complex linear from $\mathbb{C}$ to $\mathbb{C}$, nor does a real linear S in Hom $\mathbb{R}^4$ become a complex linear mapping in Hom $\mathbb{C}^2$ when $\mathbb{R}^4$ is viewed as $\mathbb{C}^2$. We shall study this question in the exercises.

The complex differentiability of a mapping F between complex vector spaces has the obvious definition $\Delta F_\alpha = T + \mathfrak{o}$, where T is complex linear, and then F is also real differentiable, in view of the above remarks. But F may be real differentiable without being complex differentiable. It follows from the discussion at the end of Section 8 that if $\{a_n\} \subset \mathbb{C}$ and $\{|a_n|\delta^n\}$ is bounded, then the series $\sum a_n\zeta^n$ converges on the ball $B_\delta(0)$ in the (real) Banach algebra $\mathbb{C}$, and $F(\zeta) = \sum_0^\infty a_n\zeta^n$ is real differentiable on this ball, with $dF_\beta(\zeta) = (\sum_1^\infty na_n\beta^n)\zeta = F'(\beta)\cdot\zeta$. But multiplication by $F'(\beta)$ is obviously a complex linear operation on the one-dimensional complex vector space $\mathbb{C}$. Therefore, complex-valued functions defined by convergent complex power series are automatically complex differentiable. But we can go even further. In this case, if $\zeta \neq 0$, we can divide by ζ in the defining equation

$$\Delta F_\beta(\zeta) = F'(\beta)\cdot\zeta + \mathfrak{o}(\zeta)$$

to get the result that

$$\frac{\Delta F_\beta(\zeta)}{\zeta} \to F'(\beta) \qquad \text{as} \qquad \zeta \to 0.$$

That is, $F'(\beta)$ is now an honest derivative again, with the complex infinitesimal ζ in the denominator of the difference quotient.

The consequences of complex differentiability are incalculable, and we shall mostly leave them as future pleasures to be experienced in a course on functions of complex variables. See, however, the problems on the residue calculus at the end of Chapter 12 and the proof in Chapter 11, Exercise 4.3, of the following fundamental theorem of algebra.

Theorem. Every polynomial with complex coefficients is a product of linear factors.

A weaker but equivalent statement is that every polynomial has at least one (complex) root. The crux of the matter is that $x^2 + 1$ cannot be factored over $\mathbb{R}$ (i.e., it has no real root), but over $\mathbb{C}$ we have $x^2 + 1 = (x + i)(x - i)$, with the two roots $\pm i$.

For later use we add a few more words about the complex exponential function $\exp \zeta = e^\zeta = \sum_0^\infty \zeta^n/n!$. If $\zeta = x + iy$, we have $e^\zeta = e^{x+iy} = e^x e^{iy}$, and $e^{iy} = \sum_0^\infty (iy)^n/n! = (1 - y^2/2! + y^4/4! - \cdots) + i(y - y^3/3! + y^5/5! - \cdots) = \cos y + i \sin y$. Thus $e^{x+iy} = e^x(\cos y + i \sin y)$. That is, the real and imaginary parts of the complex-valued function $\exp(x + iy)$ are $e^x \cos y$ and $e^x \sin y$, respectively.

EXERCISES

11.1 Prove the associativity of complex multiplication directly from its definition.

11.2 Prove the distributive law,

$$\alpha(\xi + \eta) = \alpha\xi + \alpha\eta,$$

for complex numbers.

11.3 Show that scalar multiplication by a real number a, $a\langle x, y\rangle = \langle ax, ay\rangle$, in $\mathbb{C} = \mathbb{R}^2$ is consistent with the interpretation of a as the complex number $\langle a, 0\rangle$ and the definition of complex multiplication.

11.4 Let θ be an automorphism of the complex number field leaving the real numbers fixed. Prove that θ is either the identity or complex conjugation. [*Hint:* $(\theta(i))^2 = \theta(i^2) = \theta(-1) = -1$. Show that the only complex numbers $x + iy$ whose squares are -1 are $\pm i$, and then finish up.]

11.5 If we remember that $\mathbb{C}$ is in particular the two-dimensional real vector space $\mathbb{R}^2$, we see that multiplying the elements of $\mathbb{C}$ by the complex number $a + ib$ must define a linear transformation on $\mathbb{R}^2$. Show that its matrix is

$$\begin{bmatrix} a & -b \\ b & a \end{bmatrix}.$$

11.6 The above exercise suggests that the complex number system may be like the set A of all 2×2 real matrices of the form

$$\begin{bmatrix} a & -b \\ b & a \end{bmatrix}.$$

Prove that A is a subalgebra of the matrix algebra $\mathbb{R}^{2\times 2}$ (that is, A is closed under multiplication, addition, and scalar multiplication) and that the mapping

$$\begin{bmatrix} a & -b \\ b & a \end{bmatrix} \to a + ib$$

is a bijection from A to $\mathbb{C}$ that preserves all algebra operations. We therefore can conclude that the laws of an algebra automatically hold for $\mathbb{C}$. Why?

11.7 In the above matrix model of the complex number system show that the absolute value identity $|\zeta\gamma| = |\zeta|\,|\gamma|$ is a determinant property.

11.8 Let W be a real vector space, and let V be the real vector space $W \times W$. Show that there is a θ in Hom V such that $\theta^2 = -I$. (Think of $\mathbb{C}$ as being the real vector space $\mathbb{R}^2 = \mathbb{R} \times \mathbb{R}$ under multiplication by i.)

11.9 Let V be a real vector space, and let θ in Hom V satisfy $\theta^2 = -I$. Show that V becomes a complex vector space if $i\alpha$ is defined as $\theta(\alpha)$. If the complex vector space V is made from the real vector space W as in this and the above exercise, we shall call V the *complexification* of W. We shall regard W itself as being a real subspace of V (actually $W \times \{0\}$), and then $V = W \oplus iW$.

11.10 Show that the complex vector space $\mathbb{C}^n$ is the complexification of $\mathbb{R}^n$. Show more generally that for any set A the complex vector space $\mathbb{C}^A$ is the complexification of the real vector space $\mathbb{R}^A$.

11.11 Let V be the complexification of the real vector space W. Define the operation of complex conjugation on V. That is, show that there is a real linear mapping φ such that $\varphi^2 = I$ and $\varphi(i\alpha) = -i\varphi(\alpha)$. Show, conversely, that if V is a complex vector space and φ is a conjugation on V [a real linear mapping φ such that $\varphi^2 = I$ and $\varphi(i\alpha) = -i\varphi(\alpha)$], then V is (isomorphic to) the complexification of a real linear space W. (Apply Theorem 5.5 of Chapter 1 to the identity $\varphi^2 - I = 0$.)

11.12 Let W be a real vector space, and let V be its complexification. Show that every T in Hom W "extends" to a complex linear S in Hom V which commutes with the conjugation φ. By S extending T we mean, of course, that $S \restriction (W \times \{0\}) = T$. Show, conversely, that if S in Hom V commutes with conjugation, then S is the extension of a T in Hom W.

11.13 In this situation we naturally call S the complexification of T. Show finally that if S is the complexification of T, then its null space X in V is the direct sum $X = N \oplus iN$, where N is the null space of T in W. Remember that we are viewing V as $W \oplus iW$.

11.14 On a *complex* normed linear space V the norm is required to be *complex* homogeneous:

$$\|\lambda\alpha\| = |\lambda| \cdot \|\alpha\|$$

for all complex numbers λ. Show that the natural definitions of $\| \ \|_1$, $\| \ \|_2$, and $\| \ \|_\infty$ on $\mathbb{C}^n$ have this property.

11.15 If a real normed linear space W is complexified to $V = W \oplus iW$, there is no trivial formula which converts the real norm for W into a complex norm for V. Show that, nevertheless, any product norm on V (which really is $W \times W$) can be used to generate an equivalent complex norm. [*Hint:* Given $\langle \xi, \eta \rangle \in V$, consider the set of numbers $\{\|(x + iy)\langle \xi, \eta \rangle\| : |x + iy| = 1\}$, and try to obtain from this set a single number that works.]

11.16 Show that every nonzero complex number has a logarithm. That is, show that if $u + iv \neq 0$, then there exists an $x + iy$ such that $e^{x+iy} = u + iv$. (Write the equation $e^x(\cos y + i \sin y) = u + iv$, and solve by being slightly clever.)

11.17 The fundamental theorem of algebra and Theorem 5.5 of Chapter 1 imply that if V is a complex vector space and T in Hom V satisfies $p(T) = 0$ for a polynomial

p, then there are subspaces $\{V_i\}_1^n$ of V, complex numbers $\{\lambda_i\}_1^n$, and integers $\{m_i\}_1^n$ such that $V = \bigoplus_1^n V_i$, V_i is T-invariant for each, and $(T - \lambda_i T)^{m_i} = 0$ on V_i for each i. Show that this is so. Show also that if V is finite-dimensional, then every T in Hom V must satisfy some polynomial equation $p(t) = 0$. (Consider the linear independence or dependence of the vector $I, T, T^2, \ldots, T^{n^2}, \ldots$, in the vector space Hom V.)

11.18 Suppose that the polynomial p in the above exercise has real coefficients. Use the fact that complex conjugation is an automorphism of $\mathbb{C}$ to prove that if λ is a root of p, then so is $\bar{\lambda}$.

Show that if V is the complexification of a real space W and T is the complexification of $R \in \text{Hom } W$, then there exists a real polynomial p such that $p(T) = 0$.

11.19 Show that if W is a finite-dimensional real vector space and $R \in \text{Hom } W$ is an isomorphism, then there exists an $A \in \text{Hom } W$ such that $R = e^A$ (that is, log R exists). This is a hard exercise, but it can be proved from Exercises 8.19 through 8.23, 11.12, 11.17, and 11.18.

*12. WEAK METHODS

Our theorem that all norms are equivalent on a finite-dimensional space suggests that the limit theory of such spaces should be accessible independently of norms, and our earlier theorem that every linear transformation with a finite-dimensional domain is automatically bounded reinforces this impression. We shall look into this question in this section. In a sense this effort is irrelevant, since we can't do without norms completely, and since they are so handy that we use them even when we don't have to.

Roughly speaking, what we are going to do is to study a vector-valued map F by studying the whole collection of real-valued maps $\{l \circ F : l \in V^*\}$.

Theorem 12.1. If V is finite-dimensional, then $\xi_n \to \xi$ in V (with respect to any, and so every, norm) if and only if $l(\xi_n) \to l(\xi)$ in $\mathbb{R}$ for each l in V^*.

Proof. If $\xi_n \to \xi$ and $l \in V^*$, then $l(\xi_n) \to l(\xi)$, since l is automatically continuous. Conversely, if $l(\xi_n) \to l(\xi)$ for every l in V^*, then, choosing a basis $\{\beta_i\}_1^n$ for V, we have $\epsilon_i(\xi_n) \to \epsilon_i(\xi)$ for each functional ϵ_i in the dual basis, and this implies that $\xi_n \to \xi$ in the associated one-norm, since $\|\xi_n - \xi\|_1 = \sum_1^n |\epsilon_i(\xi_n) - \epsilon_i(\xi)| \to 0$. □

Remark. If V is an arbitrary normed linear space, so that $V^* = \text{Hom}(V, \mathbb{R})$ is the set of *bounded* linear functionals, then we say that $\xi_n \to \xi$ *weakly* if $l(\xi_n) \to l(\xi)$ for each $l \in V^*$. The above theorem can therefore be rephrased to say that *in a finite-dimensional space, weak convergence and norm convergence are equivalent notions.*

We shall now see that in a similar way the integration and differentiation of parametrized arcs can all be thrown back to the standard calculus of real-valued functions of a real variable by applying functionals from V^* and using the natural isomorphism of V^{**} with V. Thus, if $f \in \mathcal{C}([a, b], V)$ and $\lambda \in V^*$, then

$\lambda \circ f \in \mathcal{C}([a, b], \mathbb{R})$, and so the integral $\int_a^b \lambda \circ f$ exists from standard calculus. If we vary λ, we can check that the map $\lambda \mapsto \int_a^b \lambda \circ f$ is linear, hence is in V^{**}, and therefore is given by a uniquely determined vector $\alpha \in V$ (by duality; see Chapter 2, Theorem 3.2.). That is, there exists a unique $\alpha \in V$ such that $\lambda(\alpha) = \int_a^b \lambda \circ f$ for every $\lambda \in V^*$, and we *define* this α to be $\int_a^b f$. Thus integration is defined so as to commute with the application of linear functionals: $\int_a^b f$ is that vector such that

$$\lambda\left(\int_a^b f\right) = \int_a^b \lambda(f(t))\, dt \qquad \text{for all} \quad \lambda \in V^*.$$

Similarly, if all the real-valued functions $\{\lambda \circ f : \lambda \in V^*\}$ are differentiable at x_0, then the mapping $\lambda \mapsto (\lambda \circ f)'(x_0)$ is linear by the linearity of the derivative in the standard calculus:

$$((c_1\lambda_1 + c_2\lambda_2) \circ f)' = (c_1(\lambda_1 \circ f) + c_2(\lambda_2 \circ f))' = c_1(\lambda_1 \circ f)' + c_2(\lambda_2 \circ f)'.$$

Therefore, there is again a unique $\alpha \in V$ such that

$$(\lambda \circ f)'(x_0) = \lambda(\alpha) \qquad \text{for all} \quad \lambda \in V^*,$$

and if we define this α to be the derivative $f'(x_0)$, we have again defined an operation of the calculus by commutativity with linear functionals:

$$(\lambda \circ f')(x_0) = (\lambda \circ f)'(x_0).$$

Now the fundamental theorem of the calculus appears as follows.

If $F(x) = \int_a^x f$, then $(\lambda \circ F)(x) = \int_a^x \lambda \circ f$ by the weak definition of the integral. The fundamental theorem of the standard calculus then says that $(\lambda \circ F)'$ exists and $(\lambda \circ F)'(x) = (\lambda \circ f)(x) = \lambda(f(x))$. By the weak definition of the derivative we then have that F' exists and $F'(x) = f(x)$.

The one conclusion that we don't get so easily by weak methods is the norm inequality $\|\int_a^b f\| \leq (b - a)\|f\|_\infty$. This requires a theorem about norms on finite-dimensional spaces that we shall not prove in this course.

Theorem 12.2. $\|\alpha^{**}\| = \|\alpha\|$ for each $\alpha \in V$.

What is being asserted is that lub $|\alpha^{**}(\lambda)|/\|\lambda\| = \|\alpha\|$. Since $\alpha^{**}(\lambda) = \lambda(\alpha)$, and since $|\lambda(\alpha)| \leq \|\lambda\| \cdot \|\alpha\|$ by the definition of $\|\lambda\|$, we see that

$$\text{lub } |\alpha^{**}(\lambda)|/\|\lambda\| \leq \|\alpha\|.$$

Our problem is therefore to find $\lambda \in V^*$ with $\|\lambda\| = 1$ and $|\lambda(\alpha)| = \|\alpha\|$. If we multiply through by a suitable constant (replacing α by $c\alpha$, where $c = 1/\|\alpha\|$), we can suppose that $\|\alpha\| = 1$. Then α is on the unit spherical surface, and the problem is to find a functional $\lambda \in V^*$ such that the affine subspace (hyperplane) where $\lambda = 1$ touches the unit sphere at α (so that $\lambda(\alpha) = 1$) and otherwise lies outside the unit sphere (so that $|\lambda(\xi)| \leq 1$ when $\|\xi\| = 1$, and hence $\|\lambda\| \leq 1$). It is clear geometrically that such "tangent planes" must exist, but we shall drop the matter here.

If we assume this theorem, then, since

$$\begin{aligned}\left|\lambda\left(\int_a^b f\right)\right| = \left|\int_a^b \lambda(f(t))\,dt\right| &\le (b-a)\max\{|\lambda(f(t))| : t \in [a, b]\}\\ &\le (b-a)\|\lambda\|\max\{\|f(t)\|\} \qquad (\text{from } |\lambda(\alpha)| \le \|\lambda\|\cdot\|\alpha\|)\\ &= (b-a)\|\lambda\|\cdot\|f\|_\infty,\end{aligned}$$

we get

$$\left\|\int_a^b f\right\| = \left\|\left(\int_a^b f\right)^{**}\right\| = \operatorname{lub}\left|\lambda\left(\int_a^b f\right)\right|/\|\lambda\| \le (b-a)\|f\|_\infty,$$

the extreme members of which form the desired inequality.

CHAPTER 5

SCALAR PRODUCT SPACES

In this short chapter we shall look into what is going on behind two-norms, and we shall find that a wholly new branch of linear analysis is opened up. These norms can be characterized abstractly as those arising from scalar products. They are the finite and infinite-dimensional analogues of ordinary geometric length, and they carry with them practically all the concepts of Euclidean geometry, such as the notion of the angle between two vectors, perpendicularity (orthogonality) and the Pythagorean theorem, and the existence of many rigid motions.

The impact of this extra structure is particularly dramatic for infinite-dimensional spaces. Infinite orthogonal bases exist in great profusion and can be handled about as easily as bases in finite-dimensional spaces, although the basis expansion of a vector is now a convergent infinite series, $\xi = \sum_1^\infty x_i\alpha_i$. Many of the most important series expansions in mathematics are examples of such orthogonal basis expansions. For example, we shall see in the next chapter that the Fourier series expansion of a continuous function f on $[0, \pi]$ is the basis expansion of f under the two-norm $\|f\|_2 = (\int_0^\pi f^2)^{1/2}$ for the particular orthogonal basis $\{\alpha_n\}_1^\infty = \{\sin nt\}_1^\infty$. If a vector space is *complete* under a scalar product norm, it is called a *Hilbert space*. The more advanced theory of such spaces is one of the most beautiful parts of mathematics.

1. SCALAR PRODUCTS

A *scalar product* on a real vector space V is a real-valued function from $V \times V$ to $\mathbb{R}$, its value at the pair $\prec \xi, \eta \succ$ ordinarily being designated (ξ, η), such that

a) (ξ, η) is linear in ξ when η is held fixed;

b) $(\xi, \eta) = (\eta, \xi)$ (symmetry);

c) $(\xi, \eta) > 0$ if $\xi \neq 0$ (positive definiteness).

If (c) is replaced by the weaker condition

c′) $(\xi, \xi) \geq 0$ for all $\xi \in V$,

then (ξ, η) is called a *semiscalar product*.

Two important examples of scalar products are

$$(\mathbf{x}, \mathbf{y}) = \sum_1^n x_i y_i \qquad \text{when} \qquad V = \mathbb{R}^n$$

and

$$(f, g) = \int_a^b f(t)\, g(t)\, dt \qquad \text{when} \qquad V = \mathcal{C}([a, b]).$$

On a complex vector space (b) must be replaced by

b′) $(\xi, \eta) = \overline{(\eta, \xi)}$ (Hermitian symmetry),

where the bar denotes complex conjugation. The corresponding examples are $(\mathbf{z}, \mathbf{w}) = \sum_1^n z_i \overline{w_i}$ when $V = \mathbb{C}^n$ and $(f, g) = \int_a^b f\overline{g}$ when V is the space of continuous complex-valued functions on $[a, b]$. We shall study only the real case.

It follows from (a) and (b) that a semiscalar product is also linear in the second variable when the first variable is held fixed, and therefore is a symmetric bilinear functional whose associated quadratic form $q(\xi) = (\xi, \xi)$ is positive definite or positive semidefinite [(c) or (c′); see the last section in Chapter 2]. The definiteness of the form q has far-reaching consequences, as we shall begin to see at once.

Theorem 1.1. The Schwarz inequality

$$|(\xi, \eta)| \leq (\xi, \xi)^{1/2}(\eta, \eta)^{1/2}$$

is valid for any semiscalar product.

Proof. We have $0 \leq (\xi - t\eta, \xi - t\eta) = (\xi, \xi) - 2t(\xi, \eta) + t^2(\eta, \eta)$ for every $t \in \mathbb{R}$. Since this quadratic in t is never negative, it cannot have distinct roots, and the usual $(b^2 - 4ac)$-formula implies that $4(\xi, \eta)^2 - 4(\xi, \xi)(\eta, \eta) \leq 0$, which is equivalent to the Schwarz inequality. □

We can also proceed directly. If $(\eta, \eta) > 0$, and if we set $t = (\xi, \eta)/(\eta, \eta)$ in the quadratic inequality in the first line of the proof, then the resulting expression simplifies to the Schwarz inequality. If $(\eta, \eta) = 0$, then (ξ, η) must also be 0 (or else the beginning inequality is clearly false for some t), and now the Schwarz inequality holds trivially.

Corollary. If (ξ, η) is a scalar product, then $\|\xi\| = (\xi, \xi)^{1/2}$ is a norm.

Proof

$$\begin{aligned} \|\xi + \eta\|^2 &= (\xi + \eta, \xi + \eta) \\ &= \|\xi\|^2 + 2(\xi, \eta) + \|\eta\|^2 \leq \|\xi\|^2 + 2\|\xi\|\,\|\eta\| + \|\eta\|^2 \qquad \text{(by Schwarz)} \\ &= (\|\xi\| + \|\eta\|)^2, \end{aligned}$$

proving the triangle inequality. Also, $\|c\xi\| = (c\xi, c\xi)^{1/2} = (c^2(\xi, \xi))^{1/2} = |c|\,\|\xi\|$. □

Note that the Schwarz inequality $|(\xi, \eta)| \leq \|\xi\|\,\|\eta\|$ is now just the statement that the bilinear functional (ξ, η) is bounded by one with respect to the scalar product norm.

A normed linear space V in which the norm is a scalar product norm is called a *pre-Hilbert space*. If V is complete in this norm, it is a *Hilbert space*.

The two examples of scalar products mentioned earlier give us the real explanation of our two-norms for the first time:

$$\|\mathbf{x}\|_2 = \left(\sum_1^n x_i^2\right)^{1/2} \qquad \text{for} \quad \mathbf{x} \in \mathbb{R}^n$$

and

$$\|f\|_2 = \left(\int_a^b f^2\right)^{1/2} \qquad \text{for} \quad f \in \mathcal{C}([a, b])$$

are scalar product norms.

Since the scalar product norm on $\mathbb{R}^n$ becomes Euclidean length under a Cartesian coordinate correspondence with Euclidean n-space, it is conventional to call $\mathbb{R}^n$ itself Euclidean n-space $\mathbb{E}^n$ when we want it understood that the scalar product norm is being used.

Any finite-dimensional space V is a Hilbert space with respect to any scalar product norm, because its finite dimensionality guarantees its completeness. On the other hand, we shall see in Exercise 1.10 that $\mathcal{C}([a, b])$ is incomplete in the two-norm, and is therefore a pre-Hilbert space but not a Hilbert space in this norm. (Remember, however, that $\mathcal{C}([a, b])$ *is* complete in the uniform norm $\|f\|_\infty$.) It is important to the real uses of Hilbert spaces in mathematics that any pre-Hilbert space can be completed to a Hilbert space, but the theory of infinite-dimensional Hilbert spaces is for the most part beyond the scope of this book.

Scalar product norms have in some sense the smoothest possible unit spheres, because these spheres are quadratic surfaces.

It is orthogonality that gives the theory of pre-Hilbert spaces its special flavor. Two vectors α and β are said to be *orthogonal*, written $\alpha \perp \beta$, if $(\alpha, \beta) = 0$. This definition gets its inspiration from geometry; we noted in Chapter 1 that two geometric vectors are perpendicular if and only if their coordinate triples $\mathbf{x}$ and $\mathbf{y}$ satisfy $(\mathbf{x}, \mathbf{y}) = 0$. It is an interesting problem to go further and to show from the law of cosines ($c^2 = a^2 + b^2 - 2ab \cos \theta$) that the angle θ between two geometric vectors is given by $(\mathbf{x}, \mathbf{y}) = \|\mathbf{x}\| \, \|\mathbf{y}\| \cos \theta$. This would motivate us to define the angle θ between two vectors ξ and η in a pre-Hilbert space by $(\xi, \eta) = \|\xi\| \, \|\eta\| \cos \theta$, but we shall have no use for this more general formulation.

We say that two subsets A and B are orthogonal, and we write $A \perp B$, if $\alpha \perp \beta$ for every α in A and β in B; for any subset A we set $A^\perp = \{\beta \in V : \beta \perp A\}$.

Lemma 1.1. If β is orthogonal to the set A, then β is orthogonal to $\overline{L(A)}$, the closure of the linear span of A. It follows that $B^\perp$ is a closed subspace for every subset B.

Proof. The first assertion depends on the linearity and continuity of the scalar product in one of its variables; it will be left to the reader. As for $A = B^\perp$, it includes the closure of its own linear span, by the first part, and so is a closed subspace. □

Lemma 1.2. In any pre-Hilbert space we have the parallelogram law,

$$\|\alpha + \beta\|^2 + \|\alpha - \beta\|^2 = 2(\|\alpha\|^2 + \|\beta\|^2),$$

and the Pythagorean theorem,

$$\alpha \perp \beta \qquad \text{if and only if} \qquad \|\alpha + \beta\|^2 = \|\alpha\|^2 + \|\beta\|^2.$$

If $\{\alpha_i\}_1^n$ is a (pairwise) orthogonal collection of vectors, then

$$\left\|\sum_1^n \alpha_i\right\|^2 = \sum_1^n \|\alpha_i\|^2.$$

Proof. Since $\|\alpha + \beta\|^2 = \|\alpha\|^2 + 2(\alpha, \beta) + \|\beta\|^2$, by the bilinearity of the scalar product $\|\alpha + \beta\|^2$, we see that $\|\alpha + \beta\|^2 = \|\alpha\|^2 + \|\beta\|^2$ if and only if $(\alpha, \beta) = 0$, which is the Pythagorean theorem. Writing down the similar expansion of $\|\alpha - \beta\|^2$ and adding, we have the parallelogram law. The last statement follows from the Pythagorean theorem and Lemma 1.1 by induction. Or we can obtain this statement directly by expanding the scalar product on the left and noticing that all "mixed terms" drop out by orthogonality. □

The reader will notice that the Schwarz inequality has not been used in this lemma, but it would have been silly to state the lemma before proving that $\|\xi\| = (\xi, \xi)^{1/2}$ is a norm.

If $\{\alpha_i\}_1^n$ are orthogonal and nonzero, then the identity $\|\sum_1^n x_i\alpha_i\|^2 = \sum_1^n x_i^2\|\alpha_i\|^2$ shows that $\sum_1^n x_i\alpha_i$ can be zero only if all the coefficients x_i are zero. Thus,

Corollary. A finite collection of (pairwise) orthogonal nonzero vectors is independent. Similarly, a finite collection of orthogonal subspaces is independent.

EXERCISES

1.1 Complete the second proof of Theorem 1.1.

1.2 Reexamine the proof of Theorem 1.1 and show that if ξ and η are independent, then the Schwarz inequality is strict.

1.3 Continuing the above exercise, now show that the triangle inequality is strict if ξ and η are independent.

1.4 a) Show that the sum of two semiscalar products is a semiscalar product.

b) Show that if (μ, ν) is a semiscalar product on a vector space W and if T is a linear transformation from a vector space V to W, then $[\xi, \eta] = (T\mu, T\nu)$ is a semiscalar product on V.

c) Deduce from (a) and (b) that

$$(f, g) = f(a)g(a) + \int_a^b f'(t)g'(t)\,dt$$

is a semiscalar product on $V = \mathcal{C}^1([a, b])$. Prove that it is a scalar product.

1.5 If α is held fixed, we know that $f(\xi) = (\xi, \alpha)$ is continuous. Why? Prove more generally that (ξ, η) is continuous as a map from $V \times V$ to $\mathbb{R}$.

1.6 Let V be a two-dimensional Hilbert space, and let $\{\alpha_1, \alpha_2\}$ be any basis for V. Show that a scalar product (ξ, η) has the form

$$(\xi, \eta) = ax_1y_1 + b(x_1y_2 + x_2y_1) + cx_2y_2,$$

where $b^2 < ac$. Here, of course, $\xi = x_1\alpha_1 + x_2\alpha_2$, $\eta = y_1\alpha_1 + y_2\alpha_2$.

1.7 Prove that if $\omega(\mathbf{x}, \mathbf{y}) = ax_1y_1 + b(x_1y_2 + x_2y_1) + cx_2y_2$ and $b^2 < ac$, then ω is a scalar product on $\mathbb{R}^2$.

1.8 Let $\omega(\xi, \eta)$ be any symmetric bilinear functional on a finite-dimensional vector space V, and let $q(\xi) = \omega(\xi, \xi)$ be its associated quadratic form. Show that for any choice of a basis for V the equation $q(\xi) = 1$ becomes a quadratic equation in the coordinates $\{x_i\}$ of ξ.

1.9 Prove in detail that if a vector β is orthogonal to a set A in a pre-Hilbert space, then β is orthogonal to $\overline{L(A)}$.

1.10 We know from the last chapter that the Riemann integral is defined for the set $\overline{\mathcal{E}}$ of uniform limits of real-valued step functions on $[0, 1]$ and that $\overline{\mathcal{E}}$ includes all the continuous functions. Given that k is the step function whose value is 1 on $[0, \frac{1}{2}]$ and 0 on $[\frac{1}{2}, 1]$, show that $\|f - k\|_2 > 0$ for any continuous function f. Show, however, that there is a sequence of continuous functions $\{f_n\}$ such that $\|f_n - k\|_2 \to 0$. Show, therefore, that $\mathcal{C}([0, 1])$ is incomplete in the two-norm, by showing that the above sequence $\{f_n\}$ is Cauchy but not convergent in $\mathcal{C}([0, 1])$.

2. ORTHOGONAL PROJECTION

One of the most important devices in geometric reasoning is "dropping a perpendicular" from a point to a line or a plane and then using right triangle arguments. This device is equally important in pre-Hilbert space theory. If M is a subspace and α is any element in V, then by "the foot of the perpendicular dropped from α to M" we mean that vector μ in M such that $(\alpha - \mu) \perp M$, if such a μ exists. (See Fig. 5.1.) Writing α as $\mu + (\alpha - \mu)$, we see that the existence of the "foot" μ in M for each α in V is equivalent to the direct sum decomposition $V = M \oplus M^{\perp}$. Now it is precisely this direct sum decomposition that the completeness of a Hilbert space guarantees, as we shall shortly see. We start by proving the geometrically intuitive fact that μ is the foot of the perpendicular dropped from α to M if and only if μ is the point in M closest to α.

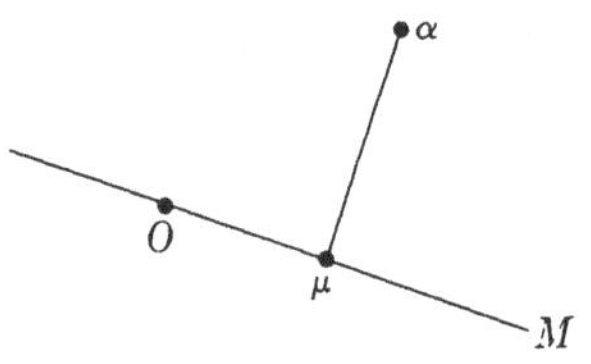

Fig. 5.1

Lemma 2.1. If μ is in the subspace M, then $(\alpha - \mu) \perp M$ if and only if μ is the unique point in M closest to α, that is, μ is the "best approximation" to α in M.

Proof. If $(\alpha - \mu) \perp M$ and ξ is any other point in M, then $\|\alpha - \xi\|^2 = \|(\alpha - \mu) + (\mu - \xi)\|^2 = \|\alpha - \mu\|^2 + \|\mu - \xi\|^2 > \|\alpha - \mu\|^2$. Thus μ is the unique point in M closest to α. Conversely, suppose that μ is a point in M closest to α, and let ξ be any nonzero vector in M. Then $\|\alpha - \mu\|^2 \leq \|(\alpha - \mu) + t\xi\|^2$, which becomes $0 \leq 2t(\alpha - \mu, \xi) + t^2\|\xi\|^2$ when the right-hand scalar product is expanded. This can hold for *all* t only if $(\alpha - \mu, \xi) = 0$ (otherwise let $t = ?$). Therefore, $(\alpha - \mu) \perp M$. □

On the basis of this lemma it is clear that a way to look for μ is to take a sequence μ_n in M such that $\|\alpha - \mu_n\| \to \rho(\alpha, M)$ and to hope to define μ as its limit. Here is the crux of the matter: We can prove that such a sequence $\{\mu_n\}$ is always Cauchy, but its limit may not exist if M is not complete!

Lemma 2.2. If $\{\mu_n\}$ is a sequence in the subspace M whose distance from some vector α converges to the distance ρ from α to M, then $\{\mu_n\}$ is Cauchy.

Proof. By the parallelogram law, $\|\mu_n - \mu_m\|^2 = \|(\alpha - \mu_n) - (\alpha - \mu_m)\|^2 = 2(\|\alpha - \mu_n\|^2 + \|\alpha - \mu_m\|^2) - \|2\alpha - (\mu_n + \mu_m)\|^2$. Since the first term on the right converges to $4\rho^2$ as $n, m \to \infty$, and since the second term is always $\leq -4\rho^2$ (factor out the 2), we see that $\|\mu_n - \mu_m\|^2 \to 0$ as $n,m \to \infty$. □

Theorem 2.1. If M is a complete subspace of a pre-Hilbert space V, then $V = M \oplus M^\perp$. In particular, this is true for any finite-dimensional subspace of a pre-Hilbert space and for any closed subspace of a Hilbert space.

Proof. This follows at once from the last two lemmas, since now $\mu = \lim \mu_n$ exists, $\|\alpha - \mu\| = \rho(\alpha, M)$, and so $(\alpha - \mu) \perp M$. □

If $V = M \oplus M^\perp$, then the projection on M along $M^\perp$ is called *the orthogonal projection on M*, or simply *the projection on M*, since among all the projections on M associated with the various complements of M, the orthogonal projection is distinguished. Thus, if M is a complete subspace of V, and if P is the projection on M, then $P(\xi)$ is at once the foot of the perpendicular dropped from ξ to M (which is where the word "projection" comes from) and also the best approximation to ξ in M (Lemma 2.1).

Lemma 2.3. If $\{M_i\}_1^n$ is a finite collection of complete, pairwise orthogonal subspaces, and if for a vector α in V, α_i is the projection of α on M_i for $i = 1, \ldots, n$, then $\sum_1^n \alpha_i$ is the projection of α on $\bigoplus_1^n M_i$.

Proof. We have to show that $\alpha - \sum_1^n \alpha_i$ is orthogonal to $\bigoplus_1^n M_j$, and it is sufficient to show it orthogonal to each M_j separately. But if $\xi \in M_j$, then $(\alpha - \sum_1^n \alpha_i, \xi) = (\alpha - \alpha_j, \xi)$, since $(\alpha_i, \xi) = 0$ for $i \neq j$, and $(\alpha - \alpha_j, \xi) = 0$ because α_j is the projection of α on M_j. Thus $(\alpha - \sum_1^n \alpha_i, \xi) = 0$. □

Lemma 2.4. The projection of ξ on the one-dimensional span of a single nonzero vector η is $((\xi, \eta)/\|\eta\|^2)\eta$.

Proof. Here μ must be of the form $x\eta$. But $(\xi - x\eta) \perp \eta$ if and only if

$$0 = (\xi - x\eta, \eta) = (\xi, \eta) - x\|\eta\|^2, \qquad \text{or} \qquad x = (\xi, \eta)/\|\eta\|^2. \ \square$$

We call the number $(\xi, \eta)/\|\eta\|^2$ the *η-Fourier coefficient* of ξ. If η is a unit (normalized) vector, then this Fourier coefficient is just (ξ, η). It follows from Lemma 2.3 that if $\{\varphi_i\}_1^n$ is an orthogonal collection of nonzero vectors, and if $\{x_i\}_1^n$ are the corresponding Fourier coefficients of a vector ξ, then $\sum_1^n x_i\varphi_i$ is the projection of ξ on the subspace M spanned by $\{\varphi_i\}_1^n$. Therefore, $\xi - \sum_1^n x_i\varphi_i \perp M$, and (Lemma 2.1) $\sum_1^n x_i\varphi_i$ is the best approximation to ξ in M. If ξ is in M, then both of these statements say that $\xi = \sum_1^n x_i\varphi_i$. (This can of course be verified directly, by letting $\xi = \sum_1^n a_i\varphi_i$ be the basis expansion of ξ and computing $(\xi, \varphi_j) = \sum_1^n a_i(\varphi_i, \varphi_j) = a_j\|\varphi_j\|^2$.)

If an orthogonal set of vectors $\{\varphi_i\}$ is also normalized ($\|\varphi_i\| = 1$), then we call the set *orthonormal*.

Theorem 2.2. If $\{\varphi_i\}_1^\infty$ is an infinite orthonormal sequence, and if $\{x_i\}_1^\infty$ are the corresponding Fourier coefficients of a vector ξ, then

$$\sum_1^\infty x_i^2 \leq \|\xi\|^2 \qquad \text{(Bessel's inequality)},$$

and $\xi = \sum_1^\infty x_i\varphi_i$ if and only if $\sum_1^\infty x_i^2 = \|\xi\|^2$ (Parseval's equation).

Proof. Setting $\sigma_n = \sum_1^n x_i\varphi_i$ and $\xi = (\xi - \sigma_n) + \sigma_n$, and remembering that $\xi - \sigma_n \perp \sigma_n$, we have

$$\|\xi\|^2 = \|\xi - \sigma_n\|^2 + \sum_1^n x_i^2.$$

Therefore, $\sum_1^n x_i^2 \leq \|\xi\|^2$ for all n, proving Bessel's inequality, and $\sigma_n \to \xi$ (that is, $\|\xi - \sigma_n\| \to 0$) if and only if $\sum_1^n x_i^2 \to \|\xi\|^2$, proving Parseval's identity. □

We call the formal series $\sum x_i\varphi_i$ the *Fourier series* of ξ (with respect to the orthonormal set $\{\varphi_i\}$). The Parseval condition says that the Fourier series of ξ converges to ξ if and only if $\|\xi\|^2 = \sum_1^\infty x_i^2$.

An infinite orthonormal sequence $\{\varphi_i\}_1^\infty$ is called a *basis* for a pre-Hilbert space V if every element in V is the sum of its Fourier series.

Theorem 2.3. An infinite orthonormal sequence $\{\varphi_i\}_1^\infty$ is a basis for a pre-Hilbert space V if (and only if) its linear span is dense in V.

Proof. Let ξ be any element of V, and let $\{x_i\}$ be its sequence of Fourier coefficients. Since the linear span of $\{\varphi_i\}$ is dense in V, given any ϵ, there is a finite linear combination $\sum_1^n y_i\varphi_i$ which approximates ξ to within ϵ. But $\sum_1^m x_i\varphi_i$ is the *best* approximation to ξ in the span of $\{\varphi_i\}_1^m$, by Lemmas 2.3 and 2.1, and so

$$\left\|\xi - \sum_1^m x_i\varphi_i\right\| \leq \epsilon \qquad \text{for any} \quad m \geq n.$$

That is, $\xi = \sum_1^\infty x_i\varphi_i$. □

Corollary. If V is a Hilbert space, then the orthonormal sequence $\{\varphi_i\}_1^\infty$ is a basis if and only if $\{\varphi_i\}^\perp = \{0\}$.

Proof. Let M be the closure of the linear span of $\{\varphi_i\}_1^\infty$. Since $V = M + M^\perp$, and since $M^\perp = \{\varphi_i\}^\perp$, by Lemma 1.1, we see that $\{\varphi_i\}^\perp = \{0\}$ if and only if $V = M$, and, by the theorem, this holds if and only if $\{\varphi_i\}$ is a basis. □

Note that when orthogonal bases only are being used, the coefficient of a vector ξ at a basis element β is always the Fourier coefficient $(\xi, \beta)/\|\beta\|^2$. Thus the β-coefficient of ξ depends only on β and is independent of the choice of the rest of the basis. However, we know from Chapter 2 that when an arbitrary basis containing β is being used, then the β-coefficient of ξ varies with the basis. This partly explains the favored position of orthogonal bases.

We often obtain an orthonormal sequence by "orthogonalizing" some given sequence.

Lemma 2.5. If $\{\alpha_i\}$ is a finite or infinite sequence of independent vectors, then there is an orthonormal sequence $\{\varphi_i\}$ such that $\{\alpha_i\}_1^n$ and $\{\varphi_i\}_1^n$ have the same linear span for all n.

Proof. Since normalizing is trivial, we shall only orthogonalize. Suppose, to be definite, that the sequence is infinite, and let M_n be the linear span of $\{\alpha_1, \ldots, \alpha_n\}$. Let μ_n be the orthogonal projection of α_n on M_{n-1}, and set $\varphi_n = \alpha_n - \mu_n$ (and $\varphi_1 = \alpha_1$). This is our sequence. We have $\varphi_i \in M_i \subset M_{n-1}$ if $i < n$, and $\varphi_n \perp M_{n-1}$, so that the vectors φ_i are mutually orthogonal. Also, $\varphi_n \neq 0$, since α_n is not in M_{n-1}. Thus $\{\varphi_i\}_1^n$ is an independent subset of the n-dimensional vector space M_n, by the corollary of Lemma 1.2, and so $\{\varphi_i\}_1^n$ spans M_n. □

The actual calculation of the orthogonalized sequence $\{\varphi_n\}$ can be carried out recursively, starting with $\varphi_1 = \alpha_1$, by noticing that since μ_n is the projection of α_n on the span of $\varphi_1, \ldots, \varphi_{n-1}$, it must be the vector $\sum_1^{n-1} c_i\varphi_i$, where c_i is the Fourier coefficient of α_n with respect to φ_i.

Consider, for example, the sequence $\{x^n\}_0^\infty$ in $\mathcal{C}([0, 1])$. We have $\varphi_1 = \alpha_1 = 1$. Next, $\varphi_2 = \alpha_2 - \mu_2 = x - c \cdot 1$, where

$$c = (\alpha_2, \varphi_1)/\|\varphi_1\|^2 = \int_0^1 x \cdot 1/\int_0^1 (1)^2 = \tfrac{1}{2}.$$

Then $\varphi_3 = \alpha_3 - (c_2\varphi_2 + c_1\varphi_1)$, where

$$c_1 = \int_0^1 x^2 \cdot 1/\int_0^1 (1)^2 = \tfrac{1}{3}$$

and

$$c_2 = \int_0^1 x^2(x - \tfrac{1}{2})/\int_0^1 (x - \tfrac{1}{2})^2 = (\tfrac{1}{4} - \tfrac{1}{6})/(\tfrac{2}{24}) = 1.$$

Thus the first three terms in the orthogonalization of $\{x^n\}_0^\infty$ in $\mathcal{C}([0, 1])$ are 1, $x - \frac{1}{2}$, $x^2 - (x - \frac{1}{2}) - \frac{1}{3} = x^2 - x + \frac{1}{6}$. This process is completely elementary, but the calculations obviously become burdensome after only a few terms.

We remember from general bilinear theory that if for β in V we define $\theta_\beta \colon V \to \mathbb{R}$ by $\theta_\beta(\xi) = (\xi, \beta)$, then $\theta_\beta \in V^*$ and $\theta \colon \beta \mapsto \theta_\beta$ is a linear mapping from V to V^*. If (ξ, η) is a scalar product, then $\theta_\beta(\beta) = \|\beta\|^2 > 0$ if $\beta \neq 0$, and so θ is injective. Actually, θ is an *isometry*, as we shall ask the reader to show in an exercise. If V is finite-dimensional, the injectivity of θ implies that θ is an isomorphism. But we have a much more startling result:

Theorem 2.4. θ is an isomorphism if and only if V is a Hilbert space.

Proof. Suppose first that V is a Hilbert space. We have to show that θ is surjective, i.e., that every nonzero F in V^* is of the form θ_β. Given such an F, let N be its null space, let α be a vector orthogonal to N (Theorem 2.1), and consider $\beta = c\alpha$, where c is to be determined later. Every vector ξ in V is uniquely a sum $\xi = x\beta + \eta$, where η is in N. [This only says that V/N is one-dimensional, which presumably we know, but we can check it directly by applying F and seeing that $F(\xi - x\beta) = 0$ if and only if $x = F(\xi)/F(\beta)$.] But now the equations

$$F(\xi) = F(x\beta + \eta) = xF(\beta) = xcF(\alpha)$$

and

$$\theta_\beta(\xi) = (\xi, \beta) = (x\beta + \eta, \beta) = x\|\beta\|^2 = xc^2\|\alpha\|^2$$

show that $\theta_\beta = F$ if we take $c = F(\alpha)/\|\alpha\|^2$.

Conversely, if θ is surjective (and assuming that it is an isometry), then it is an isomorphism in $\text{Hom}(V, V^*)$, and since V^* is complete by Theorem 7.6, Chapter 4, it follows that V is complete by Theorem 7.3 of the same chapter. We are finished. □

EXERCISES

2.1 In the proof of Lemma 2.1, if $(\alpha - \mu, \xi) \neq 0$, what value of t will contradict the inequality $0 \leq 2t(\alpha - \mu, \xi) + t^2\|\xi\|^2$?

2.2 Prove the "only if" part of Theorem 2.3.

2.3 Let $\{M_i\}$ be an orthogonal sequence of complete subspaces of a pre-Hilbert space V, and let P_i be the (orthogonal) projection on M_i. Prove that $\{P_i\xi\}$ is Cauchy for any ξ in V.

2.4 Show that the functions $\{\sin nt\}_{n=1}^{\infty}$ form an orthogonal collection of elements in the pre-Hilbert space $\mathcal{C}([0, \pi])$ with respect to the standard scalar product $(f, g) = \int_0^\pi f(t)\, g(t)\, dt$. Show also that $\|\sin nt\|_2 = \sqrt{\pi/2}$.

2.5 Compute the Fourier coefficients of the function $f(t) = t$ in $\mathcal{C}([0, \pi])$ with respect to the above orthogonal set. What then is the best two-norm approximation to t in the two-dimensional space spanned by $\sin t$ and $\sin 2t$? Sketch the graph of this approximating function, indicating its salient features in the usual manner of calculus curve sketching.

2.6 The "step" function f defined by $f(t) = \pi/2$ on $[0, \pi/2]$ and $f(t) = 0$ on $(\pi/2, \pi]$ is of course discontinuous at $\pi/2$. Nevertheless, calculate its Fourier coefficients with respect to $\{\sin nt\}_{n=1}^{\infty}$ in $\mathcal{C}([0, \pi])$ and graph its best approximation in the span of $\{\sin nt\}_1^3$.

2.7 Show that the functions $\{\sin nt\}_{n=1}^{\infty} \cup \{\cos nt\}_{n=0}^{\infty}$ form an orthogonal collection of elements in the pre-Hilbert space $\mathcal{C}([-\pi, \pi])$ with respect to the standard scalar product $(f, g) = \int_{-\pi}^{\pi} f(t)\, g(t)\, dt$.

2.8 Calculate the first three terms in the orthogonalization of $\{x^n\}_0^\infty$ in $\mathcal{C}([-1, 1])$.

2.9 Use the definition of the norm of a bounded linear transformation and the Schwarz inequality to show that $\|\theta_\beta\| \leq \|\beta\|$ [where $\theta_\beta(\xi) = (\xi, \beta)$]. In order to conclude that $\beta \mapsto \theta_\beta$ is an isometry, we also need the opposite inequality, $\|\theta_\beta\| \geq \|\beta\|$. Prove this by using a special value of ξ.

2.10 Show that if V is an incomplete pre-Hilbert space, then V has a proper closed subspace M such that $M^\perp = \{0\}$. [*Hint:* There must exist $F \in V^*$ not of the form $F(\xi) = (\xi, \alpha)$.] Together with Theorem 2.1, this shows that a pre-Hilbert space V is a Hilbert space if and only if $V = M \oplus M^\perp$ for every closed subspace M.

2.11 The isometry $\theta: \alpha \mapsto \theta_\alpha$ [where $\theta_\alpha(\xi) = (\xi, \alpha)$] imbeds the pre-Hilbert space V in its conjugate space V^*. We know that V^* is complete. Why? The closure of V as a subspace of V^* is therefore complete, and we can hence complete V as a Banach space. Let H be its completion. It is a Banach space including (the isometric image of) V as a dense subspace. Show that the scalar product on V extends uniquely to H and that the norm on H is the extended scalar product norm, so that H is a Hilbert space.

2.12 Show that under the isometric imbedding $\alpha \mapsto \theta_\alpha$ of a pre-Hilbert space V into V^* orthogonality is equivalent to annihilation as discussed in Section 2.3. Discuss the connection between the properties of the annihilator A° and Lemma 1.1 of this chapter.

2.13 Prove that if C is a nonempty complete convex subset of a pre-Hilbert space V, and if α is any vector not in C, then there is a unique $\mu \in C$ closest to α. (Examine the proof of Lemma 2.2.)

3. SELF-ADJOINT TRANSFORMATIONS

Definition. If V is a pre-Hilbert space, then T in Hom V is *self-adjoint* if $(T\alpha, \beta) = (\alpha, T\beta)$ for every $\alpha, \beta \in V$. The set of all self-adjoint transformations will be designated SA.

Self-adjointness suggests that T ought to become its own adjoint under the injection θ of V into V^*. We check this now. Since $(\alpha, \beta) = \theta_\beta(\alpha)$, we can rewrite the equation $(T\alpha, \beta) = (\alpha, T\beta)$ as $\theta_\beta(T\alpha) = \theta_{T\beta}(\alpha)$, and again as $\big(T^*(\theta_\beta)\big)(\alpha) = \theta_{T\beta}(\alpha)$ by the definition of T^*. This holds for all α and β if and only if $T^*(\theta_\beta) = \theta_{T\beta}$ for all $\beta \in V$, or $T^* \circ \theta = \theta \circ T$, which is the asserted identification.

Lemma 3.1. If V is a finite-dimensional Hilbert space and $\{\varphi_i\}_1^n$ is an orthonormal basis for V, then $T \in \text{Hom}(V)$ is self-adjoint if and only if the matrix $\{t_{ij}\}$ of T with respect to $\{\varphi_i\}$ is symmetric ($\mathbf{t} = \mathbf{t}^*$).

Proof. If we substitute the basis expansions of α and β and expand, we see that $(\alpha, T\beta) = (T\alpha, \beta)$ for all α and β if and only if $(\varphi_i, T\varphi_j) = (T\varphi_i, \varphi_j)$ for all i and j. But $T\varphi_l = \sum_{k=1}^n t_{kl}\varphi_k$, and when this is substituted in these last scalar products, the equation becomes $t_{ij} = t_{ji}$. That is, T is self-adjoint if and only if $\mathbf{t} = \mathbf{t}^*$. □

A self-adjoint T is said to be *nonnegative* if $(T\xi, \xi) \geq 0$ for all ξ. Then $[\xi, \eta] = (T\xi, \eta)$ is a semiscalar product!

Lemma 3.2. If T is a nonnegative self-adjoint transformation, then $\|T(\xi)\| \leq \|T\|^{1/2}(T\xi, \xi)^{1/2}$ for all ξ. Therefore, if $(T\xi, \xi) = 0$, then $T\xi = 0$, and, more generally, if $(T\xi_n, \xi_n) \to 0$, then $T(\xi_n) \to 0$.

Proof. If T is nonnegative as well as self-adjoint, then $[\xi, \eta] = (T\xi, \eta)$ is a semiscalar product, and so, by Schwarz's inequality,

$$|(T\xi, \eta)| = [\xi, \eta] \leq [\xi, \xi]^{1/2}[\eta, \eta]^{1/2} = (T\xi, \xi)^{1/2}(T\eta, \eta)^{1/2}.$$

Taking $\eta = T\xi$, the factor on the right becomes $(T(T\xi), T\xi)^{1/2}$, which is less than or equal to $\|T\|^{1/2}\|T\xi\|$, by Schwarz and the definition of $\|T\|$. Dividing by $\|T\xi\|$, we get the inequality of the lemma. □

If $\alpha \neq 0$ and $T(\alpha) = c\alpha$ for some c, then α is called an *eigenvector* (proper vector, characteristic vector) of T, and c is the associated *eigenvalue* (proper value, characteristic value).

Theorem 3.1. If V is a finite-dimensional Hilbert space and T is a self-adjoint element of Hom V, then V has an orthonormal basis consisting entirely of eigenvectors of T.

Proof. Consider the function $(T\xi, \xi)$. It is a continuous real-valued function of ξ, and on the unit sphere $S = \{\xi : \|\xi\| = 1\}$ it is bounded above by $\|T\|$ (by Schwarz). Set $m = \text{lub}\,\{(T\xi, \xi) : \|\xi\| = 1\}$. Since S is compact (being bounded and closed), $(T\xi, \xi)$ assumes the value m at some point α on S. Now $m - T$ is a nonnegative self-adjoint transformation (Check this!), and $(T\alpha, \alpha) = m$ is equivalent to $((m - T)\alpha, \alpha) = 0$. Therefore, $(m - T)\alpha = 0$ by Lemma 3.2, and $T\alpha = m\alpha$. We have thus found one eigenvector for T. Now set $V_1 = V$, $\alpha_1 = \alpha$, and $m_1 = m$, and let V_2 be $\{\alpha_1\}^\perp$. Then $T[V_2] \subset V_2$, for if $\xi \perp \alpha_1$, then $(T\xi, \alpha_1) = (\xi, T\alpha_1) = m(\xi, \alpha_1) = 0$.

We can therefore repeat the above argument for the restriction of T to the Hilbert space V_2 and find α_2 in V_2 such that $\|\alpha_2\| = 1$ and $T(\alpha_2) = m_2\alpha_2$, where $m_2 = \text{lub}\,\{(T\xi, \xi) : \|\xi\| = 1 \text{ and } \xi \in V_2\}$. Clearly, $m_2 \leq m_1$. We then set $V_3 = \{\alpha_1, \alpha_2\}^\perp$ and continue, arriving finally at an orthonormal basis $\{\alpha_i\}_1^n$ of eigenvectors of T. □

Now let $\lambda_1, \ldots, \lambda_r$ be the *distinct* values in the list $m_1, \ldots, m_n$, and let M_j be the linear span of those basis vectors α_i for which $m_i = \lambda_j$. Then the subspaces M_j are orthogonal to each other, $V = \bigoplus_1^r M_j$, each M_j is T-invariant, and the restriction of T to M_j is λ_j times the identity. Since *all* the nonzero vectors in M_j are eigenvectors with eigenvalue λ_j, if the α_i's spanning M_j are replaced by *any* other orthonormal basis for M_j, then we still have an orthonormal basis of eigenvectors. The α_i's are therefore not in general uniquely determined. *But the subspaces M_j and the eigenvalues λ_j are unique.* This will follow if we show that every eigenvector is in an M_j.

Lemma 3.3. In the context of the above discussion, if $\xi \neq 0$ and $T(\xi) = x\xi$ for some x in $\mathbb{R}$, then $\xi \in M_j$ (and so $x = \lambda_j$) for some j.

Proof. Since $V = \bigoplus_1^r M_j$, we have $\xi = \sum_1^r \xi_i$ with $\xi_i \in M_i$. Then

$$\sum_1^r x\xi_i = x\xi = T(\xi) = \sum_1^r T(\xi_i) = \sum_1^r \lambda_i \xi_i \quad \text{and} \quad \sum_1^r (x - \lambda_i)\xi_i = 0.$$

Since the subspaces M_i are independent, every component $(x - \lambda_i)\xi_i$ is 0. But some $\xi_j \neq 0$, since $\xi \neq 0$. Therefore, $x = \lambda_j$, $\xi_i = 0$ for $i \neq j$, and

$$\xi = \xi_j \in M_j. \ \square$$

We have thus proved the following theorem.

Theorem 3.2. If V is a finite-dimensional Hilbert space and T is a self-adjoint element of Hom V, then there are uniquely determined subspaces $\{V_i\}_1^r$ of V, and distinct scalars $\{\lambda_i\}_1^r$, such that $\{V_i\}$ is an orthogonal family whose sum is V and the restriction of T to V_i is λ_i times the identity.

If V is a finite-dimensional vector space and we are given $T \in$ Hom V, then we know how to compute related mappings such as T^2 and T^{-1} (if it exists) and vectors $T\alpha$, $T^{-1}\alpha$, etc., by choosing a basis for V and then computing matrix products, inverses (when they exist), and so on. Some of these computations, particularly those related to inverses, can be quite arduous. One enormous advantage of a basis consisting of eigenvectors for T is that it trivializes all of these calculations.

To see this, let $\{\beta_n\}$ be a basis of V consisting entirely of eigenvectors for T, and let $\{r_n\}$ be the corresponding eigenvalues. To compute $T\xi$, we write down the basis expansion for ξ, $\xi = \sum_1^n x_i\beta_i$, and then $T\xi = \sum_1^n r_i x_i \beta_i$. T^2 has the same eigenvectors, but with eigenvalues $\{r_i^2\}$. Thus $T^2\alpha = \sum_1^n r_i^2 x_i \beta_i$. T^{-1} exists if and only if no $r_i = 0$, in which case it has the same eigenvectors with eigenvalues $\{1/r_i\}$. Thus $T^{-1}\xi = \sum_1^n (x_i/r_i)\beta_i$. If $P(t) = \sum_0^m a_n t^n$ is any polynomial, then $P(T)$ takes β_i into $P(r_i)\beta_i$. Thus $P(T)\xi = \sum_0^m P(r_i)x_i\beta_i$. By now the point should be amply clear.

The additional value of orthonormality in a basis is already clear fom the last section. Basically, it enables us to compute the coefficients $\{x_i\}$ of ξ by scalar products: $x_i = (\xi, \beta_i)$.

This is a good place to say a few words about the general eigenvalue problem in finite-dimensional theory. Our complete analysis above was made possible by the self-adjointness of T (or the symmetry of the matrix $\mathbf{t}$). What we can say about an arbitrary T in Hom V is much less satisfactory.

We first note that the eigenvalues of T can be determined algebraically, for λ is an eigenvalue if and only if $T - \lambda I$ is not injective, or, equivalently, is singular, and we know that $T - \lambda I$ is singular if and only if its determinant $\Delta(T - \lambda I)$ is 0. If we choose any basis for V, the determinant of $T - \lambda I$ is the determinant of its matrix $\mathbf{t} - \lambda\mathbf{e}$, and our later formula in Chapter 7 shows that this is a polynomial of degree n in λ. It is easy to see that this polynomial is independent of the basis; it is called the *characteristic polynomial of T*. Thus the eigenvalues of T are exactly the roots of the characteristic polynomial of T.

However, T need not have any eigenvectors! Consider, for example, a 90° rotation in the Cartesian plane. This is the map $T\colon \langle x, y\rangle \mapsto \langle -y, x\rangle$. Thus $T(\delta^1) = \delta^2$ and $T(\delta^2) = -\delta^1$, so the matrix of T is

$$\begin{bmatrix} 0 & -1 \\ 1 & 0 \end{bmatrix}.$$

Then the matrix of $T - \lambda$ is

$$\begin{bmatrix} \lambda & -1 \\ 1 & \lambda \end{bmatrix},$$

and the characteristic polynomial of T is the determinant of this matrix: $\lambda^2 + 1$. Since this polynomial is irreducible over $\mathbb{R}$, there are no eigenvalues.

Note how different the outcome is if we consider the transformation with the same matrix on *complex* 2-space $\mathbb{C}^2$. Here the scalar field is the complex number system, and T is the map $\langle z_1, z_2\rangle \mapsto \langle -z_2, z_1\rangle$ from $\mathbb{C}^2$ to $\mathbb{C}^2$. But now $\lambda^2 + 1 = (\lambda + i)(\lambda - i)$, and T has eigenvalues $\pm i$! To find the eigenvectors for i, we solve $T(\mathbf{z}) = i\mathbf{z}$, which is the equation $\langle -z_2, z_1\rangle = \langle iz_1, iz_2\rangle$, or $z_2 = -iz_1$. Thus $\langle 1, -i\rangle$ (or $i\langle 1, -i\rangle = \langle i, 1\rangle$) is the unique eigenvector for i to within a scalar multiple.

We return to our real theory. If $T \in \operatorname{Hom} V$ and $n = d(V)$, so that $d(\operatorname{Hom} V) = n^2$, then the set of $n^2 + 1$ vectors $\{T^i\}_0^{n^2}$ in $\operatorname{Hom} V$ is dependent. But this is exactly the same as saying that $p(T) = 0$ for some polynomial p of degree $\leq n^2$. That is, any T in $\operatorname{Hom} V$ satisfies a polynomial identity $p(T) = 0$. Now suppose that r is an eigenvalue of T and that $T(\xi) = r\xi$. Then $p(T)(\xi) = p(r)\xi = 0$, and so $p(r) = 0$. That is, every eigenvalue of T is a root of the polynomial p. Conversely, if $p(r) = 0$, then we know from the remainder theorem of algebra that $t - r$ is a factor of the polynomial $p(t)$, and therefore $(t - r)^m$ will be one of the relatively prime factors of p. Now suppose that p is the minimal polynomial of T (see Exercise 3.5). Theorem 5.5, Chapter 1, tells us that $(T - rI)^m$ is zero on a corresponding subspace N of V and therefore, in particular, that $T - rI$ is not injective when restricted to N. That is, r is an eigenvalue. We have proved:

Theorem 3.3. The eigenvalues of T are zeros (roots) of every polynomial $p(t)$ such that $p(T) = 0$, and are exactly the roots of the minimal polynomial.

EXERCISES

3.1 Use the defining identity $(T\xi, \eta) = (\xi, T\eta)$ to show that the set SA of all self-adjoint elements of $\operatorname{Hom} V$ is a subspace. Prove similarly that if S and T are self-adjoint, then ST is self-adjoint if and only if $ST = TS$. Conclude that if T is self-adjoint, then so is $p(T)$ for any polynomial p.

3.2 Show that if T is self-adjoint, then $S = T^2$ is nonnegative. Show, therefore, that if T is self-adjoint and $a \geq 0$, then $T^2 + aI$ cannot be the zero transformation.

3.3 Let $p(t) = t^2 + bt + c$ be an irreducible quadratic polynomial ($b^2 < 4c$), and let T be a self-adjoint transformation. Show that $p(T) \neq 0$. (Complete the square and apply earlier exercises.)

3.4 Let T be self-adjoint and nilpotent ($T^n = 0$ for some n). Prove that $T = 0$. This can be done in various ways. One method is to show it first for $n = 2$ and then for $n = 2^m$ by induction. Finally, any n can be bracketed by powers of 2, $2^m \leq n < 2^{m+1}$.

3.5 Let V be any vector space, and let T be an element of Hom V. Suppose that there is a polynomial q such that $q(T) = 0$, and let p be such a polynomial of minimum degree. Show that p is unique (to within a constant multiple). It is called the *minimal polynomial* of T. Show that if we apply Theorem 5.5 of Chapter 1 to the minimal polynomial p of T, then the subspaces N_i must both be nontrivial.

3.6 It is a corollary of the fundamental theorem of algebra that a polynomial with real coefficients can be factored into a product of linear factors $(t - r)$ and irreducible quadratic factors $(t^2 + bt + c)$. Let T be a self-adjoint transformation on a finite-dimensional Hilbert space, and let $p(t)$ be its minimal polynomial. Deduce a new proof of Theorem 3.1 by applying to $p(t)$ the above remark, Theorem 5.5 of Chapter 1, and Exercises 3.1 through 3.4.

3.7 Prove that if T is a self-adjoint transformation on a pre-Hilbert space V, then its null space is the orthogonal complement of its range: $N(T) = (R(T))^{\perp}$. Conclude that if V is a Hilbert space, then a self-adjoint T is injective if and only if its range is dense (in V).

3.8 Assuming the above exercise, show that if V is a Hilbert space and T is a self-adjoint element of Hom V that is *bounded below* (as well as bounded), then T is surjective.

3.9 Let T be self-adjoint and nonnegative, and set $m = \operatorname{lub} \{(T\xi, \xi) : \|\xi\| = 1\}$. Use the Schwarz inequality and the inequality of Lemma 3.2 to show that $m = \|T\|$.

3.10 Let V be a Hilbert space, let T be a self-adjoint element of Hom V, and set $m = \operatorname{lub} \{(T\xi, \xi) : \|\xi\| = 1\}$. Show that if $a > m$, then $a - T$ $(= aI - T)$ is invertible and $\|(a - T)^{-1}\| \leq 1/(a - m)$. (Apply the Schwarz inequality, the definition of m, and Exercise 3.8.)

3.11 Let P be a bounded linear transformation on a pre-Hilbert space V that is a projection in the sense of Chapter 1. Prove that if P is self-adjoint, then P is an orthogonal *projection*. Now prove the converse.

3.12 Let V be a finite-dimensional Hilbert space, let T in Hom V be self-adjoint, and suppose that S in Hom V commutes with T. Show that the subspaces M_j of Theorem 3.1 and Lemma 3.3 are invariant under S.

3.13 A self-adjoint transformation T on a finite-dimensional Hilbert space V is said to have a *simple* spectrum if all its eigenvalues are distinct. By this we mean that all the subspaces M_j are one-dimensional. Suppose that T is a self-adjoint transformation with a simple spectrum, and suppose that S commutes with T. Show that S is also self-adjoint. (Apply the above exercise.)

3.14 Let H be a Hilbert space, and let $\omega[\xi, \eta]$ be a bounded bilinear form on $H \times H$. That is, there is a constant b such that

$$|\omega[\xi, \eta]| \leq b\|\xi\| \, \|\eta\| \qquad \text{for all} \quad \xi, \eta \in H.$$

Show that there is a unique T in Hom V such that $\omega[\xi, \eta] = (\xi, T\eta)$. Show that T is self-adjoint if and only if ω is symmetric.

4. ORTHOGONAL TRANSFORMATIONS

Assuming that V is a Hilbert space and that therefore $\theta: V \to V^*$ is an isomorphism, we can of course replace the adjoint $T^* \in \text{Hom } V^*$ of any $T \in \text{Hom } V$ by the corresponding transformation $\theta^{-1} \circ T^* \circ \theta \in \text{Hom } V$. In Hilbert space theory it is *this* mapping that is called the adjoint of T and is designated T^*. Then, exactly as in our discussion of a self-adjoint T, we see that

$$(T\alpha, \beta) = (\alpha, T^*\beta) \qquad \text{for all} \quad \alpha, \beta \in V$$

and that T^* is uniquely defined by this identity. Finally, T is self-adjoint if and only if $T = T^*$.

Although it really amounts to the above way of introducing T^* into Hom V, we can make a direct definition as follows. For each η the mapping $\xi \mapsto (T\xi, \eta)$ is linear and bounded, and so is an element of V^*, which, by Theorem 2.4, is given by a unique element β_η in V according to the formula $(T\xi, \eta) = (\xi, \beta_\eta)$. Now we check that $\eta \mapsto \beta_n$ is linear and bounded and is therefore an element of Hom V which we call T^*, etc.

The matrix calculations of Lemma 3.1 generalize verbatim to show that the matrix of T^* in Hom V is the transpose $\mathbf{t}^*$ of the matrix $\mathbf{t}$ of T.

Another very important type of transformation on a Hilbert space is one that preserves the scalar product.

Definition. A transformation $T \in \text{Hom } V$ is *orthogonal* if $(T\alpha, T\beta) = (\alpha, \beta)$ for all $\alpha, \beta \in V$.

By the basic adjoint identity above this is entirely equivalent to $(\alpha, T^*T\beta) = (\alpha, \beta)$, for all α, β, and hence to $T^*T = I$. An orthogonal T is injective, since $\|T\alpha\|^2 = \|\alpha\|^2$, and is therefore invertible if V is finite-dimensional. Whether V is finite-dimensional or not, if T *is* invertible, then the above condition becomes $T^* = T^{-1}$.

If $T \in \text{Hom } \mathbb{R}^n$, the matrix form of the equation $T^*T = I$ is of course $\mathbf{t}^*\mathbf{t} = \mathbf{e}$, and if this is written out, it becomes

$$\sum_{k=1}^{n} t_{ki}t_{kj} = \delta_j^i \qquad \text{for all} \quad i, j,$$

which simply says that the columns of t form an orthonormal set (and hence a basis) in $\mathbb{R}^n$. We thus have:

Theorem 4.1. A transformation $T \in \text{Hom } \mathbb{R}^n$ is orthogonal if and only if the image of the standard basis $\{\delta^i\}_1^n$ under T is another orthonormal basis (with respect to the standard scalar product).

The necessity of this condition is, of course, obvious from the scalar-product-preserving definition of orthogonality, and the sufficiency can also be checked directly using the basis expansions of α and β.

We can now state the eigenbasis theorem in different terms. By a *diagonal* matrix we mean a matrix which is zero everywhere except on the main diagonal.

Theorem 4.2. Let $\mathbf{t} = \{t_{ij}\}$ be a symmetric $n \times n$ matrix. Then there exists an orthogonal $n \times n$ matrix $\mathbf{b}$ such that $\mathbf{b}^{-1}\mathbf{t}\mathbf{b}$ is a diagonal matrix.

Proof. Since the transformation $T \in \operatorname{Hom} \mathbb{R}^n$ defined by $\mathbf{t}$ is self-adjoint, there exists an orthonormal basis $\{\mathbf{b}^i\}_1^n$ of eigenvectors of T, with corresponding eigenvalues $\{r_i\}_1^n$. Let B be the orthogonal transformation defined by $B(\delta^j) = \mathbf{b}^j$, $j = 1, \ldots, n$. (The n-tuples $\mathbf{b}^j$ are the columns of the matrix $\mathbf{b} = \{b_{ij}\}$ of B.) Then $(B^{-1} \circ T \circ B)(\delta^j) = r_j\delta^j$. Since $(B^{-1} \circ T \circ B)(\delta^j)$ is the jth column of $\mathbf{b}^{-1}\mathbf{t}\mathbf{b}$, we see that $\mathbf{s} = \mathbf{b}^{-1}\mathbf{t}\mathbf{b}$ is diagonal, with $s_{jj} = r_j$. □

For later applications we are also going to want the following result.

Theorem 4.3. Any invertible $T \in \operatorname{Hom} V$ on a finite-dimensional Hilbert space V can be expressed in the form $T = RS$, where R is orthogonal and S is self-adjoint and positive.

Proof. For any T, T^*T is self-adjoint, since $(T^*T)^* = T^*T^{**} = T^*T$. Let $\{\varphi_i\}_1^n$ be an orthonormal eigenbasis, and let $\{r_i\}_1^n$ be the corresponding eigenvalues of T^*T. Then $0 < \|T\varphi_i\|^2 = (T^*T\varphi_i, \varphi_i) = (r_i\varphi_i, \varphi_i) = r_i$ for each i. Since all the eigenvalues of T^*T are thus positive, we can define a positive square root $S = (T^*T)^{1/2}$ by $S\varphi_i = (r_i)^{1/2}\varphi_i$, $i = 1, 2, \ldots, n$. It is clear that $S^2 = T^*T$ and that S is self-adjoint.

Then $A = ST^{-1}$ is orthogonal, for $(ST^{-1}\alpha, ST^{-1}\beta) = (T^{-1}\alpha, S^2T^{-1}\beta) = (T^{-1}\alpha, T^*TT^{-1}\beta) = (T^{-1}\alpha, T^*\beta) = (TT^{-1}\alpha, \beta) = (\alpha, \beta)$. Since $T = A^{-1}S$, we set $R = A^{-1}$ and have the theorem. □

It is not hard to see that the above factorization of T is unique. Also, by starting with TT^*, we can express T in the form $T = SR$, where S is self-adjoint and positive and R is orthogonal.

We call these factorizations the polar decompositions of T, since they function somewhat like the polar coordinate factorization $z = re^{i\theta}$ of a complex number.

Corollary. Any nonsingular $n \times n$ matrix $\mathbf{t}$ can be expressed as $\mathbf{t} = \mathbf{u}\mathbf{d}\mathbf{v}$, where $\mathbf{u}$ and $\mathbf{v}$ are orthogonal and $\mathbf{d}$ is diagonal.

Proof. From the theorem we have $\mathbf{t} = \mathbf{r}\mathbf{s}$, where $\mathbf{r}$ is orthogonal and $\mathbf{s}$ is symmetric. By Theorem 4.2, $\mathbf{s} = \mathbf{b}\mathbf{d}\mathbf{b}^{-1}$, where $\mathbf{d}$ is diagonal and $\mathbf{b}$ is orthogonal. Thus $\mathbf{t} = \mathbf{r}\mathbf{s} = (\mathbf{r}\mathbf{b})\mathbf{d}\mathbf{b}^{-1} = \mathbf{u}\mathbf{d}\mathbf{v}$, where $\mathbf{u} = \mathbf{r}\mathbf{b}$ and $\mathbf{v} = \mathbf{b}^{-1}$ are both orthogonal. □

EXERCISES

4.1 Let V be a Hilbert space, and suppose that S and T in $\operatorname{Hom} V$ satisfy

$$(T\xi, \eta) = (\xi, S\eta) \qquad \text{for all} \quad \xi, \eta.$$

Write out the proof of the identity $S = \theta^{-1} \circ T^* \circ \theta$.

4.2 Write out the analogue of the proof of Lemma 3.1 which shows that the matrix of T^* is the transpose of the matrix of T.

4.3 Once again show that if $(\xi, \eta) = (\xi, \zeta)$ for all ξ, then $\eta = \zeta$. Conclude that if S, T in Hom V are such that $(\xi, T\eta) = (\xi, S\eta)$ for all η, then $T = S$.

4.4 Let $\{\mathbf{a}, \mathbf{b}\}$ be an orthonormal basis for $\mathbb{R}^2$, and let $\mathbf{t}$ be the 2×2 matrix whose columns are $\mathbf{a}$ and $\mathbf{b}$. Show by direct calculation that the rows of $\mathbf{t}$ are also orthonormal.

4.5 State again why it is that if V is finite-dimensional, and if S and T in Hom V satisfy $S \circ T = I$, then T is invertible and $S = T^{-1}$. Now let V be a finite-dimensional Hilbert space, and let T be an orthogonal transformation in Hom V. Show that T^* is also orthogonal.

4.6 Let $\mathbf{t}$ be an $n \times n$ matrix whose columns form an orthonormal basis for $\mathbb{R}^n$. Prove that the rows of $\mathbf{t}$ also form an orthonormal basis. (Apply the above exercise.)

4.7 Show that a nonnegative self-adjoint transformation S on a finite-dimensional Hilbert space has a uniquely determined nonnegative self-adjoint square root.

4.8 Prove that if V is a finite-dimensional Hilbert space and $T \in$ Hom V, then the "polar decomposition" of T, $T = RS$, of Theorem 4.3 is unique. (Apply the above exercise.)

5. COMPACT TRANSFORMATIONS

Theorem 3.1 breaks down when V is an infinite-dimensional Hilbert space. A self-adjoint transformation T does not in general have enough eigenvectors to form a basis for V, and a more sophisticated analysis, allowing for a "continuous spectrum" as well as a "discrete spectrum", is necessary. This enriched situation is the reason for the need for further study of Hilbert space theory at the graduate level, and is one of the sources of complexity in the mathematical structure of quantum mechanics.

However, there is one very important special case in which the eigenbasis theorem is available, and which will have a startling application in the next chapter.

Definition. Let V and W be any normed linear spaces, and let S be the unit ball in V. A transformation T in Hom(V, W) is *compact* if the closure of $T[S]$ in W is sequentially compact.

Theorem 5.1. Let V be any pre-Hilbert space, and let $T \in$ Hom V be self-adjoint and compact. Then the pre-Hilbert space $R =$ range (T) has an orthonormal basis $\{\varphi_i\}$ consisting entirely of eigenvectors of T, and the corresponding sequence of eigenvalues $\{r_n\}$ converges to 0 (or is finite).

Proof. The proof is just like that of Theorem 3.1 except that we have to start a little differently. Set $m = \|T\| = \text{lub}\,\{\|T(\xi)\| : \|\xi\| = 1\}$, and choose a sequence $\{\xi_n\}$ such that $\|\xi_n\| = 1$ for all n and $\|T(\xi_n)\| \to m$. Then

$$((m^2 - T^2)\xi_n, \xi_n) = m^2 - \|T(\xi_n)\|^2 \to 0,$$

and since $m^2 - T^2$ is a nonnegative self-adjoint transformation, Lemma 3.2 tells that $(m^2 - T^2)(\xi_n) \to 0$. But since T is compact, we can suppose (passing to a subsequence if necessary) that $\{T\xi_n\}$ converges, say to β. Then $T^2\xi_n \to T\beta$, and so $m^2\xi_n \to T\beta$ also. Thus $\xi_n \to T\beta/m^2$ and $\beta = \lim T\xi_n = T^2(\beta)/m^2$ Since $\|\beta\| = \lim \|T(\xi_n)\| = m$, we have a nonzero vector β such that $T^2(\beta) = m^2\beta$. Set $\alpha = \beta/\|\beta\|$.

We have thus found a vector α such that $\|\alpha\| = 1$ and $0 = (m^2 - T^2)(\alpha) = (m - T)(m + T)(\alpha)$. Then either $(m + T)(\alpha) = 0$, in which case $T(\alpha) = -m\alpha$, or $(m + T)(\alpha) = \gamma \neq 0$ and $(m - T)\gamma = 0$, in which case $T\gamma = m\gamma$. Thus there exists a vector φ_1 (α or $\gamma/\|\gamma\|$) such that $\|\varphi_1\| = 1$ and $T(\varphi_1) = r_1\varphi_1$, where $|r_1| = m$. We now proceed just as in Theorem 3.1.

For notational consistency we set $m_1 = m$, $V_1 = V$, and now set $V_2 = \{\varphi_1\}^\perp$. Then $T[V_2] \subset V_2$, since if $\alpha \perp \varphi_1$, then $(T\alpha, \varphi_1) = (\alpha, T\varphi_2) = r_1(\alpha, \varphi_1) = 0$. Thus $T \restriction V_2$ is compact and self-adjoint, and if $m_2 = \|T \restriction V_2\|$, there exists φ_2 with $\|\varphi_2\| = 1$ and $T(\varphi_2) = r_2\varphi_2$, where $|r_2| = m_2$. We continue inductively, obtaining an orthonormal sequence $\{\varphi_n\} \subset V$ and a sequence $\{r_n\} \subset \mathbb{R}$ such that $T\varphi_n = r_n\varphi_n$ and $|r_n| = \|T \restriction V_n\|$, where

$$V_n = \{\varphi_1, \ldots, \varphi_{n-1}\}^\perp.$$

We suppose for the moment, since this is the most interesting case, that $r_n \neq 0$ for all n. Then we claim that $|r_n| \to 0$. For $|r_n|$ is decreasing in any case, and if it does not converge to 0, then there exists a $b > 0$ such that $|r_n| \geq b$ for all n. Then $\|T(\varphi_i) - T(\varphi_j)\|^2 = \|r_i\varphi_i - r_j\varphi_j\|^2 = \|r_i\varphi_i\|^2 + \|r_j\varphi_j\|^2 = r_i^2 + r_j^2 \geq 2b^2$ for all $i \neq j$, and the sequence $\{T(\varphi_i)\}$ can have no convergent subsequence, contradicting the compactness of T. Therefore $|r_n| \downarrow 0$.

Finally, we have to show that the orthonormal sequence $\{\varphi_i\}$ is a basis for R. If $\beta = T(\alpha)$, and if $\{b_n\}$ and $\{a_n\}$ are the Fourier coefficients of β and α, then we expect that $b_n = r_na_n$, and this is easy to check: $b_n = (\beta, \varphi_n) = (T(\alpha), \varphi_n) = (\alpha, T(\varphi_n)) = (\alpha, r_n\varphi_n) = r_n(\alpha, \varphi_n) = r_na_n$. This is just saying that $T(a_n\varphi_n) = b_n\varphi_n$, and therefore $\beta - \sum_1^n b_i\varphi_i = T(\alpha - \sum_1^n a_i\varphi_i)$. Now $\alpha - \sum_1^n a_i\varphi_i$ is orthogonal to $\{\varphi_i\}_1^n$ and therefore is an element of V_{n+1}, and the norm of T on V_{n+1} is $|r_{n+1}|$. Moreover, $\|\alpha - \sum_1^n a_i\varphi_i\| \leq \|\alpha\|$, by the Pythagorean theorem. Altogether we can conclude that

$$\left\|\beta - \sum_1^n b_i\varphi_i\right\| \leq |r_{n+1}| \cdot \|\alpha\|,$$

and since $r_{n+1} \to 0$, this implies that $\beta = \sum_1^\infty b_i\varphi_i$. Thus $\{\varphi_i\}$ is a basis for $R(T)$. Also, since T is self-adjoint, $N(T) = R(T)^\perp = \{\varphi_i\}^\perp = \bigcap_1^\infty V_i$.

If some $r_i = 0$, then there is a first n such that $r_n = 0$. In this case $\|T \restriction V_n\| = |r_n| = 0$, so that $V_n \subset N(T)$. But $\varphi_i \in R(T)$ if $i < n$, since then $\varphi_i = T(\varphi_i)/r_i$, and so $N(T) = R(T)^\perp \subset \{\varphi_1, \ldots, \varphi_{n-1}\}^\perp = V_n$. Therefore, $N(T) = V_n$ and $R(T)$ is the span of $\{\varphi_i\}_1^{n-1}$. □

CHAPTER 6

DIFFERENTIAL EQUATIONS

This chapter is not a small differential equations textbook; we leave out far too much. We are principally concerned with some of the theory of the subject, although we shall say one or two practical things. Our first goal is the fundamental existence and uniqueness theorem of ordinary differential equations, which we prove as an elegant application of the fixed-point theorem. Next we look at the linear theory, where we make vital use of material from the first two chapters and get quite specific about the process of actually finding solutions. So far our development is linked to the *initial-value* problem, concerning the existence of, and in some cases the ways of finding, a unique solution passing through some initially prescribed point in the space containing the solution curves. However, some of the most important aspects of the subject relate to what are called *boundary-value* problems, and our last and most sophisticated effort will be directed toward making a first step into this large area. This will involve us in the theory of Chapter 5, for we shall find ourselves studying self-adjoint operators. In fact, the basic theorem about Fourier series expansions will come out of recognizing a certain right inverse of a differential operator to be a compact self-adjoint operator.

1. THE FUNDAMENTAL THEOREM

Let A be an open subset of a Banach space W, let I be an open interval in $\mathbb{R}$, and let $F: I \times A \to W$ be continuous. We want to study the differential equation

$$d\alpha/dt = F(t, \alpha).$$

A solution of this equation is a function $f: J \to A$, where J is an open subinterval of I, such that $f'(t)$ exists and

$$f'(t) = F(t, f(t))$$

for every t in J. Note that a solution f has to be continuously differentiable, for the existence of f' implies the continuity of f, and then $f'(t) = F(t, f(t))$ is continuous by the continuity of F.

We are going to see that if F has a continuous second partial differential, then there exists a uniquely determined "local" solution through any point

$$\langle t_0, \alpha_0 \rangle \in I \times A.$$

In saying that the solution f goes through $\langle t_0, \alpha_0 \rangle$, we mean, of course, that $\alpha_0 = f(t_0)$. The requirement that the solution f have the value α_0 when $t = t_0$ is called an *initial condition.*

The existence and continuity of $dF^2_{\langle t,\alpha \rangle}$ implies, via the mean-value theorem, that $F(t, \alpha)$ is *locally uniformly Lipschitz* in α. By this we mean that for any point $\langle t_0, \alpha_0 \rangle$ in $I \times A$ there is a neighborhood $M \times N$ and a constant b such that $\|F(t, \xi) - F(t, \eta)\| \leq b\|\xi - \eta\|$ for all t in M and all ξ, η in N. To see this we simply choose balls M and N about t_0 and α_0 such that $dF^2_{\langle t,\alpha \rangle}$ is bounded, say by b, on $M \times N$, and apply Theorem 7.4 of Chapter 3. This is the condition that we actually use below.

Theorem 1.1. Let A be an open subset of a Banach space W, let I be an open interval in $\mathbb{R}$, and let F be a continuous mapping from $I \times A$ to W which is locally uniformly Lipschitz in its second variable. Then for any point $\langle t_0, \alpha_0 \rangle$ in $I \times A$, for some neighborhood U of α_0 and for any sufficiently small interval J containing t_0, there is a unique function f from J to U which is a solution of the differential equation passing through the point $\langle t_0, \alpha_0 \rangle$.

Proof. If f is a solution on J through $\langle t_0, \alpha_0 \rangle$, then an integration gives

$$f(t) - f(t_0) = \int_{t_0}^{t} F(s, f(s))\, ds,$$

so that

$$f(t) = \alpha_0 + \int_{t_0}^{t} F(s, f(s))\, ds$$

for $t \in J$. Conversely, if f satisfies this "integral equation", then the fundamental theorem of the calculus implies that $f'(t)$ exists and equals $F(t, f(t))$ on J, so that f is a solution of the differential equation which clearly goes through $\langle t_0, \alpha_0 \rangle$. Now for *any* continuous $f: J \to A$ we can define $g: J \to W$ by

$$g(t) = \alpha_0 + \int_{t_0}^{t} F(s, f(s))\, ds,$$

and our argument above shows that f is a solution of the differential equation if and only if f is a fixed point of the mapping $K: f \mapsto g$. This suggests that we try to show that K is a contraction, so that we can apply the fixed-point theorem.

We start by choosing a neighborhood $L \times U$ of $\langle t_0, \alpha_0 \rangle$ on which $F(t, \alpha)$ is bounded and Lipschitz in α uniformly over t. Let J be some open subinterval of L containing t_0, and let V be the Banach space $\mathcal{BC}(J, W)$ of bounded continuous functions from J to W. Our later calculation will show how small we have to take J. We assume that the neighborhood U is a ball about α_0 of radius r, and we consider the ball of functions $\mathcal{U} = B_r(\bar{\alpha}_0)$ in V, where $\bar{\alpha}_0$ is the constant function with value α_0. Then any f in $\mathcal{U}$ has its range in U, so that $F(t, f(t))$ is defined, bounded, and continuous. That is, K as defined earlier maps the ball $\mathcal{U}$ into V.

We now calculate. Let F be bounded by m on $L \times U$ and let δ be the length of J. Then

$$\|K(\bar{\alpha}_0) - \bar{\alpha}_0\|_\infty = \mathrm{lub}\left\{\left\|\int_{t_0}^{t} F(s, \alpha_0)\, ds\right\| : t \in J\right\} \leq \delta m \tag{1}$$

by the norm inequality for integrals (see Section 10 of Chapter 4). Also, if f_1 and f_2 are in $\mathfrak{U}$, and if c is a Lipschitz constant for F on $L \times U$, then

$$\begin{aligned}\|K(f_1) - K(f_2)\|_\infty &= \mathrm{lub}\left\{\left\|\int_{t_0}^{t} F(s, f_1(s)) - F(s, f_2(s))\right\|\right\} \\ &\leq \delta\, \mathrm{lub}\,\{\|F(s, f_1(s)) - F(s, f_2(s))\|\} \\ &\leq \delta c\, \mathrm{lub}\,\{\|f_1(s) - f_2(s)\|\} \\ &= \delta c\|f_1 - f_2\|_\infty. \end{aligned} \tag{2}$$

From (2) we see that K is a contraction with constant $C = \delta c$ if $\delta c < 1$, and from (1) we see that K moves the center $\bar{\alpha}_0$ of the ball $\mathfrak{U}$ a distance less than $(1 - C)r$ if $\delta m < (1 - \delta c)r$. This double requirement on δ is equivalent to

$$\delta < \frac{r}{m + cr},$$

and with *any* such δ the theorem follows from a corollary of the fixed-point theorem (Corollary 2 of Theorem 9.1, Chapter 4). □

Corollary. The theorem holds if $F: I \times A \to W$ is continuous and has a continuous second partial differential.

We next show that *any* two solutions through $\langle t_0, \alpha_0 \rangle$ must agree on the intersection of their domains (under the hypotheses of Theorem 1.1).

Lemma 1.1. Let g_1 and g_2 be any two solutions of $d\alpha/dt = F(t, \alpha)$ through $\langle t_0, \alpha_0 \rangle$. Then $g_1(t) = g_2(t)$ for all t in the intersection $J = J_1 \cap J_2$ of their domains.

Proof. Otherwise there is a point s in J such that $g_1(s) \neq g_2(s)$. Suppose that $s > t_0$, and set $C = \{t : t > t_0 \text{ and } g_1(t) \neq g_2(t)\}$ and $x = \mathrm{glb}\, C$. The set C is open, since g_1 and g_2 are continuous, and therefore x is not in C. That is, $g_1(x) = g_2(x)$. Call this common value α and apply the theorem to $\langle x, \alpha \rangle$. With r such that $B_r(\alpha) \subset A$, we choose δ small enough so that the differential equation has a unique solution g from $(x - \delta, x + \delta)$ to $B_r(\alpha)$ passing through $\langle x, \alpha \rangle$, and we also take δ small enough so that the restrictions of g_1 and g_2 to $(x - \delta, x + \delta)$ have ranges in $B_r(\alpha)$. But then $g_1 = g_2 = g$ on this interval by the uniqueness of g, and this contradicts the definition of x. Therefore, $g_1 = g_2$ on the intersection of their domains. □

This lemma allows us to remove the restriction on the range of f in the theorem.

Theorem 1.2. Let A, I, and F be as in Theorem 1.1. Then for any point $\langle t_0, \alpha_0 \rangle$ in $I \times A$ and any sufficiently small interval neighborhood J of t_0, there is a unique solution from J to A passing through $\langle t_0, \alpha_0 \rangle$.

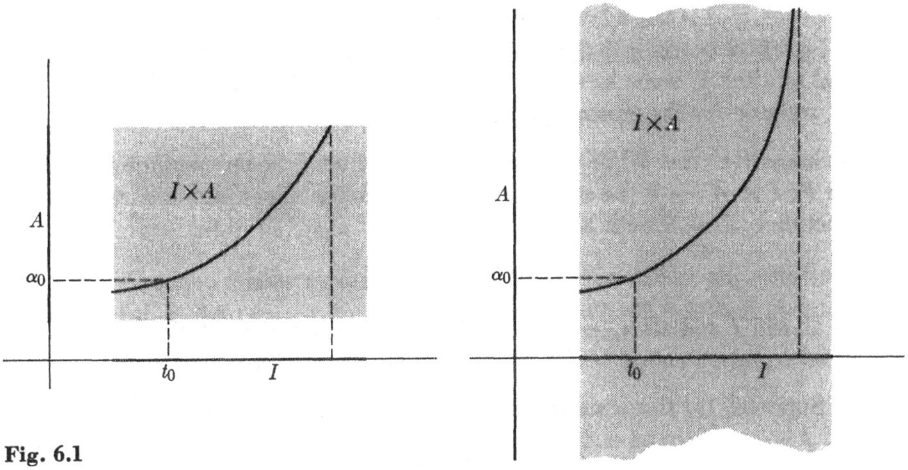

Fig. 6.1

Global solutions. The solutions we have found for the differential equation $d\alpha/dt = F(t, \alpha)$ are defined only in sufficiently small neighborhoods of the initial point t_0 and are accordingly called local solutions. Now if we run along to a point $\langle t_1, \alpha_1 \rangle$ near the end of such a local solution and then consider the local solution about $\langle t_1, \alpha_1 \rangle$, first of all it will have to agree with our first solution on the intersection of the two domains, and secondly it will in general extend farther beyond t_1 than the first solution, so the two local solutions will fit together to make a solution on a larger t-interval than either gives separately. We can continue in this way to extend our original solution to what might be called a global solution, made up of a patchwork of matching local solutions. These notions are somewhat vague as described above, and we now turn to a more precise construction of a global solution.

Given $\langle t_0, \alpha_0 \rangle \subset I \times A$, let $\mathfrak{F}$ be the family of all solutions through $\langle t_0, \alpha_0 \rangle$. Thus $g \in \mathfrak{F}$ if and only if g is a solution on an interval $J \subset I$, $t_0 \in J$, and $g(t_0) = \alpha_0$. Lemma 1.1 shows exactly that the union† f of all the functions g in $\mathfrak{F}$ is itself a function, for if $\langle t_1, \alpha_1 \rangle \in g_1$ and $\langle t_1, \alpha_2 \rangle \in g_2$, then $\alpha_1 = g_1(t) = g_2(t) = \alpha_2$.

Moreover, f is a solution, because around any x in its domain f agrees with some $g \in \mathfrak{F}$. By the way f was defined we see that f is the unique maximum solution through $\langle t_0, \alpha_0 \rangle$. We have thus proved the following theorem.

Theorem 1.3. Let $F: I \times A \to V$ be a function satisfying the hypotheses of Theorem 1.1. Then through each $\langle t_0, \alpha_0 \rangle$ in $I \times A$ there is a uniquely determined maximal solution to the differential equation $d\alpha/dt = F(t, \alpha)$.

In general, we would have to expect a maximal solution to "run into the boundary of A" and therefore to have a domain interval J properly included in I, as Fig. 6.1 suggests.

---

† Remember that we are taking a function to be a set of ordered pairs, so that the union of a family of functions makes precise sense.

However, if A is the whole space W, and if $F(t, \alpha)$ is Lipschitz in α for each t, with a Lipschitz bound $c(t)$ that is continuous in t, then we can show that each maximal solution is over the whole of I. We shall shortly see that this condition is a natural one for the linear equation.

Theorem 1.4. Let W be a Banach space, and let I be an open interval in $\mathbb{R}$. Let $F: I \times W \to W$ be continuous, and suppose that there is a continuous function $c: I \to \mathbb{R}$ such that

$$\|F(t, \alpha_1) - F(t, \alpha_2)\| \leq c(t)\|\alpha_1 - \alpha_2\|$$

for all t in I and all α_1, α_2 in W. Then each maximal solution to the differential equation $d\alpha/dt = F(t, \alpha)$ has the whole of I for its domain.

Proof. Suppose, on the contrary, that g is a maximal solution whose domain interval J has right-hand endpoint b less than that of I. We choose a finite open interval L containing b and such that $\overline{L} \subset I$ (see Fig. 6.2). Since $\overline{L}$ is compact, the continuous function $c(t)$ has a maximum value c on $\overline{L}$. We choose any t_1 in $L \cap J$ close enough to b so that $b - t_1 < 1/c$, and we set $\alpha_1 = g(t_1)$ and $m = \max \|F(t, \alpha_1)\|$ on $\overline{L}$. With these values of c and m, and with any r, the proof of Theorem 1.1 gives us a local solution f through $\langle t_1, \alpha_1 \rangle$ with domain $(t_1 - \delta, t_1 + \delta)$ for any δ less than $r/(m + rc) = 1/(c + (m/r))$. Since we now have no restriction on r (because $A = W$), this bound on δ becomes $1/c$, and since we chose t_1 so that $t_1 + (1/c) > b$, we can now choose δ so that $t_1 + \delta > b$. But this gives us a contradiction; the maximal solution g through $\langle t_1, \alpha_1 \rangle$ includes the local solution f, so that, in particular, $t_1 + \delta \leq b$. We have thus proved the theorem. □

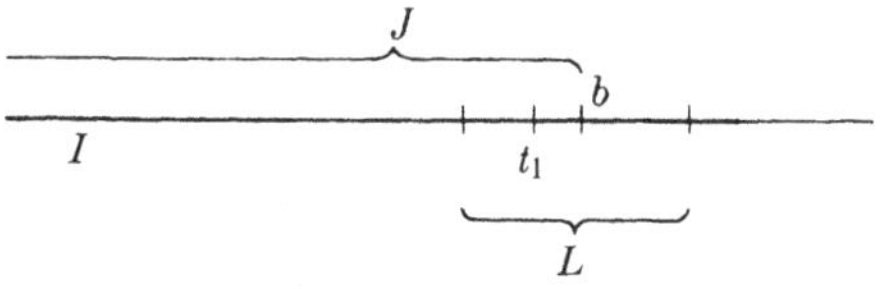

Fig. 6.2

Going back to our original situation, we can conclude that if the Lipschitz control of F is of the stronger type assumed above, and if the domain J of some maximal solution g is less than I, then the open set A cannot be the whole of W. It is in fact true that the distance from $g(t)$ to the boundary of A approaches zero as t approaches an endpoint b of J which is interior to I. That is, it is now a theorem that $\rho(f(t), A') \to 0$ as $t \to b$. The proof is more complicated than our argument above, and we leave it as a set of exercises for the interested reader.

The *n*th-order equation. Let $A_1, A_2, \ldots, A_n$ be open subsets of a Banach space W, let I be an open interval in $\mathbb{R}$, and let $G: I \times A_1 \times A_2 \times \cdots \times A_n \to W$ be continuous. We consider the differential equation

$$d^n\alpha/dt^n = G(t, \alpha, d\alpha/dt, \ldots, d^{n-1}\alpha/dt^{n-1}).$$

A function $f: J \to W$ is a solution to this equation if J is an open subinterval of I, f has continuous derivatives on J up to the nth order, $f^{(i-1)}[J] \subset A_i$, $i = 1, \ldots, n$, and

$$f^{(n)}(t) = G(t, f(t), f'(t), \ldots, f^{(n-1)}(t))$$

for $t \in J$. An initial condition is now given by a point

$$\langle t_0, \beta_1, \beta_2, \ldots, \beta_n \rangle \in I \times A_1 \times \cdots \times A_n.$$

The basic theorem is almost the same as before. To simplify our notation, let $\boldsymbol{\alpha}$ be the n-tuple $\langle \alpha_1, \alpha_2, \ldots, \alpha_n \rangle$ in $W^n = V$, and set $\mathbf{A} = \prod_1^n A_i$. Also let ψ be the mapping $f \mapsto \langle f, f', \ldots, f^{(n-1)} \rangle$. Then the solution equation becomes $f^{(n)}(t) = G(t, \psi f(t))$.

Theorem 1.5. Let $G: I \times \mathbf{A} \to W$ be as above and suppose, in addition, that $G(t, \boldsymbol{\alpha})$ is locally uniformly Lipschitz in $\boldsymbol{\alpha}$. Then for any $\langle t_0, \boldsymbol{\beta} \rangle$ in $I \times \mathbf{A}$ and for any sufficiently small open interval J containing t_0, there is a unique function f from J to W such that f is a solution to the above nth-order equation satisfying the initial condition $\psi f(t_0) = \boldsymbol{\beta}$.

Proof. There is an ancient and standard device for reducing a single nth-order equation to a system of first-order equations. The idea is to replace the single equation

$$d^n\alpha/dt^n = G(t, \alpha, d\alpha/dt, \ldots, d^{n-1}\alpha/dt^{n-1})$$

by the system of equations

$$\begin{aligned} d\alpha_1/dt &= \alpha_2, \\ d\alpha_2/dt &= \alpha_3, \\ &\vdots \\ d\alpha_{n-1}/dt &= \alpha_n, \\ d\alpha_n/dt &= G(t, \alpha_1, \alpha_2, \ldots, \alpha_n), \end{aligned}$$

and then to recognize this system as equivalent to a single first-order equation on a different space. In fact, if we define the mapping $F = \langle F^1, \ldots, F^n \rangle$ from $I \times \mathbf{A}$ to $V = W^n$ by setting $F^i(t, \boldsymbol{\alpha}) = \alpha_{i+1}$ for $i = 1, \ldots, n-1$, and $F^n(t, \boldsymbol{\alpha}) = G(t, \boldsymbol{\alpha})$, then the above system becomes the single equation

$$d\boldsymbol{\alpha}/dt = F(t, \boldsymbol{\alpha}),$$

where F is clearly locally uniformly Lipschitz in $\boldsymbol{\alpha}$. Now a function $\mathbf{f} = \langle f_1, \ldots, f_n \rangle$ from J to V is a solution of this equation if and only if

$$\begin{aligned} f_1' &= f_2, \\ f_2' &= f_3, \\ &\vdots \\ f_{n-1}' &= f_n, \\ f_n' &= G(t, f_1, \ldots, f_n), \end{aligned}$$

that is, if and only if f_1 has derivatives up to order n, $\psi(f_1) = \mathbf{f}$ and $f_1^{(n)}(t) = G(t, \psi f_1(t))$. The n-tuplet initial condition $\psi f(t_0) = \boldsymbol{\beta}$ is now just $\mathbf{f}(t_0) = \boldsymbol{\beta}$. Thus the nth-order theorem for G has turned into the first-order theorem for F, and so follows from Theorems 1.1 and 1.2 □

The local solution through $\langle t_0, \boldsymbol{\beta} \rangle$ extends to a unique maximal solution by Theorem 1.3 applied to our first-order problem $d\boldsymbol{\alpha}/dt = F(t, \boldsymbol{\alpha})$, and the domain of the maximal solution is the whole of I if $G(t, \boldsymbol{\alpha})$ is Lipschitz in $\boldsymbol{\alpha}$ with a bound $c(t)$ that is continuous and if $\mathbf{A} = W^n$, as in Theorem 1.4.

EXERCISES

1.1 Consider the equation $d\alpha/dt = F(t, \alpha)$ in the special case where $W = \mathbb{R}^2$. Write out the equation as a pair of equations involving real-valued functions and real variables.

1.2 Consider the system of differential equations

$$dx/dt = t + x^2 + y^3, \qquad dy/dt = \cos xy.$$

Define the function $F: \mathbb{R}^3 \to \mathbb{R}^2$ so that the above system becomes

$$d\alpha/dt = F(t, \alpha),$$

where $\alpha = \langle x, y \rangle$.

1.3 In the above exercise show that F is uniformly Lipschitz in α on $\mathbb{R} \times A$, where A is any bounded open set in $\mathbb{R}^2$. Is F uniformly Lipschitz on $\mathbb{R} \times \mathbb{R}^2$?

1.4 Write out the above system in terms of a solution function $f = \langle f_1, f_2 \rangle$. Write out for this system the integrated form used in proving Theorem 1.1.

1.5 The fixed-point theorem iteration sequence that we used in proving Theorem 1.1 starts off with f_0 as the constant function $\bar{\alpha}_0$ and then proceeds by

$$f_n(t) = \alpha_0 + \int_0^t F(s, f_{n-1}(s))\, ds.$$

Compute this sequence as far as f_4 for the differential equation

$$dx/dt = t + x \qquad [f'(t) = t + f(t)]$$

with the initial condition $f(0) = 0$. That is, take $f_0 = 0$ and compute f_1, f_2, f_3, and f_4 from the formula. Now guess the solution f and verify it.

1.6 Compute the iterates f_0, f_1, f_2, and f_3 for the initial-value problem

$$dy/dx = x + y^2, \qquad y(0) = 0.$$

Supposing that the solution f has a power series expansion about 0, what are its first three nonzero terms?

1.7 Make the computation in the above exercise for the initial condition $f(0) = -1$.

1.8 Do the same for $f(0) = +1$.

1.9 Suppose that W is a Banach space and that F and G are functions from $\mathbb{R} \times W^4$ to W satisfying suitable Lipschitz conditions. Show how the second-order system

$$\eta'' = F(t, \xi, \eta, \xi', \eta'), \qquad \xi'' = G(t, \xi, \eta, \xi', \eta')$$

would be brought under our standard theory by making it into a single second-order equation.

1.10 Answer the above exercise by converting it to a first-order system and then to a single first-order equation.

1.11 Let θ be a nonnegative, continuous, real-valued function defined on an interval $[0, a] \subset \mathbb{R}$, and suppose that there are constants b and $c > 0$ such that

$$\theta(x) \leq c\int_0^x \theta(t)\, dt + bx \qquad \text{for all} \quad x \in [0, a].$$

a) Prove by induction that if $m = \|\theta\|_\infty$, then

$$\theta(x) \leq \frac{m(cx)^n}{n!} + \frac{b}{c}\sum_{j=1}^{n} \frac{(cx)^j}{j!} \qquad \text{for every} \quad n.$$

b) Then prove that

$$\theta(x) \leq \frac{b}{c}(e^{cx} - 1) \qquad \text{for all} \quad x.$$

1.12 Let W be a Banach space, let I be an interval in $\mathbb{R}$, and let F be a continuous mapping from $I \times W$ to W. Suppose that $\|F(t, \alpha_0)\| \leq b$ for all $t \in I$ and that

$$\|F(t, \alpha) - F(t, \beta)\| \leq c\|\alpha - \beta\|$$

for all t in I and all α, β in W. Let f be the global solution through $\langle t_0, \alpha_0 \rangle$, and set $\theta(x) = \|f(t_0 + x) - \alpha_0\|$. Prove that

$$\theta(x) \leq \int_0^x \theta(t)\, dt + bx$$

for $x > 0$ and $t_0 + x$ in I. Then use the result in the above exercise to derive a much stronger bound than we have in the text on the growth of the solution $f(t)$ as t goes away from t_0.

1.13 With the hypotheses on F as in the above exercise, show that the iteration sequence for the solution through $\langle t_0, \alpha_0 \rangle$ converges on the *whole* of I by showing inductively that if $f_0 = \bar{\alpha}_0$ and

$$f_n(t) = \alpha_0 + \int_0^t F\big((s), f_{n-1}(s)\big)\, ds,$$

then

$$|f_n(t) - f_{n-1}(t)| \leq \frac{b}{c}\frac{(ct)^n}{n!}.$$

From these inequalities prove directly that the solution f through $\langle t_0, \alpha_0 \rangle$ satisfies

$$\|f(t) - \alpha_0\| \leq \frac{b}{c}(e^{c|t-t_0|} - 1).$$

2. DIFFERENTIABLE DEPENDENCE ON PARAMETERS

It is exceedingly important in some applications to know how the solution to the system

$$f'(t) = G(t, f(t)), \qquad f(t_1) = \alpha_1$$

varies with the initial point $\langle t_1, \alpha_1 \rangle$. In order to state the problem precisely, we fix an open interval J, set $\mathcal{U} = B_r(\bar{\alpha}_0) \subset V = \mathcal{BC}(J, W)$ as in the previous section, and require a solution in $\mathcal{U}$ passing through $\langle t_1, \alpha_1 \rangle$, where $\langle t_1, \alpha_1 \rangle$ is near $\langle t_0, \alpha_0 \rangle$. Supposing that a unique solution f exists, we then have a mapping $\langle t_1, \alpha_1 \rangle \mapsto f$, and it is the continuity and differentiability of this map that we wish to study.

Theorem 2.1. Let $L \times U$ be a neighborhood of $\langle t_0, \alpha_0 \rangle$ in the Banach space $\mathbb{R} \times W$, and let $F(t, \alpha)$ be a bounded continuous mapping from $L \times U$ to W which is Lipschitz in α uniformly over t. Then there is a neighborhood $J \times N$ of $\langle t_0, \alpha_0 \rangle$ with the following property. For any $\langle t_1, \alpha_1 \rangle$ in $J \times N$ there is a unique function f from J to U which is a solution of the differential equation $d\alpha/dt = F(t, \alpha)$ passing through $\langle t_1, \alpha_1 \rangle$, and the mapping $\langle t_1, \alpha_1 \rangle \mapsto f$ from $J \times N$ to V is continuous.

Proof. We simply reexamine the calculation of Theorem 1.1 and take δ a little smaller. Let $K(t_1, \alpha_1, f)$ be the mapping of that theorem but with initial point $\langle t_1, \alpha_1 \rangle$, so that $g = K(t_1, \alpha_1, f)$ if and only if $g(t) = \alpha_1 + \int_{t_1}^{t} F(s, f(s))\, ds$ for all t in J. Clearly K is continuous in $\langle t_1, \alpha_1 \rangle$ for each fixed f.

If N is the ball $B_{r/2}(\alpha_0)$, then the inequality (1) in the proof shows that $\|K(t_1, \alpha_1, \bar{\alpha}_0) - \bar{\alpha}_0\| \leq \|\alpha_1 - \alpha_0\| + \delta m \leq r/2 + \delta m$. The second inequality remains unchanged. Therefore, $f \mapsto K(t_1, \alpha_1, f)$ is a map from $\mathcal{U}$ to V which is a contraction with constant $C = \delta c$ if $\delta c < 1$, and which moves the center $\bar{\alpha}_0$ of $\mathcal{U}$ a distance less than $(1 - C)r$ if $r/2 + \delta m < (1 - \delta c)r$. This new double requirement on δ is equivalent to

$$\delta < \frac{r}{2(m + cr)},$$

which is just half the old value. With J of length δ, we can now apply Theorem 9.2 of Chapter 4 to the map $K(t_1, \alpha_1, f)$ from $(J \times N) \times \mathcal{U}$ to V, and so have our theorem. □

If we want the map $\langle t_1, \alpha_1 \rangle \mapsto f$ to be differentiable, it is sufficient, by Theorem 9.4 of Chapter 4, to know in addition to the above that

$$K\colon (J \times N) \times \mathcal{U} \to V$$

is continuously differentiable. And to deduce this, it is sufficient to suppose that dF exists and is *uniformly* continuous on $L \times U$.

Theorem 2.2. Let $L \times U$ be a neighborhood of $\langle t_0, \alpha_0 \rangle$ in the Banach space $\mathbb{R} \times W$, and let $F(t, \alpha)$ be a bounded mapping from $L \times U$ to W such that dF exists, is bounded, and is uniformly continuous on $L \times U$. Then, in the context of the above theorem, the solution f is a continuously differentiable function of the initial value $\langle t_1, \alpha_1 \rangle$.

Proof. We have to show that the map $K(t_1, \alpha_1, f)$ from $(J \times N) \times \mathfrak{U}$ to V is continuously differentiable, after which we can apply Theorem 9.4 of Chapter 4, as we remarked above. Now the mapping $h \mapsto k$ defined by $k(t) = \int_{t_1}^t h(s)\, ds$ is a bounded linear mapping from V to V which clearly depends continuously on t_1, and by Theorem 14.3 of Chapter 3 the integrand map $f \mapsto h$ defined by $h(s) = F(s, f(s))$ is continuously differentiable on $\mathfrak{U}$. Composing these two maps we see that $dK^3_{\langle t_1, \alpha_1, f \rangle}$ exists and is continuous on $J \times N \times \mathfrak{U}$. Now

$$\Delta K^2_{\langle t_1, \alpha_1, f \rangle}(\xi) = \xi,$$

so that $dK^2 = I$, and $\Delta K^1_{\langle t_1, \alpha_1, f \rangle}(h) = -\int_{t_1}^{t_1+h} F(s, f(s))\, ds$, from which it follows easily that $dK^1_{\langle t_1, \alpha_1, f \rangle}(h) = -hF(t_1, f(t))$. The three partial differentials dK^1, dK^2, and dK^3 thus exist and are continuous on $J \times N \times \mathfrak{U}$, and it follows from Theorem 8.3 of Chapter 3 that $K(t_1, \alpha_1, f)$ is continuously differentiable there. □

Corollary. If s is any point in J, then the value $f(s)$ of a solution at s is a differentiable function of its value at t_0.

Proof. Let f_α be the solution through $\langle t_0, \alpha \rangle$. By the theorem, $\alpha \mapsto f_\alpha$ is a continuously differentiable map from N to the function space $V = \mathcal{BC}(J, W)$. But $\pi_s : f \mapsto f(s)$ is a bounded linear mapping and thus trivially continuously differentiable. Composing these two maps, we see that $\alpha \mapsto f_\alpha(s)$ is continuously differentiable on N. □

It is also possible to make the continuous and differentiable dependence of the solution on its initial value $\langle t_0, \alpha_0 \rangle$ into a global affair. The following is the theorem. We shall not go into its proof here.

Theorem 2.3. Let f be the maximal solution through $\langle t_0, \alpha_0 \rangle$ with domain J, and let $[a, b]$ be any finite closed subinterval of J containing t_0. Then there exists an $\epsilon > 0$ such that for every $\langle t_1, \alpha_1 \rangle \in B_\epsilon(\langle t_0, \alpha_0 \rangle)$ the domain of the global solution through $\langle t_1, \alpha_1 \rangle$ includes $[a, b]$, and the restriction of this solution to $[a, b]$ is a continuous function of $\langle t_1, \alpha_1 \rangle$. If F satisfies the hypotheses of Theorem 2.2, then this dependence is continuously differentiable.

Finally, suppose that F depends continuously (or continuously differentiably) on a parameter λ, so that we have $F(\lambda, t, \alpha)$ on $M \times I \times A$. Now the solution f to the initial-value problem

$$f'(t) = F(t, f(t)), \qquad f(t_1) = \alpha_1$$

depends on the parameter λ as well as on the initial condition $f(t_1) = \alpha_1$, and if the reader has fully understood our arguments above, he will see that we can show in the same way that the dependence of f on λ is also continuous (continuously differentiable). We shall not go into these details here.

3. THE LINEAR EQUATION

We now suppose that the function F of Section 1 is from $I \times W$ to W and continuous, and that $F(t, \alpha)$ is linear in α for each fixed t. It is not hard to see that we then automatically have the strong Lipschitz hypothesis of Theorem 1.4, which we shall in any case now assume. Here this is a boundedness condition on a linear map: we are assuming that $F(t, \alpha) = T_t(\alpha)$, where $T_t \in \text{Hom } W$, and that $\|T_t\| \leq c(t)$ for all t, where $c(t)$ is continuous on I.

As one might expect, in this situation the existence and uniqueness theory of Section 1 makes contact with general linear theory. Let X_0 be the vector space $\mathcal{C}(I, W)$ of all continuous functions from I to W, and let X_1 be its subspace $\mathcal{C}^1(I, W)$ of all functions having continuous first derivatives. Norms will play no role in our theorem.

Theorem 3.1. The mapping $S: X_1 \to X_0$ defined by setting $g = Sf$ if $g(t) = f'(t) - F(t, f(t))$ is a surjective linear mapping. The set N of global solutions of the differential equation $d\alpha/dt = F(t, \alpha)$ is the null space of S, and is therefore, in particular, a vector space. For each $t_0 \in I$ the restriction to N of the coordinate (evaluation) mapping $\pi_{t_0}: f \mapsto f(t_0)$ is an isomorphism from N to W. The null space M of π_{t_0} is therefore a complement of N in X_1, and so determines a right inverse R of S. The mapping $f \mapsto \langle Sf, f(t_0) \rangle$ is an isomorphism from X_1 to $X_0 \times W$, and this fact is equivalent to all the above assertions.

Proof. For any fixed g in X_0 we set $G(t, \alpha) = F(t, \alpha) + g(t)$ and consider the (nonlinear) equation $d\alpha/dt = G(t, \alpha)$. By Theorems 1.3 and 1.4 it has a unique maximal solution f through any initial point $\langle t_0, \alpha_0 \rangle$, and the domain of f is the whole of I. That is, for each pair $\langle g, \alpha \rangle$ in $X_0 \times W$ there is a unique f in X_1 such that $\langle Sf, f(t_0) \rangle = \langle g, \alpha \rangle$. The mapping

$$\langle S, \pi_{t_0} \rangle : f \mapsto \langle Sf, f(t_0) \rangle$$

is thus bijective, and since it is clearly linear, it is an isomorphism. In particular, S is surjective. The null space N of S is the inverse image of $\{0\} \times W$ under the above isomorphism; that is, $\pi_{t_0} \restriction N$ is an isomorphism from N to W.

Finally, the null space M of π_{t_0} is the inverse image of $X_0 \times \{0\}$ under $\langle S, \pi_{t_0} \rangle$, and the direct sum decomposition $X_1 = M \oplus N$ simply reflects the decomposition $X_0 \times W = (X_0 \times \{0\}) \oplus (\{0\} \times W)$ under the inverse isomorphism. This finishes the proof of the theorem. □

The problem of finding, for a given g in X_0 and a given α_0 in W, the unique f in X_1 such that $S(f) = g$ and $f(t_0) = \alpha_0$ is called the *initial-value problem.* At the theoretical level, the problem is solved by the above theorem, which states that the uniquely determined f exists. At the practical level of computation, the problem remains important.

The fact that $M = M_{t_0}$ is a complement of N breaks down the initial-value problem into two independent subproblems. The right inverse R associated with

M_{t_0} finds h in X_1 such that $S(h) = g$ and $h(t_0) = 0$. The inverse of the isomorphism $f \mapsto f(t_0)$ from N to W selects that k in X_1 such that $S(k) = 0$ and $k(t_0) = \alpha_0$. Then $f = h + k$. The first subproblem is the problem of "solving the inhomogeneous equation with homogeneous initial data", and the second is the problem of "solving the homogeneous equation with inhomogeneous initial data". In a certain sense the initial-value problem is the "direct sum" of these two independent problems.

We shall now study the homogeneous equation $d\alpha/dt = T_t(\alpha)$ more closely. As we saw above, its solution space N is isomorphic to W under each projection map $\pi_t = f \mapsto f(t)$. Let φ_t be this isomorphism (so that $\varphi_t = \pi_t \restriction N$). We now choose some fixed t_0 in I—we may as well suppose that I contains 0 and take $t_0 = 0$—and set $K_t = \varphi_t \circ \varphi_0^{-1}$. Then $\{K_t\}$ is a one-parameter family of linear isomorphisms of W with itself, and if we set $f_\beta(t) = K_t(\beta)$, then f_β is the solution of $d\alpha/dt = T_t(\alpha)$ passing through $\langle 0, \beta \rangle$. We call K_t a *fundamental solution* of the homogeneous equation $d\alpha/dt = T_t(\alpha)$.

Since $f'_\beta(t) = T_t(f_\beta(t))$, we see that $d(K_t)/dt = T_t \circ K_t$ in the sense that the equation is true at each β in W. However, the derivative $d(K_t)/dt$ does not necessarily exist as a norm limit in Hom W. This is because our hypotheses on T_t do not imply that the mapping $t \mapsto T_t$ is continuous from I to Hom W. If this mapping *is* continuous, then the mapping $\langle t, A \rangle \mapsto T_t \circ A$ is continuous from $I \times$ Hom W to Hom W, and the initial-value problem

$$dA/dt = T_t \circ A, \qquad A_0 = I$$

has a unique solution A_t in $\mathcal{C}^1(I, \text{Hom } W)$. Because evaluation at β is a bounded linear mapping from Hom W to W, $A_t(\beta)$ is a differentiable function of t and

$$dA_t(\beta)/dt = (dA_t/dt)(\beta) = T_t(A_t(\beta)).$$

This implies that $A_t(\beta) = K_t(\beta)$ for all β, so $K_t = A_t$. In particular, the fundamental solution $t \mapsto K_t$ is now a differentiable map into Hom W, and $dK_t/dt = T_t \circ K_t$. We have proved the following theorem.

Theorem 3.2. Let $t \mapsto T_t$ be a continuous map from an interval neighborhood I of 0 to Hom W. Then the fundamental solution $t \mapsto K_t$ of the differential equation $d\alpha/dt = T_t(\alpha)$ is the parametrized arc from I to Hom W that is the solution of the initial-value problem $dA/dt = T_t \circ A$, $A_0 = I$.

In terms of the isomorphisms $K_t = K(t)$, we can now obtain an explicit solution for the inhomogeneous equation $d\alpha/dt = T_t(\alpha) + g(t)$. We want f such that

$$f'(t) - T_t(f(t)) = g(t).$$

Now $K'(t) = T_t \circ K(t)$, so that $T_t = K'(t) \circ K(t)^{-1}$, and it follows from Exercise 8.12 of Chapter 4 and the general product rule for differentiation

(Theorem 8.4 of Chapter 3) that the left side of the equation above is exactly

$$K(t)\left(\frac{d}{dt}[K(t)^{-1}(f(t))]\right).$$

The equation we have to solve can thus be rewritten as

$$\frac{d}{dt}[K(t)^{-1}(f(t))] = K(t)^{-1}(g(t)).$$

We therefore have an obvious solution, and even if the reader has found our motivating argument too technical, he should be able to check the solution by differentiating.

Theorem 3.3. In the context of Theorem 3.2, the function

$$f(t) = K_t\left[\int_0^t K_s^{-1}(g(s))\,ds\right]$$

is the solution of the inhomogeneous initial-value problem

$$d\alpha/dt = T_t(\alpha) + g(t), \qquad f(0) = 0.$$

This therefore is a formula for the right inverse R of S determined by the complement M_0 of the null space N of S.

The special case of the *constant coefficient* equation, where the "coefficient" operator T_t is a fixed T in Hom W, is extremely important. The first new fact to be observed is that if f is a solution of $d\alpha/dt = T(\alpha)$, then so is f'. For the equation $f'(t) = T(f(t))$ has a differentiable right-hand side, and differentiating, we get $f''(t) = T(f'(t))$. That is:

Lemma 3.1. The solution space N of the constant coefficient equation $d\alpha/dt = T(\alpha)$ is invariant under the derivative operator D.

Moreover, we see from the differential equation that the operator D on N is just composition with T. More precisely, the equation $f'(t) = T(f(t))$ can be rewritten $\pi_t \circ D = T \circ \pi_t$, and since the restriction of π_t to N is the isomorphism φ_t from N to W, this equation can be solved for T. We thus have the following lemma.

Lemma 3.2. For each fixed t the isomorphism φ_t from N to W takes the derivative operator D on N to the operator T on W. That is,

$$T = \varphi_t \circ D \circ \varphi_t^{-1}.$$

The equation for the fundamental solution K_t is now $dS/dt = TS$. In the elementary calculus this is the equation for the exponential function, which leads us to expect and immediately check that $K_t = e^{tT}$. (See the end of Section 8 of Chapter 4.) The solution of $d\alpha/dt = T(\alpha)$ through $\langle 0, \beta \rangle$ is thus the function

$$e^{tT}\beta = \sum_0^\infty t^j \frac{T^j(\beta)}{j!}.$$

If T satisfies a polynomial equation $p(T) = 0$, as we know it must if W is finite-dimensional, then our analysis can be carried significantly further. Suppose for now that p has only real roots, so that its relatively prime factorization is $p(x) = \prod_1^k (x - \lambda_i)^{m_i}$. Then we know from Theorem 5.5 of Chapter 1 that W is the direct sum $W = \bigoplus_1^k W_i$ of the null spaces W_i of the transformations $(T - \lambda_i)^{m_i}$, and that each W_i is invariant under T. This gives us a much simpler form for the solution curve $e^{tT}\alpha$ if the point α is in one of the null spaces W_i. Taking such a subspace W_i itself as W for the moment, we have $(T - \lambda I)^m = 0$, so that $T = \lambda I + R$, where $R^m = 0$, and the factorization $e^{tT} = e^{t\lambda}e^{tR}$, together with the now finite series expansion of e^{tR}, gives us

$$e^{tT}\alpha = e^{t\lambda}\left[\alpha + tR(\alpha) + \cdots + t^{m-1}\frac{R^{m-1}(\alpha)}{(m-1)!}\right].$$

Note that the number of terms on the right is the degree of the factor $(t - \lambda)^m$ in the polynomial $p(t)$.

In the general situation where $W = \bigoplus_1^k W_i$, we have $\alpha = \sum_1^k a_i$, $e^{tT}(\alpha) = \sum_1^k e^{tT}(\alpha_i)$, and each $e^{tT}(\alpha_i)$ of the above form. The solution of $f'(t) = T(f(t))$ through the general point $\langle 0, \alpha \rangle$ is thus a finite sum of terms of the form $t^j e^{t\lambda_i}\beta_{ij}$, the number of terms being the degree of the polynomial p.

If W is a complex Banach space, then the restriction that p have only real roots is superfluous. We get exactly the same formula but with complex values of λ. This introduces more variety into the behavior of the solution curves since an outside exponential factor $e^{t\lambda} = e^{t\mu}e^{it\nu}$ now has a periodic factor if $\nu \neq 0$.

Altogether we have proved the following theorem.

Theorem 3.4. If W is a real or complex Banach space and $T \in \operatorname{Hom} W$, then the solution curve in W of the initial-value problem $f'(t) = T(f(t))$, $f(0) = \beta$, is

$$f(t) = e^{tT}\beta = \sum_0^\infty \frac{t^j}{j!} T^j(\beta).$$

If T satisfies a polynomial equation $(T - \lambda)^m = 0$, then

$$f(t) = e^{t\lambda}\left[\beta + tR(\beta) + \cdots + \frac{t^{m-1}}{(m-1)!} R^{m-1}(\beta)\right],$$

where $R = T - \lambda I$. If T satisfies a polynomial equation $p(T) = 0$ and p has the relatively prime factorization $p(x) = \prod_1^k (x - \lambda_i)^{m_i}$, then $f(t)$ is a sum of k terms of the above type, and so has the form

$$f(t) = \sum_{i,j} t^j e^{t\lambda_i}\beta_{ij},$$

where the number of terms on the right is the degree of the polynomial p, and each β_{ij} is a fixed (constant) vector.

It is important to notice how the asymptotic behavior of $f(t)$ as $t \to +\infty$ is controlled by the polynomial roots λ_i. We first restrict ourselves to the solution through a vector α in one of the subspaces W_i, which amounts to supposing that $(T - \lambda)^m = 0$. Then if λ has a positive real part, so that $e^{t\lambda} = e^{t\mu}e^{it\nu}$ with $\mu > 0$, then $\|f(t)\| \to \infty$ in exponential fashion. If λ has a negative real part, then $f(t)$ approaches zero as $t \to \infty$ (but its norm becomes infinite exponentially fast as $t \to -\infty$). If the real part of λ is zero, then $\|f(t)\| \to \infty$ like t^{m-1} if $m > 1$. Thus the only way for f to be bounded on the whole of $\mathbb{R}$ is for the real part of λ to be zero and $m = 1$, in which case f is periodic. Similarly, in the general case where $p(T) = \prod_1^k (T - \lambda_n)^{m_n} = 0$, it will be true that all the solution curves are bounded on the whole of $\mathbb{R}$ if and only if the roots λ_n are all pure imaginary and all the multiplicities m_n are 1.

EXERCISES

3.1 Let I be an open interval in $\mathbb{R}$, and let W be a normed linear space. Let $F(t, \alpha)$ be a continuous function from $I \times W$ to W which is linear in α for each fixed t. Prove that there is a function $c(t)$ which is bounded on every closed interval $[a, b]$ included in I and such that $\|F(t, \alpha)\| \leq c(t)\|\alpha\|$ for all α and t. Then show that c can be made continuous. (You may want to use the Heine-Borel property: If $[a, b]$ is covered by a collection of open intervals, then some finite subcollection already covers $[a, b]$.)

3.2 In the text we omitted checking that $f \mapsto f^{(n)} - G(t, f, f', \ldots, f^{(n-1)})$ is surjective from X_n to X_0. Prove that this is so by tracking down the surjectivity through the reduction to a first-order system.

3.3 Suppose that the coefficients $a_i(t)$ in the operator

$$Tf = \sum_0^n a_i f^{(i)}$$

are all themselves in $\mathcal{C}^1$. Show that the null space N of T is a subspace of $\mathcal{C}^{n+1}$. State a generalization of this theorem and indicate roughly why it is true.

3.4 Suppose that W is a Banach space, $T \in \operatorname{Hom} W$, and β is an eigenvector of T with eigenvalue r. Show that the solution of the constant coefficient equation $d\alpha/dt = T(\alpha)$ through $\langle 0, \beta \rangle$ is $f(t) = e^{tr}\beta$.

3.5 Suppose next that W is finite-dimensional and has a basis $\{\beta_i\}_1^n$ consisting of eigenvectors of T, with corresponding eigenvalues r_i. Find a formula for the solution through $\langle 0, \alpha \rangle$ in terms of the basis expansion of α.

3.6 A very important special case of the linear equation $d\alpha/dt = T_t(\alpha)$ is when the operator function T_t is periodic. Suppose, for example, that $T_{t+1} = T_t$ for all t. Show that then $K_{t+n} = K_t(K_1)^n$ for all t and n.

Assume next that K_1 has a logarithm, and so can be written $K_1 = e^A$ for some A in Hom W. (We know from Exercise 11.19 of Chapter 4 that this is always possible if W is finite-dimensional.) Show that now K_t can be written in the form

$$K_t = B(t)e^{tA},$$

where $B(t)$ is periodic with period 1.

3.7 Continuing the above exercise, suppose now that W is a finite-dimensional complex vector space. Using the analysis of $e^{tA}\beta$ given in the text, show that the differential equation $d\alpha/dt = T_t(\alpha)$ has a periodic solution (with any period) only if K_1 has an eigenvalue of absolute value 1. Show also that if K_1 has an nth root of unity as an eigenvalue, then the differential equation has a periodic solution with period n.

3.8 Write out the special form that the formula of Theorem 3.3 takes in the constant coefficient situation.

3.9 It is interesting to look at the facts of Theorem 3.1 from the point of view of Theorem 5.3 of Chapter 1. Assume that $S: X_1 \to X_0$ is surjective and that its null space N is isomorphic to W under the coordinate (evaluation) map π_{t_0}. Prove that if M is the nullspace of π_{t_0} in X_1, then $S \restriction M$ is an isomorphism onto X_0 by applying this theorem.

4. THE nTH-ORDER LINEAR EQUATION

The nth-order linear differential equation is the equation

$$d^n\alpha/dt^n = G(t, \alpha, d\alpha/dt, \ldots, d^{n-1}\alpha/dt^{n-1}),$$

where $G(t, \boldsymbol{\alpha}) = G(t, \alpha_1, \ldots, \alpha_n)$ is now linear from $V = W^n$ to W for each t in I. We convert this to a first-order equation $d\boldsymbol{\alpha}/dt = F(t, \boldsymbol{\alpha})$ just as before, where now F is a map from $I \times V$ to V that is linear in its second variable $\boldsymbol{\alpha}$, $F(t, \boldsymbol{\alpha}) = T_t(\boldsymbol{\alpha})$.

Our proof of Theorem 1.5 showed that a function f in $\mathcal{C}^{(n)}(I, W)$ is a solution of the nth-order equation $d^n\alpha/dt^n = G(t, \alpha, \ldots, d^{n-1}\alpha/dt^{n-1})$ if and only if the n-tuplet $\psi f = \langle f, f', \ldots, f^{(n-1)} \rangle$ is a solution of the first-order equation $d\boldsymbol{\alpha}/dt = F(t, \boldsymbol{\alpha}) = T_t(\boldsymbol{\alpha})$. We know that the latter solutions form a vector subspace $\mathbf{N}$ of $\mathcal{C}^1(I, W^n)$, and since the map $\psi: f \mapsto \langle f, f', \ldots, f^{(n-1)} \rangle$ is linear from $\mathcal{C}^n(I, W)$ to $\mathcal{C}^1(I, W^n)$, it follows that the set N of solutions of the nth-order equation is a subspace of $\mathcal{C}^n(I, W)$ and $\psi \restriction N$ is an isomorphism from N to $\mathbf{N}$. Since the coordinate evaluation $\varphi_t = \pi_t \restriction \mathbf{N}$ is an isomorphism from $\mathbf{N}$ to W^n for each t (Theorem 3.1), it follows that the map

$$\pi_t \circ \psi: f \mapsto \langle f(t), f'(t), \ldots, f^{(n-1)}(t) \rangle$$

takes N isomorphically to W^n. Its null space M_t is a complement of N in $\mathcal{C}^n$, as before. Here M_t is the set of functions f in $\mathcal{C}^n(I, W)$ such that $f(t) = \cdots = f^{(n-1)}(t) = 0$.

We now consider the special case $W = \mathbb{R}$. For each fixed t, G is now a linear map from $\mathbb{R}^n$ to $\mathbb{R}$, that is, an element of $(\mathbb{R}^n)^*$, and its coordinate set with respect to the standard basis is an n-tuple $\mathbf{k} = \langle k_1, \ldots, k_n \rangle$. Since the linear map varies continuously with t, the n-tuple $\mathbf{k}$ varies continuously with t. Thus, when we take t into account, we have an n-tuple $k(t) = \langle k_1(t), \ldots, k_n(t) \rangle$ of continuous real-valued functions on I such that

$$G(t, x_1, \ldots, x_n) = \sum_{i=1}^{n} k_i(t)x_i.$$

The solution space N of the nth-order differential equation

$$d^n\alpha/dt^n = G(\alpha, \ldots, d^{n-1}\alpha/dt^{n-1}, t)$$

is just the null space of the linear transformation $L: \mathcal{C}^n(I, \mathbb{R}) \to \mathcal{C}^0(I, \mathbb{R})$ defined by

$$(Lf)(t) = f^{(n)}(t) - k_n(t)f^{(n-1)}(t) - \cdots - k_1(t)f(t).$$

If we shift indices to coincide with the order of the derivative, and if we let $f^{(n)}$ also have a coefficient function, then our nth-order linear differential operator L appears as

$$(Lf)(t) = a_n(t)f^{(n)}(t) + \cdots + a_0(t)f(t).$$

Giving $f^{(n)}$ a coefficient function a_n changes nothing provided $a_n(t)$ is never zero, since then it can be divided out to give the form we have studied. This is called the *regular* case. The *singular* case, where $a_n(t)$ is zero for some t, requires further study, and we shall not go into it here.

We recapitulate what our general linear theory tells us about this situation.

Theorem 4.1. L is a surjective linear transformation from the space $\mathcal{C}^n(I)$ of all real-valued functions on I having continuous derivatives through order n to the space $\mathcal{C}^0(I) = \mathcal{C}(I)$ of continuous functions on I. Its null space N is the solution space of our original differential equation. For each t_0 in I the restriction to N of the mapping $\varphi_{t_0} \circ \psi: f \mapsto \langle f(t_0), \ldots, f^{(n-1)}(t_0)\rangle$ is an isomorphism from N to $\mathbb{R}^n$, and the set M_{t_0} of functions f in $\mathcal{C}^n$ such that $f(t_0) = \cdots = f^{(n-1)}(t_0) = 0$ is therefore a complement of N in $\mathcal{C}^n(I)$, and determines a linear right inverse of L.

The practical problem of "solving" the differential equation $L(f) = g$ for f when g is given falls into two parts. First we have to find the null space N of L, that is, we have to solve the homogeneous equation $L(f) = 0$. Since N is an n-dimensional vector space, the problem of delineating it is equivalent to finding a basis, and this is clearly the efficient way to proceed. Our first problem therefore is to find n linearly independent solutions $\{u_i\}_1^n$ of $L(f) = 0$. Our second problem is to find a right inverse of L, that is, a linear way of picking *one* f such that $L(f) = g$ for each g. Here the obvious thing to do is to try to make the formula of Theorem 3.3 into a practical computation. If v is one solution of $L(f) = g$, then of course the set of all solutions is the affine subspace $N + v$.

We shall start with the first problem, that of finding a basis $\{u_i\}_1^n$ of solutions to $L(f) = 0$. Unfortunately, there is no general method available, and we have to be content with partial success. We shall see that we can easily solve the first-order equation directly, and that if we can find one solution of the nth-order equation, then we can reduce the problem to solving an equation of order $n - 1$. Moreover, in the very important special case of an operator L with constant coefficients, Theorem 3.4 gives a complete explicit solution.

The first-order homogeneous linear equation can be written in the form $y' + a(t)y = 0$, where the coefficient of y' has been divided out. Dividing by y

and remembering that $y'/y = (\log y)'$, we see that, formally at least, a solution is given by $\log y = -\int a(t)\,dt$ or $y = e^{-\int a(t)\,dt}$, and we can check it by inspection. Thus the equation $y' + y/t = 0$ has a solution $y = e^{-\log t} = 1/t$, as the reader might have noticed directly.

Suppose now that L is an nth-order operator and that we know one solution u of $Lf = 0$. Our problem then is to find $n - 1$ solutions $v_1, \ldots, v_{n-1}$ independent of each other and of u. It might even be reasonable to guess that these could be determined as solutions of an equation of order $n - 1$. We try to find a second solution $v(t)$ in the form $c(t)u(t)$, where $c(t)$ is an unknown function. Our motivation, in part, is that such a solution would automatically be independent of u unless $c(t)$ turns out to be a constant.

Now if $v(t) = c(t)u(t)$, then $v' = cu' + c'u$, and generally

$$v^{(j)} = \sum_{i=0}^{j} \binom{j}{i} c^{(i)} u^{(j-i)}.$$

If we write down $L(v) = \sum_0^n a_j(t)v^{(j)}(t)$ and collect those terms involving $c(t)$, we get

$$\begin{aligned} L(v) &= c(t)\sum_0^n a_j u^{(j)} + \text{terms involving } c', \ldots, c^{(n)} \\ &= cL(u) + S(c') = S(c'), \end{aligned}$$

where S is a certain linear differential operator of order $n - 1$ which can be explicitly computed from the above formulas. We claim that solving $S(f) = 0$ solves our original problem. For suppose that $\{g_i\}_1^{n-1}$ is a basis for the null space of S, and set $c_i(t) = \int_0^t g_i$. Then $L(c_i u) = S(c_i') = S(g_i) = 0$ for $i = 1, \ldots, n - 1$. Moreover, $u, c_1 u, \ldots, c_{n-1}u$ are independent, for if $u = \sum_1^{n-1} k_i c_i u$, then $1 = \sum_1^{n-1} k_i c_i(t)$ and $0 = \sum_1^{n-1} k_i c_i'(t) = \sum_1^{n-1} k_i g_i(t)$, contradicting the independence of the set $\{g_i\}$.

We have thus shown that if we can find one solution u of the nth-order equation $Lf = 0$, then its complete solution is reduced to solving an equation $Sf = 0$ of order $n - 1$ (although our independence argument was a little sketchy).

This reduction procedure does not combine with the solution of the first-order equation to build up a sequence of independent solutions of the nth-order equation because, roughly speaking, it "works off the top instead of off the bottom". For the combination to be successful, we would have to be able to find from a given nth-order operator a first-order operator S such that $N(S) \subset N(L)$, and we can't do this in general. However, we can do it when the coefficient functions in L are all constants, although we shall in fact proceed differently.

Meanwhile it is valuable to note that a *second*-order equation $Lf = 0$ can be solved completely if we can find *one* solution u, since the above argument reduces the remaining problem to a first-order equation which can then be solved by an integration, as we saw earlier. Consider, for instance, the equation $y'' - 2y/t^2 = 0$ over any interval I not containing 0, so that $a_0(t) = 1/t^2$ is continuous on I. We see by inspection that $u(t) = t^2$ is one solution. Then we *know* that we can find a solution $v(t)$ independent of $u(t)$ in the form $v(t) = t^2c(t)$ and that the problem will become a first-order problem for c'. We have, in fact, $v' = t^2c' + 2tc$ and $v'' = t^2c'' + 4tc' + 2c$, so that $L(v) = v'' - 2v/t^2 = t^2c'' + 4tc'$, and $L(v) = 0$ if and only if $(c')' + (4/t)c' = 0$. Thus

$$c' = e^{-\int 4dt/t} = e^{-4\log t} = 1/t^4, \qquad c = 1/t^3$$

(to within a scalar multiple; we only want a basis!), and $v = t^2c(t) = 1/t$. (The reader may wish to check that this *is* the promised solution.) The null space of the operator $L(f) = f'' - 2f/t^2$ is thus the linear span of $\{t^2, 1/t\}$.

We now turn to an important tractable case, the differential operator

$$Lf = a_nf^{(n)} + a_{n-1}f^{(n-1)} + \cdots + a_0f,$$

where the coefficients a_i are *constants* and a_n might as well be taken to 1. What makes this case accessible is that now L is a polynomial in the derivative operator D. That is, if $Df = f'$, so that $D^jf = f^{(j)}$, then $L = p(D)$, where $p(x) = \sum_0^n a_ix^i$.

The most elegant, but not the most elementary, way to handle this equation is to go over to the equivalent first-order system $d\mathbf{x}/dt = T(\mathbf{x})$ on $\mathbb{R}^n$ and to apply the relevant theory from the last section.

Theorem 4.2. If $p(t) = (t - b)^n$, then the solution space N of the constant coefficient nth-order equation $p(D)f = 0$ has the basis

$$\{e^{bt}, te^{bt}, \ldots, t^{n-1}e^{bt}\}.$$

If $p(t)$ is a polynomial which has a relatively prime factorization $p(t) = \prod_1^k p_i(t)$ with each $p_i(t)$ of the above form, then the solution space of the constant coefficient equation $p(D)f = 0$ has the basis $\bigcup B_i$, where B_i is the above basis for the solution space N_i of $p_i(D)f = 0$.

Proof. We know that the mapping $\psi: f \mapsto \langle f, f', \ldots, f^{(n-1)}\rangle$ is an isomorphism from the null space N of $p(D)$ to the null space $\mathbf{N}$ of $d\mathbf{x}/dt - T(\mathbf{x})$. It is clear that ψ commutes with differentiation, $\psi(Df) = \langle f', \ldots, f^{(n)}\rangle = D\psi(f)$, and since we know that $\mathbf{N}$ is invariant under D by Lemma 3.1, it follows (and can easily be checked directly) that N is invariant under D. By Lemma 3.2 we have $T = \varphi_t \circ D \circ \varphi_t^{-1}$, which simply says that the isomorphism $\varphi_t: \mathbf{N} \to \mathbb{R}^n$ takes the operator D on $\mathbf{N}$ into the operator T on $\mathbb{R}^n$. Altogether $\varphi_t \circ \psi$ takes D on N into T on $\mathbb{R}^n$, and since $p(D) = 0$ on N, it follows that $p(T) = 0$ on $\mathbb{R}^n$.

We saw in Theorem 3.4 that if $p(T) = 0$ and $p = (t - b)^n$, then the solution space $\mathbf{N}$ of $d\mathbf{x}/dt = T(\mathbf{x})$ is spanned by vectors of the form

$$e^{bt}\mathbf{x}, \ldots, t^{n-1}e^{bt}\mathbf{x}.$$

The first coordinates of the n-tuple-valued functions $\mathbf{g}$ in $\mathbf{N}$ form the space N (under the isomorphism $f = \psi^{-1}\mathbf{g}$), and we therefore see that N is spanned by the functions $e^{bt}, \ldots, t^{n-1}e^{bt}$. Since N is n-dimensional, and since there are n of these functions, the spanning set forms a basis.

The remainder of the theorem can be viewed as the combination of the above and the direct application of Theorem 5.5 of Chapter 1 to the equation $p(D) = 0$ on N, or as the carry-over to N under the isomorphism ψ^{-1} of the facts already established for $\mathbf{N}$ in the last section. □

If the roots of the polynomial p are not all real, then we have to resort to the complexification theory that we developed in the exercises of Section 11, Chapter 4. Except for one final step, the results are the same. The one extra fact that has to be applied is that the null space of a real operator T acting on a real vector space Y is exactly the intersection with Y of the null space of the complexification S of T acting on the complexification $Z = Y \oplus iY$ of Y. This implies that if $p(t)$ is a polynomial with real coefficients, then we get the real solutions of $p(D)f = 0$ as the real parts of the complex solutions. In order to see exactly what this means, suppose that $q(x) = (x^2 - 2bx + c)^m$ is one of the relatively prime factors of $p(x)$ over $\mathbb{R}$, with $x^2 - 2bx + c$ irreducible over $\mathbb{R}$. Over $\mathbb{C}$, $q(x)$ factors into $(x - \lambda)^m(x - \bar{\lambda})^m$, where $\lambda = b + i\omega$ and $\omega^2 = c - b^2$. It follows from our general theory above that the complex $2m$-dimensional null space of $q(D)$ is the complex span of

$$\{e^{\lambda t}, te^{\lambda t}, \ldots, t^{m-1}e^{\lambda t}, e^{\bar{\lambda} t}, te^{\bar{\lambda} t}, \ldots, t^{m-1}e^{\bar{\lambda} t}\}.$$

The real parts of the complex linear combinations of these $2m$ functions is a $2m$-dimensional real vector space spanned by the real parts of the above functions and the real parts of i times the above functions. That is, the null space of the real operator $q(D)$ is a $2m$-dimensional real space spanned by

$$\{e^{bt}\cos \omega t, te^{bt}\cos \omega t, \ldots, t^{m-1}e^{bt}\cos \omega t;\ e^{bt}\sin \omega t, \ldots, t^{m-1}e^{bt}\sin \omega t\}.$$

Since there are $2m$ of these functions, they must be independent and must form a basis for the real solution space of $q(D)f = 0$. Thus,

Theorem 4.3. If $p(t) = (t^2 + 2bt + c)^m$ and $b^2 < c$, then the solution space of the constant coefficient $2m$th-order equation $p(D)f = 0$ has the basis

$$\{t^i e^{bt}\cos \omega t\}_{i=0}^{m-1} \cup \{t^i e^{bt}\sin \omega t\}_{i=0}^{m-1},$$

where $\omega^2 = c - b^2$. For any polynomial $p(t)$ with real coefficients, if $p(t) = \prod_1^k p_i(t)$ is its relatively prime factorization into powers of linear factors and powers of irreducible quadratic factors, then the solution space N of $p(D)f = 0$ has the basis $\bigcup_1^k B_i$, where B_i is the basis for the null space of $p_i(D)$ that we displayed above if $p_i(t)$ is a power of an irreducible quadratic, and B_i is the basis of Theorem 4.2 if $p_i(t)$ is a power of a linear factor.

Suppose, for example, that we want to find a basis for the null space of $D^4 - 1 = 0$. Here $p(x) = x^4 - 1 = (x - 1)(x + 1)(x - i)(x + i)$. The basis for the complex solution space is therefore $\{e^t, e^{-t}, e^{it}, e^{-it}\}$. Since $e^{it} = \cos t + i \sin t$, the basis for the real solution space is $\{e^t, e^{-t}, \cos t, \sin t\}$.

The same problem for $D^3 - 1 = 0$ gives us

$$p(x) = x^3 - 1 = (x-1)(x^2+x+1)$$
$$= (x-1)\left(x + \frac{1+i\sqrt{3}}{2}\right)\left(x + \frac{1-i\sqrt{3}}{2}\right),$$

so that the basis for the complex solution space is

$$\{e^t, e^{-[(1+i\sqrt{3})/2]t}, e^{-[(1-i\sqrt{3}/2]t}\}$$

and the basis for the real solution space is

$$\{e^t, e^{-t/2}\cos(\sqrt{3}t/2), e^{-t/2}\sin(\sqrt{3}t/2)\}.$$

*Our results above suggest that the collection $\mathcal{A}$ of all real-valued solutions of constant coefficient homogeneous linear differential equations contains the functions t^i, e^{rt}, $\cos\omega t$, $\sin\omega t$ for all i, r, and ω, *and is closed under addition and multiplication*, and is in fact the *algebra* generated by these functions.

We can easily prove this conjecture. We first consider sums. Suppose that $T(f) = 0$ and that $S(g) = 0$, where S and T are two such constant coefficient operators. Then $f + g$ is in the null space of $S \circ T$ *because S and T commute:* $(S \circ T)(f+g) = (S \circ T)(f) + (S \circ T)(g) = S(Tf) + T(Sg) = 0 + 0 = 0$. We know that S and T commute because they are both polynomials in D.

In order to treat products, we first have to recognize that the linear span of all the trigonometric functions $\sin at$, $\cos bt$ is an algebra. In other words, any finite product of such functions is a linear combination of such functions. This is the role of a certain class of trigonometric identities, such as $2\sin x\cos y = \sin(x+y) + \sin(x-y)$, which the reader has undoubtedly had to struggle with. (And again the mystery disappears when we are allowed to treat them as complex exponentials.) Then we observe that any function in the algebra $\mathcal{A}$ is a finite sum of terms each of which is of the form $t^ie^{rt}\sin\omega t$ or $t^ie^{rt}\cos\omega t$ for some i, r, and ω. We can exhibit an operator T having such a function in its null space, and our finite sum of such terms will then be in the null space of the composition of these operators T by our first argument.

We are tempted to say one more thing. The functions t^i, e^{rt}, $\sin\omega t$, $\cos\omega t$, and sums of their products can be shown to be exactly the continuous functions $f\colon \mathbb{R} \to \mathbb{R}$ such that the set of *translates* of f has a finite-dimensional span. That is, if we define translation through x, K_x, by $(K_xf)(t) = f(t-x)$, then for exactly the above functions f the linear span of $\{K_xf, x \in \mathbb{R}\}$ is finite-dimensional. This second characterization of exactly the same class of functions cannot be accidental. Part of the secret lies in the fact that the constant coefficient operators T are exactly those linear differential operators that commute with translation. That is, if T is a linear differential operator, then $T \circ K_x = K_x \circ T$ for all x if and only if T has constant coefficients. Now we have noted in an early chapter that if $T \circ S = S \circ T$, then the null space of T is invariant under S. Therefore, the null space N of a constant coefficient operator T is invariant under all translations: $K_x[N] \subset N$ for all x. Now we know that N is finite-dimensional from our differential equation theory. Therefore, the functions in N are such that their translates have a finite-dimensional span!

This device of gaining additional information about the null space N of a linear operator T by finding operators S that commute with T, so that N is S-invariant, is much used in advanced mathematics. It is especially important when we have a *group* of commuting operators S, as we do in the above case with the operators $S = K_x$.

What we have not shown is that if a continuous function f is such that its translation generates a finite-dimensional vector space, then f is in the null space of some constant coefficient operator $p(D)$. This is delicate, and it depends on showing that if $\{K_t\}$ is a one-parameter family of linear transformations on a finite-dimensional space such that $K_{s+t} = K_s \circ K_t$ and $K_t \to I$ as $t \to 0$, then there is an S in Hom V such that $K_t = e^{tS}$.*

EXERCISES

Find solutions for the following equations.

4.1 $x'' - 3x' + 2x = 0$

4.2 $x'' + 2x' - 3x = 0$

4.3 $x'' + 2x' + 3x = 0$

4.4 $x'' + 2x' + x = 0$

4.5 $x''' - 3x'' + 3x' - x = 0$

4.6 $x''' - x = 0$

4.7 $x^{(6)} - x'' = 0$

4.8 $x''' = 0$

4.9 $x''' - x'' = 0$

4.10 Solve the initial-value problem $x'' + 4x' - 5x = 0,\ x(0) = 1,\ x'(0) = 2$.

4.11 Solve the initial-value problem $x''' + x' = 0,\ x(0) = 0,\ x'(0) = -1,\ x''(0) = 1$.

4.12 Find one solution u of the equation $4t^2x'' + x = 0$ by trying $u(t) = t^n$, and then find a second solution as in the text by setting $v(t) = c(t)u(t)$.

4.13 Solve $t^3x''' - 3tx' + 3x = 0$ by trying $u(t) = t^n$.

4.14 Solve $tx'' + x' = 0$.

4.15 Solve $t(x''' + x') + 2(x'' + x) = 0$.

4.16 Knowing that $e^{-bt}\cos \omega t$ and $e^{-bt}\sin \omega t$ are solutions of a second-order linear differential equation, and observing that their values at 0 are 1 and 0, we know that they are independent. Why?

4.17 Find constant coefficient differential equations of which the following functions are solutions: t^2, $\sin t$, $t^2 \sin t$.

4.18 If f and g are independent solutions of a second-order linear differential equation $u'' + a_1u' + a_2u = 0$ with continuous coefficient functions, then we know that the vectors $\langle f(x), f'(x)\rangle$ and $\langle g(x), g'(x)\rangle$ are independent at every point x. Show conversely that if two functions have this latter property, then they are solutions of a second-order differential equation.

4.19 Solve the equation $(D - a)^3 f = 0$ by applying the order-reducing procedure discussed in the text starting with the obvious solution e^{at}.

5. SOLVING THE INHOMOGENEOUS EQUATION

We come now to the problem of solving the inhomogeneous equation $L(f) = g$. We shall briefly describe a practical method which works *easily some* of the time and a theoretical method which works *all* the time, but which may be hard to apply. The latter is just the translation of Theorem 3.3 into matrix language.

We first consider the constant coefficient equation $L(f) = g$ in the special case where g itself is in the null space of a constant coefficient operator S. A simple example is $y' - ay = e^{bt}$ (or $y' - ay = \sin bt$), where $g(t) = e^{bt}$ is in the null space of $S = (D - b)$. In such a situation a solution f must be in the null space of $S \circ L$, for $S \circ L(f) = S(g) = 0$. We know what all these functions are, and our problem is to select f among them such that $L(f)$ is the given g.

For the moment suppose that the polynomials L and S (polynomials in D) have no factors in common. Then we know that L is an isomorphism on the null space N_S of S and therefore that there exists an f in N_S such that $Lf = g$. Since we have a basis for N_S, we could construct the matrix for the action of L on N_S and find f by solving a matrix equation, but the simplest thing to do is take a general linear combination of the basis, with unknown coefficients, let L act on it, and see what the coefficients must be to give g.

For example, to solve $y' - ay = e^{bt}$, we try $f(t) = ce^{bt}$ and apply

$$L : (D - a)(ce^{bt}) = (b - a)ce^{bt} \stackrel{?}{=} e^{bt},$$

and we see that $c = 1/(b - a)$.

Again, to solve $y' - ay = \cos bt$, we observe that $\cos bt$ is in the null space of $S = D^2 + b^2$ and that this null space has the basis $\{\sin bt, \cos bt\}$. We therefore set $f(t) = c_1 \sin bt + c_2 \cos bt$ and solve $(D - a)f = \cos bt$, getting

$$(-ac_1 - bc_2) \sin bt + (bc_1 - ac_2) \cos bt = \cos bt,$$

$$-ac_1 - bc_2 = 0,$$

$$bc_1 - ac_2 = 1,$$

and

$$f(t) = \frac{b}{a^2 + b^2} \sin bt - \frac{a}{a^2 + b^2} \cos bt.$$

When L and S do have factors in common, the situation is more complicated, but a similar procedure can be proved to work. Now an extra factor t^i must be introduced, where i is the number of occurrences of the common factor in L. For example, in solving $(D - r)^2 f = e^{rt}$, we have $S \circ L = (D - r)^3$, and so we must set $f(t) = ct^2 e^{rt}$. Our equation then becomes

$$(D - r)^2 ct^2 e^{rt} = 2ce^{rt} \stackrel{?}{=} e^{rt},$$

and so $c = \frac{1}{2}$.

For $(D^2 + 1)f = \sin t$ we have to set $f(t) = t(c_1 \sin t + c_2 \cos t)$, and after we work it out we find that $c_1 = 0$ and $c_2 = -\frac{1}{2}$, so that $f = -\frac{1}{2}t \cos t$.

This procedure, called, naturally, the method of *undetermined coefficients*, violates our philosophy about a solution process being a linear right inverse. Indeed, it is not a single process, applicable to any g occurring on the right, but varies with the operator S. However, when it is available, it is the easiest way to compute explicit solutions.

We describe next a general theoretical method, called *variation of parameters*, that *is* a right inverse to L and *does* therefore apply to every g. Moreover, it inverts the general (variable coefficient) linear nth-order operator L:

$$(Lf)(t) = \sum_0^n a_i(t) f^{(i)}(t).$$

We are assuming that we know the null space N of L; that is, we assume known n linearly independent solutions $\{u_i\}_1^n$ of the homogeneous equation $Lf = 0$. What we are going to do is to translate into this context our formula $K_t \int_0^t K_s^{-1}(g(s))\, ds$ for the solution to $d\alpha/dt = T_t(\alpha) + g(t)$. Since

$$\psi : f \mapsto \langle f, f', \ldots, f^{(n-1)} \rangle$$

is an isomorphism from the solution space N of the nth-order equation $L(f) = 0$ to the solution space $\mathbf{N}$ of the equivalent first-order system $d\mathbf{x}/dt = T_t(\mathbf{x})$, it follows that if we have a basis $\{u_i\}_1^n$ for N, then the columns of the matrix $w_{ij} = u_j^{(i-1)}$ form a basis for $\mathbf{N}$.

Let $\mathbf{w}(t)$ be the matrix $w_{ij}(t) = u_j^{(i-1)}(t)$. Since evaluation at t is the isomorphism φ_t from $\mathbf{N}$ to $\mathbb{R}^n$, the columns of $\mathbf{w}(t)$ form a basis for $\mathbb{R}^n$, for each t. But $K_t(\boldsymbol{\alpha})$ is the value at t of the solution of $d\mathbf{x}/dt = T_t(\mathbf{x})$ through the initial point $\langle 0, \boldsymbol{\alpha} \rangle$, and it follows that the linear transformation K_t takes the columns of the matrix $\mathbf{w}(0)$ to the corresponding columns of $\mathbf{w}(t)$. The matrix for K_t is therefore $\mathbf{w}(t) \cdot \mathbf{w}(0)^{-1}$, and the matrix form of our formula

$$\mathbf{f}(t) = K_t \int_0^t (K_s)^{-1}(\mathbf{g}(s))\, ds$$

is therefore

$$\mathbf{f}(t) = \mathbf{w}(t) \cdot \mathbf{w}(0)^{-1} \cdot \int_0^t \mathbf{w}(0) \cdot \mathbf{w}(s)^{-1} \cdot \mathbf{g}(s)\, ds.$$

Moreover, since integration commutes with the application of a constant linear transformation (here multiplication by a constant matrix), the middle $\mathbf{w}(0)$ factors cancel, and we have the result that

$$\mathbf{f}(t) = \mathbf{w}(t) \cdot \int_0^t \mathbf{w}(s)^{-1} \cdot \mathbf{g}(s)\, ds$$

is the solution of $d\mathbf{x}/dt = T_t(\mathbf{x}) + \mathbf{g}(t)$ which passes through $\langle 0, \mathbf{0} \rangle$. Finally, set $\mathbf{k}(s) = \mathbf{w}(s)^{-1} \cdot \mathbf{g}(s)$, so that this solution formula splits into the pair

$$\mathbf{f}(t) = \mathbf{w}(t) \cdot \int_0^t \mathbf{k}(s)\, ds, \qquad \mathbf{w}(s) \cdot \mathbf{k}(s) = \mathbf{g}(s).$$

Now we want to solve the inhomogeneous nth-order equation $L(f) = g$, and this means solving the first-order system with $\mathbf{g} = \langle 0, \ldots, 0, g \rangle$. Therefore, the second equation above is equivalent to

$$\sum_j w_{ij}(s)k_j(s) = 0, \qquad i < n,$$

$$\sum_j w_{nj}(s)k_j(s) = g(s).$$

Moreover, the solution f of the nth-order equation is the first component of the n-tuple $\mathbf{f}$ (that is, $f = \psi^{-1}\mathbf{f}$), and so we end up with

$$f(t) = \sum_{j=1}^{n} w_{1j}(t) \int_0^t k_j(s)\, ds = \sum_1^n u_j(t)c_j(t),$$

where $c_i(t)$ is the antiderivative $\int_0^t k_i(s)\, ds$. Any other antiderivative would do as well, since the difference between the two resulting formulas is of the form $\sum_1^n a_i u_i(t)$, a solution of the homogeneous equation $L(f) = 0$. We have proved the following theorem.

Theorem 5.1. If $\{u_i(t)\}_1^n$ is a basis for the solution space of the homogeneous equation $L(h) = 0$, and if $f(t) = \sum_1^n c_i(t)u_i(t)$, where the derivatives $c_i'(t)$ are determined as the solutions of the equations

$$\sum_i c_i'(t)u_i^{(j)}(t) = 0, \qquad j = 0, \ldots, n-2,$$

$$\sum_i c_i'(t)u_i^{(n-1)}(t) = g(t),$$

then $L(f) = g$.

We now consider a simple example of this method. The equation $y'' + y = \sec x$ has constant coefficients, and we can therefore easily find the null space of the homogeneous equation $y'' + y$. A basis for it is $\{\sin x, \cos x\}$. But we can't use the method of undetermined coefficients, because $\sec x$ is not a solution of a constant coefficient equation. We therefore try for a solution

$$v(x) = c_1(x) \sin x + c_2(x) \cos x.$$

Our system of equations to be solved is

$$c_1' \sin x + c_2' \cos x = 0,$$

$$c_1' \cos x - c_2' \sin x = \sec x.$$

Thus $c_2' = -c_1' \tan x$ and $c_1'(\cos x + \sin x \tan x) = \sec x$, giving

$$c_1' = 1, \qquad c_2' = -\tan x,$$

$$c_1 = x, \qquad c_2 = \log \cos x,$$

and

$$v(x) = x \sin x + (\log \cos x) \cos x.$$

(Check it!)

This is all we shall say about the process of finding solutions. In cases where everything works we have complete control of the solutions of $L(f) = g$, and we can then solve the initial-value problem. If L has order n, then we know that the null space N is n-dimensional, and if for a given g the function v is one solution of the inhomogeneous equation $L(f) = g$, then the set of all solutions is the n-dimensional plane (affine subspace) $M = N + v$. If we have found a basis $\{u_i\}_1^n$ for N, then every solution of $L(f) = g$ is of the form $f = \sum_1^n c_i u_i + v$. The initial-value problem is the problem of finding f such that $L(f) = g$ and $f(t_0) = a_1^0, f'(t_0) = a_2^0, \ldots, f^{(n-1)}(t_0) = a_n^0$, where $\langle a_1^0, \ldots, a_n^0 \rangle = \mathbf{a}^0$ is the given initial value. We can now find this unique f by using these n conditions to determine the n coefficients c_i in $f = \sum c_i u_i + v$. We get n equations in the n unknowns c_i. Our ability to solve this problem uniquely again comes back to the fact that the matrix $w_{ij}(t_0) = u_j^{(i-1)}(t_0)$ is nonsingular, as did our success in carrying out the variation of parameters process.

We conclude this section by discussing a very simple and important example. When a perfectly elastic spring is stretched or compressed, it resists with a "restoring" force proportional to its deformation. If we picture a coiled spring lying along the x-axis, with one end fixed and the free end at the origin when undisturbed (Fig. 6.3), then when the coil is stretched a distance x (compression being negative stretching), the force it exerts is $-cx$, where c is a constant representing the stiffness, or elasticity, of the spring, and the minus sign shows that the force is in the direction opposite to the displacement. This is *Hooke's law*.

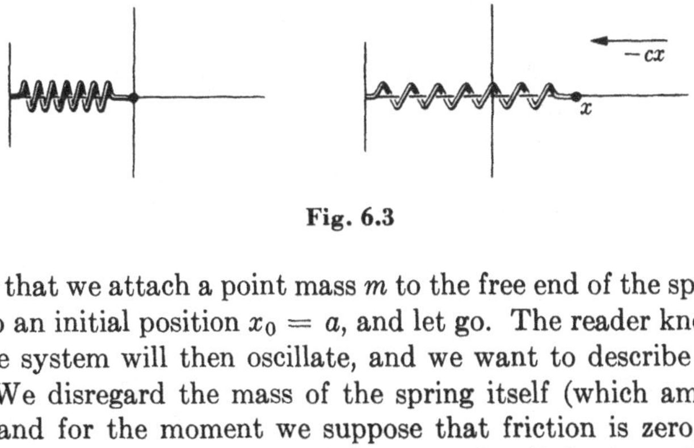

Fig. 6.3

Suppose that we attach a point mass m to the free end of the spring, pull the spring out to an initial position $x_0 = a$, and let go. The reader knows perfectly well that the system will then oscillate, and we want to describe its vibration explicitly. We disregard the mass of the spring itself (which amounts to adjusting m), and for the moment we suppose that friction is zero, so that the system will oscillate forever. Newton's law says that if the force F is applied to the mass m, then the particle will accelerate according to the equation

$$m \frac{d^2x}{dt^2} = F.$$

Here $F = -cx$, so the equation combining the laws of Newton and Hooke is

$$m \frac{d^2x}{dt^2} + cx = 0.$$

This is almost the simplest constant coefficient equation, and we know that the general solution is

$$x = c_1 \sin \Omega t + c_2 \cos \Omega t,$$

where $\Omega = \sqrt{c/m}$. Our initial condition was that $x = a$ and $x' = 0$ when $t = 0$. Thus $c_2 = a$ and $c_1 = 0$, so $x = a \cos \Omega t$. The particle oscillates forever between $x = -a$ and $x = a$. The maximum displacement a is called the *amplitude* A of the oscillation. The number of complete oscillations per unit time is called the *frequency* f, so $f = \Omega/2\pi = \sqrt{c}/2\pi\sqrt{m}$. This is the quantitative expression of the intuitively clear fact that the frequency will increase with the stiffness c and decrease as the mass m increases. Other initial conditions are equally reasonable. We might consider the system originally at rest and strike it, so that we start with an initial velocity v and an initial displacement 0 at time $t = 0$. Now $c_2 = 0$ and $x = c_1 \sin \Omega t$. In order to evaluate c_1, we remember that $dx/dt = v$ at $t = 0$, and since $dx/dt = c_1\Omega \cos \Omega t$, we have $v = c_1\Omega$ and $c_1 = v/\Omega$, the amplitude for this motion. In general, the initial condition would be $x = a$ and $x' = v$ when $t = 0$, and the unique solution thus determined would involve both terms of the general solution, with amplitude to be calculated.

The situation is both more realistic and more interesting when friction is taken into account. Frictional resistance is ideally a force proportional to the velocity dx/dt but again with a negative sign, since its direction is opposite to that of the motion. Our new equation is thus

$$m \frac{d^2x}{dt^2} + k \frac{dx}{dt} + cx = 0,$$

and we know that the system will act in quite different ways depending on the relationship among the constants m, k, and c. The reader will be asked to explore these equations further in the exercises.

It is extraordinary that exactly the same equation governs a freely oscillating electric circuit. It is now written

$$L \frac{d^2x}{dt^2} + R \frac{dx}{dt} + \frac{1}{C} x = 0,$$

where L, R, and C are the inductance, resistance, and capacitance of the circuit, respectively, and dx/dt is the current. However, the ordinary operation of such a circuit involves forced rather than free oscillation. An alternating (sinusoidal) voltage is applied as an extra, external, "force" term, and the equation is now

$$L \frac{d^2x}{dt^2} + R \frac{dx}{dt} + \frac{x}{C} = a \sin \omega t.$$

This shows the most interesting behavior of all. Using the method of undetermined coefficients, we find that the solution contains transient terms that die away, contributed by the homogeneous equation, and a permanent part of frequency $\omega/2\pi$, arising from the inhomogeneous term $a \sin \omega t$. New phenomena called *phase* and *resonance* now appear, as the reader will discover in the exercises.

EXERCISES

Find particular solutions of the following equations.

5.1 $x'' - x = t^4$ **5.2** $x'' - x = \sin t$ **5.3** $x'' - x = \sin t + t^4$

5.4 $x'' + x = \sin t$ **5.5** $y'' - y' = x^2$ (Here $y' = dy/dx$.)

5.6 $y'' - y' = e^x$

5.7 Consider the equation $y'' + y = \sec x$ that was solved in the text. To what interval I must we limit our discussion? Check that the particular solution found in the text is correct. Solve the initial-value problem for

$$f''(x) + f(x) = \sec x, \qquad f(0) = 1, \qquad f'(0) = -1.$$

Solve the following equations by variation of parameters.

5.8 $x'' + x = \tan t$ **5.9** $x''' + x' = t$ **5.10** $y'' + y = 1$

5.11 $y^{(4)} - y = \cos x$ **5.12** $y'' + 4y = \sec 2x$ **5.13** $y'' + 4y = \sec x$

5.14 Show that the general solution

$$C_1 \sin \Omega t + C_2 \cos \Omega t$$

of the frictionless elastic equation $m(d^2x/dt^2) + cx = 0$ can be rewritten in the form

$$A \sin (\Omega t - \alpha).$$

(Remember that $\sin (x - y) = \sin x \cos y - \cos x \sin y$.) This type of motion along a line is called *simple harmonic motion.*

5.15 In the above exercise express A and α in terms of the initial values $dx/dt = v$ and $x = a$ when $t = 0$.

5.16 Consider now the freely vibrating system with friction taken into account, and therefore having the equation

$$m(d^2x/dt^2) + k(dx/dt) + cx = 0,$$

all coefficients being positive. Show that if $k^2 < 4mc$, then the system oscillates forever, but with amplitude decreasing exponentially. Determine the frequency of oscillation. Use Exercise 5.14 to simplify the solution, and sketch its graph.

5.17 Show that if the frictional force is sufficiently large ($k^2 \geq 4mc$), then a freely vibrating system does not in fact vibrate. Taking the simplest case $k^2 = 4mc$, sketch the behavior of the system for the initial condition $dx/dt = 0$ and $x = a$ when $t = 0$. Do the same for the initial condition $dx/dt = v$ and $x = 0$ when $t = 0$.

5.18 Use the method of undetermined coefficients to find a particular solution of the equation of the driven electric circuit

$$L\frac{d^2x}{dt^2} + R\frac{dx}{dt} + \frac{x}{c} = a \sin \omega t.$$

Assuming that $R > 0$, show by a general argument that your particular solution is in fact the steady-state part (the part without exponential decay) of the general solution.

5.19 In the above exercise show that the "current" dx/dt for your solution can be written in the form

$$\frac{dx}{dt} = \frac{a}{\sqrt{R^2 + X^2}} \sin(\omega t - \alpha),$$

where $X = L\omega - 1/\omega C$. Here α is called the *phase* angle.

5.20 Continuing our discussion, show that the current flowing in the circuit will have a maximum amplitude when the frequency of the "impressed voltage" $a \sin \omega t$ is $1/2\pi\sqrt{LC}$. This is the phenomenon of *resonance*. Show also that the current is in phase with the impressed voltage (i.e., that $\alpha = 0$) if and only if $L = C = 0$.

5.21 What is the condition that the phase α be approximately 90°? −90°?

5.22 In the theory of a stable equilibrium point in a dynamical system we end up with two scalar products (ξ, η) and $((\xi, \eta))$ on a finite-dimensional vector space V, the quadratic form $q(\xi) = \frac{1}{2}((\xi, \xi))$ being the potential energy and $p(\xi') = \frac{1}{2}(\xi', \xi')$ being the kinetic energy. Now we know that $dq_\alpha(\xi) = ((\alpha, \xi))$ and similarly for p, and because of this fact it can be shown that the Lagrangian equations can be written

$$\frac{d}{dt}\left(\frac{d\xi}{dt}, \eta\right) = ((\xi, \eta)).$$

Prove that a basis $\{\beta_i\}_1^n$ can be found for V such that this vector equation becomes the system of second-order equations

$$\frac{d^2x_i}{dt^2} = \lambda_i x_i, \qquad i = 1, \ldots, n,$$

where the constants λ_i are positive. Show therefore that the motion of the system is the sum of n linearly independent simple harmonic motions.

6. THE BOUNDARY-VALUE PROBLEM

We now turn to a problem which seems to be like the initial-value problem but which turns out to be of a wholly different character. Suppose that T is a second-order operator, which we consider over a closed interval $[a, b]$. Some of the most important problems in physics require us to find solutions to $T(f) = g$ such that f has given values at a and b, instead of f and f' having given values at a single point t_0. This new problem is called a *boundary-value* problem, because $\{a, b\}$ is the boundary of the domain $I = [a, b]$. The boundary-value problem, like the initial-value problem, breaks neatly into two subproblems if the set

$$M = \{f \in \mathcal{C}^2([a, b]) : f(a) = f(b) = 0\}$$

turns out to be a complement of the null space N of T. However, if the reader will consider this general question for a moment, he will realize that he doesn't have a clue to it from our initial-value development, and, in fact, wholly new tools have to be devised.

Our procedure will be to forget that we are trying to solve the boundary-value problem and instead to speculate on the nature of a linear differential operator T from the point of view of scalar products and the theory of self-adjoint operators. That is, our present study of T will be by means of the scalar product $(f, g) = \int_a^b f(t)g(t)\,dt$, the general problem being the usual one of solving $Tf = g$ by finding a right inverse S of T. Also, as usual, S may be determined by finding a complement M of $N(T)$. Now, however, it turns out that if T is "formally self-adjoint", then suitable choices of M will make the associated right inverses S *self-adjoint* and *compact*, and the eigenvectors of S, computed as those solutions of the homogeneous equation $Tf - rf = 0$ which lie in M, then allow (relatively) the same easy handling of S, by virtue of Theorem 5.1 of Chapter 5, that they gave us earlier in the finite-dimensional situation.

We first consider the notion of "formal adjoint" for an nth-order linear differential operator T. The ordinary formula for integration by parts,

$$\int_a^b f'g = fg\Big]_a^b - \int_a^b fg',$$

allows the derivatives of f occurring in the scalar product (Tf, g) to be shifted one at a time to g. At the end, f is undifferentiated and g is acted on by a certain nth-order linear differential operator R. The endpoint evaluations, like the above $fg]_a^b$, that accumulate step by step can be described as

$$B(f, g)]_a^b = \sum_{0 \le i+j < n} k_{ij}(x) f^{(i)}(x) g^{(j)}(x)]_a^b,$$

where the coefficient functions $k_{ij}(x)$ are linear combinations of the coefficient functions $a_i(x)$ and their derivatives. Thus

$$(Tf, g) = (f, Rg) + B(f, g)]_a^b.$$

The operator R is called the *formal adjoint* of T, and if $R = T$, we say that T is *formally self-adjoint.*

Every application of the integration by parts formula introduces a sign change, and the reader may be able to see that the leading coefficient of R is $(-1)^n$ times the leading coefficient of T. Assuming this, we see that a necessary condition for formal self-adjointness is that n be even, so that R and T have the same first terms.

Supposing that T is formally self-adjoint, we seek a complement M of the null space N of T in $\mathcal{C}^n([a, b])$ with the further property that S, the associated right inverse of T, is self-adjoint as a mapping from the pre-Hilbert space $\mathcal{C}^0([a, b])$ to itself. Let us see what this further requirement amounts to. For any $u, v \in \mathcal{C}^0$, set $f = Su$ and $g = Sv$, so that f and g are in M and $u = Tf$, $v = Tg$. Then $(u, Sv) = (Tf, g) = (f, Tg) + B(f, g)]_a^b = (Su, v) + B(f, g)]_a^b$. We thus have:

Lemma 6.1. If T is a formally self-adjoint differential operator and M is a complement of the null space of T, then the right inverse of T determined by M is self-adjoint if and only if

$$f, g \in M \Rightarrow B(f, g)]_a^b = 0.$$

From now on we shall consider only the second-order case. However, almost everything that we are going to do works perfectly well for the general case, the price of generality being only additional notational complexity.

We start by computing the formal adjoint of the second-order operator $Tf = c_2f'' + c_1f' + c_0f$. We have

$$(Tf, g) = \int_a^b c_2f''g + \int_a^b c_1f'g + \int_a^b c_0fg,$$

$$\int_a^b c_1f'g = c_1fg\Big]_a^b - \int_a^b f(c_1g)',$$

$$\int_a^b c_2f''g = c_2f'g\Big]_a^b - \int_a^b f'(c_2g)'$$

$$= \big(c_2f'g - f(c_2g)'\big)\Big]_a^b + \int_a^b f(c_2g)'',$$

giving

$$(f, Rg) = \int f(c_2g)'' - (c_1g)' + (c_0g),$$

and

$$B(f, g) = c_2(f'g - g'f) + (c_1 - c_2')fg.$$

Thus $Rg = c_2g'' + (2c_2' - c_1)g' + (c_2'' - c_1' + c_0)g$ and $R = T$ if and only if $2c_2' - c_1 = c_1$ (and $c_2'' - c_1' = 0$), that is, $c_2' = c_1$. We have proved:

Lemma 6.2. The second-order differential operator T is formally self-adjoint if and only if

$$Tf = c_2f'' + c_2'f' + c_0f = (c_2f')' + c_0f,$$

in which case

$$B(f, g) = c_2(f'g - g'f).$$

A constant coefficient operator is thus formally self-adjoint if and only if $c_1 = 0$.

Supposing that T is formally self-adjoint, we now try to find a complement M of its null space N such that $f, g \in M \Rightarrow B(f, g)]_a^b = 0$. Since N is two-dimensional, any complement M can be described as the intersection of the null space of two linear functionals l_1 and l_2 on $X_2 = \mathcal{C}^2([a, b])$. For example, the "one point" complement M_{t_0} that we had earlier in connection with the initial-value problem is the intersection of the null spaces of the two functionals $l_1(f) = f(t_0)$ and $l_2(f) = f'(t_0)$. Here, however, the vanishing of l_1 and l_2 for two functions f and g must imply that $B(f, g)]_a^b = c_2(f'g - g'f)]_a^b = 0$, and the functionals $l_i(f)$ must therefore involve the values of f and f' at a and at b. We would naturally guess, and it can be proved, that each of l_1 and l_2 must be of the form $l(f) = k_1f(a) + k_2f'(a) + k_3f(b) + k_4f'(b)$.

Our problem can therefore be restated as follows. We must find two linear functionals l_1 and l_2 of the above general form such that if M is the intersection of their null spaces, then

a) M is a complement of N, and

b) $f, g \in M \Rightarrow c_2(f'g - g'f)]_a^b = 0$,

in which case we call the boundary condition $l_1(f) = l_2(f) = 0$ self-adjoint.

Lemma 6.3. We can replace (a) by

a') T is injective on M.

Proof. If T is injective on M, then $M \cap N = \{0\}$, so that the map

$$f \to \langle l_1(f), l_2(f) \rangle$$

is injective on N, and therefore, because N is two-dimensional, is an isomorphism from N to $\mathbb{R}^2$ (by the corollary of Theorem 2.4, Chapter 2). Then M is a complement of N by Theorem 5.3 of Chapter 1. □

Now we can easily write down various pairs l_1 and l_2 that form a self-adjoint boundary condition. We list some below.

1) $f \in M \Leftrightarrow f(a) = f(b) = 0$ [that is, $l_1(f) = f(a)$ and $l_2(f) = f(b)$].

2) $f \in M \Leftrightarrow f'(a) = f'(b) = 0$.

3) More generally, $f'(a) = kf(a)$, $f'(b) = cf(b)$. (In fact, l_1 can be any l that depends only on the values at a, and l_2 can be any l that depends only on b. Thus $l_1(f) = k_1f(a) + k_2f'(a)$, and if $l_1(f) = l_1(g) = 0$, then the pairs $\langle f(a), f'(a) \rangle$ and $\langle g(a), g'(a) \rangle$ are dependent, since both lie in the one-dimensional null space of l_1, and so $f'g - g'f|_a = 0$. The same holds for l_2 and b, so that this split pair of endpoint conditions makes $b(f, g)]_a^b = 0$ by making the values of B at a and at b separately 0.)

4) If $c_2(a) = c_2(b)$, then take $f \in M \Leftrightarrow f(a) = f(b)$ and $f'(a) = f'(b)$. That is, $l_1(f) = f(a) - f(b)$ and $l_2(f) = f'(a) - f'(b)$.

We now show that in every case but (3) the condition (a') also holds if we replace T by $T - \lambda$ for a suitable λ. This is true also for case (3), but the proof is harder, and we shall omit it.

Lemma 6.4. Suppose that M is defined by one of the self-adjoint boundary conditions (1), (2), or (4) above, that $c_2(t) \geq m > 0$ on $[a, b]$, and that $\lambda \geq c_0(t) + 1$ on $[a, b]$. Then

$$|((T - \lambda)f, f)| \geq m\|f'\|_2^2 + \|f\|_2^2$$

for all $f \in M$. In particular, M is a complement of the null space of $T - \lambda$ and hence defines a self-adjoint right inverse of $T - \lambda$.

Proof. We have

$$((\lambda - T)f, f) = -\int_a^b (c_2f')'f + \int_a^b (\lambda - c_0)f^2$$

$$= -c_2f'f\Big]_a^b + \int_a^b c_2(f')^2 + \int_a^b (\lambda - c_0)f^2.$$

Under any of conditions (1), (2), or (4), $c_2 f'f]_a^b = 0$, and the two integral terms are clearly bounded below by $m\|f'\|_2^2$ and $\|f\|_2^2$, respectively. Lemma 6.3 then implies that M is a complement of the null space of $T - \lambda$. □

We come now to our main theorem. It says that the right inverse S of $T - \lambda$ determined by the subspace M above is a compact self-adjoint mapping of the pre-Hilbert space $\mathcal{C}^0([a, b])$ into itself, and is therefore endowed with all the rich eigenvalue structures of Theorem 5.1 of the last chapter. First we present some classical terminology. A *Sturm-Liouville system* on $[a, b]$ is a formally self-adjoint second-order differential operator $Tf = (c_2 f')' + c_0 f$ defined over the closed interval $[a, b]$, together with a self-adjoint boundary condition $l_1(f) = l_2(f) = 0$ for that interval. If $c_2(t)$ is never zero on $[a, b]$, the system is called *regular*. If $c_2(a)$ or $c_2(b)$ is zero, or if the interval $[a, b]$ is replaced by an infinite interval such as $[a, \infty]$, then the system is called *singular*.

Theorem 6.1. If $T\colon l_1, l_2$ is a regular Sturm-Liouville system on $[a, b]$, with c_2 positive, then the subspace M defined by the homogeneous boundary condition is a complement of $N(T - \lambda)$ if λ is taken sufficiently large, and the right inverse of $T - \lambda$ thus determined by M is a compact self-adjoint mapping of the pre-Hilbert space $\mathcal{C}^0([a, b])$ into itself.

Proof. The proof depends on the inequality of the above lemma. Since we have proved this inequality only for boundary conditions (1), (2), and (4), our proof will be complete only for those cases.

Set $g = (T - \lambda)f$. Since $\|g\|_2\|f\|_2 \geq |((T - \lambda)f, f)|$ by the Schwarz inequality, we see from the lemma first that $\|f\|_2^2 \leq \|g\|_2\|f\|_2$, so that

$$\|f\|_2 \leq \|g\|_2,$$

and then that $m\|f'\|_2^2 \leq \|g\|_2\|f\|_2 \leq \|g\|_2^2$, so that

$$\|f'\|_2 \leq \|g\|_2/\sqrt{m}.$$

We have already checked that the right inverse S of the formally self-adjoint $T - \lambda$ defined by M is self-adjoint, and it remains for us to show that the set $S[U] = \{f : \|g\|_2 \leq 1\}$ has compact closure. For any such f the Schwarz inequality and the above inequality imply that

$$|f(y) - f(x)| \leq \int_x^y |f'| = \int_x^y |f'| \cdot 1 \leq \|f'\|_2 |y - x|^{1/2} \leq \frac{|y - x|^{1/2}}{\sqrt{m}}.$$

Thus $S[U]$ is uniformly equicontinuous. Since the common domain of the functions in $S[U]$ is the compact set $[a, b]$, we will be able to conclude from Theorem 6.1 of Chapter 4 that the set $S[U]$ is totally bounded if we can show that there is a constant C such that all the functions in $S[U]$ have their ranges in $[-C, C]$. Taking y and x in the last inequality where $|f|$ assumes its maximum and minimum values, we have $\|f\|_\infty - \min |f| \leq (b - a)^{1/2}/\sqrt{m}$. But

$(\min|f|)(b-a)^{1/2} \leq \|f\|_2 \leq \|g\|_2 \leq 1$, and therefore

$$\|f\|_\infty \leq C = 1/(b-a)^{1/2} + (b-a)^{1/2}/\sqrt{m}.$$

Thus $S[U]$ is a uniformly equicontinuous set of functions mapping the compact set $[a, b]$ into the compact set $[-C, C]$, and is therefore totally bounded in the *uniform norm*. Since $\mathcal{C}([a, b])$ is complete in the uniform norm, every sequence in $S[U]$ has a subsequence uniformly converging to some $f \in \mathcal{C}$, and since $\|f\|_2 \leq (b-a)^{1/2}\|f\|_\infty$, this subsequence also converges to f in the two-norm.

We have thus shown that if H is the pre-Hilbert space $\mathcal{C}([a, b])$ under the standard scalar product, then the image $S[U]$ of the unit ball $U \subset H$ under S has the property that every sequence in $S[U]$ has a subsequence converging in H. This is the property we actually used in proving Theorem 5.1 of Chapter 5, but it is not quite the definition of the compactness of S, which requires us to show that the closure $\overline{S[U]}$ is compact in H. However, if $\{\xi_n\}$ is any sequence in this closure, then we can choose $\{\zeta_n\}$ in $S[U]$ so that $\|\xi_n - \zeta_n\| < 1/n$. The sequence $\{\zeta_n\}$ has a convergent subsequence $\{\zeta_{n(m)}\}_m$ as above, and then $\{\xi_{n(m)}\}_m$ converges to the same limit. Thus S is a compact operator. □

Theorem 6.2. There exists an orthonormal sequence $\{\varphi_n\}$ consisting entirely of eigenvectors of T and forming a basis for M. Moreover, the Fourier expansion of any $f \in M$ with respect to the basis $\{\varphi_n\}$ converges *uniformly* to f (as well as in the two-norm).

Proof. By Theorem 5.1 of Chapter 5 there exist an eigenbasis for the range of S, which is M. Since $S\varphi_n = r_n\varphi_n$ for some nonzero r_n, we have $(T - \lambda)(r_n\varphi_n) = \varphi_n$ and $T\varphi_n = ((1 + \lambda r_n)/r_n)\varphi_n$. The *uniformity* of the series convergence comes out of the following general consideration.

Lemma 6.5. Suppose that T is a self-adjoint operator on a pre-Hilbert space V and that T is compact as a mapping from V to $\langle V, q\rangle$, where q is a second norm on V that dominates the scalar product norm p $(q \geq cp)$. Then T is compact (from p to p), and the eigenbasis expansion $\sum b_n\varphi_n$ of an element β in the range of T converges to β in both norms.

Proof. Let U be the unit ball of V in the scalar product norm. By the hypothesis of the lemma, the q-closure B of $T[U]$ is compact. B is then also p-compact, for any sequence in it has a q-convergent subsequence which also p-converges to the same limit, because $p \leq cq$. We can therefore apply the eigenbasis theorem.

Now let α and $\beta = T(\alpha)$ have the Fourier series $\sum a_i\varphi_i$ and $\sum b_i\varphi_i$, and let $T(\varphi_i) = r_i\varphi_i$. Then $b_i = r_ia_i$, because $b_i = (T(\alpha), \varphi_i) = (\alpha, T(\varphi_i)) = (\alpha, r_i\varphi_i) = r_i(\alpha, \varphi_i) = r_ia_i$. Since the sequence of partial sums $\sum_1^n a_i\varphi_i$ is p-bounded (Bessel's inequality), the sequence $\{\sum_1^n b_i\varphi_i\} = \{T(\sum_1^n a_i\varphi_i)\}$ is totally q-bounded. Any subsequence of it therefore has a subsubsequence q-converging to some element γ in V. Since it then p-converges to γ, γ must be β. Thus every subsequence has a subsubsequence q-converging to β, and so $\{\sum_1^n b_i\varphi_i\}$ itself q-converges to β by Lemma 4.1 of Chapter 4. □

EXERCISES

6.1 Given that $Tf(x) = xf''(x) + f(x)$ and $Sf(x) = f'(x)$, compute $T \circ S$ and $S \circ T$.

6.2 Show that the differential operators $T = aD$ and $S = bD$ commute if and only if the functions $a(x)$ and $b(x)$ are proportional.

6.3 Show that the differential operators $T = aD^2$ and $S = bD$ commute if and only if $b(x)$ is a first-degree polynomial $b(x) = cx + d$ and $a(x) = k(b(x))^2$.

6.4 Compute the formal adjoint S of T if

a) $Tf = f'$, b) $Tf = f''$, c) $Tf = f'''$, d) $(Tf)(x) = xf'(x)$,
e) $(Tf)(x) = x^3 f''(x)$.

6.5 Let S and T be linear differential operators of orders m and n, respectively. What are the coefficient conditions for $S \circ T$ to be a linear differential operator of order $m + n$?

6.6 Let T be the second-order linear differential operator

$$(Tf)(t) = a_2(t)f''(t) + a_1(t)f'(t) + a_0(t)f(t).$$

What are the conditions on its coefficient functions for its formal adjoint to exist? What are these conditions for T of order n?

6.7 Let S and T be linear differential operators of order m and n, respectively, and suppose that all coefficients are $\mathcal{C}^\infty$-functions (infinitely differentiable). Prove that $S \circ T - T \circ S$ is of order $\leq m + n - 1$.

6.8 A δ-blip is a continuous nonnegative function φ such that $\varphi = 0$ outside of $[-\delta, \delta]$ and $\int_{-\delta}^{\delta} \varphi = 1$ (Fig. 6.4). We assume that there exists an infinitely differentiable 1-blip φ. Show that there exists an infinitely differentiable δ-blip for every $\delta > 0$. Define what you would mean by a δ-blip centered at x, and show that one exists.

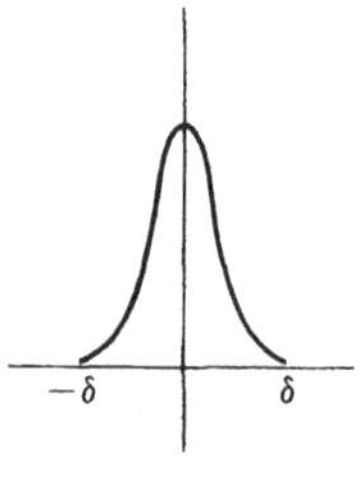

Fig. 6.4

6.9 Let f be a continuous function on $[a, b]$ such that $(f, g) = \int_a^b fg = 0$ whenever g is an infinitely differentiable function which vanishes near a and b. Show that $f = 0$. (Use the above exercise.)

6.10 Let $\mathcal{C}^\infty([a, b])$ be the vector space of infinitely differentiable functions on $[a, b]$, and let T be a second-order linear differential operator with coefficients in $\mathcal{C}^\infty$:

$$(Tf)(t) = a_2(t)f''(t) + a_1(t)f'(t) + a_0(t)f(t).$$

Let S be a linear operator on $\mathcal{C}^\infty([a, b])$ such that

$$(Tf, g) - (f, Sg) = K(f, g)$$

is a bilinear functional depending only on the values of f, g, f', and g' at a and b. Prove that S is the formal adjoint of T. [*Hint:* Take f to be a δ-blip centered at x. Then $K(f, g) = 0$. Now try to work the assertion to be proved into a form to which the above exercise can be applied.]

6.11 Prove an nth-order generalization of the above exercise.

6.12 Let X be the space of linear differential operators with $\mathcal{C}^\infty$-coefficients, and let A_T be the formal adjoint of T. Prove that $T \to A_T$ is an isomorphism from X to X. Prove that $A_{(T\circ S)} = A_S \circ A_T$.

7. FOURIER SERIES

There are not many regular Sturm-Liouville systems whose associated orthonormal eigenbases have proved to be important in actual calculations. Most orthonormal bases that are used, such as those due to Bessel, Legendre, Hermite, and Laguerre, arise from *singular* Sturm-Liouville systems and are therefore beyond the limitations we have set for this discussion. However, the most well-known example, Fourier series, is available to us.

We shall consider the constant coefficient operator $Tf = D^2 f$, which is clearly both formally self-adjoint and regular, and either the boundary condition $f(0) = f(\pi) = 0$ on $[0, \pi]$ (type 1) or the periodic boundary condition $f(-\pi) = f(\pi)$, $f'(-\pi) = f'(\pi)$ on $[-\pi, \pi]$ (type 4).

To solve the first problem, we have to find the solutions of $f'' - \lambda f = 0$ which satisfy $f(0) = f(\pi) = 0$. If $\lambda > 0$, then we know that the two-dimensional solution space is spanned by $\{e^{rx}, e^{-rx}\}$, where $r = \lambda^{1/2}$. But if $c_1 e^{rx} + c_2 e^{-rx}$ is 0 at both 0 and π, then $c_1 = c_2 = 0$ (because the pairs $\langle 1, 1\rangle$ and $\langle e^{r\pi}, e^{-r\pi}\rangle$ are independent). Therefore, there are no solutions satisfying the boundary conditions when $\lambda > 0$. If $\lambda = 0$, then $f(x) = c_1 x + c_0$ and again $c_1 = c_0 = 0$.

If $\lambda < 0$, then the solution space is spanned by $\{\sin rx, \cos rx\}$, where $r = (-\lambda)^{1/2}$. Now if $c_1 \sin rx + c_2 \cos rx$ is 0 at $x = 0$ and $x = \pi$, we get, first, that $c_2 = 0$ and, second, that $r\pi = n\pi$ for some integer n. Thus the eigenfunctions for the first system form the set $\{\sin nx\}_1^\infty$, and the corresponding eigenvalues of D^2 are $\{-n^2\}_1^\infty$.

At the end of this section we shall prove that the functions in $\mathcal{C}^2([a, b])$ that are zero near a and b are dense in $\mathcal{C}([a, b])$ in the two-norm. Assuming this, it follows from Theorem 2.3 of Chapter 5 that a basis for M is a basis for $\mathcal{C}^0$, and we now have the following corollary of the Sturm-Liouville theorem.

Theorem 7.1. The sequence $\{\sin nx\}_1^\infty$ is an orthogonal basis for the pre-Hilbert space $\mathcal{C}^0([0, \pi])$. If $f \in \mathcal{C}^2([0, \pi])$ and $f(0) = f(\pi) = 0$, then the Fourier series for f converges uniformly to f.

We now consider the second boundary problem. The computations are a little more complicated, but again if $f(x) = c_1 e^{rx} + c_2 e^{-rx}$, and if $f(-\pi) = f(\pi)$ and $f'(-\pi) = f'(\pi)$, then $f = 0$. For now we have

$$c_1 e^{-r\pi} + c_2 e^{r\pi} = c_1 e^{r\pi} + c_2 e^{-r\pi},$$

giving $c_1 = c_2$, and

$$c_1 r e^{-r\pi} - c_2 r e^{r\pi} = c_1 r e^{r\pi} - c_2 r e^{-r\pi},$$

giving $c_1(e^{r\pi} - e^{-r\pi}) = 0$, and so $c_1 = 0$. Again $f(x) = c_1x + c_0$ is ruled out Finally, if $f(x) = c_1 \sin rx + c_2 \cos rx$, our boundary conditions become

$$2c_1 \sin r\pi = 0 \qquad \text{and} \qquad 2rc_2 \sin r\pi = 0,$$

so that again $r = n$, but this time the full solution space of $(D^2 + n^2)f = 0$ satisfies the boundary condition.

Theorem 7.2. The set $\{\sin nx\}_1^\infty \cup \{\cos nx\}_0^\infty$ forms an orthogonal basis for the pre-Hilbert space $\mathcal{C}^0([-\pi, \pi])$. If $f \in \mathcal{C}^2([-\pi, \pi])$ and $f(-\pi) = f(\pi)$, $f'(-\pi) = f'(\pi)$, then the Fourier series for f converges to f uniformly on $[-\pi, \pi]$.

Remaining proof. This theorem follows from our general Sturm-Liouville discussion except for the orthogonality of $\sin nx$ and $\cos nx$. We have

$$\begin{aligned}(\sin nx, \cos nx) &= \int_{-\pi}^{\pi} \sin nt \cos nt \, dt \\ &= \tfrac{1}{2}\int_{-\pi}^{\pi} \sin 2nt \, dt \\ &= -(1/4n) \cos 2nx]_{-\pi}^{\pi} \\ &= 0.\end{aligned}$$

Or we can simply remark that the first integrand is an odd function and therefore its integral over any symmetric interval $[-a, a]$ is necessarily zero.

The orthogonality of eigenvectors having different eigenvalues follows of course, as in the proof of Theorem 3.1 of Chapter 5. □

Finally, we prove the density theorem we needed above. There are very slick ways of doing this, but they require more machinery than we have available, and rather than taking the time to make the machines, we shall prove the theorem with our bare hands.

It is standard notation to let a subscript zero on a symbol denoting a class of functions pick out those functions in the class that are zero "on the boundary" in some suitable sense. Here $\mathcal{C}_0([a, b])$ will denote the functions in $\mathcal{C}([a, b])$ that are zero in neighborhoods of a and b, and similarly for $\mathcal{C}_0^2([a, b])$.

Theorem 7.3. $\mathcal{C}^2([a, b])$ is dense in $\mathcal{C}([a, b])$ in the uniform norm, and $\mathcal{C}_0^2([a, b])$ is dense in $\mathcal{C}([a, b])$ in the two-norm.

Proof. We first approximate $f \in \mathcal{C}([a, b])$ to within ϵ by a piecewise "linear" function g by drawing straight line segments between the adjacent points on the graph of f lying over a subdivision $a = x_0 < x_1 < \cdots < x_n = b$ of $[a, b]$. If f varies by less than ϵ on each interval (x_{i-1}, x_i), then $\|f - g\|_\infty \leq \epsilon$. Now $g'(t)$ is a step function which is constant on the intervals of the above subdivision. We now alter $g'(t)$ slightly near each jump in such a way that the new function $h(t)$ is continuous there. If we do it as sketched in Fig. 6.5, the total

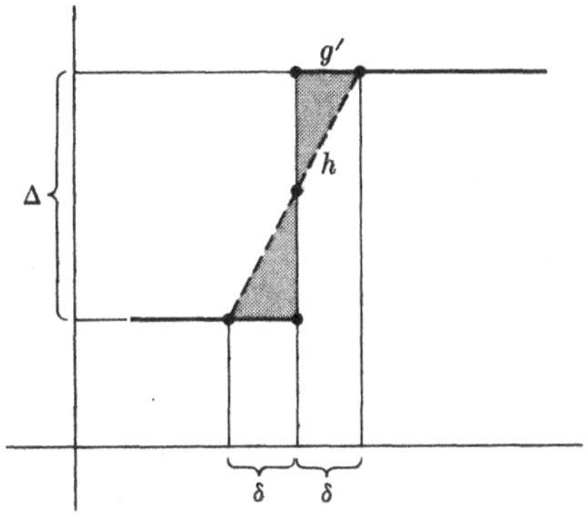

Fig. 6.5

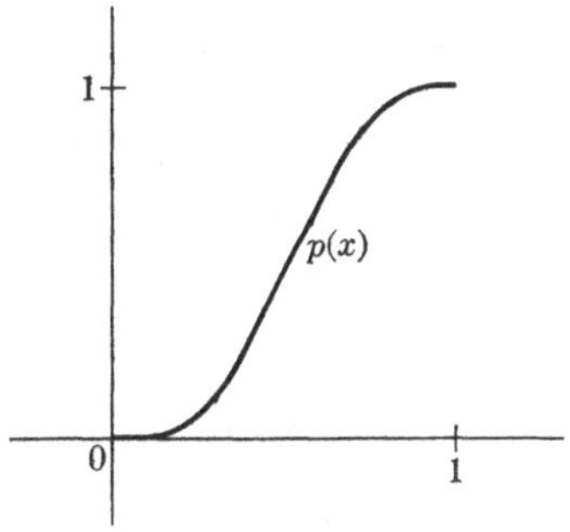

Fig. 6.6

integral error at the jump is zero, $\int_{x_i-\delta}^{x_i+\delta} (h - g') = 0$, and the maximum error $\int_{x_i-\delta}^{x_i}$ is $\delta\Delta/4$. This will be less than ϵ if we take $\delta = \epsilon/\|g'\|_\infty$, since $\Delta \leq 2\|g'\|_\infty$. We now have a continuous function h such that $|\int_a^x h(t)\,dt - (f(x) - f(a))| < 2\epsilon$. In other words, we have approximated f uniformly by a continuously differentiable function.

Now choose g and h in $\mathcal{C}^1([a, b])$ so that first $\|f - g\|_\infty < \epsilon/2$ and then $\|g' - h\|_\infty < \epsilon/2(b - a)$. Then

$$\left|g(x) - g(0) - \int_0^x h\right| < \epsilon/2,$$

and so $H(x) = \int_0^x h + g(0)$ is a twice continuously differentiable function such that $\|f - H\|_\infty < \epsilon$. In other words, $\mathcal{C}^2([a, b])$ is dense in $\mathcal{C}([a, b])$ in the uniform norm. It is then also dense in the two-norm, since

$$\|f\|_2 = \left(\int_a^b f^2\right)^{1/2} \leq \|f\|_\infty\left(\int_a^b 1\right)^{1/2} = (b - a)^{1/2}\|f\|_\infty.$$

But now we can do something which we couldn't do for the uniform norm: we can alter the approximating function to one that is zero on neighborhoods of a and b, and keep the two-norm approximation good. Given δ, let $e(t)$ be a nonnegative function on $[a, b]$ such that $e(t) = 1$ on $[a + 2\delta, b - 2\delta]$, $e(t) = 0$ on $[a, a + \delta]$ and on $[b - \delta, b]$, e'' is continuous, and $\|e\|_\infty = 1$. Such an $e(t)$ clearly exists, since we can draw it. We leave it as an interesting exercise to actually define $e(t)$. Here is a hint: Show somehow that there is a fifth-degree polynomial $p(t)$ having a graph between 0 and 1 as shown in Fig. 6.6, with a zero second derivative at 0 and at 1, and then use a piece of the graph, suitably translated, compressed, rotated, etc., to help patch together $e(t)$.

Anyway, then $\|g - eg\|_2 \leq \|g\|_\infty(4\delta)^{1/2}$ for any g on $\mathcal{C}([a, b])$, and if g has continuous derivatives up to order 2, then so does eg. Thus, if we start with f in $\mathcal{C}$ and approximate it by g in $\mathcal{C}^2$, and then approximate g by eg, we have altogether the second approximation of the theorem. □

EXERCISES

7.1 Convert the orthogonal basis $\{\sin nx\}_1^\infty$ for the pre-Hilbert space $\mathcal{C}([0, \pi])$ to an orthonormal basis.

7.2 Do the same for the orthogonal basis $\{\sin nx\}_1^\infty \cup \{\cos nx\}_0^\infty$ for $\mathcal{C}([-\pi, \pi])$.

7.3 Show that $\{\sin nx\}_1^\infty$ is an orthogonal basis for the vector space V of all *odd* continuous functions on $[-\pi, \pi]$. (Be clever. Do not calculate from scratch.) Normalize the above basis.

7.4 State and prove the corresponding theorem for the *even* functions on $[-\pi, \pi]$.

7.5 Prove that the derivative of an odd function is even, and conversely.

7.6 We now want to prove the following stronger theorem about the uniform convergence of Fourier series.

> **Theorem.** Let f have a continuous derivative on $[-\pi, \pi]$, and suppose that $f(-\pi) = f(\pi)$. Then the Fourier series for f converges to f uniformly.

Assume for convenience that f is even. (This only cuts down the number of calculations.) Show first that the Fourier series for f' is the series obtained from the Fourier series for f by term-by-term differentiation. Apply the above exercises here. Next show from the two-norm convergence of its Fourier series to f' and the Schwarz inequality that the Fourier series for f converges uniformly.

7.7 Prove that $\{\cos nx\}_0^\infty$ is an orthonormal basis for the space M of $\mathcal{C}^2$-functions on $[0, \pi]$ such that $f'(0) = f'(\pi) = 0$.

7.8 Find a fifth-degree polynomial $p(x)$ such that

$$p(0) = p'(0) = p''(0) = 0, \qquad p'(1) = p''(1) = 0, \qquad p(1) = 1.$$

(Forget the last condition until the end.) Sketch the graph of p.

7.9 Use a "piece" of the above polynomial p to construct a function $e(x)$ such that e' and e'' exist and are continuous, $e(x) = 0$ when $x < a + \delta$ and $x > b - \delta$, $e(x) = 1$ on $[a + 2\delta, b - 2\delta]$, and $\|e\|_\infty = 1$.

7.10 Prove the Weierstrass theorem given below on $[0, \pi]$ in the following steps. We know that f can be uniformly approximated by a $\mathcal{C}^2$-function g.

1) Show that c and d can be found and that $g(t) - c(t) - d$ is 0 at 0 and π.
2) Use the Fourier series expansion of this function and the Maclaurin series for the functions $\sin nx$ to show that the polynomial $p(x)$ can be found.

> **Theorem** (*The Weierstrass approximation theorem*). The polynomials are dense in $\mathcal{C}([a, b])$ in the uniform norm. That is, given any continuous function f on $[a, b]$ and any ϵ, there is a polynomial p such that $|f(x) - p(x)| < \epsilon$ for all x in $[a, b]$.

CHAPTER 7

MULTILINEAR FUNCTIONALS

This chapter is principally for reference. Although most of the proofs will be included, the reader is not expected to study them. Our goal is a collection of basic theorems about alternating multilinear functionals, or exterior forms, and the determinant function is one of our rewards.

1. BILINEAR FUNCTIONALS

We have already studied various aspects of bilinear functionals. We looked at their duality implications in Section 6, Chapter 1, we considered the "canonical forms" of symmetric bilinear functionals and their equivalent quadratic forms in Section 7, Chapter 2, and, of course, the whole scalar product theory of Chapter 5 is the theory of a still more special kind of bilinear functional. In this chapter we shall restrict ourselves to bilinear and multilinear functionals over finite-dimensional spaces, and our concerns are purely algebraic.

We begin with some material related to our earlier algebra. If V and W are finite-dimensional vector spaces, then the set of all bilinear functionals on $V \times W$ is pretty clearly a vector space. We designate it $V^* \otimes W^*$ and call it the *tensor product* of V^* and W^*. Our first theorem simply states something that was implicit in Theorem 6.1 of Chapter 1.

Theorem 1.1. The vector spaces $V^* \otimes W^*$, $\mathrm{Hom}(V, W^*)$, and $\mathrm{Hom}(W, V^*)$ are naturally isomorphic.

Proof. We saw in Theorem 6.1 of Chapter 1 that each f in $V^* \otimes W^*$ determines a linear mapping $\alpha \mapsto f_\alpha$ from W to V^*, where $f_\alpha(\xi) = f(\xi, \alpha)$, and we also noted that this correspondence from $V^* \otimes W^*$ to $\mathrm{Hom}(W, V^*)$ is bijective. All that the present theorem adds is that this bijective correspondence is linear and so constitutes a natural isomorphism, as does the similar one from $V^* \otimes W^*$ to $\mathrm{Hom}(V, W^*)$. To see this, let f_T be the bilinear functional corresponding to T in $\mathrm{Hom}(V, W^*)$. Then $f_{(T+S)} = f_T + f_S$, for $f_{(T+S)}(\alpha, \beta) = \big((T + S)(\alpha)\big)(\beta) = \big(T(\alpha) + S(\alpha)\big)(\beta) = \big(T(\alpha)\big)(\beta) + \big(S(\alpha)\big)(\beta) = f_T(\alpha, \beta) + f_S(\alpha, \beta)$. We can do the same for homogeneity.

The isomorphism of $V^* \otimes W^*$ with $\mathrm{Hom}(W, V^*)$ follows in exactly the same way by reversing the roles of the variables. We are thus finished with the proof. □

Before looking for bases in $V^* \otimes W^*$, we define a bilinear functional $\gamma \otimes \lambda$ for any two functionals $\gamma \in V^*$ and $\lambda \in W^*$ by $(\gamma \otimes \lambda)(\xi, \eta) = \gamma(\xi)\lambda(\eta)$. We call $\gamma \otimes \lambda$ the tensor product of the functionals γ and λ and call any bilinear functional having this form *elementary*. It is not too hard to see that $f \in V^* \otimes W^*$ is elementary if and only if the corresponding $T \in \mathrm{Hom}(V, W^*)$ is a dyad.

If V and W are finite-dimensional, with dimensions m and n, respectively, then the above isomorphism of $V^* \otimes W^*$ with $\mathrm{Hom}(V, W^*)$ shows that the dimension of $V^* \otimes W^*$ is mn. We now describe the basis determined by given bases in V and W.

Theorem 1.2. Let $\{\alpha_i\}_1^m$ and $\{\beta_j\}_1^n$ be any bases for V and W, and let their dual bases in V^* and W^* be $\{\mu_i\}_1^m$ and $\{\nu_j\}_1^n$. Then the mn elementary bilinear functionals $\{\mu_i \otimes \nu_j\}$ form the corresponding basis for $V^* \otimes W^*$.

Proof. Since $\mu_i \otimes \nu_j(\xi, \eta) = \mu_i(\xi)\nu_j(\eta) = x_i y_j$, the matrix expansion $f(\xi, \eta) = \sum_{i,j} t_{ij} x_i y_j$ becomes $f(\xi, \eta) = \sum_{i,j} t_{ij}(\mu_i \otimes \nu_j)(\xi, \eta)$ or

$$f = \sum_{i,j} t_{ij}(\mu_i \otimes \nu_j).$$

The set $\{\mu_i \otimes \nu_j\}$ thus spans $V^* \otimes W^*$. Since it contains the same number of elements (mn) as the dimension of $V^* \otimes W^*$, it forms a basis. □

Of course, independence can also be checked directly. If $\sum_{i,j} t_{ij}(\mu_i \otimes \nu_i) = 0$, then for every pair $\langle k, l \rangle$, $t_{kl} = \sum_{i,j} t_{ij}(\mu_i \otimes \nu_j)(\alpha_k, \beta_l) = 0$.

We should also remark that this theorem is entirely equivalent to our discussion of the basis for $\mathrm{Hom}(V, W)$ at the end of Section 4, Chapter 2.

2. MULTILINEAR FUNCTIONALS

All the above considerations generalize to multilinear functionals

$$f: V_1 \times \cdots \times V_n \to \mathbb{R}.$$

We change notation, just as we do in replacing the traditional $\langle x, y \rangle \in \mathbb{R}^2$ by $\mathbf{x} = \langle x_1, \ldots, x_n \rangle \in \mathbb{R}^n$. Thus we write $f(\alpha_1, \ldots, \alpha_n) = f(\boldsymbol{\alpha})$, where $\boldsymbol{\alpha} = \langle \alpha_1, \ldots, \alpha_n \rangle \in V_1 \times \cdots \times V_n$. Our requirement now is that

$$f(\alpha_1, \ldots, \alpha_n)$$

be a linear functional of α_j when α_i is held fixed for all $i \neq j$. The set of all such functionals is a vector space, called the tensor product of the dual spaces $V_1^*, \ldots, V_n^*$, and is designated $V_1^* \otimes \cdots \otimes V_n^*$.

As before, there are natural isomorphims between these tensor product spaces and various Hom spaces. For example,

$$V_1^* \otimes \cdots \otimes V_n^* \qquad \text{and} \qquad \mathrm{Hom}(V_1, V_2^* \otimes \cdots \otimes V_n^*)$$

are naturally isomorphic. Also, there are additional isomorphisms of a variety not encountered in the bilinear case. However, it will not be necessary for us to look into these questions.

We define *elementary* multilinear functionals as before. If $\lambda_i \in V_i^*$, $i = 1, \ldots, n$, and $\xi = \langle \xi_1, \ldots, \xi_n \rangle$, then

$$(\lambda_1 \otimes \cdots \otimes \lambda_n)(\xi) = \lambda_1(\xi_1) \cdots \lambda_n(\xi_n).$$

To keep our notation as simple as possible, and also because it is the case of most interest to us, we shall consider the question of bases only when $V_1 = V_2 = \cdots = V_n = V$. In this case $(V^*)^{ⓝ} = V^* \otimes \cdots \otimes V^*$ (in factors) is called the space of *covariant tensors* of order n (over V).

If $\{\alpha_j\}_1^m$ is a basis for V and $f \in (V^*)^{ⓝ}$, then we can expand the value $f(\xi) = f(\xi_1, \ldots, \xi_n)$ with respect to the basis expansions of the vectors ξ_i just as we did when f was bilinear, but now the result is notationally more complex. If we set $\xi_i = \sum_{j=1}^m x_j^i \alpha_j$ for $i = 1, \ldots, n$ (so that the coordinate set of ξ_i is $\mathbf{x}^i = \{x_j^i\}_j$) and use the linearity of the $f(\xi_1, \ldots, \xi_n)$ in its separate variables one variable at a time, we get

$$f(\xi_1, \ldots, \xi_n) = \sum x_{p_1}^1 x_{p_2}^2 \cdots x_{p_n}^n f(\alpha_{p_1}, \alpha_{p_2}, \ldots, \alpha_{p_n}),$$

where the sum is taken over all n-tuples $\mathbf{p} = \langle p_1, \ldots, p_n \rangle$ such that $1 \le p_i \le m$ for each i from 1 to m. The set of all these n-tuples is just the set of all functions from $\{1, \ldots, n\}$ to $\{1, \ldots, m\}$. We have designated this set $\overline{m}^{\overline{n}}$, using the notation $\overline{n} = \{1, \ldots, n\}$, and the scope of the above sum can thus be indicated in the formula as follows:

$$f(\xi_1, \ldots, \xi_n) = \sum_{\mathbf{p} \in \overline{m}^{\overline{n}}} x_{p_1}^1 \cdots x_{p_n}^n f(\alpha_{p_1}, \ldots, \alpha_{p_n}).$$

A strict proof of this formula would require an induction on n, and is left to the interested reader. At the inductive step he will have to rewrite a double sum $\sum_{\mathbf{p} \in \overline{m}^{\overline{n}}} \sum_{j \in \overline{m}}$ as the single sum $\sum_{\mathbf{q} \in \overline{m}^{\overline{n+1}}}$ using the fact that an ordered pair $\langle \mathbf{p}, j \rangle$ in $\overline{m}^{\overline{n}} \times \overline{m}$ is equivalent to an $(n+1)$-tuplet $\mathbf{q} \in \overline{m}^{\overline{n+1}}$, where $q_i = p_i$ for $i = 1, \ldots, n$ and $q_{n+1} = j$.

If $\{\mu_i\}_1^m$ is the dual basis for V^* and $\mathbf{q} \in \overline{m}^{\overline{n}}$, let $\mu_{\mathbf{q}}$ be the elementary functional $\mu_{q_1} \otimes \cdots \otimes \mu_{q_n}$. Thus $\mu_{\mathbf{q}}(\alpha_{p_1}, \ldots, \alpha_{p_n}) = \prod_1^n \mu_{q_i}(\alpha_{p_i}) = 0$ unless $\mathbf{p} = \mathbf{q}$, in which case its value is 1. More generally,

$$\mu_{\mathbf{q}}(\xi_1, \ldots, \xi_n) = \mu_{q_1}(\xi_1) \cdots \mu_{q_n}(\xi_n) = x_{q_1}^1 \cdots x_{q_n}^n.$$

Therefore, if we set $c_{\mathbf{q}} = f(\alpha_{q_1}, \ldots, \alpha_{q_n})$, the general expansion now appears as

$$f(\xi_1, \ldots, \xi_n) = \sum_{\mathbf{p} \in \overline{m}^{\overline{n}}} c_{\mathbf{p}} \mu_{\mathbf{p}}(\xi_1, \ldots, \xi_n)$$

or $f = \sum c_{\mathbf{p}} \mu_{\mathbf{p}}$, which is the same formula we obtained in the bilinear case, but with more sophisticated notation. The functionals $\{\mu_{\mathbf{p}} : \mathbf{p} \in \overline{m}^{\overline{n}}\}$ thus span $(V^*)^{ⓝ}$. They are also independent. For, if $\sum c_{\mathbf{p}} \mu_{\mathbf{p}} = 0$, then for each $\mathbf{q}$, $c_{\mathbf{q}} = \sum c_{\mathbf{p}} \mu_{\mathbf{p}}(\alpha_{q_1}, \ldots, \alpha_{q_n}) = 0$. We have proved the following theorem.

Theorem 2.1. The set $\{\mu_{\mathbf{p}} : \mathbf{p} \in \overline{m}^{\overline{n}}\}$ is a basis for $(V^*)^{\circledn}$. For any f in $(V^*)^{\circledn}$ its coordinate function $\{c_{\mathbf{p}}\}$ is defined by $c_{\mathbf{p}} = f(\alpha_{p_1}, \ldots, \alpha_{p_n})$. Thus $f = \sum c_{\mathbf{p}}\mu_{\mathbf{p}}$ and $f(\xi_1, \ldots, \xi_n) = \sum c_{\mathbf{p}}\mu_{\mathbf{p}}(\xi_1, \ldots, \xi_n) = \sum c_{\mathbf{p}}\, x^1_{p_1} \cdots x^n_{p_n}$ for any $f \in (V^*)^{\circledn}$ and any $\langle \xi_1, \ldots, \xi_n \rangle \in V^n$.

Corollary. The dimension of $(V^*)^{\circledn}$ is m^n.

Proof. There are m^n functions in $\overline{m}^{\overline{n}}$, so the basis $\{\mu_{\mathbf{p}} : \mathbf{p} \in \overline{m}^{\overline{n}}\}$ has m^n elements. □

3. PERMUTATIONS

A permutation on a set S is a bijection $f: S \to S$. If $\mathcal{S}(S)$ is the set of all permutations on S, then $\mathcal{S} = \mathcal{S}(S)$ is closed under composition ($\sigma, \rho \in \mathcal{S} \Rightarrow \sigma \circ \rho \in \mathcal{S}$) and inversion ($\sigma \in \mathcal{S} \Rightarrow \sigma^{-1} \in \mathcal{S}$). Also, the identity map I is in $\mathcal{S}$, and, of course, the composition operation is associative. Together these statements say exactly that $\mathcal{S}$ is a *group* under composition. The simplest kind of permutation other than I is one which interchanges a pair of elements of S and leaves every other element fixed. Such a permuation is called a *transposition.*

We now take S to be the finite set $\overline{n} = \{1, \ldots, n\}$ and set $\mathcal{S}_n = \mathcal{S}(\overline{n})$. It is not hard to see that then any permutation can be expressed as a product of transpositions, and in more than one way.

A more elementary fact that we shall need is that if ρ is a fixed element of $\mathcal{S}_n$, then the mapping $\sigma \mapsto \sigma \circ \rho$ is a bijection $\mathcal{S}_n \mapsto \mathcal{S}_n$. It is surjective because any σ' can be written $\sigma' = (\sigma' \circ \rho^{-1}) \circ \rho$, and it is injective because $\sigma_1 \circ \rho = \sigma_2 \circ \rho \Rightarrow (\sigma_1 \circ \rho) \circ \rho^{-1} = (\sigma_2 \circ \rho) \circ \rho^{-1} \Rightarrow \sigma_1 = \sigma_2$. Similarly, the mapping $\sigma \mapsto \rho \circ \sigma$ (ρ fixed) is bijective.

We also need the fact that there are $n!$ elements in $\mathcal{S}_n$. This is the elementary count from secondary school algebra. In defining an element $\sigma \in \mathcal{S}_n$, $\sigma(1)$ can be chosen in n ways. For each of these choices $\sigma(2)$ can be chosen in $n - 1$ ways, so that $\langle\sigma(1), \sigma(2)\rangle$ can be chosen in $n(n - 1)$ ways. For each of these choices $\sigma(3)$ can be chosen in $n - 2$ ways, etc. Altogether σ can be chosen in $n(n - 1)(n - 2) \cdots 1 = n!$ ways.

In the sequel we shall often write '$\rho\sigma$' instead of '$\rho \circ \sigma$', just as we occasionally wrote 'ST' instead of '$S \circ T$' for the composition of linear maps.

If $\boldsymbol{\xi} = \langle \xi_1, \ldots, \xi_n \rangle \in V^n$ and $\sigma \in \mathcal{S}_n$, then we can "apply σ to $\boldsymbol{\xi}$", or "permute the elements of $\langle \xi_1, \ldots, \xi_n \rangle$ through σ". We mean, of course, that we can replace $\langle \xi_1, \ldots, \xi_n \rangle$ by $\langle \xi_{\sigma(1)}, \ldots, \xi_{\sigma(n)} \rangle$, that is, we can replace $\boldsymbol{\xi}$ by $\boldsymbol{\xi} \circ \sigma$.

Permuting the variables changes a functional $f \in (V^*)^{\circledn}$ into a new such functional. Specifically, given $f \in (V^*)^{\circledn}$ and $\sigma \in \mathcal{S}_n$, we define f^σ by

$$f^\sigma(\boldsymbol{\xi}) = f(\boldsymbol{\xi} \circ \sigma^{-1}) = f(\xi_{\sigma^{-1}(1)}, \ldots, \xi_{\sigma^{-1}(n)}).$$

The reason for using σ^{-1} instead of σ is, in part, that it gives us the following formula.

Lemma 3.1. $f^{(\sigma_1\sigma_2)} = (f^{\sigma_1})^{\sigma_2}$.

Proof. $f^{(\sigma_1\sigma_2)}(\xi) = f(\xi \circ (\sigma_1 \circ \sigma_2)^{-1}) = f(\xi \circ (\sigma_2^{-1} \circ \sigma_1^{-1})) = f((\xi \circ \sigma_2^{-1}) \circ \sigma_1^{-1}) = f^{\sigma_1}(\xi \circ \sigma_2^{-1}) = (f^{\sigma_1})^{\sigma_2}(\xi)$. □

Theorem 3.1. For each σ in $\mathcal{S}_n$ the mapping T_σ defined by $f \mapsto f^\sigma$ is a linear isomorphism of $(V^*)^{\textcircled{n}}$ onto itself. The mapping $\sigma \mapsto T_\sigma$ is an antihomomorphism from the group $\mathcal{S}_n$ to the group of nonsingular elements of $\mathrm{Hom}((V^*)^{\textcircled{n}})$.

Proof. Permuting the variables does not alter the property of multilinearity, so T_σ maps $(V^*)^{\textcircled{n}}$ into itself. It is linear, since $(af + bg)^\sigma = af^\sigma + bg^\sigma$. And $T_{\rho\sigma} = T_\sigma \circ T_\rho$, because $f^{\rho\sigma} = (f^\rho)^\sigma$. Thus $\sigma \mapsto T_\sigma$ preserves products, but in the reverse order. This is why it is called an *anti*homomorphism. Finally,

$$T_{(\sigma^{-1})} \circ T_\sigma = T_{(\sigma\sigma^{-1})} = T_I = I,$$

so that T_σ is invertible (nonsingular, an isomorphism). □

The mapping $\sigma \mapsto T_\sigma$ is a *representation* (really an *anti*representation) of the group $\mathcal{S}_n$ by linear transformations on $(V^*)^{\textcircled{n}}$.

Lemma 3.2. Each T_σ carries the basis $\{\mu_{\mathbf{p}}\}$ into itself, and so is a permutation on the basis.

Proof. We have $(\mu_{\mathbf{p}})^\sigma(\xi) = \mu_{\mathbf{p}}(\xi \circ \sigma^{-1}) = \prod_{i=1}^n \mu_{p_i}(\xi_{\sigma^{-1}(i)})$. Setting $j = \sigma^{-1}(i)$ and so having $i = \sigma(j)$, this product can be rewritten $\prod_{j=1}^n \mu_{p_{\sigma(j)}}(\xi_j) = \mu_{\mathbf{p}\circ\sigma}(\xi)$. Thus

$$(\mu_{\mathbf{p}})^\sigma = \mu_{\mathbf{p}\circ\sigma},$$

and since $\mathbf{p} \mapsto \mathbf{p} \circ \sigma$ is a permutation on $\overline{m}^{\overline{n}}$, we are done. □

4. THE SIGN OF A PERMUTATION

We consider now the special polynomial E on $\mathbb{R}^n$ defined by

$$E(\mathbf{x}) = E(x_1, \ldots, x_n) = \prod_{1 \le i < j \le n} (x_i - x_j).$$

This is the product over all pairs $\langle i, j\rangle \in \overline{n} \times \overline{n}$ such that $i < j$. This set of ordered pairs is in one-to-one correspondence with the collection P of all pair sets $\{i, j\} \subset \overline{n}$ such that $i \neq j$, the ordered pair being obtained from the unordered pair by putting it in its natural order. Now it is clear that for any permutation $\sigma \in \mathcal{S}_n$, the mapping $\{i, j\} \mapsto \{\sigma(i), \sigma(j)\}$ is a permutation of P. This means that the factors in the polynomial $E^\sigma(\mathbf{x}) = E(\mathbf{x} \circ \sigma)$ are exactly the same as in the polynomial $E(\mathbf{x})$ except for the changes of sign that occur when σ reverses the order of a pair. Therefore, if n is the number of these reversals, we have $E^\sigma = (-1)^n E$. The mapping $\sigma \mapsto (-1)^n$ is designated 'sgn' (and called "sign"). Thus sgn is a function from $\mathcal{S}_n$ to $\{1, -1\}$ such that $E^\sigma = (\operatorname{sgn} \sigma)E$, for all $\sigma \in \mathcal{S}_n$. It follows that

$$\operatorname{sgn} \rho\sigma = (\operatorname{sgn} \rho)\,(\operatorname{sgn} \sigma),$$

for $(\operatorname{sgn} \rho\sigma)E = E^{\rho\sigma} = (E^\rho)^\sigma = (\operatorname{sgn} \sigma)E^\rho = (\operatorname{sgn} \rho)\,(\operatorname{sgn} \sigma)E$, and we can evaluate E at any n-tuple $\mathbf{x}$ such that $E(\mathbf{x}) \neq 0$ and cancel the factor $E(\mathbf{x})$. Also

$$\operatorname{sgn} \sigma = -1 \qquad \text{if} \quad \sigma \text{ is a transposition.}$$

This is clear if σ interchanges adjacent numbers because it then changes the sign of just one factor in $E(\mathbf{x})$; we leave the general case as an exercise for the interested reader.

5. THE SUBSPACE $\mathcal{A}^n$ OF ALTERNATING TENSORS

Definition. A covariant tensor $f \in (V^*)^{ⓝ}$ is *symmetric* if $f^\sigma = f$ for all $\sigma \in \mathcal{S}_n$.

If f is bilinear [$f \in (V^*)^{②}$], this is just the condition $f(\xi, \eta) = f(\eta, \xi)$ for all $\xi, \eta \in V$.

Definition. A covariant tensor $f \in (V^*)^{ⓝ}$ is *antisymmetric* or *alternating* if $f^\sigma = (\operatorname{sgn} \sigma) f$ for all $\sigma \in \mathcal{S}_n$.

Since each σ is a product of transpositions, this can also be expressed as the fact that f just changes sign if two of its arguments are interchanged. In the case of a bilinear functional it is the condition $f(\xi, \eta) = -f(\eta, \xi)$ for all $\xi, \eta \in V$. It is important to note that if f is alternating, then $f(\boldsymbol{\xi}) = 0$ whenever the n-tuple $\boldsymbol{\xi} = \langle \xi_1, \ldots, \xi_n \rangle$ is not injective ($\xi_i = \xi_j$ for some $i \neq j$). The set of all symmetric elements of $(V^*)^{ⓝ}$ is clearly a subspace, as is also the (for us) more important set $\mathcal{A}^n$ of all alternating elements. There is an important linear projection from $(V^*)^{ⓝ}$ to $\mathcal{A}^n$ which we now describe.

Theorem 5.1. The mapping $f \mapsto (1/n!)\sum_{\sigma \in \mathcal{S}_n} (\operatorname{sgn} \sigma) f^\sigma$ is a projection Ω from $(V^*)^{ⓝ}$ to $\mathcal{A}^n$.

Proof. We first check that $\Omega f \in \mathcal{A}^n$ for every f in $(V^*)^{ⓝ}$. We have $(\Omega f)^\rho = (1/n!)\sum_\sigma (\operatorname{sgn} \sigma) f^{\sigma\rho}$. Now $\operatorname{sgn} \sigma = (\operatorname{sgn} \sigma\rho)(\operatorname{sgn} \rho)$. Setting $\sigma' = \sigma \circ \rho$ and remembering that $\sigma \mapsto \sigma'$ is a bijection, we thus have

$$(\Omega f)^\rho = \frac{(\operatorname{sgn} \rho)}{n!} \sum_{\sigma'} (\operatorname{sgn} \sigma') f^{\sigma'} = (\operatorname{sgn} \rho)(\Omega f).$$

Hence $\Omega f \in \mathcal{A}^n$.

If f is already in $\mathcal{A}^n$, then $f^\sigma = (\operatorname{sgn} \sigma) f$ and $\Omega f = (1/n!)\sum_{\sigma \in \mathcal{S}_n} f$. Since $\mathcal{S}_n$ has $n!$ elements, $\Omega f = f$. Thus Ω is a projection from $(V^*)^{ⓝ}$ to $\mathcal{A}^n$. □

Lemma 5.1. $\Omega(f^\rho) = (\operatorname{sgn} \rho)\Omega f$.

Proof. The formula for $\Omega(f^\rho)$ is the same as that for $(\Omega f)^\rho$ except that $\rho\sigma$ replaces $\sigma\rho$. The proof is thus the same as the one for the theorem above. □

Theorem 5.2. The vector space $\mathcal{A}^n$ of alternating n-linear functionals over the m-dimensional vector space V has dimension $\binom{m}{n}$.

Proof. If $f \in \mathcal{A}^n$ and $f = \sum_{\mathbf{p}} c_{\mathbf{p}}\mu_{\mathbf{p}}$, then since $f^\sigma = (\operatorname{sgn}\sigma)f$, we have $\sum_{\mathbf{p}} c_{\mathbf{p}}\mu_{\mathbf{p}\circ\sigma} = \sum_{\mathbf{p}} (\operatorname{sgn}\sigma)c_{\mathbf{p}}\mu_{\mathbf{p}}$ for any σ in $\mathcal{S}_n$. Setting $\mathbf{p}\circ\sigma = \mathbf{q}$, the left sum becomes $\sum_{\mathbf{q}} c_{\mathbf{q}\circ\sigma^{-1}}\mu_{\mathbf{q}}$, and since the basis expansion is unique, we must have $c_{\mathbf{q}\circ\sigma^{-1}} = \operatorname{sgn}\sigma c_{\mathbf{q}}$ or $c_{\mathbf{p}} = (\operatorname{sgn}\sigma)c_{\mathbf{p}\circ\sigma}$ for all $\mathbf{p} \in \overline{m}^{\overline{n}}$. Working backward, we see, conversely, that this condition implies that $f^\sigma = (\operatorname{sgn}\sigma)f$. Thus $f \in \mathcal{A}^n$ if and only if its coordinate function $c_{\mathbf{p}}$ satisfies the identity

$$c_{\mathbf{p}} = (\operatorname{sgn}\sigma)c_{\mathbf{p}\circ\sigma} \qquad \text{for all} \quad \mathbf{p} \in \overline{m}^{\overline{n}} \quad \text{and all} \quad \sigma \in \mathcal{S}_n.$$

This has many consequences. For one thing, $c_{\mathbf{p}} = 0$ unless $\mathbf{p}$ is one-to-one (injective). For if $p_i = p_j$ and σ is the transposition interchanging i and j, then $\mathbf{p}\circ\sigma = \mathbf{p}$, $c_{\mathbf{p}} = (\operatorname{sgn}\sigma)c_{\mathbf{p}\circ\sigma} = -c_{\mathbf{p}}$, and so $c_{\mathbf{p}} = 0$. Since no $\mathbf{p}$ can be injective if $n > m$, we see that in this case the only element of $\mathcal{A}^n$ is the zero functional. Thus $n > m \Rightarrow \dim \mathcal{A}^n = 0$.

Now suppose that $n \le m$. For any injective $\mathbf{p}$, the set $\{\mathbf{p}\circ\sigma : \sigma \in \mathcal{S}_n\}$ consists of all the (injective) n-tuples with the same range set as $\mathbf{p}$. There are clearly $n!$ of them. Exactly one $\mathbf{q} = \mathbf{p}\circ\sigma$ counts off the range set in its natural order, i.e., satisfies $q_1 < q_2 < \cdots < q_n$. We select this unique $\mathbf{q}$ as the representative of all the elements $\mathbf{p}\circ\sigma$ having this range. The collection C of these canonical (representative) q's is thus in one-to-one correspondence with the collection of all (range) subsets of $\overline{m} = \{1, \ldots, m\}$ of size n.

Each injective $\mathbf{p} \in \overline{m}^{\overline{n}}$ is uniquely expressible as $\mathbf{p} = \mathbf{q}\circ\sigma$ for some $\mathbf{q} \in C$, $\sigma \in \mathcal{S}_n$. Thus each f in $\mathcal{A}^n$ is the sum $\sum_{\mathbf{q}\in C}\sum_{\sigma\in\mathcal{S}_n} t_{\mathbf{q}\circ\sigma}\mu_{\mathbf{q}\circ\sigma}$. Since $t_{\mathbf{q}\circ\sigma} = (\operatorname{sgn}\sigma)t_{\mathbf{q}}$, this sum can be rewritten $\sum_{\mathbf{q}\in C} t_{\mathbf{q}}\sum_\sigma (\operatorname{sgn}\sigma)\mu_{\mathbf{q}\circ\sigma} = \sum_{\mathbf{q}\in C} t_{\mathbf{q}}\nu_{\mathbf{q}}$, where we have set $\nu_{\mathbf{q}} = \sum_\sigma (\operatorname{sgn}\sigma)\mu_{\mathbf{q}\circ\sigma} = n!\Omega(\mu_{\mathbf{q}})$.

We are just about done. Each $\nu_{\mathbf{q}}$ is alternating, since it is in the range of Ω, and the expansion

$$f = \sum_{\mathbf{q}\in C} t_{\mathbf{q}}\nu_{\mathbf{q}}$$

which we have just found to be valid for every $f \in \mathcal{A}^n$ shows that the set $\{\nu_{\mathbf{q}} : \mathbf{q} \in C\}$ spans $\mathcal{A}^n$. It is also independent, since $\sum_{\mathbf{q}\in C} t_{\mathbf{q}}\nu_{\mathbf{q}} = \sum_{\mathbf{p}\in\overline{m}^{\overline{n}}} t_{\mathbf{p}}\mu_{\mathbf{p}}$ and the set $\{\mu_{\mathbf{p}}\}$ is independent. It is therefore a basis for $\mathcal{A}^n$.

Now the total number of injective mappings $\mathbf{p}$ from $\overline{n} = \{1, \ldots, n\}$ to $\overline{m} = \{1, \ldots, m\}$ is $m(m-1)\cdots(m-n+1)$, for the first element can be chosen in m ways, the second in $m-1$ ways, and so on down through n choices, the last element having $m-(n-1) = m-n+1$ possibilities. We have seen above that the number of these $\mathbf{p}$'s with a given range is $n!$. Therefore, the number of different range sets is

$$m(m-1)\cdots\frac{m-n+1}{n!} = \frac{m!}{n!(m-n)!} = \binom{m}{n}.$$

And this is the number of elements $\mathbf{q} \in C$. □

The case $n = m$ is very important. Now C contains only one element, the identity I in $\mathcal{S}_m$, so that

$$f = c_I \nu_I = c_I \sum_\sigma (\operatorname{sgn} \sigma)\mu_\sigma$$

and

$$f(\xi_1, \ldots, \xi_m) = c_I \sum_\sigma (\operatorname{sgn} \sigma)\mu_{\sigma(1)}(\xi_1) \cdots \mu_{\sigma(m)}(\xi_m)$$
$$= c_I \sum_\sigma (\operatorname{sgn} \sigma) x^1_{\sigma(1)} \cdots x^m_{\sigma(m)}.$$

This is essentially the formula for the determinant, as we shall see.

6. THE DETERMINANT

We saw in Section 5 that the dimension of the space $\mathfrak{A}^m$ of alternating m-forms over an m-dimensional V is $\binom{m}{m} = 1$. Thus, to within scalar multiples there is only one *alternating m-linear* functional D over $V = \mathbb{R}^m$, and we can adjust the constant so that $D(\delta^1, \ldots, \delta^m) = 1$. This uniquely determined m-form is the *determinant* functional, and its value $D(\mathbf{x}^1, \ldots, \mathbf{x}^m)$ at the m-tuple $\langle \mathbf{x}^1, \ldots, \mathbf{x}^m \rangle$ is the *determinant of the matrix* $\mathbf{x} = \{x_{ij}\}$ whose jth column is $\mathbf{x}^j$ for $j = 1, \ldots, m$.

Lemma 6.1. $D(\mathbf{t}^1, \ldots, \mathbf{t}^m) = \sum_{\sigma \in \mathcal{S}_m} (\operatorname{sgn} \sigma) t_{\sigma(1),1} \cdots t_{\sigma(m),m}$.

Proof. This is just the last remark of the last section, with the constant $c_I = 1$, since $D(\delta^1, \ldots, \delta^m) = 1$, and with the notation changed to the usual matrix form t_{ij}. □

Corollary 1. $D(\mathbf{t}^*) = D(\mathbf{t})$.

Proof. If we reorder the factors of the product $t_{\sigma_1,1} \cdots t_{\sigma_m,m}$ in the order of the values σ_i, the product becomes $t_{1,\rho_1} \cdots t_{m,\rho_m}$, where $\rho = \sigma^{-1}$. Since

$$\sigma \mapsto \rho = \sigma^{-1}$$

is a bijection from $\mathcal{S}_n$ to $\mathcal{S}_n$, and since $\operatorname{sgn}(\sigma^{-1}) = \operatorname{sgn} \sigma$, the sum in the lemma can be rewritten as $\sum_{\rho \in \mathcal{S}_m} (\operatorname{sgn} \rho)\, t_{1,\rho_1} \cdots t_{m,\rho_m}$. But this is

$$\sum (\operatorname{sgn} \rho) t^*_{\rho_1,1} \cdots t_{\rho_m,m} = D(t^*). \ \square$$

Corollary 2. $D(\mathbf{t})$ is an alternating m-linear functional of the rows of $\mathbf{t}$.

Now let $\dim V = m$, and let f be any nonzero alternating m-form on V. For any T in Hom V the functional f_T defined by $f_T(\xi_1, \ldots, \xi_n) = f(T\xi_1, \ldots, T\xi_n)$ also belongs to $\mathfrak{A}^m$. Since $\mathfrak{A}^m$ is one-dimensional, $f_T = k_T f$ for some constant k_T. Moreover, k_T is independent of f, since if $g_T = k_T' g$ and $g = cf$, we must have $cf_T = k_T' cf$ and $k_T' = k_T$. This unique constant is called the *determinant of* T; we shall designate it $\Delta(T)$. Note that $\Delta(T)$ is defined independently of any basis for V.

Theorem 6.1. $\Delta(S \circ T) = \Delta(S)\,\Delta(T)$.

Proof

$$\begin{aligned}\Delta(S\circ T)f(\xi_1,\ldots,\xi_m) &= f((S\circ T)(\xi_1),\ldots,(S\circ T)(\xi_m))\\ &= f(S(T(\xi_1)),\ldots,S(T(\xi_m)))\\ &= \Delta(S)f(T(\xi_1),\ldots,T(\xi_m)) = \Delta(T)\,\Delta(S)f(\xi_1,\ldots,\xi_m).\end{aligned}$$

Now divide out f. □

Theorem 6.2. If θ is an isomorphism from V to W, and if $T \in \text{Hom } V$ and $S = \theta \circ T \circ \theta^{-1}$, then $\Delta(S) = \Delta(T)$.

Proof. If f is any nonzero alternating m-form on W, and if we define g by $g(\xi_1, \ldots, \xi_n) = f(\theta\xi_1, \ldots, \theta\xi_n)$. Then g is a nonzero alternating m-form on V. Now $f(S\circ\theta\xi_1, \ldots, S\circ\theta\xi_n) = \Delta(S)f(\theta\xi, \ldots, \theta\xi_n) = \Delta(S)g(\xi_1, \ldots, \xi_n)$, and also $f(S\circ\theta\xi_1, \ldots, S\circ\theta\xi_n) = f(\theta\circ T\xi_1, \ldots, \theta\circ T\xi_n) = g(T\xi_1, \ldots, T\xi_n) = \Delta(T)g(\xi_1, \ldots, \xi_n)$. Thus $\Delta(S)g = \Delta(T)g$ and $\Delta(S) = \Delta(T)$. □

The reader will expect the two notions of determinant we have introduced to agree; we prove this now.

Corollary 1. If $\mathbf{t}$ is the matrix of T with respect to some basis in V, then $D(\mathbf{t}) = \Delta(T)$.

Proof. If θ is the coordinate isomorphism, then $\overline{T} = \theta\circ T\circ\theta^{-1}$ is in Hom $\mathbb{R}^m$ and $\Delta(T) = \Delta(\overline{T})$ by the theorem. Also, the columns of $\mathbf{t}$ are the m-tuple $\overline{T}(\delta^1), \ldots, \overline{T}(\delta^m)$. Thus $D(\mathbf{t}) = D(\mathbf{t}^1, \ldots, \mathbf{t}^m) = D(\overline{T}(\delta^1), \ldots, \overline{T}(\delta^m)) = \Delta(\overline{T})\,D(\delta^1, \ldots, \delta^m) = \Delta(\overline{T})$. Altogether we have $D(\mathbf{t}) = \Delta(T)$. □

Corollary 2. If $\mathbf{s}$ and $\mathbf{t}$ are $m \times m$ matrices, then $D(\mathbf{s}\cdot\mathbf{t}) = D(\mathbf{s})\,D(\mathbf{t})$.

Proof. $D(\mathbf{s}\cdot\mathbf{t}) = \Delta(S\circ T) = \Delta(S)\,\Delta(T) = D(\mathbf{s})\,D(\mathbf{t})$. □

Corollary 3. $D(\mathbf{t}) = 0$ if and only if $\mathbf{t}$ is singular.

Proof. If $\mathbf{t}$ is nonsingular, then $\mathbf{t}^{-1}$ exists and $D(\mathbf{t})\,D(\mathbf{t}^{-1}) = D(\mathbf{t}\mathbf{t}^{-1}) = D(I) = 1$. In particular, $D(\mathbf{t}) \neq 0$. If $\mathbf{t}$ is singular, some column, say $\mathbf{t}_1$, is a linear combination of the others, $\mathbf{t}_1 = \sum_2^m c_i\mathbf{t}_i$, and $D(\mathbf{t}_1, \ldots, \mathbf{t}_m) = \sum_2^m c_iD(\mathbf{t}_i, \mathbf{t}_2, \ldots, \mathbf{t}_m) = 0$, since each term in the sum evaluates D at an m-tuple having two identical elements, and so is 0 by the alternating property. □

We still have to show that Δ has all the properties we ascribed to it in Chapter 2. Some of them are in hand. We know that $\Delta(S\circ T) = \Delta(S)\,\Delta(T)$, and the one- and two-dimensional properties are trivial. Thus, if T interchanges independent vectors α_1 and α_2 in a two-dimensional space, then its matrix with respect to them as a basis is $\mathbf{t} = \left[\begin{smallmatrix}0 & 1\\ 1 & 0\end{smallmatrix}\right]$, and so $\Delta(T) = D(\mathbf{t}) = -1$.

The following lemma will complete the job.

Lemma 6.2. Consider $D(\mathbf{t}) = D(\mathbf{t}^1, \ldots, \mathbf{t}^m)$ under the special assumption that $\mathbf{t}^m = \delta^m$. If $\mathbf{s}$ is the $(m-1)\times(m-1)$ matrix obtained from the $m\times m$ matrix $\mathbf{t}$ by deleting its last row and last column, then $D(\mathbf{s}) = D(\mathbf{t})$.

Proof. This can be made to follow from an inspection of the formula of Lemma 6.1, but we shall argue directly.

If $\mathbf{t}$ has δ^m also as its jth column for some $j \neq m$, then of course $D(\mathbf{t}) = 0$ by the alternating property. This means that $D(\mathbf{t})$ is unchanged if the jth column is altered in the mth place, and therefore $D(\mathbf{t})$ depends only on the values t_{ij} in the rows $i \neq m$. That is, $D(\mathbf{t})$ depends only on $\mathbf{s}$. Now $\mathbf{t} \mapsto \mathbf{s}$ is clearly a surjective mapping to $\mathbb{R}^{\overline{m-1} \times \overline{m-1}}$, and, as a function of $\mathbf{s}$, $D(\mathbf{t})$ is alternating $(m-1)$-linear. It therefore is a constant multiple of the determinant D on $\mathbb{R}^{(m-1)\times(m-1)}$. To see what the constant is, we evaluate at

$$\mathbf{s} = \langle \delta^1, \ldots, \delta^{m-1} \rangle.$$

Then $D(\mathbf{s}) = 1 = D(\mathbf{t})$ for this special choice, and so $D(\mathbf{s}) = D(\mathbf{t})$ in general. □

In order to get a hold on the remaining two properties, we consider an $m \times m$ matrix $\mathbf{t}$ whose last $m - n$ columns are $\delta^{n+1}, \ldots, \delta^m$, and we apply the above lemma repeatedly. We have, first, $D(\mathbf{t}) = D((\mathbf{t})^{mm})$, where $(\mathbf{t})^{mm}$ is the $(m-1) \times (m-1)$ matrix obtained from $\mathbf{t}$ by deleting the last row and the last column. Since this matrix has δ^{m-1} as its last column (δ^{m-1} being now an $(m-1)$-tuple), the same argument shows that its determinant is the same as that of the $(m-2) \times (m-2)$ matrix obtained from it in the same way. We can keep on going as long as the δ-columns last, and thus see that $D(\mathbf{t})$ is the determinant of the $n \times n$ matrix that is the upper left corner of $\mathbf{t}$. If we interpret this in terms of transformations, we have the following lemma.

Lemma 6.3. Suppose that V is m-dimensional and that T in Hom V is the identity on an $(m-n)$-dimensional subspace X. Let Y be a complement of X, and let p be the projection on Y along X. Then $p \circ (T \restriction Y)$ can be considered an element of Hom Y and $\Delta(T) = \Delta_Y(p \circ (T \restriction Y))$.

Proof. Let $\alpha_1, \ldots, \alpha_n$ be a basis for Y, and let $\alpha_{n+1}, \ldots, \alpha_m$ be a basis for X. Then $\{\alpha_i\}_1^m$ is a basis for V, and since $T(\alpha_i) = \alpha_i$ for $i = n+1, \ldots, m$, the matrix for T has δ^i as its ith column for $i = n+1, \ldots, m$. The lemma will therefore follow from our above discussion if we can show that the matrix of $p \circ (T \restriction Y)$ in Hom Y is the $n \times n$ upper left corner of $\mathbf{t}$. The student should be able to see this if he visualizes what vector $(p \circ T)(\alpha_i)$ is for $i \leq n$. □

Corollary. In the above situation if Y is also invariant under T, then $\Delta(T) = \Delta_Y(T \restriction Y)$.

Proof. The proof follows immediately since now $p \circ (T \restriction Y) = T \restriction Y$. □

If the roles of X and Y are interchanged, both being invariant under T and T being the identity on Y, then this same lemma tells us that $\Delta(T) = \Delta_X(T \restriction X)$. If we only know that X and Y are T-invariant, then we can factor T into a commuting product $T = T_1 \circ T_2 = T_2 \circ T_1$, where T_1 and T_2 are of the two more special types discussed above, and so have the rule $\Delta(T) = \Delta(T_1)\,\Delta(T_2) = \Delta_X(T \restriction X)\,\Delta_Y(T \restriction Y)$, another of our properties listed in Chapter 2.

The final rule is also a consequence of the above lemma. If T is the identity on X and also on V/X, then it isn't too hard to see that $p \circ (T \upharpoonright Y)$ is the identity, as an element of Hom Y, and so $\Delta(T) = 1$ by the lemma.

We now prove the theorem concerning "expansion by minors (or cofactors)". Let $\mathbf{t}$ be an $m \times m$ matrix, and let $(\mathbf{t})^{pr}$ be the $(m-1) \times (m-1)$ submatrix obtained from $\mathbf{t}$ by deleting the pth row and rth column. Then,

Theorem 6.3. $D(\mathbf{t}) = \sum_{i=1}^{m} (-1)^{i+r} t_{ir}\, D((\mathbf{t})^{ir})$. That is, we can evaluate $D(\mathbf{t})$ by going down the rth column, multiplying each element by the determinant of the $(m-1) \times (m-1)$ matrix associated with it, and adding. The two occurrences of 'D' in the theorem are of course over dimensions m and $m-1$, respectively.

Proof. Consider $D(\mathbf{t}) = D(\mathbf{t}^1, \ldots, \mathbf{t}^m)$ under the special assumption that $\mathbf{t}^r = \delta^p$. Since $D(\mathbf{t})$ is an alternating linear functional both of the columns of $\mathbf{t}$ and of the rows of $\mathbf{t}$, we can move the rth column and pth row to the right-bottom border, and apply Lemma 6.2. Thus

$$D(\mathbf{t}) = (-1)^{m-r}(-1)^{m-p}\, D(\mathbf{t}^{pr}) = (-1)^{p+r} D(\mathbf{t}^{pr}),$$

assuming that the rth column of $\mathbf{t}$ is δ^r. In general, $\mathbf{t}^r = \sum_{i=1}^{m} t_{ir}\, \delta^i$, and if we expand $D(\mathbf{t}^1, \ldots, \mathbf{t}^m)$ with respect to this sum in the rth place, and if we use the above evaluation of the separate terms of the resulting sum, we get $D(\mathbf{t}) = \sum_{i=1}^{m} (-1)^{i+r}\, t_{ir}\, D(\mathbf{t})^{ir}$. □

Corollary 1. If $s \neq r$, then $\sum_{i=1}^{m} (-1)^{i+r} t_{is}\, D((\mathbf{t})^{ir}) = 0$.

Proof. We now have the expansion of the theorem for a matrix with identical sth and rth columns, and the determinant of this matrix is zero by the alternating property. □

For simpler notation, set $c_{ij} = (-1)^{i+j}\, D((\mathbf{t})^{ij})$. This is called the *cofactor* of the element t_{ij} in $\mathbf{t}$. Our two results together say that

$$\sum_{i=1}^{m} c_{ir} t_{is} = \delta_r^s\, D(\mathbf{t}).$$

In particular, if $D(\mathbf{t}) \neq 0$, then the matrix $\mathbf{s}$ whose entries are $s_{ri} = c_{ir}/D(\mathbf{t})$ is the inverse of $\mathbf{t}$. This observation gives us a neat way to express the solution of a system of linear equations. We want to solve $\mathbf{t} \cdot \mathbf{x} = \mathbf{y}$ for $\mathbf{x}$ in terms of $\mathbf{y}$, supposing that $D(\mathbf{t}) \neq 0$. Since $\mathbf{s}$ is the inverse of $\mathbf{t}$, we have $\mathbf{x} = \mathbf{s} \cdot \mathbf{y}$. That is, $x_j = \sum_{i=1}^{m} s_{ji} y_i = (\sum_{i=1}^{m} y_i c_{ij})/D(\mathbf{t})$ for $j = 1, \ldots, m$. According to our expansion theorem, the numerator in this expression is exactly the determinant d_j of the matrix obtained from $\mathbf{t}$ by replacing its jth column by the m-tuple $\mathbf{y}$. Hence, with d_j defined this way, the solution to $\mathbf{t} \cdot \mathbf{x} = \mathbf{y}$ is the m-tuple

$$\mathbf{x} = \left\langle \frac{d_1}{D(\mathbf{t})}, \ldots, \frac{d_m}{D(\mathbf{t})} \right\rangle.$$

This is *Cramer's rule*. It was stated in slightly different notation in Section 2.5.

7. THE EXTERIOR ALGEBRA

Our final job is to introduce a multiplication operation between alternating n-linear functionals (also now called *exterior n-forms*). We first extend the tensor product operation that we have used to fashion elementary covariant tensors out of functionals.

Definition. If $f \in (V^*)^{\textcircled{n}}$ and $g \in (V^*)^{\textcircled{l}}$, then $f \otimes g$ is that element of $(V^*)^{\textcircled{n+l}}$ defined as follows:

$$f \otimes g(\xi_1, \ldots, \xi_{n+l}) = f(\xi_1, \ldots, \xi_n)g(\xi_{n+1}, \ldots, \xi_{n+l}).$$

We naturally ask how this operation combines with the projection Ω of $(V^*)^{\textcircled{n+l}}$ onto $\mathfrak{a}^{n+l}$.

Theorem 7.1. $\Omega(f \otimes g) = \Omega(f \otimes \Omega g) = \Omega(\Omega f \otimes g)$.

Proof. We have

$$\begin{aligned}\Omega(f \otimes \Omega g) &= \frac{1}{(n+l)!} \sum_{\sigma} (\operatorname{sgn} \sigma)(f \otimes \Omega g)^{\sigma} \\ &= \frac{1}{(n+l)!} \sum_{\sigma} (\operatorname{sgn} \sigma) \left(f \otimes \frac{1}{l!} \sum_{\rho} (\operatorname{sgn} \rho) g^{\rho}\right)^{\sigma} \\ &= \frac{1}{(n+l)!\,l!} \sum_{\sigma,\rho} (\operatorname{sgn} \sigma)(\operatorname{sgn} \rho)(f \otimes g^{\rho})^{\sigma}.\end{aligned}$$

We can regard ρ as acting on the full $n + l$ places of $f \otimes g$ by taking it as the identity on the first n places. Then $(f \otimes g^{\rho})^{\sigma} = (f \otimes g)^{\rho\sigma}$. Set $\rho\sigma = \sigma'$. For each σ' there are exactly $l!$ pairs $\langle \rho, \sigma \rangle$ with $\rho\sigma = \sigma'$, namely, the pairs $\{\langle \rho, \rho^{-1}\sigma' \rangle : \rho \in \mathcal{S}_l\}$. Thus the above sum is

$$\frac{1}{(n+l)!} \sum_{\sigma'} (\operatorname{sgn} \sigma')(f \otimes g)^{\sigma'} = \Omega(f \otimes g).$$

The proof for $\Omega(\Omega f \otimes g)$ is essentially the same. □

Definition. If $f \in \mathfrak{a}^n$ and $g \in \mathfrak{a}^l$, then $f \wedge g = \binom{n+l}{n}\Omega(f \otimes g)$.

Lemma 7.1. $f_1 \wedge f_2 \wedge \cdots \wedge f_k = (n!/n_1!n_2! \cdots n_k!)\Omega(f_1 \otimes \cdots \otimes f_k)$, where n_i is the order of f_i, $i = 1, \ldots, k$, and $n = \sum_1^k n_i$.

Proof. This is simply an induction, using the definition of the wedge operation $\wedge$ and the above theorem. □

Corollary. If $\lambda_i \in V^*$, $i = 1, \ldots, n$, then

$$\lambda_1 \wedge \cdots \wedge \lambda_n = n!\Omega(\lambda_1 \otimes \cdots \otimes \lambda_n).$$

In particular, if $q_1 < \cdots < q_n$ and $\{\mu_i\}_1^m$ is a basis for V^*, then

$$\mu_{q_1} \wedge \cdots \wedge \mu_{q_n} = n!\Omega(\mu_{\mathbf{q}}) = \text{the basis element } \nu_{\mathbf{q}} \text{ of } \mathfrak{a}^n.$$

Theorem 7.2. If $f \in \mathcal{A}^n$ and $g \in \mathcal{A}^l$, then $g \wedge f = (-1)^{ln} f \wedge g$. In particular, $\lambda \wedge \lambda = 0$ for $\lambda \in V^*$.

Proof. We have $g \otimes f = (f \otimes g)^\sigma$, where σ is the permutation moving each of the last l places over each of the first n places. Thus σ is the product of ln transpositions, $(\operatorname{sgn} \sigma) = (-1)^{ln}$, and

$$\Omega(g \otimes f) = \Omega(f \otimes g)^\sigma = (\operatorname{sgn} \sigma)\Omega(f \otimes g) = (-1)^{ln}\Omega(f \otimes g).$$

We multiply by $\binom{n+l}{n}$ and have the theorem. □

Corollary. If $\{\lambda_i\}_1^n \subset V^*$, then $\lambda_1 \wedge \cdots \wedge \lambda_n = 0$ if and only if the sequence $\{\lambda_i\}_1^n$ is dependent.

Proof. If $\{\lambda_i\}$ is independent, it can be extended to a basis for V^*, and then $\lambda_1 \wedge \cdots \wedge \lambda_n$ is some basis vector $\nu_\mathbf{q}$ of $\mathcal{A}^n$ by the above corollary. In particular, $\lambda_1 \wedge \cdots \wedge \lambda_n \neq 0$.

If $\{\lambda_i\}$ is dependent, then one of its elements, say λ_1, is a linear combination of the rest, $\lambda_1 = \sum_2^n c_i\lambda_i$ and $\lambda_1 \wedge \lambda_2 \wedge \cdots \wedge \lambda_n = \sum_{i=2}^n c_i\lambda_i \wedge (\lambda_2 \wedge \cdots \wedge \lambda_n)$. The ith of these terms repeats λ_i, and so is 0 by the lemma and the above corollary. □

Lemma 7.2. The mapping $\langle f, g \rangle \mapsto f \wedge g$ is a bilinear mapping from $\mathcal{A}^n \times \mathcal{A}^l$ to $\mathcal{A}^{n+l}$.

Proof. This follows at once from the obvious bilinearity of $f \otimes g$. □

We conclude with an important extension theorem.

Theorem 7.3. Let θ be the alternating n-linear map

$$\langle \lambda_1, \ldots, \lambda_n \rangle \mapsto \lambda_1 \wedge \cdots \wedge \lambda_n$$

from $(V^*)^n$ to $\mathcal{A}^n$. Then for any alternating n-linear functional $F(\lambda_1, \ldots, \lambda_n)$ on $(V^*)^n$, there is a uniquely determined *linear* functional G on $\mathcal{A}^n$ such that $F = G \circ \theta$. The mapping $G \mapsto F$ is thus a canonical isomorphism from $(\mathcal{A}^n)^*$ to $\mathcal{A}^n(V^*)$.

Proof. The straightforward way to prove this is to define G by establishing its necessary values on a basis, using the equation $F = G \circ \theta$, and then to show from the linearity of G, the alternating multilinearity of

$$\langle \lambda_1, \ldots, \lambda_n \rangle \mapsto \lambda_1 \wedge \cdots \wedge \lambda_n,$$

and the alternating multilinearity of F that the identity $F = G \circ \theta$ holds everywhere. This computation becomes notationally complex. Instead, we shall be devious. We shall see that by proving more than the theorem asserts we get a shorter proof of the theorem.

We consider the space $\mathcal{A}^n(V^*)$ of all alternating n-linear functions on $(V^*)^n$. We know from Theorem 5.2 that $d(\mathcal{A}^n(V^*)) = \binom{m}{n}$, since $d(V^*) = d(V) = m$.

Now for each functional G in $(\mathcal{A}^n)^*$, the functional $G \circ \theta$ is alternating and n-linear, and so $G \mapsto F = G \circ \theta$ is a mapping from $(\mathcal{A}^n)^*$ to $\mathcal{A}^n(V^*)$ which is clearly linear. Moreover it is injective, for if $G \neq 0$, then $F(\mu_{q(1)}, \ldots, \mu_{q(n)}) = G(\nu_{\mathbf{q}}) \neq 0$ for some basis vector $\nu_{\mathbf{q}} = \mu_{q(1)} \wedge \cdots \wedge \mu_{q(n)}$ of $\mathcal{A}^n(V^*)$. Since $d(\mathcal{A}^n(V^*)) = \binom{m}{n} = d(\mathcal{A}^n) = d((\mathcal{A}^n)^*)$, the mapping is an isomorphism (by the corollary of Theorem 2.4, Chapter 2). In particular, every F in $\mathcal{A}^n(V^*)$ is of the form $G \circ \theta$. □

It can be shown further that the property asserted in the above theorem is an abstract characterization of $\mathcal{A}^n$. By this we mean the following. Suppose that a vector space X and an alternating mapping φ from $(V^*)^n$ to X are given, and suppose that every alternating functional F on $(V^*)^n$ extends uniquely to a linear functional G on X (that is, $F = G \circ \varphi$). Then X is isomorphic to $\mathcal{A}^n$, and in such a way that φ becomes θ.

To see this we simply note that the hypothesis of unique extensibility is exactly the hypothesis that $\Phi\colon G \mapsto F = G \circ \varphi$ is an isomorphism from X^* to $\mathcal{A}^n(V^*)$. The theorem gave an isomorphism Θ from $(\mathcal{A}^n)^*$ to $\mathcal{A}^n(V^*)$, and the adjoint $(\Phi^{-1} \circ \Theta)^*$ is thus an isomorphism from X^{**} to $(\mathcal{A}^n)^{**}$, that is, from X to $\mathcal{A}^n$. We won't check that φ "becomes" θ.

By virtue of Corollary 1 of Theorem 6.2, the identity $D(\mathbf{t}) = D(\mathbf{t}^*)$ is the matrix form of the more general identity $\Delta(T) = \Delta(T^*)$, and it is interesting to note the "coordinate free" proof of this equation. Here, of course, $T \in \operatorname{Hom} V$.

We first note that the identity $(T^*\lambda)(\xi) = \lambda(T\xi)$ carries through the definitions of $\otimes$ and $\wedge$ to give

$$T^*\lambda_1 \wedge \cdots \wedge T^*\lambda_n(\xi_1, \ldots, \xi_n) = \lambda_1 \wedge \cdots \wedge \lambda_n(T\xi_1, \ldots, T\xi_n). \qquad (*)$$

Also, $\mathrm{ev}_{\boldsymbol{\xi}}\colon \lambda_1 \wedge \cdots \wedge \lambda_n \mapsto \lambda_1 \wedge \cdots \wedge \lambda_n(\xi_1, \ldots, \xi_n)$ is an alternating n-linear functional on $\mathcal{A}_n(V^*)$ for each $\boldsymbol{\xi} \in V^n$. The left member of $(*)$ is thus $\mathrm{ev}_{\boldsymbol{\xi}}(T^*\lambda_1, \ldots, T^*\lambda_n)$, and, if $n = \dim V$, this is $\Delta(T^*)\mathrm{ev}_{\boldsymbol{\xi}}(\lambda_1, \ldots, \lambda_n)$ by the definition of Δ. By the same definition the right side of $(*)$ becomes

$$\Delta(T)[\lambda_1 \wedge \cdots \wedge \lambda_n(\xi_1, \ldots, \xi_n)] = \Delta(T)\mathrm{ev}_{\boldsymbol{\xi}}(\lambda_1, \ldots, \lambda_n).$$

Thus $(*)$ implies the identity $\Delta(T^*)\mathrm{ev}_{\boldsymbol{\xi}} = \Delta(T)\mathrm{ev}_{\boldsymbol{\xi}}$. Since $\mathrm{ev}_{\boldsymbol{\xi}} \neq 0$ if $\boldsymbol{\xi} = \{\xi_i\}_1^n$ is independent, we have proved that $\Delta(T^*) = \Delta(T)$.

We call a wedge product $\lambda_1 \wedge \cdots \wedge \lambda_n$ of functionals $\lambda_i \in V^*$ a *multivector*. We saw above that $\lambda_1 \wedge \cdots \wedge \lambda_n \neq 0$ if and only if $\{\lambda_i\}_1^n$ is independent, in which case $\{\lambda_i\}_1^n$ spans an n-dimensional subspace of V^*. The following lemma shows that this geometric connection is not accidental.

Lemma 7.3. Two independent n-tuples $\{\lambda_i\}_1^n$ and $\{\mu_i\}_1^n$ in V^* have the same linear span if and only if $\mu_1 \wedge \cdots \wedge \mu_n = k(\lambda_1 \wedge \cdots \wedge \lambda_n)$ for some k.

Proof. If $\{\mu_j\}_1^n \subset L(\{\lambda_i\}_1^n)$, then each μ_j is a linear combination of the λ_i's, and if we expand $\mu_1 \wedge \cdots \wedge \mu_n$ according to these basis expansions, we get $k(\lambda_1 \wedge \cdots \wedge \lambda_n)$. If, furthermore, $\{\mu_i\}_1^n$ is independent, then k cannot be zero.

Now suppose, conversely, that $\mu_1 \wedge \cdots \wedge \mu_n = k(\lambda_1 \wedge \cdots \wedge \lambda_n)$, where $k \neq 0$. This implies first that $\{\mu_i\}_1^n$ is independent, and then that

$$\mu_j \wedge (\lambda_1 \wedge \cdots \wedge \lambda_n) = 0$$

for each j, so that each μ_j is dependent on $\{\lambda_i\}_1^n$. Together, these two consequences imply that the set $\{\mu_i\}_1^n$ has the same linear span as $\{\lambda_i\}_1^n$. □

This lemma shows that a multivector has a relationship to the subspace it determines like that of a single vector to its span.

8. EXTERIOR POWERS OF SCALAR PRODUCT SPACES

Let V be a finite-dimensional vector space, and let $(\ ,\ )$ be a nondegenerate (nonsingular) symmetric bilinear form on V. In this and the next section we shall call any such bilinear form a scalar product, even though it may not be positive definite. We know that the bilinear form $(\ ,\ )$ induces an isomorphism of V with V^* sending $y \in V$ into $\overline{y} \in V^*$, where $\overline{y}(x) = \langle x, \overline{y}\rangle = (x, y)$ for all $x \in V$. We then get a nondegenerate form (scalar product), which we shall continue to denote by $(\ ,\ )$, on V^* by setting $(\overline{u}, \overline{v}) = (u, v)$. We also obtain a nondegenerate scalar product on $\mathcal{A}^q$ by setting

$$(\overline{u}_1 \wedge \cdots \wedge \overline{u}_q, \overline{v}_1 \wedge \cdots \wedge \overline{v}_q) = \det\left((\overline{u}_i, \overline{v}_j)\right). \tag{8.1}$$

To check that (8.1) makes sense, we first remark that for fixed $\overline{v}_1, \ldots, \overline{v}_q \in V^*$, the right-hand side of (8.1) is an antisymmetric multilinear function of the vectors $\overline{u}_1, \ldots, \overline{u}_q$, and therefore extends to a linear function on $\mathcal{A}^q(V^*)$ by Theorem 7.3. Similarly, holding the $\overline{u}$'s fixed determines a linear function on $\mathcal{A}^q(V^*)$, and (8.1) is well defined and extends to a bilinear function on $\mathcal{A}^q(V^*)$. The right-hand side of (8.1) is clearly symmetric in u and v, so that the bilinear form we get is indeed symmetric. To see that it is nondegenerate, let us choose a basis $u_1, \ldots, u_n$ so that

$$(u_i, u_j) = \pm\delta_{ij}. \tag{8.2}$$

(We can always find such a basis by Theorem 7.1 of Chapter 2.) We know that $\{\overline{u}_{\mathbf{i}}\} = \{\overline{u}_{i_1} \wedge \cdots \wedge \overline{u}_{i_q}\}$ forms a basis for $\mathcal{A}^q$, where $\mathbf{i} = \langle i_1, \ldots, i_q\rangle$ ranges over all q-tuplets of integers such that $1 \leq i_1 < \cdots < i_q \leq n$, and we claim that

$$(\overline{u}_{\mathbf{i}}, \overline{u}_{\mathbf{j}}) = \pm\delta_{\mathbf{ij}}. \tag{8.3}$$

In fact, if $\mathbf{i} \neq \mathbf{j}$, then $i_r \neq j_s$ for some value of r between 1 and q and for all s. In this case one whole row of the matrix (u_{i_l}, u_{j_m}) vanishes, namely, the rth row. Thus (8.1) gives zero in this case. If $\mathbf{i} = \mathbf{j}$, then (8.2) says that the matrix has ± 1 down the diagonal and zeros elsewhere, establishing (8.3), and thus the fact that $(\ ,\ )$ is nondegenerate on $\mathcal{A}^q$. In particular, we have

$$(u_1 \wedge \cdots \wedge u_n, u_1 \wedge \cdots \wedge u_n) = (-1)^{\#}, \tag{8.4}$$

where # is the number of minus signs occurring in (8.3).

9. THE STAR OPERATOR

Let V be a finite-dimensional vector space endowed with a nondegenerate scalar product as in Section 8. The space $\mathfrak{A}^n$ is one-dimensional if n is the dimension of V. The induced scalar product on $\mathfrak{A}^n$ is nondegenerate, so that $(\overline{u}, \overline{u})$ is either always positive or always negative for all nonzero $\overline{u} \in \mathfrak{A}^n$. In particular, there are exactly two $\overline{u}$'s in $\mathfrak{A}^n$ with $(\overline{u}, \overline{u}) = \pm 1$. Let us choose one of them and hold it fixed for the remainder of this section. Geometrically, this amounts to choosing an orientation on V. We thus have picked a

$$\overline{u} \in \mathfrak{A}^n \qquad \text{with} \qquad (\overline{u}, \overline{u}) = (-1)^{\#}. \tag{9.1}$$

Let $\overline{v}$ be some fixed element of $\mathfrak{A}^q$. Then for any $\overline{y} \in \mathfrak{A}^{n-q}$, $\overline{v} \wedge \overline{y} \in \mathfrak{A}^n$, and so we can write $\overline{v} \wedge \overline{y} = f_{\overline{v}}(\overline{y})\overline{u}$, where $f_{\overline{v}}(\overline{y})$ depends linearly on $\overline{y}$. Since the induced scalar product $(\ ,\ )$ on $\mathfrak{A}^{n-q}$ is nondegenerate, there is a unique element $*\overline{v} \in \mathfrak{A}^{n-q}$ such that $(\overline{y}, *\overline{v}) = f_{\overline{v}}(\overline{y})$. To repeat, we have assigned a $*\overline{v} \in \mathfrak{A}^{n-q}$ to each $\overline{v} \in \mathfrak{A}^q$ by setting

$$(\overline{y}, *\overline{v})\overline{u} = \overline{v} \wedge \overline{y}. \tag{9.2}$$

We have thus defined a map, $*$, from $\mathfrak{A}^q$ to $\mathfrak{A}^{n-q}$. It is clear from (9.2) that this map is linear. Let $u_1, \ldots, u_n$ be a basis for V satisfying (8.2) and also $\overline{u} = u_1 \wedge \cdots \wedge u_n$, and construct the corresponding bases for the spaces $\mathfrak{A}^q$ and $\mathfrak{A}^{n-q}$. Then $\overline{u}_{\mathbf{i}} \wedge \overline{u}_{\mathbf{j}} = 0$ if any i_l occurring in the q-tuplet $\mathbf{i}$ also occurs in $\mathbf{j}$. If no i_l occurs in $\mathbf{j}$ then

$$\overline{u}_{\mathbf{i}} \wedge \overline{u}_{\mathbf{j}} = \epsilon_{i_1 \ldots i_q j_1 \ldots j_{n-q}} \overline{u},$$

where $\epsilon_{\mathbf{k}} = \operatorname{sgn} \mathbf{k}$.

If we compare this with (9.2) and (8.3), we see that

$$*\overline{u}_{\mathbf{i}} = \pm \epsilon_{i_1 \ldots i_q j_1 \ldots j_{n-q}} \overline{u}_{\mathbf{j}}, \tag{9.3}$$

where the sign is the same as that occurring in (8.3), i.e., the sign is positive or negative according as the number of j_l with $(\overline{u}_{j_l}, \overline{u}_{j_l}) = -1$ which appear in $\mathbf{j}$ is even or odd. Applying $*$ to (9.3), we see that

$$**\overline{v} = (-1)^{q(n-q)+\#} v \qquad \text{for all} \quad v \in \mathfrak{A}^q. \tag{9.4}$$

Let v and w be elements of $\mathfrak{A}^q$. Then

$$(*v, *w)u = v \wedge *w = (-1)^{q(n-q)} *w \wedge v = (-1)^{q(n-q)}(**w, v)u.$$

If we apply (9.4), we see that

$$(*v, *w) = (-1)^{\#}(v, w). \tag{9.5}$$

CHAPTER 8

INTEGRATION

1. INTRODUCTION

In this chapter we shall present a theory of integration in n-dimensional Euclidean space $\mathbb{E}^n$, which the reader will remember is simply Cartesian n-space $\mathbb{R}^n$ together with the standard scalar product. Our main item of business is to introduce a notion of size for subsets of $\mathbb{E}^n$ (area in two dimensions, volume in three . . .). Before proceeding to the formal definitions, let us see what properties we would like our notion of size to have. We are looking for a function μ which assigns a number $\mu(A)$ to bounded subsets $A \subset \mathbb{E}^n$.

i) We would like $\mu(A)$ to be a nonnegative real number.

ii) If $A \subset B$, we would expect to have $\mu(A) \leq \mu(B)$.

iii) If A and B are disjoint (that is, $A \cap B = \emptyset$), then we would expect to have $\mu(A \cup B) = \mu(A) + \mu(B)$.

iv) Let T be any Euclidean motion.* For any set A let TA be the set of all points of the form Tx, where $x \in A$. We then would expect to have $\mu(TA) = \mu(A)$. (Thus we want "congruent" sets to have the same size.)

v) We would expect a "lower-dimensional set" (where this is suitably defined) to have zero size. Thus points in the line, curves in the plane, surfaces in three-space, etc., should all have zero size.

vi) By the same token, we would expect open sets to have positive size.

In the above discussion we did not specify what kind of sets we were talking about. One might be ambitious and try to assign a size to every subset of $\mathbb{E}^n$. This proves to be impossible, however, for the following reason: Let U and V be any two bounded open subsets of $\mathbb{E}^3$. It can be shown† that we can find

* Recall that a Euclidean motion is an isometry of $\mathbb{E}^n$ and can thus be represented as the composition of the translation and an orthogonal transformation.

† S. Banach and A. Tarski, Sur la décomposition des ensembles de pointes en partie respectivement congruentes, *Fund. Math.* **6**, 244–277 (1924). R. M. Robinson, On the decomposition of spheres, *Fund. Math.* **34**, 246–260 (1947).

decompositions

$$U = \bigcup_{i=1}^{k} U_i \qquad \text{and} \qquad V = \bigcup_{i=1}^{k} V_i$$

with $U_i \cap U_j = \emptyset = V_i \cap V_j$ for $i \neq j$, and Euclidean motions T_i with $T_iU_i = V_i$. In other words, we can break up U into finitely many pieces, move these pieces around, and then recombine them to get V. Needless to say, the sets U_i will have to look very bad. A moment's reflection shows that if we wish to assign a size to *all* subsets (including those like U_i), we cannot satisfy (ii), (iii), (iv), and (vi). In fact, (iii) [repeated $(k - 1)$ times] implies that

$$\mu(U) = \sum_{i=1}^{k} \mu(U_i),$$

and (iv) implies that $\mu(U_i) = \mu(V_i)$. Thus $\mu(U) = \mu(V)$, or the size of any two open sets would coincide. Since any open set contains two disjoint open sets, this implies, by (ii), that $\mu(U) \geq 2\mu(U)$, so $\mu(U) = 0$.

We are thus faced with a choice. Either we dispense with some of requirements (i) through (vi) above, or we do not assign a size to every subset of $\mathbb{E}^n$. Since our requirements are reasonable, we prefer the second alternative. This means, of course, that now, in addition to introducing a notion of size, we must describe the class of "good" sets we wish to admit.

We shall proceed axiomatically, listing some "reasonable" axioms for a class of subsets and a function μ.

2. AXIOMS

Our axioms will concern a class $\mathfrak{D}$ of subsets of $\mathbb{E}^n$ and a function μ defined on $\mathfrak{D}$. (That is, $\mu(A)$ is defined if A is a subset of $\mathbb{E}^n$ belonging to our collection $\mathfrak{D}$.)

I. $\mathfrak{D}$ is a collection of subsets of $\mathbb{E}^n$ such that:

$\mathfrak{D}$1. If $A \in \mathfrak{D}$ and $B \in \mathfrak{D}$, then $A \cup B \in \mathfrak{D}$, $A \cap B \in \mathfrak{D}$, and $A - B \in \mathfrak{D}$.

$\mathfrak{D}$2. If $A \in \mathfrak{D}$ and T is a translation, then $TA \in \mathfrak{D}$.

$\mathfrak{D}$3. The set $\square_0^1 = \{x : 0 \leq x^i < 1\}$ belongs to $\mathfrak{D}$.

II. The real-valued function μ has the following properties:

μ1. $\mu(A) \geq 0$ for all $A \in \mathfrak{D}$.

μ2. If $A \in \mathfrak{D}$, $B \in \mathfrak{D}$, and $A \cap B = \emptyset$, then $\mu(A \cup B) = \mu(A) + \mu(B)$.

μ3. For any $A \in \mathfrak{D}$ and any translation T, we have $\mu(TA) = \mu(A)$.

μ4. $\mu(\square_0^1) = 1$.

Before proceeding, some remarks about our axioms are in order. Axiom $\mathfrak{D}$1 will allow us to perform elementary set-theoretical operations with the elements of $\mathfrak{D}$. Note that in Axioms $\mathfrak{D}$2 and μ3 we are only allowing *translations*, but in our list of desired properties we wanted proper behavior with respect to *all* Euclidean motions in (iv). The reason for this is that we shall show that for "good" choices of $\mathfrak{D}$, the axioms, as they stand, uniquely determine μ. It will then turn out that μ actually satisfies the stronger condition (iv), while we assume the weaker condition $\mu 3$ as an axiom.

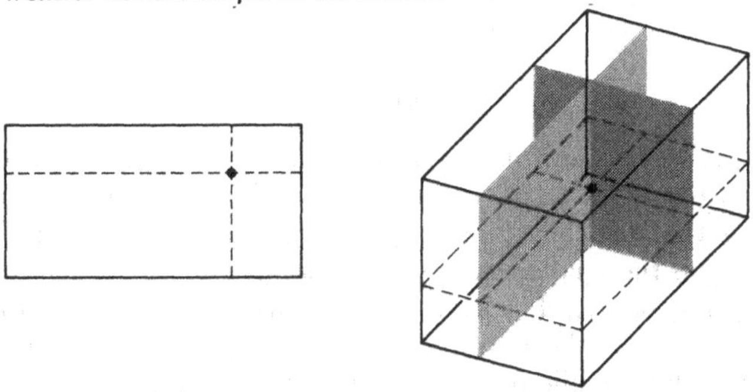

Fig. 8.1

Axiom $\mathfrak{D}3$ guarantees that our theory is not completely trivial, i.e., the collection $\mathfrak{D}$ is not empty. Axiom $\mu 4$ has the effect of normalizing μ. Without it, any μ satisfying $\mu 1$, $\mu 2$, and $\mu 3$ could be multiplied by any nonnegative real number, and the new function μ' so obtained would still satisfy our axioms. In particular, $\mu 4$ guarantees that we do not choose μ to be the trivial function assigning to each A the value zero.

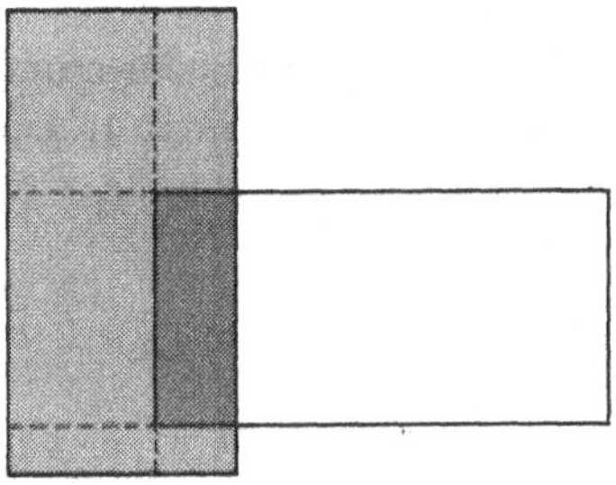

Fig. 8.2

Our program for the next few sections is to make some reasonable choices for $\mathfrak{D}$ and to show that for the given $\mathfrak{D}$ there exists a unique μ satisfying $\mu 1$ through $\mu 4$.

An important elementary consequence of the $\mathfrak{D}, \mu$-axioms that we shall frequently use without comment is:

$\mu 5$. If $A \subset \bigcup_1^n A_i$ and all the sets are in $\mathfrak{D}$, then $\mu(A) \leq \sum_1^n \mu(A_i)$.

Our beginning work will be largely combinational. We will first consider (generalized) rectangles, which are just Cartesian products of intervals, and the way in which a point inside a rectangle determines a splitting of the rectangle into a collection of smaller rectangles, as indicated in Fig. 8.1. This is associated with the fact that the intersection of any two rectangles is a rectangle and the difference of two rectangles is a finite disjoint union of rectangles (see Fig. 8.2).

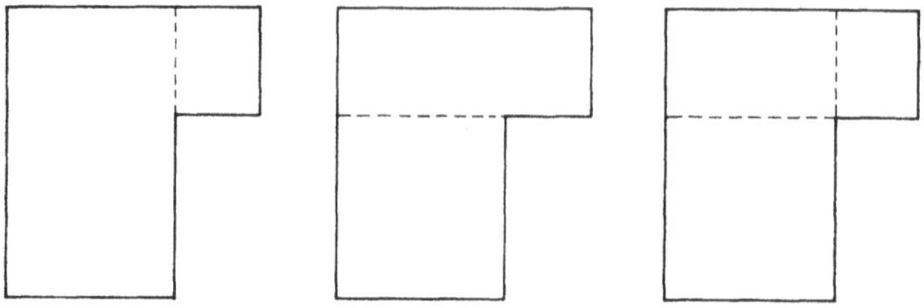

Fig. 8.3

We call a set A *paved* if it can be expressed as the union of a finite disjoint collection p of rectangles (a paving of A). It will follow from our combinational considerations that the collection $\mathfrak{D}_{\min}$ of all the paved sets satisfies Axioms $\mathfrak{D}1$ through $\mathfrak{D}3$ and is the smallest family that does: any other collection $\mathfrak{D}$ satisfying the axioms includes $\mathfrak{D}_{\min}$. It will then follow that if μ satisfies $\mu 1$ through $\mu 4$ on $\mathfrak{D}_{\min}$, then it must have the natural value (the product of the lengths of the sides) for a rectangle. This implies that μ is uniquely defined on $\mathfrak{D}_{\min}$ by requirements $\mu 1$ through $\mu 4$, since the value $\mu(A)$ for any paved set A must be the sum of the natural values for the rectangles in a paving of A. The existence of μ on $\mathfrak{D}_{\min}$ thus depends on the crucial lemma that two different pavings of the set A give the same sum. (See Fig. 8.3.)

This comes down to the fact that the "intersection" of two pavings of A is a third paving "finer" than either, and the fact that when a single rectangle is broken up, the natural values of μ for the pieces add up to μ for the fragmented rectangle.

All these considerations are elementary but exceedingly messy in detail. We give the proofs below for the reader to refer to in case of doubt, but he may prefer to study only the definitions and statements of results and then to proceed to Section 6.

3. RECTANGLES AND PAVED SETS

We first introduce some notation and terminology. Let $\mathbf{a} = \langle a^1, \ldots, a^n \rangle$ and $\mathbf{b} = \langle b^1, \ldots, b^n \rangle$ be elements of $\mathbb{E}^n$. By the *rectangle* $\square_{\mathbf{a}}^{\mathbf{b}}$ we shall mean the set of all $\mathbf{x} = \langle x^1, \ldots, x^n \rangle$ in $\mathbb{E}^n$ with $a^i \leq x^i < b^i$. Thus

$$\square_{\mathbf{a}}^{\mathbf{b}} = \{\mathbf{x} : a^i \leq x^i < b^i, \ i = 1, \ldots, n\}. \tag{3.1}$$

Note that in order for $\square_{\mathbf{a}}^{\mathbf{b}}$ to be nonempty, we must have $a^i < b^i$ for all i. In other words,

$$\square_{\mathbf{a}}^{\mathbf{b}} = \emptyset \qquad \text{if} \quad a^i \geq b^i \text{ for some } i. \tag{3.2}$$

In the plane ($n = 2$) for instance, our rectangles $\square_{\mathbf{a}}^{\mathbf{b}}$ correspond to ordinary Euclidean rectangles whose *sides are parallel to the axes.* (We should perhaps use an additional adjective and call our sets level rectangles, braced rectangles, or something else, but for simplicity we shall just call them rectangles.) Note that in the plane our rectangles include the left-hand and lower edges but not the right-hand and upper ones (see Fig. 8.4).

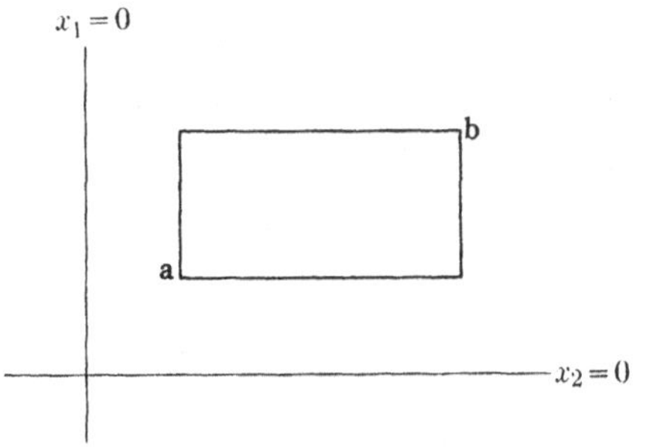

Fig. 8.4

For general n, if we set $\mathbf{1} = \langle 1, \ldots, 1 \rangle$, then our notation coincides with that of $\mathfrak{D}3$.

We now collect some elementary facts about rectangles. It follows immediately from the definition (3.1) that if $\mathbf{a} = \langle a^1, \ldots, a^k \rangle$, $\mathbf{b} = \langle b^1, \ldots, b^n \rangle$, etc., then

$$\square_{\mathbf{a}}^{\mathbf{b}} \cap \square_{\mathbf{c}}^{\mathbf{d}} = \square_{\mathbf{e}}^{\mathbf{f}}, \tag{3.3}$$

where

$$e^i = \max(a^i, c^i) \qquad \text{and} \qquad f^i = \min(b^i, d^i), \qquad i = 1, \ldots, n.$$

(The reader should draw various different instances of this equation in the plane to get the correct geometrical feeling.) Note that the case where $\square_{\mathbf{a}}^{\mathbf{b}} \cap \square_{\mathbf{c}}^{\mathbf{d}} = \emptyset$ is included in (3.3) by (3.2). Another immediate consequence of the definition (3.1) is

$$T\,\square_{\mathbf{a}}^{\mathbf{b}} = \square_{T\mathbf{a}}^{T\mathbf{b}} \qquad \textit{for any translation } T. \tag{3.4}$$

We will now establish some elementary results which will imply that any $\mathfrak{D}$ satisfying Axioms $\mathfrak{D}1$ through $\mathfrak{D}3$ must contain all rectangles.

Lemma 3.1. Any rectangle $\square_{\mathbf{a}}^{\mathbf{b}}$ can be written as the disjoint union

$$\square_{\mathbf{a}}^{\mathbf{b}} = \bigcup_{r=1}^{k} \square_{\mathbf{a}_r}^{\mathbf{b}_r}, \qquad \text{where} \qquad \mathbf{b}_r - \mathbf{a}_r \in \square_{\mathbf{0}}^{\mathbf{1}}.$$

(What this says is that any "big" rectangle can be written as a finite union of "small" ones.)

Proof. We may assume that $\square_{\mathbf{a}}^{\mathbf{b}} \neq \emptyset$ (otherwise take $k = 0$ in the union). Thus $b^i > a^i$. In particular, if we choose the integer m sufficiently large, $(1/2^m)(\mathbf{b} - \mathbf{a})$ will lie in $\square_{\mathbf{0}}^{\mathbf{1}}$.

By induction, it therefore suffices to prove that we can decompose $\square_{\mathbf{a}}^{\mathbf{b}}$ into the disjoint union

$$\square_{\mathbf{a}}^{\mathbf{b}} = \bigcup_{s=1}^{2^n} \square_{\mathbf{c}_s}^{\mathbf{d}_s} \qquad \text{with} \qquad \mathbf{d}_s - \mathbf{c}_s = \tfrac{1}{2}(\mathbf{b} - \mathbf{a}). \tag{3.5}$$

(For then we can continue to subdivide until the rectangles we get are small enough.)

We get this subdivision in the obvious way by choosing the vertex "in the middle" of the rectangle and considering all rectangles obtained by cutting $\square_{\mathbf{a}}^{\mathbf{b}}$ through this point by coordinate hyperplanes. To write down an explicit

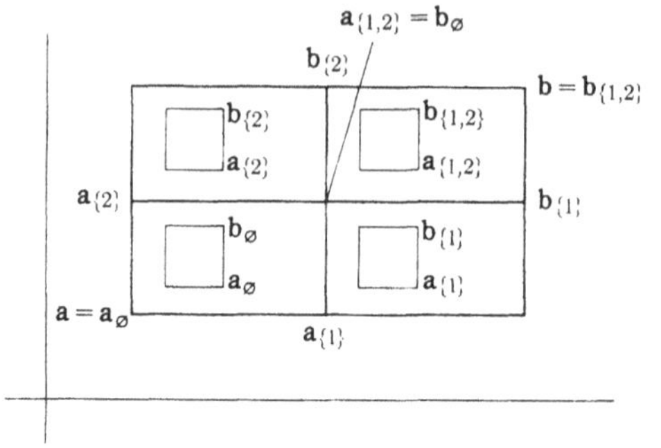

Fig. 8.5

formula, it will be convenient to use the set of all subsets of $\{1, \ldots, n\}$ as an indexing set, rather than the integers $1, \ldots, 2^n$. Let J denote an arbitrary subset of $\{1, 2, \ldots, n\}$. Let $\mathbf{a}_J = \langle a_J^1, \ldots, a_J^n \rangle$ and $\mathbf{b}_J = \langle b_J^1, \ldots, b_J^n \rangle$ be given by

$$a_J^i = \begin{cases} a^i + \dfrac{b^i - a^i}{2} & \text{if } i \in J, \\ a^i & \text{if } i \notin J \end{cases} \quad \text{and} \quad b_J^i = \begin{cases} b^i & \text{if } i \in J, \\ b^i - \dfrac{b^i - a^i}{2} & \text{if } i \notin J. \end{cases}$$

Then any $x \in \Box_{\mathbf{a}}^{\mathbf{b}}$ lies in one and only one $\Box_{\mathbf{a}_J}^{\mathbf{b}_J}$. In other words, $\Box_{\mathbf{a}_J}^{\mathbf{b}_J} \cap \Box_{\mathbf{a}_K}^{\mathbf{b}_K} = \varnothing$ if $J \neq K$ and $\bigcup_{\text{all } J} \Box_{\mathbf{a}_J}^{\mathbf{b}_J} = \Box_{\mathbf{a}}^{\mathbf{b}}$. (The case where $n = 2$ is shown in Fig. 8.5.) Since $\mathbf{b}_J - \mathbf{a}_J = \frac{1}{2}(\mathbf{b} - \mathbf{a})$ for all J, we have proved the lemma. □

We now observe that for any $\mathbf{c} \in \Box_{\mathbf{0}}^{\mathbf{1}}$ we have, by (3.3),

$$\Box_{\mathbf{0}}^{\mathbf{c}} = \Box_{\mathbf{0}}^{\mathbf{1}} \cap \Box_{\mathbf{c}-\mathbf{1}}^{\mathbf{c}}. \tag{3.6}$$

Let T_v denote translation through the vector v. Then $\Box_{\mathbf{c}-\mathbf{1}}^{\mathbf{c}} = T_{\mathbf{c}-\mathbf{1}} \Box_{\mathbf{0}}^{\mathbf{1}}$ by (3.4). Thus by Axioms $\mathfrak{D}2$ and $\mathfrak{D}3$ the rectangle $\Box_{\mathbf{c}-\mathbf{1}}^{\mathbf{c}}$ must belong to $\mathfrak{D}$. By (3.6) and Axiom $\mathfrak{D}1$ we conclude that $\Box_{\mathbf{0}}^{\mathbf{c}} \in \mathfrak{D}$ for any $\mathbf{c} \in \Box_{\mathbf{0}}^{\mathbf{1}}$.

Observe that $T_{\mathbf{a}} \Box_{\mathbf{0}}^{\mathbf{b}-\mathbf{a}} = \Box_{\mathbf{a}}^{\mathbf{b}}$ by (3.4). Thus

$$\Box_{\mathbf{a}}^{\mathbf{b}} \in \mathfrak{D} \qquad \text{whenever} \qquad \mathbf{b} - \mathbf{a} \in \Box_{\mathbf{0}}^{\mathbf{1}}.$$

If we now apply Lemma 3.1 we conclude that

$$\Box_{\mathbf{a}}^{\mathbf{b}} \in \mathfrak{D} \qquad \text{for all} \quad \mathbf{a} \text{ and } \mathbf{b}. \tag{3.7}$$

We make the following definition.

Definition 3.1. A subset $S \subset \mathbb{E}^n$ will be called a *paved set* if S is the disjoint union of finitely many rectangles.

We can then assert:

Proposition 3.1. Any $\mathfrak{D}$ satisfying Axioms $\mathfrak{D}1$ through $\mathfrak{D}3$ must contain all paved sets. Let $\mathfrak{D}_{\min}$ denote the collection of all finite unions of rectangles; then $\mathfrak{D}_{\min}$ satisfies Axioms $\mathfrak{D}1$ through $\mathfrak{D}3$.

Proof. We have already proved the first part of this proposition. We leave the second part as an exercise for the reader.

4. THE MINIMAL THEORY

We are now going to see how far μ is determined by Axioms $\mu 1$ through $\mu 3$. In fact, we are going to show the $\mu(\square_{\mathbf{a}}^{\mathbf{b}})$ is what it should be; i.e., if

$$\mathbf{a} = \langle a^1, \ldots, a^n \rangle \quad \text{and} \quad \mathbf{b} = \langle b^1, \ldots, b^n \rangle,$$

then we must have

$$\mu(\square_{\mathbf{a}}^{\mathbf{b}}) = \begin{cases} 0 & \text{if } \square_{\mathbf{a}}^{\mathbf{b}} = \emptyset, \\ (b_1 - a_1) \cdots (b_n - a_n) & \text{if } \square_{\mathbf{a}}^{\mathbf{b}} \neq \emptyset. \end{cases} \tag{4.1}$$

Axiom $\mu 4$ says that (4.1) holds for the special case $\mathbf{a} = \mathbf{0}$, $\mathbf{b} = \mathbf{1}$. Examining the proof of Lemma 3.1 shows that $\square_{\mathbf{0}}^{\mathbf{1}}$ can be written as the disjoint union of 2^n rectangles, all congruent (via translation) to $\square_{\mathbf{0}}^{\mathbf{1/2}}$, where $\frac{\mathbf{1}}{\mathbf{2}} = (\frac{1}{2}, \ldots, \frac{1}{2})$. Axioms $\mu 2$ and $\mu 3$ then imply that

$$\mu(\square_{\mathbf{0}}^{\mathbf{1/2}}) = \frac{1}{2^n}.$$

Repeating this argument inductively shows that

$$\mu(\square_{\mathbf{0}}^{\mathbf{1/2^r}}) = \frac{1}{2^{nr}} \quad \text{if} \quad \frac{\mathbf{1}}{\mathbf{2^r}} = \left\langle \frac{1}{2^r}, \ldots, \frac{1}{2^r} \right\rangle. \tag{4.2}$$

We shall now use (4.2) to verify (4.1). The idea is to approximate any rectangle by unions of translates of cubes

$$\square_{\mathbf{0}}^{\mathbf{1/2^r}}.$$

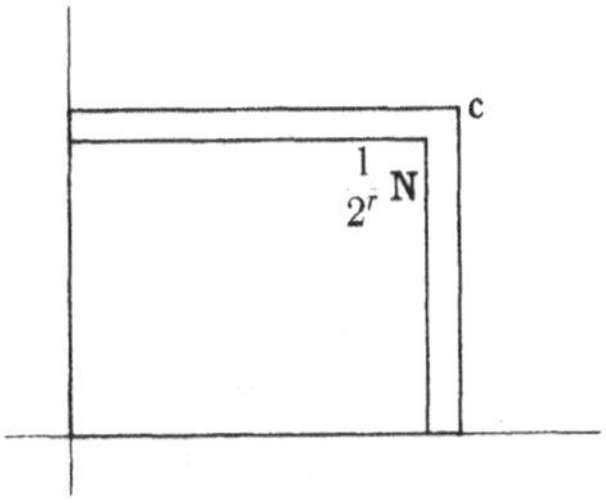

Fig. 8.6

Observe that in proving (4.1) we need to consider only rectangles of the form $\square_{\mathbf{0}}^{\mathbf{c}}$. In fact, we take $\mathbf{c} = \mathbf{b} - \mathbf{a}$ and observe that

$$T_{-\mathbf{a}}(\square_{\mathbf{a}}^{\mathbf{b}}) = \square_{\mathbf{0}}^{\mathbf{c}},$$

so Axiom $\mu 3$ implies that $\mu(\square_{\mathbf{a}}^{\mathbf{b}}) = \mu(\square_{\mathbf{0}}^{\mathbf{c}})$, and by definition $c^1 \cdots c^n = (b^1 - a^1) \cdots (b^n - a^n)$. If $\square_{\mathbf{0}}^{\mathbf{c}} = \emptyset$, then (4.1) is trivially true (from Axiom $\mu 2$). Suppose that $\square_{\mathbf{0}}^{\mathbf{c}} \neq \emptyset$. Then $\mathbf{c} = \langle c^1, \ldots, c^n \rangle$ with $c^i > 0$ for all i. For each r there are n integers $N^1, \ldots, N^n$ such that (Fig. 8.6)

$$N^i/2^r \leq c^i \leq (N^i + 1)/2^r. \tag{4.3}$$

In what follows, let $\mathbf{k} = \langle k^1, \ldots, k^n \rangle$, $\mathbf{l} = \langle l^1, \ldots, l^n \rangle$, etc., denote vectors with integral coordinates (i.e., the k_i's are integers). Let us write $\mathbf{k} < \mathbf{l}$ if $k_i < l_i$ for all i. If $\mathbf{N} = \langle N_1, \ldots, N_n \rangle$, then it follows from (4.3) and the definitions that

$$\square_{(1/2^r)\mathbf{k}}^{(1/2^r)\mathbf{k}+\mathbf{1/2^r}} \subset \square_{\mathbf{0}}^{\mathbf{c}} \quad \text{whenever} \quad \mathbf{k} < \mathbf{N}.$$

For any $\mathbf{k}$ and $\mathbf{l}$,

$$\square_{(1/2^r)\mathbf{k}}^{(1/2^r)\mathbf{k}+\mathbf{1/2^r}} \cap \square_{(1/2^r)\mathbf{l}}^{(1/2^r)\mathbf{l}+\mathbf{1/2^r}} = \emptyset \quad \text{if} \quad \mathbf{k} \neq \mathbf{l}.$$

Since

$$\mu \square_{(1/2^r)\mathbf{k}}^{(1/2^r)\mathbf{k}+\mathbf{1/2^r}} = \frac{1}{2^{nr}}$$

by (4.2) (and Axiom $\mu 2$) and

$$\bigcup_{\mathbf{k}<\mathbf{N}} \square_{(1/2^r)\mathbf{k}}^{(1/2^r)\mathbf{k}+1/2^{\mathbf{r}}} \subset \square_{\mathbf{0}}^{\mathbf{c}},$$

we conclude that

$$\mu(\square_{\mathbf{0}}^{\mathbf{c}}) \geq \frac{1}{2^{nr}} \times \text{(the number of } \mathbf{k} \text{ satisfying } \mathbf{0} \leq \mathbf{k} < \mathbf{N}).$$

It is easy to see that there are $N_1 \cdot N_2 \cdot \ldots \cdot N_n$ such $\mathbf{k}$, so that

$$\mu(\square_{\mathbf{0}}^{\mathbf{c}}) \geq \frac{1}{2^{nr}} \times (N_1 \cdots N_n) = \left(\frac{N_1}{2^r}\right) \cdots \left(\frac{N_n}{2^r}\right).$$

According to (4.3), $N_i/2^r \geq \mathbf{c}_i - 1/2^r$, so we have

$$\mu(\square_{\mathbf{0}}^{\mathbf{c}}) \geq \left(\mathbf{c}^1 - \frac{1}{2^r}\right) \cdots \left(\mathbf{c}^n - \frac{1}{2^r}\right). \tag{4.4}$$

Similarly,

$$\square_{\mathbf{0}}^{\mathbf{c}} \subset \bigcup_{\mathbf{k}<\mathbf{N}+2} \square_{(1/2^r)\mathbf{k}}^{(1/2^r)\mathbf{k}+1/2^{\mathbf{r}}},$$

and we conclude that

$$\mu(\square_{\mathbf{0}}^{\mathbf{c}}) \leq \left(\mathbf{c}^1 + \frac{2}{2^r}\right) \cdots \left(\mathbf{c}^n + \frac{2}{2^r}\right). \tag{4.5}$$

Letting $r \to \infty$ in (4.4) and (4.5) proves (4.1).

In deriving (4.1) we made use of Axiom $\mu 4$. Examining our argument shows that if μ' satisfied $\mu 2$ and $\mu 3$ but *not* $\mu 4$, we could argue in the same manner except that we would have to multiply everything by the fixed constant $\mu'(\square_{\mathbf{0}}^{\mathbf{1}})$. To sum up, we have proved:

Proposition 4.1. If μ satisfies Axioms $\mu 1$ through $\mu 3$, then the value of μ for any rectangle is uniquely determined and is given by (4.1). If μ' satisfies $\mu 1$ and $\mu 2$, then for any rectangle $\square_{\mathbf{a}}^{\mathbf{b}}$,

$$\mu'(\square_{\mathbf{a}}^{\mathbf{b}}) = K\mu(\square_{\mathbf{a}}^{\mathbf{b}}) \qquad \text{where} \qquad K = \mu'(\square_{\mathbf{0}}^{\mathbf{1}}).$$

5. THE MINIMAL THEORY (Continued)

We will now show that formula (4.1) extends to give a unique μ defined on $\mathfrak{D}_{\min}$ so as to satisfy Axioms $\mu 1$ through $\mu 4$. We must establish essentially two facts.

1) *Every union of rectangles can be written as a disjoint union of rectangles.*

This will then allow us to use Axiom $\mu 1$ to determine $\mu(A)$ for every $A \in \mathfrak{D}_{\min}$, by setting

$$\mu(A) = \sum \mu(\square_{\mathbf{a}_i}^{\mathbf{b}_i})$$

if A is the disjoint union of the $\square_{\mathbf{a}_i}^{\mathbf{b}_i}$. Since A might be written in another way as a disjoint union of rectangles, this formula is not well defined until we establish that:

2) *If* $A = \bigcup \square_{\mathbf{a}_i}^{\mathbf{b}_i} = \bigcup \square_{\mathbf{c}_j}^{\mathbf{d}_j}$ *are two representations of* A *as a disjoint union of rectangles, then*

$$\sum \mu(\square_{\mathbf{a}_i}^{\mathbf{b}_i}) = \sum \mu(\square_{\mathbf{c}_j}^{\mathbf{d}_j}).$$

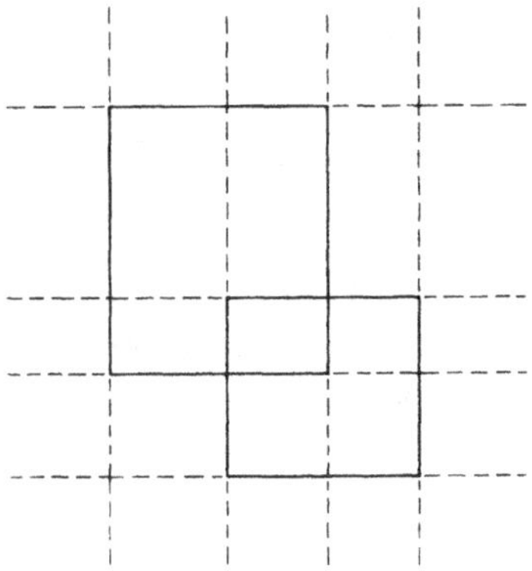

Fig. 8.7

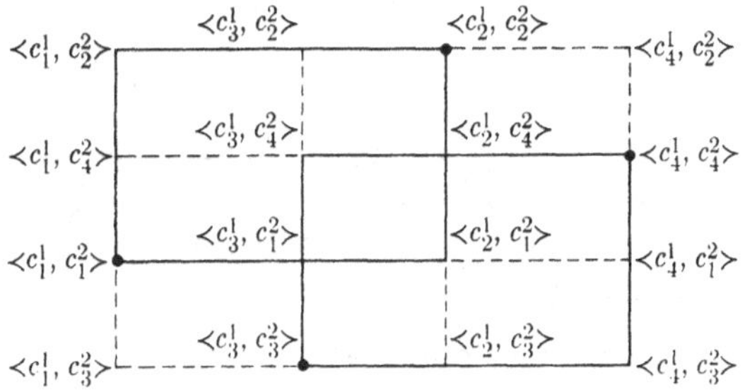

Fig. 8.8

We first introduce some notation.

Definition 5.1. A paving $\mathfrak{p}$ of $\mathbb{E}^n$ is a finite collection of mutually disjoint rectangles. The *floor* of this paving, denoted by $|\mathfrak{p}|$, is the union of all rectangles belonging to $\mathfrak{p}$.

If $\mathfrak{p} = \{\square_{\mathbf{a}_i}^{\mathbf{b}_i}\}$ and T is a translation, we set $T\mathfrak{p} = \{T\square_{\mathbf{a}_i}^{\mathbf{b}_i}\}$.

If $\mathfrak{p}$ and $\mathfrak{z}$ are two pavings, we say that $\mathfrak{z}$ is finer than $\mathfrak{p}$ (and write $\mathfrak{z} \prec \mathfrak{p}$) if every rectangle of $\mathfrak{p}$ is a union of rectangles of $\mathfrak{z}$. It is clear that if $\mathfrak{p} \prec \mathfrak{v}$ and $\mathfrak{z} \prec \mathfrak{p}$, then $\mathfrak{z} \prec \mathfrak{v}$. Note also that $\mathfrak{z} \prec \mathfrak{p}$ implies $|\mathfrak{p}| \subset |\mathfrak{z}|$.

Proposition 5.1. Let $\mathfrak{p}$ and $\mathfrak{z}$ be any two pavings. There exists a third paving $\mathfrak{v}$ such that $\mathfrak{v} \prec \mathfrak{p}$ and $\mathfrak{v} \prec \mathfrak{z}$.

Proof. The idea of the proof is very simple. Each rectangle in $\mathfrak{p}$ or in $\mathfrak{z}$ determines 2^n hyperplanes (each hyperplane containing a face of the rectangle). If we collect all these hyperplanes, they will "enclose" a number of rectangles. We let $\mathfrak{v}$ consist of those rectangles in this collection which do not contain any smaller rectangle. Figure 8.7 shows the case (for $n = 2$) where $\mathfrak{p}$ and $\mathfrak{z}$ each contain one rectangle. Here $\mathfrak{v}$ contains nine rectangles.

We now fill in the details of this argument. Let $\mathbf{c}_1 = \langle c_1^1, \ldots, c_1^n \rangle, \ldots, \mathbf{c}_k = \langle c_k^1, \ldots, c_k^n \rangle$ be all the vectors that occur in the description of the rectangles of $\mathfrak{p}$ and $\mathfrak{z}$. (In other words, if $\square_{\mathbf{a}}^{\mathbf{b}} \in \mathfrak{p}$ or $\in \mathfrak{z}$, then $\mathbf{a}$ and $\mathbf{b}$ are among the $\mathbf{c}$'s.) Let $\mathbf{d}_1, \ldots, \mathbf{d}_{k^n}$ be the vectors of the form $\langle c_{i_1}^1, \ldots, c_{i_n}^n \rangle$, where the i_j's range independently from 1 to k, (so that there are k^n of them). (See Fig. 8.8 for the case where $n = 2$ and $\mathfrak{p}$ and $\mathfrak{z}$ consist of one rectangle each.) For each $\mathbf{d}_i$ there is at most one smallest $\mathbf{d}_{j(i)}$ such that $\mathbf{d}_i < \mathbf{d}_{j(i)}$. In fact, if

$$\mathbf{d}_i = \langle c_{i_1}^1, \ldots, c_{i_n}^n \rangle,$$

then set $\mathbf{d}_{j(i)} = \langle c_{j_1}^1, \ldots, c_{j_n}^n \rangle$, where

$$c_{j_l}^l = \min_{c_m^l > c_{i_l}^l} c_m^l.$$

Let $\mathscr{v} = \{\square_{\mathbf{d}_i}^{\mathbf{d}_{j(i)}}\}$. Then $\mathscr{v}$ is finer than $\mathscr{p}$ and $\mathscr{z}$. In fact, if $\square_{\mathbf{a}}^{\mathbf{b}} \in \mathscr{p}$, say, then $\square_{\mathbf{a}}^{\mathbf{b}} = \square_{\mathbf{d}_\beta}^{\mathbf{d}_\alpha}$ for suitable α and β and

$$\square_{\mathbf{d}_\alpha}^{\mathbf{d}_\beta} = \bigcup_{\substack{\mathbf{d}_\alpha \leq \mathbf{d}_i \\ \mathbf{d}_{j(i)} \leq \mathbf{d}_\beta}} \square_{\mathbf{d}_i}^{\mathbf{d}_{j(i)}}. \tag{5.1}$$

To see this observe that if $\mathbf{x} \in \square_{\mathbf{d}_\alpha}^{\mathbf{d}_\beta}$, then $\mathbf{d}_\alpha \leq \mathbf{x} < \mathbf{d}_\beta$. Choose a largest $\mathbf{d}_i \leq \mathbf{x}$. Then $\mathbf{d}_i \leq \mathbf{x} < \mathbf{d}_{j(i)}$, so $\mathbf{x} \in \square_{\mathbf{d}_i}^{\mathbf{d}_{j(i)}}$. This proves the proposition. We will later want to use the particular form of the $\mathscr{v}$ we constructed to find additional information. □

We can now prove (1) and (2).

Lemma 5.1. Let $\mathscr{p}_1, \ldots, \mathscr{p}_l$ be pavings. Then there exists a paving $\mathscr{z}$ such that $|\mathscr{z}| = |\mathscr{p}_1| \cup \cdots \cup |\mathscr{p}_l|$.

Proof. By repeated applications of Proposition 5.1 we can choose a paving $\mathscr{v}$ which is finer than all the $\mathscr{p}_i$'s. Then each $|\mathscr{p}_i|$ is the union of suitable rectangles of $\mathscr{v}$. Let $\mathscr{z}$ be the collection of all these rectangles occurring in all the $\mathscr{p}_i$'s. Then $|\mathscr{z}| = |\mathscr{p}_1| \cup \cdots \cup |\mathscr{p}_k|$. □

In particular, we have proved (1). More generally, we have shown that every $A \in \mathfrak{D}_{\min}$ is of the form $A = |\mathscr{p}|$ for a suitable paving $\mathscr{p}$. We now wish to turn our attention to (2).

Lemma 5.2. Let $c_1^1 < \cdots < c_{r_1}^1, c_1^2 < \cdots < c_{r_2}^2, \ldots, c_1^n < \cdots < c_{r_n}^n$ be n sequences of numbers. Then

$$\mu \square_{<c_1^1, \ldots, c_1^n>}^{<c_{r_1}^1, \ldots, c_{r_n}^n>} = \sum_{\substack{1 \leq i_1 < r_1 \\ \vdots \\ 1 \leq i_n < r_n}} \mu \square_{<c_{i_1}^1, \ldots, c_{i_n}^n>}^{<c_{i_1+1}^1, \ldots, c_{i_n+1}^n>}$$

Proof. In fact, $c_{r_i}^i - c_1^i = c_2^i - c_1^i + c_3^i - c_2^i + \cdots + c_{r_i}^i - c_{r_i-1}^i$, so that the lemma follows from (4.1) when we multiply out all the factors. □

We now prove (2). Let $\mathscr{p} = \{\square_{\mathbf{a}_i}^{\mathbf{b}_i}\}$ and $\mathscr{z} = \{\square_{\mathbf{e}k}^{\mathbf{f}_k}\}$, where $A = |\mathscr{p}| = |\mathscr{z}|$. Let $\mathscr{v}$ be the paving we constructed in the proof of Proposition 5.1. Let $\mathscr{s} = \{\square_{\mathbf{d}_l}^{\mathbf{d}_m}\}$ be the collection of those rectangles $\square_{\mathbf{d}_\alpha}^{\mathbf{d}_\beta}$ of $\mathscr{v}$ such that $\square_{\mathbf{d}_\alpha}^{\mathbf{d}_\beta} \subset |\mathscr{p}| = |\mathscr{z}|$. Then to prove (2) it suffices to show that

$$\sum \mu(\square_{\mathbf{d}_l}^{\mathbf{d}_m}) = \sum \mu(\square_{\mathbf{a}_i}^{\mathbf{b}_i}) = \sum \mu(\square_{\mathbf{e}k}^{\mathbf{f}_k}). \tag{5.2}$$

Now each rectangle $\square_{\mathbf{a}_i}^{\mathbf{b}_i}$ is decomposed into rectangles $\square_{\mathbf{d}_l}^{\mathbf{d}_{j(l)}}$ according to (5.1), that is, $\mathbf{a}_i = \mathbf{d}_\alpha$, $\mathbf{b}_i = \mathbf{d}_\beta$, etc.

By construction of the $\mathbf{d}$'s, this is exactly a decomposition of the type described in Lemma 5.2. Thus (5.1) implies that

$$\mu(\square_{\mathbf{d}_\alpha}^{\mathbf{d}_\beta}) = \sum_{\substack{\mathbf{d}_\alpha \leq \mathbf{d}_i \\ \mathbf{d}_{j(i)} < \mathbf{d}_p}} \mu(\square_{\mathbf{d}_i}^{\mathbf{d}_{j(i)}}).$$

Summing over all $\square_{\mathbf{a}_i}^{\mathbf{b}_i}$ (and doing the same for $\square_{\mathbf{c}_j}^{\mathbf{d}_j}$) proves (5.2). We can thus state:

Theorem 5.1. Every $A \in \mathfrak{D}_{\min}$ can be written as $A = |p|$. The number $\mu(A) = \sum_{\square \in p} \mu(\square)$ does not depend on the choice of p. We thus get a well-defined function μ on $\mathfrak{D}_{\min}$. It satisfies Axioms $\mu 1$ through $\mu 4$. If μ' is any other function on $\mathfrak{D}_{\min}$ satisfying $\mu 2$ and $\mu 3$, then $\mu'(A) = K\mu(A)$, where $K = \mu'(\square_{\mathbf{0}}^{\mathbf{1}})$.

Proof. The proof of the last two assertions of the theorem is easy and is left as an exercise for the reader.

6. CONTENTED SETS

Theorem 5.1 shows that our axioms are not vacuous. It does not provide us with a satisfactory theory, however, because $\mathfrak{D}_{\min}$ contains far too few sets. In particular, it does *not* fulfill requirement (iii), since $\mathfrak{D}_{\min}$ is not invariant under rotations, except under very special ones. We are now going to remedy this by repeating the arguments of Section 4; we are going to try to approximate more general sets by sets whose μ's we know, i.e., by sets contained in $\mathfrak{D}_{\min}$. This idea goes back to Archimedes, who used it to find the areas of figures in the plane.

Definition 6.1. Let A be any subset of $\mathbb{E}^n$. We say that p is an *inner paving* of A if $|p| \subset A$. We say that $\mathfrak{z}$ is an *outer paving* of A if $A \subset |\mathfrak{z}|$.

We list several obvious facts.

$$\text{If } |p| \subset A \subset |\mathfrak{z}|, \text{ then } \mu(p) \leq \mu(\mathfrak{z}). \tag{6.1}$$

$$\text{If } |p| \subset A \subset |\mathfrak{z}|, \text{ then } |Tp| \subset TA \subset |T\mathfrak{z}|. \tag{6.2}$$

$$\text{If } A_1 \cap A_2 = \emptyset \text{ and } |p_1| \subset A_1,\ |p_2| \subset A_2, \text{ then } p_1 \cup p_2 \text{ is an inner paving of } A_1 \cup A_2. \tag{6.3}$$

Definition 6.2. For any bounded subset A of $\mathbb{E}^n$ let

$$\mu^*(A) = \operatorname*{lub}_{|p| \subset A} \mu(|p|)$$

be called the *inner content* of A and let

$$\bar{\mu}(A) = \operatorname*{glb}_{A \subset |\mathfrak{z}|} \mu(|\mathfrak{z}|)$$

be called the *outer content* of A.

Note that since A is bounded, there exists a $\mathfrak{z}$ with $A \subset |\mathfrak{z}|$. This shows that $\bar{\mu}(A)$ is defined. This together with (6.1) shows that $\mu^*(A)$ is defined and that

$$\mu^*(A) \leq \bar{\mu}(A). \tag{6.4}$$

Definition 6.3. A set A will be called *contented* if $\mu^*(A) = \bar{\mu}(A)$. We call $\mu^*(A) = \bar{\mu}(A)$ the *content* of A and denote it by $\mu(A)$.

Observe that every $A \in \mathfrak{D}_{\min}$ is contented. In fact, if $A = |\mathscr{v}|$, then $\mathscr{v}$ is both an inner and an outer paving of A. Thus $\mu^*(A) = \bar{\mu}(A) = \mu(|\mathscr{v}|)$, and the new definition of $\mu(A)$ coincides with the old one.

Our next immediate objective is to show that the collection of all contented sets fulfills Axioms $\mathfrak{D}1$ through $\mathfrak{D}3$.

Proposition 6.1. A set A is contented if and only if its boundary is contented and has content zero.

Proof. Suppose A is contented. For any $\delta > 0$ we can find an inner paving $\mathscr{p}$ and an outer paving $\mathscr{z}$ such that $\mu(\mathscr{z}) - \mu(\mathscr{p}) < \delta/2$. We want to replace $\mathscr{p}$ by a close paving $\mathscr{p}'$ with $|\mathscr{p}'| \subset \operatorname{int} A$. To do this, we choose a small number η and replace each rectangle $\square_{\mathbf{a}}^{\mathbf{b}}$ of $\mathscr{p}$ by $\square_{\mathbf{a}+\eta(\mathbf{b}-\mathbf{a})}^{\mathbf{b}-\eta(\mathbf{b}-\mathbf{a})}$. We let $\mathscr{p}_\eta$ be the collection of all these rectangles. Then $|\mathscr{p}_\eta| \subset \operatorname{int} |\mathscr{p}|$, so $|\mathscr{p}_\eta| \subset \operatorname{int} A$. Furthermore, $\mu(|\mathscr{p}_\eta|) = (1 - 2\eta)^n \mu(|\mathscr{p}|)$, since the factor $(1 - 2\eta)$ is the decrease of each side of each rectangle of $\mathscr{p}$. Similarly, we replace $\mathscr{z}$ by a slightly larger $\mathscr{z}_\eta$, with $A \subset \operatorname{int} \mathscr{z}_\eta$ and $\mu(\mathscr{z}_\eta) \leq (1 + 2\eta)^n \mu(\mathscr{z})$. By choosing η sufficiently small, we can thus arrange that $\mu(\mathscr{z}_\eta) - \mu(\mathscr{p}_\eta) < \delta$. Let $\mathscr{v}$ be a paving which is finer than $\mathscr{z}_\eta$ and $\mathscr{p}_\eta$, with $|\mathscr{v}| = |\mathscr{z}_\eta|$. Let $\mathscr{s} \subset \mathscr{v}$ consist of those rectangles of $\mathscr{v}$ lying in $\operatorname{int} A$. Then $|\mathscr{v}| = |\mathscr{z}_\eta| \supset |\mathscr{s}| \supset |\mathscr{p}_\eta|$, so $\mu(|\mathscr{v}|) - \mu(|\mathscr{s}|) \leq \delta$. But $\partial A \subset |\mathscr{v} - \mathscr{s}|$, so that $\bar{\mu}(\partial A) \leq \mu(|\mathscr{v} - \mathscr{s}|) = \mu(|\mathscr{v}|) - \mu(|\mathscr{s}|) < \delta$. In other words, $\bar{\mu}(\partial A) = 0$.

Conversely, suppose that ∂A has content zero. Let $\mathscr{s}$ be an outer paving of ∂A with $\mu(|\mathscr{s}|) < \epsilon$. Let $\mathscr{v}$ be a paving finer than $\mathscr{s}$ and such that $A \subset |\mathscr{v}|$. Let $\mathscr{p} \subset \mathscr{v}$ consist of those rectangles contained in A. Let $\mathscr{z} \subset \mathscr{v}$ consist of those rectangles lying in $|\mathscr{p}| \cup |\mathscr{s}|$. Then $\mu(|\mathscr{z}|) \leq \mu(|\mathscr{p}|) + \mu(|\mathscr{s}|) < \mu(|\mathscr{p}|) + \epsilon$. Furthermore, $A \subset |\mathscr{z}|$. In fact, let $x \in A$. Then $x \in \square$ for some $\square \in \mathscr{v}$. If $\square \cap \partial A \neq \emptyset$, then $\square \cap |\mathscr{s}| \neq \emptyset$, so $\square \subset |\mathscr{s}|$, since $\mathscr{v}$ is a refinement of $\mathscr{s}$. If $\square \cap \partial A = \emptyset$, then every point of $\square$ must lie in A, so that $\square \subset |\mathscr{p}|$. We have thus constructed $\mathscr{p}$ and $\mathscr{z}$ with $|\mathscr{p}| \subset A \subset |\mathscr{z}|$ and $\mu(\mathscr{z}) - \mu(\mathscr{p}) < \epsilon$. Since we can do this for any ϵ, this implies that A is contented. □

Proposition 6.2. The union of any finite number of sets with content zero has content zero. If $A \subset B$ and B has content zero, then so does A.

Proof. The proof is obvious.

Theorem 6.1. Let $\mathfrak{D}_{\text{con}}$ denote the collection of all contented sets. Then $\mathfrak{D}_{\text{con}}$ satisfies Axioms $\mathfrak{D}1$ through $\mathfrak{D}3$, and the μ given in Definition 6.3 satisfies $\mu1$ through $\mu4$. If μ' is any other function on $\mathfrak{D}_{\text{con}}$ satisfying $\mu1$ through $\mu3$, then $\mu' = K\mu$, where $K = \mu'(\square_0^1)$.

Proof. Let us verify the axioms.

$\mathfrak{D}1$. For any A and B,

$$\partial(A \cup B) \subset \partial A \cup \partial B \quad \text{and} \quad \partial(A \cap B) \subset \partial A \cup \partial B.$$

By Proposition 6.1, if A and B are contented, then ∂A and ∂B have content zero. Thus so do $\partial A \cup \partial B$, $\partial(A \cup B)$, and $\partial(A \cap B)$, by Proposition 6.2. Hence $A \cup B$ and $A \cap B$ are contented.

$\mathfrak{D}$2. Follows immediately from (6.1).

$\mathfrak{D}$3. Is obvious.

μ2. If A_1 and A_2 are contented, we can find inner pavings $\mathfrak{p}_1$ and $\mathfrak{p}_2$ such that $\mu(A_1) - \mu(|\mathfrak{p}_1|) < \epsilon/2$ and $\mu(A_2) - \mu(|\mathfrak{p}_2|) < \epsilon/2$. If $A_1 \cap A_2 = \emptyset$, then $\mathfrak{p}_1 \cup \mathfrak{p}_2$ is an inner paving of $A_1 \cup A_2$, and so

$$\mu(A_1 \cup A_2) \geq \mu(A_1) + \mu(A_2).$$

On the other hand, let $\mathfrak{z}_1$ and $\mathfrak{z}_2$ be outer pavings of A_1 and A_2, respectively, with $\mu(\mathfrak{z}_1) < \mu(A_1) + \epsilon/2$ and $\mu(\mathfrak{z}_2) < \mu(A_2) + \epsilon/2$.

Let $\mathfrak{v}$ be a paving with $|\mathfrak{v}| = |\mathfrak{z}_1| \cup |\mathfrak{z}_2|$. Then $\mathfrak{v}$ is an outer paving of $A_1 \cup A_2$ and $\mu(|\mathfrak{v}|) \leq \mu(|\mathfrak{z}_1|) + \mu(|\mathfrak{z}_2|)$. Thus $\mu(A_1 \cup A_2) \leq \mu(|\mathfrak{v}|) \leq \mu(A_1) + \mu(A_2) + \epsilon$, or $\mu(A_1 \cup A_2) \leq \mu(A_1) + \mu(A_2)$. These two inequalities together give μ2.

μ1. Is obvious.

μ3. Follows from (6.2) and Definition 6.3.

μ4. We already know.

The second part of the theorem follows from Theorem 5.1 and Definition 6.3. In fact, we know that $\mu'(|\mathfrak{p}|) = K\mu(|\mathfrak{p}|)$, and (6.1) together with Axiom μ2 implies that $\mu'(|\mathfrak{p}|) \leq \mu'(A) \leq \mu'(|\mathfrak{z}|)$. Since we can choose $\mathfrak{p}$ and $\mathfrak{z}$ to be arbitrarily close approximations to A (relative to μ), we are done. □

Remark. It is useful to note that we have actually proved a little more than what is stated in Theorem 6.1. We have proved, namely, that if $\mathfrak{D}$ is any collection of sets satisfying $\mathfrak{D}$1 through $\mathfrak{D}$3, such that $\mathfrak{D}_{\min} \subset \mathfrak{D} \subset \mathfrak{D}_{\text{con}}$, and if $\mu' : \mathfrak{D} \to \mathbb{R}$ satisfies μ1 through μ3, then $\mu'(A) = K\mu(A)$ for all A in $\mathfrak{D}$, where $K = \mu'(\square_0^1)$.

7. WHEN IS A SET CONTENTED?

We will now establish some useful criteria for deciding whether a given set is contented.

Recall that a closed ball $B_{\mathbf{x}}^r$ with center $\mathbf{x}$ and radius r is given by

$$B_{\mathbf{x}}^r = \{\mathbf{y} : \|\mathbf{y} - \mathbf{x}\| \leq r\}. \tag{7.1}$$

Note that

$$B_{\mathbf{x}}^r \subset \square_{\mathbf{x}-r\mathbf{1}}^{\mathbf{x}+(r+\epsilon)\mathbf{1}} \quad \text{for any} \quad \epsilon > 0, \tag{7.2}$$

and

$$\square_{\mathbf{x}-r\mathbf{1}}^{\mathbf{x}+r\mathbf{1}} \subset B_{\mathbf{x}}^{\sqrt{n}\,r}. \tag{7.3}$$

(See Fig. 8.9.)

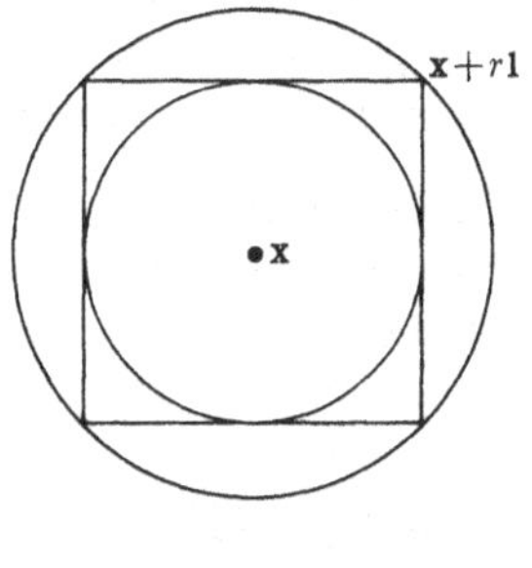

Fig. 8.9

If we combine (7.2) and (7.3), we see that any cube C lies in a ball B such that $\bar{\mu}(B) \leq 2^n(\sqrt{n})^n\mu(C)$ and that any ball B lies in a cube C such that $\mu(C) < 3^n(\sqrt{n})^n\bar{\mu}(B)$.

Lemma 7.1. Let A be a subset of $\mathbb{E}^n$. Then A has content zero if and only if for every $\epsilon > 0$ there exist a finite number of balls $\{B_i\}$ covering A with $\sum \bar{\mu}(B_i) < \epsilon$.

Proof. If we have such a collection of covering balls, then by the above remark we can enlarge each ball to a rectangle to get a paving p such that $A \subset |p|$ and $\mu(|p|) < 3^n(\sqrt{n})^n\epsilon$. Therefore, $\bar{\mu}(A) = 0$ if we can always find the $\{B_i\}$.

Conversely, suppose A has content 0. Then for any δ we can find an outer paving p with $\mu(|p|) < \delta$. For each rectangle $\square$ in the paving we can, by the arguments of Section 4, find a finite number of *cubes* which cover $\square$ and whose total content is as close as we like to $\mu(\square)$, say $<2\mu(\square)$. By doing this for each $\square \in p$, we have a finite number of cubes $\{\square_i\}$ covering A with total content less than 2δ. Then by our remark before the lemma each cube $\square_i$ lies in a ball B_i such that $\mu(\square_i) < 2^n(\sqrt{n})^n\bar{\mu}(B_i)$, and so we have a covering of A by balls B_i such that $\sum \bar{\mu}(B_i) < 2^{n+1}(\sqrt{n})^n\delta$. If we take $\delta = \epsilon/2^{n+1}(\sqrt{n})^n$, we have the desired collection of balls, proving the lemma. $\square$

Recall that a map φ of $U \subset \mathbb{E}^n \to \mathbb{E}^n$ is said to satisfy a Lipschitz condition if there is a constant K (called the *Lipschitz constant*) such that

$$\|\varphi(y) - \varphi(x)\| < K\|y - x\|. \tag{7.4}$$

Proposition 7.1. Let A be a set of content zero with $\bar{A} \subset U$, and let $\varphi: U \to \mathbb{E}^n$ satisfy a Lipschitz condition. Then $\varphi(A)$ has content zero.

Proof. The proof consists of applying both parts of Lemma 7.1. Since A has content zero, for any $\epsilon > 0$ we can find a finite number of balls covering A whose total outer content is less than ϵ/K^n. By (7.4), $\varphi(B_{\mathbf{x}}^r) \subset B_{\varphi(\mathbf{x})}^{Kr}$, so that the images of the balls covering A cover $\varphi(A)$ and have a total volume less than ϵ. $\square$

Recall that if φ is a (continuously) differentiable map of an open set U into $\mathbb{E}^n$, then φ satisfies a Lipschitz condition on any compact subset of U.

As a consequence of Proposition 7.1, we can thus state:

Proposition 7.2. Let φ be a continuously differentiable map defined on an open set U, and let A be a bounded set of content zero with $\bar{A} \subset U$. Then $\varphi(A)$ has content zero.

Let A be any compact subset of $\mathbb{E}^n$ lying entirely in the subspace given by $x^n = 0$. Then A has content zero. In fact, for some sufficiently large fixed r, the set A is contained in the rectangle

$$\square_{\langle -r,\ldots,-r,0\rangle}^{\langle r,\ldots,r,\epsilon\rangle} \qquad \text{for any} \quad \epsilon > 0,$$

which has arbitrarily small volume.

Now let $\psi \colon V \subset \mathbb{E}^{n-1} \to \mathbb{E}^n$ be a continuously differentiable map given by

$$\langle y^1, \ldots, y^{n-1} \rangle \to \langle \psi^1(y^1, \ldots, y^{n-1}), \ldots, \psi^n(y^1, \ldots, y^{n-1}) \rangle.$$

Let B be any bounded subset of $\mathbb{E}^{n-1}$ with $\overline{B} \subset V$. We can then write $\psi(B) = \varphi(A)$, where A is the set of points in $\mathbb{E}^n$ of the form $(y, 0)$, where $y \in B$, and where φ is a differentiable map such that

$$\varphi(x^1, \ldots, x^n) = \langle \psi^1(x^1, \ldots, x^{n-1}), \ldots, \psi^n(x^1, \ldots, x^{n-1}) \rangle.$$

By Proposition 7.2 we see that $\mu(\psi(B)) = 0$. Thus,

Proposition 7.3. Let ψ be a differentiable map of $V \subset \mathbb{E}^{n-1}$ into $\mathbb{E}^n$, and let B be a bounded set such that $\overline{B} \subset V$. Then $\psi(B)$ has content zero.

We have thus recovered requirement (v) of Section 1.

An immediate consequence of Propositions 7.3 and 6.1 is:

Proposition 7.4. Let $A \subset \mathbb{E}^n$ be such that $\partial A \subset \bigcup \psi_i(B_i)$ where each ψ_i and B_i is as in Proposition 7.3. Then A is contented.

This shows that every set "we can draw" is contented.

Exercise. Show that every ball is contented.

8. BEHAVIOR UNDER LINEAR DISTORTIONS

We shall continue to derive consequences of Proposition 7.1.

Proposition 8.1. Let φ be a one-to-one map of $U \to \mathbb{E}^n$ which satisfies a Lipschitz condition and is such that φ^{-1} is continuous. If $\overline{A} \subset U$ is contented, then so is $\varphi(A)$.

Proof. Since A is contented, ∂A has content zero. By the conditions on φ, we know that $\partial\varphi(A) = \varphi(\partial A)$. Thus $\partial\varphi(A)$ has content zero, and so $\varphi(A)$ is contented. □

An immediate consequence of Proposition 8.1 is:

Proposition 8.2. Let L be a linear transformation of $\mathbb{E}^n$. Then LA is contented whenever A is contented.

Proof. If L is nonsingular, Proposition 8.1 applies. If L is singular, it maps all of $\mathbb{E}^n$ onto a proper subspace. Any such subspace is contained in the image of $\{\mathbf{x} : x^n = 0\}$ by a suitable linear transformation, and so $\mu(LA) = 0$ for any contented A. □

Theorem 8.1. Let L be a linear transformation of $\mathbb{E}^n$. Then for any contented A we have

$$\mu(LA) = |\det L|\mu(A). \tag{8.1}$$

Proof. We can restrict our attention to nonsingular L, since we have already checked Eq. (8.1) for $\det L = 0$. If L is nonsingular, then L carries the class of contented sets into itself. Let us define μ' by $\mu'(A) = \mu(LA)$ for each $A \in \mathfrak{D}_{\text{con}}$. We claim that μ' satisfies Axioms $\mu 1$ through $\mu 3$ on $\mathfrak{D}_{\text{con}}$.

In fact, $\mu 1$ and $\mu 2$ are obviously true; $\mu 3$ follows from the fact that for any translation T_v, we have $T_{Lv}L = LT_v$, so that

$$\mu'(T_vA) = \mu(LT_vA) = \mu(T_{Lv}LA) = \mu(LA) = \mu'(A).$$

By Theorem 5.2 we thus conclude that

$$\mu' = k_L\mu,$$

where k_L is some constant depending on L. We must show that $k_L = |\det L|$.

We first observe that if O is an orthogonal transformation, then

$$\mu(OA) = \mu(A).$$

In fact, we know that $\mu(OA) = k_O\mu(A)$. If we take A to be the unit ball $B_{\mathbf{0}}^1$, then $OB_{\mathbf{0}}^1 = B_{\mathbf{0}}^1$, so $k_O = 1$.

Next we observe that $\mu(L_1L_2A) = k_{L_1}\mu(L_2A) = k_{L_1}k_{L_2}\mu(A)$, so that

$$k_{L_1L_2} = k_{L_1}k_{L_2}.$$

Now we recall that any nonsingular L can be written as $L = PO$, where P is a positive self-adjoint operator and O is orthogonal. Thus $k_L = k_P$ and $|\det L| = |\det P|\,|\det O| = |\det P|$, so we need only verify (8.1) for positive self-adjoint linear transformations. Any such P can be written as $P = O_1DO_1^{-1}$, where O_1 is orthogonal and D is diagonal. Since P is positive, all the eigenvalues of D are positive. Since $\det P = \det D$ and $k_P = k_D$, we need only verify (8.1) for the case where L is given by a diagonal matrix with positive eigenvalues $\lambda_1, \ldots, \lambda_n$. But then $L\square_{\mathbf{0}}^1 = \square_{\mathbf{0}}^{\langle\lambda_1,\ldots,\lambda_n\rangle}$, so that

$$\mu'(\square_{\mathbf{0}}^1) = \mu(\square_{\mathbf{0}}^{\langle\lambda_1,\ldots,\lambda_n\rangle}) = \lambda_1\cdots\lambda_n = |\det L|,$$

verifying (8.1). □

Exercise. Let $\mathbf{v}_1, \ldots, \mathbf{v}_n$ be vectors of $\mathbb{E}^n$. By the parallelepiped spanned by $\mathbf{v}_1, \ldots, \mathbf{v}_n$ we mean the set of all vectors of the form $\sum_{i=1}^n x^i\mathbf{v}_i$, where $0 \le x^i \le 1$. Show that its content is $|\det((\mathbf{v}_i, \mathbf{v}_j))|^{1/2}$.

9. AXIOMS FOR INTEGRATION

So far we have shown that there is a unique μ defined for a large collection of sets in $\mathbb{E}^n$. However, we do not have an effective way to compute μ, except in very special cases. To remedy this we must introduce a theory of integration. We first introduce some notation.

Definition 9.1. Let f be any real-valued function on $\mathbb{E}^n$. By the *support* of f, denoted by supp f, we shall mean the closure of the set where f is not zero; that is,

$$\operatorname{supp} f = \overline{\{\mathbf{x} : f(\mathbf{x}) \neq 0\}}.$$

Observe that

$$\operatorname{supp}(f+g) \subset \operatorname{supp} f \cup \operatorname{supp} g \tag{9.1}$$

and

$$\operatorname{supp} fg \subset \operatorname{supp} f \cap \operatorname{supp} g. \tag{9.2}$$

We shall say that f has compact support if $\operatorname{supp} f$ is compact. Equation (9.1) [and Eq. (9.2) applied to constant g] shows that the set of all functions with compact support form a vector space.

Let T be any one-to-one transformation of $\mathbb{E}^n$ onto itself. For any function f we denote by Tf the functions given by

$$(Tf)(\mathbf{x}) = f(T^{-1}\mathbf{x}). \tag{9.3}$$

Observe that if T and T^{-1} are *continuous*, then

$$\operatorname{supp} Tf = T \operatorname{supp} f. \tag{9.4}$$

Definition 9.2. Let A be a subset of $\mathbb{E}^n$. By the *characteristic function* of A, denoted by e_A, we shall mean the function given by

$$e_A(\mathbf{x}) = \begin{cases} 1 & \text{if } \mathbf{x} \in A, \\ 0 & \text{if } \mathbf{x} \notin A. \end{cases} \tag{9.5}$$

Note that

$$e_{A_1 \cap A_2} = e_{A_1} \cdot e_{A_2}, \tag{9.6}$$

$$e_{A_1 \cup A_2} = e_{A_1} + e_{A_2} - e_{A_1 \cap A_2}, \tag{9.7}$$

$$\operatorname{supp} e_A = \overline{A}, \tag{9.8}$$

and

$$Te_A = e_{TA} \tag{9.9}$$

for any one-to-one map T of $\mathbb{E}^n$ onto itself.

By a theory of integration on $\mathbb{E}^n$ we shall mean a collection $\mathfrak{F}$ of functions and a rule $\int$ which assigns a real number $\int f$ to each $f \in \mathfrak{F}$, subject to the following axioms:

$\mathfrak{F}$1. $\mathfrak{F}$ is a vector subspace of the space of all bounded functions of compact support.

$\mathfrak{F}$2. If $f \in \mathfrak{F}$ and T is a translation, then $Tf \in \mathfrak{F}$.

$\mathfrak{F}$3. $e_\square$ belongs to $\mathfrak{F}$ for any rectangle $\square$.

$\int$1. $\int$ is a linear function on $\mathfrak{F}$.

$\int$2. $\int Tf = \int f$ for any translation T.

$\int$3. If $f \geq 0$, then $\int f \geq 0$.

$\int$4. $\int e_{\square_0^1} = 1$.

Note that the axioms imply that $\mathfrak{F}$ contains all functions of the form $e_{\square_1} + e_{\square_2} + \cdots + e_{\square_k}$ for any rectangles $\square_1, \ldots, \square_k$. In particular, for any paving p, the function $e_{|p|}$ must belong to $\mathfrak{F}$.

Also note that from $\int 3$ we have at once the stronger version:

$\int 3'$. $f \leq g \Rightarrow \int f \leq \int g$, since then $g - f \geq 0$.

Proposition 9.1. Let $\mathfrak{F}$, $\int$ be a system satisfying Axioms $\mathfrak{F}1$ through $\mathfrak{F}3$ and $\int 1$ through $\int 4$. Then

$$\int e_A = \mu(A) \tag{9.10}$$

for every contented set A such that $e_A \in \mathfrak{F}$, and

$\int 5$. $|\int f| \leq \|f\|_\infty \mu(\operatorname{supp} f)$ for every $f \in \mathfrak{F}$.

Proof. The axioms guarantee that $e_A \in \mathfrak{F}$ for every $A \in \mathfrak{D}_{\min}$ and that $\nu(A) = \int e_A$ satisfies $\mu 1$ through $\mu 4$. Therefore, $\int e_A = \mu(A)$ for every $A \in \mathfrak{D}_{\min}$ by the uniqueness of μ (Proposition 4.1). It follows that if A is a contented set such that $e_A \in \mathfrak{F}$, and if $\mathfrak{p}$ and $\mathfrak{z}$ are inner and outer pavings of A, then

$$\mu(|\mathfrak{p}|) = \int e_{|\mathfrak{p}|} \leq \int e_A \leq \int e_{|\mathfrak{z}|} = \mu(|\mathfrak{z}|).$$

Therefore, $\int e_A$ lies between $\mu^*(A)$ and $\bar{\mu}(A)$, and so equals $\mu(A)$. For any $f \in \mathfrak{F}$ and any $A \in \mathfrak{D}_{\min}$ such that $\operatorname{supp} f \subseteq A$, we have $-\|f\|_\infty e_A \leq f \leq \|f\|_\infty e_A$, and therefore $|\int f| \leq \|f\|_\infty \mu(A)$ by $\int 3'$ and (9.10). Taking the greatest lower bound of the right side over all such sets A, we have $\int 5$. □

10. INTEGRATION OF CONTENTED FUNCTIONS

We will now proceed to deal with Axioms $\mathfrak{F}$ and $\int$ in the same way we dealt with Axioms $\mathfrak{D}$ and μ. We will construct a "minimal" theory and then get a "big" one by approximating. According to Proposition 9.1, the class $\mathfrak{F}$ must contain the function $e_{|\mathfrak{p}|}$ for any paving $\mathfrak{p}$. By $\mathfrak{F}1$ it must therefore contain all linear combinations of such.

Definition 10.1. By a *paved function* we shall mean a function $f = f_\mathfrak{p}$ given by

$$f = \sum_{\square_i \in \mathfrak{p}} c_i e_{\square_i} \tag{10.1}$$

for some paving $\mathfrak{p}$.

It is easy to see that the collection of all paved functions satisfies Axioms $\mathfrak{F}1$ through $\mathfrak{F}3$. Furthermore, by Proposition 9.1 and Axiom $\int 1$ the integral, $\int$, is uniquely determined on the class of all paved functions by

$$\int f = \sum c_i \mu(\square_i) \tag{10.2}$$

if f is given by (10.1).

The reader should verify that if we let $\mathfrak{F}_P$ be the class of all paved functions and let $\int$ be given by (10.2), then all our axioms are satisfied. Don't forget to show that $\int$ is well defined: if f is expressed as in (10.1) in two ways, then the sums given by (10.2) are equal.

The paved functions obviously form too small a collection of functions. We would like to have an $\mathfrak{F}$ including all continuous functions with compact support and all characteristic functions of the form e_A with A contented, for example.

Definition 10.2. A bounded function f with compact support is said to be *contented* if for any $\epsilon > 0$ and $\delta > 0$ there exists a paved function $g = g_{\epsilon,\delta}$ and a contented set $A = A_{\epsilon,\delta}$ such that

$$|f(\mathbf{x}) - g(\mathbf{x})| < \epsilon \qquad \text{for all} \quad \mathbf{x} \notin A \tag{10.3}$$

and

$$\mu(A) < \delta. \tag{10.4}$$

The pair $\langle g, A \rangle$ will be called a paved ϵ,δ-approximation to f.

Let us verify that the collection of all contented functions, $\mathfrak{F}_{\text{con}}$, satisfies Axioms $\mathfrak{F}1$ through $\mathfrak{F}3$. It is clear that if f is contented, so is af for any constant a. If f_1 and f_2 are contented, let $\langle g_1, A_1 \rangle$ and $\langle g_2, A_2 \rangle$ be paved ϵ,δ-approximations to f_1 and f_2, respectively. Then

$$|(f_1 + f_2)(\mathbf{x}) - (g_1 + g_2)(\mathbf{x})| < 2\epsilon \qquad \text{for all} \quad \mathbf{x} \notin A_1 \cup A_2,$$

and

$$\mu(A_1 \cup A_2) < 2\delta.$$

Thus $\langle g_1 + g_2, A_1 \cup A_2 \rangle$ gives a paved $2\epsilon, 2\delta$-approximation to $f_1 + f_2$.

To verify $\mathfrak{F}2$ we simply observe that if $\langle g, A \rangle$ is a paved ϵ,δ-approximation to f, then $\langle Tg, TA \rangle$ is one to Tf.

A similar argument establishes the analogous result for multiplication:

Proposition 10.1. Let f_1 and f_2 be two contented functions. Then $f_1 f_2$ is contented.

Proof. Let M be such that $|f_1(\mathbf{x}) < M$ and $|f_2(\mathbf{x})| < M$ for all $\mathbf{x}$. Recall that the product of two paved functions is a paved function. Using the same notation as before, we have

$$\begin{aligned} |f_1 f_2(\mathbf{x}) - g_1(\mathbf{x}) g_2(\mathbf{x})| &\leq |f_1(\mathbf{x})|\, |f_2(\mathbf{x}) - g_2(\mathbf{x})| + |g_2(\mathbf{x})|\, |f_1(\mathbf{x}) - g_1(\mathbf{x})| \\ &< M\epsilon + (M + \epsilon)\epsilon \qquad \text{for all} \quad \mathbf{x} \notin A_1 \cup A_2. \end{aligned}$$

Thus $\langle g_1 g_2, A_1 \cup A_2 \rangle$ is a paved $(2M + \epsilon)\epsilon, 2\delta$-approximation to $f_1 f_2$. □

As for $\mathfrak{F}3$, it is immediate that a stronger statement is true:

Proposition 10.2. If B is a contented set, then e_B is a contented function.

Proof. In fact, let p be an inner paving of B with $\mu(B) - \mu(|p|) < \delta$. Then

$$e_B(\mathbf{x}) - e_{|p|}(\mathbf{x}) = 0 \qquad \text{if} \quad \mathbf{x} \notin B - |p|,$$

and

$$\mu(B - |p|) < \delta,$$

so $e_{|p|}, |p|$ is a paved ϵ,δ-approximation to e_B for any $\epsilon > 0$. □

We now establish a useful alternative characterization of a contented function.

Proposition 10.3. A function f is contented if and only if for every ϵ there are paved functions h and k such that $h \leq f \leq k$ and $\int (k - h) < \epsilon$.

Proof. If f is contented, let R be a rectangle including supp f. Let $\langle g, A \rangle$ be an ϵ, δ-approximation to f. Let P be a paved set including $A = A_{\epsilon,\delta}$ such that $\mu(P) < \delta$, and let m be a bound of $|f|$. Then $g - \epsilon(e_R) - me_P \leq f \leq g + \epsilon(e_R) + me_P$, where the outside functions are clearly paved and the differences of their integrals is less than $2\epsilon\mu(R) + 2m\delta$. Since ϵ and δ are arbitrary, we have our h and k. Conversely, if h and k are paved functions such that $h \leq f \leq k$ and $\int (k - h) < a$, then the set where $k - h \geq a^{1/2}$ is a paved set A. Furthermore, $a^{1/2}\mu(A) \leq \int e_A(k - h) \leq \int (k - h) \leq a$, so that $\mu(A) \leq a^{1/2}$. Given ϵ and δ, we only have to choose $a \leq \min(\epsilon^2, \delta^2)$ and take g as either k or h to see that f is contented. □

Corollary. A function f is contented if for every ϵ there are contented functions f_1 and f_2 such that $f_1 \leq f \leq f_2$ and $\int (f_2 - f_1) < \epsilon$.

Proof. For then we can find paved functions $h \leq f_1$ and $k \geq f_2$ such that $\int (f_1 - h) < \epsilon$ and $\int (k - f_2) < \epsilon$ and end up with $h \leq f \leq k$ and $\int (k - h) < 3\epsilon$. □

Theorem 10.1. Let $\mathfrak{F}$ be a class of functions satisfying Axioms $\mathfrak{F}1$ through $\mathfrak{F}3$ and such that $\mathfrak{F}_P \subset \mathfrak{F} \subset \mathfrak{F}_{\text{con}}$. Then there exists a unique $\int$ satisfying Axioms $\int 1$ through $\int 4$ on $\mathfrak{F}$.

Proof. If $\int$ is any integral on $\mathfrak{F}$ satisfying Axioms $\int 1$ through $\int 4$, then we must have $\int f$ simultaneously equal to lub $\int h$ for h paved and $\leq f$ and equal to glb $\int k$ for k paved and $\geq f$, by Proposition 10.3. The integral is thus uniquely determined on $\mathfrak{F}$. Moreover, it is easy to see that if the integral on $\mathfrak{F}$ is defined by $\int f =$ lub $\int h =$ glb $\int k$, then Axioms $\int 1$ through $\int 4$ follow from the fact that they hold for the uniquely determined integral on the paved functions. □

Exercise 10.1. Let f and g be contented functions such that $f(x) = g(x)$ for $x \notin A$, where $\mu(A) = 0$. Then $\int f = \int g$. (This shows that for the purpose of integration we need to know a function only as to a set of content zero.)

Definition 10.3. Let f be a contented function and A a contented set. We call $\int e_A f$ the *integral of f over A* and denote it by $\int_A f$. Thus

$$\int_A f = \int e_A f. \tag{10.5}$$

An immediate consequence of Axiom $\int 1$ and (9.7) is

$$\int_{A_1 \cup A_2} f = \int_{A_1} f + \int_{A_2} f - \int_{A_1 \cap A_2} f. \tag{10.6}$$

An immediate consequence of Exercise 10.1 is

$$\left| \int_A f \right| \leq \sup_{\mathbf{x} \in A} |f(\mathbf{x})| \mu(A).$$

We close this section by giving another useful characterization of contented functions.

Proposition 10.4. Let f be a bounded function with compact support. Then f is contented if and only if to every $\epsilon > 0$ and $\delta > 0$ we can find an $\eta > 0$ and a contented set A_δ such that $\mu(A_\delta) < \delta$ and

$$|f(\mathbf{x}) - f(\mathbf{y})| < \epsilon \quad \text{whenever} \quad \|\mathbf{x} - \mathbf{y}\| < \eta \quad \text{and} \quad \mathbf{x}, \mathbf{y} \notin A_\delta. \tag{10.7}$$

Proof. Suppose that for every ϵ, δ we can find η and A_δ. Let $p = \{\square_i\}$ be a paving such that

i) $\operatorname{supp} f \subset |p|$;

ii) if $\mathbf{x}, \mathbf{y} \in \square_i$, then $\|\mathbf{x} - \mathbf{y}\| < \eta$;

iii) if $\mathfrak{z} = \{\square_i \in p : \square_i \cap A_\delta \neq \emptyset\}$, then $\mu(|\mathfrak{z}|) < 2\delta$.

Then let $f_{\epsilon,2\delta}(\mathbf{x}) = f(\mathbf{x}_i)$ when $\mathbf{x} \in \square_i$, where $\mathbf{x}_i$ is some point of $\square_i$. By (ii) and (iii), we see that $f_{\epsilon,2\delta}, |\mathfrak{z}|$ is a paved $\epsilon, 2\delta$-approximation to f. Thus f is contented.

Conversely, suppose that f is contented, and let $f_{\epsilon/2,\delta/2}$, $A_{\epsilon/2,\delta/2}$ be a paved approximation to f.

Let $p = \{\square_i\}$ be the paving associated with $f_{\epsilon/2,\delta/2}$. Replace each $\square_i$ by the rectangle $\square_i'$ obtained by contracting $\square_i$ about its center by a factor $(1 - \xi)$. (See Fig. 8.10.) Thus $\mu(\square_i') = (1 - \xi)^n \mu(\square_i)$. For any $\mathbf{x}$, $\mathbf{y} \in \bigcup \square_i'$, if

$$\|\mathbf{x} - \mathbf{y}\| < \eta,$$

where η is sufficiently small, then $\mathbf{x}$ and $\mathbf{y}$ belong to the same $\square_i'$. If

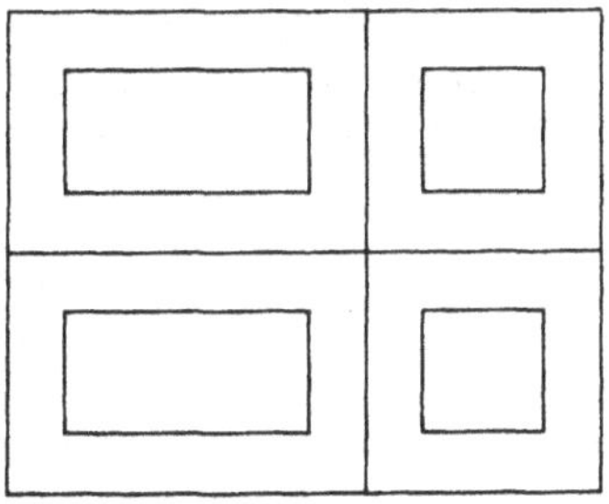

Fig. 8.10

$$\mathbf{x}, \mathbf{y} \in \bigcup \square_i', \qquad \mathbf{x}, \mathbf{y} \notin A_{\epsilon/2,\delta/2}, \qquad \text{and} \qquad \|\mathbf{x} - \mathbf{y}\| < \eta,$$

then

$$|f(\mathbf{x}) - f(\mathbf{y})| \le |f(\mathbf{x}) - f_{\epsilon/2,\delta/2}(\mathbf{x})| + |f(\mathbf{y}) - f_{\epsilon/2,\delta/2}(\mathbf{y})| + |f_{\epsilon/2,\delta/2}(\mathbf{x}) - f_{\epsilon/2,\delta/2}(\mathbf{y})|.$$

But the third term vanishes, so that $|f(\mathbf{x}) - f(\mathbf{y})| < \epsilon$. Now by first choosing ξ sufficiently small, we can arrange that $\mu(|p| - \bigcup \square_i') < \delta/2$. Then we can choose η so small that $\|x - y\| < \eta$ implies that $\mathbf{x}, \mathbf{y}$ belong to the same $\square_i'$ if $\mathbf{x}, \mathbf{y} \in \bigcup \square_i'$. For this η and for $A_\delta = A_{\epsilon/2,\delta/2} \cup (|p| - \bigcup \square_i')$, Eq. (10.7) holds, and $\mu(A_\delta) < \delta$. □

In particular, a bounded function which is continuous except at a set of content zero and has compact support is contented.

EXERCISES

10.2 Show that for any bounded set A, e_A is a contented function if and only if A is a contented set.

10.3 Let f be a contented function whose support is contained in a cube $\square$. For each δ let $p_\delta = \{\square_{i,\delta}\}_{i\in I_\delta}$ be a paving with $|p_\delta| = \square$ and whose cubes have diameter less than δ. Let $\mathbf{x}_{i,\delta}$ be some point of $\square_{i,\delta}$. The expression

$$\sum_{i\in I_\delta} f(\mathbf{x}_{i,\delta})\mu(\square_{i,\delta})$$

is called a Riemann δ-approximating sum for f. Show that for any $\epsilon > 0$ there exists a $\delta_0\,[=\delta_0(f)] > 0$ such that

$$\left|\int f - \sum_{i\in I_\delta} f(\mathbf{x}_{i,\delta})\mu(\square_{i,\delta})\right| < \epsilon$$

whenever $\delta < \delta_0$.

11. THE CHANGE OF VARIABLES FORMULA

This section will be devoted to the proof of the following theorem, which is of fundamental importance.

Theorem 11.1. Let U and V be bounded *open* sets in $\mathbb{R}^n$, and let φ be a differentiable one-to-one map of U onto V with φ^{-1} differentiable. Let f be a contented function with $\operatorname{supp} f \subset V$. Then $(f\circ\varphi)$ is a contented function, and

$$\int_V f = \int_U (f\circ\varphi)|\det J_\varphi|. \tag{11.1}$$

Recall that if the map φ is given by $y^i = \varphi^i(x^1, \ldots, x^n)$, then J_φ is the linear transformation whose matrix is $[\partial\varphi_i/\partial x_j]$.

Note that if φ is a nonsingular linear transformation (so that J_φ is just φ), then Theorem 11.1 is an easy consequence of Theorem 8.1. In fact, for functions of the form e_A we observe that $e_A\circ\varphi = e_{\varphi^{-1}A}$, and Eq. (11.1) reduces, in this case, to (8.1). By linearity, (11.1) is valid for all paved functions.

Furthermore, $f\circ\varphi$ is contented. Suppose $|f(\mathbf{x}) - f(\mathbf{y})| < \epsilon$ when $\|\mathbf{x}-\mathbf{y}\| < \varphi$ and $\mathbf{x}, \mathbf{y} \notin A$, with $\mu(A) < \delta$. Then $|f\circ\varphi(\mathbf{u}) - f\circ\varphi(\mathbf{v})| < \epsilon$ when

$$\|\mathbf{u}-\mathbf{v}\| < \eta/\|\varphi\| \qquad \text{and} \qquad \mathbf{u}, \mathbf{v} \notin \varphi^{-1}(A),$$

with $\mu(\varphi^{-1}A) < \delta/|\det\varphi|$.

Now let $g_{\epsilon,\delta}$, $A_{\epsilon,\delta}$ be an approximating family of paved functions for f. Then $|f\circ\varphi(\mathbf{x}) - g_{\epsilon,\delta}\circ\varphi(\mathbf{x})| < \epsilon$ for $\mathbf{x}\notin\varphi^{-1}(A_{\epsilon,\delta})$ and $\mu(\varphi^{-1}A_{\epsilon,\delta}) < \delta/|\det\varphi|$. Thus $\int(g_{\epsilon,\delta}\circ\varphi)|\det\varphi| \to \int(f\circ\varphi)|\det\varphi|$, and Eq. (11.1) is valid for all contented f.

The proof of Theorem 11.1 for nonlinear maps is a bit more tricky. It consists essentially of approximating φ locally by linear maps, and we shall do it in several steps. We shall use the uniform norm $\|\mathbf{x}\|_\infty = \max|x^i|$ on $\mathbb{R}^n$. This is convenient because a ball in this norm is actually a cube, although this nicety isn't really necessary.

Let ψ be a (continuously) differentiable map defined on a convex open set U. If the cube $\Box = \Box_{\mathbf{p}-r\mathbf{1}}^{\mathbf{p}+r\mathbf{1}}$ lies in U, then the mean-value theorem (Section 7, Chapter 3) implies that for any $\mathbf{y} \in \Box$,

$$\|\psi(\mathbf{y}) - \psi(\mathbf{p})\|_\infty < \|\mathbf{y} - \mathbf{p}\|_\infty \sup_{\mathbf{z}\in\Box} \|J_\psi(\mathbf{z})\|.$$

Thus

$$\psi(\Box) \subset \Box_{\psi(\mathbf{p})-Kr\mathbf{1}}^{\psi(\mathbf{p})+Kr\mathbf{1}}, \qquad \text{where} \qquad K = \sup_{\mathbf{z}\in\Box} \|J_\psi(\mathbf{z})\|.$$

Thus

$$\mu(\psi(\Box)) < (\sup_{\mathbf{z}\in\Box} \|J_\psi(\mathbf{z})\|)^n \mu(\Box). \tag{11.2}$$

Lemma 11.1. Let φ be as in Theorem 11.1. Then for any contented set A with $\bar{A} \subset U$ we have

$$\mu(\varphi(A)) \le \int_A |\det J_\varphi|. \tag{11.3}$$

Proof. Let us apply Eq. (11.2) to the map $\psi = L^{-1}\varphi$, where L is a linear transformation. Then

$$|\det L|^{-1}\mu(\varphi(\Box)) = \mu(L^{-1}\varphi(\Box)) \le (\sup_{\mathbf{z}\in\Box} \|J_{L^{-1}\varphi}(\mathbf{z})\|)^n \mu(\Box).$$

Since $J_{L^{-1}\varphi} = L^{-1}J_\varphi$, we get

$$\mu(\varphi(\Box)) \le |\det L| (\sup_{\mathbf{z}\in\Box} \|L^{-1}J_\varphi(\mathbf{z})\|)^n \mu(\Box) \tag{11.4}$$

for any $\Box$ contained in the domain of the definition of φ and for any linear transformation L.

For any $\epsilon > 0$, let δ be so small that $\|J_\varphi(\mathbf{x})^{-1}J_\varphi(\mathbf{y})\| < 1 + \epsilon$ for

$$\|\mathbf{x} - \mathbf{y}\|_\infty < \delta$$

for all $\mathbf{x}$, $\mathbf{y}$ in a compact neighborhood of A. (It is possible to choose such a δ, since $J(\mathbf{x})$ is a uniformly continuous function of $\mathbf{x}$, so that $J_\varphi(\mathbf{x})^{-1}J_\varphi(\mathbf{y})$ is close to the identity matrix when $\mathbf{x}$ is close to $\mathbf{y}$; see Section 8, Chapter 4.)

Choose an outer paving $\mathfrak{z} = \{\Box_i\}$ of A, where the $\Box_i$ are cubes all having edges of length less than δ. Let $\mathbf{x}_i$ be a point of $\Box_i$. Then applying (11.4) to each $\Box_i$ taking $L = J_\varphi(\mathbf{x}_i)$, we get

$$\mu(\varphi(A)) < \mu(\varphi(|\mathfrak{z}|)) = \sum \mu\varphi(\Box_i) < \sum |\det J_\varphi(\mathbf{x}_i)|(1+\epsilon)^n\mu(\Box_i).$$

We can also suppose δ to have been taken small enough so that

$$|\det J_\varphi(\mathbf{z})| > (1-\epsilon)|\det J_\varphi(\mathbf{x}_i)| \qquad \text{for all} \quad \mathbf{z} \in \Box_i \quad \text{and all} \quad i.$$

Then we have

$$\int_{\Box_i} |\det J_\varphi| > (1-\epsilon)|\det J_\varphi(\mathbf{x}_i)|\mu(\Box_i),$$

and so

$$\mu(\varphi(A)) < \frac{1}{1-\epsilon}(1+\epsilon)^n \int_{|\mathfrak{z}|} |\det J_\varphi|.$$

Since ϵ is arbitrary and $\mathfrak{z}$ is an arbitrary outer paving of A, we get (11.3). $\Box$

We can now conclude that $f \circ \varphi$ is contented for any contented f with $\operatorname{supp} f \subset V$. In fact, let K be chosen so large that it is a Lipschitz constant for φ on $\varphi^{-1}(\operatorname{supp} f)$, and so large that $K > |\det J_{\varphi^{-1}}(\mathbf{u})|$ for $\mathbf{u} \in \operatorname{supp} f$. Now given ϵ and δ, we can find an η such that

$$|f(\mathbf{u}) - f(\mathbf{v})| < \epsilon \quad \text{if} \quad \|\mathbf{u} - \mathbf{v}\| < \eta \text{ and } \mathbf{u}, \mathbf{v} \notin A_\delta \text{ with } \mu(A_\delta) < \delta.$$

But this implies that

$$|f \circ \varphi(\mathbf{x}) - f \circ \varphi(\mathbf{y})| < \epsilon \quad \text{if} \quad \|\mathbf{x} - \mathbf{y}\| < \eta/K \text{ and } \mathbf{x}, \mathbf{y} \notin \varphi^{-1}(A_\delta),$$

where $\mu(\varphi^{-1}(A_\delta)) < K_\delta$, by (11.3). Since K was chosen independently of ϵ and δ, this shows that $f \circ \varphi$ is contented.

Lemma 11.2. Let φ, U, and V be as in Theorem 11.1. Let f be a nonnegative contented function with $\operatorname{supp} f \subset V$. Then

$$\int_V f \le \int_U (f \circ \varphi)|\det J_\varphi|. \tag{11.5}$$

Proof. Let $\langle g, A \rangle$ be a paved ϵ, δ-approximation to f with $g(\mathbf{u}) \le f(\mathbf{u})$ for all $\mathbf{u}$. If $p = \{\square_i\}$ is the paving associated with g, we may assume that $\operatorname{supp} f \subset |p|$. Then

$$\begin{aligned}\int g &= \sum_{\substack{i \\ u_i \in \square_i}} g(\mathbf{u}_i)\mu(\square_i) \le \sum g(\mathbf{u}_i) \int_{\varphi^{-1}(\square_i)} |\det J_\varphi| \le \sum_i \int_{\varphi^{-1}(\square_i)} (f \circ \varphi)|\det J_\varphi| \\ &= \int_{\bigcup \varphi^{-1}(\square_i)} (f \circ \varphi)|\det J_\varphi| = \int (f \circ \varphi)|\det J_\varphi|.\end{aligned}$$

Since we can choose g so that $\int g \to \int f$, we obtain (11.5). $\square$

Lemma 11.3. Let φ, U, V, and f be as in Theorem 11.1. Let f be a nonnegative function. Then Eq. (11.5) holds.

Proof. Let us apply (11.4) to the map φ^{-1} and the function $(f \circ \varphi)|\det J_\varphi|$. Since $J_\varphi(\mathbf{x}) \circ J_{\varphi^{-1}}(\varphi(\mathbf{x})) = \text{id}$, we obtain

$$\begin{aligned}\int (f \circ \varphi)|\det J_\varphi| &\le \int [(f \circ \varphi) \circ \varphi^{-1}](|\det J_\varphi| \circ \varphi^{-1})|\det J_{\varphi^{-1}}| \\ &= \int f.\end{aligned}$$

Combining this with (11.5) proves the lemma. $\square$

Completion of the proof of Theorem 11.1. Any real-valued contented function can be written as the difference of two positive contented functions. If for all $\mathbf{x}$, $f(\mathbf{x}) > -M$ for some large M, we write $f = (f + Me_\square) - Me_\square$, where $\operatorname{supp} f \subset \square$. Since we have verified Eq. (11.1) for nonnegative functions, and since both sides of (11.1) are linear in f, we are done. Similarly, any bounded complex-valued contented function f can be written as $f = f_1 + if_2$, where f_1 and f_2 are bounded real-valued contented functions. $\square$

In practice, we sometimes may apply Eq. (11.1) to a situation where the hypotheses of Theorem 11.1 are not, strictly speaking, verified. For instance, in $\mathbb{R}^2$ we may want to introduce "polar coordinates". That is, we let r, θ be coordinates on $\mathbb{R}^2$; if S is the set $0 \le \theta < 2\pi$, $0 \le r$, we consider the map $\varphi: S \to \mathbb{R}^2$ given by $x = r\cos\theta$, $y = r\sin\theta$, where x, y are coordinates on a second copy of $\mathbb{R}^2$. Now this map is one-to-one and has positive Jacobian for $r > 0$. If we consider the open sets $U \subset S$ given by $0 < r$, $0 < \theta \le 2\pi$ and $V \subset \mathbb{R}^2$ given by $V = \mathbb{R}^2 - \{x, y : y = 0, x \ge 0\}$, the hypotheses of Theorem 11.1 are fulfilled, and we can write (since $\det J_\varphi = r$)

$$\int f = \int (f \circ \varphi) r \tag{11.6}$$

if $\operatorname{supp} f \subset V$. However, Eq. (11.6) is valid without the restriction $\operatorname{supp} f \subset V$. In fact, if D_ϵ is a strip of width ϵ about the ray $y = 0$, $x \ge 0$, then $f = fe_{D_\epsilon} + fe_{\mathbb{R}^n - D_\epsilon}$ and $\int fe_{D_\epsilon} \to 0$ as $\epsilon \to 0$ (Fig. 8.11). Similarly, $\int (f \circ \varphi)(r \circ \varphi) e_{D_\epsilon} \circ \varphi \to 0$, so that (11.6) is valid for all contented f by this simple limit argument.

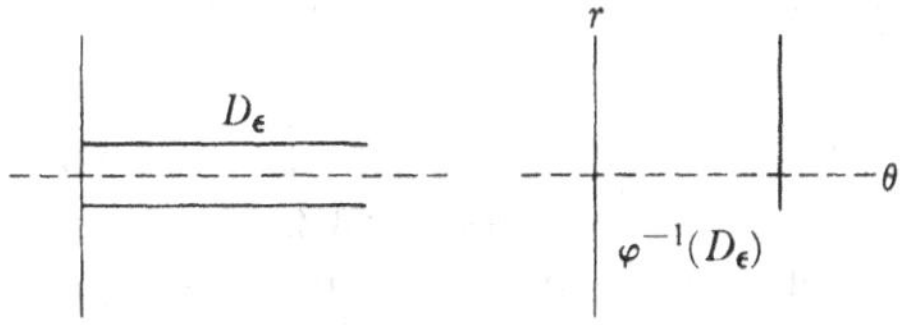

Fig. 8.11

We will not state a general theorem covering all such useful extensions of Theorem 11.1. In each case the limit argument is usually quite straightforward and will be left to the reader.

EXERCISES

11.1 By the parallelepiped spanned by $\mathbf{v}^1, \ldots, \mathbf{v}^n$ we mean the set of all $x = \sum \xi^1 \mathbf{v}^1 + \cdots + \xi^n \mathbf{v}^n$, where $0 \le \xi^i < 1$. Show that the content of this parallelepiped is given by

$$|\det((\mathbf{v}_i, \mathbf{v}_j))|^{1/2}.$$

11.2 Express the content of the ellipsoid

$$\left\{\mathbf{x} : \frac{(x^1)^2}{(a^1)^2} + \cdots + \frac{(x^n)^2}{(a^n)^2} \le 1\right\}$$

in terms of the content of the unit ball.

11.3 Compute the Jacobian determinant of the map $\langle r, \theta \rangle \mapsto \langle x, y \rangle$, where $x = r\cos\theta$, $y = r\sin\theta$.

11.4 Compute the Jacobian determinant of the map $\langle r, \theta, \varphi \rangle \mapsto \langle x, y, z \rangle$, where $x = r\cos\varphi\sin\theta$, $y = r\sin\varphi\sin\theta$, $z = r\cos\theta$.

11.5 Compute the Jacobian determinant of the map $\langle r, \theta, z \rangle \mapsto \langle x, y, z \rangle$, where $x = r\cos\theta$, $y = r\sin\theta$, $z = z$.

12. SUCCESSIVE INTEGRATION

In the case of one variable, i.e., the theory of integration on $\mathbb{R}^1$, the fundamental theorem of the calculus reduces the computation of the integral of a function to the computation of its antiderivative. The generalization of this theorem to n dimensions will be presented in a later chapter. In this section we will show how, in many cases, the computation of an n-dimensional integral can be reduced to n successive one-dimensional integrations.

Suppose we regard $\mathbb{R}^n$, in some fixed way, as the direct product $\mathbb{R}^n = \mathbb{R}^k \times \mathbb{R}^l$. We shall write every $\mathbf{z} \in \mathbb{R}^n$ as $\mathbf{z} = \langle \mathbf{x}, \mathbf{y} \rangle$, where $\mathbf{x} \in \mathbb{R}^k$ and $\mathbf{y} \in \mathbb{R}^l$.

Definition 12.1. We say that a contented function f is contented relative to the decomposition $\mathbb{R}^n = \mathbb{R}^k \times \mathbb{R}^l$ if there exists a set $A_f \subset \mathbb{R}^k$ of content zero (in $\mathbb{R}^k$) such that

i) for each fixed $\mathbf{x} \in \mathbb{R}^k$, $\mathbf{x} \notin A_f$, the function $f(\mathbf{x}, \cdot)$ is a contented function on $\mathbb{R}^l$;

ii) the function $\int_{\mathbb{R}^l} f$ which assigns to x the number $\int_{\mathbb{R}^l} f(\mathbf{x}, \cdot)$ is a contented function on $\mathbb{R}^k$.

It is easy to see that the set of all such functions satisfies Axioms $\mathfrak{F}1$ through $\mathfrak{F}3$. (The only axiom that is not immediate is $\mathfrak{F}2$. But this is an easy consequence of the fact that any translation T can be rewritten as $T_1 T_2$,where T_1 is a translation in $\mathbb{R}^k$ and T_2 is a translation in $\mathbb{R}^l$.)

It is equally easy to verify that the rule which assigns to any such f the number

$$\int_{\mathbb{R}^k} \left(\int_{\mathbb{R}^l} f(\cdot, \cdot) \right)$$

satisfies Axioms $\int 1$ through $\int 4$. The only one which isn't immediately obvious is $\int 3$. However, if p is any paving with $\operatorname{supp} f \subset |p|$, then

$$f \leq \|f\| e_{|p|}$$

and

$$\int_{\mathbb{R}^k} \left(\int_{\mathbb{R}^l} \|f\| e_{|p|} \right) = \|f\| \int_{\mathbb{R}^k} \int_{\mathbb{R}^l} e_{|p|} = \|f\| \mu(e_{|p|}),$$

since

$$\int_{\mathbb{R}^k} \int_{\mathbb{R}^l} e_{\square} = \mu(e_{\square})$$

for any rectangle (direct verification). Thus, by the uniqueness part of Theorem 10.1, we have

$$\int_{\mathbb{R}^n} f = \int_{\mathbb{R}^k} \left(\int_{\mathbb{R}^l} f(\cdot, \cdot) \right). \tag{12.1}$$

Note, in particular, that if f is also contented relative to the decomposition $\mathbb{R}^n = \mathbb{R}^l \times \mathbb{R}^k$, then

$$\int_{\mathbb{R}^n} f = \int_{\mathbb{R}^l} \left(\int_{\mathbb{R}^k} f(\cdot, \cdot) \right).$$

In particular, for such f the double integration is independent of the order.

In practice, all the functions that we shall come across will be contented relative to any decomposition of $\mathbb{R}^n$. In particular, writing $\mathbb{R}^n = \mathbb{R}^1 \times \cdots \times \mathbb{R}^1$, we have

$$\int_{\mathbb{R}^n} f = \int\left(\cdots\left(\int f(\cdot, \ldots, \cdot)\right)\right). \tag{12.2}$$

In terms of the rectangular coordinates $x^1, \ldots, x^n$, this last expression is usually written as

$$\int\left(\cdots\left(\int f(x^1, \ldots, x^n)\, dx^1\right)\cdots\right) dx^n.$$

For this reason, the expression on the left-hand side of (12.2) is frequently written as

$$\int_{\mathbb{R}^n} f = \int \cdots \int f(x^1, \ldots, x^n)\, dx^1 \cdots dx^n.$$

Let us work out some simple examples illustrating the methods of integration given in the previous sections.

Example 1. *Compute the volume of the intersection of the solid cone with vertex angle α (vertex at O) with the spherical shell* $1 \leq r \leq 2$ (Fig. 8.12). By a Euclidean motion we may assume that the axis of the cone is the z-axis. If we introduce polar coordinates, we see that the set in question is the image of the set

$$\square_{\langle 1,0,0\rangle}^{\langle 2,2\pi,\alpha/2\rangle}, \qquad 1 \leq r < 2, \qquad 0 \leq \varphi < 2\pi, \qquad 0 \leq \theta \leq \alpha/2$$

in the $\langle r, \varphi, \theta\rangle$-space (Fig. 8.13).

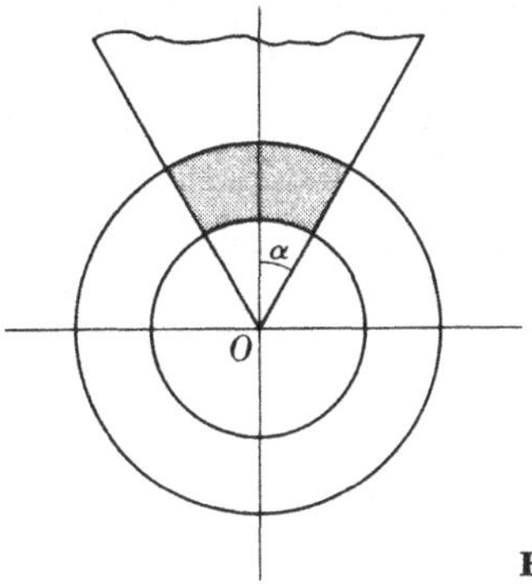

Fig. 8.12

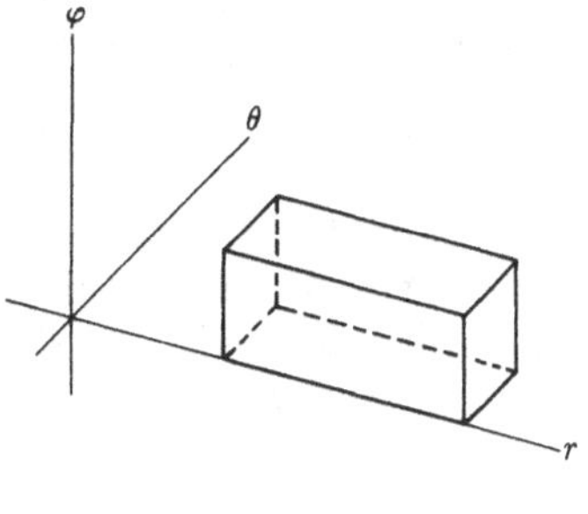

Fig. 8.13

By the change of variables formula and Exercise 11.4 we see that the volume in question is given by

$$\begin{aligned}\int r^2 \sin\theta &= \int_1^2 \int_0^{2\pi} \int_0^{\alpha/2} r^2 \sin\theta\, d\theta\, d\varphi\, dr \\ &= 2\pi \int_1^2 \int_0^{\alpha/2} r^2 \sin\theta\, d\theta\, dr \\ &= 2\pi \int_1^2 [1 - \cos(\alpha/2)] r^2\, dr \\ &= 2\pi[1 - \cos(\alpha/2)](\tfrac{8}{3} - \tfrac{1}{3}).\end{aligned}$$

Example 2. Let B be a contented set in the plane, and let f_1 and f_2 be two contented functions defined on B. Let A be the set of all $\langle x, y, z \rangle \in \mathbb{E}^3$ such that $\langle x, y \rangle \in B$ and $f_1(x, y) \le z \le f_2(x, y)$. If G is any contented function on A, we can express the integral $\int_A G$ as

$$\int_A G = \int_A \left\{ \int_{f_1(x,y)}^{f_2(x,y)} G(x, y, z)\, dz \right\} dx\, dy.$$

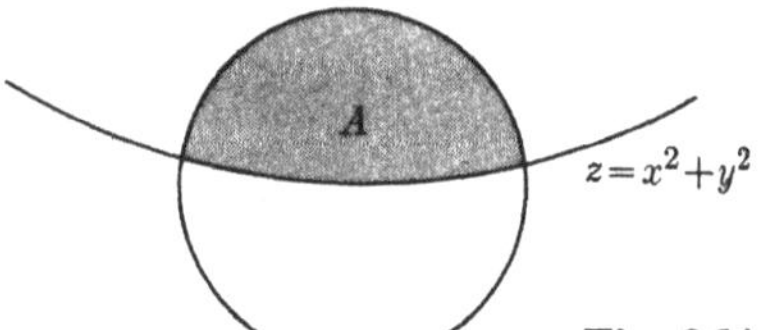

Fig. 8.14

For example, *compute the integral* $\int_A z$, *where* A *is the set of all points in the unit ball lying above the surface* $z = x^2 + y^2$ (Fig. 8.14). Thus

$$A = \{\langle x, y, z \rangle : x^2 + y^2 + z^2 \le 1, z \ge x^2 + y^2\}.$$

We must have $x^2 + y^2 \le a$, where $a^2 + a = 1$ [so that $a = (\sqrt{5} - 1)/2$], in order for $\langle x, y, z \rangle$ to belong to A. Then $f_1(x, y) = x^2 + y^2$, $f_2(x, y) = \sqrt{1 - (x^2 + y^2)}$, and

$$\int_{f_1(x,y)}^{f_2(x,y)} z\, dz = \tfrac{1}{2}[1 - (x^2 + y^2) - (x^2 + y^2)^2],$$

so that, using polar coordinates in the plane (and Exercise 11.3),

$$\int_A z = \tfrac{1}{2} \int_{x^2+y^2 \le a} [1 - (x^2 + y^2) - (x^2 + y^2)^2] = \pi \int_0^{\sqrt{a}} r(1 - r^2 - r^4)\, dr.$$

As we saw in the last example, part of the problem of computing an integral as an iterated integral is to determine a good description of the domain of integration in terms of the decomposition of the vector space. It is usually a great help in visualizing the situation to draw a figure.

Example 3. *Compute the volume enclosed by a surface of revolution.* Here we are given a function f of one variable, and we consider the surface obtained by rotating the curve $x = f(z)$, $z_1 \le z \le z_2$, around the z-axis (Fig. 8.15). We thus wish to compute $\mu(A)$, where

$$A = \{\langle x, y, z \rangle : x^2 + y^2 \le f(z), z_1 \le z \le z_2\}.$$

Here it is obviously convenient to use cylindrical coordinates, and we see that A is the image of the set

$$B = \{\langle r, \theta, z \rangle : r \le f(z), 0 \le \theta < 2\pi\}$$

in the $\langle r, \theta, z \rangle$-space. By Exercise 11.5, we wish to compute

$$\int_B r = \int_0^{2\pi} \int_{z_1}^{z_2} \int_0^{f(z)} r\, dr\, dz\, d\theta = 2\pi \int_{z_1}^{z_2} \left(\int_0^{f(z)} r\, dr \right) dz = 2\pi \int_{z_1}^{z_2} \frac{f(z)^2}{2}\, dz.$$

Thus

$$\mu(A) = \pi \int_{z_1}^{z_2} f(z)^2\, dz.$$

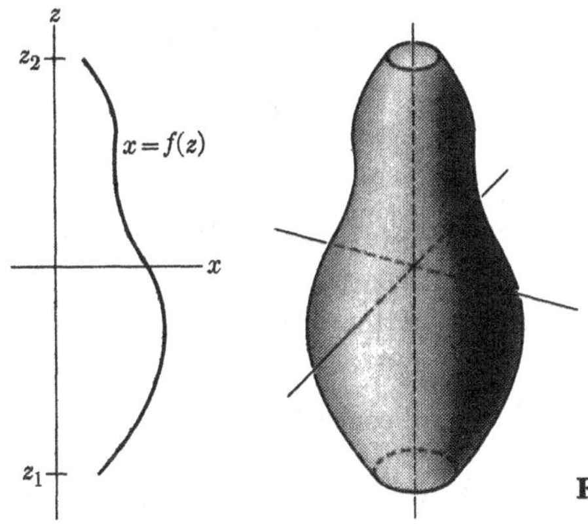

Fig. 8.15

EXERCISES

12.1 Compute the volume of the region between the surfaces $z = x^2 + y^2$ and $z = x + y$.

12.2 Find the volume of the region in $\mathbb{E}^3$ bounded by the plane $z = 0$, the cylinder $x^2 + y^2 = 2x$, and the cone $z = +\sqrt{x^2 + y^2}$.

12.3 Compute $\int_A (x^2 + y^2)^2 \, dx \, dy \, dz$, where A is the region bounded by the plane $z = 2$ and the surface $x^2 + y^2 = 2z$.

12.4 Compute $\int_A x$, where

$$A = \{\langle x, y, z\rangle : x^2 + y^2 + z^2 \leq a^2, x \geq 0, y \geq 0, z \geq 0\}.$$

12.5 Compute

$$\int_A \left(\frac{x^2}{a^2} + \frac{y^2}{b^2} + \frac{z^2}{c^2}\right)^{1/2},$$

where A is the region bounded by the ellipsoid

$$\frac{x^2}{a^2} + \frac{y^2}{b^2} + \frac{z^2}{c^2} = 1.$$

Let ρ be a nonnegative function (to be called the density of mass in the following discussion) defined on a domain D in $\mathbb{E}^3$. The total mass of $\langle D, \rho\rangle$ is defined as

$$M = \int_D \rho(x) \, dx.$$

If $M \neq 0$, the center of gravity of $\langle D, \rho\rangle$ is the point $C = \langle C_1, C_2, C_3\rangle$, where

$$C_1 = \frac{1}{M} \int_D x_1 \rho(x) \, dx,$$

$$C_2 = \frac{1}{M} \int_D x_2 \rho(x) \, dx,$$

$$C_3 = \frac{1}{M} \int_D x_3 \rho(x) \, dx.$$

12.6 A homogeneous solid (where ρ is constant) is given by $x_1 \geq 0, x_2 \geq 0, x_3 \geq 0$, and

$$\frac{x_1^2}{a^2} + \frac{x_2^2}{b^2} + \frac{x_3^2}{c^2} \leq 1.$$

Find its center of gravity.

12.7 The unit cube has density $\rho(\mathbf{x}) = x_1x_3$. Find its total mass and its center of gravity.

12.8 Find the center of mass of the homogeneous body bounded by the surfaces $x^2 + y^2 + z^2 = a^2$ and $x^2 + y^2 = ax$.

The notion of center of mass can, of course, be defined for a region in a Euclidean space of any dimension. Thus, for a region D in the plane with density ρ, the center of mass will be the point $\langle x_0, y_0 \rangle$, where

$$x_0 = \frac{\int_D x\rho}{\int_D \rho} \qquad \text{and} \qquad y_0 = \frac{\int_D y\rho}{\int_D \rho}.$$

12.9 Let D be a region in the xz-plane which lies entirely in the half-plane $x > 0$. Let A be the solid in $\mathbb{E}^3$ obtained by rotating D about the z-axis. Show that $\mu(A) = 2\pi\, d\mu(D)$, where d is the distance of the center of mass of the region D (with uniform density) from the z-axis. (Use cylindrical coordinates.) This is known as *Guldin's rule.*

Observe that in the definition of center of gravity we obtain a vector (i.e., a point in $\mathbb{E}^3$) as the answer by integrating each of its coordinates. This suggests the following definition: Let V be a finite-dimensional vector space, and let $e_1, \ldots, e_k$ be a basis for V. Call a map f from $\mathbb{E}^n$ to V (f is a vector-valued function on $\mathbb{E}^n$ with values in V) contented if when we write $f(\mathbf{x}) = \sum f_i(\mathbf{x})e_i$, each of the (real-valued) functions f_i is contented. Define the integral of f over D by

$$\int_D f = \sum \left(\int_D f_i\right) e_i.$$

12.10 Show that the condition that a function be contented and the value of its integral are independent of the choice of basis $e_1, \ldots, e_k$.

Let ξ be a point not in the closed domain D, which has a mass distribution ρ. The gravitational force on a particle of unit mass situated at ξ is defined to be the vector

$$\int_D \frac{\rho(\mathbf{x})(\mathbf{x} - \xi)}{\|\mathbf{x} - \xi\|^3}\, dx$$

(here $\mathbf{x} - \xi$ is an $\mathbb{E}^3$-valued function on $\mathbb{E}^3$).

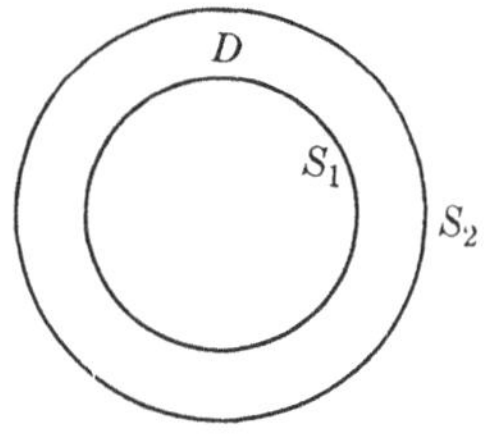

Fig. 8.16

12.11 Let D be the spherical shell bounded by two concentric spheres S_1 and S_2 (Fig. 8.16), with center at the origin. Let ρ be a mass distribution on D which depends only on the distance from the center, that is, $\rho(\mathbf{x}) = f(\|\mathbf{x}\|)$. Show that the gravitational force vanishes at any ξ inside S_1.

12.12 $\langle D, \rho \rangle$ is as in Exercise 12.5. Show that the gravitational force on a point outside S_2 is the same as that due to a particle situated at the origin and whose mass is the total mass of D.

13. ABSOLUTELY INTEGRABLE FUNCTIONS

Thus far we have been dealing with bounded functions of compact support. In practice, we would like to be able to integrate functions which neither are bounded nor have compact support. Let f be a function defined on $\mathbb{E}^n$, let M be a nonnegative real number, and let A be a (bounded) contented subset of $\mathbb{E}^n$. Let f_A^M be the function

$$f_A^M(x) = \begin{cases} 0 & \text{if } x \notin A, \\ M & \text{if } x \in A \text{ and } |f(x)| > M, \\ f(x) & \text{if } x \in A \text{ and } |f(x)| \leq M. \end{cases}$$

Thus f_A^M is a bounded function of compact support. It is obtained from f by cutting f back to zero outside A and cutting f back to M when $|f(x)| > M$.

We say that a function f is *absolutely integrable* if

i) f_A^M is a contented function for all $M > 0$ and contented sets A; and

ii) for any $\epsilon > 0$ there is a bounded contented set A_ϵ such that $e_{A_\epsilon} \cdot f$ is bounded and for all $M > 0$ and all B with $B \cap A_\epsilon = \emptyset$,

$$\int |f_B^M| < \epsilon.$$

It is easy to check that the sum of two absolutely integrable functions is again absolutely integrable. Thus the set of absolutely integrable functions forms a vector space. Note that if f satisfies condition (i) and $|f(x)| < |g(x)|$ for all x, where g is absolutely integrable, then f is absolutely integrable.

Let f be an absolutely integrable function. Given any ϵ, choose a corresponding A_ϵ. Then for any numbers M_1 and $M_2 \geq \max_{x \in A_\epsilon} |f(x)|$ and for any sets $A_1 \supset A_\epsilon$ and $A_2 \supset A_\epsilon$,

$$\left| \int f_{A_1}^{M_1} - \int f_{A_2}^{M_2} \right| \leq \int \left| f_{A_1 - A_\epsilon}^{M_1} \right| + \int \left| f_{A_2 - A_\epsilon}^{M_2} \right| < 2\epsilon.$$

If we let $\epsilon \to 0$ and choose a corresponding family of A_ϵ, then the above inequality implies that the lim $\int f_{A_\epsilon}$ is independent of the choice of the A_ϵ. We define this limit to be $\int f$.

We now list some very crude sufficient criteria for a function to be absolutely integrable. We will consider the two different causes of trouble—nonboundedness and lack of compact support.

Let f be a bounded function with f_A contented for any contented set A. Suppose $|f(x)| \leq C\|x\|^{-k}$ for large values of $\|x\|$. Let B_r be the ball of radius r centered at the origin. If r_1 is large enough so that the inequality holds for $\|x\| \geq r_1$, then for $r_2 \geq r_1$ we have

$$\int |f_{B_{r_2} - B_{r_1}}| \leq C \int_{B_{r_2} - B_{r_1}} \|x\|^{-k} = C\Omega_n \int_{r_1}^{r_2} r^{n-1-k},$$

where Ω_n is some constant depending on n (in fact, it is the "surface area" of the unit sphere in $\mathbb{E}^n$). If $k > n$, this last integral becomes

$$\frac{C\Omega_n}{n-k}(r_2^{n-k} - r_1^{n-k}),$$

which is $\leq [Cl_n/(k-n)]r_1^{n-k}$, which tends to zero as $r_1 \to \infty$ if $k > n$. Thus we can assert:

Let f be a bounded function such that f_A is contented for any contented set A. Suppose that $|f(x)| \to 0$ as $\|x\| \to \infty$ in such a way that $\|x\|^k|f(x)|$ is bounded for some $k > n$. Then f is absolutely integrable.

Now let us examine the situation when f is of compact support but unbounded. Suppose first that there is a point x_0 such that f is bounded in the complement of any neighborhood of x_0. Suppose, furthermore, that

$$|f(x)| \leq C^k\|x - x_0\|^{-k}$$

for some constants C and k. Then if $|f(x)| > M$, $\|x - x_0\|^{-k} > M/C$ or $\|x - x_0\| < C/M^{1/k}$.

Let B_1 be the ball of radius $CM_1^{-1/k}$ centered at x_0. Then $|f(x)| > M_1$ implies that $x \in B_1$. Furthermore, for $M_2 > M_1$ we have

$$\int_{B_1} |f^{M_2}| < C^k \int_{B_1 - B_2} \|x - x_0\|^k + M_2\mu(B_2),$$

where B_2 is the ball of radius $CM_2^{-1/k}$ centered at x_0. Thus

$$\int_{B_1} |f^{M_2}| \leq C^k\Omega_n \int_{CM_2^{-1/k}}^{CM_1^{-1/k}} r^{n-1-k}\, dr + M_2C^{n-1}V_nM_2^{-n/k},$$

where Ω_n and V_n depend only on n. If $k < n$, the integral on the right becomes

$$\frac{C^{n-k}}{n-k}(M_1^{(k-n)/k} - M_2^{(k-n)/k}) < \frac{C^{n-k}}{n-k}M_1^{(k-n)/k}.$$

Thus

$$\int_{B_1} |f^{M_2}| < \text{const}\,(M_1^{(k-n)/k} + M_2^{(k-n)/n}),$$

which can be made arbitrarily small by choosing M_1 large.

Thus if f has compact support and is such that f^M is contented for all M and $|f(x)| < C\|x - x_0\|^{-k}$ with $k < n$, then f is absolutely integrable.

More generally, let S be a bounded subset of an l-dimensional subspace of $\mathbb{E}^n$. Let $d(x)$ denote the distance from x to S. Let f be a function of compact support with f^M contented for all M. If $|f(x)| < C\,d(x)^{-k}$ *with* $k < n - l$, *then f is absolutely integrable.* The proof is similar to that given above and is left to the reader.

Let $\{f_k\}$ be a sequence of absolutely integrable functions. Under what conditions will the sequence $\int f_n \to \int f$ if the sequence $f_k(x) \to f(x)$? Even if the sequence converges uniformly, there is no guarantee that the integrals converge. For instance, if $f_k = (1/k^n)e_{\square_0^k}$, then $|f_k(x)| \leq 1/k^n$, so that f_k approaches zero uniformly. On the other hand, $\int f_k = 1$ for all k.

We say that a set of functions $\{f_k\}$ is *uniformly* absolutely integrable if for any $\epsilon > 0$ there is an A_ϵ which can be chosen *independently* of k such that

$$\int_B \left|f_k^M\right| < \epsilon \qquad \text{for all } M$$

wherever $B \cap A_\epsilon = \emptyset$.

We frequently verify that $\{f_k\}$ is uniformly absolutely integrable by showing that there is an absolutely integrable function g such that $|f_k(x)| \leq |g(x)|$ for all k and x.

Let $\{f_k\}$ be a uniformly absolutely integrable sequence of functions. Suppose that $f_k \to f$ uniformly. Suppose in addition that f is absolutely integrable. Then $\int f_k \to \int f$. In fact, for any $\delta > 0$ we can find a k_0 such that $|f_k(x) - f(x)| < \delta$ for all $k > k_0$ and all x. We can also find A_ϵ and M_ϵ such that

$$\begin{aligned}\left|\int f_k - \int f\right| &\leq \left|\int f_{k,A_\epsilon}^M - \int f_k\right| + \left|\int f_{A_\epsilon}^M - \int f\right| + \left|\int f_{k,A_\epsilon}^M - \int f_{A_\epsilon}^M\right| \\ &\leq \epsilon + \epsilon + \delta\mu(A_\epsilon),\end{aligned}$$

which can be made arbitrarily small by first choosing ϵ small (which then gives an A_ϵ) and then choosing δ small (which means choosing k_0 large).

The main applications that we shall make of the preceding ideas will be to the problems of computing iterated integrals and of differentiating under the integral sign.

Proposition 13.1. Let f be a function on $\mathbb{R}^k \times \mathbb{R}^l$. Suppose that the set of functions $\{f(x, \cdot)\}$ is uniformly absolutely integrable, where x is restricted to lie in a bounded contented set $K \subset \mathbb{R}^k$. Then the function $e_{K\times\mathbb{R}^l} \cdot f$ is absolutely integrable, and

$$\int_{K\times\mathbb{R}^l} f = \int_K \int_{\mathbb{R}^l} f(x, y)\, dy\, dx = \int_{\mathbb{R}^l} \int_K f(x, y)\, dx\, dy.$$

Proof. By assumption, for any $\epsilon > 0$ we can find M and $A_\epsilon \subset \mathbb{R}^l$ such that

$$\int_{\mathbb{R}^l} |f_A^M(x, \cdot)| < \epsilon \qquad \text{if} \quad A \cap A_\epsilon = \emptyset. \tag{13.1}$$

Now for any set B in $\mathbb{R}^n$,

$$\begin{aligned}\int e_{K\times\mathbb{R}^l}|f_B^M| &= \int e_K(x) \int_{\mathbb{R}^l} |f_B^M(x, \cdot)| \\ &\leq \mu(K)\epsilon \qquad \text{if} \quad B \cap K \times A_\epsilon = \emptyset.\end{aligned}$$

This shows that $e_{K\times\mathbb{R}^l} f$ is absolutely integrable on $\mathbb{R}^n$. Now choose a sufficiently large $\square = \square_1 \times \square_2$ and an M such that

$$\left|\int_{K\times\mathbb{R}^l} f - \int_{K\times\mathbb{R}^l} f_\square^M\right| < \epsilon,$$

and also such that

$$\left|\int f(x, \cdot) - \int f_\square^M(x, \cdot)\right| < \epsilon \qquad \text{for all} \quad x \in K.$$

Then we have

$$\left|\int_{K\times\mathbb{R}^l} f - \int_{K\times\mathbb{R}^l} f_{\square}^M\right| < \epsilon$$

and

$$\int_{K\times\mathbb{R}^l} f_{\square}^M = \int_K \int_{\mathbb{R}^l} f_B^M(x, y)\, dy\, dx.$$

Thus

$$\left|\int_{K\times\mathbb{R}^l} f - \int_K \int_{\mathbb{R}^l} f(x, y)\, dy\, dx\right| < \epsilon + \mu(K)\epsilon,$$

so that

$$\int_{K\times\mathbb{R}^l} f = \int_K \int_{\mathbb{R}^l} f(x, y)\, dy\, dx.$$

Finally Eq. (13.1) shows that the function $F(y) = \int_K f(\cdot, y)$ is absolutely integrable. In fact, using the same A and M as in (13.1), we get

$$\int |F_B^{M_1}| \leq \mu(K)\epsilon.$$

Thus we get

$$\int_{K\times\mathbb{R}^l} f = \int_K \int_{\mathbb{R}^l} f(x, y)\, dy\, dx = \int_{\mathbb{R}^l} \int_K f(x, y)\, dx\, dy. \;\square$$

An extension of the same argument shows the following.

Proposition 13.2. Let f be absolutely integrable on $\mathbb{R}^n$ and such that the functions $f(x, \cdot)$ are uniformly absolutely integrable for each $x \in \mathbb{R}^k$. Then

$$\int f = \iint f(x, y)\, dy\, dx.$$

We now turn our attention to the problem of differentiating under the integral sign.

Proposition 13.3. Let $(t, x) \mapsto F(t, x)$ be a function on $I \times \mathbb{R}^n$, where $I = [a, b] \subset \mathbb{R}$. Suppose that

i) F and $\partial F/\partial t$ are continuous functions on $I \times \mathbb{R}^n$;

ii) $(\partial F/\partial t)(t, \cdot)$ is a uniformly absolutely integrable family of functions;

iii) $F(t, \cdot)$ is absolutely integrable for all $t \in I$.

Let $f(t) = \int F(t, \cdot)$. Then f is a differentiable function of t and

$$f'(t) = \int_{\mathbb{R}^n} (\partial F/\partial t)(t, \cdot).$$

Proof. Let $G(t) = \int_{\mathbb{R}^n} (\partial F/\partial t)(t, \cdot)$. Then $G(t)$ is continuous; hence we can pass to the limit under the integral sign of a family of absolutely integrable functions. Furthermore,

$$\int_a^t G(s)\, ds = \int_{\mathbb{R}^n} \int_a^t (\partial F/\partial t)(s, \cdot)\, ds$$

by Proposition 13.1. Thus

$$\int_a^t G(s) = \int_{\mathbb{R}^n} (F(t,\cdot) - F(a,\cdot)) = \int_{\mathbb{R}^n} F(t,\cdot) - \int_{\mathbb{R}^n} F(a,\cdot) = f(t) - f(a).$$

Differentiating this equation with respect to t gives the desired result. □

Finally, let us state the change of variables formula for absolutely integrable functions.

Let $\varphi\colon U \to V$ be a differentiable one-to-one map with differentiable inverse, where U and V are two open sets in $\mathbb{R}^n$. Let f be an absolutely integrable function defined on V. Then $(f \circ \varphi)|\det J_\varphi|$ is an absolutely integrable function on U and

$$\int_V f = \int_U (f \circ \varphi)|\det J_\varphi|.$$

Proof. To show that $(f \circ \varphi)|\det J_\varphi|$ is absolutely integrable, let $\epsilon > 0$ and choose an $A_\epsilon \subset V$ such that (ii) holds. Then $\overline{A}_\epsilon$ is compact, and therefore so is $\varphi^{-1}(\overline{A}_\epsilon)$. In particular, $\varphi^{-1}(\overline{A}_\epsilon)$ is a bounded contented set and $|\det J_\varphi|$ is bounded on it. If $B \cap \varphi^{-1}(\overline{A}_\epsilon) = \varnothing$, where $B \subset U$ is bounded and contented, then

$$\int_B |\det J_\varphi|^M |(f \circ \varphi)^M| \le \int_{\varphi(B)} |f^M| < \epsilon.$$

This shows that $(f \circ \varphi)|\det J_\varphi|$ is absolutely integrable. The rest of the proposition then follows from

$$\int_{A_\epsilon} f = \int_{\varphi^{-1}(A_\epsilon)} (f \circ \varphi)|\det J_\varphi|$$

by letting $\epsilon \to 0$. □

EXERCISES

13.1 Evaluate the integral $\int_{-\infty}^{\infty} e^{-x^2}\,dx$. [*Hint:* Compute its square.]

13.2 Evaluate the integral $\int_0^\infty e^{-x^2}x^{2k}\,dx$.

13.3 Evaluate the volume of the unit ball in an odd-dimensional space. [*Hint:* Observe that the Jacobian determinant for "polar" coordinates is of the form $r^{n-1} \times f$, where f is a function of the "angular variables". Thus the volume of the unit ball is of the form $C\int_0^1 r^{n-1}\,dr$, where C is determined by integrating f over the "angular variables". Evaluate C by computing $(\int_{-\infty}^{\infty} e^{-x^2}\,dx)^n$.]

14. PROBLEM SET: THE FOURIER TRANSFORM

Let $\alpha = \langle \alpha_1, \ldots, \alpha_n \rangle$ be an n-tuple whose entries are nonnegative integers. By D^α we shall mean the differential operator

$$D^\alpha = \frac{\partial^{\alpha_1 + \cdots + \alpha_n}}{\partial x_1^{\alpha_1} \cdots \partial x_n^{\alpha_n}}.$$

Let $|\alpha| = \alpha_1 + \cdots + \alpha_n$. Let $Q(x, D) = \sum_{|\alpha|\leq k} a_\alpha(x)D^\alpha$ be the differential operator where each a_α is a polynomial in x. Thus if f is a C^k-function on $\mathbb{R}^n$, we have

$$(Qf)(x) = \sum_{|\alpha|\leq k} a_\alpha(x)D^\alpha f(x).$$

For any f which is C^∞ on $\mathbb{R}^n$ we set

$$\|f\|_Q = \sup_{x\in\mathbf{R}^n} |Qf(x)|.$$

We denote by $\mathcal{S}$ the space of all $f \in C^\infty$ such that

$$\|f\|_Q < \infty \tag{14.1}$$

for all Q. To see what this means, let us consider those Q with $k = 0$. Then (14.1) says that for any polynomial $a(\cdot)$ the function $a \cdot f$ is bounded. In other words, f vanishes at infinity faster than the inverse of any polynomial; that is,

$$\lim_{\|x\|\to\infty} \|x\|^p f(x) = 0$$

for all p. To say that (14.1) holds means that the same is true for any derivative of f as well.

If f is a C^∞-function of compact support, then (14.1) obviously holds, so $f \in \mathcal{S}$. A more instructive example is provided by the function n given by

$$n(x) = e^{-\|x\|^2}.$$

Since $\lim_{r\to\infty} r^p e^{-r^2} = 0$ for any p, it follows that $\lim_{\|x\|\to\infty} a(x)n(x) = 0$. On the other hand, it is easy to see (by induction) that $D^\alpha n(x) = P_\alpha(x)n(x)$ for some polynomial P_α. Thus $Qn(x) = P_Q(x)n(x)$, where P_Q is a polynomial. Thus $n \in \mathcal{S}$.

It is easy to see that the space $\mathcal{S}$ is a vector space. We shall introduce a notion of convergence on this space by saying that $f_n \to f$ if for every fixed Q,

$$\|f_n - f\|_Q \to 0.$$

(Note that the space $\mathcal{S}$ is *not* a Banach space in that convergence depends on an infinity of different norms.)

EXERCISES

14.1 Let φ be a C^∞-function which grows slowly at infinity. That is, suppose that for every α there is a polynomial P_α such that

$$|D^\alpha\varphi(x)| < P_\alpha(x) \qquad \text{for all} \quad x.$$

Show that if $f \in \mathcal{S}$, then $\varphi f \in \mathcal{S}$. Furthermore, the map of $\mathcal{S}$ into itself sending $f \to \varphi f$ is continuous, that is, if $f_n \to f$, then $\varphi f_n \to \varphi f$.

For $x = \langle x^1, \ldots, x^n \rangle \in \mathbb{R}^n$ and $\xi = \langle \xi^1, \ldots, \xi^n \rangle \in \mathbb{R}^{n*}$ we denote the value of ξ at x by

$$\langle x, \xi \rangle = x^1\xi^1 + \cdots + x^n\xi^n.$$

Also for any $\alpha = \langle \alpha^1, \ldots, \alpha^n \rangle$ and any $x \in \mathbb{R}^n$ we let

$$x^\alpha = (x^1)^{\alpha_1} \cdots (x^n)^{\alpha_n},$$

and similarly $\xi^\alpha = (\xi^1)^{\alpha_1} \cdots (\xi^n)^{\alpha_n}$, etc.

For any $f \in \mathcal{S}$ we define its Fourier transform $\hat{f}$, which is a function on $\mathbb{R}^{n*}$, by

$$\hat{f}(\xi) = \int e^{-i\langle x,\xi \rangle} f(x)\, dx.$$

We note that

$$\hat{f}(0) = \int f \quad \text{and} \quad |\hat{f}(\xi)| \le \int |f|.$$

14.2 Show that $\hat{f}$ possesses derivatives of all orders with respect to ξ and that

$$D_\xi^\alpha \hat{f}(\xi) = (-i)^{|\alpha|} \int e^{-i\langle x,\xi\rangle} x^\alpha f(x)\, dx;$$

in other words,

$$D_\xi^\alpha \hat{f}(\xi) = \hat{g}(\xi),$$

where $g(x) = (-i)^{|\alpha|} x^\alpha f(x)$.

14.3 Show that

$$\widehat{\frac{\partial f}{\partial x^j}}(\xi) = i\xi^j \hat{f}(\xi).$$

[*Hint:* Write the integral as an iterated integral and use integration by parts with respect to the jth variable.]

14.4 Conclude that the map $f \mapsto \hat{f}$ sends $\mathcal{S}(\mathbb{R}^n)$ into $\mathcal{S}(\mathbb{R}^{n*})$ and that if $f_n \to 0$ in $\mathcal{S}$, then $\hat{f}_n \to 0$ in $\mathcal{S}(\mathbb{R}^{n*})$.

14.5 Show that

$$\widehat{T_\omega f}(\xi) = e^{-i\langle \omega,\xi\rangle} \hat{f}(\xi) \quad \text{for any} \quad \omega \in \mathbb{R}^n.$$

Recall that $T_\omega f(x) = f(x - \omega)$.

14.6 For any $f \in \mathcal{S}$ define $\tilde{f}$ by

$$\tilde{f}(x) = \overline{f(-x)},$$

where $\overline{}$ denotes complex conjugation. Show that

$$\hat{\tilde{f}}(\xi) = \overline{\hat{f}(\xi)}.$$

14.7 Let $n = 1$, and let f be an even real-valued function of x. Show that

$$\hat{f}(\xi) = \int \cos \langle x, \xi \rangle\, f(x)\, dx.$$

14.8 Let $n(x) = e^{-(1/2)x^2}$ where $x \in \mathbb{R}^1$. Show that

$$\frac{d\hat{n}}{d\xi}(\xi) = -\xi\hat{n}(\xi),$$

and conclude that

$$\log \hat{n}(\xi) = -\tfrac{1}{2}\xi^2 + \text{const},$$

so that

$$\hat{n}(\xi) = \text{const} \times e^{-(1/2)\xi^2}.$$

Evaluate this constant as $\sqrt{2\pi}$ by setting $\xi = 0$ and using Exercise 13.1. Thus

$$\hat{n}(\xi) = \sqrt{2\pi}\, e^{-(1/2)\xi^2}.$$

14.9 Show that the limit $\lim_{\epsilon \to 0} \int_\epsilon^{1/\epsilon} (\sin x)/x \, dx$ exists. Let us call this limit d. Show that for any $R > 0$, $\lim_{\epsilon \to 0} \int_\epsilon^{1/\epsilon} (\sin Rx)/x \, dx = d$.

If $f \in \mathcal{S}$, we have seen that $\hat{f} \in \mathcal{S}(\mathbb{R}^{n*})$. We can therefore consider the function

$$\int e^{i\langle y, \xi \rangle} \hat{f}(\xi) \, d\xi.$$

The purpose of the next few exercises is to show that

$$f(y) = \frac{1}{(2\pi)^n} \int e^{i\langle y, \xi \rangle} \hat{f}(\xi) \, d\xi. \tag{14.2}$$

We first remark that since all integrals involved are absolutely convergent, it suffices to show that

$$f(y) = \lim_{R_1 \to \infty} \cdots \lim_{R_n \to \infty} \frac{1}{(2\pi)^n} \int_{-R_n}^{R_n} \int_{-R_1}^{R_1} \hat{f}(\xi^1, \ldots, \xi^n) e^{i(y^1\xi^1 + \cdots + y^n\xi^n)} \, d\xi^1 \cdots d\xi^n.$$

Substituting the definition of f into this formula and interchanging the order of integration with respect to x and ξ, we get

$$\lim_{R_1 \to \infty} \cdots \lim_{R_n \to \infty} \left(\frac{1}{2\pi}\right)^n \iint_{-R_n}^{R_n} \cdots \int_{-R_1}^{R_1} f(x^1, \ldots, x^n) e^{i[(y^1 - x^1)\xi^1 + \cdots + (y^n - x^n)\xi^n]} \, d\xi^1 \cdots d\xi^n dx.$$

It therefore suffices to evaluate this limit one variable at a time (provided the convergence is uniform, which will be clear from the proof). We have thus reduced the problem to functions of one variable. We must show that if $f \in \mathcal{S}(\mathbb{R}^1)$, then

$$f(y) = \lim_{R \to \infty} \frac{1}{2\pi} \iint_{-R}^{R} f(x) e^{i(y-x)\xi} \, d\xi \, dx.$$

We shall first show that

$$f(y) = \lim_{R \to \infty} \frac{1}{4d} \iint_{-R}^{R} f(x) e^{i(y-x)\xi} \, d\xi \, dx,$$

where d is given in Exercise 14.9.

14.10 Show that this last integral can be written as

$$\frac{1}{2d} \int_{-\infty}^{\infty} f(x) \frac{\sin R(y-x)}{y-x} \, dx = \frac{1}{d} \int_0^{\infty} \frac{f(y-u) + f(y+u)}{2} \sin \frac{Ru}{u} \, du.$$

14.11 Let

$$g(u) = \frac{f(y-u) + f(y+u)}{2} - f(y).$$

Show that $g(0) = 0$ and conclude that $g(x) = xh(x)$ for $0 \leq x \leq 1$, where $h \in C^1$. By integrating by parts, show that

$$\left| \int_1^{\epsilon} g(u) \sin \frac{Ru}{u} + \int_{1/\epsilon}^1 g(u) \sin \frac{Ru}{u} \right| < \text{const} \frac{1}{R}.$$

Conclude that

$$\lim_{R \to \infty} \frac{1}{d} \int_0^{\infty} \frac{f(y-u) + f(y+u)}{2} \sin \frac{Ru}{u} \, du = f(y).$$

This proves that

$$f(y) = \frac{1}{4d} \int e^{iy\xi} f(\xi) \, d\xi.$$

14.12 Using Exercise 14.8, conclude that $d = \pi/2$.

Let $f_1 \in \mathcal{S}$ and $f_2 \in \mathcal{S}$. Define the function $f_1 \star f_2$ by setting

$$f_1 \star f_2(x) = \int f_1(x-y) f_2(y) \, dy.$$

Note that this makes good sense, since the integrand on the right clearly converges for each fixed value of x. We can be more precise. Since $f_i \in \mathcal{S}$, we can, for any integer p, find a K_p such that

$$|f_i(y)| < \frac{K_p}{1 + \|y\|^p},$$

so that

$$\int_{\|y\| > R} |f_2(y)| < \frac{L_p R^n}{1 + R^p}.$$

Then

$$\int (1 + \|x\|^q) f_1(x-y) f_2(y) \, dy = \int_{\|y\| < (1/2)\|x\|} (1 + \|x\|^q) f_1(x-y) f_2(y) \, dy$$
$$+ \int_{\|y\| > (1/2)\|x\|} (1 + \|x\|^q) f_1(x-y) f_2(y) \, dy.$$

The first integral is at most

$$C_n(\tfrac{1}{2}\|x\|)^n (1 + \|x\|)^q \max_z |f_2(z)| \frac{K_p}{1 + (\frac{1}{2}\|x\|)^p},$$

while the second is at most

$$(1 + \|x\|^q) \max_u |f_1(u)| \frac{L_p(\frac{1}{2}\|x\|)^n}{1 + (\frac{1}{2}\|x\|)^p}.$$

By choosing $p > q + n$, we see that both terms go to zero. Thus

$$\lim_{\|x\| \to \infty} (1 + \|x\|^q) f_1 \star f_2(x) = 0.$$

14.13 Show that

$$\frac{\partial}{\partial x^i}(f_1 \star f_2) = \left(\frac{\partial f_1}{\partial x^i}\right) \star f_2 = f_1 \star \left(\frac{\partial f_2}{\partial x^i}\right).$$

Conclude that $f_1 \star f_2 \in \mathcal{S}$.

14.14 Show that if φ is any bounded continuous function on $\mathbb{R}^n$, then

$$\iint \varphi(x+y)f\,(x)f_2(y)\,dx\,dy = \int \varphi(u)(f_1 \star f_2)(u)\,du.$$

14.15 Conclude that

$$\widehat{f_1 \star f_2}(\xi) = \hat{f}_1(\xi)\hat{f}_2(\xi).$$

14.16 Show that

$$f \star \tilde{f}(y) = \left(\frac{1}{2\pi}\right)^n \int |\hat{f}(\xi)|^2 e^{i\langle y,\xi\rangle}\,d\xi.$$

14.17 Conclude that for any $f \in \mathcal{S}$,

$$\int |f|^2 = \left(\frac{1}{2\pi}\right)^n \int |\hat{f}|^2. \tag{14.3}$$

[*Hint:* Set $y = 0$ in Exercise 14.16.]

The following exercises use the Fourier transform to develop facts which are useful in the study of partial differential equations. We will make use of these facts at the end of the last chapter. The reader may prefer to postpone his study of these problems until then.

On the space $\mathcal{S}$, define the norm $\|\;\|_s$ by setting

$$\|f\|_s^2 = (2\pi)^{-n}\int (1+\|\xi\|^2)^s|\hat{f}(\xi)|^2\,d\xi,$$

and the scalar product $(f, g)_s$ by

$$(f, g)_s = \int (1+\|\xi\|^2)^s\hat{f}(\xi)\bar{\hat{g}}(\xi)\,d\xi.$$

14.18 Let $s = R$ be a nonnegative integer. Show that

$$\|f\|_R^2 = \sum_{|\alpha|\le R} \frac{R}{\alpha!(R-|\alpha|)!}\int |D^\alpha f(x)|^2\,dx,$$

where $\alpha! = \alpha_1!\cdots\alpha_n!$. [Use the multinomial theorem, a repeated application of Exercise 14.3, and Eq. (15.3).]

We thus see that $\|f\|_R$ measures the size of f and its derivatives to order R in the square integral norm. It is helpful to think of $\|\;\|_s$ as a generalization of this notion of size, where now s can be an arbitrary real number.

Note that

$$\|f\|_s \le \|f\|_t \quad \text{if} \quad s \le t.$$

For any real s define the operator K^s by setting

$$\widehat{K^sf}(\xi) = (1+\|\xi\|^2)^s\hat{f}(\xi).$$

14.19 Show that the operator $K = K^1$ is given by

$$Kf = f - \sum \frac{\partial^2 f}{\partial x_i^2}.$$

14.20 Show that for any real numbers s and t,

$$\|K^s f\|_t = \|f\|_{t+2s}$$

and

$$(K^s f, g)_t = (f, K^s g)_t = (f, g)_{s+t}.$$

14.21 Show that $K^{s+t} = K^s \circ K^t$, so that, in particular, K^s in invertible for all s.

We now define the space H_s to be the completion of $\mathcal{S}$ under the norm $\| \ \|_s$. The space H_s is a Hilbert space with a scalar product $(\ , \ )_s$. We can think of the elements of H_s as "generalized functions with generalized derivatives up to order s". By construction, the space $\mathcal{S}$ is a dense subspace of H_s in the norm $\| \ \|_s$. We note that Exercise 14.20 implies that the operator K^s can be extended to an isometric map of H_t into H_{t-2s}. We shall also denote this extended map by K^s. By Exercise 14.21,

$$K^{-s}\colon H_{t-2s} \to H_t$$

is the inverse of K^s, so that K^s is a norm-preserving isomorphism of H_t onto H_{t-2s}.

14.22 Let $u \in H_s$ and $v \in H_{-s}$. Show that

$$|(u, v)_0| \leq \|u\|_s \|v\|_{-s}.$$

Thus we can extend $\langle u, v \rangle \to (u, v)_0$ to a function on $H_s \times H_{-s}$ which is linear in u and antilinear in v [that is, $(u, av_1 + b_2v_2)_0 = \bar{a}(u, v_1) + \bar{b}(u, v_2)$] and satisfies the above inequality. Thus any $v \in H_{-s}$ defines a bounded linear function, l, on H_s by $l(u) = (u, v)_0$.

14.23 Conversely, let l be a bounded linear function on H_s. Show that there is a $v \in H_{-s}$ with $l(u) = (u, v)_0$ for all $u \in H_s$. [*Hint:* Consider the linear form $v = K^s w$, where w is a suitable element of H_s, using Theorem 2.4 of Chapter 5.]

14.24 Show that

$$\|v\|_{-s} = \sup_{\substack{u \in H_s \\ u \neq 0}} \frac{|(u, v)_0|}{\|u\|_s}.$$

(Exercise 14.22 gives an inequality. If $v \neq 0$, take $u = K^{-s}v$ to get

$$\|v\|_{-s}\|u\|_s = (K^{-s/2}v, K^{-s/2}v) = \|v\|_s^2,$$

in order to get an equality.)

14.25 Let $2s > n$ (where our functions are defined on $\mathbb{R}^n$). Show that for any $f \in \mathcal{S}$ we have

$$\sup_{\mathbb{R}^n} |f(x)| \leq \|f\|_s \left[\int_{\mathbb{R}^n} (1 + \|\xi\|^2)^{-s}\right]^{1/2} d\xi \qquad \text{(Sobolev's inequality)}.$$

(Use Eq. (14.2), Schwarz's inequality, and the fact that the integral on the right of the inequality is absolutely convergent.)

Sobolev's inequality shows that the injection of $\mathcal{S}$ into $C(\mathbb{R}^n)$ extends to a continuous injection of H_s into $C(\mathbb{R}^n)$, where $C(\mathbb{R}^n)$ is given the uniform norm. We can thus regard the elements of H_s as actual functions on $\mathbb{R}^n$ if $s > n/2$.

By induction on $|\alpha|$ we can assert that for $s > n/2$, any $f \in H_{|\alpha|+s}$ has $|\alpha|$ continuous derivatives and

$$\sup_{\mathbb{R}^n} |D^\alpha f(x)| \leq C\alpha \|f\|_{|\alpha|+s}. \tag{14.4}$$

14.26 Let Ω be a bounded open subset of $\mathbb{R}^n$. Let $\varphi \in \mathcal{S}$ satisfy $\operatorname{supp} \varphi \subset \Omega$. Show that

$$|\hat{\varphi}(\xi)| \leq \mu(\Omega)^{1/2} \|\varphi\|_0 \qquad \text{for all} \quad \xi.$$

14.27 Let $d^i = \operatorname{lub}_{x \in \Omega} |x^i|$, and let $d^\alpha = (d^1)^{\alpha_1} \cdots (d^n)^{\alpha_n}$. Show that

$$|D^\alpha \hat{\varphi}(\xi)| \leq d^\alpha \mu(\Omega)^{1/2} \|\varphi\|_0.$$

14.28 Show that

$$|\xi^\beta \hat{\varphi}(\xi)| \leq \mu(\Omega)^{1/2} \|D^\beta \varphi\|_0,$$

and conclude that

$$|(1 + \|\xi\|^2)^k \hat{\varphi}^2(\xi)| \leq \mu(\Omega) \|\varphi\|_k^2.$$

14.29 More generally, let ψ be a function in $\mathcal{S}$ which satisfies $\psi(x) = 1$ for all $x \in \Omega$, and let $\varphi \in \mathcal{S}$ satisfy $\operatorname{supp} \varphi \subset \Omega$. Show that

$$|\hat{\varphi}(\xi)| = |(\varphi, \psi_\xi)_0| \leq \|\varphi\|_s \|\psi_\xi\|_{-s},$$

where $\psi_\xi(x) = \psi(x)\, e^{-i(x,\xi)}$, and that

$$|D_\xi^\alpha \hat{\varphi}(\xi)| \leq \|\varphi\|_s \|\psi_\xi^\alpha\|_{-s},$$

where $\psi_\xi^\alpha(x) = x^\alpha \psi(x) e^{-i(x,\xi)}$.

Let us denote by H_s^Ω the completion under $\| \;\|_s$ of the space of those functions in $\mathcal{S}$ whose supports lie in Ω. According to Exercise 14.29, any $\varphi \in H_s^\Omega$ defines an actual function $\hat{\varphi}$ of ξ which is differentiable and satisfies

$$|D_\xi^\alpha \hat{\varphi}(\xi)| \leq \|\varphi\|_s \|\psi_\xi^\alpha(x)\|_{-s},$$

where $\|\psi_\xi^\alpha(x)\|_{-s}$ depends only on Ω, α, ξ, and $-s$, and is independent of φ. Furthermore, $\|\varphi\|_s^2 = \int (1 + \|\xi\|^2)^s |\hat{\varphi}(\xi)|^2 \, d\xi$.

14.30 Let $s < t$. Then the injection $H_t \to H_s$ is a compact mapping. That is, if $\{\varphi_i\}$ is a sequence of elements of H_t^Ω such that $\|\varphi_i\|_t \leq 1$ for all i, then we can select a subsequence $\{\varphi_{i_j}\}$ which converges in $\| \;\|_s$. [*Hint:* By Exercise 14.29, the sequence of functions $\varphi_i(\xi)$ is bounded and equicontinuous on $\{\xi : \|\xi\| \leq r\}$ for only fixed r. We can thus choose a subsequence which converges uniformly and therefore a subsubsequence which converges on $\{\xi : \|\xi\| < r\}$ for all r (the uniformity possibly depending on r). Then if $\{\varphi_{i_j}\}$ is this subsubsequence,

$$\begin{aligned}
\|\varphi_{i_j} - \varphi_{i_k}\|_s^2 &= \int (1 + \|\xi\|^2)^s |\varphi_{i_j}(\xi) - \varphi_{i_k}(\xi)|^2 \, d\xi \\
&= \int_{\|\xi\| \leq r} (1 + \|\xi\|^2)^s |\varphi_{i_j}(\xi) - \varphi_{i_k}(\xi)|^2 \, d\xi \\
&\quad + \int_{\|\xi\| > r} (1 + \|\xi\|^2)^s |\varphi_{i_j}(\xi) - \varphi_{i_k}(\xi)|^2 \, d\xi \\
&\leq \int_{\|\xi\| \leq r} (1 + \|\xi\|^2)^s |\varphi_{i_j}(\xi) - \varphi_{i_k}(\xi)|^2 \, d\xi \\
&\quad + (1 + \|\xi\|^2)^{s-t} \{\|\varphi_{i_j}\|_t^2 + \|\varphi_{i_k}\|_t^2\}.]
\end{aligned}$$

CHAPTER 9

DIFFERENTIABLE MANIFOLDS

Thus far our study of the calculus has been devoted to the study of properties of and operations on functions defined on (subsets of) a vector space. One of the ideas used was the approximation of possibly nonlinear functions at each point by linear functions. In this chapter we shall generalize our notion of space to include spaces which cannot, in any natural way, be regarded as open subsets of a vector space. One of the tools we shall use is the "approximation" of such a space at each point by a linear space.

Suppose we are interested in studying functions on (the surface of) the unit sphere in $\mathbb{E}^3$. The sphere is a two-dimensional object in the sense that we can describe a neighborhood of every point of the sphere in a bicontinuous way by two coordinates. On the other hand, we cannot map the sphere in a bicontinuous one-to-one way onto an open subset of the plane (since the sphere is compact and an open subset of $\mathbb{E}^2$ is not). Thus pieces of the sphere can be described by open subsets of $\mathbb{E}^2$, but the whole sphere cannot. Therefore, if we want to do calculus on the whole sphere at once, we must introduce a more general class of spaces and study functions on them.

Even if a space can be regarded as a subset of a vector space, it is conceivable that it cannot be so regarded in any canonical way. Thus the state of a (homogeneous ideal) gas in equilibrium is specified when one gives any two of the three parameters: temperature, pressure, or volume. There is no reason to prefer any two to the third. The transition from one set of parameters to the other is given by a one-to-one bidifferentiable map. Thus any function of the states of the gas which is a differentiable function in terms of one choice of parameters is differentiable in terms of any other. Thus it makes sense to talk of *differentiable* functions on the states of the gas. However, a function which is linear in terms of one choice of parameters need not be linear in terms of the other. Thus it doesn't really make sense to talk of *linear* functions on the states of the gas. In such a situation we would like to know what properties of functions and what operations make sense in the space and are not artifacts of the description we give of the space.

Finally, even in a vector space it is sometimes convenient to introduce "nonlinear coordinates" for the solution of specific problems: for example, polar coordinates in Exercises 11.3 and 11.4, Chapter 8. We would therefore like to know how various objects change when we change coordinates and, if possible, to introduce notation which is independent of the coordinate system.

We will begin our formal discussion with the definition of differentiable manifolds. The basic idea is similar to the one that is used in everyday life to describe the surface of the earth. One gives a collection of charts describing small overlapping portions of the globe. We can piece the whole picture together by seeing how the charts match up.

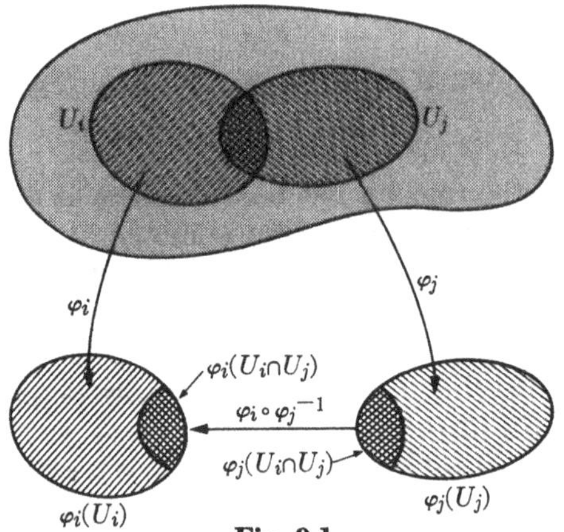

Fig. 9.1

1. ATLASES

Let M be a set. Let V be a Banach space. (For almost all our applications we shall take V to be $\mathbb{R}^n$ for some integer n.) A *V-atlas of class C^k* on M is a collection $\mathcal{A}$ of pairs (U_i, φ_i) called *charts*, where U_i is a subset of M and φ_i is a bijective map of U_i onto an open subset of V subject to the following conditions (Fig. 9.1):

A1. For any $(U_i, \varphi_i) \in \mathcal{A}$ and $(U_j, \varphi_j) \in \mathcal{A}$ the sets $\varphi_i(U_i \cap U_j)$ and $\varphi_j(U_i \cap U_j)$ are open subsets of V, and the maps

$$\varphi_i \circ \varphi_j^{-1}: \varphi_j(U_i \cap U_j) \to \varphi_i(U_i \cap U_j)$$

are differentiable of class C^k.

A2. $\bigcup U_i = M$.

The functions $\varphi_i \circ \varphi_j^{-1}$ are called the *transition functions* of the atlas $\mathcal{A}$. The following are examples of sets with atlases.

Example 1. *The trivial example.* Let M be an open subset of V. If we take $\mathcal{A}$ to consist of the single element (U, φ), where $U = M$ and $\varphi: U \to V$ is the identity map, then Axioms A1 and A2 are trivially fulfilled.

Example 2. *The sphere.* Let $M = S^n$ denote the subset of $\mathbb{R}^{n+1}$ given by $(x^1)^2 + \cdots + (x^{n+1})^2 = 1$. Let the set U_1 consist of those points for which $x^{n+1} > -1$, and let U_2 consist of those points for which $x^{n+1} < 1$. Let

$$\varphi_1: U_1 \to \mathbb{R}^n$$

be given by

$$y^i \circ \varphi_1(x^1, \ldots, x^{n+1}) = \frac{x^i}{1 + x^{n+1}}, \qquad i = 1, \ldots, n,$$

where $y^1, \ldots, y^n$ are coordinates on $\mathbb{R}^n$. Thus the map φ_1 is given by the projection from the "south pole", $\langle 0, \ldots, 0, -1 \rangle$, to $\mathbb{R}^n$ regarded as the equatorial plane (see Fig. 9.2). Similarly, define φ_2 by

$$y^i \circ \varphi_2(x^1, \ldots, x^{n+1}) = \frac{x^i}{1 - x^{n+1}}.$$

Then $\varphi_1(U_1 \cap U_2) = \varphi_2(U_1 \cap U_2) = \{y \in \mathbb{R}^n : y \neq 0\}$. Now

$$\sum (y^i \circ \varphi_1)^2(x^1, \ldots, x^{n+1}) = \frac{(x^1)^2 + \cdots + (x^n)^2}{(1 + x^{n+1})^2}$$

$$= \frac{1 - (x^{n+1})^2}{(1 + x^{n+1})^2} = \frac{1 - x^{n+1}}{1 + x^{n+1}}.$$

Thus

$$y^i \circ \varphi_2(x^1, \ldots, x^{n+1}) = \frac{y^i \circ \varphi_1(x^1, \ldots, x^{n+1})}{\sum [y^i \circ \varphi_1(x^1, \ldots, x^{n+1})]^2},$$

or

$$\varphi_2(x) = \frac{\varphi_1(x)}{\|\varphi_1(x)\|^2}.$$

In other words, the map $\varphi_2 \circ \varphi_1^{-1}$, defined for all $y \neq 0$, is given by

$$\varphi_2 \circ \varphi_1^{-1}(y) = \frac{y}{\|y\|^2}.$$

Thus conditions A1 and A2 are fulfilled.

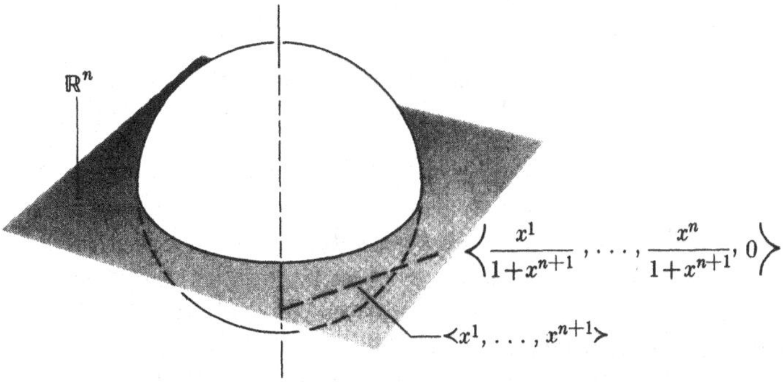

Fig. 9.2

Note that the atlas we gave for the sphere contains only two charts (each given by polar projection). An atlas of the earth usually contains many more charts. In other words, many different atlases can be used to describe the same set. We shall return to this point later.

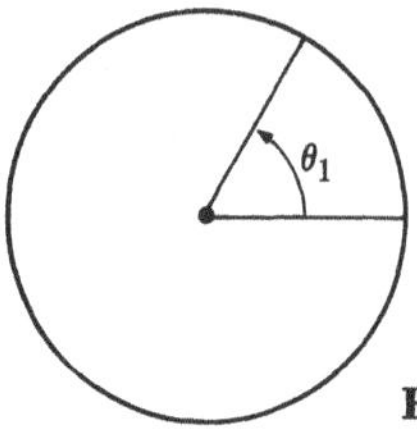

Fig. 9.3

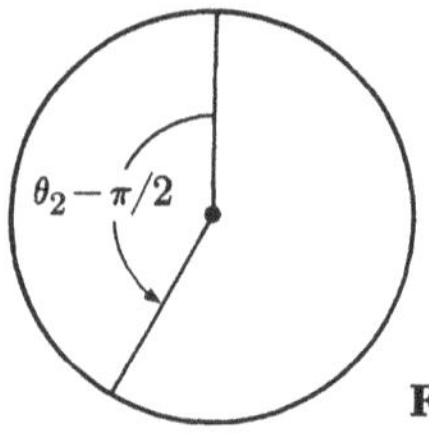

Fig. 9.4

Example 3. *The circle.* The circle S^1 is a "one-dimensional sphere" and therefore has an atlas as described in Example 2. We wish to describe a different atlas on S^1. Regard S^1 as the unit circle $x_1^2 + x_2^2 = 1$, and consider the function θ_1, defined in a neighborhood of $\langle 1, 0 \rangle$ on the upper semicircle of S^1, which gives the angle from the point on S^1 to $\langle 1, 0 \rangle$ (see Fig. 9.3). As we move counterclockwise around the circle, this function is well defined until we hit $\langle 1, 0 \rangle$ again. We will take, as the first chart in our atlas, (U_1, θ_1), where $U_1 = S^1 - \{\langle 1, 0 \rangle\}$ and θ_1 is the function defined above. Let $U_2 = S^1 - \{\langle 0, 1 \rangle\}$, and define θ_2 to be $\pi/2$ plus the angle (measured counterclockwise) from $\langle 0, 1 \rangle$ (see Fig. 9.4). Now $U_1 \cap U_2 = S^1 - \{\langle 1, 0 \rangle, \langle 0, 1 \rangle\}$, and $\theta_1(U_1 \cap U_2) = (0, 2\pi) - \{\pi/2\}$.

Also, $\theta_2(U_1 \cap U_2) = (\pi/2, 2\pi + \pi/2) - \{2\pi\}$.

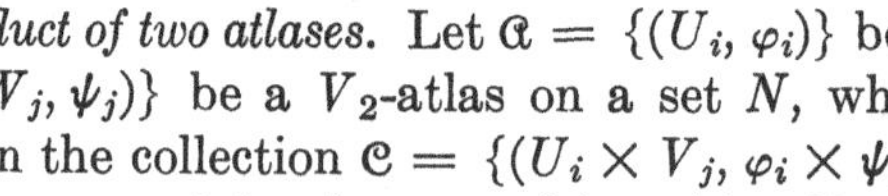

The map $\theta_2 \circ \theta_1^{-1}$ is given by

$$\theta_2 \circ \theta_1^{-1}(x) = \begin{cases} x + 2\pi & \text{if } 0 < x < \pi/2, \\ x & \text{if } \pi/2 < x < 2\pi. \end{cases}$$

Example 4. *The product of two atlases.* Let $\mathcal{A} = \{(U_i, \varphi_i)\}$ be a V_1-atlas on a set M, and let $\mathcal{B} = \{(W_j, \psi_j)\}$ be a V_2-atlas on a set N, where V_1 and V_2 are Banach spaces. Then the collection $\mathcal{C} = \{(U_i \times V_j, \varphi_i \times \psi_j)\}$ is a $(V_1 \times V_2)$-atlas on $M \times N$. Here $\varphi_i \times \psi_j(p, q) = \langle \varphi_i(p), \psi_j(q) \rangle$ if $\langle p, q \rangle \in U_i \times W_j$. It is easy to check that $\mathcal{C}$ satisfies conditions A1 and A2. We shall call $\mathcal{C}$ the product of $\mathcal{A}$ and $\mathcal{B}$ and write $\mathcal{C} = \mathcal{A} \times \mathcal{B}$.

For instance, let $M = (0, 1) \subset \mathbb{R}^1$ and $N = S^1$. Then we can regard $M \times N$ as a cylinder or an annulus. If $M = N = S^1$, then $M \times N$ is a torus.

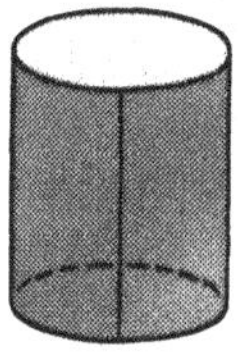

Cylinder

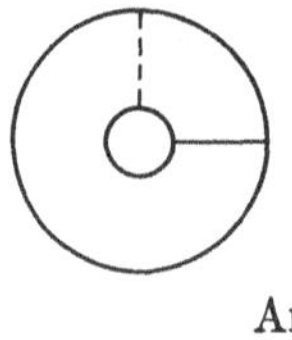

Annulus

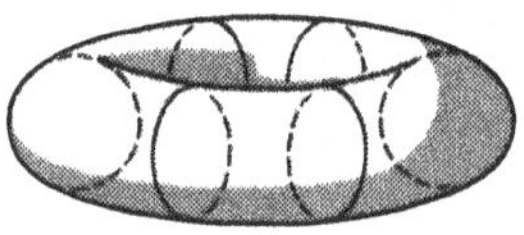

Torus

It is an instructive exercise to write down the atlases and transition functions explicitly in these cases.

Example 5. As a generalization of our first example, let S be a submanifold of an $(n+m)$-dimensional vector space X, as defined in Section 12 of Chapter 3. For each neighborhood N defined there, the set $S \cap N$, together with the map φ which is defined as the projection π_1 restricted to S, provides a chart with values in V (where X is viewed as $V \times W$). In such a neighborhood N the set S is presented as a graph of function F. In other words,

$$S \cap N = \{\langle x, F(x)\rangle \in V \times W : x \in \pi_1(S)\},$$

where F is a smooth map of $A = \pi_1(S \cap N)$ into W. Let N' be another such neighborhood with corresponding projection π_1' (where now X is identified with $V \times W$ in some other way). Then $\varphi' \circ \varphi^{-1}(x) = \pi_1'(x, F(x))$, which shows that $\varphi' \circ \varphi^{-1}$ is a smooth map. Thus every submanifold in the sense of Chapter 3 possesses an atlas.

Exercise. Let $\mathbb{P}^n$ (projective n-space) denote the space of all lines through the origin in $\mathbb{R}^{n+1}$. Any such line is determined by a nonzero vector lying on the line. Two such vectors, $\langle x^1, \ldots, x^{n+1}\rangle$ and $\langle y^1, \ldots, y^{n+1}\rangle$, determine the same line if and only if they differ by a factor, that is, $y^i = \lambda x^i$ for all i, where λ is some (nonzero) real number. We can thus regard an element of $\mathbb{P}^n$ as an equivalence class of nonzero vectors. For each i between 1 and $n+1$, let $U_i \subset \mathbb{P}^n$ be the set of those elements coming from vectors with $x^i \neq 0$. Map

$$U_i \xrightarrow{\alpha_i} \mathbb{R}^n$$

by sending

$$\langle x^1, \ldots, x^{n+1}\rangle \mapsto \left\langle \frac{x^1}{x^i}, \ldots, \frac{x^{i-1}}{x^i}, \frac{x^{i+1}}{x^i}, \ldots, \frac{x^{n+1}}{x^i}\right\rangle.$$

Show that the map α_i is well defined and that $\{(U_i, \alpha_i)\}$ is an atlas on $\mathbb{P}^n$.

2. FUNCTIONS, CONVERGENCE

Let $\mathcal{A}$ be a V-atlas of class C^k on a set M. Let f be a real-valued function defined on M. For a chart (U_i, φ_i) we obtain a function f_i defined on $\varphi_i(U_i)$ by setting

$$f_i = f \circ \varphi_i^{-1}. \tag{2.1}$$

The function f_i can be regarded as the "local expression of f" in terms of the chart (U_i, φ_i). In general, the functions f_i will look quite different from one another. For example, let $M = S^n$, let $\mathcal{A}$ be the atlas described, and let f be the function on the sphere assigned to the point $\langle x^1, \ldots, x^{n+1}\rangle$ the value x^{n+1}. Then

$$f_1(y) = f \circ \varphi_1^{-1}(y) = \frac{2}{1 + \|y\|^2} - 1,$$

while

$$f_2(y) = f \circ \varphi_2^{-1}(y) = 1 - \frac{2}{1 + \|y\|^2},$$

as one can check by solving the equations.

Returning to the general discussion, we observe that the functions f_i are not completely independent of one another. In fact, it follows from the definition (2.1) that we have

$$f_i \circ \varphi_i \circ \varphi_j^{-1} = f_j \quad \text{on} \quad \varphi_j(U_i \cap U_j). \tag{2.2}$$

[Thus in the example cited above we indeed have $f_2(y) = f_1(y/\|y\|^2)$, as is required by (2.2).]

We now come to a simple but important observation. Suppose we start with a collection of functions $\{f_i\}$, each f_i defined on $\varphi_i(U_i)$, and such that (2.2) holds. Then there exists a unique function f on M such that $f_i = f \circ \varphi_i^{-1}$. In fact, define f by setting $f(p) = f_i(\varphi_i(p))$ if $p \in U_i$. For f to be well defined, we must be sure that this definition is consistent, i.e., that if p is also in U_j, then $f_i(\varphi_i(p)) = f_j(\varphi_j(p))$, but this is exactly what (2.2) says.

We can thus think of a real-valued function in two ways: as either

i) an object defined invariantly on M, i.e., a map from M to $\mathbb{R}$, or

ii) a collection of objects (in this case functions) one defined for each chart and satisfying certain "transition laws", namely (2.2).

This dual way of looking at objects on M will recur quite frequently in what follows.

Let M be a set with an atlas of class C^k. We will say that a function f is of class C^l $(l \leq k)$ if each of the functions f_i defined by (2.1) is of class C^l. Note that since $l \leq k$, this can happen without any interference from (2.2). If $f_i \in C^l$ and $\varphi_i^{-1} \circ \varphi_j \in C^k$ $(k \geq l)$, then $f_i \circ (\varphi_i^{-1} \circ \varphi_j) \in C^l$. If l were larger than k, then in general f_i would not be of class C^l if f_j were, and there would be very few functions of class C^l.

Since we will not wish to constantly specify degrees of differentiability of our atlas, *from now on when we speak of an atlas we shall mean an atlas of class* C^∞.

Let M be a set with an atlas $\mathcal{A}$. We shall say that a sequence of points $\{x_i \in M\}$ converges to $x \in M$ if

i) there exists a chart $(U_i, \varphi_i) \in \mathcal{A}$ and an integer N_i such that $x \in U_i$ and for all $k > N$, $x_k \in U_i$;

ii) $\varphi_i(x_k)_{k>N}$ converges to $\varphi(x)$.

Note that if (U_j, φ_j) is any other chart with $x \in U_j$, then there exists an N_j such that $\varphi_j(x_k) \in U_j$ for $k > N_j$ and $\varphi_j(x_k) \to \varphi_j(x)$. In fact, choose N_j so that $\varphi_i(x_k) \in \varphi_i(U_i \cap U_j)$ for all $k \geq N_j$. (This is possible since $\varphi_j(U_i \cap U_j)$ is open by A1.) The fact that the $\varphi_j(x_k)$ converge to $\varphi_j(x)$ follows from the continuity of $\varphi_j \circ \varphi_i^{-1}$. It thus makes good sense to say that $\{x_k\}$ converges to x.

Warning. It does *not* make sense to say that a sequence $\{x_k\}$ is a Cauchy sequence. Thus, for example, let $M = S^n$ with the atlas described above. If $\{x_k\}$ is a sequence of points converging to the north pole in S^n, then $\varphi_1(x_k) \to 0$, while $\varphi_2(x_k) \to \infty$. This example becomes even more sticky if we remove the north pole, i.e., let $M = S^n - \{\langle 0, \ldots, 0, 1\rangle\}$ and define the charts as before.

Then $\{x_k\}$ has no limit (in M). Clearly, $\{\varphi_1(x_k)\}$ is a Cauchy sequence, while $\{\varphi_2(x_k)\}$ is not.

Once we have a notion of convergence, we can talk about such things as open sets and closed sets. We could also define them directly. For instance, a set U is *open* if $\varphi_i(U \cap U_i)$ is an open subset of $\varphi_i(U_i)$ for all charts (U_i, φ_i), and so on.

EXERCISES

2.1 Show that the above definition of a set's being open is consistent, i.e., that there exist nonempty open sets. (In fact, each of the U_i's is open.)

2.2 Show that a sequence $\{x_\alpha\}$ converges to x if and only if for every open set U containing x there is an N_U with $x_\alpha \in U$ for $\alpha > N_U$.

Let $\mathfrak{A} = \{(U_i, \varphi_i)\}$ be an atlas on M, and let U be an open subset of M relative to this atlas. Let $\mathfrak{A} \upharpoonright U$ be the collection of all pairs $(U_i \cap U, \varphi_i \upharpoonright U)$. It is easy to check that $\mathfrak{A} \upharpoonright U$ is an atlas on U. We shall call it the restriction of $\mathfrak{A}$ to U.

Let f be a function defined on the open set U. We say that f is of class C^l on U if it is of class C^l relative to the atlas $\mathfrak{A} \upharpoonright U$ on U. For later convenience we shall say that a function f defined on a subset of M is of class C^l if

i) the domain of f is some open set U of M, and

ii) f is of class C^l on U.

3. DIFFERENTIABLE MANIFOLDS

In our discussion of the examples in Section 1, the particular choice of atlas that we made in each case was rather arbitrary. We could equally well have introduced a different atlas in each case without changing the class of differentiable functions, or the class of open sets, or convergent sequences, and so on. We therefore introduce an equivalence relation between atlases on M:

> Let $\mathfrak{A}_1$ and $\mathfrak{A}_2$ be atlases on M. We say that they are equivalent if their union $\mathfrak{A}_1 \cup \mathfrak{A}_2$ is again an atlas on M.

The crucial condition is that A1 still hold for the union. This means that for any charts $(U_i, \varphi_i) \in \mathfrak{A}_1$ and $(W_j, \psi_j) \in \mathfrak{A}_2$ the sets $\varphi_i(U_i \cap W_j)$ and $\psi_j(U_i \cap W_j)$ are open and $\varphi_i \circ \psi_j^{-1}$ is a differentiable map of $\psi_j(U_i \cap W_j)$ onto $\varphi_i(U_i \cap W_j)$ with a differentiable inverse.

It is clear that the relation introduced is an equivalence relation. Furthermore, it is an easy exercise to check that if f is a function of class C^l with respect to a given atlas, it is of class C^l with respect to any equivalent one. The same is true for the notions of open set and convergence.

Definition 3.1. A set M together with an equivalence class of atlases on M is called a *differentiable manifold* if it satisfies the "Hausdorff property": For any two points $x_1 \neq x_2$ of M there are open sets U_1 and U_2 with $x_1 \in U_1$ and $x_2 \in U_2$ with $U_1 \cap U_2 = \emptyset$.

In what follows we shall (by abuse of the language) denote a differentiable manifold by M, where the equivalence class of atlases is understood. By *an atlas of* M we shall then mean an atlas belonging to the given equivalence class, and by *a chart of* M we shall mean a chart belonging to some atlas of M.

We shall also adopt the notational convention that V is the Banach space where the charts on M take their values (and shall say that M is a V-manifold). If there are several manifolds, M_1, M_2, etc., under discussion, we shall denote the corresponding vector spaces by V_1, V_2, etc. If $V = \mathbb{R}^n$, we say that M is an *n-dimensional* manifold.

Let M_1 and M_2 be differentiable manifolds. A map $\varphi: M_1 \to M_2$ is called *continuous* if for any open set $U_2 \subset M_2$ the set $\varphi^{-1}(U_2)$ is an open subset of M_1. Let $x_2 \in M_2$, and let U_2 be any open set containing x_2. If $\varphi(x_1) = x_2$, then $\varphi^{-1}(U_2)$ is an open set containing x_1. If (W, α) is a chart about x_1, then $W \cap \varphi^{-1}(U_2)$ is an open subset of W, and $\alpha(W \cap \varphi^{-1}(U_2))$ is an open set in V_1 containing $\alpha(x_1)$. Therefore, there exists an $\epsilon > 0$ such that $\varphi(x) \in U_2$ for all $x \in W$, with $\|\alpha(x) - \alpha(x_1)\| < \epsilon$. In this sense, all points "close to x_1" are mapped "close to x_2". Note that the *choice of ϵ will depend on the chart* (W, α) *as well as on* x_1, x_2, U_2, and φ.

If M_1, M_2, and M_3 are differentiable manifolds, and if $\varphi: M_1 \to M_2$ and $\psi: M_2 \to M_3$ are continuous maps, it is easy to see that their composition $\psi \circ \varphi$ is a continuous map from M_1 to M_3.

Let φ be a continuous map from M_1 to M_2. Let (W_1, α_1) be a chart on M_1 and (W_2, α_2) a chart on M_2. We say that these charts are *compatible* (under φ) if $\varphi(W_1) \subset W_2$. If $\mathcal{A}_2$ is an atlas on M_2 and $\mathcal{A}_1$ is an atlas on M_1, we say that $\mathcal{A}_1$ and $\mathcal{A}_2$ are compatible under φ if for every $(W_1, \alpha_1) \in \mathcal{A}_1$ there exists a $(W_2, \alpha_2) \in \mathcal{A}_2$ compatible with it, i.e., such that $\varphi(W_1) \subset W_2$. (Note that the map $\alpha_2 \circ (\varphi \restriction W_1) \circ \alpha_1^{-1}$ is then a continuous map of an open subset of V_1 into V_2.) Given $\mathcal{A}_2$ and φ, we can always find an $\mathcal{A}_1$ compatible with $\mathcal{A}_2$ under φ. In fact, let $\mathcal{A}'_1$ be any atlas on M_1, and set

$$\mathcal{A}_1 = \{(W_1 \cap \varphi^{-1}(W_2)), \alpha \restriction (W_1 \cap \varphi^{-1}(W_2))\},$$

where (W_1, α) ranges over all charts of $\mathcal{A}'_1$ and (W_2, β) ranges over all charts of $\mathcal{A}_2$.

Definition 3.2. Let M_1 and M_2 be differentiable manifolds, and let φ be a map: $M_1 \xrightarrow{\varphi} M_2$. We say that φ is differentiable if the following hold:

i) φ is continuous.

ii) Let $\mathcal{A}_1$ and $\mathcal{A}_2$ be compatible atlases under φ. Then for any compatible $(W_1, \alpha_1) \in \mathcal{A}_1$ and $(W_2, \alpha_2) \in \mathcal{A}_2$, the map

$$\alpha_2 \circ \varphi \circ \alpha_1^{-1}: \alpha_1(W_1) \to \alpha_2(W_2)$$

is differentiable (as a map of an open subset of a Banach space into a Banach space). (See Fig. 9.5.)

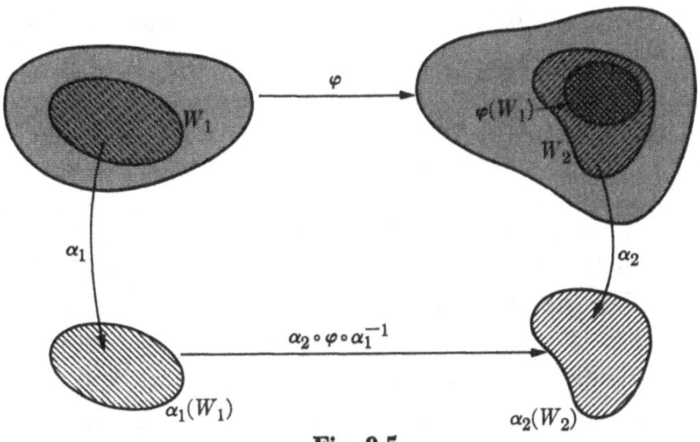

Fig. 9.5

In order to check that a continuous map φ is differentiable, it suffices to check much less than (ii). Condition (ii) relates to *any* pair of compatible atlases and *any* pair of compatible charts. In fact, we can assert:

Proposition 3.1. Let $\varphi\colon M_1 \to M_2$ be continuous, and let $\mathcal{A}_1$ and $\mathcal{A}_2$ be compatible atlases under φ. Suppose that for every $(W_1, \alpha_1) \in \mathcal{A}_1$ there exists a $(W_2, \alpha_2) \in \mathcal{A}_2$ with $\varphi(W_1) \subset W_2$ and $\alpha_2 \circ \varphi \circ \alpha_1^{-1}$ differentiable. Then φ is differentiable.

Proof. Let (U_1, β_1) and (U_2, β_2) be any charts on M_1 and M_2 with $\varphi(U_1) \subset U_2$. We must show that $\beta_2 \circ \varphi \circ \beta_1^{-1}$ is differentiable. It suffices to show that it is differentiable in the neighborhood of every point $\beta(x_1)$, where $x_1 \in U_1$. Choose $(W_1, \alpha_1) \in \mathcal{A}_1$ with $x \in W_1$, and choose $(W_2, \alpha_1) \in \mathcal{A}_2$ with $\varphi(W_1) \subset W_2$. Then on $\beta_1(W_1 \cap U_1)$, we have

$$\beta_2 \circ \varphi \circ \beta_1^{-1} = (\beta_2 \circ \alpha_2^{-1}) \circ (\alpha_2 \circ \varphi \circ \alpha_1^{-1}) \circ (\alpha_1 \circ \beta_1^{-1}),$$

so that the left-hand side is differentiable. □

In other words, it suffices to verify differentiability with one pair of atlases. We have as a consequence:

Proposition 3.2. Let $\varphi\colon M_1 \to M_2$ and $\psi\colon M_2 \to M_3$ be differentiable. Then $\psi \circ \varphi$ is differentiable.

Proof. Let $\mathcal{A}_3$ be an atlas on M_3. Choose $\mathcal{A}_2$ compatible with $\mathcal{A}_3$ under ψ, and then choose an atlas $\mathcal{A}_1$ on M_1 compatible with $\mathcal{A}_2$ under φ. For any $(W_1, \alpha_1) \in \mathcal{A}_1$ choose $(W_2, \alpha_2) \in \mathcal{A}_2$ and $(W_3, \alpha_3) \in \mathcal{A}_3$ with $\varphi(W_1) \subset W_2$ and $\psi(W_2) \subset W_3$. Then $\alpha_3 \circ \psi \circ \varphi \circ \alpha_1^{-1} = (\alpha_3 \circ \psi \circ \alpha_2^{-1}) \circ (\alpha_2 \circ \varphi \circ \alpha_1^{-1})$ is differentiable. □

Exercise 3.1. Let $M_1 = S^n$, let $M_2 = \mathbb{P}^n$, and let $\varphi\colon M_1 \to M_2$ be the map sending each point of the unit sphere into the line it determines. (Note that two antipodal points of S^n go into the same point of $\mathbb{P}^n$.) Construct compatible atlases for φ and show that φ is differentiable.

Note that if f is any function on M with values in a Banach space, then f is differentiable as a function (in the sense of Section 2) if and only if it is differentiable as a map of manifolds. In particular, let $\varphi: M_1 \to M_2$ be a differentiable map, and let f be a differentiable function on M_2 (defined on some open subset, say U_2). Then $f \circ \varphi$ is a differentiable function on M_1 [defined on the open set $\varphi^{-1}(U_2)$]. Thus φ "pulls back" a differentiable function on M_2 to M_1. From this point of view we can say that φ induces a map from the collection of differentiable functions on M_2 to the collection of differentiable functions on M_1. We shall denote this induced map by φ^*. Thus

$$\text{differentiable functions on } M_2 \xrightarrow{\varphi^*} \text{differentiable functions on } M_1$$

is given by

$$\varphi^*[f] = f \circ \varphi.$$

If $\psi: M_2 \to M_3$ is a second differentiable map, then $(\psi \circ \varphi)^*$ goes from functions on M_3 to functions on M_1, and we have

$$(\psi \circ \varphi)^* = \varphi^* \circ \psi^* \tag{3.1}$$

(note the change of order). In fact, for g on M_3,

$$(\psi \circ \varphi)^* g = g \circ (\psi \circ \varphi) = (g \circ \psi) \circ \varphi = \varphi^*[\psi^*[g]].$$

Observe that if φ is *any* map from $M_1 \to M_2$ and f is any function defined on a subset S_2 of M_2, then the "pullback" $\varphi^*[f] = f \circ \varphi$ is a function defined on $\varphi^{-1}(S_2)$ of M_1. The fact that φ is continuous allows us to conclude that if S_2 is open, then so is $\varphi^{-1}(S_2)$. The fact that φ is differentiable implies that $\varphi^*[f]$ is differentiable whenever f is.

The map φ^* commutes with all algebraic operations whenever they are defined. More precisely, suppose f and g take values in the same vector space and have domains of definition U_1 and U_2. Then $f + g$ is defined on $U_1 \cap U_2$ and $\varphi^*[f] + \varphi^*[g]$ is defined on $\varphi^{-1}(U_1 \cap U_2)$, and we clearly have

$$\varphi^*[f + g] = \varphi^*[f] + \varphi^*[g].$$

EXERCISES

3.2 Let M_2 be a finite-dimensional manifold, and let $\varphi: M_1 \to M_2$ be continuous. Suppose that $\varphi^*[f]$ is differentiable for any (locally defined) differentiable real-valued function f. Conclude that φ is differentiable.

3.3 Show that if φ is a bounded linear map between Banach spaces, then φ^* as defined above is an extension of φ^* as defined in Section 3, Chapter 2.

4. THE TANGENT SPACE

In this section we are going to construct an "approximating vector space" to a differentiable manifold at each point of the manifold. This will allow us to formulate most of the notions of the differential calculus on manifolds.

Let M be a differentiable manifold, and let x be a point of M (Fig. 9.6). Let $I \subset \mathbb{R}$ be an interval containing the origin. Let φ be a differentiable map of I into M such that $\varphi(0) = x$. We will call φ a (differentiable) curve through x.

Let f be any differentiable real-valued function on M defined in a neighborhood of x. Then $\varphi^*[f]$ is differentiable on $\mathbb{R}$ and we can consider its derivative at the origin. Define the operator D_φ by

$$D_\varphi(f) = \left.\frac{d\varphi^*[f]}{dt}\right|_{t=0}.$$

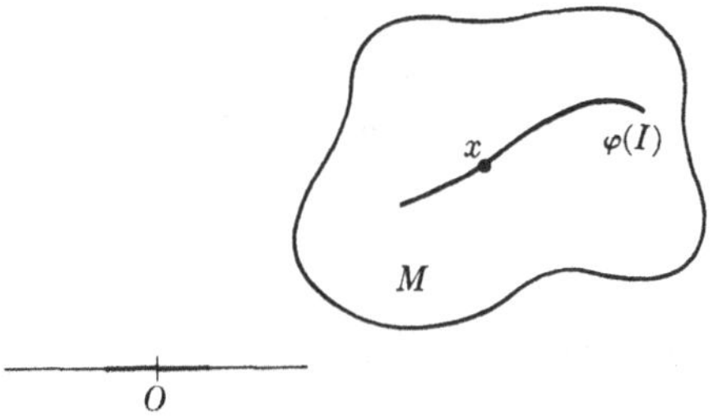

Fig. 9.6

In view of the linearity of φ^*, the map $f \to D_\varphi(f)$ is linear:

$$D_\varphi(af + bg) = aD_\varphi(f) + bD_\varphi(g).$$

Similarly, we have Leibnitz's rule:

$$D_\varphi(fg) = f(x)D_\varphi(g) + g(x)D_\varphi(f),$$

which can easily be checked. The functional D_φ depends on the curve φ. If ψ is a second curve, then, in general, $D_\varphi \neq D_\psi$. If, however, $D_\psi = D_\varphi$, then we say that the curves φ and ψ are tangent at x, and we write $\varphi \sim \psi$. Thus

$$\varphi \sim \psi \quad \text{if and only if} \quad D_\varphi(f) = D_\psi(f) \quad \text{for all differentiable functions } f.$$

It is easy to check that $\sim$ is an equivalence relation. An equivalence class of curves through x will be called a *tangent vector* at x. If ξ is a tangent vector at x and $\varphi \in \xi$, we say that ξ is tangent to φ at x.

For any differentiable function f defined about x and any tangent vector ξ, we set

$$\xi(f) = D_\varphi(f),$$

where $\varphi \in \xi$. Thus ξ gives us a functional on differentiable functions defined about x. We have

$$\xi(af + bg) = a\xi(f) + b\xi(g), \tag{4.1}$$

$$\xi(fg) = f(x)\xi(g) + g(x)\xi(f). \tag{4.2}$$

Let us examine what the equivalence relation $\sim$ says in terms of a chart (W, α) about x. The functional $D_\varphi(f)$ can be written as

$$\left.\frac{df \circ \varphi}{dt}\right|_{t=0} = \left.\frac{d(f \circ \alpha^{-1}) \circ (\alpha \circ \varphi)}{dt}\right|_{t=0}.$$

If we set $\Phi = \alpha \circ \varphi$ and $F = f \circ \alpha^{-1}$, then Φ is a parametrized curve in a Banach space and F is a differentiable function there. We can thus write

$$D_\varphi[f] = dF(\Phi'(0)) = D_{\Phi'(0)}F.$$

From this expression we see (setting $\Psi = \alpha \circ \psi$) that $\psi \sim \varphi$ if and only if $\Phi'(0) = \Psi'(0)$. We thus see that in terms of a chart (W, α), every tangent vector ξ at x corresponds to a unique vector $\xi_\alpha \in V$ given by

$$\xi_\alpha = (\alpha \circ \varphi)'(0),$$

where $\varphi \in \xi$.

Conversely, given any $v \in V$, there is a tangent vector ξ with $\xi_\alpha = v$. In fact, define φ by setting $\varphi(t) = \alpha^{-1}(\alpha(x) + tv)$. Then φ is defined in a small enough interval about 0, and $(\alpha \circ \varphi)' = v$.

In short, a choice of chart allows us to identify the set of all tangent vectors at x with V. Let (U, β) be a second chart about x. Then

$$\xi_\beta = (\beta \circ \varphi)'(0) = (\beta \circ \alpha^{-1}) \circ (\alpha \circ \varphi)'(0).$$

By the chain rule we thus have

$$\xi_\beta = J_{\beta\circ\alpha^{-1}}(\alpha(x))\, \xi_\alpha, \tag{4.3}$$

where $J_\gamma(p)$ is the differential $d\gamma_p$ of γ at p.

Since $J_{\beta\circ\alpha^{-1}}(\alpha(x))$ is a linear map of V into itself, Eq. (4.3) says that the set of all tangent vectors at x can be identified with V, the identification being determined up to an automorphism of V. In particular, we can make the set of all tangent vectors at x into a vector space by defining

$$a\xi + b\eta = \zeta,$$

where ζ is determined by

$$a\xi_\alpha + b\eta_\alpha = \zeta_\alpha$$

for some chart α. Equation (4.3) shows that this definition is independent of α.

We shall denote the space of tangent vectors at x by $T_x(M)$ and shall call it the tangent space (to M) at x.

Let ψ be a differentiable map of M_1 to M_2, and let φ be a curve passing through $x \in M_1$ (see Fig. 9.7). Then $\psi \circ \varphi$ is a curve passing through $\psi(x) \in M_2$. It is easy to check that if $\varphi \sim \bar{\varphi}$, then $\psi \circ \varphi \sim \psi \circ \bar{\varphi}$. Thus the map ψ induces a mapping of $T_x(M_1)$ into $T_{\psi(x)}(M_2)$, which we shall denote by ψ_{*x}. To repeat,

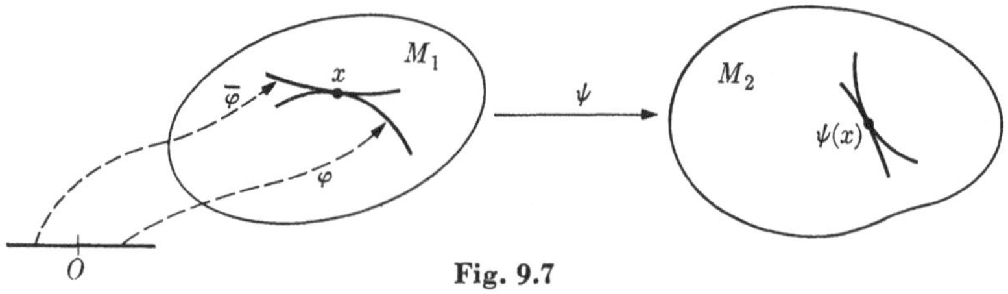

Fig. 9.7

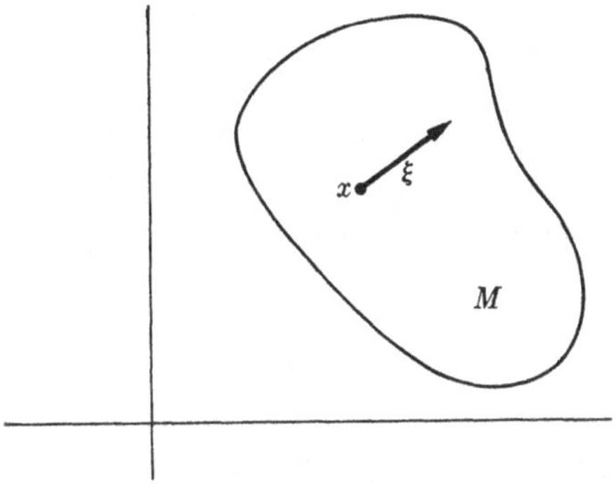

Fig. 9.8

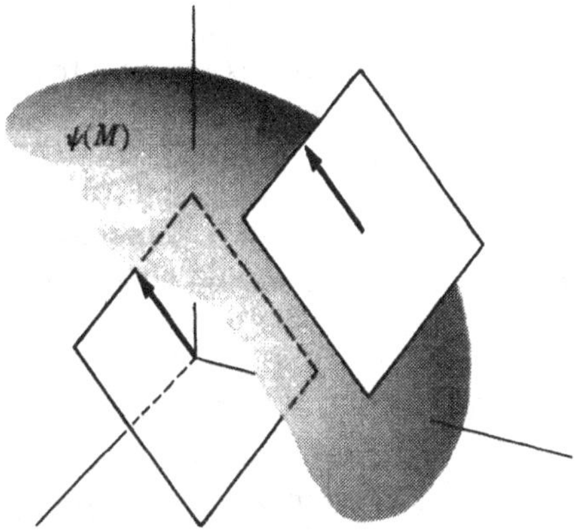

Fig. 9.9

if $\xi \in T_x(M_1)$, then $\psi_{*x}(\xi) = \eta$ is determined by

$$\psi \circ \varphi \in \eta \qquad \text{for all} \quad \varphi \in \xi.$$

Let (U, α) be a chart about x, and let (W, β) be a chart about $\psi(x)$. Then

$$\xi_\alpha = (\alpha \circ \varphi)'(0)$$

and

$$\eta_\beta = (\beta \circ \psi \circ \varphi)'(0) = (\beta \circ \psi \circ \alpha^{-1}) \circ (\alpha \circ \varphi)'(0).$$

By the chain rule we can thus write

$$\eta_\beta = J_{\beta \circ \psi \circ \alpha^{-1}}\big(\alpha(x)\big)\, \xi_\alpha.$$

This says that if we identify $T_x(M_1)$ with V_1 via α and identify $T_{\psi(x)}(M_2)$ with V_2 via β, then ψ_{*x} becomes identified with the linear map $J_{\beta \circ \psi \circ \alpha^{-1}}\big(\alpha(x)\big)$. In particular, *the map ψ_{*x} is a continuous linear mapping* from $T_x(M_1)$ to $T_{\psi(x)}(M_2)$. If $\varphi \colon M_1 \to M_2$ and $\psi \colon M_2 \to M_3$ are two differentiable mappings, then it follows immediately from the definitions that

$$(\psi \circ \varphi)_{*x} = \psi_{*\varphi(x)} \circ \varphi_{*x}. \tag{4.4}$$

We have seen that the choice of chart (U, α) identifies $T_x(M)$ with V. Now suppose that M is actually V itself (or an open subset of V) regarded as a differentiable manifold. Then M has a distinguished chart, namely (M, id). Thus on an open subset of V the identity chart gives us a distinguished way of identifying $T_x(M)$ with V. It is sometimes convenient to picture $T_x(M)$ as a copy of V whose origin has been translated to x. We would then draw a tangent vector at x as an arrow originating at x. (See Fig. 9.8.)

Now suppose that M is a general manifold and that ψ is a differentiable map of M into a vector space V_1. Then $\psi_*\big(T_x(M)\big)$ is a subspace of $T_{\psi(x)}(V_1)$. If we regard $\psi_*\big(T_x(M)\big)$ as a subspace of V_1 and consider the corresponding hyperplane through x, we get the "plane tangent to $\psi(M)$ at x" in the intuitive sense (Fig. 9.9).

It is very convenient to think of tangent vectors in this way, that is, to regard them as vectors tangent to M if M were mapped into a vector space.

If f is a real-valued differentiable function defined in a neighborhood U of of $x \in M$, then we can regard it as a map of the manifold U to the manifold $\mathbb{R}^1$. We therefore get a map $f_{*x}\colon T_x(M) \to T_{f(x)}(\mathbb{R}^1)$. Recall that we identify $T_y(\mathbb{R}^1)$ with $\mathbb{R}^1$ for any $y \in \mathbb{R}^1$. Therefore, f_{*x} can be viewed as a map from $T_{*x}(M)$ to $\mathbb{R}^1$. The reader should check that this map is indeed given by

$$f_{*x}(\xi) = \xi(f) \qquad \text{for} \quad \xi \in T_x(M). \tag{4.5}$$

In particular, if we take $M_3 = \mathbb{R}$ and $\psi = f$ in (4.4), we can assert:

Let ψ be a differentiable map of M_1 to M_2, and let f be a differentiable function on M_2 defined in a neighborhood of $\psi(x)$. Then for any $\xi \in T_x(M_1)$,

$$\xi(\psi^*(f)) = \psi_{*x}(\xi)(f). \tag{4.6}$$

From now on, we shall frequently drop the subscript x in ψ_{*x} when it can be understood from the context. Thus we would write (4.4) as $(\psi \circ \varphi)_* = \psi_* \circ \varphi_*$. Some authors call the mapping ψ_{*x} the differential of ψ at x and designate it $d\psi_x$. If M_1 and M_2 are open subsets of Banach spaces V_1 and V_2 (and hence are differentiable manifolds under their identity charts), then ψ_{*x} as defined above does reduce to the differential $d\psi_x$ when $T_x(M_i)$ is identified with V_i. This reduction does depend on the identification, however.

5. FLOWS AND VECTOR FIELDS

Let M_1 and M_2 be differentiable manifolds. A map g from $M_1 \to M_2$ is called a *diffeomorphism* if g is a differentiable one-to-one map of M_1 onto M_2 such that g^{-1} is also differentiable.

Let M be a differentiable manifold. A map $\varphi\colon M \times \mathbb{R} \to M$ is called a *one-parameter group* if

i) φ is differentiable;

ii) $\varphi(x, 0) = x$ for all $x \in M$;

iii) $\varphi(\varphi(x, s), t) = \varphi(x, s + t)$ for all $x \in M$ and $s, t \in \mathbb{R}$.

We can express conditions (ii) and (iii) a little differently. Let $\varphi_t\colon M \to M$ be given by

$$\varphi_t(x) = \varphi(x, t).$$

For each $t \in \mathbb{R}$ the map φ_t is differentiable. In fact,

$$\varphi_t = \varphi \circ \iota_t,$$

where ι_t is the differentiable map of $M \to M \times \mathbb{R}$ given by $\iota_t(x) = (x, t)$.

Then condition (ii) says that $\varphi_0 = \text{id}$. Condition (iii) says that

$$\varphi_t \circ \varphi_s = \varphi_{t+s}.$$

If we take $t = -s$ in this equation, we get $\varphi_t \circ \varphi_{-t} = \text{id}$. Thus for each t the map φ_t is a diffeomorphism and $(\varphi_t)^{-1} = \varphi_{-t}$.

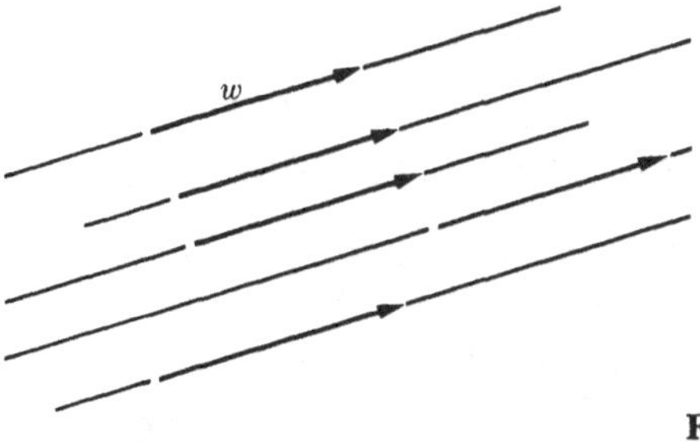

Fig. 9.10

We now give some examples of one-parameter groups.

Example 1. Let $M = V$ be a vector space, and let $w \in M$. Let $\varphi\colon V \times \mathbb{R} \to V$ be given by

$$\varphi(v, t) = v + tw.$$

It is easy to check that (i), (ii), and (iii) are satisfied. (See Fig. 9.10.)

Example 2. Let $M = V$ be a finite-dimensional vector space, and let A be a linear transformation $A\colon V \to V$. Recall that the linear transformation e^{tA} is defined by

$$e^{tA} = 1 + tA + \frac{t^2A^2}{2!} + \frac{t^3A^3}{3!} + \cdots;$$

i.e., for any $v \in V$,

$$e^{tA}v = \sum_{j=0}^{\infty} \frac{t^j}{j!} A^j v.$$

(See Figs. 9.11 and 9.12.) Since the convergence of the series is uniform on any

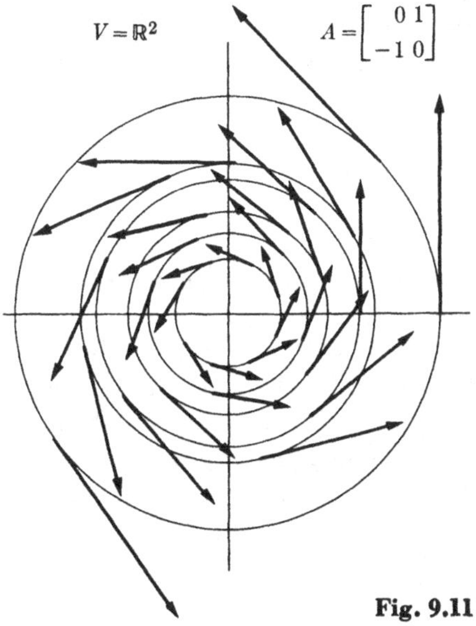

Fig. 9.11

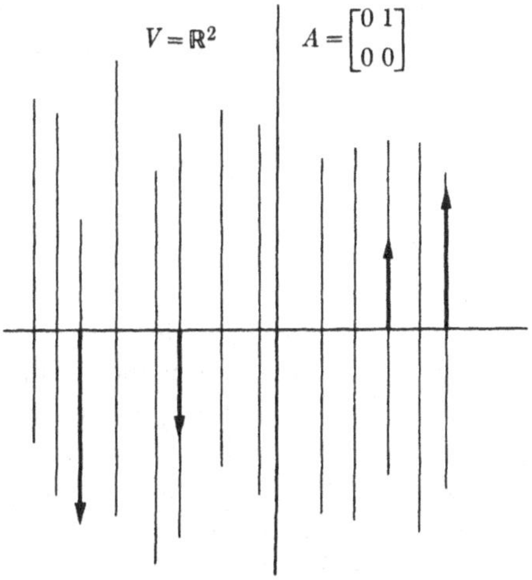

Fig. 9.12

compact set of $\langle v, t\rangle$, the map $\varphi: M \times \mathbb{R} \to M$ given by

$$\varphi(v, t) = e^{tA}v$$

is easily seen to be differentiable and to satisfy (ii) and (iii) as well.

Example 3. Let M be the circle S^1, and let a be any real number. Let φ_t^a be the diffeomorphism consisting of rotation through angle ta. In terms of the atlas $\mathcal{A} = \{(U_1, \theta_1), (U_2, \theta_2)\}$, the map φ is given by

$$\begin{aligned} \theta_1(\varphi(x, t)) &= \theta_1(x) + ta, && x \in U_1, \quad \theta_1(x) < 2\pi - ta, \\ &= \theta_1(x) + ta - 2\pi, && x \in U_1, \quad \theta_1(x) > 2\pi - ta, \\ \theta_2(\varphi(x, t)) &= \theta_2(x) + ta, && x \in U_2, \quad \theta_2(x) < 2\pi + \pi/2 - ta, \\ &= \theta_2(x) + ta - 2\pi, && x \in U_2, \quad \theta_2(x) > 2\pi + \pi/2 - ta. \end{aligned}$$

(Strictly speaking, this doesn't quite define φ for all values of $\langle x, t\rangle$. If $x = \langle 1, 0\rangle$ and $ta = \pi/2$, then $x \notin U_1$ and $\varphi(x, \pi/2) \notin U_2$. This is easily remedied by the introduction of a third chart.) It is easy to see that φ is a one-parameter group.

Example 4. Let $M = S^1 \times S^1$ be the torus, and let a and b be real numbers. Write $x \in M$ as $x = \langle x_1, x_2\rangle$, where $x_i \in S^1$. Define $\varphi^{\langle a,b\rangle}$ by

$$\varphi^{\langle a,b\rangle}(x_1, x_2, t) = \langle \varphi_t^a(x_1), \varphi_t^b(x_2)\rangle,$$

where φ^a and φ^b are given in Example 3. Then $\varphi^{\langle a,b\rangle}$ is a one-parameter group and indeed a rather instructive one. The reader should check to see that essentially different behavior arises according to whether b/a is rational or irrational.

[The construction of Example 4 from Example 3 can be generalized as follows. If $\varphi: M \times \mathbb{R} \to M$ and $\psi: N \times \mathbb{R} \to N$ are one-parameter groups, then we can construct a one-parameter group on $M \times N$ given by $\varphi_t \times \psi_t$. The map of $M \times N \times \mathbb{R} \to M \times N$ sending $\langle x, y, t\rangle \to \langle \varphi_t(x), \psi_t(y)\rangle$ is differentiable because it can be written as the composite $(\varphi \times \psi) \circ \Delta$, where

$$\varphi \times \psi: M \times \mathbb{R} \times N \times \mathbb{R} \to M \times N,$$

and

$$\Delta: M \times N \times \mathbb{R} \to M \times \mathbb{R} \times N \times \mathbb{R}$$

is given by $\Delta(x, y, t) = \langle x, t, y, t\rangle$.]

In each of the four preceding examples we started out with an "infinitesimal generator" to construct the one-parameter group, namely, the vector w in Example 1, the linear transformation A in Example 2, the real number a in Example 3, and the pair $\langle a, b\rangle$ in Example 4. We will now show that associated with any one-parameter group on a manifold, there is a nice object which we can regard as the infinitesimal generator of the one-parameter group.

Let $\varphi: M \times \mathbb{R} \to M$ be a one-parameter group. For each $x \in M$ consider the map φ_x of $\mathbb{R} \to M$ given by

$$\varphi_x(t) = \varphi(x, t).$$

In view of condition (ii), we know that $\varphi_x(0) = x$. Thus φ_x is a curve passing through x (see Fig. 9.13). Let us denote the tangent to this curve by $X(x)$. We thus get a mapping X which assigns to each $x \in M$ a vector $X(x) \in T_x(M)$. Any such map, i.e., any rule assigning to each $x \in M$ a vector in $T_x(M)$, will be called a *vector field*. We have seen that every one-parameter group gives rise to a vector field which we shall call the *infinitesimal generator* of the one-parameter group.

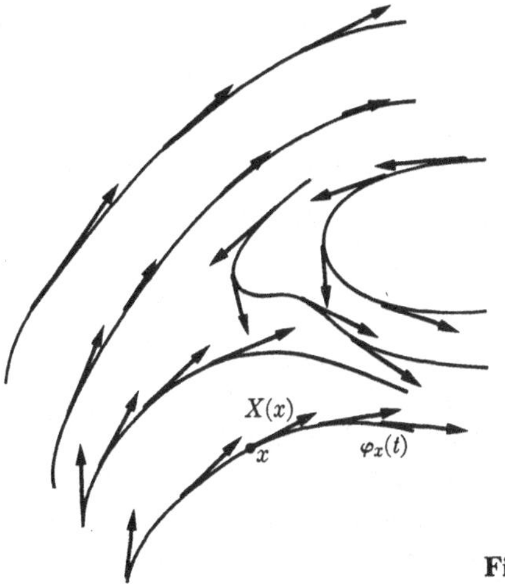

Fig. 9.13

Let Y be a vector field on M, and let (U, α) be a chart on M. For each $x \in U$ we get a vector $Y(x)_\alpha \in V$. We can regard this as defining a V-valued function Y_α on $\alpha(U)$:

$$Y_\alpha(v) = Y\big(\alpha^{-1}(v)\big)_\alpha \qquad \text{for} \quad v \in \alpha(U). \tag{5.1}$$

Let (W, β) be a second chart, and let Y_β be the corresponding V-valued function on $\beta(W)$. If we compare (5.1) with (4.3), we see that

$$Y_\beta\big(\beta \circ \alpha^{-1}(v)\big) = J_{\beta\circ\alpha^{-1}}(v) \circ Y_\alpha(v) \qquad \text{if} \quad v \in \alpha(U \cap W). \tag{5.2}$$

Equation (5.1) gives the "local expression" of a vector field with respect to a chart, and Eq. (5.2) describes the "transition law" from one chart to another.

Conversely, let $\mathcal{A}$ be an atlas of M, and let Y_α be a V-valued function defined on $\alpha(U)$ for each chart $(U, \alpha) \in \mathcal{A}$. Suppose that the Y_α satisfy (5.2). Then for each $x \in M$ we can let $Y(x) \in T_x(M)$ be defined by setting

$$Y(x)_\alpha = Y_\alpha\big(\alpha(x)\big)$$

for some chart (U, α) about x. It follows from the transition law given by (4.3) and (5.2) that this definition does not depend on the choice of (U, α).

Observe that $J_{\beta\circ\alpha^{-1}}$ is a C^∞-function (linear transformation-valued function) on $\alpha(U \cap W)$. Therefore, if Y is a vector field and Y_α is a V-valued C^∞-function on $\alpha(U)$, the function Y_β will be C^∞ on $\beta(U \cap V)$. In other words, it is consistent to require that the functions Y_α be of class C^∞. We shall therefore say that Y

is a C^∞-vector field if the function Y_α is C^∞ for every chart (U, α). As in the case of functions and mappings, in order to verify that Y is C^∞, it suffices to check that the Y_α are C^∞ for all charts (U, α) belonging to some atlas of M.

Let us check that the infinitesimal generator X of a one-parameter group φ is a C^∞-vector field. In fact, if (U, α) is a chart, then

$$X_\alpha(v) = (\alpha \circ \varphi_x)'(0),$$

where $\varphi_x(t) = \varphi(x, t)$. We can write $\alpha \circ \varphi_x(t) = \Phi(v, t)$, where

$$\Phi_\alpha(v, t) = \alpha \circ \varphi(\alpha^{-1}(v), t).$$

Let $U' \subset U$ be a neighborhood of x such that $\varphi(y, t) \in U$ for $y \in U'$ and $|t| < \epsilon$. Then Φ is a differentiable map of $\alpha(U') \times I \to \alpha(U)$, where $I = \{t : |t| < \epsilon\}$. In terms of this representation, we can write

$$X_\alpha(v) = \frac{\partial \Phi_\alpha}{\partial t}(v, 0). \tag{5.3}$$

This shows that X is a C^∞-vector field.

If we evaluate (5.3) in the case of Example 1, we get $\Phi_{\text{id}}(v, t) = v + tw$, so that $X_{\text{id}} = w$. In the case of Example 2 we get $X_{\text{id}}(v) = Av$.

There are various algebraic operations that can be performed with vector fields. The set of all vector fields on M forms a vector space in the obvious way. If X and Y are C^∞-vector fields, then so is $aX + bY$ (a and b are constants), where

$$(aX + bY)(x) = aX(x) + bY(x), \qquad x \in M.$$

Similarly, we can multiply a vector field by a function. If f is a function and X is a vector field, we define fX by

$$(fX)(x) = f(x)X(x), \qquad x \in M.$$

It is easy to see that if f and X are differentiable, then so is fX. It is also easy to check the various associative laws for this multiplication.

We have seen that any one-parameter group defines a smooth vector field. Let us examine the converse. Does any C^∞-vector field define a one-parameter group? The answer to the question as stated is "no".

In fact, let $X = \partial/\partial x^1$ be the vector field corresponding to translation in the x^1-direction in $\mathbb{R}^n$. Let $M = \mathbb{R}^2 - C$, where C is some nonempty closed set of $\mathbb{R}^n$. Then if p is any point of M that lies on a line parallel to that x^1-axis which intersects C (Fig. 9.14), then $\varphi_t(p)$ will not be defined (will not lie in M) for every t.

The reader may object that M "has more points missing" and that is why X does not generate a one-parameter group. But we can construct a counterexample on $\mathbb{R}^2$ itself. In fact, if we consider the vector field X on $\mathbb{R}^2$ given by

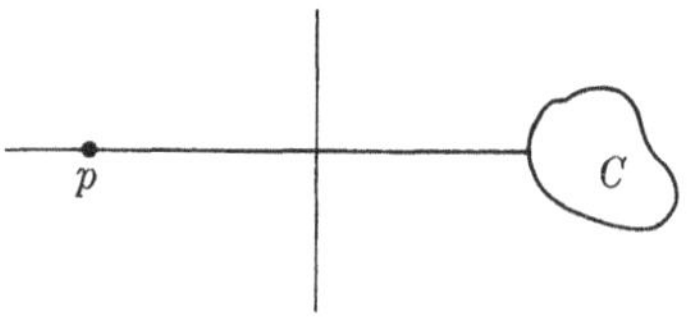

Fig. 9.14

$$X_{\text{id}}(x^1, x^2) = (1, -(x^2)^2),$$

then (5.3) shows that φ, if defined, satisfies

$$\frac{d\Phi}{dt}(x, t) = \frac{d\Phi}{dt}(\Phi(x, t), 0) = X(\Phi(x, t)),$$

where $\Phi = \Phi_{\text{id}}$. If we let $y^i(t, x) = x^i \circ \Phi(x, t)$, then

$$\frac{dy^1}{dt} = 1, \qquad y^1(0) = x^1,$$

$$\frac{dy^2}{dt} = -(y^2)^2, \qquad y^2(0) = x^2.$$

If $x^2 \neq 0$, then the unique solution of the second equation is given by

$$y^2(t) = \frac{1}{t + 1/x^2},$$

which is not defined for all values of t. Of course, the trouble is that we only have a *local* existence theorem for differential equations.

We must therefore give up on the requirement that φ be defined on all of $M \times \mathbb{R}$.

Definition 5.1. A *flow* on M is a map φ of an open set $U \subset M \times \mathbb{R} \to M$ such that

i) $M \times \{0\} \subset U$;
ii) φ is differentiable;
iii) $\varphi(x, 0) = x$;
iv) $\varphi(\varphi(x, s), t) = \varphi(x, s + t)$ whenever both sides of this equation are defined.

For x fixed, $\varphi_x(t) = \varphi(x, t)$ is defined for sufficiently small t, so that φ gives rise to a vector field X as before. We shall call X the *infinitesimal generator* of the flow φ.

As the previous examples show, there may be no $t \neq 0$ such that $\varphi(x, t)$ is defined for all x, and there may be no x such that $\varphi(x, t)$ is defined for all t.

Proposition 5.1. Let X be a smooth vector field on M. Then there exists a neighborhood U of $M \times \{0\}$ in $M \times \mathbb{R}$ and a flow $\varphi\colon U \to M$ having X as its infinitesimal generator.

Proof. We shall first construct the curve $\varphi_x(t)$ for any $x \in M$, and shall then verify that $\langle x, t\rangle \mapsto \varphi(x, t)$ is indeed a flow.

Let x be a point of M, and let (U, α) be a chart about x. Then X_α gives us an ordinary differential equation in $\alpha(U)$, namely,

$$\frac{dv}{dt} = X_\alpha(v), \qquad v \in \alpha(U).$$

By the fundamental existence theorem for ordinary differential equations, there exists an $\epsilon > 0$, an open set O containing $\alpha(x)$, and a map

$$\Phi_\alpha\colon O \times \{t : |t| < \epsilon\} \to \alpha(U)$$

such that

$$\Phi_\alpha \text{ is } C^\infty, \qquad \Phi_\alpha(v, 0) = v,$$

and

$$\frac{d\Phi_\alpha(v, t)}{dt} = X_\alpha\big(\Phi_\alpha(v, t)\big).$$

Here the choice of the open set O and of ϵ depends on $\alpha(x)$ and $\alpha(U)$. The uniqueness part of the theorem asserts that Φ_α is uniquely determined up to the domain of definition; i.e., if Φ_v is any curve defined for $|t| < \epsilon'$ with $\Phi_v(0) = v$ and

$$\frac{d\Phi_v(t)}{dt} = X_\alpha\big(\Phi_v(t)\big), \tag{5.4}$$

then $\Phi_v(t) = \Phi_\alpha(v, t)$.

This implies that

$$\Phi_\alpha(v, t + s) = \Phi_\alpha\big(\Phi_\alpha(v, s), t\big)$$

whenever both sides are defined. (Just hold s fixed in the equation.)

Consider the curve $\phi^\alpha{}_x(\cdot)$ defined by

$$\phi^\alpha{}_x(t) = \alpha^{-1}(\Phi_\alpha(\alpha(x),t)). \tag{5.5}$$

It is defined for $|t| < \epsilon$, and is a continuous, in fact differentiable map of this interval into M. Furthermore, if we write $\psi = \phi^\alpha{}_x(\cdot)$ then (5.4) asserts that the tangent vector to the curve $\psi(t + \cdot)$ is $X(\psi(t))$, the value of the vector field at the point $\psi(t)$. We will write this condition as

$$\psi'(t) = X(\psi(t)). \tag{5.6}$$

Equation (5.6) is the way we would write the "first order differential equation" on M corresponding to the vector field X. A differentiable curve ψ satisfying (5.6) is called an *integral curve* of X. We now can formulate a manifold version of the uniqueness theorem of differential equations:

Lemma 5.1. Let $\psi_1{:}I \to M$ and $\psi_2{:}I \to M$ be two integral curves of X defined on the same interval I. If $\psi_1(s) = \psi_2(s)$ at some point $s \in I$ then $\psi_1 = \psi_2$, i.e. $\psi_1(t) = \psi_2(t)$ for all $t \in I$.

Proof. We wish to show that the set where $\psi_1(t) \neq \psi_2(t)$ is empty. Let

$$A = \{t{:}t \geq s \text{ and } \psi_1(t) \neq \psi_2(t)\}.$$

We wish to show that A is empty, and similarly that the set $B = \{t{:}t \leq s$ and $\psi_1(t) \neq \psi_2(t)\}$ is empty. Suppose that A is not empty, and let

$$t_+ = \text{glb}\, A = \text{glb}\,\{t{:}t \geq s \text{ and } \psi_1(t) \neq \psi_2(t)\}.$$

We will derive a contradiction by

i) using the uniqueness theorem for differential equations to show that $\psi_1(t_+) \neq \psi_2(t_+)$, and

ii) using the Hausdorff property of manifolds to show that $\psi_1(t_+) = \psi_2(t_+)$.

Details: i). Suppose that $\psi_1(t_+) = \psi_2(t_+) = y \in M$. We can find a coordinate chart (β, W) about y, and then $\beta \circ \psi_1$ and $\beta \circ \psi_2$ are solutions of the same system of first order ordinary differential equations, and they both take on the value $\beta(y)$ at $t = t_+$. Hence, by uniqueness for differential equations, $\beta \circ \psi_1$ and $\beta \circ \psi_2$ must be equal in some interval about t_+, and hence $\psi_1(t) = \psi_2(t)$ for all t in this interval. This is impossible since there must be points arbitrarily close to t_+ where $\psi_1(t) \neq \psi_2(t)$ by the glb property of t_+. This proves i). Now suppose that $\psi_1(t_+) \neq \psi_2(t_+)$. We can find neighborhoods U_1 of $\psi_1(t_+)$ and U_2 of $\psi_2(t_+)$ such that $U_1 \cap U_2 = \varnothing$. But then the continuity of the ψ_1 imply that $\psi_1(t) \in U_1$ and $\psi_2(t) \in U_2$ for t close enough to t_+, and hence that $\psi_1(t) \neq \psi_2(t)$ for t in some interval about t_+. This once again contradicts the glb property of t_+, proving ii). The same argument with glb replaced by lub shows that B is empty proving Lemma 5.1. The above argument is typical of a "connectedness argument." We showed that the set where $\psi_1(t) = \psi_2(t)$ is both open and closed, and hence must be the whole interval I.

Lemma 5.1 shows that (5.5) defines a solution curve of X passing through x at time $t = 0$, and is independent of the choice of chart in any common interval of definition about 0. In other words it is legitimate to define the curve $\phi_x(\cdot)$ by

$$\phi_x(t) = \alpha^{-1}(\Phi_\alpha(\alpha(\ ),t))$$

which defines $\phi_x(t)$ for $|t| < \epsilon$. Unfortunately the ϵ depends not only on x but also on the choice of chart. We use Lemma 5.1 and extend the definition of $\phi_x(\cdot)$ as far as possible, much as we did for ordinary differential equations on a vector space in Chapter 6: For any s with $|s| < \epsilon$ we let $y = \phi_x(s)$ and obtain a curve $\phi_y(\cdot)$ defined for $|t| < \epsilon'$. By Lemma 5.1

$$\phi_y(t) = \phi_x(s+t) \text{ if } |s+t| < \epsilon. \tag{5.7}$$

It may happen that $|s| + \epsilon' > \epsilon$. Then there will exist a t with $|t| < \epsilon'$ and $|s+t| > \epsilon$. Then the right hand side of (5.7) is not defined, but the left is. We then take (5.7) as the *definition* of $\phi_x(s+t)$, extending the domain of definition of $\phi_x(\cdot)$. We then continue: Let $I_x{}^+$ denote the set of all $s > 0$ for which there exists a finite sequence of real numbers $s_0 = 0 < s_1 < \ldots < s_k = s$ and points $x_0, \ldots x_{k-1} \in M$ with $x_0 = x$, s_1 in the domain of definition of $\phi_x(\cdot)$, $x_2 = \phi_x(s_1)$ and, inductively,

s_{i+1} in the domain of definition of $\phi_{x_i}(\cdot)$ and $x_{i+1} = \phi_{x_i}(s_{i+1})$.

If $s \in I_x^+$, so is s' for $0 < s' < s$, and so is $s + \eta$ for sufficiently small positive η. Thus I_x^+ is an interval, half open on the right. By repeated use of (5.4) we define $\varphi_x(s)$ for $s \in I_x^+$. We construct I_x^- in a similar fashion and set $I_x = I_x^+ \cup I_x^-$. Then $\varphi_x(s)$ is defined for $s \in I_x$, and I is the maximal interval for which our construction defines φ_x. For each $x \in M$ we obtain an open interval I_x in which the curve $\varphi_x(\cdot)$ is defined.

Let $U = \bigcup_{x \in M}\{x\} \times I_x$. Then U is an open subset of $M \times I$. To verify this, let $(\bar{x}, \bar{s}) \in U$. We must show that there is a neighborhood W of $\bar{x}$ and an $\epsilon > 0$ such that $s \in I_x$ for all $|s - \bar{s}| < \epsilon$ and $x \in W$. By definition, there is a finite sequence of points $\bar{x} = \bar{x}_0, \bar{x}_1, \ldots, \bar{x}_k$ and charts $(U_1, \alpha_1), \ldots, (U_k, \alpha_k)$ with $x_{i-1} \in U_i$ and $x_i \in U_i$ and such that

$$\alpha_i(\bar{x}_i) = \Phi_{\alpha_i}(\alpha_i(\bar{x}_{i-1}), t_i),$$

where $t_1 + \cdots + t_k = s$. It is now clear from the continuity properties of the Φ_α that if we choose x_0 such that $\alpha_1(x_0)$ is close enough to $\alpha_1(\bar{x}_0)$, then the points x_i defined inductively by

$$\alpha_i(x_i) = \Phi_{\alpha_i}(\alpha_i(x_{i-1}), t_i)$$

will be well defined. [That is, $\alpha_j(x_{j-1})$ will be in the domain of the definition of $\Phi_{\alpha_i}(\cdot, t_i)$.] This means that $\bar{s} \in I_{x_0}$ for all such points x_0. The same argument shows that $\bar{s} + \eta \in I_{x_0}$ for η sufficiently small and x sufficiently close to $\bar{x}$. This shows that U is open.

Now define φ by setting

$$\varphi(x, t) = \varphi_x(t) \qquad \text{for} \quad (x, t) \in U.$$

That φ is differentiable near $M \times \{0\}$ follows from the fact that φ is given (in terms of a chart) as the solution of an ordinary differential equation. The fundamental existence theorem then guarantees the differentiability. Near the point $(\bar{x}, \bar{t})$ we can write

$$\varphi(x, t) = \varphi(\varphi(\cdots(\varphi(x, t_1), t_2)\cdots)t_k), \qquad t = t_1 + t_2 + \cdots + t_k,$$

and so φ is differentiable because it is the composite of differentiable maps. □

6. LIE DERIVATIVES

Let φ be a one-parameter group on a manifold M, and let f be a differentiable function on M. Then for each t the function $\varphi_t^*[f]$ is differentiable, and for $t \neq 0$ we can form the function

$$\frac{\varphi_t^*[f] - f}{t}. \tag{6.1}$$

We claim that the limit of this expression as $t \to 0$ exists. In fact, for any $x \in M$, $\varphi_t^*[f](x) = f \circ \varphi_t(x) = f \circ \varphi_x(t)$ and, therefore,

$$\lim_{t \to 0} \frac{\varphi_t^*[f] - f}{t}(x) = \lim_{t \to 0} \frac{f \circ \varphi_x(t) - f \circ \varphi_x(0)}{t} = D_{\varphi_x} f = X(x)f. \tag{6.2}$$

Here $X(x)$ is a tangent vector at x and we are using the notation introduced in Section 4. We shall call the limit of (6.1) the derivative of f with respect to the one-parameter group φ, and shall denote it by $D_X f$. More generally, for any smooth vector field X and differentiable function f we define $D_X f$ by

$$D_X f(x) = X(x)f \qquad \text{for all} \quad x \in M, \tag{6.3}$$

and call it the *Lie derivative* of f with respect to X. In terms of the flow generated by X, we can, near any $x \in M$, represent $D_X f$ as the limit of (6.1), where, in general, (6.1) will only be defined for a sufficiently small neighborhood of x and for sufficiently small $|t|$.

Our notation generalizes the notation in Chapter 3 for directional derivative. In fact, if M is an open subset of V and X is the "constant vector field" of Example 1

$$X_{\mathrm{id}} = w \in V,$$

then

$$(D_X f)_{\mathrm{id}} = D_w f_{\mathrm{id}},$$

where D_w is the directional derivative with respect to w.

Note that $D_X f$ is linear in X; that is,

$$D_{aX+bY} f = aD_X f + bD_Y g$$

if X and Y are vector fields and a and b are constants.

Let ψ be a diffeomorphism of M_1 onto M_2, and let X be a vector field on M_2. We define the "pullback" vector field $\psi^*[X]$ on M_1 by setting

$$\psi^*[X](x) = \psi_*^{-1} X(\psi(x)) \qquad \text{for all} \quad x \in M. \tag{6.4}$$

Note that ψ must be a diffeomorphism for (6.4) to make sense, since ψ^{-1} enters into the definition. This is in contrast to the "pullback" for functions, which made sense for any differentiable map. Equation (6.4) does indeed define a vector field, since

$$\psi_*^{-1} = \psi_{*x}^{-1} : T_{\psi(x)}(M_2) \to T_x(M_1) \qquad \text{and} \qquad X(\psi(x)) \in T_{\psi(x)}(M_2).$$

Let us check that $\psi^*[X]$ is a smooth vector field if X is. To this effect, let $\mathcal{A}_1$ and $\mathcal{A}_2$ be compatible atlases on M_1 and M_2, and let $(U, \alpha) \in \mathcal{A}_1$ and $(W, \beta) \in \mathcal{A}_2$ be compatible charts. Then (6.4) says that

$$\psi^*[X]_\alpha(v) = J_{\alpha\circ\psi^{-1}\circ\beta^{-1}}(\beta \circ \psi \circ \alpha^{-1}(v)) \cdot X_\beta(\beta \circ \psi \circ \alpha^{-1}(v)) \qquad \text{for} \quad v \in \alpha(U),$$

which is a differentiable function of v. Since, by the chain rule,

$$J_{\alpha\circ\psi^{-1}\circ\beta^{-1}}(\beta \circ \psi \circ \alpha^{-1}(v)) \cdot J_{\beta\circ\psi\circ\alpha^{-1}}(v) = 1,$$

we can rewrite the last expression more simply as

$$\psi^*[X]_\alpha(v) = (J_{\beta\circ\psi\circ\alpha^{-1}}(v))^{-1} X_\beta(\beta \circ \psi \circ \alpha^{-1}(v)) \qquad \text{for} \quad v \in \alpha(U). \tag{6.5}$$

Thus $\psi^*[X]_\alpha$ is the product of a smooth $\mathrm{Hom}(V_2, V_1)$-valued function and a smooth V_2-valued function, which shows that $\psi^*[X]$ is a smooth vector field.

Exercise. Let φ be the flow generated by X on M_2. Show that the flow generated by $\psi^*[X]$ is given by

$$\langle x, t \rangle \mapsto \psi^{-1}\varphi(\psi(x), t). \tag{6.6}$$

If φ is a one-parameter group, then we can write (6.6) as

$$\langle x, t \rangle \mapsto \psi^{-1} \circ \varphi_t \circ \psi(x). \tag{6.6$'$}$$

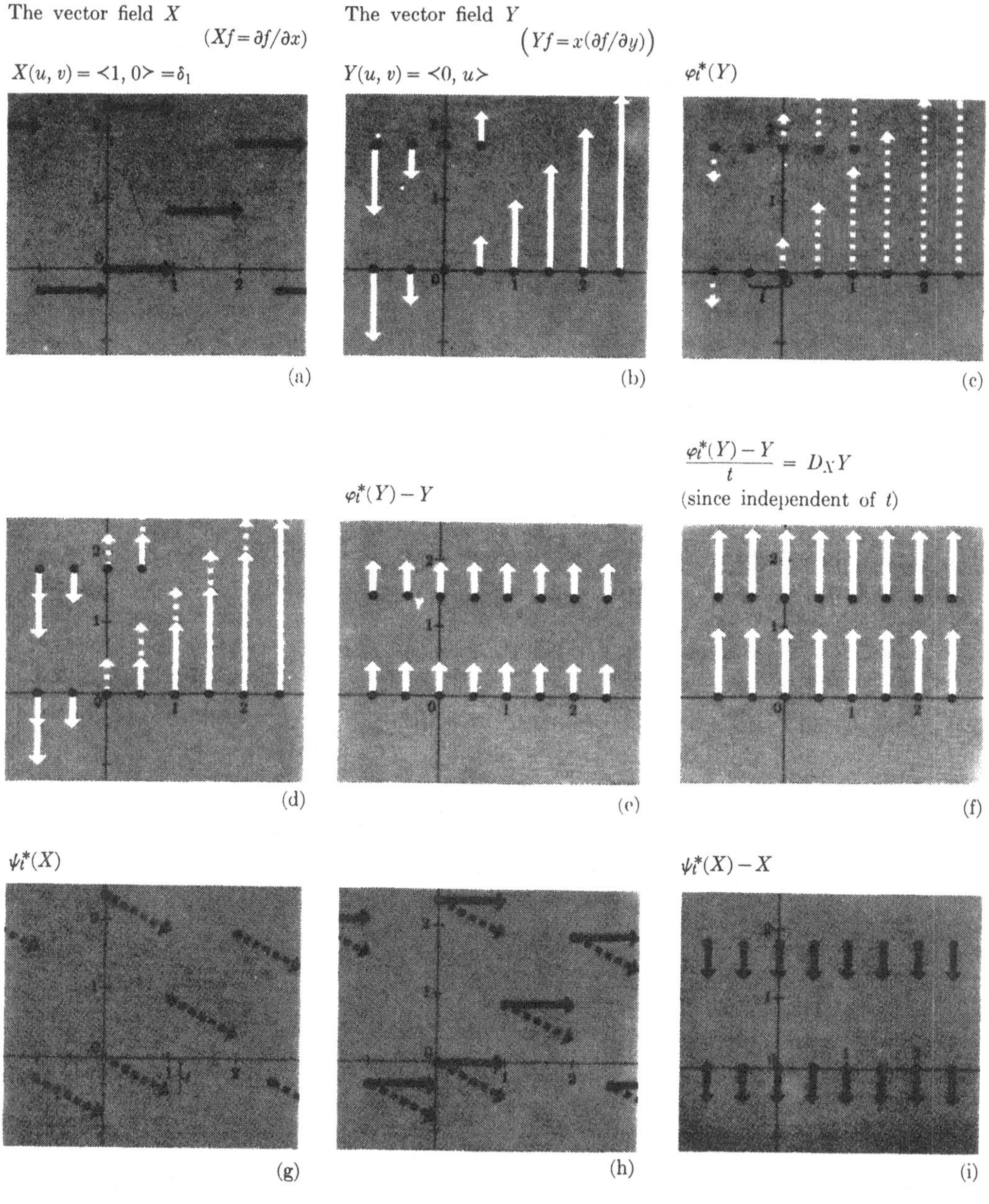

Fig. 9.15

It is easy to check that if $\psi_1 \colon M_1 \to M_2$ and $\psi_2 \colon M_2 \to M_3$ are diffeomorphisms and Y is a vector field on M_3, then

$$(\psi_2 \circ \psi_1)^* Y = \psi_1^* \psi_2^* Y.$$

$\frac{\psi_t^*(X) - X}{t} = D_Y X$
(since independent of t)

(j)

Fig. 9.15 *(cont.)*

If f is a differentiable function on M_2, then

$$D_{\psi^*[X]}(\psi^*[f]) = \psi^*(D_X f). \tag{6.7}$$

In fact, by (6.3) and (4.6) we have, for $x \in M_1$,

$$\begin{aligned} D_{\psi^*[X]}\psi^*[f](x) &= \psi^*[X](x)\psi^*[f] && \text{by (6.3)} \\ &= \psi_*^{-1}X(\psi(x))\psi^*[f] && \text{by (6.4)} \\ &= (\psi_*\psi_*^{-1}X(\psi(x)))f && \text{by (4.6)} \\ &= X(\psi(x))f \\ &= (D_X f)(\psi(x)) \\ &= \psi^*(D_X f)(x). \end{aligned}$$

Let φ be a one-parameter group on M with infinitesimal generator X, and let Y be another smooth vector field on M. For $t \neq 0$ we can form the vector field

$$\frac{\varphi_t^*[Y] - Y}{t} \tag{6.8}$$

and investigate its limit as $t \to 0$, which we shall call $D_X Y$. In Fig. 9.15 we have shown the calculation of $D_Y X$ and $D_X Y$ for two very simple fields on the Cartesian plane $\mathbb{R}^2$. The field X is the constant field $X_{\text{id}} = \delta_1$, so that $Xf = \partial f/\partial x$ in terms of Cartesian coordinates x, y. The corresponding flow is given by $\varphi_t(x, y) = \langle x + t, y\rangle$. Thus $\varphi_{t*} = \text{id}$ if we identify the tangent space at each point of the plane with the plane itself. Then $Y \mapsto \varphi_t^* Y$ consists of "moving" the vector field Y to the left by t units. Here we have taken $Y = x\delta_2$, so that $Yf = x(\partial f/\partial y)$. In Fig. 9.15(c) we have pictured $\varphi_t^* Y$, and have superimposed Y and $\varphi_t^* Y$ in Fig. 9.15(d). Figure 9.15(e) represents $\varphi_t^* Y - Y$ and Fig. 9.15(f) is $(1/t)\{\varphi_t^* Y - Y\}$, which coincides with its limit, $D_X Y$, since the expression is independent of t. The one-parameter group generated by Y is ψ_t where $\psi_t(x, y) = \langle x, y + tx\rangle$. Here at any $p \in \mathbb{R}^2$ we have $\psi_{t*}\delta_1 = \delta_1 + t\delta_2$, so that $\psi_t^* X = \psi_{-t*}X(\psi(x)) = \delta_1 - t\delta_2$. In Fig. 9.15(g) we have drawn $\psi_t^* X$ and in Fig. 9.15(h) we have superimposed it on X. Note that $D_X Y = -D_Y X$. However, these two derivatives are nonzero for quite different reasons. The field $\varphi_t^* Y$ varies with t because the field Y is not constant. The field $\psi_t^* X$ varies with t because of "distortion" in the flow ψ_t. See Fig. 9.15(g) and (h). In the general case, $D_X Y$ will result from a superposition of these two effects. We now make the general calculation.

Let (U, α) be a chart on M, and for $v \in \alpha(U)$ let O be a sufficiently small open set containing v, and let $\epsilon > 0$ be sufficiently small, so that Φ_α given by

$$\Phi_\alpha(w, t) = \alpha \circ \varphi(\alpha^{-1}(w), t)$$

is defined for $w \in O$ and $|t| < \epsilon$. Then, for $|t| < \epsilon$, Eq. (6.5) implies that

$$\varphi_t^*[Y]_\alpha(v) = (J_{\Phi_\alpha(v,t)}(v))^{-1} Y_\alpha(\Phi_\alpha(v, t)). \tag{6.9}$$

The right-hand side of this equation is of the form $A_t^{-1}z_t$, where A_t and z_t are differentiable functions of t with $A_0 = I$. Therefore, its derivative with respect to t exists and

$$\begin{aligned}\left.\frac{d(A_t^{-1}z_t)}{dt}\right|_{t=0} &= \lim_{t\to 0}\frac{A_t^{-1}z_t - z_0}{t}\\ &= \lim_{t\to 0} A_t\frac{(A_t^{-1}z_t - z_0)}{t}\\ &= \lim_{t\to 0}\frac{z_t - A_tz_0}{t}\\ &= \lim_{t\to 0}\left(\frac{z_t - z_0}{t} - \frac{A_tz_0 - z_0}{t}\right)\\ &= z_0' - A_0'z_0.\end{aligned}$$

Now in (6.9) $z_t = Y_\alpha(\Phi_\alpha(v, t))$, so

$$\begin{aligned} z_0' &= dY_\alpha\left(\frac{\partial \Phi_\alpha}{\partial t}(v, 0)\right)\\ &= dY_\alpha(X_\alpha(v)).\end{aligned}$$

Here Y_α is a V-valued function, so dY_α is its differential at the point $\Phi_\alpha(v, 0)$. Hence $dY_\alpha(X_\alpha(v))$ is the value of this differential at $X_\alpha(v)$. The transformation $A_t = J_{\Phi_\alpha(v,t)} = d(\Phi_\alpha)_{(v,t)}$, so

$$\begin{aligned}\left.\frac{dA}{dt}\right|_{t=0} &= \left.\frac{\partial\, d\Phi_\alpha}{\partial t}\right|_{t=0}\\ &= d\,\frac{\partial \Phi_\alpha}{\partial t}\\ &= d(X_\alpha)_v.\end{aligned}$$

Thus the derivative of (6.9) at $t = 0$ can be written as

$$d(Y_\alpha)_v(X_\alpha(v)) - d(X_\alpha)_v(Y_\alpha(v)) = D_{X_{\alpha}(v)}Y_\alpha - D_{Y_{\alpha}(v)}X_\alpha.$$

We have thus shown that the limit in (6.8) exists. If we denote it by D_XY, we can write

$$(D_XY)_\alpha(v) = D_{X_{\alpha}(v)}Y_\alpha - D_{Y_{\alpha}(v)}X_\alpha. \tag{6.10}$$

As before, we can use (6.10) as the definition of D_XY for arbitrary vector fields X and Y. Again, this represents the derivative of Y with respect to the flow generated by X, that is, the limit of (6.9) where now (6.8) is only locally defined.

From (6.10) we derive the surprising result that $D_XY = -D_YX$. For this reason it is convenient to introduce a notation which expresses the antisymmetry more clearly, and we shall write

$$D_XY = [X, Y].$$

The expression on the right-hand side is called the *Lie bracket* of X and Y. We have

$$[X, Y] = -[Y, X]. \tag{6.11}$$

Let us evaluate the Lie bracket for some of the examples listed in the beginning of Section 5. Let $M = \mathbb{R}^n$.

Example 1. If $X_{\mathrm{id}} = w_1$ and $Y_{\mathrm{id}} = w_2$ are "constant" vector fields, then (6.10) shows that $[X, Y] = 0$.

Example 2. Let $X_{\mathrm{id}}(v) = Av$, where A is a linear transformation, and let $Y_{\mathrm{id}} = w$. Then (6.10) says that

$$[X, Y]_{\mathrm{id}}(v) = -Aw,$$

since the directional derivative of the linear function Av with respect to w is Aw.

Example 3. Let $X_{\mathrm{id}}(v) = Av$ and $Y_{\mathrm{id}}(v) = Bv$, where A and B are linear transformations. Then by (6.10),

$$[X, Y]_{\mathrm{id}}(v) = BAv - ABv = (BA - AB)v. \tag{6.12}$$

Thus in this case $[X, Y]$ again comes from a linear transformation, namely, $BA - AB$. In this case the antisymmetry in A and B is quite apparent.

We now return to the general case. Let φ be a one-parameter group on M, let Y be a smooth vector field on M, and let f be a differentiable function on M. According to (6.7),

$$D_{\varphi_t^*[Y]}(\varphi_t^*[f]) = \varphi_t^*(D_Yf).$$

Then

$$\begin{aligned}\frac{\varphi_t^*(D_Yf) - D_Yf}{t} &= \frac{D_{\varphi_t^*[Y]}(\varphi_t^*[f]) - D_Y(\varphi_t^*[f])}{t} + \frac{D_Y(\varphi_t^*f) - D_Yf}{t}\\ &= D_{\left\{\frac{\varphi_t^*[Y]-Y}{t}\right\}}\varphi_t^*[f] - D_Y\left(\frac{\varphi_t^*f - f}{t}\right).\end{aligned}$$

Since the functions $\varphi_t^*[f]$ are uniformly differentiable, we may take the limit as $t \to 0$ to obtain

$$\begin{aligned}D_X(D_Yf) &= D_{D_XY}f + D_Y(D_Xf)\\ &= D_{[X,Y]}f + D_Y(D_Xf).\end{aligned}$$

In other words,

$$D_{[X,Y]}f = D_X(D_Yf) - D_Y(D_Xf). \tag{6.13}$$

In view of its definition as a derivative, it is clear that D_XY is linear in Y:

$$D_X(aY_1 + bY_2) = aD_XY_1 + bD_XY_2$$

if a and b are constants and X and Y are vector fields. By the antisymmetry, it must therefore also be linear in X. That is,

$$D_{aX_1+bX_2}Y = [aX_1 + bX_2, Y] = a[X_1, Y] + b[X_2, Y] = aD_{X_1}Y + bD_{X_2}Y.$$

Let X and Y be vector fields on a manifold M_2, and let ψ be a diffeomorphism of M_1 onto M_2. Then

$$\psi^*[X, Y] = [\psi^*X, \psi^*Y]. \tag{6.14}$$

In fact, suppose X generates the flow φ. Then

$$\begin{aligned}\psi^*[X, Y] = \psi^* D_X Y = \psi^* \lim_{t=0}\left(\frac{\varphi_t^* Y - Y}{t}\right) &= \lim_{t=0} \frac{\psi^*\varphi_t^* Y - \psi^* Y}{t} \\ &= \lim_{t=0} \frac{\psi^*\varphi_t^*\psi^{-1*}\psi^* Y - \psi^* Y}{t} \\ &= \lim_{t=0} \frac{(\psi^{-1}\circ\varphi_t\circ\psi)^*\psi^* Y - \psi^* Y}{t}.\end{aligned}$$

Since $\psi^{-1} \circ \varphi_t \circ \psi$ is the flow generated by ψ^*X, we conclude that the last limit is $D_{\psi^*X}\psi^*Y$, which proves (6.14).

Now let Y and Z be smooth vector fields on M, and let X be the infinitesimal generator of φ. Then

$$\begin{aligned}D_X[Y, Z] &= \lim_{t=0} \frac{\varphi_t^*[Y, Z] - [Y, Z]}{t} \\ &= \lim_{t=0} \frac{[\varphi_t^* Y, \varphi_t^* Z] - [Y, Z]}{t} \\ &= \lim_{t=0} \left\{\left[\frac{\varphi_t^* Y - Y}{t}, \varphi_t^* Z\right] + \left[Y, \frac{\varphi_t^* Z - Z}{t}\right]\right\} \\ &= [D_X Y, Z] + [Y, D_X Z].\end{aligned}$$

Thus

$$[X, [Y, Z]] = [[X, Y], Z] + [Y, [X, Z]]. \tag{6.15}$$

In view of the antisymmetry of the Lie bracket, Eq. (6.15) can be rewritten as

$$[X, [Y, Z]] + [Y, [Z, X]] + [Z, [X, Y]] = 0. \tag{6.16}$$

Equation (6.15), or (6.16), is known as Jacobi's identity.

7. LINEAR DIFFERENTIAL FORMS

Let M be a differentiable manifold. We have attached to each $x \in M$ a vector space $T_x(M)$. Its dual space, $(T_x(M))^*$, is called the *cotangent space* to M at x, and will be denoted by $T_x^*(M)$. Thus an element of $T_x^*(M)$ is a continuous linear function on $T_x(M)$; it is called a *covector*.

Some explanation of the word "continuous" is in order. In the case where M [and hence $T_x(M)$] is finite-dimensional, all linear functions on $T_x(M)$ are continuous, so no further comment is necessary. We shall be concerned primarily with this case. More generally, let l be a linear function on $T_x(M)$. For any chart (U, α) about x we have identified $T_x(M)$ with V, thus identifying $\xi \in T_x(M)$ with $\xi_\alpha \in V$. Then l determines a linear function l_α on V by

$$\langle \xi_\alpha, l_\alpha \rangle = \langle \xi, l \rangle. \tag{7.1}$$

If (W, β) is a second chart, then

$$\langle \xi_\beta, l_\beta \rangle = \langle J_{\alpha \circ \beta^{-1}}(\beta(x))\, \xi_\beta, l_\alpha \rangle.$$

Since $J_{\alpha \circ \beta^{-1}}(\beta(x))$ is a continuous map of V into V, we see that l_α is continuous if and only if l_β is. We shall therefore say that l is continuous if l_α is continuous for some (and hence any) α. In this case we see that (7.1) gives us an identification of $T_x^*(M)$ with V^* sending l into l_α. The last equation says that the rule for change of charts is given by

$$l_\beta = (J_{\alpha \circ \beta^{-1}}(\beta(x)))^* l_\alpha. \tag{7.2}$$

Let f be a differentiable function on M, and let $x \in M$. Then the function on $T_x(M)$ sending each $\xi \in T_x(M)$ into $\xi(f)$ will be denoted by $df(x)$. Thus

$$\langle \xi, df(x) \rangle = \xi f.$$

It is easy to see that $df \in T_x^*(M)$. In fact, in terms of a chart (U, α) about x,

$$\langle \xi, df(x) \rangle = D_\xi(f_\alpha)(\alpha(x)).$$

Note that f assigns an element $df(x)$ of $T_x^*(M)$ to each $x \in M$. A map which assigns to each $x \in M$ an element of $T_x^*(M)$ will be called a covector field or a *linear differential form*. The linear differential form determined by the function f will be denoted simply by df.

Let ω be a linear differential form. Thus $\omega(x) \in T_x^*(M)$ for each $x \in M$. Let $\mathcal{A}$ be an atlas of M. For each $(U, \alpha) \in M$ we obtain the V^*-valued function ω_α on $\alpha(U)$ defined by

$$\omega_\alpha(v) = \{\omega(\alpha^{-1}(v))\}_\alpha \quad \text{for} \quad v \in \alpha(U). \tag{7.3}$$

If $(W, \beta) \in \mathcal{A}$, then (7.2) says that

$$\begin{aligned} \omega_\beta(\beta \circ \alpha^{-1}(v)) &= (J_{\alpha \circ \beta^{-1}}(\beta \circ \alpha^{-1}(v)))^* \omega_\alpha(v) \\ &= (J_{\beta \circ \alpha^{-1}}(v))^{-1*} \omega_\alpha(v) \qquad \text{for} \quad v \in \alpha(U \cap W). \end{aligned} \tag{7.4}$$

As before, Eq. (7.4) shows that it makes sense to require that ω be smooth. We say that ω is a C^k-differential form if ω_α is a V^*-valued C^k-function for every chart (U, α). By (7.4) it suffices to check this for all charts in an atlas. Also, if we are given V^*-valued functions ω_α, each defined on $\alpha(U)$, $(U, \alpha) \in \mathcal{A}$, and satisfying (7.4), then they define a linear differential form ω on M via (7.3).

If ω is a differential form and f is a function, we define the form $f\omega$ by $f\omega(x) = f(x)\omega(x)$. Similarly, we define $\omega_1 + \omega_2$ by

$$(\omega_1 + \omega_2)(x) = \omega_1(x) + \omega_2(x).$$

Let M_1 and M_2 be differentiable manifolds, and let $\varphi: M_1 \to M_2$ be a differentiable map. For any $x \in M_1$ we have the map $\varphi_{*x}: T_x(M_1) \to T_{\varphi(x)}(M_2)$. It therefore defines a dual map

$$(\varphi_{*x})^*: T^*_{\varphi(x)}(M_2) \to T^*_x(M_1).$$

(The reader can check that if $l \in T^*_{\varphi(x)}(M_2)$, then $\xi \to \langle \varphi_*(\xi), l\rangle$ is a continuous linear function of ξ, by verification in terms of a chart.)

Now let ω be a differential form on M_2. It assigns $\omega(\varphi(x)) \in T^*_{\varphi(x)}(M_2)$ to $\varphi(x)$, and thus assigns an element $(\varphi_{*x})^*(\omega(\varphi(x))) \in T^*_x(M_1)$ to $x \in M_1$. We thus "pull back" the form ω to obtain a form on M_1 which we shall call $\varphi^*\omega$. Thus

$$\varphi^*\omega(x) = (\varphi_{*x})^*\omega(\varphi(x)). \tag{7.5}$$

Note that φ^* is defined for any differentiable map as in the case of functions, not only for diffeomorphisms (and in contrast to the situation for vector fields).

It is easy to give the expression for φ^* in terms of compatible charts (U, α) of M_1 and (W, β) of M_2. In fact, from the local expression for φ^* we see that

$$(\varphi^*\omega)_\alpha(v) = (J_{\beta\circ\varphi\circ\alpha^{-1}}(v))^*\omega_\beta(\beta\circ\varphi\circ\alpha^{-1}(v)), \qquad v \in \alpha(U). \tag{7.6}$$

From (7.6) we see that $\varphi^*\omega$ is smooth if ω is. It is clear that φ^* preserves algebraic operations:

$$\varphi^*(\omega_1 + \omega_2) = \varphi^*\omega_1 + \varphi^*\omega_2 \tag{7.7}$$

and

$$\varphi^*(f\omega) = \varphi^*[f]\varphi^*(\omega). \tag{7.8}$$

If $\varphi: M_1 \to M_2$ and $\psi: M_2 \to M_3$ are differentiable maps, then (4.4) and (7.5) show that

$$(\psi\circ\varphi)^*\omega = \varphi^*\psi^*\omega. \tag{7.9}$$

Let $\psi: M_1 \to M_2$ be a differentiable map, and let f be a differentiable function on M_2. Then (4.6) and the definition df show that

$$d(\psi^*[f]) = \psi^*\, df. \tag{7.10}$$

Let φ be a flow on M with infinitesimal generator X, and let ω be a smooth linear differential form on M. Then the form $\varphi_t^*\omega$ is locally defined and, as in the case of functions and vector fields, the limit as $t \to 0$ of

$$\frac{\varphi_t^*\omega - \omega}{t}$$

exists. We can verify this by using (7.6) and proceeding as we did in the case of vector fields. The limit will be a smooth covector field which we shall call $D_X\omega$. We could give an expression for $D_X\omega$ in terms of a chart, just as we did for vector fields.

If f is a differentiable function, ω a smooth differential form, and X the infinitesimal generator of φ, then

$$D_X(f\omega) = (D_Xf)\omega + fD_X\omega. \tag{7.11}$$

In fact,

$$\begin{aligned} D_X(f\omega) &= \lim_{t\to 0} \frac{\varphi_t^* f\omega - f\omega}{t} \\ &= \lim_{t\to 0} \left(\frac{\varphi_t^* f - f}{t} \varphi_t^*(\omega) + f \frac{\varphi_t^* \omega - \omega}{t} \right) \\ &= (D_Xf)\omega + fD_X\omega. \end{aligned}$$

If g is a differentiable function on M, then

$$\frac{\varphi_t^*\, dg - dg}{t} = \frac{d\varphi_t^*[g] - dg}{t} = d\left(\frac{\varphi_t^*[g] - g}{t} \right).$$

An easy verification in terms of a chart shows that the limit of this last expression exists and is indeed $d(D_X\varphi)$. Thus

$$D_X(df) = d(D_Xf). \tag{7.12}$$

Equations (7.11) and (7.12) show that if

$$\omega = f_1\, dg_1 + \cdots + f_k\, dg_k,$$

then

$$D_X\omega = (D_Xf_1)\, dg_1 + \cdots + (D_Xf_k)\, dg_k + f_1\, d(D_Xg_1) + \cdots + f_k\, d(D_Xg_k). \tag{7.12$'$}$$

Let ω be a smooth linear differential form, and let X be a smooth vector field. Then $\langle X, \omega \rangle$ is a smooth function given by

$$\langle X, \omega \rangle(x) = \langle X(x), \omega(x) \rangle.$$

Note that $\langle X, \omega \rangle$ is linear in both X and ω. Also observe that for any smooth function f we have

$$\langle X, df \rangle = D_Xf. \tag{7.13}$$

8. COMPUTATIONS WITH COORDINATES

For the remainder of this chapter we shall assume that our manifolds are finite-dimensional. Let M be a differentiable manifold whose $V = \mathbb{R}^n$. If (U, α) is a chart of M, then we define the function x_α^i on U by setting

$$x_\alpha^i(x) = i\text{th coordinate of } \alpha(x). \tag{8.1}$$

If f is any differentiable function on U, then we can write Eq. (2.1) as

$$f(x) = f_\alpha\big(x_\alpha^1(x), \ldots, x_\alpha^n(x)\big),$$

which we shall write as

$$f = f_\alpha(x_\alpha^1, \ldots, x_\alpha^n). \tag{8.2}$$

We define the vector field $\partial/\partial x_\alpha^i$ on U by

$$\left(\frac{\partial}{\partial x_\alpha^i}\right)_\alpha (v) = \delta_i(= \langle 0, \ldots, \underset{i\text{th position}}{1}, \ldots, 0 \rangle). \tag{8.3}$$

If X is any vector field on U, then we have

$$X = X_\alpha^1 \frac{\partial}{\partial x_\alpha^1} + \cdots + X_\alpha^n \frac{\partial}{\partial x_\alpha^n}, \tag{8.4}$$

where the functions X_α^i are defined by

$$(X)_\alpha(\alpha(x)) = \langle X_\alpha^1(x), \ldots, X_\alpha^n(x) \rangle. \tag{8.5}$$

Equation (8.4) allows us to regard the vector field X as a "differential operator". In fact, it follows from the definitions that

$$D_X f = X_\alpha^1 \frac{\partial f_\alpha}{\partial x_\alpha^1} + \cdots + X_\alpha^n \frac{\partial f_\alpha}{\partial x_\alpha^n}. \tag{8.6}$$

Since x_α^i is a differentiable function on U, dx_α^i is a differential form on U and

$$(dx_\alpha^i)_\alpha(v) = \delta_i \qquad \text{for all} \quad v \in U. \tag{8.7}$$

In particular,

$$\left\langle \frac{\partial}{\partial x_\alpha^i}, dx_\alpha^j \right\rangle = \delta_i^j. \tag{8.8}$$

If ω is a differential form on U, then

$$\omega = a_{1\alpha}\, dx_\alpha^1 + \cdots + a_{n\alpha}\, dx_\alpha^n, \tag{8.9}$$

where the functions $a_{i\alpha}$ are defined by

$$\omega_\alpha(\alpha(x)) = \langle a_{1\alpha}(x), \ldots, a_{n\alpha}(x) \rangle \in \mathbb{R}^{n*}. \tag{8.10}$$

It then follows from the definitions that

$$df = \frac{\partial f_\alpha}{\partial x_\alpha^i}\, dx_\alpha^1 + \cdots + \frac{\partial f_\alpha}{\partial x_\alpha^n}\, dx_\alpha^n. \tag{8.11}$$

Equation (8.11) has built into it the transition law for differential forms under a change of charts. In fact, if (W, β) is a second chart, then on $U \cap W$ we have, by (8.11),

$$dx_\beta^i = \frac{\partial x_\beta^i}{\partial x_\alpha^1}\, dx_\alpha^1 + \cdots + \frac{\partial x_\beta^i}{\partial x_\alpha^n}\, dx_\alpha^n. \tag{8.12}$$

If we write $\omega = a_{1\beta}\,dx^1_\beta + \cdots + a_{n\beta}\,dx^n_\beta$ and substitute (8.12), we get

$$a_{j\alpha} = \sum \frac{\partial x^i_\beta}{\partial x^j_\alpha} a_{i\beta}.$$

Now

$$\left[\frac{\partial x^i_\beta}{\partial x^j_\alpha}\right]$$

is the matrix $J_{\beta\circ\alpha^{-1}}$. If we compare with (8.10), we see that we have recovered (7.4).

Since the subscripts α, β, etc., clutter up the formulas, we shall frequently use the following notational conventions: Instead of writing x^i_α we shall write x^i, and instead of writing x^j_β we shall write y^j. Thus

$$x^i = x^i_\alpha, \qquad y^j = x^j_\beta, \qquad z^k = x^k_\gamma, \quad \text{etc.}$$

Similarly, we shall write X^i for X^i_α, Y^i for X^i_β, a_i for $a_{i\alpha}$, b_i for $a_{i\beta}$, and so on.

Then Eqs. (8.1) through (8.12) can be written as

$$x^i(x) = i\text{th coordinate of } \alpha(x), \tag{8.1$'$}$$

$$f = f_\alpha(x^1, \ldots, x^n), \tag{8.2$'$}$$

$$\left(\frac{\partial}{\partial x^i}\right)_\alpha (v) = \delta_i, \tag{8.3$'$}$$

$$X = X^1 \frac{\partial}{\partial x^1} + \cdots + X^n \frac{\partial}{\partial x^n}, \tag{8.4$'$}$$

$$(X)_\alpha(\alpha(x)) = \langle X^1(x), \ldots, X^n(x)\rangle, \tag{8.5$'$}$$

$$D_X f = X^1 \frac{\partial f_\alpha}{\partial x^1} + \cdots + X^n \frac{\partial f_\alpha}{\partial x^n}, \tag{8.6$'$}$$

$$(dx^i)_\alpha(v) = \delta_i, \tag{8.7$'$}$$

$$\left\langle \frac{\partial}{\partial x^i}, dx^j \right\rangle = \delta^j_i, \tag{8.8$'$}$$

$$\omega = a_1\,dx^1 + \cdots + a_n\,dx^n, \tag{8.9$'$}$$

$$\omega_\alpha(\alpha(x)) = \langle a_1(x), \ldots, a_n(x)\rangle, \tag{8.10$'$}$$

$$df = \frac{\partial f_\alpha}{\partial x^1} dx^1 + \cdots + \frac{\partial f_\alpha}{\partial x^j} dx^j, \tag{8.11$'$}$$

$$dy^i = \frac{\partial y^i}{\partial x^1} dx^1 + \cdots + \frac{\partial y^i}{\partial x^n} dx^n. \tag{8.12$'$}$$

The formulas for "pullback" also take a simple form. Let $\psi: M_1 \to M_2$ be a differentiable map, and suppose that M_1 is m-dimensional and M_2 is n-dimensional. Let (U, α) and (W, β) be compatible charts. Then the map

$$\beta \circ \psi \circ \alpha^{-1}: \alpha(U) \to \beta(W)$$

is given by

$$y^i(\psi(x)) = y^i(x^1, \ldots, x^m), \qquad i = 1, \ldots, n, \tag{8.13}$$

that is, by n functions of m real variables. If f is a function on M_2 with

$$f = f_\beta(y^1, \ldots, y^n) \qquad \text{on} \quad W,$$

then

$$\psi^*[f] = f_\alpha(x^1, \ldots, x^m) \qquad \text{on} \quad U,$$

where

$$f_\alpha(x^1, \ldots, x^m) = f_\beta(y^1(x^1, \ldots, x^m), \ldots, y^n(x^1, \ldots, x^m)). \tag{8.14}$$

The rule for "pulling back" a differential form is also very easy. In fact, if

$$\omega = a_1\, dy^1 + \cdots + a_n\, dy^n \qquad \text{on} \quad W,$$

then $\psi^*\omega$ has the same form on U, where we now regard the a's and y's as functions of the x's and expand by using (8.12). Thus

$$\psi^*\omega = \sum a_i \frac{\partial y^i}{\partial x^j}\, dx^j,$$

where $a_i = a_i(y^1(x^1, \ldots, x^m), \ldots, y^n(x^1, \ldots, x^m))$.

Let $x \in U$. Then

$$\psi_*\left(\frac{\partial}{\partial x^i}\right)(x) = \sum_j \frac{\partial y^j}{\partial x^i}(x) \frac{\partial}{\partial y^j}(\psi(x)) \tag{8.15}$$

gives the formula for ψ_{*x}.

EXERCISES

8.1 Let x and y be rectangular coordinates on $\mathbb{E}^2$, and let (r, θ) be polar "coordinates" on $\mathbb{E}^2 - \{0\}$. Express the vector fields $\partial/\partial r$ and $\partial/\partial\theta$ in terms of rectangular coordinates. Express $\partial/\partial x$ and $\partial/\partial y$ in terms of polar coordinates.

8.2 Let x, y, z be rectangular coordinates on $\mathbb{E}^3$. Let

$$X = y\frac{\partial}{\partial z} - z\frac{\partial}{\partial y}, \qquad Y = z\frac{\partial}{\partial x} - x\frac{\partial}{\partial z}, \qquad \text{and} \qquad Z = x\frac{\partial}{\partial y} - y\frac{\partial}{\partial x}.$$

Compute $[X, Y]$, $[X, Z]$, and $[Y, Z]$. Note that X represents the infinitesimal generator of the one-parameter group of rotations about the x-axis. We sometimes call X the "infinitesimal rotation about the x-axis". We can do the same for Y and Z.

8.3 Let

$$A = y\frac{\partial}{\partial z} + z\frac{\partial}{\partial y}, \qquad B = x\frac{\partial}{\partial z} + z\frac{\partial}{\partial x}, \qquad \text{and} \qquad C = x\frac{\partial}{\partial y} - y\frac{\partial}{\partial x}.$$

Compute $[A, B]$, $[A, C]$, and $[B, C]$. Show that $Af = Bf = Cf = 0$ if $f(x, y, z) = x^2 + y^2 - z^2$. Sketch the integral curves of each of the vector fields A, B, and C.

8.4 Let

$$D = u\frac{\partial}{\partial v} + v\frac{\partial}{\partial u}, \qquad E = u\frac{\partial}{\partial v} - v\frac{\partial}{\partial u}, \qquad \text{and} \qquad F = u\frac{\partial}{\partial u} - v\frac{\partial}{\partial v}.$$

Compute $[D, E]$, $[D, F]$, and $[E, F]$.

8.5 Let $P_1, \ldots, P_n$ be polynomials in $x^1, \ldots, x^n$ with no constant term, that is,

$$P_1(0, \ldots, 0) = 0.$$

Let

$$I = x^1\frac{\partial}{\partial x^1} + \cdots + x^n\frac{\partial}{\partial x^n}$$

and

$$X = P_1\frac{\partial}{\partial x^1} + \cdots + P_n\frac{\partial}{\partial x^n}.$$

Show that

$$[I, X] = 0$$

if and only if the P_i's are linear. [*Hint:* Consider the expansion of the P_i's into homogeneous terms.]

8.6 Let X and the P_i's be as in Exercise 8.5, and suppose that the P_i's are linear. Let

$$A = \lambda_1 x^1\frac{\partial}{\partial x^1} + \cdots + \lambda_n x^n\frac{\partial}{\partial x^n},$$

and suppose that

$$\lambda_i \neq \lambda_j \qquad \text{for} \quad i \neq j.$$

Show that $[A, X] = 0$ if and only if $P_i = \mu_i x^i$, that is,

$$X = \mu_1 x^1\frac{\partial}{\partial x^1} + \cdots + \mu_n x^n\frac{\partial}{\partial x^n} \qquad \text{for some} \quad \mu^1, \ldots, \mu^n.$$

8.7 Let A be as in Exercise 8.6, and suppose, in addition, that

$$\lambda_i \neq \lambda_j + \lambda_r \qquad \text{for any} \quad i, j, r.$$

Show that if the P_i's are at most quadratic, then

$$[A, X] = 0$$

if and only if $P_i = \mu_i x^i$. Generalize this result to the case where P_i can be a polynomial of degree at most m.

9. RIEMANN METRICS

Let M be a finite-dimensional differentiable manifold. A Riemann metric, m, on M is a rule which assigns a positive definite scalar product $(\ ,\ )_{\mathsf{m},x}$ to the vector space $T_x(M)$ for each $x \in M$. We shall usually drop the subscripts m and x when they are understood from the context. Thus if m is a Riemann metric on M, $x \in M$, and $\xi, \eta \in T_x(M)$, we shall write the scalar product of ξ and η as

$$(\xi, \eta) = (\xi, \eta)_{\mathsf{m},x}.$$

Let (U, α) be a chart of M. Define the functions g_{ij} on U by setting

$$g_{ij}(x) = \left(\frac{\partial}{\partial x^i}(x), \frac{\partial}{\partial x^j}(x)\right), \tag{9.1}$$

so that $g_{ij} = g_{ji}$. If $\xi, \eta \in T_x(M)$ with

$$\xi = \sum \xi^i \frac{\partial}{\partial x^i}(x) \qquad \text{and} \qquad \eta = \sum \eta^i \frac{\partial}{\partial x^i}(x),$$

then

$$(\xi, \eta) = \sum_{i,j} g_{ij}(x)\xi^i\eta^j.$$

Since $dx^1(x), \ldots, dx^n(x)$ is the basis of $T_x^*(M)$ dual to the basis

$$\frac{\partial}{\partial x^1}(x), \ldots, \frac{\partial}{\partial x^n}(x),$$

we have

$$\xi^i = \langle \xi, dx^i(x) \rangle, \qquad \eta^j = \langle \eta, dx^j(x) \rangle,$$

so that the last equation can be written as

$$(\xi, \eta)_{\mathsf{m},x} = \sum g_{ij}(x)\langle \xi, dx^i \rangle \langle \eta, dx^j \rangle. \tag{9.2}$$

Equation (9.2) is usually written more succinctly as

$$\mathsf{m} \restriction U = \sum g_{ij}(x)\, dx^i\, dx^j. \tag{9.3}$$

[Here (9.3) is to be interpreted as a short way of writing (9.2).]

Let (W, β) be a second chart with

$$h_{kl}(x) = \left(\frac{\partial}{\partial y^k}(x), \frac{\partial}{\partial y^l}(x)\right), \qquad x \in W,$$

that is,

$$\mathsf{m} \restriction W = \sum h_{kl}\, dy^k\, dy^l. \tag{9.4}$$

Then for $x \in U \cap W$, we have

$$\frac{\partial}{\partial x^i}(x) = \sum \frac{\partial y^k}{\partial x^i}(x) \frac{\partial}{\partial y^k}(x), \qquad \frac{\partial}{\partial x^j}(x) = \sum \frac{\partial y^l}{\partial x^j}(x) \frac{\partial}{\partial y^l},$$

so

$$g_{ij}(x) = \left(\frac{\partial}{\partial x^i}(x), \frac{\partial}{\partial x^j}(x)\right) = \sum_{k,l} \frac{\partial y^k}{\partial x^i}(x) \frac{\partial y^l}{\partial x^j}(x) h_{kl}(x),$$

that is,

$$g_{ij} = \sum_{k,l} h_{kl} \frac{\partial y^k}{\partial x^i} \frac{\partial y^l}{\partial x^j}. \tag{9.5}$$

Note that (9.5) is the answer we would get if we formally substituted (8.12) for the dy's in (9.4) and collected the coefficients of $dx^i\, dx^j$.

In any event, it is clear from (9.5) that if the h_{ij} are all smooth functions on W, then the g_{ij} are smooth on $U \cap W$. In view of this we shall say that a

Riemann metric is smooth if the functions g_{ij} given by (9.3) are smooth for any chart (U, α) belonging to an atlas $\mathcal{A}$ of M. Also, if we are given functions $g_{ij} = g_{ji}$ defined for each $(U, \alpha) \in \mathcal{A}$ such that

i) $\sum g_{ij}(x)\xi^i\xi^j > 0$ unless $\xi = 0$ for all $x \in U$,

ii) the transition law (9.5) holds,

then the g_{ij} define a Riemann metric on M. In the following discussion we shall assume that our Riemann metrics are smooth.

Let $\psi: M_1 \to M_2$ be a differentiable map, and let m be a Riemann metric on M_2. For any $x \in M_1$ define $(\ ,\ )_{\psi^*(\mathsf{m}),x}$ on $T_x(M_1)$ by

$$(\xi, \eta)_{\psi^*(\mathsf{m}),x} = (\psi_*(\xi), \psi_*(\eta))_{\mathsf{m},\psi(x)}. \tag{9.6}$$

Note that this defines a symmetric bilinear function of ξ and η. It is not necessarily positive definite, however, since it is conceivable that $\psi_*(\xi) = 0$ with $\xi \neq 0$. Thus, in general, (9.6) does not define a Riemann metric on M_1. For certain ψ it does.

A differentiable map $\psi: M_1 \to M_2$ is called an *immersion* if ψ_{*x} is an injection (i.e., is one-to-one) for all $x \in M_1$.

If $\psi: M_1 \to M_2$ is an immersion and m is a Riemann metric on M_2, then we define the Riemann metric $\psi^*(\mathsf{m})$ on M_1 by (9.6).

Let (U, α) and (W, β) be compatible charts of M_1 and M_2, and let

$$\mathsf{m} \restriction W = \sum h_{kl}\, dy^k\, dy^l.$$

Then

$$\psi^*(\mathsf{m}) \restriction U = \sum g_{ij}\, dx^i\, dx^j,$$

where

$$\begin{aligned} g_{ij}(x) &= \left(\frac{\partial}{\partial x^i}(x), \frac{\partial}{\partial x^j}(x)\right)_{\psi^*(\mathsf{m}),x} \\ &= \left(\psi_*\left(\frac{\partial}{\partial x^i}(x)\right), \psi_*\left(\frac{\partial}{\partial x^j}(x)\right)\right)_{\mathsf{m},\psi(x)} \\ &= \left(\sum \frac{\partial y^k}{\partial x^i}\frac{\partial}{\partial y^k}(\psi(x)), \sum \frac{\partial y^l}{\partial x^j}\frac{\partial}{\partial y^l}(\psi(x))\right)_{\mathsf{m},\psi(x)} \\ &= \sum_{k,l} h_{kl}\frac{\partial y^k}{\partial x^j}\frac{\partial y^l}{\partial x^j}, \end{aligned}$$

which is just (9.5) again (with a different interpretation). Or, more succinctly,

$$\psi^*(\mathsf{m}) \restriction U = \sum \psi^*(h_{kl})\psi^*(dy^k)\psi^*(dy^l).$$

Let us give some examples of these formulas. If $M = \mathbb{R}^n$, then the identity chart induces a Riemann metric on $\mathbb{R}^n$ given by

$$(dx^1)^2 + \cdots + (dx^n)^2.$$

Let us see what this looks like in terms of polar coordinates in $\mathbb{R}^2$ and $\mathbb{R}^3$.

In $\mathbb{R}^2$ if we write

$$x^1 = r\cos\theta, \qquad x^2 = r\sin\theta,$$

then

$$\begin{aligned} dx^1 &= \cos\theta\, dr - r\sin\theta\, d\theta, \\ dx^2 &= \sin\theta\, dr + r\cos\theta\, d\theta, \end{aligned}$$

so

$$(dx^1)^2 + (dx^2)^2 = dr^2 + r^2\, d\theta^2. \tag{9.7}$$

Note that (9.7) holds whenever the forms dr and $d\theta$ are defined, i.e., on all of $\mathbb{R}^2 - \{0\}$. (Even though the function θ is not well defined on all of $\mathbb{R}^2 - \{0\}$, the form $d\theta$ is. In fact, we can write

$$d\theta = \frac{x^1\, dx^2 - x^2\, dx^1}{(x^1)^2 + (x^2)^2}\cdot) \tag{9.8}$$

In $\mathbb{R}^3$ we introduce

$$\begin{aligned} x^1 &= r\cos\varphi\sin\theta, \\ x^2 &= r\sin\varphi\sin\theta, \\ x^3 &= r\cos\theta. \end{aligned}$$

Then

$$\begin{aligned} dx^1 &= \cos\varphi\sin\theta\, dr - r\sin\varphi\sin\theta\, d\varphi + r\cos\varphi\cos\theta\, d\theta, \\ dx^2 &= \sin\varphi\sin\theta\, dr + r\cos\varphi\sin\theta\, d\varphi + r\sin\varphi\cos\theta\, d\theta, \\ dx^3 &= \cos\theta\, dr - r\sin\theta\, d\theta. \end{aligned}$$

Thus

$$(dx^1)^2 + (dx^2)^2 + (dx^3)^2 = dr^2 + r^2\sin^2\theta\, d\varphi^2 + r^2\, d\theta^2. \tag{9.9}$$

Again, (9.9) is valid wherever the forms on the right are defined, which this time means when $(x^1)^2 + (x^2)^2 \neq 0$.

Let us consider the map ι of the unit sphere $S^2 \to \mathbb{R}^3$, which consists of regarding a point of S^2 as a point of $\mathbb{R}^3$. We then get an induced Riemann metric on S^2.

Let us set

$$\overline{d\theta} = \iota^*\, d\theta \qquad \text{and} \qquad \overline{d\varphi} = \iota^*\, d\varphi,$$

so the forms $\overline{d\theta}$ and $\overline{d\varphi}$ are defined on $U = S^2 - \{\langle 0, 0, 1\rangle, \langle 0, 0, -1\rangle\}$. Then on U we can write (since $r = 1$ on S^2)

$$\iota^*\big((dx^1)^2 + (dx^2)^2 + (dx^3)^2\big) = \overline{d\theta}^2 + \sin^2\theta\, \overline{d\varphi}^2. \tag{9.10}$$

We now return to general considerations. Let M be a differentiable manifold and let $C\colon I \to M$ be a differentiable map, where I is an interval in $\mathbb{R}^1$. Let t denote the coordinate of the identity chart on I. We shall set

$$C'(s) = C_*\left(\frac{\partial}{\partial t}\right)(s), \qquad s \in I,$$

so that $C'(s) \in T_{C(s)}(M)$ is the tangent vector to the curve C at s. If (U, α) is a chart on M and $x^1, \ldots, x^n$ are the coordinate functions of (U, α), then if

$C(I') \subset U$ for some $I' \subset I$,

$$\alpha \circ C = \langle x^1 \circ C, \ldots, x^n \circ C \rangle,$$

so that

$$C'(t)_\alpha = \left\langle \frac{dx^1 \circ C}{dt}, \ldots, \frac{dx^n \circ C}{dt} \right\rangle (t).$$

When there is no possibility of confusion, we shall omit the $\circ\, C$ and simply write

$$\alpha \circ C(t) = \langle x^1(t), \ldots, x^n(t) \rangle \qquad \text{and} \qquad C'(t)_\alpha = \langle x^{1\prime}(t), \ldots, x^{n\prime}(t) \rangle.$$

Now let $\mathbf{m}$ be a Riemann metric on M. Then $\|C'(t)\| = (C'(t), C'(t))^{1/2}$ is a continuous function. In fact, in terms of a chart, we can write

$$\|C'(t)\| = \sqrt{\sum g_{ij}(C(t)) x^{i\prime}(t) x^{j\prime}(t)}\ .$$

The integral

$$\int_I \|C'(t)\|\, dt \tag{9.11}$$

is called the length of the curve C. It will be defined if $\|C'(t)\|$ is integrable over I. This will certainly be the case if I and $\|C'(t)\|$ are both bounded, for instance. Note that the length is independent of the parametrization. More precisely, let $\varphi\colon J \to I$ be a one-to-one differentiable map, and let $C_1 = C \circ \varphi$. Then at any $\tau \in J$ we have

$$\begin{aligned} C_1' &= C_{1*}\left(\frac{\partial}{\partial \tau}\right) = C_* \circ \varphi_* \left(\frac{\partial}{\partial \tau}\right) \\ &= \frac{dt}{d\tau} C_* \left(\frac{\partial}{\partial t}\right), \end{aligned}$$

that is,

$$C_1'(\tau) = \frac{dt}{d\tau} C'(t).$$

Thus

$$\|C_1'(\tau)\| = \|C'(\varphi(\tau))\| \left| \frac{d\varphi}{d\tau} \right| . \tag{9.12}$$

On the other hand, by the law for change of variable in an integral we have

$$\begin{aligned} \int_I \|C'(\cdot)\| &= \int_J \|C'(\varphi(\cdot))\| \left| \frac{d\varphi}{d\tau} \right| \\ &= \int_J \|C_1'(\cdot)\| \qquad \text{by } \ (9.12). \end{aligned}$$

More generally, we say that a curve C defined on an interval I is piecewise differentiable if

i) C is continuous;

ii) $I = I_1 \cup \cdots \cup I_r$ and C, on each I_j, is the restriction of a differentiable curve defined on some interval I_j' strictly containing I_j.

(Thus a piecewise differentiable curve is a curve with a finite number of "corners".) If C is piecewise differentiable, then $\|C'(t)\|$ is defined and continuous except at a finite number of t's, where it may have a jump discontinuity. In particular, the integral (9.11) will exist and the curve will have a length.

Exercise. Let C be a curve mapping I onto a straight line segment in $\mathbb{R}^n$ in a one-to-one manner. Show that the length of C is the same as the length of the segment.

Let $C: [0, 1] \to \mathbb{R}^2$ be a curve with $C(0) = 0$ and $C(1) = v \in \mathbb{R}^2$. If we use the expression (9.7), we see that

$$\begin{aligned}\int \|C'(t)\|\, dt &= \int \sqrt{(r'(t))^2 + (r(t)\theta'(t))^2}\, dt \\ &\geq \int |r'(t)|\, dt \\ &\geq \int_0^1 r'(t)\, dt \\ &= \|v\|,\end{aligned}$$

with equality holding if and only if $\theta' \equiv 0$ and $r' \geq 0$. Thus among all curves joining 0 to v, the straight line segment has shortest length.

Similarly, on the sphere, let C be any curve $C: [0, 1] \to S^2$ with $C(0) = (0, 0, 1)$ and $C(1) = p \neq (0, 0, -1)$, and let $\theta_1 = \theta(C(1))$. Then

$$\begin{aligned}\int \|C'(t)\|\, dt &= \int \sqrt{(\theta'(t))^2 + \sin^2\theta(\varphi'(t))^2}\, dt \\ &\geq \int_0^1 |\theta'(t)|\, dt.\end{aligned}$$

If we let t_1 denote the first point in [0, 1] where $\theta = \theta_1$, then

$$\int \|C'(t)\|\, dt \geq \int_0^1 |\theta'(t)|\, dt \geq \int_0^{t_1} |\theta'(t)|\, dt \geq \int_0^{t_1} \theta'(t)\, dt = \theta_1,$$

with equality only if $\varphi' \equiv 0$ and $t_1 = 1$. Thus the shortest curve joining any two points on S^2 is the great circle joining them.

In both examples above we were aided by a very fortuitous choice of coordinates (polar coordinates in the plane and a kind of polar coordinates on the sphere). We shall see in Section 11, Chapter 13, that this is not accidental. We shall see that on *any* Riemann manifold one can introduce local coordinates in terms of which it is easy to describe the curves that locally minimize length.

CHAPTER 10

THE INTEGRAL CALCULUS ON MANIFOLDS

In this chapter we shall study integration on manifolds. In order to develop the integral calculus, we shall have to restrict the class of manifolds under consideration. In this chapter we shall assume that all manifolds M that arise satisfy the following two conditions:

1) M is finite-dimensional.
2) M possesses an atlas $\mathcal{A}$ containing (at most) a countable number of charts; that is, $\mathcal{A} = \{(U_i, \alpha_i)\}_{i=1,2,\ldots}$.

Before getting down to the business of integration, there are several technical facts to be established. The first two sections will be devoted to this task.

1. COMPACTNESS

A subset A of a manifold M is said to be *compact* if it has the following property:

i) If $\{U_\iota\}$ is any collection of open sets with

$$A \subset \bigcup_\iota U_\iota,$$

there exist finitely many of the U_ι, say $U_{\iota_1}, \ldots, U_{\iota_r}$, such that

$$A \subset U_{\iota_1} \cup \cdots \cup U_{\iota_r}.$$

Alternatively, we can say:

ii) A set A is compact if and only if for any family $\{F_\iota\}$ of closed sets such that

$$A \cap \bigcap_\iota F_\iota = \varnothing,$$

there exist finitely many of the F_ι such that

$$A \cap F_{\iota_1} \cap \cdots \cap F_{\iota_r} = \varnothing.$$

The equivalence of (i) and (ii) can be seen by taking U_ι equal to the complement of F_ι.

In Section 5 of Chapter 4 we established that if $M = U$ is an open subset of $\mathbb{R}^n$, then *$A \subset U$ is compact if and only if A is a closed bounded subset of $\mathbb{R}^n$.*

We make some further trivial remarks about compactness:

iii) If $A_1, \ldots, A_r$ are compact, so is $A_1 \cup \cdots \cup A_r$.

In fact, if $\{U_\iota\}$ covers $A_1 \cup \cdots \cup A_r$, it certainly covers each A_j. We can thus choose for each j a finite subcollection which covers A_j. The union of these subcollections forms a finite subcollection covering $A_1 \cup \cdots \cup A_r$.

iv) If $\psi: M_1 \to M_2$ is continuous and $A \subset M_1$ is compact, then $\psi[A]$ is compact.

In fact, if $\{U_\iota\}$ covers $\psi[A]$, then $\{\psi^{-1}(U_\iota)\}$ covers A. If the U_ι are open, so are the $\psi^{-1}(U_\iota)$, since ψ is continuous. We can thus choose $\iota_1, \ldots, \iota_r$ so that

$$A \subset \psi^{-1}(U_{\iota_1}) \cup \cdots \cup \psi^{-1}(U_{\iota_r}),$$

which implies that $\psi[A] \subset U_{\iota_1} \cup \cdots \cup U_{\iota_r}$.

We see from this that if $A = A_1 \cup \cdots \cup A_r$, where each A_j is contained in some W_i, where (W_i, β_i) is a chart, and $\beta_i(A_j)$ is a compact subset of $\mathbb{R}^n$, then A is compact. In particular, the manifold M itself may be compact. For instance, we can write S^n as the union of the upper and lower hemispheres: $S^n = \{x : x^{n+1} \geq 0\} \cup \{x : x^{n+1} \leq 0\}$. Each hemisphere is compact. In fact, the upper hemisphere is mapped onto $\{y : \|y\| \leq 1\}$ by the map φ_1 of Section 8.1, and the lower hemisphere is mapped onto the same set by φ_2. Thus the sphere is compact.

On the other hand, an open subset of $\mathbb{R}^n$ is not compact. However, it can be written as a *countable* union of compact sets. In fact, if $U \subset \mathbb{R}^n$ is an open set, let

$$A_n = \{x \in U : \|x\| \leq n \text{ and } \rho(x, \partial U) \geq 1/n\}.$$

It is easy to check that A_n is compact and that

$$\bigcup A_n = U.$$

In view of condition (2), we can say the same for any manifold M under consideration:

Proposition 1.1. Any manifold M satisfying (1) and (2) can be written as

$$M = \bigcup_{i=1}^{\infty} A_i,$$

where each $A_i \subset M$ is compact.

Proof. In fact, by (2)

$$M = \bigcup_{j=1}^{\infty} U_j,$$

and by the preceding discussion each U_j can be written as the countable union of compact sets. Since the countable union of a countable union is still countable, we obtain Proposition 1.1. □

An immediate corollary is:

Proposition 1.2. Let M be a manifold [satisfying (1) and (2)], and let $\{U_\iota\}$ be an open covering of M. Then we can select a countable subcollection $\{U_j\}$ such that

$$\bigcup U_j = M.$$

Proof. Write $M = \bigcup A_r$, where A_r is compact. For each r we can choose finitely many $U_{r,1}, U_{r,2}, \ldots, U_{r,k_r}$ so that

$$A_r \subset U_{r,1} \cup \cdots \cup U_{r,k_r}.$$

The collection

$$\{U_{r,k}\}_{\substack{r=1,\ldots,\infty \\ k=1,\ldots,k_r}}$$

is a countable subcollection covering M. □

2. PARTITIONS OF UNITY

In the following discussion it will be convenient for us to have a method of "breaking up" functions, vector fields, etc., into "little pieces". For this purpose we introduce the following notation:

Definition 2.1. A collection $\{g_j\}$ of C^∞-functions is said to be a *partition of unity* if

i) $g_i \geq 0$ for all i;

ii) supp g_i† is compact for all i;

iii) each $x \in M$ has a neighborhood V_x such that $V_x \cap \operatorname{supp} g_i = \emptyset$ for all but a finite number of i; and

iv) $\sum g_i(x) = 1$ for all $x \in M$.

Note that in view of (iii) the sum occurring in (iv) is actually finite, since for any x all but a finite number of the $g_i(x)$ vanish. Note also that:

Proposition 2.1. If A is a compact set and $\{g_j\}$ is a partition of unity, then

$$A \cap \operatorname{supp} g_i = \emptyset$$

for all but a finite number of i.

Proof. In fact, each $x \in A$ has a neighborhood V_x given by (iii). The sets $\{V_x\}_{x\in A}$ form an open covering of A. Since A is compact, we can select a finite subcollection $\{V_1, \ldots, V_r\}$ with $A \subset V_1 \cup \cdots \cup V_r$. Since each V_k has a nonempty intersection with only finitely many of the supp g_i, so does their union, and so *a fortiori* does A. □

---

† Recall that supp g is the closure of the set $\{x : g(x) \neq 0\}$.

Definition 2.2. Let $\{U_\iota\}$ be an open covering of M, and let $\{g_j\}$ be a partition of unity. We say that $\{g_j\}$ is *subordinate* to $\{U_\iota\}$ if for every j there exists an $\iota(j)$ such that

$$\operatorname{supp} g_j \subset U_{\iota(j)}. \tag{2.1}$$

Theorem 2.1. Let $\{U_\iota\}$ be any open covering of M. There exists a partition of unity $\{g_j\}$ subordinate to $\{U_\iota\}$.

The proof that we shall present below is due to Bonic and Frampton.† First we introduce some preliminary notions.

The function f on $\mathbb{R}$ defined by

$$f(u) = \begin{cases} e^{-1/u} & \text{if } u > 0, \\ 0 & \text{if } u \le 0 \end{cases}$$

is C^∞. For $u \neq 0$ it is clear that f has derivatives of all orders. To check that f is C^∞ at 0, it suffices to show that $f^{(k)}(u) \to 0$ as $u \to 0$ from the right. But $f^{(k)}(u) = P_k(1/u)e^{-1/u}$, where P_k is a polynomial of degree k. So

$$\lim_{u \to 0} f^{(k)}(u) = \lim_{s \to \infty} P_k(s)e^{-s} = 0,$$

since e^s goes to infinity faster than any polynomial.

Note that $f(u) > 0$ if and only if $u > 0$. Now consider the function g_a^b on $\mathbb{R}$ defined by

$$g_a^b(x) = f(x-a)f(b-x).$$

Then g_a^b is C^∞ and nonnegative and

$$g_a^b(x) \neq 0 \qquad \text{if and only if} \qquad a < x < b.$$

More generally, if $\mathbf{a} = \langle a^1, \ldots, a^k \rangle$ and $\mathbf{b} = \langle b^1, \ldots, b^k \rangle$, define the function $g_{\mathbf{a}}^{\mathbf{b}}$ on $\mathbb{R}^k$ by setting

$$g_{\mathbf{a}}^{\mathbf{b}}(x) = g_{a^1}^{b^1}(x)g_{a^2}^{b^2}(x^2)\cdots g_{a^k}^{b^k}(x^k),$$

where $x = \langle x^1, \ldots, x^k \rangle$. Then $g_{\mathbf{a}}^{\mathbf{b}} \geq 0$, $g_{\mathbf{a}}^{\mathbf{b}} \in C^\infty$, and

$$g_{\mathbf{a}}^{\mathbf{b}}(x) > 0 \qquad \text{if and only if} \qquad a^1 < x^1 < b^1, \ldots, a^k < x^k < b^k. \tag{2.2}$$

Lemma. Let $f_1, \ldots, f_k$ be C^∞-functions on a manifold M, and let $W = \{x : a^1 < f_1(x) < b^1, \ldots, a^k < f_k(x) < b^k\}$. There exists a nonnegative C^∞-function g such that $W = \{x : g(x) > 0\}$.

In fact, if we define g by

$$g(x) = g_{\mathbf{a}}^{\mathbf{b}}(f_1(x), \ldots, f_k(x)),$$

then it is clear that g has the desired properties.

† Smooth functions on Banach manifolds, *J. Math and Mech.* **15,** 877–898 (1966).

We now turn to the proof of Theorem 2.1.

Proof. For each $x \in M$ choose a U_ι containing x and a chart (U, α) about x. Then $\alpha(U \cap U_\iota)$ is an open set containing $\alpha(x)$ in $\mathbb{R}^n$. Choose $\mathbf{a}$ and $\mathbf{b}$ such that

$$\alpha(x) \in \operatorname{int} \square_{\mathbf{a}}^{\mathbf{b}} \quad \text{and} \quad \overline{\square_{\mathbf{a}}^{\mathbf{b}}} \subset \alpha(U \cap U_\iota).$$

Let $W_x = \alpha^{-1}(\operatorname{int} \square_{\mathbf{a}}^{\mathbf{b}})$. Then

$$\overline{W_x} \subset U_\iota \quad \text{and} \quad \overline{W_x} \text{ is compact.} \tag{2.3}$$

Also if $x^1, \ldots, x^n$ are the coordinates given by α,

$$W_x = \{y : a^1 < x^1(y) < b^1, \ldots, a^n < x^n(y) < b^n\}.$$

By our lemma we can find a nonnegative C^∞-function f_x such that

$$W_x = \{y : f_x(y) > 0\}.$$

Since $x \in W_x$, the $\{W_x\}$ cover M. By Proposition 1.2 we can select a countable subcovering $\{W_i\}$. Let us denote the corresponding functions by f_i; that is, if $W_i = W_x$, we set $f_i = f_x$.

Let

$$\begin{aligned} V_1 &= W_1 = \{x : f_1(x) > 0\}, \\ V_2 &= \{x : f_2(x) > 0, f_1(x) < \tfrac{1}{2}\}, \\ &\vdots \\ V_r &= \{x : f_r(x) > 0, f_1(x) < 1/r, \ldots, f_{r-1}(x) < 1/r\}. \end{aligned}$$

It is clear that V_j is open and that $V_j \subset W_j$, so that, by (2.3),

$$\overline{V_j} \text{ is compact} \quad \text{and} \quad \overline{V_j} \subset U_\iota \tag{2.4}$$

for some $\iota = \iota(j)$.

For each $x \in M$ let $q(x)$ denote the first integer q for which $f_q(x) > 0$. Thus $f_p(x) = 0$ if $p < q(x)$ and $f_{q(x)}(x) > 0$.

Let $V_x = \{y : f_{q(x)}(y) > \frac{1}{2}f_{q(x)}(x)\}$. Since $f_{q(x)}(x) > 0$, it follows that $x \in V_x$ and V_x is open. Furthermore,

$$V_x \cap \overline{V_r} = \emptyset \quad \text{if} \quad r > q(x) \quad \text{and} \quad 1/r < \tfrac{1}{2}f_{q(x)}(x). \tag{2.5}$$

According to the lemma, each set V_i can be given as $V_i = \{x : \bar{g}_i(x) > 0\}$, where $\bar{g}_i$ is a suitable C^∞-function. Let $g = \sum \bar{g}_i$. In view of (2.5) this is really a finite sum in the neighborhood of any x. Thus g is C^∞. Now $\bar{g}_{q(x)}(x) > 0$, since $x \in V_{q(x)}$. Thus $g > 0$. Set

$$g_j = \frac{\bar{g}_j}{g}.$$

We claim that $\{g_j\}$ is the desired partition of unity. In fact, (i) holds by our construction, (ii) and (2.1) follow from (2.4), (iii) follows from (2.5), and (iv) holds by construction. $\square$

3. DENSITIES

If we regard $\mathbb{R}^n$ as a differentiable manifold, then the law for change of variables for an integral shows that the integrand does not have the same transition law as that of a function under change of chart. For this reason we cannot expect to integrate functions on a manifold. We now introduce the type of object that we can integrate.

Definition 3.1. A *density* ρ is a rule which assigns to each chart (U, α) of M a function ρ_α defined on $\alpha(U)$ subject to the following transition law: If (W, β) is a second chart of M, then

$$\rho_\alpha(v) = \rho_\beta(\beta \circ \alpha^{-1}(v))|\det J_{\beta \circ \alpha^{-1}}(v)| \qquad \text{for} \quad v \in \alpha(U \cap W). \tag{3.1}$$

If $\mathcal{A}$ is an atlas of M and functions ρ_{α_i} are given for all $(U_i, \alpha_i) \in \mathcal{A}$ satisfying (3.1), then the ρ_{α_i} define a density ρ on M. In fact, if (U, α) is any chart of M (not necessarily belonging to $\mathcal{A}$), define ρ_α by

$$\rho_\alpha(v) = \rho_{\alpha_i}(\alpha_i \circ \alpha^{-1}(v))|\det J_{\alpha_i \circ \alpha^{-1}}(v)| \qquad \text{if} \quad v \in \alpha(U \cap U_i).$$

This definition is consistent: If $v \in \alpha(U \cap U_i) \cap \alpha(U \cap U_j)$, then by (3.1),

$$\begin{aligned} &\rho_{\alpha_j}(\alpha_j \circ \alpha^{-1}(v))|\det J_{\alpha_j \circ \alpha^{-1}}(v)| \\ &\quad = \rho_{\alpha_i}(\alpha_i \circ \alpha_j^{-1}(\alpha_j \circ \alpha^{-1}(v)))|\det J_{\alpha_i \circ \alpha_j^{-1}}(\alpha_j \circ \alpha^{-1}(v))||\det J_{\alpha_j \circ \alpha^{-1}}(v)| \\ &\quad = \rho_{\alpha_i}(\alpha_i \circ \alpha^{-1}(v))|\det J_{\alpha_i \circ \alpha^{-1}}(v)| \end{aligned}$$

by the chain rule and the multiplicative property of determinants.

In view of (3.1) it makes sense to talk about local smoothness properties of densities. We will say that a density ρ is C^k if for any chart (U, α) the function ρ_α is C^k. As usual, it suffices to verify this for all charts (U, α) belonging to some atlas. Similarly, we say that a density ρ is *locally absolutely integrable* if for any chart (U, α) the function ρ_α is absolutely integrable. By the last proposition of Chapter 8 this is again independent of the choice of atlases.

Let ρ be a density on M, and let x be a point of M. It does not make sense to talk about the value of ρ at x. However, (3.1) shows that it does make sense to talk about the sign of ρ at x, More precisely, we say that

$$\rho > 0 \text{ at } x \qquad \text{if} \quad \rho_\alpha(\alpha(x)) > 0 \tag{3.2}$$

for a chart (U, α) about x. Equation (3.1) shows that if $\rho_\alpha(\alpha(x)) > 0$, then $\rho_\beta(\beta(x)) > 0$ for any other chart (W, β) about x. Similarly, it makes sense to say that $\rho < 0$ at x, $\rho > 0$ at x, or $\rho \neq 0$ at x.

Definition 3.2. Let ρ be a density on M. By the *support* of ρ, denoted by supp ρ, we shall mean the closure of the set of points of M at which ρ does not vanish. That is,

$$\operatorname{supp} \rho = \overline{\{x : \rho \neq 0 \text{ at } x\}}.$$

Let ρ_1 and ρ_2 be densities. We define their sum by setting

$$(\rho_1 + \rho_2)_\alpha = \rho_{1\alpha} + \rho_{2\alpha} \tag{3.3}$$

for any chart (U, α). It is immediate that the right-hand side of (3.3) satisfies the transition law (3.1), and so defines density on M.

Let ρ be a density, and let f be a function. We define the density $f\rho$ by

$$(f\rho)_\alpha = f_\alpha \rho_\alpha. \tag{3.4}$$

Again, the verification of (3.1) is immediate in view of the transition laws for functions.

It is clear that

$$\operatorname{supp}(\rho_1 + \rho_2) \subset \operatorname{supp}\rho_1 \cup \operatorname{supp}\rho_2 \tag{3.5}$$

and

$$\operatorname{supp}(f\rho) = \operatorname{supp} f \cap \operatorname{supp}\rho. \tag{3.6}$$

We shall write

$$\rho_1 \le \rho_2 \text{ at } x \qquad \text{if} \quad \rho_2 - \rho_1 \ge 0 \text{ at } x$$

and

$$\rho_1 \le \rho_2 \qquad \text{if} \quad \rho_1 \le \rho_2 \text{ at all } x \in M.$$

Let P denote the space of locally absolutely integrable densities of compact support. We observe that P is a vector space and that the product $f\rho$ belongs to P if f is a (bounded) locally contented function and $\rho \in \mathrm{P}$.

Theorem 3.1. There exists a unique linear function $\int$ on P satisfying the following condition: If $\rho \in \mathrm{P}$ is such that $\operatorname{supp}\rho \subset U$, where (U, α) is a chart of M, then

$$\int \rho = \int_{\alpha(U)} \rho_\alpha. \tag{3.7}$$

Proof. We first show that there is at most one linear function satisfying (3.7). Let $\mathcal{A}$ be an atlas of M, and let $\{g_j\}$ be a partition of unity subordinate to $\mathcal{A}$. For each j choose an $i(j)$ so that

$$\operatorname{supp} g_j \subset U_{i(j)}.$$

Write $\rho = 1 \cdot \rho = \sum g_j \cdot \rho$. Since $\operatorname{supp}\rho$ is compact, only finitely many of the terms $g_j\rho$ are not identically zero. Thus the sum is finite. Since $\int$ is linear,

$$\int \rho = \int \sum g_j\rho = \sum \int g_j\rho.$$

By (3.7),

$$\int g_j\rho = \int_{\alpha_{i(j)}(U_{i(j)})} (g_j\rho)_{\alpha_{i(j)}}.$$

Thus

$$\int \rho = \sum_j \int_{\alpha_{i(j)}(U_{i(j)})} (g_j\rho)_{\alpha_{i(j)}}. \tag{3.8}$$

Thus $\int$, if it exists, must be given by (3.8). To establish the existence of $\int$, we must show that (3.8) defines a linear function on P satisfying (3.7). The linearity is obvious; we must verify (3.7).

Suppose supp $\rho \subset U$ for some chart (U, α). We must show that

$$\int_{\alpha(U)} \rho_\alpha = \sum_j \int_{\alpha_{i(j)}(U_{i(j)})} (g_j\rho)_{\alpha_{i(j)}}.$$

Since $\rho = \sum g_j\rho$ and therefore $\rho_\alpha = \sum (g_j\rho)_\alpha$, it suffices to show that

$$\int_{\alpha(U)} (g_j\rho)_\alpha = \int_{\alpha_i(U_i)} (g_j\rho)_{\alpha_i}, \tag{3.9}$$

where supp $g_j\rho \subset U \cap U_i$. By (3.1),

$$(g_j\rho)_\alpha = (g_j\rho)_{\alpha_i} \circ (\alpha_i \circ \alpha^{-1}) \cdot |\det J_{\alpha_i \circ \alpha^{-1}}|,$$

so that (3.9) holds by the transformation law for integrals in $\mathbb{R}^n$. □

We can derive a number of useful properties of the integral from the formula (3.8):

$$\text{if} \quad \rho_1 \leq \rho_2, \quad \text{then} \quad \int \rho_1 \leq \int \rho_2. \tag{3.10}$$

In fact, since $g_j \geq 0$, we have $(g_j\rho_1)_\alpha \leq (g_j\rho_2)_\alpha$ for any chart (U, α). Thus (3.10) follows from the corresponding fact on $\mathbb{R}^n$ if we use (3.8).

Let us say that a set A has content zero if $A \subset A_1 \cup \cdots \cup A_p$ where each A_i is compact, $A_i \subset U_i$ for some chart (U_i, α_i), and $\alpha_i(A_i)$ has content zero in $\mathbb{R}^n$. It is easy to see that the union of any finite number of sets of content zero has content zero. It is also clear that the function e_A is contented.

Let us call a set $B \subset M$ contented if the function e_B is contented. For any $\rho \in$ P we define $\int_B \rho$ by

$$\int_B \rho = \int e_B\rho. \tag{3.11}$$

It follows from (3.8) that

$$\int_A \rho = 0$$

for any $\rho \in$ P if A has content zero. We can thus ignore sets of content zero for the purpose of integration. In practice, one usually takes advantage of this when computing integrals, rather than using (3.8). For instance, in computing an integral over S^n, we can "ignore" any meridian: for example, if

$$A = \{x \in S^n : x = (t, 0, \ldots, \pm\sqrt{1 - t^2}) \in \mathbb{R}^{n+1}\},$$

then

$$\int_{S^n} \rho = \int_{S^n - A} \rho \qquad \text{for any } \rho.$$

This means that we can compute $\int_{S^n} \rho$ by introducing polar coordinates (Fig. 10.1) and expressing ρ in terms of them. Thus in S^2, if $U = S^2 - A$ and α is the polar coordinate chart on U, then

$$\int_{S^2} \rho = \int_0^{2\pi} \int_0^{\pi} \rho_\alpha \, d\theta \, d\varphi.$$

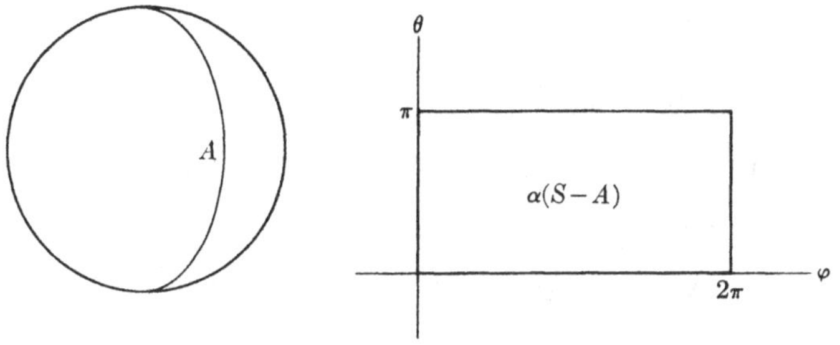

Fig. 10.1

It is worth observing that if N is a differentiable manifold of dimension less than dim M and ψ is a differentiable map of $N \to M$, then Proposition 7.3 of Chapter 8 implies that if A is any compact subset of N, then $\psi(A)$ has content zero in M. In this sense, one can ignore "lower-dimensional sets" when integrating on M.

4. VOLUME DENSITY OF A RIEMANN METRIC

Let M be a differentiable manifold with a Riemann metric $\mathfrak{a}$. We define the density σ [$=\sigma(\mathfrak{a})$] as follows. For each chart (U, α) with coordinates $x^1, \ldots, x^n$ let

$$\sigma_\alpha(\alpha(x)) = \left|\det\left[\left(\frac{\partial}{\partial x^i}(x), \frac{\partial}{\partial x^j}(x)\right)\right]\right|^{1/2} = |\det(g_{ij}(x))|^{1/2}. \tag{4.1}$$

Here

$$\left[\left(\frac{\partial}{\partial x^i}(x), \frac{\partial}{\partial x^j}(x)\right)\right]$$

is the matrix whose ijth entry is the scalar product of the vectors

$$\frac{\partial}{\partial x^i}(x) \qquad \text{and} \qquad \frac{\partial}{\partial x^j}(x),$$

so that (in view of Exercise 8.1 of Chapter 8)

$\sigma_\alpha(\alpha(x)) =$ volume of the parallelepiped spanned by $(\partial/\partial x^i)(x)$ with respect to the Euclidean metric $(\ ,\ )_{\mathfrak{a},x}$ on $T_x(M)$.

It is easy to see that (4.1) actually defines a density. Let (W, β) be a second chart about x with coordinates $y^1, \ldots, y^n$. Then

$$\frac{\partial}{\partial y^k} = \sum_i \frac{\partial x^i}{\partial y^k}\frac{\partial}{\partial x^i},$$

so that

$$\sigma_\beta(\beta(x)) = \left|\det\left[\left(\frac{\partial}{\partial y^k}(x), \frac{\partial}{\partial y^l}(x)\right)\right]\right|^{1/2}.$$

Now

$$\left(\frac{\partial}{\partial y^k}, \frac{\partial}{\partial y^l}\right) = \sum_{i,j} \frac{\partial x^i}{\partial y^k} \cdot \frac{\partial x^j}{\partial y^l} \left(\frac{\partial}{\partial x^i}, \frac{\partial}{\partial x^j}\right)$$

for all k and l. We can write this as the matrix equation

$$\left[\left(\frac{\partial}{\partial y^k}, \frac{\partial}{\partial y^l}\right)\right] = \left[\frac{\partial x^i}{\partial y^k}\right]\left[\left(\frac{\partial}{\partial x^i}, \frac{\partial}{\partial y^j}\right)\right]\left[\frac{\partial x^j}{\partial y^l}\right],$$

so that

$$\begin{aligned}\sigma_\beta(\beta(x)) &= \left|\det\left[\left(\frac{\partial}{\partial x^i}(x), \frac{\partial}{\partial x^j}(x)\right)\right]\det\left[\frac{\partial x^i}{\partial y^k}\right]\det\left[\frac{\partial x^j}{\partial y^l}\right]\right|^{1/2} \\ &= \left|\det\left[\left(\frac{\partial}{\partial x^i}(x), \frac{\partial}{\partial x^j}(x)\right)\right]\right|^{1/2}\left|\det\left[\frac{\partial x^i}{\partial y^k}\right]\right| \\ &= \sigma_\alpha(\alpha(x))\left|\det\left[\frac{\partial x^i}{\partial y^k}\right]\right|(x).\end{aligned}$$

If M is an open subset of Euclidean space with the Euclidean metric, then the volume density, when integrated over any contented set, yields the ordinary Euclidean volume of that set. In fact, if $x^1, \ldots, x^n$ are orthonormal coordinates corresponding to the identity chart, then $g_{ij}(x) = 0$ if $i \neq j$ and $g_{ii} = 1$, so that $\sigma_{\text{id}} \equiv 1$ and thus

$$\int_A \sigma = \int_A 1 = \mu(A).$$

More generally, let φ be an immersion of a k-dimensional manifold M into $\mathbb{R}^n$ such that $\varphi(M)$ is an open subset of a k-dimensional hyperplane in $\mathbb{R}^n$, and let m be the Riemann metric induced on M by φ. Then, if σ denotes the corresponding volume density, $\int_A \sigma$ is the k-dimensional Euclidean volume of $\varphi(A)$. In fact, by a Euclidean motion, we may assume that φ maps M into $\mathbb{R}^k \subset \mathbb{R}^n$. Then, since φ is an immersion and M is k-dimensional, we can use $x^1, \ldots, x^k$ as coordinates on M and conclude, as before, that σ is given by the function in terms of these coordinates, and hence that $\int_A \sigma = \mu(\varphi(A))$.

Now let φ_1 and φ_2 be two immersions of $M \to \mathbb{R}^n$. Let (U, α) be a coordinate chart on M with coordinates $y^1, \ldots, y^k$. If m_1 is the Riemann metric induced by φ_i, then

$$\left(\frac{\partial}{\partial y^i}, \frac{\partial}{\partial y^j}\right)_{\mathsf{m}_1} = \left(\frac{\partial \varphi_1}{\partial y^i}, \frac{\partial \varphi_1}{\partial y^j}\right)$$

and

$$\left(\frac{\partial}{\partial y^i}, \frac{\partial}{\partial y^j}\right)_{\mathsf{m}_2} = \left(\frac{\partial \varphi_2}{\partial y^i}, \frac{\partial \varphi_2}{\partial y^j}\right),$$

where the scalar product on the right is the Euclidean scalar product. Let σ_1

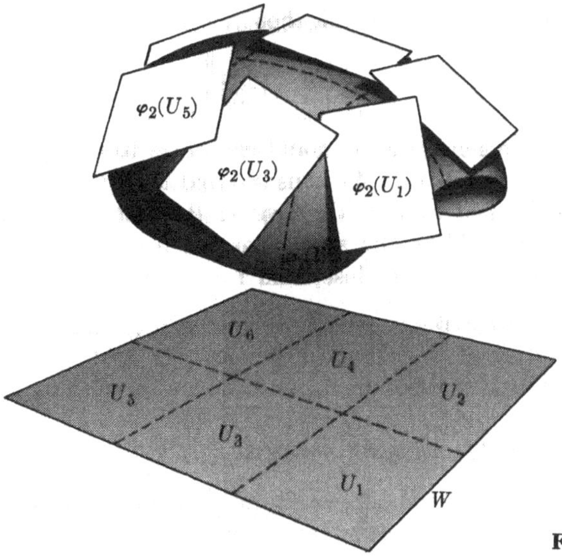

Fig. 10.2

and σ_2 be the volume densities corresponding to m_1 and m_2. Then

$$\sigma_{1\alpha} = \left|\det\left[\left(\frac{\partial\varphi_1}{\partial y^i}, \frac{\partial\varphi_1}{\partial y^j}\right)\right]\right|^{1/2}$$

and

$$\sigma_{2\alpha} = \left|\det\left[\left(\frac{\partial\varphi_2}{\partial y^i}, \frac{\partial\varphi_2}{\partial y^j}\right)\right]\right|^{1/2}.$$

In particular, given an $L > 0$, there is a $K = K(k, n, L)$ such that if

$$\left\|\frac{\partial\varphi_1}{\partial y^i}\right\| < L \quad \text{and} \quad \left\|\frac{\partial\varphi_2}{\partial y^i}\right\| < L \quad \text{for all} \quad i = 1, \ldots, k,$$

then, by the mean-value theorem,

$$|\sigma_{1\alpha} - \sigma_{2\alpha}| \leq K\left(\left\|\frac{\partial\varphi_2}{\partial y^1} - \frac{\partial\varphi_1}{\partial y^1}\right\| + \cdots + \left\|\frac{\partial\varphi_2}{\partial y^k} - \frac{\partial\varphi_1}{\partial y^k}\right\|\right).$$

Roughly speaking, this means that if φ_1 and φ_2 are close, in the sense that their derivatives are close, then the densities they induce are close.

We apply this remark to the following situation. We let φ_1 be an immersion of M into $\mathbb{R}^n$ and let (W, α) be some chart of M with coordinates $y^1, \ldots, y^k$. We let $U = W - C = \bigcup U_l$, where C is some closed set of content zero and such that $U_l \cap U_{l'} = \emptyset$ if $l \neq l'$. For each l let z_l be a point of U_l whose coordinates are $\langle y_l^1, \ldots, y_l^k \rangle$, and for $z = \langle y^1, \ldots, y^k \rangle$ define φ_2 by setting,

$$\varphi_2(y^1, \ldots, y^k) = \varphi_1(z_l) + \sum (y^i - y_l^i)\frac{\partial\varphi_1}{\partial y^i}(z_l)$$

if $z \in U_l$. (See Fig. 10.2.)

If the U_l's are sufficiently small, then

$$\left\| \frac{\partial \varphi_2}{\partial y^i} - \frac{\partial \varphi_1}{\partial y^i} \right\|$$

will be small. More generally, we could choose φ_2 to be any affine linear map approximating φ_1 on each U_l. We thus see that *the volume of W in terms of the Riemann metric induced by φ is the limit of the (surface) volume of polyhedra approximating $\varphi(W)$*. Here the approximation must be in the *sense of slope* (i.e., the derivatives must be close) and not merely in the sense of position.

The construction of the volume density can be generalized and suggests an alternative definition of the notion of density. In fact, let ρ be a rule which assigns to each x in M a function, ρ_x, on n tangent vectors in $T_x(M)$ subject to the rule

$$\rho_x(A\xi_1, \ldots, A\xi_n) = |\det A| \rho_x(\xi_1, \ldots, \xi_n), \tag{4.2}$$

where $\xi_i \in T_x(M)$ and $A\colon T_x(M) \to T_x(M)$ is a linear transformation. Then we see that ρ determines a density by setting

$$\rho_\alpha(\alpha(x)) = \rho\left(\frac{\partial}{\partial u^1}(x), \ldots, \frac{\partial}{\partial u^n}(x)\right) \tag{4.3}$$

if (U, α) is a chart with coordinates $u^1, \ldots, u^n$. The fact that (4.3) defines a density follows immediately from (4.2) and the transformation law for the $\partial/\partial u^i$ under change of coordinates.

Conversely, given a density ρ in terms of the ρ_α, define $\rho(\partial/\partial u^1, \ldots, \partial/\partial u^n)$ by (4.3). Since the vectors $\{\partial/\partial u^i\}_{i=1,\ldots,n}$ form a basis at each x in U, any $\xi_1, \ldots, \xi_n$ in $T_x(M)$ can be written as

$$\xi_i = B\frac{\partial}{\partial u^i}(x),$$

where B is a linear transformation of $T_x(M)$ into itself. Then (4.2) determines $\rho(\xi_1, \ldots, \xi_n)$ as

$$\rho(\xi_1, \ldots, \xi_n) = |\det B| \rho_\alpha(\alpha(x)). \tag{4.4}$$

That this definition is consistent (i.e., doesn't depend on α) follows from (4.2) and the transformation law (3.1) for densities.

EXERCISES

4.1 Let $M = S^1 \times S^1$ be the torus, and let $\varphi\colon M \to \mathbb{R}^4$ be given by

$$\begin{aligned} x^1 \circ \varphi(\theta_1, \theta_2) &= \cos\theta_1, \\ x^2 \circ \varphi(\theta_1, \theta_2) &= \sin\theta_1, \\ x^3 \circ \varphi(\theta_1, \theta_2) &= 2\cos\theta_2, \\ x^4 \circ \varphi(\theta_1, \theta_2) &= 2\sin\theta_2, \end{aligned}$$

where $x^1, \ldots, x^4$ are the rectangular coordinates on $\mathbb{R}^4$ and θ^1, θ^2 are angular coordinates on M.

a) Express the Riemann metric induced on M by φ (from the Euclidean metric on $\mathbb{R}^4$) in terms of the coordinates θ^1, θ^2. [That is, compute the $g_{ij}(\theta^1, \theta^2)$.]

b) What is the volume of M relative to this Riemann metric?

4.2 Consider the Riemann metric induced on $S^1 \times S^1$ by the immersion φ into $\mathbb{E}^3$ by

$$x \circ \varphi(u, v) = (a - \cos u) \cos v,$$
$$y \circ \varphi(u, v) = (a - \cos u) \sin v,$$
$$z \circ \varphi(u, v) = \sin u,$$

where u and v are angular coordinates and $a > 2$. What is the total surface area of $S^1 \times S^1$ under this metric?

4.3 Let φ map a region U of the xy-plane into $\mathbb{E}^3$ by the formula

$$\varphi(x, y) = (x, y, F(x, y)),$$

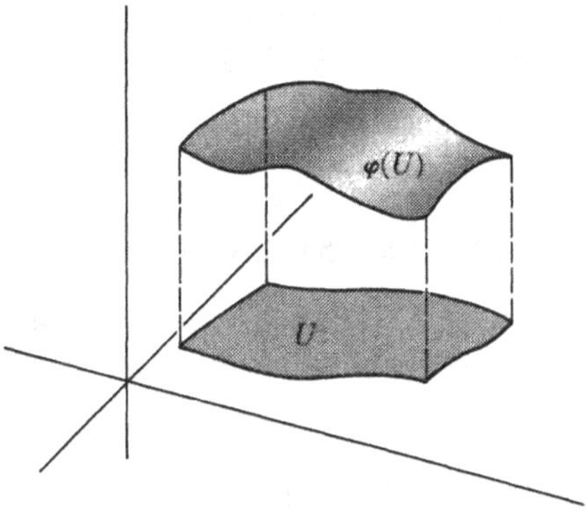

Fig. 10.3

so that $\varphi(U)$ is the surface $z = F(x, y)$. (See Fig. 10.3.) Show that the area of this surface is given by

$$\int_U \sqrt{1 + \left(\frac{\partial F}{\partial x}\right)^2 + \left(\frac{\partial F}{\partial y}\right)^2}.$$

4.4 Find the area of the paraboloid

$$z = x^2 + y^2 \quad \text{for} \quad x^2 + y^2 \leq 1.$$

4.5 Let $U \subset \mathbb{R}^2$, and let $\varphi\colon U \to \mathbb{E}^3$ be given by

$$\varphi(u, v) = (x(u, v), y(u, v), z(u, v)),$$

where x, y, z are rectangular coordinates on $\mathbb{E}^3$. Show the area of the surface $\varphi(U)$ is given by

$$\int_U \sqrt{\left(\frac{\partial x}{\partial u}\frac{\partial y}{\partial v} - \frac{\partial x}{\partial v}\frac{\partial y}{\partial u}\right)^2 + \left(\frac{\partial y}{\partial u}\frac{\partial z}{\partial v} - \frac{\partial y}{\partial v}\frac{\partial z}{\partial u}\right)^2 + \left(\frac{\partial x}{\partial u}\frac{\partial z}{\partial v} - \frac{\partial x}{\partial v}\frac{\partial y}{\partial u}\right)^2}.$$

4.6 Compute the surface area of the unit sphere in $\mathbb{E}^3$.

4.7 Let M_1 and M_2 be differentiable manifolds, and let σ be a density on M_2 which is nowhere zero. For each density ρ on $M_1 \times M_2$, each product chart $(U_1 \times U_2, \alpha_1 \times \alpha_2)$, and each $x_2 \in U_2$, define the function $\rho_{1\alpha_1}(\cdot, x_2)$ by

$$\rho_{1\alpha_1}(v_1, x_2)\sigma_\alpha(\alpha_2(x_2)) = \rho_{\alpha_1 \times \alpha_2}(v_1, \alpha_2(x_2))$$

for all $v_1 \in \alpha_1(U_1)$.

a) Show that $\rho_{1\alpha_1}(v_1, x_2)$ is independent of the chart (U_2, α_2).

b) Show that for each fixed $x_2 \in M_2$ the functions $\rho_{1\alpha_1}(\cdot, x_2)$ define a density on M_1. We shall call this density $\rho_1(x_2)$.

c) Show that if ρ is a smooth density of compact support on $M_1 \times M_2$ and σ is smooth, then $\rho_1(x_2)$ is a smooth density of compact support on M_1.

d) Let ρ be as in (c). Define the function F_ρ on M_2 by

$$F_\rho(x_2) = \int_{M_1} \rho_1(x_2).$$

Sketch how you would prove the fact that F_ρ is a smooth function of compact support on M_2 and that

$$\int_{M_1 \times M_2} \rho = \int_{M_2} F_\rho \cdot \sigma.$$

5. PULLBACK AND LIE DERIVATIVES OF DENSITIES

Let $\varphi: M_1 \to M_2$ be a diffeomorphism, and let ρ be a density on M_2. Define the density $\varphi^*\rho$ on M_1 by

$$\varphi^*\rho(\xi_1, \ldots, \xi_n) = \rho(\varphi_*\xi_1, \ldots, \varphi_*\xi_n) \tag{5.1}$$

for $\xi_i \in T_x(M_1)$ and $\varphi_* = \varphi_{*x}$. To show that $\varphi^*\rho$ is actually a density, we must check that (4.2) holds for any linear transformation A of $T_x(M_1)$. But

$$\begin{aligned}\varphi^*\rho(A\xi_1, \ldots, A\xi_n) &= \rho(\varphi_*A\xi_1, \ldots, \varphi_*A\xi_n)\\ &= \rho(\varphi_*A\varphi_*^{-1}\varphi_*\xi_1, \ldots, \varphi_*A\varphi_*^{-1}\varphi_*\xi_n)\\ &= |\det \varphi_*A\varphi_*^{-1}|\,\rho(\varphi_*\xi_1, \ldots, \varphi_*\xi_n)\\ &= |\det A|\varphi^*\rho(\xi_1, \ldots, \xi_n),\end{aligned}$$

which is the desired identity.

Let (U, α) and (W, β) be compatible charts on M_1 and M_2 with coordinates $u^1, \ldots, u^n$ and $w^1, \ldots, w^n$, respectively. Then for all points of U we have, by (4.3),

$$\begin{aligned}(\varphi^*\rho)_\alpha(\alpha(\cdot)) &= \rho\left(\varphi_*\frac{\partial}{\partial u^1}, \ldots, \varphi_*\frac{\partial}{\partial u^n}\right) = \left|\det\left(\frac{\partial w^j}{\partial u^i}\right)\right| \rho\left(\frac{\partial}{\partial w^1}, \ldots, \frac{\partial}{\partial w^n}\right)\\ &= \left|\det\left(\frac{\partial w^j}{\partial u^i}\right)\right| \rho_\beta(\beta \circ \varphi(\cdot)).\end{aligned}$$

In other words, we have

$$(\varphi^*\rho)_\alpha = |\det J_{\beta\circ\varphi\circ\alpha^{-1}}|\rho(\beta \circ \varphi \circ \alpha^{-1}(\cdot)). \tag{5.2}$$

The density $\varphi^*\rho$ is called the pullback of ρ by φ^*. It is clear that

$$\varphi^*(\rho_1 + \rho_2) = \varphi^*(\rho_1) + \varphi^*(\rho_2)$$

and that

$$\varphi^*(f\rho) = \varphi^*(f)\varphi^*(\rho)$$

for any function f.

It follows directly from the definition that

$$\operatorname{supp} \varphi^*\rho = \varphi^{-1}[\operatorname{supp} \rho].$$

Proposition 5.1. Let $\varphi\colon M_1 \to M_2$ be a diffeomorphism, and let ρ be a locally absolutely integrable density with compact support on M_2. Then

$$\int \varphi^*\rho = \int \rho. \tag{5.3}$$

Proof. It suffices to prove (5.3) for the case

$$\operatorname{supp} \rho \subset \varphi(U)$$

for some chart (U, α) of M_1 with $\varphi(U) \subset W$, where (W, β) is a chart of M_2. In fact, the set of all such $\varphi(U)$ is an open covering of M_2, and we can therefore choose a partition of unity $\{g_j\}$ subordinate to it. If we write $\rho = \sum g_j\rho$, then the sum is finite and each $g_j\rho$ has the observed property. Since both sides of (5.3) are linear, we conclude that it suffices to prove (5.3) for each term.

Now if $\operatorname{supp} \rho \subset \varphi(U)$, then

$$\int \rho = \int_{\beta(W)} \rho_\beta = \int_{\beta\circ\varphi(U)} \rho_\beta$$

and

$$\begin{aligned}\int \varphi^*\rho &= \int_{\alpha(U)} (\varphi^*\rho)_\alpha \\ &= \int_{\alpha(U)} \rho_\beta(\beta\circ\varphi\circ\alpha^{-1})|\det J_{\beta\circ\varphi\circ\alpha^{-1}}| \\ &= \int_{\beta\circ\varphi(U)} \rho_\beta,\end{aligned}$$

thus establishing (5.3). □

Now let φ_t be a one-parameter group on M with infinitesimal generator X. Let ρ be a density on M, let (U, α) be a chart, and let W be an open subset of U such that $\varphi_t(W) \subset U$ for all $|t| < \epsilon$. Then

$$(\varphi_t^*\rho)_\alpha(v) = \rho_\alpha(\Phi_\alpha(v, t)) \left|\det\left(\frac{\partial\Phi_\alpha}{\partial v}\right)_{(v,t)}\right| \qquad \text{for} \quad v \in \alpha(W),$$

where $\Phi_\alpha(v, t) = \alpha\circ\varphi_t\circ\alpha^{-1}(v)$ and $(\partial\Phi_\alpha/\partial v)_{(v,t)}$ is the Jacobian of $v \mapsto \Phi_\alpha(v, t)$. We would like to compute the derivative of this expression with respect to t at $t = 0$. Now $\Phi_\alpha(v, 0) = v$, and so

$$\det\left(\frac{\partial\Phi_\alpha}{\partial v}\right)_{(v,0)} = 1.$$

Consequently, we can conclude that

$$\det\left(\frac{\partial\Phi_\alpha}{\partial v}\right)_{(v,t)} > 0$$

for t close to zero. We can therefore omit the absolute-value sign and write

$$\left.\frac{d(\varphi_t^*\rho)_\alpha}{dt}\right|_{t=0} = \left.\frac{d\rho_\alpha(\Phi_\alpha)}{dt}\right|_{t=0} + \rho_\alpha(v)\left.\frac{d}{dt}\left(\det\frac{\partial\Phi_\alpha}{\partial v}\right)\right|_{t=0}.$$

We simply evaluate the first derivative on the right by the chain rule, and get

$$d\rho_\alpha\left(\frac{\partial\Phi_\alpha}{\partial t}\right) = d\rho_\alpha(X_\alpha(v)).$$

In terms of coordinates $x^1, \ldots, x^n$, we can write

$$\frac{d\rho_\alpha(\Phi_\alpha(v, t))}{dt} = \sum \frac{\partial\rho_\alpha}{\partial x^i} X^i_\alpha$$

if $X_\alpha = \langle X^1_\alpha, \ldots, X^n_\alpha\rangle$.

To evaluate the second term on the right, we need to make a preliminary observation. Let $A(t) = (a_{ij}(t))$ be a differentiable matrix-valued function of t with $A(0) = \mathrm{id} = (\delta^j_i)$. Then

$$\frac{d(\det A(t))}{dt} = \lim \frac{1}{t}(\det A(t) - 1).$$

Now $a_{ii}(0) = 1$ and $a_{ij}(0) = 0$ $(i \neq j)$. To say that A is differentiable means that each of the functions $a_{ij}(t)$ is differentiable. We can therefore find a constant K such that $|a_{ij}(t)| \leq K|t|$ $(i \neq j)$ and $|a_{ii}(t) - 1| \leq K|t|$. In the expansion of $\det A(t)$, the only term which will not vanish at least as t^2 is the diagonal product $a_{11}(t) \cdots a_{nn}(t)$. In fact, any other term in $\sum \pm a_{1i_1}(t) \cdots a_{ni_n}(t)$ involves at least two off-diagonal terms and thus vanishes at least as t^2. Thus

$$\begin{aligned}\frac{d}{dt}(\det A(t)) &= \lim_{t=0} \frac{1}{t}(a_{11}(t) \cdots a_{nn}(t) - 1)\\ &= a'_{11}(0) + \cdots + a'_{nn}(0)\\ &= \operatorname{tr} A'(0).\end{aligned}$$

If we take $A = \partial\Phi_\alpha/\partial v$, we conclude that

$$\frac{d}{dt}\left(\det \frac{\partial\Phi_\alpha}{\partial v}\right) = \operatorname{tr} \frac{\partial X_\alpha}{\partial v} = \sum \frac{\partial X^i_\alpha}{\partial x^i}.$$

Thus

$$\frac{d(\overset{*}{\varphi_t}\rho)_\alpha}{dt} = \sum \frac{\partial\rho_\alpha}{\partial x^i} X^i_\alpha + \rho_\alpha \frac{\partial X^i_\alpha}{\partial x^i} = \sum \frac{\partial}{\partial x^i}(\rho_\alpha X^i_\alpha).$$

We repeat:

Proposition 5.2. Let φ_t be a one-parameter group of diffeomorphisms of M with infinitesimal generator X, and let ρ be a differentiable density on M. Then

$$D_X\rho = \lim_{t=0} \frac{\overset{*}{\varphi_t}\rho - \rho}{t}$$

exists and is given locally by

$$(D_X\rho)_\alpha = \sum \frac{\partial(\rho_\alpha X^i_\alpha)}{\partial x^i}$$

if $X = \langle X^1_\alpha, \ldots, X^n_\alpha\rangle$ on the chart (U, α).

The density $D_X\rho$ is sometimes called the *divergence* of $\langle X, \rho \rangle$ and is denoted by div $\langle X, \rho \rangle$. Thus div $\langle X, \rho \rangle = D_X\rho$ is the density given by

$$(\text{div} \langle X, \rho \rangle)_\alpha = \sum \frac{\partial}{\partial x^i} (X^i_\alpha \rho_\alpha) \qquad \text{on} \quad (U, \alpha).$$

Now let ρ be a differentiable density, and let A be a compact contented set. Then

$$\begin{aligned} \int_{\varphi_t(A)} \rho &= \int_M e_{\varphi_t(A)}\rho \\ &= \int_M \varphi_t^*(e_{\varphi_t(A)}\rho) \\ &= \int (\varphi_t^* e_{\varphi_t(A)})(\varphi_t^*\rho) \\ &= \int e_A \varphi_t^*(\rho) \\ &= \int_A \varphi_t^*\rho. \end{aligned}$$

Thus

$$\frac{1}{t}\left(\int_{\varphi_t(A)} \rho - \int_A \rho\right) = \int_A \frac{1}{t}(\varphi_t^*\rho - \rho).$$

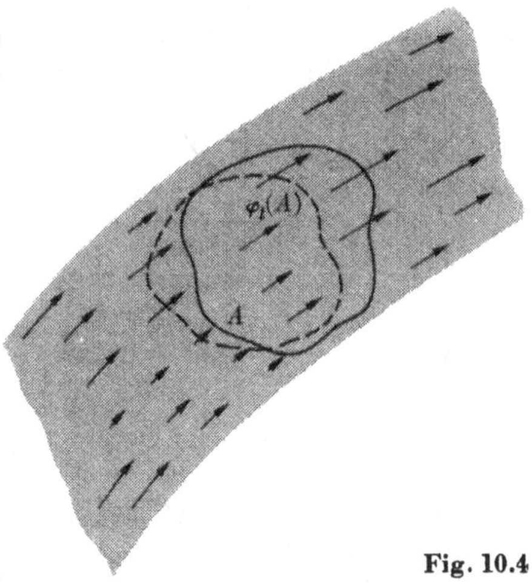

Fig. 10.4

Using a partition of unity, we can easily see that the limit under the integral sign is uniform, and we thus have the formula

$$\frac{d}{dt}\left(\int_{\varphi_t(A)} \rho\right)\Bigg|_{t=0} = \int_A D_X\rho = \int_A \text{div} \langle X, \rho \rangle.$$

6. THE DIVERGENCE THEOREM

Let φ be a flow on a differentiable manifold M with infinitesimal generator X. Let ρ be a density belonging to P, and let A be a contented subset of M. Then for small values of t, we would expect the difference $\int_{\varphi_t(A)} \rho - \int_A \rho$ to depend only on what is happening near the boundary of A (Fig. 10.4). In the limit, we would expect the derivative of $\int_{\varphi_t(A)} \rho$ at $t = 0$ (which is given by $\int_A \text{div} \langle X, \rho \rangle$) to be given by some integral over ∂A. In order to formulate such a result, we must first single out a class of sets whose boundaries are sufficiently nice to allow us to integrate over them. We therefore make the following definition:

Definition. Let M be a differentiable manifold, and let D be a subset of M. We say that D is a *domain with regular boundary* if for every $x \in M$ there is a chart (U, α) about x, with coordinates $x^1_\alpha, \ldots, x^n_\alpha$, such that one of the following three possibilities holds:

i) $U \cap D = \emptyset$;

ii) $U \subset D$;

iii) $\alpha(U \cap D) = \alpha(U) \cap \{\mathbf{v} = \langle v^1, \ldots, v^n \rangle \in \mathbb{R}^n : v^n \geq 0\}$.

Note that if $x \notin \overline{D}$, we can always find a (U, α) about x such that (i) holds. If $x \in \operatorname{int} D$, we can always find a chart (U, α) about x such that (ii) holds. This imposes no restrictions on D. The crucial condition is imposed when $x \in \partial D$. Then we cannot find charts about x satisfying (i) or (ii). In this case, (iii) implies that $\alpha(U \cap \partial D)$ is an open subset of $\mathbb{R}^{n-1}$ (Fig. 10.5). In fact, $\alpha(U \cap \partial D) = \{\mathbf{v} \in \alpha(U) : v^n = 0\} = \alpha(U) \cap \mathbb{R}^{n-1}$, where we regard $\mathbb{R}^{n-1}$ as the subspace of $\mathbb{R}^n$ consisting of those vectors with last component zero.

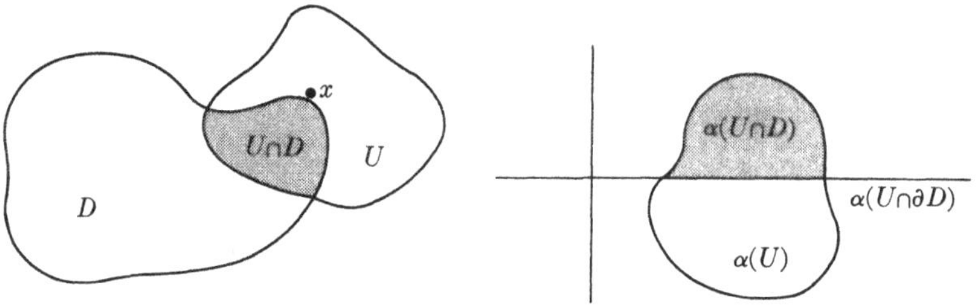

Fig. 10.5

Let $\mathfrak{a}$ be an atlas of M such that each chart of $\mathfrak{a}$ satisfies either (i), (ii), or (iii). For each $(U, \alpha) \in \mathfrak{a}$ consider the map $\alpha \restriction \partial D: U \cap \partial D \to \mathbb{R}^{n-1} \subset \mathbb{R}^n$. [Of course, the maps $\alpha \restriction \partial D$ will have a nonempty domain of definition only for charts of type (iii).] We claim that $\{(U \cap \partial D, \alpha \restriction \partial D)\}$ is an atlas on ∂D. In fact, let (U, α) and (W, β) be two charts in $\mathfrak{a}$ such that $U \cap W \cap \partial D \neq \emptyset$. Let $x^1, \ldots, x^n$ be the coordinates of (U, α), and let $y^1, \ldots, y^n$ be those of (W, β). The map $\beta \circ \alpha^{-1}$ is given by

$$\langle x^1, \ldots, x^n \rangle \mapsto \langle y^1(x^1, \ldots, x^n), \ldots, y^n(x^1, \ldots, x^n) \rangle.$$

On $\alpha(U \cap W \cap \partial D)$, we have $x^n = 0$ and $y^n = 0$. In particular,

$$y^n(x^1, \ldots, x^{n-1}, 0) \equiv 0,$$

and the functions $y^1(x^1, \ldots, x^{n-1}, 0), \ldots, y^{n-1}(x^1, \ldots, x^{n-1}, 0)$ are differentiable. This shows that $(\beta \restriction \partial D) \circ (\alpha \restriction \partial D)^{-1}$ is differentiable on $\alpha(U \cap \partial D)$. We thus get a manifold structure on ∂D.

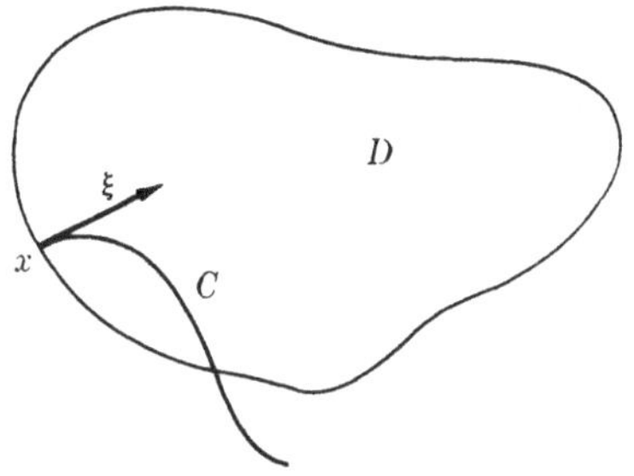

Fig. 10.6

It is easy to see that this manifold structure is independent of the particular atlas of M that was chosen. We shall denote by ι the map of $\partial D \to M$ which sends each $x \in \partial D$, regarded as an element of M, into itself. It is clear that ι is a differentiable map. (In fact, $(U \cap \partial D, \alpha \restriction \partial D)$ and (U, α) are compatible charts in terms of which $\alpha \circ \iota \circ (\alpha \restriction \partial D)^{-1}$ is just the map of $\mathbb{R}^{n-1} \to \mathbb{R}^n$.)

Let x be a point of ∂D regarded as a point of M, and let ξ be an element of $T_x(M)$. We say that ξ points into D if for every curve C with $C'(0) = \xi$, we have $C(t) \in D$ for sufficiently small positive t (Fig. 10.6). In

terms of a chart (U, α) of type (iii), let $\xi_\alpha = \langle \xi^1, \ldots, \xi^n \rangle$. Then it is clear that ξ points into D if and only if $\xi^n > 0$. Similarly, a tangent vector ξ points out of D (obvious definition) if and only if $\xi^n < 0$. If $\xi^n = 0$, then ξ is tangent to the boundary—it lies in $\iota_* T_x(\partial D)$.

Let ρ be a density on M and X a vector field on M. Define the density ρ_X on ∂D by

$$\rho_X(\xi_1, \ldots, \xi_{n-1}) = \rho(\iota_*\xi_1, \ldots, \iota_*\xi_{n-1}, X(x)) \qquad \text{for} \quad \xi_i \in T_x(\partial D). \tag{6.1}$$

It is easy to check that (6.1) defines a density. (This is left as an exercise for the reader.) If (U, α) is a chart of type (iii) about x and $X_\alpha = \langle X^1, \ldots, X^n \rangle$, then applying (4.3) to the chart $(U \cap \partial D, \alpha \restriction \partial D)$ and the density ρ_X, we see that

$$(\rho_X)_{\alpha \restriction \partial D} = \rho\left(\frac{\partial}{\partial x^1}, \ldots, \frac{\partial}{\partial x^{n-1}}, X\right).$$

Let A be the linear transformation of $T_x(M)$ given by

$$A\frac{\partial}{\partial x^1} = \frac{\partial}{\partial x^1}, \qquad A\frac{\partial}{\partial x^{n-1}} = \frac{\partial}{\partial x^{n-1}}, \qquad A\frac{\partial}{\partial x^n} = X.$$

The matrix of A is

$$\begin{bmatrix} 1 & 0 & & \cdots & X^1 \\ 0 & 1 & 0 & \cdots & X^2 \\ \vdots & & & & \vdots \\ 0 & \cdots & \cdots & 1 & 0 \\ 0 & \cdots & \cdots & & X^n \end{bmatrix},$$

and therefore $|\det A| = |X^n|$. Thus we have

$$(\rho_X)_{\alpha \restriction \partial D} = |X^n|\rho_\alpha \qquad \text{at all points of} \quad \alpha(U \cap \partial D). \tag{6.2}$$

We can now state our results.

Theorem 6.1 (*The divergence theorem*).† Let D be a domain with regular boundary, let $\rho \in \mathrm{P}$, and let X be a smooth vector field on M. Define the function ϵ_X on ∂D by

$$\epsilon_X(x) = \begin{cases} 1 & \text{if} \quad X(x) \text{ points out of } D, \\ 0 & \text{if} \quad X(x) \text{ is tangent to } \partial D, \\ -1 & \text{if} \quad X(x) \text{ points into } D. \end{cases}$$

Then

$$\int_D \operatorname{div} \langle X, \rho \rangle = \int_{\partial D} \epsilon_X \rho_X. \tag{6.3}$$

Remark. In terms of a chart of type (iii), the function ϵ_X is given by

$$\epsilon_X = -\operatorname{sgn} X^n. \tag{6.4}$$

---

† This formulation and proof of the divergence theorem was suggested to us by Richard Rasala.

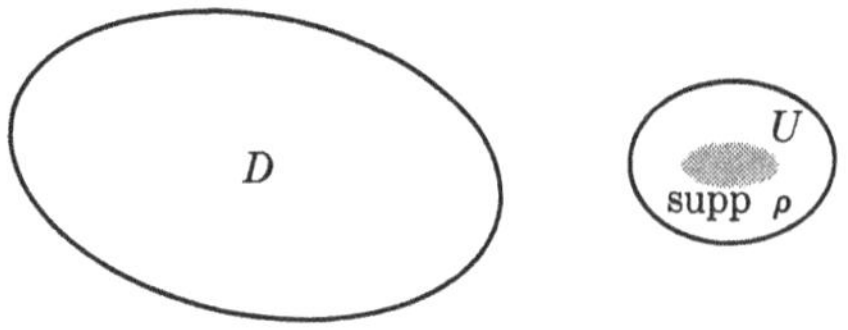

Fig. 10.7

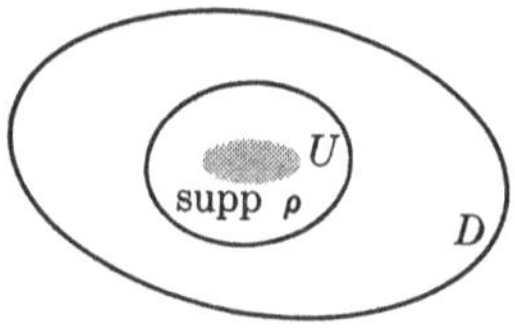

Fig. 10.8

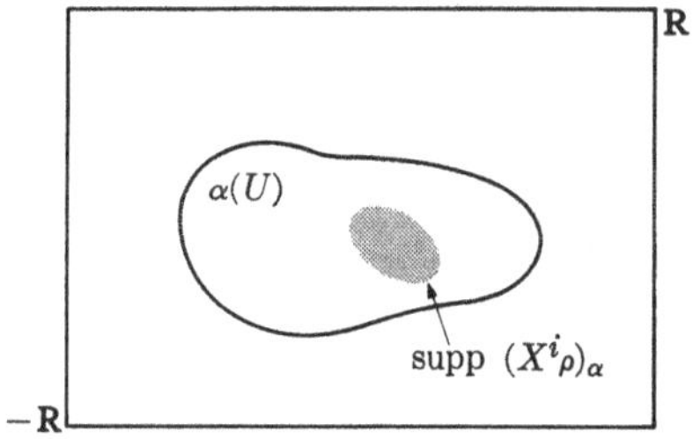

Fig. 10.9

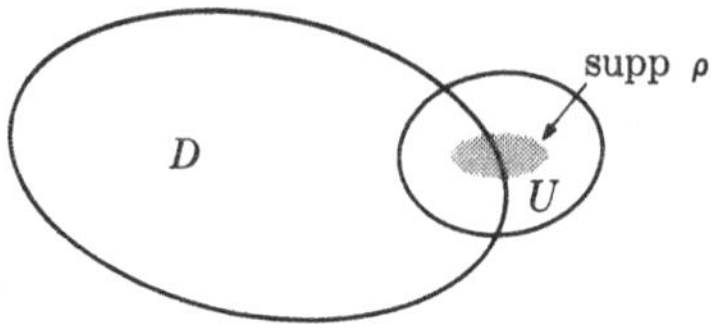

Fig. 10.10

Proof. Let $\mathfrak{A}$ be an atlas of M each of whose charts is one of the three types. Let $\{g_i\}$ be a partition of unity subordinate to $\mathfrak{A}$. Write $\rho = \sum g_i\rho$. This is a finite sum. Since both sides of (6.3) are linear functions of ρ it suffices to verify (6.3) for each of the summands $g_i\rho$. Changing our notation (replacing $g_i\rho$ by ρ), we reduce the problem to proving (6.3) under the additional assumption $\operatorname{supp}\rho \subset U$, where (U, α) is a chart of type (i), (ii), or (iii). There are therefore three cases to consider.

CASE I. $\operatorname{supp}\rho \subset U$ and $U \cap \overline{D} = \varnothing$. (See Fig. 10.7.) Then both sides of (6.3) vanish, and so (6.3) is correct.

CASE II. $\operatorname{supp}\rho \subset U$ with $U \subset \operatorname{int} D$. (See Fig. 10.8.) Then the right-hand side of (6.3) vanishes. We must show that the left-hand side does also. But

$$\int_D \operatorname{div} \langle X, \rho\rangle = \int_U \operatorname{div} \langle X, \rho\rangle = \int_{\alpha(U)} \sum \frac{\partial(X^i\rho_\alpha)}{\partial x^i} = \sum \int_{\alpha(U)} \frac{\partial(X^i\rho_\alpha)}{\partial x^i}.$$

Now each of the functions $X^i\rho_\alpha$ has its support lying inside $\alpha(U)$. Choose some large R so that $\alpha(U) \subset \Box^{\mathbf{R}}_{-\mathbf{R}}$. We can then replace $\int_{\alpha(U)}$ by $\int_{\Box^{\mathbf{R}}_{-\mathbf{R}}}$. We extend its domain of definition to all of $\mathbb{R}^n$ by setting it equal to zero outside $\alpha(U)$. (See Fig. 10.9.) Writing the integral as an iterated integral and integrating with respect to x^i first, we see that

$$\int_{\alpha(U)} \frac{\partial X^i\rho_\alpha}{\partial x^i}$$

$$= \int X^i\rho_\alpha(\ldots, R, \ldots) - X^i\rho_\alpha(\ldots, -R, \ldots)\, dx^1\, dx^2\, dx^{i-1}\, dx^i \cdots dx^n = 0.$$

This last integral vanishes, because the function $X^i\rho_\alpha$ vanishes outside $\alpha(U)$.

CASE III. supp ρ is contained in a chart of type (iii). (See Fig. 10.10.) Then

$$\int_D \operatorname{div} \langle X, \rho\rangle = \int_{D\cap U} \operatorname{div} \langle X, \rho\rangle = \sum \int_{\alpha(D\cap U)} \frac{\partial X^i \rho_\alpha}{\partial x^i}.$$

Now

$$\alpha(U \cap D) = \alpha(U) \cap \{\mathbf{v} : v^n \geq 0\}.$$

We can therefore replace the domain of integration by the rectangle $\square^{\langle R,\ldots,R\rangle}_{\langle -R,\ldots,-R,0\rangle}$. (See Fig. 10.11.) For $1 \leq i < n$ all the integrals in the sum vanish as before. For $i = n$ we obtain

$$\int_D \operatorname{div} \langle X, \rho\rangle = -\int_{\mathbb{R}^{n-1}} X^n \rho_\alpha.$$

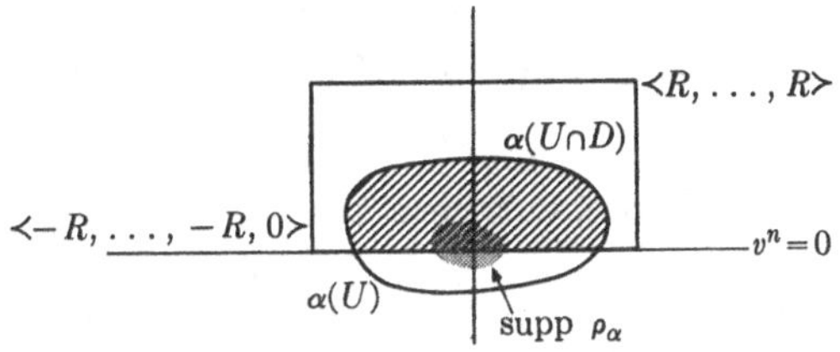

Fig. 10.11

If we compare this with (6.2) and (6.4), we see that this is exactly the assertion of (6.3). □

If the manifold M is given a Riemann metric, then we can give an alternative version of the divergence theorem. Let dV be the volume density of the Riemann metric, so that

$$dV(\xi_1, \ldots, \xi_n) = |\det((\xi_i, \xi_j))|^{1/2}, \quad \xi_i \in T_x(M),$$

is the volume of the parallelepiped spanned by the ξ_i in the tangent space (with respect to the Euclidean metric given by the scalar product on the tangent space).

Now the map ι is an immersion, and therefore we get an induced Riemann metric on ∂D. Let dS be the corresponding volume density on ∂D. Thus, if $\{\xi_i\}_{i=1,\ldots,n-1}$ are $n-1$ vectors in $T_x(\partial D)$, $dS(\xi_1, \ldots, \xi_{n-1})$ is the $(n-1)$-dimensional volume of the parallelepiped spanned by $\iota_*\xi_1, \ldots, \iota_*\xi_{n-1}$ in $\iota_*T_x(\partial D) \subset T_x(M)$. For any $x \in \partial D$ let $\mathbf{n} \in T_x(M)$ be the vector of unit length which is orthogonal to $\iota_*T_x(\partial D)$ and which points out of D (Fig. 10.12). We

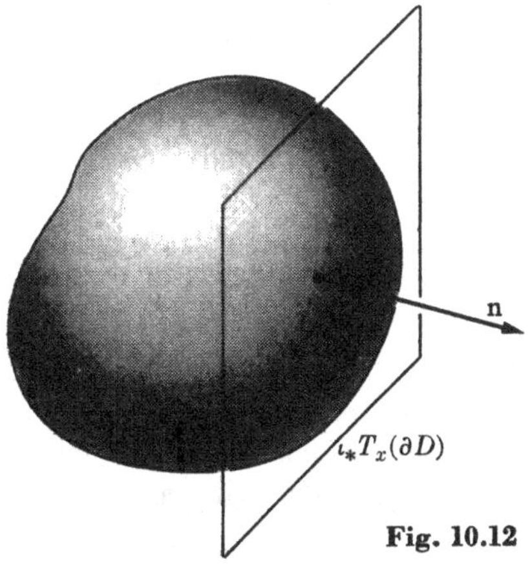

Fig. 10.12

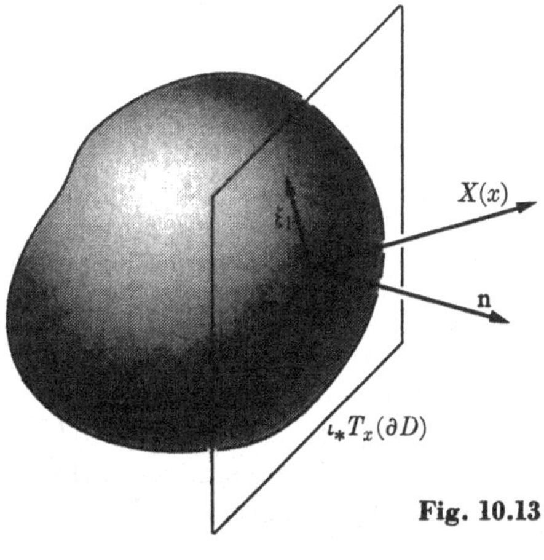

Fig. 10.13

clearly have

$$dS(\xi_1, \ldots, \xi_{n-1}) = dV(\iota_*\xi_1, \ldots, \iota_*\xi_{n-1}, n).$$

For any vector $X(x) \in T_x(M)$ (Fig. 10.13) the volume of the parallelepiped spanned by $\xi_1, \ldots, \xi_{n-1}, X(x)$ is $|(X(x), \mathbf{n})|dS(\xi_1, \ldots, \xi_{n-1})$. [In fact, write

$$X(x) = (X(x), \mathbf{n})\mathbf{n} + \mathbf{m},$$

where $\mathbf{m} \in \iota_*T(\partial D)$.] If we compare this with (6.1), we see that

$$dV_X = |(X, \mathbf{n})|dS. \tag{6.5}$$

Furthermore, it is clear that

$$\epsilon(x) = \operatorname{sgn}\,(X(x), \mathbf{n}).$$

Let ρ be any density on M. Then we can write

$$\rho = f\,dV,$$

where f is a function. Furthermore, we clearly have $\rho_X = f\,dV_X$ and

$$\operatorname{div} \langle X, \rho \rangle = \operatorname{div} \langle X, f\,dV \rangle.$$

We can then rewrite (6.3) as

$$\int_D \operatorname{div} \langle X, f\,dV \rangle = \int_{\partial D} f \cdot (X, \mathbf{n})\, dS. \tag{6.6}$$

7. MORE COMPLICATED DOMAINS

For many purposes, Theorem 6.1 is not quite sufficiently broad. The trouble is that we would like to apply (6.3) to domains whose boundaries are not completely smooth. For instance, we would like to apply it to a rectangle in $\mathbb{R}^n$. Now the boundary of a rectangle is regular at all points except those lying on an edge (i.e., the intersection of two faces). Since the edges form a set "of dimension $n - 2$", we would expect that their presence does not invalidate (6.3). This is in fact the case.

Let M be a differentiable manifold, and let D be a subset of M. We say that D is a domain with *almost regular boundary* if to every $x \in M$ there is a chart (U, α) about x, with coordinates $x^1, \ldots, x^n_\alpha$, such that one of the following four possibilities holds:

i) $U \cap D = \emptyset$;

ii) $U \subset D$;

iii) $\alpha(U \cap D) = \alpha(U) \cap \{\mathbf{v} = \langle v^1, \ldots, v^n \rangle \in \mathbb{R}^n : v^n \geq 0\}$;

iv) $\alpha(U \cap D) = \alpha(U) \cap \{\mathbf{v} = \langle v^1, \ldots, v^n \rangle \in \mathbb{R}^n : v^k \geq 0, \ldots, v^n \geq 0\}$.

The novel point is that we are now allowing for possibility (iv) where $k < n$. This, of course, is a new possibility only if $n > 1$. Let us assume $n > 1$ and see what (iv) allows. We can write $\alpha(U \cap \partial D)$ as the union of certain open subsets lying in $(n - 1)$-dimensional subspaces of $\mathbb{R}^{n-1}$, together with a union of portions lying in subspaces of dimension $n - 2$.

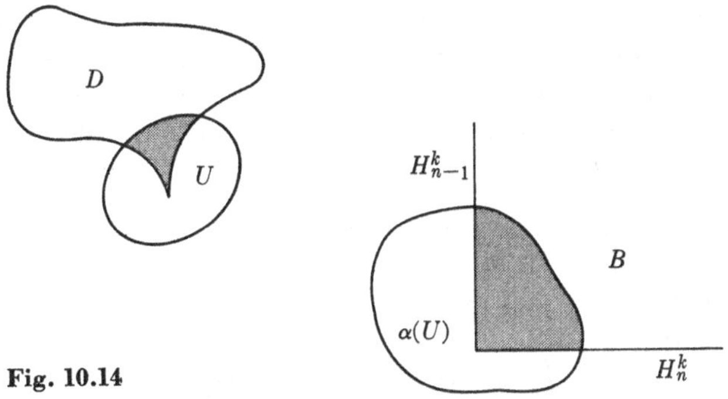

Fig. 10.14

In fact, for $k \leq p \leq n$ let

$$H^k_p = \{\mathbf{v} : v^k > 0, \ldots, v^p = 0, v^{p+1} > 0, \ldots, v^n > 0\}.$$

Thus H^k_p is an open subset of the $(n - 1)$-dimensional subspace given by $v^p = 0$. (See Fig. 10.14.) We can write

$$\alpha(U \cap \partial D) \subset \alpha(U) \cap \{(H^k_k \cup H^k_{k+1} \cup \cdots \cup H^k_n) \cup S\},$$

where S is the union of the subspaces (of dimension $n - 2$) where at least two of the v^p vanish.

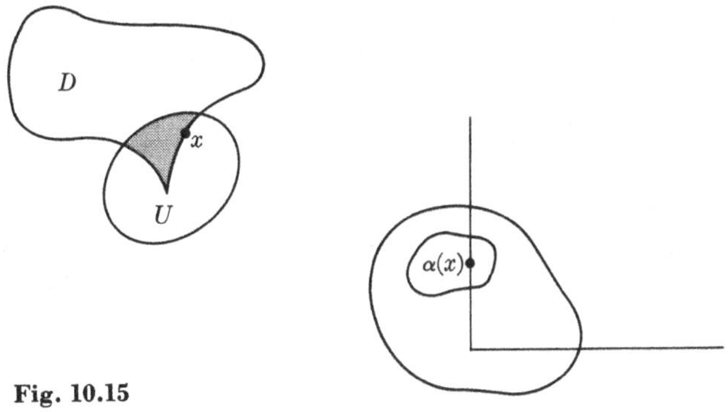

Fig. 10.15

Observe that if $x \in U \cap \partial D$ is such that $\alpha(x) \in H^k_p$ for some p, then there is a chart about x of type (iii). In fact, simply renumber the coordinates so that v^p becomes v^n, that is, map $\mathbb{R}^n \xrightarrow{\varphi} \mathbb{R}^n$ by sending $\langle v^1, \ldots, v^n \rangle \to \langle w^1, \ldots, w^n \rangle$, where

$$\begin{aligned} w^i &= v^i && \text{for } i < p, \\ w^i &= v^{i+1} && \text{for } p \leq i < n, \\ w^n &= v^p. \end{aligned}$$

Then in a sufficiently small neighborhood U^1 of x the chart $(U^1, \varphi \circ \alpha)$ is of type (iii). (See Fig. 10.15.)

We next observe the set of $x \in \partial D$ having a neighborhood of type (iii) forms a differentiable manifold. The argument is just as before. The only difference is that this time these points do not exhaust all of ∂D. We shall denote this manifold by $\breve{\partial D}$. Thus $\breve{\partial D}$ is a manifold which, as a set, is not ∂D but only the "regular" points of ∂D, that is, those having charts of type (iii).

Theorem 7.1 (*The divergence theorem*). Let M be an n-dimensional manifold, and let $D \subset M$ be a domain with almost regular boundary. Let $\breve{\partial D}$ be as above, and let i be the injection of $\breve{\partial D} \to M$. Then for any $\rho \in \mathrm{P}$ we have

$$\int_D \operatorname{div} \langle X, \rho \rangle = \int_{\breve{\partial D}} \epsilon_X \rho_X. \tag{7.1}$$

Proof. The proof proceeds as before. We choose a connecting atlas of charts of types (i) through (iv) and a partition of unity $\{g_j\}$ subordinate to the atlas. We write $\rho = \sum g_j\rho$ and now have four cases to consider. The first three cases have already been handled.

The new case arises when ρ has its support in U, where (U, α) is a chart of type (iv). We must evaluate

$$\int_{\alpha(U \cap D)} \sum \frac{\partial X^i \rho_\alpha}{\partial x^i}.$$

The terms in the sum corresponding to $i < k$ make no contribution to the integral, as before. Let us extend $X^i\rho_\alpha$ to be defined on all of $\mathbb{R}^n$ by setting it equal to zero outside $\alpha(U)$, just as before. Then, for $k \le i \le n$ we have

$$\int_{\alpha(U \cap D)} \frac{\partial X^i \rho_\alpha}{\partial x^i} = \int_B \frac{\partial X^i \rho_\alpha}{\partial x^i},$$

where $B = \{\mathbf{v} : v^k \ge 0, \ldots, v^n \ge 0\}$. Writing this as an iterated integral and integrating first with respect to x^i, we obtain

$$\int_B \frac{\partial X^i \rho_\alpha}{\partial x^i} = \int_{A_i} X^i \rho_\alpha,$$

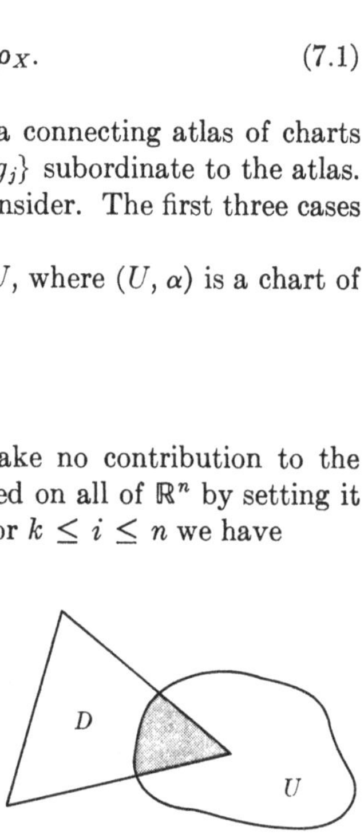

Fig. 10.16

where the set $A_i \subset \mathbb{R}^{n-1}$ is given by

$$A_i = \{\langle v^1, \ldots, v^{i-1}, v^{i+1}, \ldots, v^n \rangle : v^k \ge 0, \ldots, v^n \ge 0\}.$$

Note that A_i differs from H_i^k by a set of content zero in $\mathbb{R}^{n-1}$ (namely, where at least one of the $v^l = 0$ for $k = l \le n$). Thus we can replace the A_i by the H_i^k in the integral. Summing over $k \le i \le n$, we get

$$\int_{\alpha(D \cap U)} \sum \frac{\partial X^i \rho_\alpha}{\partial x^i} = \sum_{i=k}^{n} \int_{H_i^k} X^i \rho_\alpha,$$

which is exactly the assertion of Theorem 7.1 for case (iv). □

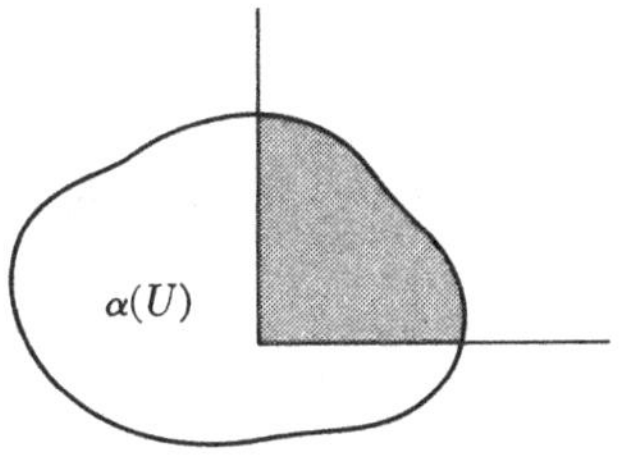

Fig. 10.17

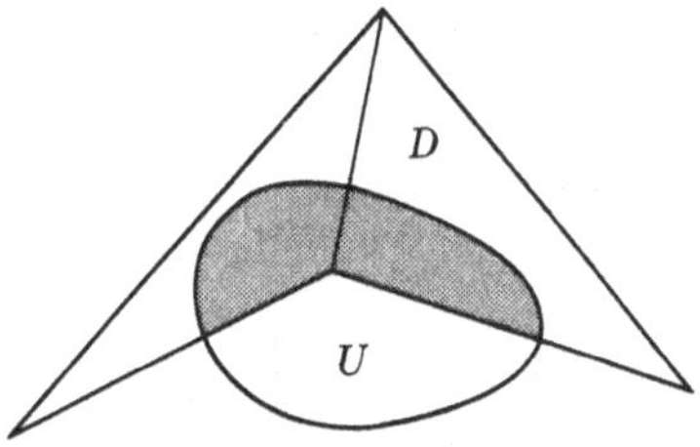

Fig. 10.18

We should point out that even Theorem 7.1 does not cover all cases for which it is useful to have a divergence theorem. For instance, in the plane, Theorem 7.1 does apply to the case where D is a triangle. (See Fig. 10.16.) This is because we can "stretch" each angle to a right angle (in fact, we can do this by a linear change of variables of $\mathbb{R}^2$). (See Fig. 10.17.)

However Theorem 7.1 does not apply to a quadrilateral such as the one in Fig. 10.18, since there is no C^1-transformation that will convert an angle greater than π into one smaller than π (since its Jacobian at the corner must carry lines into lines). Thus Theorem 7.1 doesn't apply directly. However, we can write the quadrilateral as the union of two triangles, apply Theorem 7.1 to each triangle, and note that the contributions of each triangle coming from the common boundary cancel each other out. Thus the divergence theorem does apply to our quadrilateral.

This procedure works in a quite general context. In fact, it works for all cases where we shall need the divergence theorem in this book, whether Theorem 7.1 applies directly or we can reduce to it by a finite subdivision of our domain, followed by a limiting argument. We shall not, however, formulate a general theorem covering all such cases; it is clear in each instance how to proceed.

EXERCISES

In Euclidean space we shall write div X instead of div $\langle X, \rho \rangle$ when ρ is taken to be the Euclidean volume density.

7.1 Let x, y, z be rectangular coordinates on $\mathbb{E}^3$. Let the vector field X be given by

$$X = r^2\left(x\frac{\partial}{\partial x} + y\frac{\partial}{\partial y} + z\frac{\partial}{\partial z}\right),$$

where $r^2 = x^2 + y^2 + z^2$. Show directly that

$$\int_S (X, n)\, dA = \int_B \operatorname{div} X$$

by integrating both sides. Here B is a ball centered at the origin and S is its boundary.

7.2 Let the vector field Y be given by

$$Y = Y_r n_r + Y_\theta n_\theta + Y_\varphi n_\varphi$$

in terms of polar "coordinates" r, θ, φ on $\mathbb{E}^3$, where n_r, n_θ and n_φ are the unit vectors in the directions $\partial/\partial r$, $\partial/\partial\theta$ and $\partial/\partial\varphi$ respectively. Show that

$$\operatorname{div} Y = \frac{1}{r^2 \sin\varphi}\left\{\frac{\partial}{\partial r}(r^2 \sin\varphi\, Y_r) + \frac{\partial}{\partial\theta}(rY_\theta) + \frac{\partial}{\partial\varphi}(r \sin\varphi\, Y_\varphi)\right\}.$$

7.3 Compute the divergence of a vector field in terms of polar coodrinates in the plane.

7.4 Compute the divergence of a vector field in terms of cylindrical coordinates in $\mathbb{E}^3$.

7.5 Let σ be the volume (area) density on the unit sphere S^2. Compute $\operatorname{div} \sigma X$ in terms of the coordinates θ, φ (polar coordinates) on the sphere.

CHAPTER 11

EXTERIOR CALCULUS

Let M be a differentiable manifold and let ω be a linear differential form in M. For any differentiable curve $C\colon [a, b] \to M$ we can consider the integral $\int_a^b \langle C'(t), \omega_{C(t)}\rangle\, dt$. Let $[c, d] \to [a, b]$ be a differentiable map given by $s \to t(s)$. The curve $B\colon [c, d] \to M$ given by $B(s) = C(t(s))$ satisfies

$$B'(s) = t'(s)C'(t(s)).$$

Thus if $t'(s) > 0$ for all s,

$$\int_c^d \langle B'(s), \omega_{B(s)}\rangle\, ds = \int_a^b \langle C'(t), \omega_{C(t)}\rangle\, dt.$$

Thus a linear differential form is something we can integrate over "oriented" curves of M and is independent of the parametrization. In this chapter we shall introduce objects which can be integrated over "oriented k-dimensional surfaces" of M and study their properties.

1. EXTERIOR DIFFERENTIAL FORMS

We defined a linear differential form to be a rule which assigns an element of $T_x^*(M)$ to each $x \in M$. We can regard $T_x^*(M)$ as $\mathfrak{A}^1(T_x(M))$. In view of this, we make the following generalization of this definition. By an exterior differential form of degree q on M we mean a rule which assigns an element of $\mathfrak{A}^q(T_x(M))$ to each $x \in M$. If ω is an exterior form of degree q and (U, α) is a chart, then, since α identifies each $T_x(M)$ with V for $x \in U$, we obtain an $\mathfrak{A}^q(V)$-valued function, ω_α, on $\alpha(U)$ defined by

$$\omega_\alpha(v)(\xi_\alpha^1, \ldots, \xi_\alpha^q) = \omega(x)(\xi^1, \ldots, \xi^q) \qquad \text{if } \; v = \alpha(x) \text{ and } \xi^1, \ldots, \xi^q \in T_x(M).$$

It is easy to write down the transition laws. In fact, if (W, β) is a second chart, we have

$$\omega_\beta(\beta(x))(\xi_\beta^1, \ldots, \xi_\beta^q) = \omega(x)(\xi^1, \ldots, \xi^q) = \omega_\alpha(\alpha(x))(\xi_\alpha^1, \ldots, \xi_\alpha^q)$$

or, since $\xi_\beta = J_{\beta\circ\alpha^{-1}}(\alpha(x))(\xi_\alpha)$ for $\xi \in T_x(M)$, we see that

$$\omega_\alpha(v)(\xi_\alpha^1, \ldots, \xi_\alpha^q) = \omega_\beta(\beta\circ\alpha^{-1}(v))(J_{\beta\circ\alpha^{-1}}(v)\xi_\alpha^1, \ldots, J_{\beta\circ\alpha^{-1}}(v)\xi_\alpha^q). \tag{1.1}$$

In order to write (1.1) in a less cumbersome form, we introduce the following notation. Let V_1 and V_2 be vector spaces, and let $l\colon V_1 \to V_2$ be a linear map.

We define $\mathcal{A}^p(l)$ to be the linear map of $\mathcal{A}^p(V_2) \to \mathcal{A}^p(V_1)$ given by

$$\mathcal{A}^p(l)(w)(v_1, \ldots, v_p) = w(l(v_1), \ldots, l(v_p))$$

for all $w \in \mathcal{A}^p(V_2)$ and $v_1, \ldots, v_p \in V_1$. Note that under the identification of $\mathcal{A}^1(V)$ with V^* the map $\mathcal{A}^1(l)$ coincides with the map $l^*: V_2^* \to V_1^*$. Note also that if $w_1 \in \mathcal{A}^p(V_2)$ and $w_2 \in \mathcal{A}^q(V_2)$, then

$$\mathcal{A}^p(l)w_1 \wedge \mathcal{A}^q(l)w_2 = \mathcal{A}^{p+q}(l)(w_1 \wedge w_2). \tag{1.2}$$

This follows directly from the definitions. Also, if $l_1: V_1 \to V_2$ and $l_2: V_2 \to V_3$, then

$$\mathcal{A}^p(l_2 \circ l_1) = \mathcal{A}^p(l_1) \circ \mathcal{A}^p(l_2). \tag{1.3}$$

It is clear that if l depends differentiably on some parameters, then so does $\mathcal{A}^p(l)$ for any p.

We can now write (1.1) as

$$\omega_\alpha(v) = \mathcal{A}^q(J_{\beta\circ\alpha^{-1}}(v))\omega_\beta(\beta \circ \alpha^{-1}(v)). \tag{1.1$'$}$$

It is clear from (1.1$'$) that it is consistent to require that ω_α be a smooth function. We therefore say that ω is a smooth differential form if all the functions ω_α are C^∞ on $\alpha(U)$ for all charts (U, α). As usual, it suffices to verify this for all charts in an atlas. We let $\bigwedge^q(M)$ denote the space of all smooth exterior forms of degree q.

Let $\omega_1 \in \bigwedge^p(M)$ and $\omega_2 \in \bigwedge^q(M)$. We define the exterior $(p+q)$-form $\omega_1 \wedge \omega_2$ by

$$(\omega_1 \wedge \omega_2)(x) = \omega_1(x) \wedge \omega_2(x) \qquad \text{for all} \quad x \in M.$$

It is easy to check that $\omega_1 \wedge \omega_2$ is a smooth $(p+q)$-form. We thus get a multiplication on exterior forms. To make the formalism complete, it is convenient to denote the space of differentiable functions on M by $\bigwedge^0(M)$ and to denote the product of a function f and a p-form ω by $f\omega$ or $f \wedge \omega$. This product is given by

$$(f \wedge \omega)(x) = (f\omega)(x) = f(x)\omega(x) \qquad \text{for all} \quad x \in M.$$

We have thus defined, for all $0 \leq p \leq n$ and $0 \leq q \leq n$, a multiplication sending $\omega_1 \in \bigwedge^p(M)$ and $\omega_2 \in \bigwedge^q(M)$ into $\omega_1 \wedge \omega_2 \in \bigwedge^{p+q}(M)$ (where $\omega_1 \wedge \omega_2 \equiv 0$ if $p + q > n = \dim M$). The rules for the $\wedge$-product on antisymmetric tensors carry over and thus, for instance,

$$\begin{aligned}
\omega_1 \wedge \omega_2 &= (-1)^{pq}\omega_2 \wedge \omega_1 \qquad \text{if} \quad \omega_1 \in \textstyle\bigwedge^p(M) \text{ and } \omega_2 \in \bigwedge^q(M),\\
\omega_1 \wedge (\omega_2 \wedge \omega_2) &= (\omega_1 \wedge \omega_2) \wedge \omega_3,\\
\omega_1 \wedge (\omega_2 + \omega_3) &= \omega_1 \wedge \omega_2 + \omega_1 \wedge \omega_3,
\end{aligned}$$

and so on.

Let M_1 and M_2 be differentiable manifolds, and let $\varphi: M_1 \to M_2$ be a differentiable map. For each $\omega \in \bigwedge^q(M_2)$ we define the form $\varphi^*\omega \in \bigwedge^q(M_1)$ by

$$\varphi^*\omega(x) = \mathcal{A}^q(\varphi_{*x})(\omega(\varphi(x))). \tag{1.4}$$

It is easy to check that $\varphi^*\omega$ is indeed an element of $\bigwedge^q(M_1)$, that is, it is a smooth q-form. Note also that (7.5) of Chapter 9 is a special case of (1.4)—the case $q = 1$. (If we make the convention that $\mathfrak{A}^0(l) = \text{id}$, then the case $q = 0$ of (1.4) is the rule for pullback of functions.)

It follows from (1.4) that φ^* is linear, that is,

$$\varphi^*(\omega_1 + \omega_2) = \varphi^*(\omega_1) + \varphi^*(\omega_2), \tag{1.5}$$

and from (1.2) that

$$\varphi^*(\omega_1 \wedge \omega_2) = \varphi^*(\omega_1) \wedge \varphi^*(\omega_2). \tag{1.6}$$

If φ is a one-parameter group on a manifold M with infinitesimal generator X, then we can show that the

$$\lim_{t \to 0} \frac{\overset{*}{\varphi_t}\omega - \omega}{t} = D_X\omega$$

exists for any $\omega \in \bigwedge^q(M)$. The proof of the existence of this limit is straightforward and will be omitted. We shall derive a useful formula allowing a simple calculation of $D_X\omega$ in Section 3.

Let us now see how to compute with the $\bigwedge^q(M)$ in terms of local coordinates. Let (U, α) be a chart of M with coordinates $x^1, \ldots, x^n$. Then $dx^i \in \bigwedge^1(U)$ (where by $\bigwedge^q(U)$ we mean the set of differentiable q-forms defined on U). For any $i_1, \ldots, i_q$ the form $dx^{i_1} \wedge \cdots \wedge dx^{i_q}$ belongs to $\bigwedge^q(U)$, and for every $x \in U$ the forms

$$\{(dx^{i_1} \wedge \cdots \wedge dx^{i_q})(x)\}_{i_1 < \cdots < i_q}$$

form a basis for $\mathfrak{A}^q(T_x(M))$. From this it follows that every exterior form ω of degree q which is defined on U can be written as

$$\omega = \sum_{i_1 < \cdots < i_q} a_{i_1, \ldots, i_q}\, dx^{i_1} \wedge \cdots \wedge dx^{i_q}, \tag{1.7}$$

where the a's are functions; that is,

$$\omega(x) = \sum_{i_1 < \cdots < i_q} a_{i_1, \ldots, i_q}(x)(dx^{i_1} \wedge \cdots \wedge dx^{i_q})(x)$$

for all $x \in U$. It is easy to see that $\omega \in \bigwedge^q(U)$ if and only if all the functions $a_{i_1, \ldots, i_q}$ are C^∞-functions on U.

If (W, β) is a second chart with coordinates $y^1, \ldots, y^n$ and

$$\omega = \sum b_{j_1, \ldots, j_q}\, dy^{j_1} \wedge \cdots \wedge dy^{j_q}, \tag{1.8}$$

then it is easy to compute the transition law relating the b's to the a's on $U \cap W$.

In fact, on $U \cap W$ we have

$$dy^j = \sum \frac{\partial y^j}{\partial x^i}\, dx^i, \tag{1.9}$$

where $y^j = y^j(x^1, \ldots, x^n)$. Then all we have to do is to substitute (1.9) into (1.8) and collect the coefficients of $dx^{i_1} \wedge \cdots \wedge dx^{i_q}$. For instance, if $q = 2$, then we have

$$\begin{aligned}\omega &= \sum_{j_1<j_2} b_{j_1 j_2}\, dy^{j_1} \wedge dy^{j_2}\\ &= \sum_{j_1<j_2} b_{j_1 j_2}\left(\frac{\partial y^{j_1}}{\partial x^1}\, dx^1 + \cdots + \frac{\partial y^{j_1}}{\partial x^n}\, dx^n\right) \wedge \left(\frac{\partial y^{j_2}}{\partial x^1}\, dx^1 + \cdots + \frac{\partial y^{j_2}}{\partial x^n}\, dx^n\right).\end{aligned}$$

If we collect the coefficients of $dx^{i_1} \wedge dx^{i_2}$ (remember the $\wedge$-multiplication is anticommutative), we get

$$\omega = \sum_{i_1<i_y}\left[\sum_{j_1<j_2} b_{j_1 j_2}\left(\frac{\partial y^{j_1}}{\partial x^{i_1}}\frac{\partial y^{j_2}}{\partial x^{i_2}} - \frac{\partial y^{j_2}}{\partial x^{i_1}}\frac{\partial y^{j_1}}{\partial x^{i_2}}\right)\right] dx^{i_1} \wedge dx^{i_2}.$$

Thus

$$a_{i_1 i_2} = \sum_{j_1<j_2} b_{j_1 j_2} \det \begin{bmatrix} \dfrac{\partial y^{j_1}}{\partial x^{i_1}} & \dfrac{\partial y^{j_2}}{\partial x^{i_1}} \\[2ex] \dfrac{\partial y^{j_1}}{\partial x^{i_2}} & \dfrac{\partial y^{j_2}}{\partial y^{i_2}} \end{bmatrix}. \tag{1.10}$$

Although (1.10) looks a little formidable, the point is that all one has to remember is (1.9) and the law for $\wedge$-multiplication. For general q the same argument gives

$$a_{i_1,\ldots,i_q} = \sum_{j_1<\cdots<j_q} b_{j_1,\ldots,j_q} \det \begin{bmatrix} \dfrac{\partial y^{j_1}}{\partial x^{i_1}} & \cdots & \dfrac{\partial y^{j_q}}{\partial x^{i_1}} \\ \vdots & & \vdots \\ \dfrac{\partial y^{j_1}}{\partial x^{i_q}} & \cdots & \dfrac{\partial y^{j_q}}{\partial x^{i_q}} \end{bmatrix}. \tag{1.11}$$

The formula for pullback takes exactly the same form. Let $\varphi\colon M_1 \to M_2$ be a differentiable map, and suppose that (U, α) and (W, β) are compatible charts, where $x^1, \ldots, x^m$ are the coordinates of (U, α) and $y^1, \ldots, y^n$ are those of (W, β). Then we get that $y^i \circ \varphi$ are functions on U and can thus be written as

$$y^j \circ \varphi = y^j(x^1, \ldots, x^m).$$

Since $\varphi^* dy^j = d(y^j \circ \varphi)$, we have

$$\varphi^*(dy^j) = \sum \frac{\partial y^j}{\partial x^i}\, dx^i. \tag{1.12}$$

If

$$\omega = \sum_{j_1<\cdots<j_q} b_{j_1,\ldots,j_q}\, dy^j \wedge \cdots \wedge dy^{j_q} \in \textstyle\bigwedge^q(W_2)$$

then, by (1.5) and (1.6),

$$\varphi^*(\omega) = \sum_{j_1<\cdots<j_q} (b_{j_1,\ldots,j_q} \circ \varphi)(\varphi^* dy^{j_1}) \wedge \cdots \wedge (\varphi^* dy^{j_q}). \tag{1.13}$$

The expression for (1.13) in terms of the dx's can be computed by substituting (1.12) into (1.13) and collecting coefficients. The answer, of course, will look just like it did before. If

$$\varphi^*(\omega) = \sum_{i_1<\cdots<i_q} a_{i_1,\ldots,i_q}\, dx^{i_1} \wedge \cdots \wedge dx^{i_q},$$

then the a's are given by

$$a_{i_1,\ldots,i_q} = \sum_{j_1<\cdots<j_q} (b_{j_1,\ldots,j_q} \circ \varphi) \det \begin{bmatrix} \dfrac{\partial y^{j_1}}{\partial x^{i_1}} & \cdots & \dfrac{\partial y^{j_q}}{\partial x^{i_1}} \\ \vdots & & \vdots \\ \dfrac{\partial y^{j_1}}{\partial x^{i_q}} & \cdots & \dfrac{\partial y^{j_q}}{\partial x^{i_q}} \end{bmatrix}. \tag{1.14}$$

Again, we emphasize that there is no need to remember a complicated looking formula like (1.14); Eqs. (1.5), (1.6), and (1.12) (and of course the rules for $\wedge$-multiplication) are sufficient. In many cases, it is much more convenient to do the substitutions directly than to use (1.14).

2. ORIENTED MANIFOLDS AND THE INTEGRATION OF EXTERIOR DIFFERENTIAL FORMS

Let M be an n-dimensional manifold. Let (U, α) and (W, β) be two charts on M with coordinates $x^1, \ldots, x^n$ and $y^1, \ldots, y^n$. Let ω be an exterior differential form of degree n. Then we can write

$$\omega = a\, dx^1 \wedge \cdots \wedge dx^n \qquad \text{on } U$$

and

$$\omega = b\, dy^1 \wedge \cdots \wedge dy^n \qquad \text{on } W,$$

where the functions a and b are related on $U \cap W$ by (1.11), which, in this case $(q = n)$, becomes

$$a = b \det \begin{bmatrix} \dfrac{\partial y^1}{\partial x^1} & \cdots & \dfrac{\partial y^n}{\partial x^1} \\ \vdots & & \vdots \\ \dfrac{\partial y^1}{\partial x^n} & \cdots & \dfrac{\partial y^n}{\partial x^n} \end{bmatrix}$$

or

$$a_\alpha(\alpha(x)) = b_\beta(\beta \circ \alpha^{-1}(\alpha(x))) \det J_{\beta\circ\alpha^{-1}}(\alpha(x))$$

or, finally,

$$a_\alpha(v) = b_\beta(\beta \circ \alpha^{-1}(v)) \det J_{\beta\circ\alpha^{-1}}(v) \qquad \text{for} \quad v \in \alpha(U \cap W). \tag{2.1}$$

If ρ is a density on M, then the transition laws for ρ_α are given by

$$\rho_\alpha(v) = \rho_\beta(\beta \circ \alpha^{-1}(v))|\det J_{\beta\circ\alpha^{-1}}(v)|. \tag{2.2}$$

Note that (2.2) and (2.1) look almost the same; the difference is the absolute-value sign that occurs in (2.2) but not in (2.1). In particular, if (U, α) and (W, β) were such that $\det J_{\beta\circ\alpha^{-1}} > 0$, then (2.2) and (2.1) would agree for this pair of charts.

This leads us to the following definition: An atlas $\mathcal{A}$ of M is said to be *oriented* if for any pair of charts (U, α) and (W, β) of $\mathcal{A}$ we have

$$\det J_{\beta\circ\alpha^{-1}}(\alpha(x)) > 0 \qquad \text{for all} \quad x \in U \cap W.$$

There is no guarantee that there exists an oriented atlas on a given manifold M. In fact, it is not difficult to show that there does not exist an oriented atlas on certain manifolds. (An example of a manifold possessing no oriented atlas is the Möbius strip.)

We say that a manifold M is *orientable* if it has an oriented atlas.

Let M be an orientable manifold, and let $\mathcal{A}_1$ and $\mathcal{A}_2$ be two oriented atlases. We say that $\mathcal{A}_1$ and $\mathcal{A}_2$ have the same orientation, and write $\mathcal{A}_1 \,\tilde{o}\, \mathcal{A}_2$, if $\mathcal{A}_1 \cup \mathcal{A}_2$ is again an oriented atlas. To say that $\mathcal{A}_1 \,\tilde{o}\, \mathcal{A}_2$ means that for any $(U, \alpha) \in \mathcal{A}_1$ and any $(W, \beta) \in \mathcal{A}_2$ we have

$$\det J_{\beta\circ\alpha^{-1}}(v) > 0 \qquad \text{on} \quad \alpha(U \cap W).$$

It is clear that $\tilde{o}$ is an equivalence relation. An equivalence class of oriented atlases is called an *orientation* of M. An orientable manifold, together with a choice of orientation, will be called an *oriented* manifold. We shall denote an oriented manifold by $\mathbf{M}$. That is, $\mathbf{M}$ is a manifold M together with a choice of orientation. Thus an oriented one-dimensional manifold has a preferred direction at each point (Fig. 11.1); an oriented two-dimensional manifold has a notion of clockwise versus counterclockwise direction (Fig. 11.2); and at any point of an oriented three-dimensional manifold we can distinguish between right- and left-handedness.

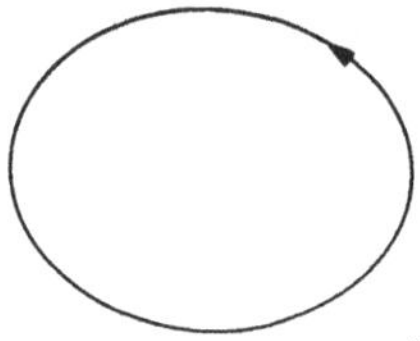

Fig. 11.1

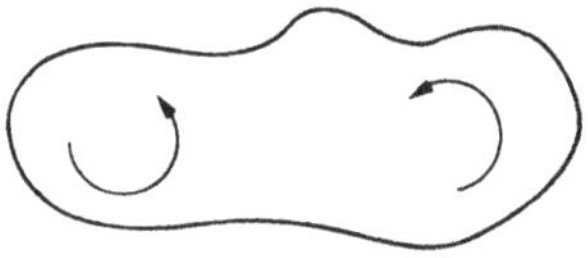

Fig. 11.2

In general, let $\mathbf{M}$ be an oriented manifold, and let (U, α) be a chart of M with coordinates $x^1, \ldots, x^n$. We say that (U, α) is a *positive* chart if $J_{\beta\circ\alpha^{-1}} > 0$ for any chart (W, β) belonging to any oriented atlas defining (i.e., belonging to) the orientation. (It suffices to check this, of course, for all (W, β) belonging to one fixed atlas defining the orientation.) Note that if U is connected, then if (U, α) is not positive, then the chart (U, α^1), where

$$\alpha^1(x) = \langle -v^1, v^2, \ldots, v^n \rangle \qquad \text{if} \quad \alpha(x) = \langle v^1, v^2, \ldots, v^n \rangle,$$

is a positive chart.

We shall say that (U, α) is a negative chart if $\det J_{\beta\circ\alpha^{-1}} < 0$ for all (W, β) belonging to an atlas defining the orientation. (Thus, if U is connected, then (U, α) must be either positive or negative.)

We now return to our initial observation comparing (2.1) with (2.2).

Proposition 2.1. Let **M** be an oriented n-dimensional manifold. We can identify exterior forms of degree n with densities by sending the form ω into the density ρ^ω, where for any positive chart (U, α) with coordinates $x^1, \ldots, x^n$, the function ρ^ω_α is determined by

$$\omega = \rho^\omega_\alpha(\alpha(\cdot))\, dx^1 \wedge \cdots \wedge dx^n \qquad \text{on} \quad U. \tag{2.3}$$

Another way of writing (2.3) is

$$\omega(\partial/\partial x^1, \ldots, \partial/\partial x^n) = \rho(\partial/\partial x^1, \ldots, \partial/\partial x^n). \tag{2.3$'$}$$

In other words, if $\omega = a\, dx^1 \wedge \cdots \wedge dx^n$ on U, then $\rho^\omega_\alpha(v) = a_\alpha$. That ρ^ω is really a density follows from the fact that (2.2) reduces to (2.1) for all pairs of charts belonging to a positive atlas.

It is clear that this identification is additive,

$$\rho^{\omega_1+\omega_2} = \rho^{\omega_1} + \rho^{\omega_2}, \tag{2.4}$$

and that for any function,

$$\rho^{f\omega} = f\rho^\omega. \tag{2.5}$$

Furthermore, if $\omega(x) = 0$, then $\rho^\omega = 0$ at x. By the support of a differential form we mean, as usual, the closure of the set of x for which $\omega(x) \neq 0$. We say that an n-form ω is locally absolutely integrable if the density ρ^ω is locally absolutely integrable. Note that to say that ω is locally absolutely integrable means that for any chart (U, α), with coordinates $x^1, \ldots, x^n$ of some atlas $\mathfrak{A}$, if

$$\omega = a\, dx^1 \wedge \cdots \wedge dx^n \qquad \text{on} \quad U,$$

then the function $a_\alpha = a \circ \alpha^{-1}$ is an absolutely integrable function on $\alpha(U)$. Let $\Gamma(M)$ denote the space of absolutely integrable n-forms of compact support. It is clear that $\Gamma(M)$ is a vector space and that $f\omega \in \Gamma(M)$ if f is a (bounded) contented function and $\omega \in \Gamma(M)$. As a consequence of Proposition 2.1 and Theorem 3.1 of Chapter 10, we can state:

Theorem 2.1. Let **M** be an oriented manifold. There exists a unique linear function $\int$ on $\Gamma(M)$ satisfying the following condition: If $\operatorname{supp} \omega \subset U$, where (U, α) is a positive chart with coordinates $x^1, \ldots, x^n$, and if $\omega = a\, dx^1 \wedge \cdots \wedge dx^n$, then

$$\int \omega = \int_{\alpha(U)} a_\alpha. \tag{2.6}$$

Observe that we can write

$$\int \omega = \int \rho^\omega \qquad \text{for all} \quad \omega \in \Gamma(M). \tag{2.7}$$

The recipe for computing $\int \omega$ is now very simple. We break ω up into small pieces such that each piece lies in some U. (We can ignore sets of content zero in the process.) If supp $\omega \subset U$, and if (U, α) *is a positive chart*, we express ω as

$$\omega = a\, dx^1 \wedge \cdots \wedge dx^n.$$

And if a is given as $a = a_\alpha(x^1, \ldots, x^n)$, we integrate the function a_α over $\mathbb{R}^n$. The computations are automatic. Thus one point that has to be checked is that the chart (U, α) is positive. If it is negative, then $\int \omega$ is given by $-\int a_\alpha$.

Let $\mathbf{M}_1$ be an oriented manifold of dimension q, let $\varphi\colon \mathbf{M}_1 \to M_2$ be a differentiable map, and let $\omega \in \bigwedge^q(M_2)$. Then for any contented compact set $A \subset \mathbf{M}_1$ the form $e_A\varphi^*(\omega)$ belongs to $\Gamma(M_1)$, so we can consider its integral. This integral is sometimes denoted by $\int_{\varphi(A)} \omega$; that is, we make the definition

$$\int_{\varphi(A)} \omega = \int_{\mathbf{M}_1} e_A\varphi^*\omega. \tag{2.8}$$

If we regard $\varphi(A)$ as an "oriented q-dimensional surface" in M_2, then we see that the elements of $\bigwedge^q(M_2)$ are objects that we can integrate over such "surfaces". (Of course, if $q = 1$, we say "curves".)

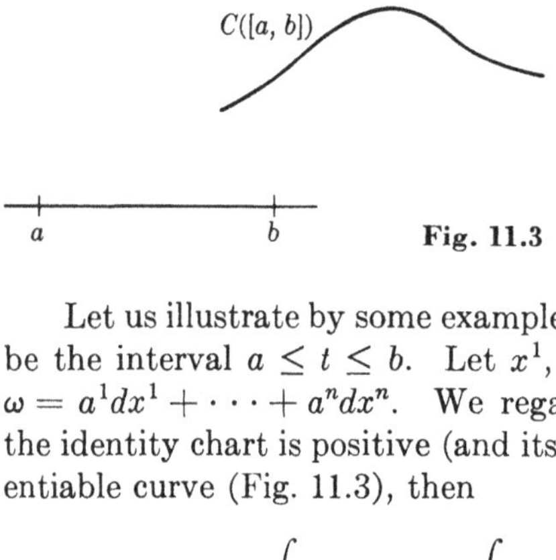

Fig. 11.3

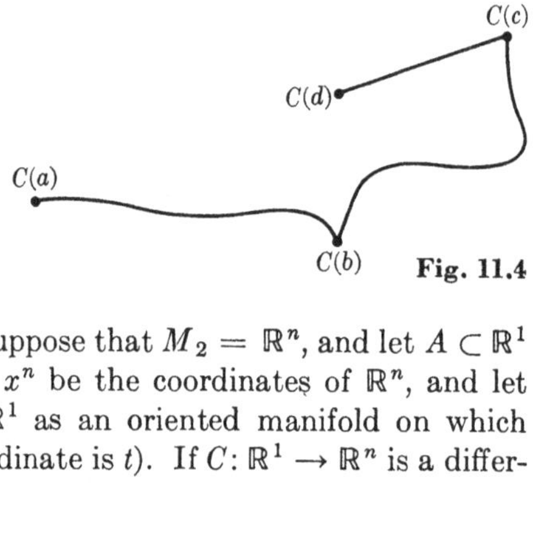

Fig. 11.4

Let us illustrate by some examples. Suppose that $M_2 = \mathbb{R}^n$, and let $A \subset \mathbb{R}^1$ be the interval $a \le t \le b$. Let $x^1, \ldots, x^n$ be the coordinates of $\mathbb{R}^n$, and let $\omega = a^1dx^1 + \cdots + a^ndx^n$. We regard $\mathbb{R}^1$ as an oriented manifold on which the identity chart is positive (and its coordinate is t). If $C\colon \mathbb{R}^1 \to \mathbb{R}^n$ is a differentiable curve (Fig. 11.3), then

$$\begin{aligned}\int_{C([a,b])} \omega &= \int e_{[a,b]}C^*(\omega)\\ &= \int_a^b \left(a^1 \frac{dx^1}{dt} + \cdots + a^n \frac{dx^n}{dt}\right) dt\\ &= \int_a^b \langle C'(t), \omega\rangle\, dt. \end{aligned}\tag{2.9}$$

From this last expression we see that C does not have to be differentiable everywhere in order for $\int_{C([a,b])} \omega$ to make sense. In fact, if C is differentiable everywhere on $\mathbb{R}$ except at a finite number of points, and if $C'(t)$ is always bounded (when regarded as an element of $\mathbb{R}^n$), then the function $\langle C'(\cdot), \omega\rangle$ is defined everywhere except for a set of content zero and is bounded. Thus

$C^*(\omega)$ is a contented density and (2.9) still makes sense. Now the curve can have corners. (See Fig. 11.4.)

It should be observed that if $\omega = df$ (and if C is continuous), then

$$\int_{C([a,b])} df = \int_{C([a,b])} d(f \circ C) = \int_a^b (f \circ C)'$$
$$= f(C(b)) - f(C(a)). \tag{2.10}$$

In this case the integral depends not on the particular curve C but on the end-points. In general, $\int_C \omega$ depends on the curve C. We will obtain conditions for it to be independent of C in Section 5.

In the next example let $M_2 = \mathbb{R}^3$ and $M_1 = U \subset \mathbb{R}^2$ where (u, v) are Euclidean coordinates on $\mathbb{R}^2$ and x, y, z are Euclidean coordinates on $\mathbb{R}^3$. Let

$$\omega = P\,dx \wedge dy + Q\,dx \wedge dz + R\,dy \wedge dz$$

be an element of $\bigwedge^2(\mathbb{R}^3)$. If $\varphi: U \to \mathbb{R}^3$ is given by the functions $x(u, v)$, $y(u, v)$, and $z(u, v)$, then for $A \subset U$,

$$\int_{\varphi(A)} \omega = \int e_A \varphi^* \omega = \int e_A \varphi^*(P\,dx \wedge dy + Q\,dx \wedge dz + R\,dy \wedge dz)$$
$$= \int_A \left[(P \circ \varphi) \left(\frac{\partial x}{\partial u} \frac{\partial y}{\partial v} - \frac{\partial y}{\partial u} \frac{\partial x}{\partial v} \right) + (Q \circ \varphi) \left(\frac{\partial x}{\partial u} \frac{\partial z}{\partial v} - \frac{\partial z}{\partial u} \frac{\partial x}{\partial v} \right) \right.$$
$$\left. + (R \circ \varphi) \left(\frac{\partial y}{\partial u} \frac{\partial z}{\partial v} - \frac{\partial z}{\partial u} \frac{\partial y}{\partial v} \right) \right].$$

We conclude this section with another look at the volume density of Riemann metrics, this time for an oriented manifold. If **M** is an oriented manifold with a Riemann metric, then the volume density σ corresponds to an n-form Ω. By our rule for this correspondence, if (U, α) is a positive chart with coordinates $x^1, \ldots, x^n$, then

$$\Omega = a\,dx^1 \wedge \cdots \wedge dx^n,$$

where, by (4.1) of Chapter 10, $a(x) = |\det(g_{ij})|^{1/2}$ is the volume in $T_x(M)$ of the parallelepiped spanned by

$$\frac{\partial}{\partial x^1}(x), \ldots, \frac{\partial}{\partial x^n}(x).$$

Let $e_1(x), \ldots, e_n(x)$ be an orthonormal basis of $T_x(M)$ (relative to the scalar product given by the Riemann metric). Then

$$|\det(g_{ij})|^{1/2} = \left| \det\left[\left(\frac{\partial}{\partial x^i}, e_j \right) \right] \right| = |\det A|,$$

where $A = (\partial/\partial x^i, e_j)$ is the matrix of the linear transformation carrying $e_j \to \partial/\partial x^j$. If $\omega^1(x), \ldots, \omega^n(x)$ is the dual basis of the e's, then

$$\omega^1(x) \wedge \cdots \wedge \omega^n(x) = \det A\, dx^1(x) \wedge \cdots \wedge dx^n(x).$$

Now $\omega^1(x), \ldots, \omega^n(x)$ can be any orthonormal basis of $T_x^*(M)$. [$T_x^*(M)$ has a scalar product, since it is the dual space of the scalar product space $T_x(M)$.] We thus get the following result: If $\omega^1, \ldots, \omega^n$ are linear differential forms such that for each $x \in M$, $\omega^1(x), \ldots, \omega^n(x)$ is an orthonormal basis of $T_x^*(M)$, then

$$\Omega = \pm\omega^1 \wedge \cdots \wedge \omega^n.$$

We can write

$$\Omega = \omega^1 \wedge \cdots \wedge \omega^n \tag{2.11}$$

if we know that $\omega^1 \wedge \cdots \wedge \omega^n$ is a positive multiple of $dx^1 \wedge \cdots \wedge dx^n$. Can we always find such forms $\omega^1, \ldots, \omega^n$ on U? The answer is "yes": we can do it by applying the orthonormalization procedure to $dx^1, \ldots, dx^n$. That is, we set

$$\omega^1 = \frac{dx^1}{\|dx^1\|}, \qquad \text{where } \|dx^1\|(x) = \|dx^1(x)\| > 0 \text{ is a } C^\infty\text{-function on } U,$$

$$\omega^2 = \frac{dx^2 - (dx^2, \omega^1)\,\omega^1}{\|dx^2 - (dx^2, \omega^1)\,\omega^1\|},$$

$$\vdots$$

The matrix which relates the dx's to the ω's is composed of C^∞-functions, so that the $\omega^i \in \bigwedge^1(U)$. Furthermore, it is a triangular matrix with positive entries on the diagonal, so its determinant is positive. We have thus constructed the desired forms $\omega^1, \ldots, \omega^n$, so (2.11) holds. For instance, it follows from Eq. (9.10), Chapter 9, that $d\theta$, $\sin\theta\, d\varphi$ form an orthonormal basis for $T_x(S^2)$ at all $x \in S^2$ (except the north and south poles). If we choose the orientation on S^2 so that θ, φ form a positive chart, then the volume form is given by

$$\Omega = \sin\theta\, d\theta \wedge d\varphi.$$

3. THE OPERATOR *d*

With every function f we have associated a linear differential form df. We can thus regard d as a map from $\bigwedge^0(M)$ to $\bigwedge^1(M)$. As such, it is linear and satisfies

$$d(f_1f_2) = f_2\, df_1 + f_1\, df_2.$$

We now seek to define a $d\colon \bigwedge^k(M) \to \bigwedge^{k+1}(M)$ for $k > 0$ as well. We shall require that d be linear and satisfy some identity with regard to multiplication, generalizing the above formula for $d(f_1f_2)$. The condition we will impose is that

$$d(\omega_1 \wedge \omega_2) = d\omega_1 \wedge \omega_2 + (-1)^p\omega_1 \wedge d\omega_2$$

if ω_1 is a form of degree p. The factor $(-1)^p$ accounts for the anticommutativity of $\wedge$. The reader should check that d is consistent with this law, at least to the extent that $d(\omega_1 \wedge \omega_2) = (-1)^{pq}\, d(\omega_2 \wedge \omega_1)$ if ω_1 is of degree p and ω_2 is of degree q.

We are going to impose one further condition on d which will uniquely determine it. This condition (which lies at the heart of the matter) requires

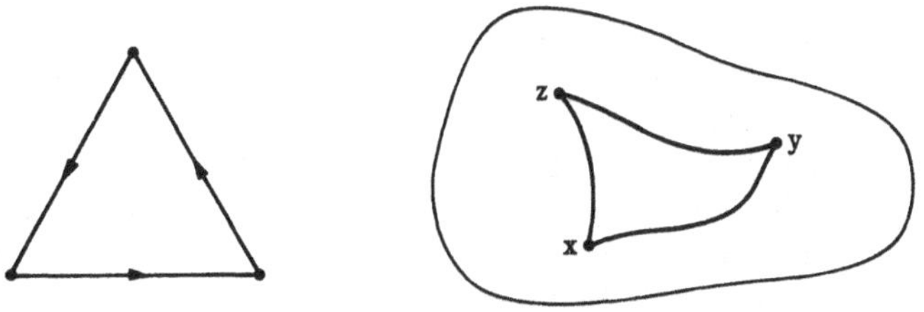

Fig. 11.5

some introduction. Let f be a differentiable function, and let $C: I \to M$ be a differentiable curve. For any points $a, b \in I$, the fundamental theorem of the calculus implies that

$$f(C(b)) - f(C(a)) = \int_a^b \frac{d(f \circ C)}{dt}\, dt = \int_a^b C^*\, df.$$

We can regard b and a (with $\pm$ signs attached) as the "oriented boundary" of the interval $[a, b]$. Let us make the convention that "integrating" an element of $\bigwedge^0(p)$ is just evaluating the function at the point p. As such, the equation above says that the integral of the "pullback" of f over the "boundary", that is, $f(b) - f(a)$, equals the integral of the "pullback" of df over $[a, b]$. In some sense, we would like to be able to say that if ω is a form of degree k, then the integral of the "pullback" of ω over the k-dimensional boundary" of a $(k+1)$-dimensional region is equal to the integral of the pullback of $d\omega$ over the $(k+1)$-dimensional region. Without trying to make this requirement precise, let us see what it says for the case where $k = 1$ and the region is a triangle in the plane. Let φ be a smooth map of some neighborhood of the triangle $\Delta \subset \mathbb{R}^2$ into M, and let the vertices of Δ be mapped by φ into $\mathbf{x}$, $\mathbf{y}$, and $\mathbf{z}$ (see Fig. 11.5). The boundary of Δ consists of three curves (segments) C_1, C_2, and C_3 (with the proper orientations). Let ω be a linear differential form on M. We would then expect that

$$\int \varphi^*\, d\omega = \int C_1^* \varphi^* \omega + \int C_2^* \varphi^* \omega + \int C_3^* \varphi^* \omega.$$

If $\omega = df$, then the three integrals on the right become (by the fundamental theorem of the calculus) $f(\mathbf{y}) - f(\mathbf{x}) + f(\mathbf{z}) - f(\mathbf{y}) + f(\mathbf{x}) - f(\mathbf{z}) = 0$. Thus $\int \varphi^*\, d(df) = 0$. Since the triangle was arbitrary, we expect that

$$d(df) = 0.$$

We now assert:

Theorem 3.1. There exists a unique linear map $d: \bigwedge^k(M) \to \bigwedge^{k+1}(M)$ such that on $\bigwedge^0$ it coincides with the old d and such that

$$d(\omega_1 \wedge \omega_2) = d\omega_1 \wedge \omega_2 + (-1)^p \omega_1 \wedge d\omega_2 \qquad \text{if} \quad \omega_1 \in \textstyle\bigwedge^p(M) \tag{3.1}$$

and

$$d(df) = 0 \qquad \text{if} \quad f \in \textstyle\bigwedge^0(M). \tag{3.2}$$

Proof. We first establish the uniqueness of d. To do this we observe that (3.1) implies that d is *local*, in the sense that if $\omega = \omega'$ on some open set U, then $d\omega = d\omega'$ on U. In fact, let W be an open set with $\overline{W} \subset U$, and let φ be a C^∞-function such that $\varphi(x) \equiv 1$ for $x \in W$ and supp $\varphi \subset U$. Then $\varphi\omega = \varphi\omega'$ everywhere on M, and thus $d(\varphi\omega) = d(\varphi\omega')$. But, by (3.1), $d(\varphi\omega) = \varphi\, d\omega + d\varphi \wedge \omega = d\omega$ on W, since $\varphi \equiv 1$ and $d\varphi = 0$ there. Thus $d\omega = d\omega'$ on W. Since W can be arbitrary, we conclude that $d\omega = d\omega'$ on U.

Let (U, α) be a chart with coordinates $x^1, \ldots, x^n$. Every $\omega \in \bigwedge^k(M)$ can be written as

$$\omega = \sum_{i_1 < \cdots < i_k} a_{i_1,\ldots,i_k}\, dx^{i_1} \wedge \cdots \wedge dx^{i_k} \qquad \text{on} \quad U.$$

Now [by induction on k, using (3.1) and (3.2)] $d(dx^{i_1} \wedge \cdots \wedge dx^{i_k}) = 0$. Thus (3.1) implies that

$$d\omega = \sum da_{i_1,\ldots,i_k} \wedge dx^{i_1} \wedge \cdots \wedge dx^{i_k} \qquad \text{on} \quad U. \tag{3.3}$$

Equation (3.3) gives a local formula for d. It also shows that d is unique. In fact, we have shown that there is at most one operator d on any open subset $O \subset M$ mapping $\bigwedge^k(O) \to \bigwedge^{k+1}(O)$ and satisfying the hypotheses of the theorem (for O). On the set $O \cap U$ it must be given by (3.3).

We now claim that in order to establish the existence of d, it suffices to show that the d given by (3.3) [in any chart (U, α)] satisfies the requirement of the theorem on $\bigwedge^k(U)$. In fact, suppose we have shown this to be so. Let $\mathcal{A}$ be an atlas of M, and for each chart $(U, \alpha) \in \mathcal{A}$ define the operator $d_\alpha\colon \bigwedge^k(U) \to \bigwedge^{k+1}(U)$ by (3.3). We would like to set $d\omega = d_\alpha\omega$ on U. For this to be consistent, we must show that $d_\alpha\omega = d_\beta\omega$ on $U \cap W$ if (W, β) is some other chart. But both d_α and d_β satisfy the hypotheses of the theorem on $U \cap W$, and they must therefore coincide there.

Thus to prove the theorem, it suffices to check that the operator d, defined by (3.3), fulfills our requirements as a map of $\bigwedge^k(U) \to \bigwedge^{k+1}(U)$. It is obviously linear. To check (3.2), we observe that

$$df = \sum \frac{\partial f}{\partial x^i}\, dx^i,$$

so

$$\begin{aligned} d(df) &= \sum d\left(\frac{\partial f}{\partial x^i}\right) \wedge dx^i = \sum_{i,j} \left(\frac{\partial^2 f}{\partial x^i\, \partial x^j}\, dx^j\right) \wedge dx^i \\ &= \sum_{i<j} \left(\frac{\partial^2 f}{\partial x^j\, \partial x^i} - \frac{\partial^2 f}{\partial x^i\, \partial x^j}\right) dx^i \wedge dx^j \\ &= 0 \end{aligned}$$

by the equality of mixed partials.

Now we turn to (3.1). Since both sides of (3.1) are linear in ω_1 and ω_2 separately, it suffices to check (3.1) for $\omega_1 = a\, dx^{i_1} \wedge \cdots \wedge dx^{i_p}$ and $\omega_2 =$

$b\,dx^{j_1} \wedge \cdots \wedge dx^{j_q}$. Now $\omega_1 \wedge \omega_2 = ab\,dx^{i_1} \wedge \cdots \wedge dx^{i_p} \wedge dx^{j_1} \wedge \cdots \wedge dx^{j_q}$ and $d(ab) = b\,da + a\,db$; therefore,

$$\begin{aligned} d(\omega_1 \wedge \omega_2) &= b\,da \wedge dx^{i_1} \wedge \cdots \wedge dx^{i_p} \wedge dx^{j_1} \wedge \cdots \wedge dx^{j_q} \\ &\quad + a\,db \wedge dx^{i_1} \wedge \cdots \wedge dx^{i_p} \wedge dx^{j_1} \wedge \cdots \wedge dx^{j_q}, \end{aligned}$$

while

$$\begin{aligned} d\omega_1 \wedge \omega_2 &= (da \wedge dx^{i_1} \wedge \cdots \wedge dx^{i_p}) \wedge (b\,dx^{j_1} \wedge \cdots \wedge dx^{j_q}) \\ &= b\,da \wedge dx^{i_1} \wedge \cdots \wedge dx^{i_p} \wedge dx^{j_1} \wedge \cdots \wedge dx^{j_q} \end{aligned}$$

and

$$\begin{aligned} \omega_1 \wedge d\omega_2 &= (a\,dx^{i_1} \wedge \cdots \wedge dx^{i_p}) \wedge (db \wedge dx^{j_1} \wedge \cdots \wedge dx^{j_q}) \\ &= (-1)^p a\,db \wedge dx^{i_1} \wedge \cdots \wedge dx^{i_p} \wedge dx^{j_1} \wedge \cdots \wedge dx^{j_q}, \end{aligned}$$

so we see that (3.1) holds. This proves the theorem. □

We can draw a number of important corollaries from Eq. (3.3).

First of all, it follows immediately that for $\omega \in \bigwedge^k(M)$, for any k, we have

$$d(d\omega) = 0. \tag{3.4}$$

(Remember we merely assumed it for $k = 0$.)

Secondly, let $\varphi\colon M_1 \to M_2$ be a differentiable map. Then for $\omega \in \bigwedge^k(M_2)$ we have

$$d\varphi^*\omega = \varphi^*\,d\omega. \tag{3.5}$$

To check (3.5), it suffices to verify it for any pair of compatible charts. But if $x^1, \ldots, x^n$ are coordinates on M_2 and, locally,

$$\omega = \sum a_{i_1,\ldots,i_k}\,dx^{i_1} \wedge \cdots \wedge dx^{i_k},$$

we have

$$\begin{aligned} d\varphi^*\omega &= d\left(\sum \varphi^*(a_{i_1,\ldots,i_k})\varphi^*\,dx^{i_1} \wedge \cdots \wedge \varphi^*\,dx^{i_k}\right) \\ &= \sum d\varphi^*(a_{i_1,\ldots,i_k}) \wedge d\varphi^*x^{i_1} \wedge \cdots \wedge d\varphi^*x^{i_k} \\ &= \sum \varphi^*\,da_{i_1,\ldots,i_k} \wedge d\varphi^*x^{i_1} \wedge \cdots \wedge d\varphi^*x^{i_k} \\ &= \varphi^*\left(\sum da_{i_1,\ldots,i_k} \wedge dx^{i_1} \wedge \cdots \wedge dx^{i_k}\right) \\ &= \varphi^*\,d\omega. \end{aligned}$$

In particular, if X is a vector field on M, we conclude that

$$D_X\,d\omega = d(D_X\omega). \tag{3.6}$$

EXERCISES

3.1 Compute d of the following differential forms.

a) $\gamma = \sum_1^n (-1)^{i-1} x_i\,dx_1 \wedge \cdots \wedge dx_{i-1} \wedge dx_{i+1} \wedge \cdots \wedge dx_n$

b) $r^{-n}\gamma$, where γ is as in (a) and $r = \{x_1^2 + \cdots + x_n^2\}^{1/2}$

c) $\sum p_i\,dq_i$

d) $\sin(x^2 + y^2 + z^2)(x\,dx + y\,dy + z\,dz)$

Let V be a vector space equipped with a nonsingular bilinear form and an orientation. Then we can define the $*$-operator as in Chapter 7. Since we identify the tangent space $T_x(V)$ with V for any $x \in V$, we can consider the $*$-operator as mapping $\bigwedge^k(V) \to \bigwedge^{n-k}(V)$. For instance, in $\mathbb{R}^2$, with the rectangular coordinates (x, y), we have

$$*dx = dy, \qquad *dy = -dx,$$

and so on.

3.2 Show that

$$d * df = \left(\frac{\partial^2 f}{\partial x^2} + \frac{\partial^2 f}{\partial y^2}\right) dx \wedge dy$$

for any function f on $\mathbb{R}^2$.

3.3 Obtain a similar expression for $d * d$ in $\mathbb{R}^n$ with its usual scalar product. (Recall that

$$*dx^1 \wedge \cdots \wedge dx^k = dx^{k+1} \wedge \cdots \wedge dx^n$$

and, more generally,

$$*dx^{i_1} \wedge \cdots \wedge dx^{i_k} = \pm\, dx^{j_1} \wedge \cdots \wedge dx^{j_{n-k}},$$

where $(i_1, \ldots, i_k, j_1, \ldots, j_{n-k})$ is a permutation of $(1, \ldots, n)$ and the $\pm$ is the sign of the permutation.)

3.4 Let x, y, z, t be coordinates on $\mathbb{R}^4$. Introduce a scalar product on the tangent space at each point so that

$$(dx, dx) = (dy, dy) = (dz, dz) = 1,$$

$$(dx, dy) = (dx, dz) = (dx, dt) = (dy, dz) = (dy, dt) = (dz, dt) = 0,$$

and

$$(c\,dt, c\,dt) = -1,$$

where c is a positive constant. Let the two-form ω be given by

$$\omega = c(E_1\, dx \wedge dt + E_2\, dy \wedge dt + E_3\, dz \wedge dt) + B_1\, dy \wedge dz + B_2\, dz \wedge dx + B_3\, dx \wedge dy.$$

Let the three-form γ be given by

$$\gamma = \rho\, dx \wedge dy \wedge dz - (J_1\, dy \wedge dz + J_2\, dz \wedge dx + J_3\, dx \wedge dy) \wedge dt.$$

Write the equations

$$d\omega = 0, \qquad d * \omega = 4\pi\gamma$$

as equations involving the various coefficients and their partial derivatives.

4. STOKES' THEOREM

In this section we shall prove a theorem which will be a far-reaching generalization of the fundamental theorem of the calculus of one variable. It should, perhaps, be called the fundamental theorem of the calculus of several variables. We first make some definitions.

Let D be a domain with regular boundary in a manifold M. We recall (page 419) that each point of M lies in a chart (U, α) which is one of three types.

Let (U, α) and (W, β) be two charts of M of type (iii). Then, as on page 420, the matrix of $J_{\beta\circ\alpha^{-1}}$ is given by

$$\begin{bmatrix} \dfrac{\partial y^1}{\partial x^1} & \cdots & \cdots & \cdots & \dfrac{\partial y^1}{\partial x^n} \\ \vdots & & & & \vdots \\ \dfrac{\partial y^{n-1}}{\partial x^1} & \cdots & \cdots & \cdots & \dfrac{\partial y^{n-1}}{\partial x^n} \\ 0 & 0 & 0 & 0 & \dfrac{\partial y^n}{\partial x^n} \end{bmatrix},$$

and so

$$\det J_{\beta\circ\alpha^{-1}} = \frac{\partial y^n}{\partial x^n} \times \det \begin{bmatrix} \dfrac{\partial y^1}{\partial x^1} & \cdots & \dfrac{\partial y^1}{\partial x^{n-1}} \\ \vdots & & \vdots \\ \dfrac{\partial y^{n-1}}{\partial x^{n-1}} & \cdots & \dfrac{\partial y^{n-1}}{\partial x^{n-1}} \end{bmatrix}$$

$$= \frac{\partial y^n}{\partial x^n} \times \det J_{(\beta\restriction\partial D)\circ(\alpha\restriction\partial D)^{-1}}. \tag{4.1}$$

Furthermore, $y^n(x^1, \ldots, x^n) > 0$ if $x^n > 0$, since $\alpha(U \cap W) \cap \{\mathbf{v} : v^n > 0\} = \alpha(U \cap W \cap \operatorname{int} D)$. Thus $\partial y^n/\partial x^n > 0$ **at a boundary point where** $x_n = 0$.

Now suppose that **M** is an oriented manifold, and let $D \subset M$ be a domain with regular boundary. We shall make ∂D into an *oriented* manifold. We say that an atlas $\mathcal{A}$ is *adjusted* if each $(U, \alpha) \in \mathcal{A}$ is of type (i), (ii), or (iii) and, in addition, if each chart of $\mathcal{A}$ is positive.

If $\dim M > 1$, we can always find an adjusted atlas. In fact, by choosing the U connected, we find that every (U, α) is either positive or negative. If (U, α) is negative, we replace it by (U, α'), where $x^1_{\alpha'} = -x^1_\alpha$.

If $\dim M = 1$, then ∂D consists of a discrete set of points (which we can regard as a "zero-dimensional manifold"). Each $x \in \partial D$ lies in a chart of type (iii) which is either positive or negative. We assign a plus sign to x if any chart (and hence all constricted charts) of type (iii) is negative. We assign a minus sign to x if its charts of type (iii) are positive. In this way we "orient" ∂D, as shown in Fig. 11.6.

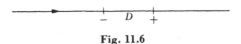

Fig. 11.6

If $\dim M > 1$, we choose an adjusted (oriented) atlas on M. It then follows from (4.1) and the fact that $\partial y^n/\partial x^n > 0$ that

$$\det J_{(\beta\restriction\partial D)\circ(\alpha\restriction\partial D)^{-1}} > 0.$$

This shows that $(U \restriction \partial D, \alpha \restriction \partial D)$ is an oriented atlas on ∂D. We thus get an orientation on ∂D. This is not quite the orientation we want on ∂D. For reasons that will soon become apparent, we choose the orientation on ∂D so that

$(U \upharpoonright \partial D, \alpha \upharpoonright \partial D)$ has the same sign as $(-1)^n$. That is, $(U \upharpoonright \partial D, \alpha \upharpoonright \partial D)$ is a positive chart if n is even, and we take the orientation opposite to that determined by $(U \upharpoonright \partial D, \alpha \upharpoonright \partial D)$ if n is odd. We can now state our main theorem.

Theorem 4.1 (*Stokes' theorem*). Let **M** be an n-dimensional oriented manifold, and let $D \subset M$ be a domain with regular boundary. Let $\partial\mathbf{D}$ denote the boundary of D regarded as an oriented manifold. Then for any $\omega \in \bigwedge^{n-1}(M)$ with compact support we have

$$\int_{\partial \mathbf{D}} \iota^* \omega = \int_{\mathbf{D}} d\omega, \tag{4.2}$$

where, as usual, ι is the injection of boundary **D** into **M**.

Proof. For $n = 1$ this is just the fundamental theorem of the calculus.

For $n > 1$ our proof is almost exactly the same as the proof of Theorem 6.1 of Chapter 10. Choose an adjusted atlas $\mathcal{A}$ and a partition of unity $\{g_j\}$ subordinate to $\mathcal{A}$. Since ω has compact support, we can write

$$\omega = \sum g_j \omega,$$

where the sum is finite. Since both sides of (4.2) are linear, it suffices to verify (4.2) for each of the summands $g_j\omega$. Since $\operatorname{supp} g_j\omega \subset U$, where (U, α), we must check the three possibilities: (U, α) satisfies (i), (ii), or (iii).

If (U, α) satisfies (i), $\iota^*\omega = 0$, since $\operatorname{supp} \omega \cap \partial D = \varnothing$, and

$$\int_{\mathbf{D}} d\omega = \int_{\mathbf{M}} e_D \, d\omega = 0,$$

since $D \cap \operatorname{supp} \omega = \varnothing$. Thus both sides of (4.2) vanish.

If (U, α) satisfies (ii), the left-hand side of (4.2) vanishes. We must show that the same holds for the right-hand side. Let $x^1, \ldots, x^n$ be the coordinates on (U, α), and write

$$\begin{aligned} g_j\omega = a_1\, dx^2 \wedge \cdots \wedge dx^n + a_2\, dx^1 \wedge dx^3 \wedge \cdots \wedge dx^n + \cdots \\ + a_n\, dx^1 \wedge \cdots \wedge dx^{n-1}. \end{aligned}$$

Then

$$dg_j\omega = \sum (-1)^{i-1} \frac{\partial a_i}{\partial x^i}\, dx^1 \wedge \cdots \wedge dx^n,$$

and thus

$$\int dg_j\omega = \sum (-1)^{i-1} \int_{\mathbb{R}^n} \frac{\partial a_i}{\partial x^i}.$$

Since $g_j\omega$ has compact support, the functions a_i have compact support, and we can replace the integral over $\mathbb{R}^n$ by the integral over $\square_{-\mathbf{R}}^{\mathbf{R}}$, where $\mathbf{R} = \langle R, \ldots, R \rangle$ and R is chosen so large that $\operatorname{supp} a_i \subset \square_{-\mathbf{R}}^{\mathbf{R}}$. But writing the multiple integral as an integral, we get

$$\int_{\square_{-\mathbf{R}}^{\mathbf{R}}} \frac{\partial a_i}{\partial x^i} = \int_{\mathbb{R}^{n-1}} a_i(\ldots, R, \ldots) - a_i(\ldots, -R, \ldots) = 0,$$

since $a_i(\ldots, R, \ldots) = a_i(\ldots, -R, \ldots) = 0$.

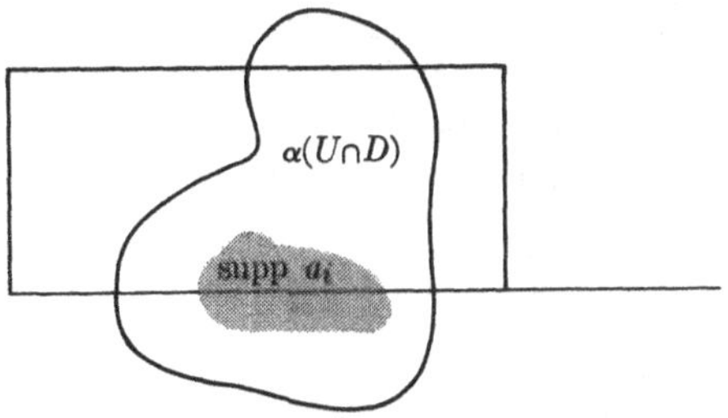

Fig. 11.7

We now examine $\int dg_j\omega$ in case (iii). The argument proceeds exactly as before, except that we must compute $\int_{\alpha(U\cap D)} \partial a_i/\partial x_i$ instead of $\int_{\alpha(U)}$. (See Fig. 11.7.)

We can now replace the region of integration by a rectangle of the form $\square^{<R,\ldots,R>}_{<-R,\ldots,-R,0>}$ for large R. If $i < n$, $\int \partial a_i/\partial x_i = 0$ as before. If $i = n$, we get

$$\int_{\alpha(U\cap D)} \frac{\partial a_n}{\partial x_n} = -\int_{\mathbf{R}^{n-1}} a_n(\cdot, \cdot, \ldots, \cdot, 0),$$

so that

$$\int d(g_j\omega) = \sum (-1)^{i-1} \int \frac{\partial a_i}{\partial x^i} = (-1)^n \int_{\mathbf{R}^{n-1}} a_n(\cdot, \cdot, \ldots, \cdot, 0).$$

Now since $x^n = 0$ on $U \cap \partial D$, we see that $\iota^*\, dx^n = 0$. Thus

$$\iota^*\omega = (\iota^* a_n)(\iota^*\, dx^1) \wedge \cdots \wedge (\iota^*\, dx^{n-1}),$$

or if (by abuse of notation) we regard $x^1, \ldots, x^{n-1}$ as the coordinates of $(U \restriction \partial D, \alpha \restriction \partial D)$, we get

$$\iota^*\omega = a(\cdot, \cdot, \ldots, \cdot, 0)\, dx^1 \wedge \cdots \wedge dx^{n-1}.$$

In view of the choice we made for the orientation of ∂D, we conclude that

$$\int_{\partial \mathbf{D}} \iota^*\omega = (-1)^n \int_{\mathbf{R}^{n-1}} a_n(\cdot, \cdot, \ldots, \cdot, 0).$$

This completes the proof of the theorem. □

Theorem 4.1, like the divergence theorem, is not sufficiently broad for us to apply to more general domains. For this purpose, we will again use the notion of a domain with almost regular boundary.

We have already seen that the set of $x \in \partial D$ having a neighborhood of type (iii) forms a differentiable manifold. (Recall that these points need not exhaust all of ∂D). Similarly, if $\mathbf{M}$ is an oriented manifold, then this collection of points becomes an oriented manifold (with $(-1)^n$ times the induced orientation, as before). By abuse of language we shall denote this oriented manifold by $\partial\mathbf{D}$. Thus $\partial\mathbf{D}$ is an oriented manifold which, as a set, is not ∂D but only the "regular" points of ∂D, that is, the points of $\breve{\partial D}$.

Theorem 4.2 (*Stokes' theorem*). Let $\mathbf{M}$ be an n-dimensional oriented manifold, and let $D \subset M$ be a domain with almost regular boundary. Let $\partial\mathbf{D}$ be as above, and let ι be the injection of $\partial\mathbf{D} \to M$. Then for any $\omega \in \bigwedge^{n-1}(M)$ with compact support we have

$$\int_{\partial\mathbf{D}} \iota^*\omega = \int_{\mathbf{D}} d\omega. \tag{4.2}$$

Proof. The proof proceeds as before. We choose an adjusted atlas and a partition of unity $\{g_j\}$ subordinate to the atlas. We write $\omega = \sum g_j\omega$ and now have four cases to consider. The first three cases have been handled already. The new case is where

$$g_j\omega = \sum_j a_j\, dx^1 \wedge \cdots \wedge \widehat{dx^j} \wedge \cdots \wedge dx^n,$$

where the $\widehat{\ }$ indicates that dx^j is to be omitted, has its support contained in U, where (U, α) is a chart of type (iv). By linearity, it suffices to verify (4.2) for each summand on the right, i.e., for

$$a_j\, dx^1 \wedge \cdots \wedge \widehat{dx^j} \wedge \cdots \wedge dx^n.$$

Now $\iota^*(a_j\, dx^1 \wedge \cdots \wedge \widehat{dx^j} \wedge \cdots \wedge dx^n) = 0$ unless $j \geq k$, since dx^p vanishes on the piece of $\partial\mathbf{D} \cap U$ whose image under α lies in H^k_p. If $j < p$, then all these dx^p occur, and thus $\iota^*(a_j\, dx^1 \wedge \cdots \wedge \widehat{dx^j} \wedge \cdots \wedge dx^n) = 0$. If $j > p$, then $\iota^*(a_j\, dx^1 \wedge \cdots \wedge \widehat{dx^j} \wedge dx^n)$ vanish everywhere except on the portion of $\partial\mathbf{D}$ which maps under α onto H^k_j.

On the other hand,

$$d(a_j\, dx^1 \wedge \cdots \wedge \widehat{dx^j} \wedge \cdots \wedge dx^n) = (-1)^{j-1}\frac{\partial a_j}{\partial x^j}\, dx^1 \wedge \cdots \wedge dx^n.$$

We can evaluate the integral $\int_D$ by integrating over the rectangle

$$\square^{\langle R, \ldots, R, \ldots, R\rangle}_{\langle -R, \ldots, -R, 0, \ldots, 0, 0\rangle}$$

(where the $-R$'s extend through the $(k-1)$th position). Integrating first with regard to x^j, we obtain

$$\int_D d(a_j\, dx^1 \wedge \cdots \wedge \widehat{dx^j} \wedge \cdots \wedge dx^n) = (-1)^j \int_{H^k_j} a_j.$$

On the other hand, the orientation on H^k_j is such that this integral has the sign necessary to make (4.2) hold. This proves Theorem 4.2. □

As before, we can apply Theorems 4.1 and 4.2 to still more general domains by using a limit argument. For instance, Theorem 4.2, as stated, does not apply to the domain D in Fig. 11.8, because the curves C_1 and C_2 are tangent at P. It does apply, however, to the approximating domain obtained by "breaking off a little piece" (Fig. 11.9), and it is clear that the values of both sides of (4.2) for D' are close to those for D. We thus obtain (4.2) for D by passing to the limit. As before, we will not state a more general theorem covering these cases. It will be clear in each instance how to apply a limit argument.

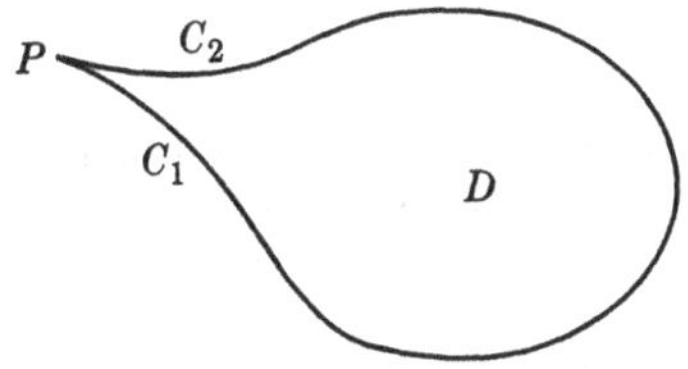

Fig. 11.8

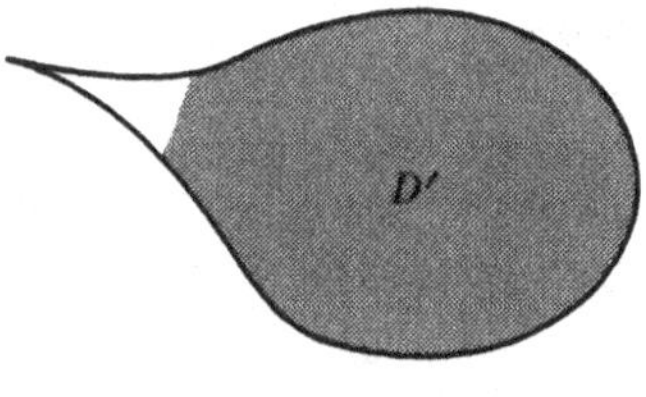

Fig. 11.9

Since the statement and proof of Stokes' theorem are so close to those of the divergence theorem, the reader might suspect that one implies the other. On an oriented manifold, the divergence theorem is, indeed, a corollary of Stokes' theorem. To see this, let Ω be an element of $\bigwedge^n(M)$ corresponding to the density ρ. If X is a vector field, then the n-form $D_X\Omega$ clearly corresponds to the density $D_X\rho = \operatorname{div} \langle X, \rho \rangle$. Anticipating some notation that we shall introduce in Section 6, let $X \lrcorner \Omega$ be the $(n-1)$-form defined by

$$X \lrcorner \Omega(\xi^1, \ldots, \xi^{n-1}) = (-1)^{n-1}\Omega(\xi^1, \ldots, \xi^{n-1}, X).$$

In terms of coordinates, if $\Omega = a\, dx^1 \wedge \cdots \wedge dx^n$, then

$$X \lrcorner \Omega = a[X^1\, dx^2 \wedge \cdots \wedge dx^n - X^2\, dx^1 \wedge dx^3 \wedge \cdots \wedge dx^n + \cdots + (-1)^{n-1}X^n\, dx^1 \wedge \cdots \wedge dx^{n-1}].$$

Note that

$$d(X \lrcorner \Omega) = \left(\sum \frac{\partial a X^i}{\partial x^i}\right) dx^1 \wedge \cdots \wedge dx^n,$$

which is exactly the n-form $D_X\Omega$, since it corresponds to the density $D_X\rho = \operatorname{div} \langle X, \rho \rangle$. Thus, by Stokes' theorem,

$$\int_{\partial \mathbf{D}} \iota^*(X \lrcorner \Omega) = \int_{\mathbf{D}} d(X \lrcorner \Omega) = \int_D \operatorname{div} \langle X, \rho \rangle.$$

We must compare $X \lrcorner \Omega$ with the density ρ_X on ∂D. By (2.2) they agree on everything up to sign. To check that the signs agree, it suffices to compare

$$\rho_X\left(\frac{\partial}{\partial x^1}, \ldots, \frac{\partial}{\partial x^{n-1}}\right) = \rho\left(\frac{\partial}{\partial x^1}, \ldots, \frac{\partial}{\partial x^{n-1}}, X\right)$$

with

$$\iota^*\left(X \lrcorner \Omega\left(\frac{\partial}{\partial x^1}, \ldots, \frac{\partial}{\partial x^{n-1}}\right)\right)$$

at any $x \in \partial D$. Now

$$\iota^*(X \lrcorner \Omega) = (-1)^{n-1}X^n\, dx^1 \wedge \cdots \wedge dx^{n-1}$$

and, according to our convention, $x^1, \ldots, x^{n-1}$ is a positive or negative coordinate system according to the sign of $(-1)^n$. Thus the two coincide if and only if X^n is negative, that is,

$$\int_{\partial \mathbf{D}} \iota^*(X \lrcorner \Omega) = \int_{\partial D} \epsilon_X \rho_X.$$

EXERCISES

4.1 Compute the following surface integrals both directly and by using Stokes' theorem. Let $\square$ denote the unit cube, and let B be the unit ball in $\mathbb{R}^3$.

a) $\int_{\partial\square} x\,dy \wedge dz + y\,dz \wedge dx + z\,dx \wedge dy$

b) $\int_{\partial B} x^3\,dy \wedge dz$

c) $\int_{\partial\square} \cos z\,dx \wedge dy$

d) $\int_{\partial U} x\,dy \wedge dz$, where

$$U = \{(x, y, z) : x \geq 0, y \geq 0, z \geq 0, x^2 + y^2 + z^2 \leq 1\}$$

4.2 Let $\omega = yz\,dx + x\,dy + dz$. Let γ be the unit circle in the plane oriented in the counterclockwise direction. Compute $\int_\gamma \omega$. Let

$$A_1 = \{(x, y, z) : z = 0, x^2 + y^2 \leq 1\},$$
$$A_2 = \{(x, y, z) : z = 1 - x^2 - y^2, x^2 + y^2 \leq 1\}.$$

Orient the surfaces A_1 and A_2 so that $\partial \mathbf{A}_1 = \partial \mathbf{A}_2 = \gamma$. Verify that $\int_{\mathbf{A}_1} d\omega = \int_{\mathbf{A}_2} d\omega = \int \omega$ by computing the integrals.

4.3 Let S^1 be the circle and define $\omega = (1/2\pi)\,d\theta$, where θ is the angular coordinate.

a) Let $\varphi: S^1 \to S^1$ be a differentiable map. Show that $\int \varphi^*\omega$ is an integer. This integer is called the *degree* of φ and is denoted by $\deg \varphi$.

b) Let φ_t be a collection of maps (one for each t) which depends differentiably on t. Show that $\deg \varphi_0 = \deg \varphi_1$.

c) Let us regard S^1 as the unit circle in the complex numbers. Let f be some function on the complex numbers, and suppose that $f(z) \neq 0$ for $|z| = r$. Define $\varphi_{r,f}$ by setting $\varphi_{r,f}(e^{i\theta}) = f(re^{i\theta})/|f(re^{i\theta})|$. Suppose $f(z) = z^n$. Compute $\deg \varphi_{r,f}$ for $r \neq 0$.

d) Let f be a polynomial of degree $n \geq 1$. Thus

$$f(z) = a_n z^n + a_{n-1} z^{n-1} + \cdots + a_0,$$

where $a_n \neq 0$. Show that there is at least one complex number z_0 at which $f(z_0) = 0$. [*Hint:* Suppose the contrary. Then $\varphi_{r,(1/a_n)f}$ is defined for all $0 \leq r < \infty$ and $\deg \varphi_{r,(1/a_n)f} = \text{const}$, by (b). Evaluate $\lim_{r=0}$ and $\lim_{r=\infty}$ of this expression.]

Let X be a vector field defined in some neighborhood U of the origin in $\mathbb{E}^2$, and suppose that $X(0) = 0$ and that $X(x) \neq 0$ for $x \neq 0$. Thus X vanishes only at the origin. Define the map $\varphi_r: S^1 \to S^1$ by

$$\varphi_r(e^{i\theta}) = \frac{X(re^{i\theta})}{\|X(re^{i\theta})\|}.$$

This map is defined for sufficiently small r. By Exercise 4.3(b) the degree of this map does not depend on r. This degree is called the *index* of the vector field X at the origin.

4.4 Compute the index of

a) $x\dfrac{\partial}{\partial x} + y\dfrac{\partial}{\partial x}$, b) $x\dfrac{\partial}{\partial x} - y\dfrac{\partial}{\partial y}$, c) $y\dfrac{\partial}{\partial x} - x\dfrac{\partial}{\partial y}$.

d) Construct a vector field with index 2.

e) Show that the index of $-X$ is the same as the index of X for any vector field X.

4.5 Let X be a vector field on an oriented two-dimensional manifold, and suppose that $X(p) = 0$ for some $p \in M$ and that X does not vanish at any other point in a small neighborhood of p. By choosing an oriented chart mapping p into zero, we get a vector field on $\mathbb{E}^2$ vanishing at the origin. Show that the index of this vector field does not depend on the choice of charts. We can thus define the index of X at p.

4.6 a) On the sphere S^2 let X be a vector field which is tangent to the meridian circles everywhere and vanishes only at the north and south poles. What is its index at each pole?

b) Let Y be a vector field which is tangent to the circles of latitude everywhere and vanishes only at the north and south poles. What is its index at each pole?

5. SOME ILLUSTRATIONS OF STOKES' THEOREM

As a simple but important corollary of Theorem 4.2, we state:

Theorem 5.1. Let $\varphi: M_1 \to M_2$ be a differentiable map of the oriented k-dimensional manifold $\mathbf{M}_1$ into the n-dimensional manifold M_2. Let ω be a form of degree $k - 1$ on M_2, and let $D \subset M_1$ be a domain with almost regular boundary on M_1. Then we have

$$\int_{\partial \mathbf{D}} \iota^* \varphi^* \omega = \int_{\mathbf{D}} \varphi^*(d\omega). \qquad (5.1)$$

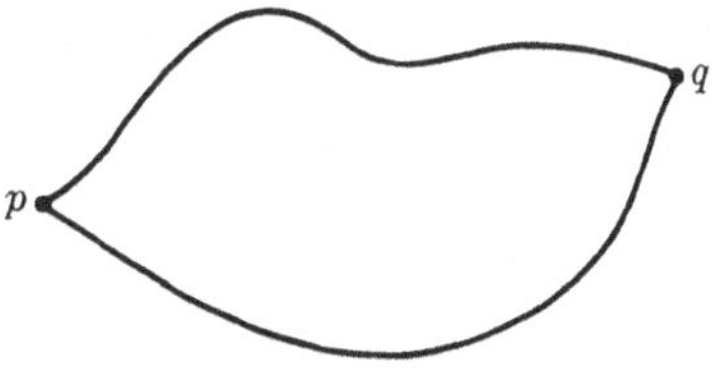

Fig. 11.10

Equation (5.1) follows directly from (4.2) and from the fact that $\varphi^* d = d\varphi^*$.

We can regard the right-hand side of (5.1) as the integral of $d\omega$ over the "oriented k-dimensional hypersurfaces" $\varphi(\mathbf{D})$. Equation (5.1) says that this integral is equal to the integral of ω over the $(k - 1)$-dimensional hypersurface $\varphi(\partial \mathbf{D})$.

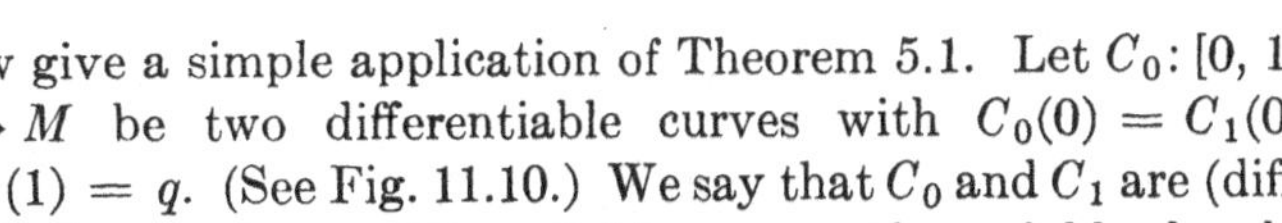

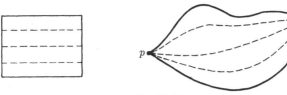

Fig. 11.11

We now give a simple application of Theorem 5.1. Let $C_0: [0, 1] \to M$ and $C_1: [0, 1] \to M$ be two differentiable curves with $C_0(0) = C_1(0) = p$ and $C_0(1) = C_1(1) = q$. (See Fig. 11.10.) We say that C_0 and C_1 are (differentiably) homotopic if there exists a differentiable map φ of a neighborhood of the unit square $[0, 1] \times [0, 1] \subset \mathbb{R}^2$ into M such that $\varphi(t, 0) = C_0(t)$, $\varphi(t, 1) = C_1(t)$, $\varphi(0, s) = p$, and $\varphi(1, s) = q$. (See Fig. 11.11.) For each value of s we get the curve C_s given by $C_s(t) = \varphi(t, s)$. We think of φ as providing a differentiable "deformation" of the curve C_0 into the curve C_1.

Proposition 5.1. Let C_0 and C_1 be differentiably homotopic curves, and let ω be a linear differential form on M with $d\omega = 0$. Then

$$\int_{C_1} \omega = \int_{C_2} \omega. \tag{5.2}$$

Proof. In fact,

$$\int_{\partial\square\substack{<1,1>\\<0,0>}} \varphi^*\omega = \int_{\square\substack{<1,1>\\<0,0>}} \varphi^* d\omega = 0.$$

But $\int_{\partial\square}$ is the sum of the four terms corresponding to the four sides of the square. The two vertical sides ($t = 0$ and $t = 1$) contribute nothing, since φ maps these curves into points. The top gives $-\int_{C_1}$ (because of the counterclockwise orientation), and the bottom gives $\int_{C_0}$. Thus $\int_{C_0} \omega - \int_{C_1} \omega = 0$, proving the proposition. □

It is easy to see that the proposition extends without difficulty to piecewise differentiable curves and piecewise differentiable homotopies. Let us say that two piecewise differentiable curves, C_0 and C_1, are (piecewise differentiably) homotopic if there is continuous map φ of $[0, 1] \times [0, 1] \to M$ such that

i) $\varphi(0, s) = p$, $\varphi(1, s) = q$;

ii) $\varphi(t, 0) = C_0(t)$, $\varphi(t, 1) = C_1(t)$;

iii) there are a finite number of points $t_0 < t_1 < \cdots < t_m$ such that φ coincides with the restriction of a differentiable map defined in some neighborhood of each rectangle $[t_i, t_{i+1}] \times [0, 1]$. (See Fig. 11.12.)

To verify that Proposition 5.1 holds for the case of piecewise differentiable homotopies, we apply Stokes' theorem to each rectangle and observe that the contribution of the interior vertical lines cancel one another.

We say that a manifold M is *connected* if every pair of points can be joined by a (piecewise differentiable) curve. Thus $\mathbb{R}^n$, for example, is connected. We say that M is *simply connected* if all (piecewise differentiable) curves joining the same two points are (piecewise differentiably) homotopic. (Note that the circle, S^1 is *not* simply connected.) Let us verify that $\mathbb{R}^n$ is simply connected. If C_0 and C_1 are two curves, let φ: $[0, 1] \times [0, 1] \to \mathbb{R}^n$ be given by

$$\varphi(t, s) = sC_0(t) + (1 - s)C_1(t).$$

It is clear that φ has all the desired properties.

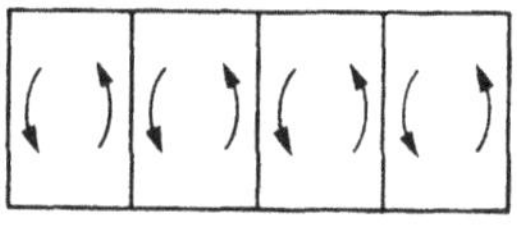

Fig. 11.12

Proposition 5.2. Let M be a connected and simply connected manifold, and let $O \in M$. Let $\omega \in \bigwedge^1(M)$ satisfy $d\omega = 0$. For any $x \in M$ let $f(x) = \int_C \omega$, where C is some piecewise differentiable curve joining O to x. The function f is well defined and differentiable, and $df = \omega$.

Proof. It follows from Proposition 5.1 that f is well defined. If C_0 and C_1 are two curves joining O to x, then they are homotopic, and so $\int_{C_0} \omega = \int_{C_1} \omega$. It is clear that f is continuous, since

$$f(x) - f(y) = \int_D \omega,$$

where D is any curve joining y to x (Fig. 11.13).

To check that f is differentiable, let (U, α) be a chart about x with coordinates $\langle x^1, \ldots, x^n \rangle$. Then

$$f(x^1, \ldots, x^i + h, \ldots, x^n) - f(x^1, \ldots, x^n) = \int_C \omega,$$

where C is any curve joining p to q, where $\alpha(p) = (x^1, \ldots, x^i, \ldots, x^n)$, and where $\alpha(q) = (x^1, \ldots, x^i + h, \ldots, x^n)$. We can take C to be the curve given by

$$\alpha \circ C(t) = (x^1, \ldots, x^i + ht, \ldots, x^n).$$

If $\omega = a_1\, dx^1 + \cdots + a_n\, dx^n$, then

$$\int_C \omega = \int_0^1 ha_i\, dt = \int_0^h a_i(x^1, \ldots, x^i + s, \ldots, x^n)\, ds.$$

(See Fig. 11.14.) Thus

$$\lim_{h \to 0} \frac{1}{h} [f(x^1, \ldots, x^i + h^1, \ldots, x^n) - f(x^1, \ldots, x^n)] = a^i,$$

that is, $\partial f / \partial x^i = a^i$. This shows that f is differentiable and that $df = \omega$, proving the proposition. □

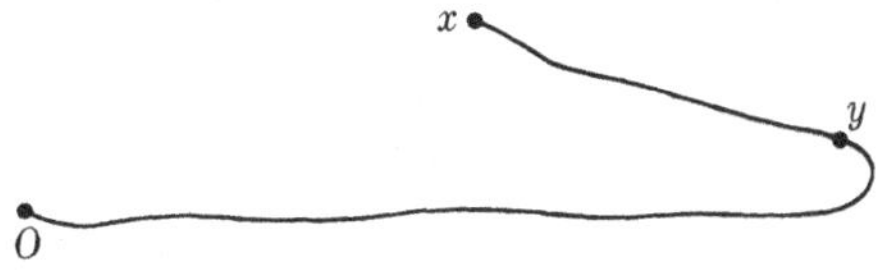

Fig. 11.13

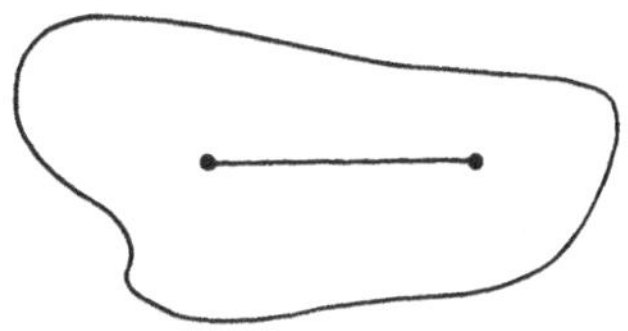

Fig. 11.14

We have thus established that every $\omega \in \bigwedge^1(\mathbb{R}^n)$ with $d\omega = 0$ is of the form df. More generally, it can be established that if $\Omega \in \bigwedge^k(\mathbb{R}^n)$ satisfies $d\Omega = 0$, then $\Omega = d\omega$ for some $\omega \in \bigwedge^{k-1}(\mathbb{R}^n)$.

This is not true for an arbitrary manifold. For instance, every $\omega \in \bigwedge^1(S^1)$ satisfies $d\omega = 0$. Yet the element of angle form (which is, unfortunately, denoted by $d\theta$) is not the d of any function. The fact that $d^2 = 0$ shows that if $\Omega = d\omega$, then $d\Omega = 0$. Thus the space $d[\bigwedge^{k-1}(M)] \subset \bigwedge^k(M)$ is a subspace of the space $\ker_k d$ of elements in $\bigwedge^k(M)$ satisfying $d\Omega = 0$. The quotient space $\ker_k d/d[\bigwedge^{k-1}]$ is denoted by $H^k(M)$ and is called the kth cohomology group of M. If M is compact, it can be shown that H^k is finite-dimensional. It measures (roughly speaking) "how many" k-dimensional holes there are in M.

6. THE LIE DERIVATIVE OF A DIFFERENTIAL FORM

Let M be a differentiable manifold, and let φ be a flow on M with infinitesimal generator X. For any $\omega \in \bigwedge^k(M)$ we can consider the expression

$$\frac{\varphi_t^* \omega - \omega}{t}.$$

It is not difficult (using local expressions) to verify that the limit as $t \to 0$ exists and is again an element of $\bigwedge^k(M)$, which we denote by $D_X\omega$. The purpose of this section is to provide an effective formula for computing $D_X\omega$. For this purpose, we first collect some properties of D_X. First of all, we have that it is linear:

$$D_X(\omega_1 + \omega_2) = D_X\omega_1 + D_X\omega_2. \tag{6.1}$$

Secondly, we have

$$\begin{aligned}\varphi_t^*(\omega_1 \wedge \omega_2) - \omega_1 \wedge \omega_2 &= (\varphi_t^*\omega_1) \wedge (\varphi_t^*\omega_2) - \omega_1 \wedge \omega_2 \\ &= (\varphi_t^*\omega_1) \wedge (\varphi_t^*\omega_2) - (\varphi_t^*\omega_1) \wedge \omega_2 \\ &\quad + (\varphi_t^*\omega_1) \wedge \omega_2 - \omega_1 \wedge \omega_2.\end{aligned}$$

Dividing by t and passing to the limit, we see that

$$D_X(\omega_1 \wedge \omega_2) = (D_X\omega_1) \wedge \omega_2 + \omega_1 \wedge D_X\omega_2. \tag{6.2}$$

Finally, since $\varphi_t^* \, d = d\varphi_t^*$, we have

$$D_X \, d\omega = d(D_X\omega). \tag{6.3}$$

Actually, these three formulas suffice for the computation. If

$$\omega = \sum a_{i_1,\ldots,i_k} \, dx^{i_1} \wedge \cdots \wedge dx^{i_k},$$

then

$$\begin{aligned}D_X\omega &= \sum D_X(a_{i_1,\ldots,i_k} \, dx^{i_1} \wedge \cdots \wedge dx^{i_k}) \qquad \text{by (6.1)} \\ &= \sum [(D_X a_{i_1,\ldots,i_k}) \, dx^{i_1} \wedge \cdots \wedge dx^{i_k} + a_{i_1,\ldots,i_k}(D_X \, dx^{i_1}) \wedge \cdots \wedge dx^{i_k} \\ &\quad + \cdots + a_{i_1,\ldots,i_k} \, dx^{i_1} \wedge \cdots \wedge (D_X \, dx^{i_k})] \quad \text{by repeated use of (6.2)} \\ &= \sum [(D_X a_{i_1,\ldots,i_k}) \, dx^{i_1} \wedge \cdots \wedge dx^{i_k} + a_{i_1,\ldots,i_k} \, d(D_X x^1) \wedge \cdots \wedge dx^k \\ &\quad + \cdots + a_{i_1,\ldots,i_k} \, dx^{i_1} \wedge \cdots \wedge d(D_X x^{i_k})] \quad \text{by (6.3).}\end{aligned}$$

Since this expression is rather cumbersome (the $d(D_X x^i)$ have to be expanded and the terms collected), we shall derive a simpler and more convenient expression for $D_X\omega$. In order to do this, we make an algebraic detour.

Recall that the operator $d\colon \bigwedge^k(M) \to \bigwedge^{k+1}(M)$ is linear and satisfies the identity

$$d(\omega_1 \wedge \omega_2) = d\omega_1 \wedge \omega_2 + (-1)^k \omega_1 \wedge d\omega_2 \tag{6.4}$$

if $\omega_1 \in \bigwedge^k(M)$. More generally, any (sequence of linear) maps θ of

$$\textstyle\bigwedge^k(M) \to \bigwedge^{k+1}(M)$$

satisfying the identity

$$\theta(\omega_1 \wedge \omega_2) = \theta\omega_1 \wedge \omega_2 + (-1)^k\omega_1 \wedge \theta\omega_2 \tag{6.4'}$$

and

$$\operatorname{supp} \theta\omega \subset \operatorname{supp} \omega^{\dagger} \tag{6.5}$$

will be called an antiderivation of the algebra $\bigwedge(M)$.

It follows from (6.5) that if $\omega_1 \equiv \omega_2$ on an open set U, then $\theta(\omega_1) \equiv \theta(\omega_2)$ on U. Now about every $x \in M$ we can find a neighborhood U and functions $x^1, \ldots, x^n$, so that $\omega \in \bigwedge^k(M)$ can be written as

$$\omega \equiv \sum a_{i_1,\ldots,i_k}\, dx^{i_1} \wedge \cdots \wedge dx^{i_k} \quad \text{on} \quad U. \tag{6.6}$$

Then by repeated use of (6.4′) we have

$$\theta(\omega) \equiv \sum [\theta(a_{i_1,\ldots,i_k}) \wedge dx^{i_1} \wedge \cdots \wedge dx^{i_k} + a_{i_1,\ldots,i_k}\theta(dx^{i_1}) \wedge \cdots \wedge dx^{i_k} + \cdots + (-1)^{k-1}a_{i_1,\ldots,i_k}\, dx^{i_1} \wedge \cdots \wedge \theta(dx^{i_k})]. \tag{6.7}$$

We thus arrive at the important conclusion:

Proposition 6.1. Any antiderivation $\theta\colon \bigwedge^k(M) \to \bigwedge^{k+1}(M)$, $k = 0, \ldots, n$, is uniquely determined by its action on $\bigwedge^0(M)$ and $\bigwedge^1(M)$. That is, if $\theta_1(\omega) = \theta_2(\omega)$ for all $\omega \in \bigwedge^0(M)$ and $\bigwedge^1(M)$, then $\theta_1(\Omega) = \theta_2(\Omega)$ for $\Omega \in \bigwedge^k(M)$ for any k.

Now suppose we are given maps

$$\theta\colon \bigwedge^0(M) \to \bigwedge^1(M) \qquad \text{and} \qquad \theta\colon \bigwedge^1(M) \to \bigwedge^2(M)$$

which satisfy (6.5) and (6.4′) where it makes sense, that is,

$$\theta(f_1f_2) = \theta(f_1)f_2 + f_1\theta(f_2) \qquad \text{and} \qquad \theta(f\omega) = \theta(f) \wedge \omega + f\theta(\omega). \tag{6.8}$$

Then any chart (U, α) defines $\theta\colon \bigwedge^k(U) \to \bigwedge^{k+1}(U)$ by (6.7). This gives an antiderivation θ_U on U, as can easily be checked by the use of the argument on pp. 440–441. By the uniqueness argument, if (W, β) is a second chart, the antiderivations θ_U and θ_W coincide on $U \cap W$. Therefore, Eq. (6.7) is consistent and yields a well-defined antiderivation on $\bigwedge(M)$. (Observe that we have just repeated about two-thirds of the proof of Theorem 3.1 for the more general context of any antiderivation.)

† This condition is actually a consequence of (6.4). In fact, let U be an open set containing $\operatorname{supp} \omega$. Since $\{U, M\text{-}\operatorname{supp} \omega\}$ is an open covering of M, we can find a partition of unity subordinate to it. In particular, we can find a C^∞-function φ which is identically one on $\operatorname{supp} \omega$ and vanishes outside U. Then $\omega = \varphi\omega$, so that

$$\theta(\omega) = \theta(\varphi\omega) = \theta(\varphi) \wedge \omega + \varphi\theta(\omega).$$

Thus $\operatorname{supp} \theta(\omega) \subset \operatorname{supp} \omega \cup \operatorname{supp} \varphi \subset U$. Since U is an arbitrary neighborhood of $\operatorname{supp} \omega$, we conclude that $\operatorname{supp} \theta(\omega) \subset \operatorname{supp} \omega$.

Also observe that in the above arguments, nothing changes if instead of $\theta\colon \bigwedge^k(M) \to \bigwedge^{k+1}(M)$ we have $\theta\colon \bigwedge^k(M) \to \bigwedge^{k-1}(M)$. [We take this to mean $\theta(f) = 0$ for $f \in \bigwedge^0(M)$.] In fact, the same argument works for

$$\theta\colon \textstyle\bigwedge^k \to \bigwedge^{k+r}$$

for any odd integer r. We can thus state:

Proposition 6.2. **Let $\theta\colon \bigwedge^0(M) \to \bigwedge^r(M)$ and $\theta\colon \bigwedge^1(M) \to \bigwedge^{r+1}(M)$ be linear maps satisfying (6.5) and (6.8), where r is odd. Then there exists one and only one way of extending θ to an antiderivation $\theta\colon \bigwedge^k(M) \to \bigwedge^{k+r}(M)$ satisfying (6.4).**

As an application of this proposition, we will attach an antiderivation $\theta(X)\colon \bigwedge^k(M) \to \bigwedge^{k-1}(M)$ to every smooth vector field X on M. Since $r = -1$, for $f \in \bigwedge^0(M)$ we set

$$\theta(X)f = 0.$$

For $\omega \in \bigwedge^1(M)$ we set

$$\theta(X)\omega = \langle X, \omega\rangle. \tag{6.9}$$

To verify (6.8) means to check that

$$\theta(X)(f\omega) = f\theta(X)\omega,$$

that is, that

$$\langle X, f\omega\rangle = f\langle X, \omega\rangle,$$

which is obvious.

If f is a function and θ is an antiderivation, we denote by $f\theta$ the map which sends $\omega \to f\theta(\omega)$. It is easy to check that this is again an antiderivation.

We can assert the following as a consequence of the uniqueness theorem: Let X and Y be smooth vector fields, and let f and g be smooth functions. Then

$$\theta(fX + gY) = f\theta(X) + g\theta(Y). \tag{6.10}$$

By the proposition, it suffices to check (6.10) on all $\omega \in \bigwedge^1(M)$. By (6.9), this is just

$$\langle fX + gY, \omega\rangle = f\langle X, \omega\rangle + g\langle Y, \omega\rangle,$$

which is obvious.

In particular, in a chart (U, h), if

$$X = X^1 \frac{\partial}{\partial x^1} + \cdots + X^n \frac{\partial}{\partial x^n},$$

then

$$\theta(X) = \sum X^i \theta\left(\frac{\partial}{\partial x^i}\right).$$

To evaluate $\theta(\partial/\partial x^i)$, we use (6.8) and the fact that

$$\theta\left(\frac{\partial}{\partial x^i}\right) f = 0, \qquad \theta\left(\frac{\partial}{\partial x^i}\right) dx^j = \begin{cases} 0 & \text{if } i \neq j, \\ 1 & \text{if } i = j. \end{cases}$$

Thus, for example, $\theta(\partial/\partial x^i)\, dx^p \wedge dx^q = 0$ if neither $p = i$ nor $q = i$, while $\theta(\partial/\partial x^i)(dx^i \wedge dx^j) = dx^j$, $\theta(\partial/\partial x^i)(dx^j \wedge dx^i) = -dx^j$, etc.

Let us call a (sequence of) map(s) $D\colon \bigwedge^k(M) \to \bigwedge^{k+s}(M)$, where s is *even*, a *derivation* if it satisfies (6.5) and

$$D(\omega_1 \wedge \omega_2) = D\omega_1 \wedge \omega_2 + \omega_1 \wedge D\omega_2. \tag{6.11}$$

Since s is even, this is consistent. The most important example is D_X where $s = 0$. Then (6.11) is just (6.2).

All the previous arguments about existence and uniqueness of extensions apply unchanged to derivations, as can easily be checked. We can therefore assert:

Proposition 6.3. Let $D\colon \bigwedge^0(M) \to \bigwedge^{0+s}(M)$ and $D\colon \bigwedge^1(M) \to \bigwedge^{1+s}(M)$, where s is even, be maps satisfying (6.5) and (6.8) (with θ replaced by D). Then there exists one and only one way of extending D to a derivation of $\bigwedge(M)$.

We need one further algebraic fact.

Proposition 6.4. Let $\theta_1\colon \bigwedge^k \to \bigwedge^{k+r_1}$ and $\theta_2\colon \bigwedge^k \to \bigwedge^{k+r_2}$ be antiderivations. Then $\theta_1\theta_2 + \theta_2\theta_1\colon \bigwedge^k \to \bigwedge^{k+r_1+r_2}$ is a derivation.

Proof. Since r_1 and r_2 are both odd, $r_1 + r_2$ is even. Equation (6.5) obviously holds. To verify (6.4′), let $\omega_1 \in \bigwedge^k(M)$. Then

$$\begin{aligned}\theta_1\theta_2(\omega_1 \wedge \omega_2) &= \theta_1[\theta_2\omega_1 \wedge \omega_2 + (-1)^k\omega_1 \wedge \theta_2\omega_2] \\ &= \theta_1\theta_2\omega_1 \wedge \omega_2 + (-1)^{k+r_2}\theta_2\omega_1 \wedge \theta_1\omega_2 \\ &\quad + (-1)^k\theta_1\omega_1 \wedge \theta_2\omega_2 + \omega_1 \wedge \theta_1\theta_2\omega_2.\end{aligned}$$

Similarly,

$$\begin{aligned}\theta_2\theta_1(\omega_1 \wedge \omega_2) &= \theta_2\theta_1\omega_1 \wedge \omega_2 + (-1)^{k+r_1}\theta_1\omega_1 \wedge \theta_2\omega_2 \\ &\quad + (-1)^k\theta_2\omega_1 \wedge \theta_1\omega_2 + \omega_1 \wedge \theta_2\theta_1\omega_2.\end{aligned}$$

Since r_1 and r_2 are both odd, the middle terms cancel when we add. Hence we get

$$(\theta_1\theta_2 + \theta_2\theta_1)(\omega_1 \wedge \omega_2) = (\theta_1\theta_2 + \theta_2\theta_1)\omega_1 \wedge \omega_2 + \omega_1 \wedge (\theta_1\theta_2 + \theta_2\theta_1)\omega_2. \square$$

As a first application of Proposition 6.3, we observe that

$$\theta(X) \circ \theta(Y) = -\theta(Y) \circ \theta(X). \tag{6.12}$$

In fact, by Proposition 6.4, $\theta(X)\theta(Y) + \theta(Y)\theta(X)$ is a derivation of degree -2, that is, it vanishes on $\bigwedge^0$ and $\bigwedge^1$. It must therefore vanish identically. We could, of course, directly verify (6.12) from the local description of $\theta(X)$ and $\theta(Y)$.

As a more serious use of Proposition 6.4, consider $\theta(X) \circ d + d \circ \theta(X)$, where X is a smooth vector field. Since $d\colon \bigwedge^k \to \bigwedge^{k+1}$ and $\theta(X)\colon \bigwedge^k \to \bigwedge^{k-1}$, we conclude that $\theta(X) \circ d + d \circ \theta(X)\colon \bigwedge^k \to \bigwedge^k$. We now assert the main formula of this section:

$$D_X = \theta(X) \circ d + d \circ \theta(X). \tag{6.13}$$

Since both sides of (6.13) are derivations, it suffices to check (6.13) for functions and linear differential forms. If $f \in \bigwedge^0(M)$, then $\theta(X)f = 0$. Thus, by (6.9), Eq. (6.13) becomes

$$D_X f = \langle X, df \rangle,$$

which we know holds. Next we must verify (6.13) for $\omega \in \bigwedge^1(M)$. By (6.5), it suffices to verify (6.13) locally. If we write $\omega = a_1\, dx^1 + \cdots + a_n\, dx^n$, it suffices, by linearity, to verify (6.13) for each term $a_i\, dx^i$. Since both sides of (6.13) are derivations, we have

$$D_X(a_i\, dx^i) = (D_X a_i)\, dx^i + a_i(D_X\, dx^i)$$

and

$$[\theta(X)\, d + d\theta(X)](a_i\, dx^i) = [\theta(X)\, d + d\theta(X)](a_i)\, dx^i + a_i[\theta(X)\, d + d\theta(X)]\, dx^i.$$

Since we have verified (6.13) for functions, we only have to check (6.13) for dx^i. Now

$$D_X\, dx^i = dD_X x^i$$

and

$$[\theta(X)\, d + d\theta(X)]\, dx^i = d\theta(X)\, dx^i = d\langle X, dx^i \rangle = dD_X x^i.$$

This completes the proof of (6.13).

In many circumstances it will be convenient to free the letter θ for other uses. We shall therefore occasionally adopt the notation

$$X \lrcorner\, \omega = \theta(X)\omega.$$

The symbol $\lrcorner$ is called the *interior product*. $X \lrcorner\, \omega$ is the interior product of the form ω with the vector field X. If $\omega \in \bigwedge^k$, then $X \lrcorner\, \omega \in \bigwedge^{k-1}$. Equation (6.13) can then be rewritten as

$$D_X\omega = X \lrcorner\, d\omega + d(X \lrcorner\, \omega). \tag{6.14}$$

Let us see what (6.14) says in some special cases in terms of local coordinates. If $\omega = a_1\, dx^1 + \cdots + a_n\, dx^n$ and

$$X = X^1 \frac{\partial}{\partial x^1} + \cdots + X^n \frac{\partial}{\partial x^n},$$

then

$$d\omega = \sum da_i \wedge dx^i.$$

Hence

$$\begin{aligned} X \lrcorner\, d\omega &= \sum [(X \lrcorner\, da_i) \wedge dx^i - da_i \wedge X \lrcorner\, dx^i] \\ &= \sum_{i,j} \left(X^j \frac{\partial a_i}{\partial x^j}\, dx^i - X^j \frac{\partial a_j}{\partial x^i}\, dx^i \right), \end{aligned}$$

while

$$X \lrcorner\, \omega = \sum a_j X^j,$$

so

$$d(X \lrcorner\, \omega) = \sum X^j \left(\frac{\partial a_j}{\partial x^i}\right) dx^i + \sum a_j \left(\frac{\partial X^j}{\partial x^i}\right) dx^i.$$

Thus

$$D_X\omega = \sum_i \left(\sum_j X^j \frac{\partial a_i}{\partial x^j} + a_j \frac{\partial X^j}{\partial x^i} \right) dx^i,$$

which agrees with Eq. (7.12′) of Chapter 9.

As a second illustration, let $\Omega = a\, dx^1 \wedge \cdots \wedge dx^n$, where $n = \dim M$. Then $d\Omega = 0$, so (6.14) reduces to

$$D_X\Omega = d(X \lrcorner \Omega).$$

If $X = \sum X^i(\partial/\partial x^i)$, then

$$\begin{aligned} X \lrcorner \Omega &= \sum X^i \left(\frac{\partial}{\partial x^i}\right) \lrcorner\, \omega \\ &= \sum aX^i \left(\frac{\partial}{\partial x^i}\right) \lrcorner\, (dx^1 \wedge \cdots \wedge dx^n) \\ &= \sum (-1)^{i-1} aX^i\, dx^1 \wedge \cdots \wedge dx^{i-1} \wedge dx^{i+1} \wedge \cdots \wedge dx^n, \end{aligned}$$

which is merely the formula introduced at the end of Section 4. Then

$$D_X\Omega = d(X \lrcorner \Omega) = \left(\sum \frac{\partial aX^i}{\partial x^i}\right) dx^1 \wedge \cdots \wedge dx^n.$$

Since we can always locally identify a density with an n-form by identifying ρ with $\rho_\alpha\, dx^1 \wedge \cdots \wedge dx^n$ on (U, α), we obtain another proof of Proposition 5.2 of Chapter 10.

Appendix I. "VECTOR ANALYSIS"

We list here the relationships between notions introduced in this chapter and various concepts found in books on "vector analysis", although we shall have no occasion to use them.

In oriented Euclidean three-space $\mathbb{E}^3$, there are a number of identifications we can make which give a special form to some of the operations we have introduced in this chapter.

First of all, in $\mathbb{E}^3$, as in any Riemann space, we can (and shall) identify vector fields with linear differential forms. Thus for any function f we can regard df as a vector field. As such, it is called grad f. Thus, in $\mathbb{E}^3$, in terms of rectangular coordinates x, y, z,

$$\operatorname{grad} f = \left\langle \frac{\partial f}{\partial x}, \frac{\partial f}{\partial y}, \frac{\partial f}{\partial z} \right\rangle,$$

where we have also identified vector fields on $\mathbb{E}^3$ with $\mathbb{E}^3$-valued functions.

Secondly, since $\mathbb{E}^3$ is oriented, we can, via the $*$-operator (acting on each T_z^*), identify $\bigwedge^2(\mathbb{E}^3)$ with $\bigwedge^1(\mathbb{E}^3)$. Recall that $*$ is given by

$$*(dx \wedge dy) = dz, \qquad *(dx \wedge dz) = -dy, \qquad *(dy \wedge dz) = dx. \tag{I.1}$$

In particular, if $\omega_1 = \langle P, Q, R \rangle = P\,dx + Q\,dy + R\,dz$ and $\omega_2 = \langle L, M, N \rangle = L\,dx + M\,dy + N\,dz$, we can introduce the so-called "vector product" of ω_1 with ω_2. It is defined by

$$\omega_1 \times \omega_2 = *(\omega_1 \wedge \omega_2)$$

and is given [in view of (I.1)] by

$$\langle P, Q, R \rangle \times \langle L, M, N \rangle = \langle QN - RM, RL - PN, PM - QL \rangle.$$

Also we introduce the operator

$$\text{curl } \omega = *d\omega.$$

Thus, if $\omega = \langle P, Q, R \rangle$, we have

$$\text{curl } \omega = \left\langle \frac{\partial R}{\partial y} - \frac{\partial Q}{\partial z}, \frac{\partial P}{\partial z} - \frac{\partial R}{\partial x}, \frac{\partial Q}{\partial x} - \frac{\partial R}{\partial y} \right\rangle.$$

Consider an oriented surface in $\mathbb{E}^3$; i.e., let $\varphi: S \to \mathbb{E}^3$. Let Ω be the volume form on S associated with the Riemann metric induced by φ. By definition, if $\xi_1, \xi_2 \in T_x(S)$, then

$$\Omega(\xi_1, \xi_2) = dV \langle \varphi_* \xi_1, \varphi_* \xi_2, n \rangle,$$

where dV is the volume element of $\mathbb{E}^3$ and n is the unit normal vector. Another way of writing this is to say that

$$\Omega(\xi_1, \xi_2) = \sigma(\varphi_* \xi_1, \varphi_* \xi_2),$$

where $\sigma = *n$ when we regard n as a differential form. Now let $\overline{w}$ be a form in $\mathbb{E}^3$, and suppose that $\varphi^* \bar{\omega} = f\Omega$ for some function f. Then

$$f(x) = (\bar{\omega}, *n)(\varphi(x)).$$

Thus

$$\int_S \varphi^*(\overline{\omega}) = \int_S f\Omega = \int_S (\overline{\omega}, *n)\Omega = \int (*\overline{\omega}, n)\Omega.$$

Applying this to $\bar{\omega} = d\omega$, where $\omega = P\,dx + Q\,dy + R\,dz$, we can rewrite Stokes' theorem as

$$\int_C \omega = \int_C P\,dx + Q\,dy + R\,dz = \int_S (\text{curl } \omega, n)\Omega,$$

where S is some surface spanning the closed curve C.

If we apply the remark to the case $\bar{\omega} = *\omega$ and $S = \partial D$, we obtain, since $** = \text{id}$ (for $n = 3$),

$$\int (\omega, n)\Omega = \int_D d * \omega$$

Note that

$$d * \omega = \left(\frac{\partial P}{\partial x} + \frac{\partial Q}{\partial y} + \frac{\partial R}{\partial z} \right) dx \wedge dy \wedge dz,$$

which we write as div ω; that is,

$$\text{div } \omega = d * \omega.$$

(It is in fact div $\{\omega, dV\}$, where dV is the volume element and we regard ω as a vector field.) Thus we get the divergence theorem again. Note that

$$\operatorname{curl}(\operatorname{grad} f) = *d\,df = 0$$

and

$$\operatorname{div}(\operatorname{curl} \omega) = d ** d\omega = d^2\omega = 0,$$

since $d^2 = 0$.

Appendix II. ELEMENTARY DIFFERENTIAL GEOMETRY OF SURFACES IN $\mathbb{E}^3$

For purposes of computation, it is convenient to introduce the notion of a *vector-valued differential form*. Let E be a vector space, and let M be a differentiable manifold. By an E-valued exterior differential form Ω of degree p we shall mean a rule which assigns an element Ω_x to each $x \in M$, where Ω_x is an antisymmetric E-valued multilinear function of degree p on $T_x(M)$. For instance, if $p = 0$, then an E-valued zero-form is just a function on M with values in E. An E-valued one-form is a rule which assigns an element of E to each tangent vector ξ at any point of M, and so on.

Suppose that E is finite-dimensional and that $\{e_1, \ldots, e_N\}$ is a basis for E. Let $\Omega_1, \ldots, \Omega_N$ be (real-valued) p-forms. We can then consider the E-valued p-form $\Omega = \Omega_1 e_1 + \cdots + \Omega_N e_N$, where, for any p vectors $\xi_1, \ldots, \xi_p$ in $T_x(M)$, we have

$$\Omega_x(\xi_1, \ldots, \xi_p) = \Omega_{1x}(\xi_1, \ldots, \xi_p)e_1 + \cdots + \Omega_{Nx}(\xi_1, \ldots, \xi_p)e_N.$$

Conversely, if Ω is an E-valued form, then real-valued forms $\Omega_1, \ldots, \Omega_N$ can be defined by the above equation. In short, once a basis for an N-dimensional vector space E has been chosen, giving an E-valued differential form Ω is the same as giving N real-valued forms, and we can write

$$\Omega = \sum_1^N \Omega_i e_i \qquad \text{or} \qquad \Omega = \langle \Omega_1, \ldots, \Omega_N \rangle .$$

The rules for local description of E-valued forms, as well as the transition laws, are similar to those of real-valued forms, so we won't describe them in detail. For the sake of simplicity, we shall restrict our attention to the case where E is finite-dimensional, although for the most part this assumption is unnecessary.

If ω is a real-valued differential form of degree p, and if Ω is an E-valued form of degree q, then we can define the form $\omega \wedge \Omega$ in the obvious way. In terms of a basis, if $\Omega = \langle \Omega_1, \ldots, \Omega_N \rangle$, then $\omega \wedge \Omega = \langle \omega \wedge \Omega_1, \ldots, \omega \wedge \Omega_N \rangle$.

More generally, let E and F be (finite-dimensional) vector spaces, and let $\#$ be a bilinear map of $E \times F \to G$, where G is a third vector space. Let

$$\{e_1, \ldots, e_N\}$$

be a basis for E, let $\{f_1, \ldots, f_M\}$ be a basis for F, and let $\{g_1, \ldots, g_L\}$ be a basis for G. Suppose that the map $\#$ is given by

$$\#\{e_i, f_j\} = \sum_k a_{ij}^k g_k.$$

Then if $\omega = \sum \omega_i e_i$ is an E-valued form and $\Omega = \sum \Omega_j f_j$ is an F-valued form, we define the G-valued form $\omega \wedge \Omega$ by

$$\omega \wedge \Omega = \sum_k \left(\sum_{i,j} a_{ij}^k \omega_i \wedge \Omega_j\right) g_k.$$

It is easy to check that this does not depend on the particular bases chosen.

We shall want to apply this notion primarily in two contexts. First of all, we will be interested in the case where $E = F$ and $G = \mathbb{R}$, so that $\#$ is a bilinear form on E. Suppose $\#$ is a scalar product and $e_1, \ldots, e_N$ is an orthonormal basis. Then we shall write $(\omega \wedge \Omega)$ to remind us of the scalar product. If

$$\omega = \langle \omega_1, \ldots, \omega_N \rangle \qquad \text{and} \qquad \Omega = \langle \Omega_1, \ldots, \Omega_N \rangle;$$

then

$$(\omega \wedge \Omega) = \sum \omega_i \wedge \Omega_i.$$

Note that in this case if ω is a p-form and Ω is a q-form, then

$$(\omega \wedge \Omega) = (-1)^{pq}(\Omega \wedge \omega),$$

as in the case of real-valued forms.

The second case we shall be interested in is where $F = G$ and $E = \text{Hom}(F)$, and $\#$ is just the evaluation map evaluating a linear transformation on a vector of F to give another element of F. This time, choosing a basis for F determines a basis for $\text{Hom}(F)$, so we can regard ω as a *matrix* of real-valued differential forms. If $\omega = (\omega_{ij})$ and $\Omega = \langle \Omega_1, \ldots, \Omega_M \rangle$, then

$$\omega \wedge \Omega = \left\langle \sum \omega_{1j} \wedge \Omega_j, \ldots, \sum \omega_{Mj} \wedge \Omega_j \right\rangle.$$

The operator d makes sense for vector-valued forms just as it did for real-valued forms, and it satisfies the same rules. Thus, if $\Omega = \langle \Omega_1, \ldots, \Omega_N \rangle$, then $d\Omega = \langle d\Omega_1, \ldots, d\Omega_N \rangle$ and

$$d(\omega \wedge \Omega) = d\omega \wedge \Omega + (-1)^p \omega \wedge d\Omega$$

if ω is an E-valued form of degree p and Ω is an F-valued form.

We shall apply the notion of vector-valued forms to develop (mostly in exercise form) some elementary facts about the geometry of oriented surfaces in $\mathbb{E}^3$. Let M be an oriented two-dimensional manifold, and let φ be a differentiable map of M into $\mathbb{E}^3$. We shall assume that φ_* is not singular at any point of M, i.e., that φ is an immersion. Thus at each point $p \in M$ the space $\varphi_*(T_p(M))$ is a two-dimensional subspace of $T_{\varphi(p)}(\mathbb{E}^3)$. Since we can identify $T_{\varphi(p)}(\mathbb{E}^3)$ with $\mathbb{E}^3$, we can regard $\varphi_*(T_p(M))$ as a two-dimensional subspace of $\mathbb{E}^3$. (See Fig. 11.15.) Since M is oriented, so is the tangent plane $\varphi_*(T_p(M))$. Therefore, there is a unique unit vector orthogonal to the tangent plane which,

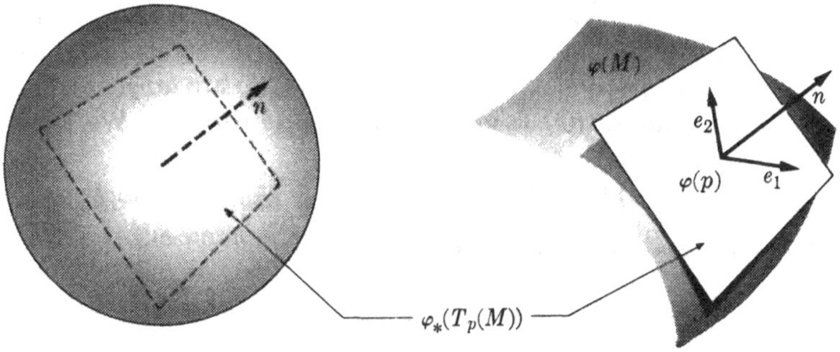

Fig. 11.15

together with an oriented basis of the tangent plane, gives an oriented basis of $\mathbb{E}^3$. This vector is called the *normal vector* and will be denoted by $n(p)$. We can consider n an $\mathbb{E}^3$-valued function on M. Since $\|n\| = 1$, we can regard n as a mapping from M to the unit sphere. Note that $\varphi(M)$ lies in a fixed plane of $\mathbb{E}^3$ if and only if $n = \text{const}$ ($n =$ the normal vector to the plane). We therefore can expect the variation of n to be useful in describing how the surface $\varphi(M)$ is "bending".

Let Ω be the (oriented) area form on M corresponding to the Riemann metric induced by φ. Let Ω_S be the (oriented) area form on the unit sphere. Then $n^*(\Omega_S)$ is a two-form on M, and therefore we can write

$$n^*\Omega_S = K\Omega.$$

The function K is called the *Gaussian curvature* of the surface $\varphi(M)$. Note that $K = 0$ if $\varphi(M)$ lies in a plane. Also, $K = 0$ if $\varphi(M)$ is a cylinder (see the exercises).

For any oriented two-dimensional manifold with a Riemann metric we let $\mathfrak{F}$ denote the set of all oriented bases in all tangent spaces of M. Thus an element of $\mathfrak{F}$ is given by $\langle f_1, f_2 \rangle$, where $\langle f_1, f_2 \rangle$ is an orthonormal basis of $T_x(M)$ for some $x \in M$. Note that f_2 is determined by f_1, because of the orientation and the fact that $f_2 \perp f_1$. Thus we can consider $\mathfrak{F}$ the space of all tangent vectors of unit length. For each $x \in M$ the set of all unit vectors is just a circle. We leave it to the reader to verify that $\mathfrak{F}$ is, in fact, a three-dimensional manifold. We denote by π the map that assigns to each $\langle f_1, f_2 \rangle$ the point x when $\langle f_1, f_2 \rangle$ is an orthonormal basis at x. Again, the reader should verify that π is a differentiable map of $\mathfrak{F}$ onto M.

In the case at hand, where the metric comes from an immersion φ, we define several vector-valued functions X, e_1, e_2, and e_3 on $\mathfrak{F}$ as follows:

$$\begin{aligned} X &= \varphi \circ \pi, \\ e_1(\langle f_1, f_2 \rangle) &= \varphi_* f_1, \\ e_2(\langle f_1, f_2 \rangle) &= \varphi_* f_2, \\ e_3 &= n \circ \pi. \end{aligned}$$

(In the middle two equations we regard $\varphi_* f_i$ as elements of $\mathbb{E}^3$ via the identification of $T_{\varphi(x)}\mathbb{E}^3$ with $\mathbb{E}^3$.) Thus at any point z of $\mathfrak{F}$, the vectors $e_1(z), e_2(z), e_3(z)$ form an orthonormal basis of $\mathbb{E}^3$, where $e_1(z)$ and $e_2(z)$ are tangent to the surface at $\varphi(\pi(z)) = X(z)$ and $e_3(z)$ is orthogonal to this surface. We can therefore write

$$(dX \wedge e_3) = (dX, e_3) = 0 \qquad \text{and} \qquad (e_i, e_j) = \delta_i^j.$$

By the first equation we can write

$$dX = \omega_1 e_1 + \omega_2 e_2, \tag{II.1}$$

where

$$\omega_1 = (dX, e_1) \qquad \text{and} \qquad \omega_2 = (dX, e_2)$$

are (real-valued) linear differential forms defined on $\mathfrak{F}$.

Similarly, let us define the forms ω_{ij} by setting

$$\omega_{ij} = (de_i, e_j).$$

Applying d to the equation $(e_i, e_j) = \delta_{ik}$ shows that

$$\omega_{ij} = -\omega_{ji}. \tag{II.2}$$

If we apply d to (II.1), we get

$$0 = d\,dX = d\omega_1 e_1 - \omega_1 \wedge de_1 + d\omega_2 e_2 - \omega_2 \wedge de_2.$$

Taking the scalar product of this equation with e_1 and e_2, respectively, shows (since $\omega_{11} = 0$ and $\omega_{22} = 0$) that

$$d\omega_1 = \omega_2 \wedge \omega_{21} \qquad \text{and} \qquad d\omega_2 = \omega_1 \wedge \omega_{12}. \tag{II.3}$$

If we apply d to the equation

$$de_i = \sum_j \omega_{ij} e_j,$$

we get

$$0 = \sum (d\omega_{ij} e_j - \omega_{ij} \wedge de_j)$$

and if we take the scalar product with e_j, we get

$$d\omega_{ij} = \sum_k \omega_{ik} \wedge \omega_{kj}. \tag{II.4}$$

If we apply d to the equation $(dX, e_3) = 0$, we get

$$0 = d(dX, e_3) = (dX, de_3) = (\omega_1 e_1 + \omega_2 e_2, \omega_{31} e_1 + \omega_{32} e_2),$$

which implies that

$$\omega_1 \wedge \omega_{31} + \omega_2 \wedge \omega_{32} = 0. \tag{II.5}$$

We will now interpret these equations. Let $z = \langle f_1, f_2 \rangle$ be a point of $\mathfrak{F}$. For any $\xi \in T_z(\mathfrak{F})$ we have

$$\langle \xi, dX \rangle = \langle \xi, d\pi^*\varphi \rangle = \langle \xi, \pi^* d\varphi \rangle = \langle \pi_* \xi, d\varphi \rangle = \varphi_*(\pi_* \xi).$$

Therefore,

$$\begin{aligned}\langle \xi, \omega_1 \rangle &= (\varphi_* \pi_* \xi, e_1) \\ &= (\varphi_* \pi_* \xi, \varphi_* f_1) \\ &= (\pi_* \xi, f_1),\end{aligned} \tag{II.6}$$

since the metric was defined to make φ_* an isometry. In other words, $\langle \xi, \omega_1 \rangle$ and $\langle \xi, \omega_2 \rangle$ are the components of $\pi_* \xi$ with respect to the basis $\langle f_1, f_2 \rangle$. If η is another tangent vector at z, then $\omega_1 \wedge \omega_2(\xi, \eta)$ is the (oriented) area of the parallelogram spanned by $\pi_* \xi$ and $\pi_* \eta$. In other words,

$$\omega_1 \wedge \omega_2 = \pi^* \Omega, \tag{II.7}$$

where Ω is the oriented area form on M.

Similarly,

$$\langle \xi, de_3 \rangle = n_* \pi_* \xi, \tag{II.8}$$

and we have

$$\begin{aligned} n_* \pi_* \xi &= \langle \xi, \omega_{31} \rangle e_1 + \langle \xi, \omega_{32} \rangle e_2 \\ &= \langle \xi, \omega_{31} \rangle \varphi_* f_1 + \langle \xi, \omega_{32} \rangle \varphi_* f_2. \end{aligned} \tag{II.9}$$

Since we can regard e_1 and e_2 as an orthonormal basis of the *tangent space to the unit sphere*, we conclude that $\omega_{31} \wedge \omega_{32}(\xi, \eta)$ is the oriented area on the unit sphere of the parallelogram spanned by $n_* \pi_* \xi$ and $n_* \pi_* \eta$. Thus

$$\begin{aligned} \omega_{31} \wedge \omega_{32} &= \pi^* n^* \Omega_S \\ &= \pi^* K \Omega \\ &= K \omega_1 \wedge \omega_2. \end{aligned}$$

Let

$$\begin{bmatrix} a & b \\ b^1 & c \end{bmatrix}$$

be the matrix of the linear transformation $n_*: T_x(M) \to T_{n(x)}(S^2)$ in terms of the basis $\langle f_1, f_2 \rangle$ of $T_x(M)$ and $\langle e_1, e_2 \rangle$ of $T_{n(x)}(S^2)$. Then comparing (II.6) with (II.9) shows that

$$\omega_{31} = a\omega_1 + b^1 \omega_2 \qquad \text{and} \qquad \omega_{32} = b\omega_1 + c\omega_2. \tag{II.10}$$

If we substitute this into (II.5), we conclude that $b = b^1$, i.e., that the matrix of n_* is symmetric. This suggests that it corresponds to a symmetric bilinear form of some geometrical significance. In other words, we want to consider the quadratic form

$$a\omega_1^2 + 2b\omega_1\omega_2 + c\omega_2^2$$

[where it is understood that this is the quadratic form on $T_z(\mathfrak{F})$ which assigns the number

$$a\langle \xi, \omega_1 \rangle^2 + 2b\langle \xi, \omega_1 \rangle \langle \xi, \omega_2 \rangle + c\langle \xi, \omega_2 \rangle^2$$

to any $\xi \in T_z(\mathfrak{F})$].

EXERCISES

II.1 Show that

$$a\langle\xi, \omega_1\rangle^2 + 2b\langle\xi, \omega_1\rangle\langle\xi, \omega_2\rangle + c\langle\xi, \omega_2\rangle^2 = (\varphi_*\pi_*\xi, n_*\pi_*\xi).$$

II.2 The quadratic form which assigns to each $\zeta \in T_x(M)$ the number $(\varphi_*\zeta, n_*\zeta)$ is called the *second fundamental form* of the surface. We shall denote it by $\mathrm{II}(\zeta)$. (What is usually called the first fundamental form is just $\|\zeta\|^2$ in our terminology.) Let C be any smooth curve with $C'(0) = \zeta$. Show that

$$\mathrm{II}(\zeta) = -\left(\frac{d^2\varphi \circ C}{dt^2}(0), n(x)\right).$$

Thus $\mathrm{II}(\zeta)$ measures how much the curve $\varphi \circ C$ is bending in the n-direction. Suppose we choose C to be such that $\varphi \circ C$ lies in the plane spanned by $\varphi_*\xi$ and $n(x)$. [Geometrically, this amounts to considering the curve obtained on the surface by intersecting the surface with the plane spanned by $\varphi_*\xi$ and $n(x)$.] Show that $\mathrm{II}(\zeta)$ is the curvature of this plane curve.

In this sense, the second fundamental form $\mathrm{II}(\zeta)$ tells us how much the surface is bending in the direction of ζ.

Note that

$$K = \det\begin{bmatrix} a & b \\ b & c \end{bmatrix}.$$

Let λ_1 and λ_2 be the eigenvalues of the matrix

$$\begin{bmatrix} a & b \\ b & c \end{bmatrix}.$$

Thus

$$\lambda_1 = \max \mathrm{II}(\zeta) \quad \text{and} \quad \lambda_2 = \min \mathrm{II}(\zeta) \quad \text{for} \quad \|\zeta\| = 1.$$

If $\lambda_1 \neq \lambda_2$, there are two orthogonal eigenvectors which are called the directions of principal curvature of the surface. (Note that they must be orthogonal, since they are eigenvectors of a symmetric matrix.)

If A is a Euclidean motion of $\mathbb{E}^3$, then $\psi = A \circ \varphi$ is another immersion of M and it is easy to check that both the Riemann metric induced by ψ and the second fundamental form associated with ψ coincide with those attached to φ. What is not so obvious is the converse: *If ψ and φ induce the same metric and the same second fundamental form, then $\psi = A \circ \varphi$ for some Euclidean motion A.* We will not prove this fact, although it is a fairly easy consequence of what we have already established.

We have seen the meaning of ω_1, ω_2, ω_{31}, and ω_{32} in geometric terms. Let us now interpret the one remaining form, ω_{12}.

Let γ be a differentiable curve on M. A differentiable family of unit vectors $f_1(\cdot)$ along γ [where $f_1(s) \in T_{\gamma(s)}(M)$] is the same as a curve C in $\mathfrak{F}$ with $\pi \circ C = \gamma$. [Here $C(s) = \prec f_1(s), f_2(s) \succ$.] Let us call the family $f_1(s)$ *parallel* if the unit vectors are all changing normally to the surface in three-space. In other words, $f_1(s)$ is parallel if the vector

$$\frac{d\varphi_{*\gamma(s)}(f_1(s))}{ds}$$

is normal to $\varphi(M)$ for all s. Let us see how to express this condition. Let ξ_s be the tangent vector to the curve C at $C(s)$. Then, by the definition of de_1,

$$\frac{d}{ds}\varphi_{*\gamma(s)}(f_1(s)) = \langle \xi_s, de_1 \rangle.$$

Note that $(\langle \xi_s, de_1\rangle, e_1(C(s))) = 0$ and $(\langle \xi_s, de_1\rangle, e_2(C(s))) = \langle \xi_s, \omega_{12}\rangle$. Now e_1 and e_2 span the tangent space to $\varphi(M)$, so saying that $f_1(\cdot)$ is parallel is the same as saying that $\langle \xi_s, \omega_{12}\rangle = 0$.

Thus $f_1(s)$ is parallel along γ if and only if $\langle \xi_s, \omega_{12}\rangle = 0$.

Let M and $\overline{M}$ be two-dimensional manifolds with Riemann metrics. Let $u: M \to \overline{M}$ be a differentiable map which is an isometry. Let $\mathfrak{F}$ be the manifold of orthonormal bases of M, and let $\overline{\mathfrak{F}}$ be the manifold of orthornormal bases of $\overline{M}$. Then u induces a map $\tilde{u}$ of $\mathfrak{F} \to \overline{\mathfrak{F}}$ by

$$\tilde{u}(\langle f_1, f_2\rangle) = \langle u_* f_1, u_* f_2\rangle.$$

Let ω_1 be the differential form on $\mathfrak{F}$ given, as in (II.6), by

$$\langle \xi, \omega_1\rangle = (\pi_*\xi, f_1) \qquad \text{for} \quad \xi \in T_z(\mathfrak{F}),$$

where $z = \langle f_1, f_2\rangle$, with the corresponding definition for $\omega_2, \bar{\omega}_1$, and $\bar{\omega}_2$. Then for any $\xi \in T_z(\mathfrak{F})$ we have, since $\bar{\pi} \circ \tilde{u} = u \circ \pi$,

$$\langle \xi, \tilde{u}_*\bar{\omega}_1\rangle = \langle \tilde{u}_*\xi, \bar{\omega}_1\rangle = (\bar{\pi}_*\tilde{u}_*\xi, u_*f_1) = (u_*\pi_*\xi, u_*f_1) = (\pi_*\xi, f_1) = \langle \xi, \omega_1\rangle.$$

In other words,

$$\tilde{u}^*\bar{\omega}_1 = \omega_1 \qquad \text{and} \qquad \tilde{u}^*\bar{\omega}_2 = \omega_2.$$

Now suppose that the metrics on M and $\overline{M}$ come from immersions φ and $\bar{\varphi}$. Then we get forms ω_{ij} and $\bar{\omega}_{ij}$. Now by (II.3) we have

$$\tilde{u}^*(\bar{\omega}_1 \wedge \bar{\omega}_{21}) = \tilde{u}^*d\bar{\omega}_1 = d(\tilde{u}^*\bar{\omega}_1) = d\omega_1 = \omega_1 \wedge \omega_{21}.$$

Thus

$$\omega_1 \wedge \tilde{u}^*\bar{\omega}_{21} = \omega_1 \wedge \omega_{21} \qquad \text{and} \qquad \omega_2 \wedge \tilde{u}^*\bar{\omega}_{12} = \omega_2 \wedge \omega_{12},$$

or

$$(\tilde{u}^*\bar{\omega}_{12} - \omega_{12}) \wedge \omega_1 = 0 \qquad \text{and} \qquad (\tilde{u}^*\bar{\omega}_{12} - \omega_{12}) \wedge \omega_2 = 0.$$

Since the differential forms ω_1 and ω_2 are linearly independent, this can only happen if

$$\tilde{u}^*\bar{\omega}_{12} = \omega_{12}.$$

In other words, if the two surfaces $\varphi(M)$ and $\bar{\varphi}(M)$ are isometric, they have the "same" ω_{12}, that is, the same notion of "parallel vector fields". Observe that a piece of a cylinder and a piece of the plane are isometric, even though they are *not* congruent by a Euclidean motion. In different terms, while the forms ω_{13} and ω_{23} depend on how the surface is immersed in $\mathbb{E}^3$, the form ω_{12} depends only on the Riemann metric induced by the immersion.

Now we have (II.4):

$$d\omega_{12} = \omega_{13} \wedge \omega_{23} = -\pi^*\Omega_s = -K\Omega.$$

From this we conclude that the Gaussian curvature K also does not depend only on the immersion, but only on the Riemann metric coming from the immersion.

Since ω_{12} does not depend on φ, we should be able to define it for an arbitrary two-dimensional manifold with a Riemann metric. Note that the preceding argument shows that ω_{12} is uniquely determined by Eq. (II.4). It therefore suffices to construct an ω_{12} on a coordinate neighborhood so as to satisfy (II.4). It will then follow from the uniqueness that any two such coincide to give a well-defined form. Let U be a coordinate neighborhood of M, and let $\psi: U \to \mathfrak{F}$ be a differentiable map such that $\pi \circ \psi = \text{id}$. Thus ψ assigns a basis $\langle f_1, f_2 \rangle$ to each $x \in U$, in a differentiable manner. (One possible way to construct ψ is to apply the orthonormalization procedure to the vector fields $\langle \partial/\partial x^1, \partial/\partial x^2 \rangle$.)

Once we have chosen ψ, any basis of T_x differs from $\psi(x)$ by a rotation. If we let τ denote the (angular) coordinate giving this rotation (so that τ is only defined mod 2π), then we can use the local coordinates on U together with τ as coordinates on $\pi^{-1}(U)$. More precisely, if x^1 and x^2 are local coordinates on U, we define y^1, y^2, τ by

$$y^1 = x^1 \circ \pi, \qquad y^2 = x^2 \circ \pi,$$

and $\tau(z)$ is given for $z = \langle e_1, e_2 \rangle$ by

$$e_1 = \cos \tau(z) f_1 + \sin \tau(z) f_2, \qquad e_2 = -\sin \tau(z) f_1 + \cos \tau(z) f_2, \tag{II.11}$$

where $\langle f_1, f_2 \rangle = \psi(x)$ when $\langle e_1, e_2 \rangle \in T_x(M)$.

Now let

$$\theta_1 = \psi^*(\omega_1) \qquad \text{and} \qquad \theta_2 = \psi^*(\omega_2),$$

so that θ_1 and θ_2 are forms defined on U and are, in fact, the dual basis for $\psi(x)$ at each $x \in M$. If we set

$$\alpha_1 = \pi^*\theta_1 \qquad \text{and} \qquad \alpha_2 = \pi^*\theta_2,$$

then (II.11) gives

$$\omega_1 = \cos \tau \alpha_1 + \sin \tau \alpha_2 \qquad \text{and} \qquad \omega_2 = -\sin \tau \alpha_1 + \cos \tau \alpha_2.$$

Note that

$$\omega_1 \wedge \omega_2 = \alpha_1 \wedge \alpha_2.$$

Define the functions l_1 and l_2 on M by

$$d\theta_1 = l_1 \theta_1 \wedge \theta_2 \qquad \text{and} \qquad d\theta_2 = l_2 \theta_1 \wedge \theta_2.$$

Let $k_1 = l_1 \circ \pi$ and $k_2 = l_2 \circ \pi$, so that

$$d\alpha_1 = k_1 \alpha_1 \wedge \alpha_2 \qquad \text{and} \qquad d\alpha_2 = k_2 \alpha_1 \wedge \alpha_2.$$

Now

$$\begin{aligned} d\omega_1 &= -\sin \tau \, d\tau \wedge \alpha_1 + \cos \tau \, d\tau \wedge \alpha_2 + (k_1 \cos \tau + k_2 \sin \tau) \alpha_1 \wedge \alpha_2, \\ d\omega_2 &= -\cos \tau \, d\tau \wedge \alpha_1 - \sin \tau \, d\tau \wedge \alpha_2 + (+k_2 \cos \tau - k_1 \sin \tau) \alpha_1 \wedge \alpha_2. \end{aligned}$$

Since $\omega_1 \wedge \omega_2 = \alpha_1 \wedge \alpha_2$, we can rewrite these equations as

$$\begin{aligned} d\omega_1 &= (d\tau + (k_1 \cos \tau + k_2 \sin \tau)\omega_1) \wedge \omega_2, \\ d\omega_2 &= -(d\tau - (+k_2 \cos \tau - k_1 \sin \tau)\omega_2) \wedge \omega_1. \end{aligned}$$

We thus see that the form

$$\begin{aligned} \omega_{12} &= d\tau + (k_1 \cos \tau + k_2 \sin \tau)\omega_1 + (-k_1 \sin \tau + k_2 \cos \tau)\omega_2 \\ &= d\tau + k_1\alpha_1 + k_2\alpha_2 \end{aligned}$$

satisfies the desired equations.

As before, on any two-dimensional Riemann manifold we will call a family of unit vectors parallel along a curve γ if $\langle \xi_s, \omega_{12} \rangle = 0$. With this definition of parallel translation we can state the following:

Theorem. Let γ be any differentiable curve on M. Given the unit vector $g_1 \in T_{\gamma(0)}(M)$, there is a unique parallel family of unit vectors $g_1(s)$ along γ, with $g_1(0) = g_1$. If $g_1'(0)$ is another unit vector of $T_{\gamma(0)}(M)$ differing from g_1 by an angle σ, then $g_1'(s)$ differs from $g_1(s)$ by the same angle σ for all s.

Proof. It is clearly sufficient (by breaking γ up into small pieces if necessary) to prove the theorem for curves γ lying entirely in a coordinate chart. Then we can use the local expression for ω_{12}.

Let us rewrite the condition for parallel translation along $\gamma(s)$. In terms of local coordinates, the unit vector $g_1(s)$ is given by a function $\tau(s)$, where

$$g_1(s) = \cos \tau(s) f_1(\gamma(s)) - \sin \tau(s) f_2(\gamma(s)).$$

Then

$$\begin{aligned} \langle \xi_s, \omega_{12} \rangle &= \langle \xi_s\, d\tau \rangle + \langle \xi_s, k_1\alpha_1 + k_2\alpha_2 \rangle \\ &= \frac{d\tau(s)}{ds} + \langle \xi_s, \pi^*(k_1\theta_1 + k_2\theta_2) \rangle \\ &= \frac{d\tau(s)}{ds} - \langle \pi_*\xi_s, k_1\theta_1 + k_2\theta_2 \rangle. \end{aligned}$$

But $\pi_*\xi_s = \zeta_s$ is the tangent vector to γ at $\gamma(s)$. Thus

$$\langle \xi_s, \omega_{12} \rangle = \frac{d\tau(s)}{ds} - F_\gamma(s),$$

where $F_\gamma(s) = \langle \zeta_s, k_1\theta_1 + k_2\theta_2 \rangle$ is a function depending only on s. In particular, $g_1(s)$ is parallel if and only if

$$\frac{d\tau(s)}{ds} = F_\gamma(s).$$

From this we see that given $g_1(0)$ there is a unique parallel family $g_1(s)$, starting with $g_1(0)$. Furthermore, if $g_1'(0)$ is a second unit vector at $\gamma(0)$, the angle between $g_1(s)$ and $g_1'(s)$ is equal to the angle between $g_1(0)$ and $g_1'(0)$. Thus parallel translation preserves angles, which proves the theorem. □

Note that if M is (locally isometric to) Euclidean space, then we can choose

$$f_1 = \frac{\partial}{\partial x^1} \quad \text{and} \quad f_2 = \frac{\partial}{\partial x^2}$$

so that $\theta_1 = dx^1$ and $\theta_2 = dx^2$. In this case, $k_1 = k_2 = 0$ and τ is just the angle that g_1 makes with $\partial/\partial x^1$, that is, with the x_1-axis. Thus $\omega_{12} = -d\tau$ in this case. Then the condition for parallel translation becomes $d\tau/ds = 0$, which coincides with the usual notion of parallelism in Euclidean geometry. Note that in Euclidean space the parallelism does not depend on the curve γ. This is not true in general.

Exercise 11.3. Let γ_1 and γ_2 be two arcs of great circles joining the north and south poles on S^2. Suppose that γ_1 and γ_2 are orthogonal at the poles. Let ζ be a tangent vector of the north pole. Compare its translates to the south pole via γ_1 and γ_2.

Let M be any two-dimensional Riemann manifold. For any curve γ on M there is an obvious way of choosing unit vectors along γ: just let $g_1(s)$ be the unit tangent vector to γ at $\gamma(s)$. Thus for every curve γ on M we get a curve, which we shall call $\tilde{\gamma}$, on $\mathfrak{F}$. [Here $\tilde{\gamma} = (\gamma(s), g_1(s), g_2(s))$ and $g_1(s)$ is the tangent to $\gamma(s)$.]

We call the form $\tilde{\gamma}^*(\omega_{12})$ the *geodesic curvature* form of γ. [In the Euclidean case this is just the ordinary curvature (see the exercises).]

Let us consider those curves whose geodesic curvatures vanish, i.e., those curves whose tangent vectors are parallel. We shall call such a curve a *geodesic* with respect to the given Riemann metric. Note that the condition that a curve be geodesic is given, in local coordinates, by a second-order differential equation. Therefore, a geodesic $C(\cdot)$ is uniquely specified by giving $C(t)$ and $C'(t)$ at any fixed value of t. In Chapter 13 we use the term "geodesic" to mean a curve which locally minimizes length. It is the purpose of the next few exercises to show that geodesics in our present sense have this property.

EXERCISES

11.4 Let x, y be local coordinates on $U \subset M$. Through each point of the curve $y = 0$ (that is, the x-axis in the local coordinates), construct the unique geodesic orthogonal to this curve. (See Fig. 11.16.) Let s be the arc-length parameter along the geodesic, so that the geodesic passing through $(u, 0)$ is given by

$$(y(u, s), x(u, s)).$$

Show that the map $(u, s) \mapsto (y(u, s), x(u, s))$ has nonzero Jacobian at $(0, 0)$ and therefore defines a coordinate system in some open subset $U' \subset U$.

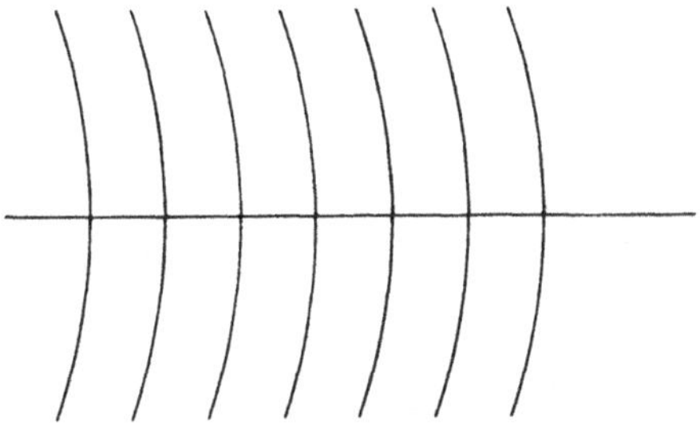

Fig. 11.16

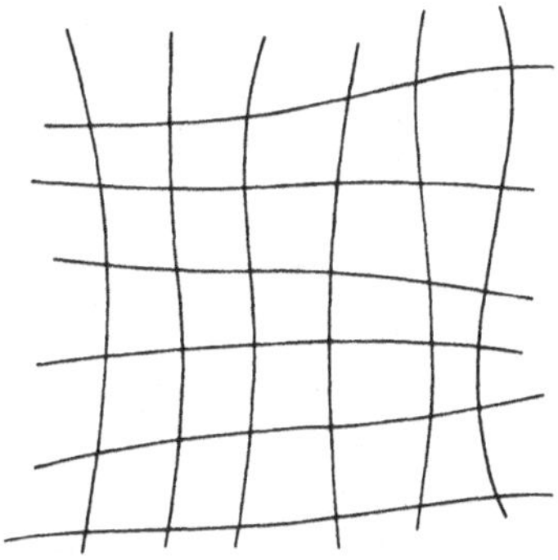

Fig. 11.17

II.5 We are going to make a further change of coordinates. Let Y be the vector field on U' defined by the properties

$$\|Y\| = 1, \qquad \left(Y, \frac{\partial}{\partial s}\right) = 0, \qquad \langle Y, du\rangle > 0.$$

Thus Y is orthogonal to the geodesics $u = \text{const}$ and points in the increasing U-direction. Let us consider the solution curves of this vector field parametrized by the initial position along the geodesic $u = 0$. That is, let v be the arc-length parameter along the geodesic $u = 0$, and consider the map

$$(u, v) \mapsto \big(u, s(u, v)\big),$$

where $s(u, v)$ is the s-coordinate of the intersection of the solution curve of Y passing through $(0, v)$ with the geodesic given by u. (See Fig. 11.17.) Again the existence theorem and smooth dependence on parameters, together with the fact that the curves $u = 0$ and $s = 0$ are already orthogonal, guarantees that we can find some neighborhood W so that (u, v) are coordinates on W. We have thus constructed coordinates such that the curves $u = \text{const}$ are geodesics and the curves $u = \text{const}$ and $v = \text{const}$ are orthogonal. Such a system of coordinates is called a *geodesic parallel coordinate system*.

II.6 Let (u, v) be a coordinate system on $U \subset M$ for which $(\partial/\partial u, \partial/\partial v) \equiv 0$, so that the metric takes the form

$$ds^2 = p\, du^2 + q\, dv^2.$$

Define the choice of frame ψ by normalizing $\partial/\partial u, \partial/\partial v$ so that $\psi(x) = \langle f_1, f_2\rangle$, where $f_1 = (\partial/\partial u)/\|(\partial/\partial u)\|$ and $f_2 = (\partial/\partial v)/\|(\partial/\partial v)\|$. Show that the forms θ_1 and θ_2 are given by

$$\theta_1 = p\, du, \qquad \theta_2 = q\, dv,$$

and

$$\omega_{12} = d\tau - \pi^*\left(\frac{1}{q}\frac{\partial p}{\partial v}\, du + \frac{1}{p}\frac{\partial q}{\partial u}\, dv\right)$$

and

$$K = -\frac{1}{pq}\left[\frac{\partial}{\partial u}\left(\frac{1}{p}\frac{\partial q}{\partial u}\right) + \frac{\partial}{\partial v}\left(\frac{1}{q}\frac{\partial p}{\partial v}\right)\right].$$

II.7 Let (u, v) be a geodesic parallel coordinate system, as in Exercise II.5. The curve C_u given by $C_u(v) = (u, v)$ is a geodesic. Thus $\langle C''(v), \omega_{12}\rangle = 0$. But in terms of our local coordinates, $\langle C''(v), d\tau\rangle = 0$, since $C'(v)$ is always parallel to one of the base vectors f_2, and

$$\langle C''(v), \bar{\pi}^*du\rangle = \langle C^1(v), du\rangle = 0,$$

since $u = $ const along C. Thus we conclude that $\partial q/\partial u = 0$, or $q = q(v)$. Let us replace the parameter v by $w = \int_0^v q(t)\, dt$. Then (u, v) is a geodesic parallel coordinate system for which we have

$$ds^2 = p\, du^2 + dw^2,$$

and now the arc length along any curve $u = $ const is $\int dw$.

II.8 Show that for $|w|$ sufficiently small, any curve joining $(0, 0)$ to $(0, w)$ must have arc length at least $|w|$. Conclude that (since the choice of our original curve $x = 0$ was arbitrary) the geodesics locally minimize length.

II.9 Let $\langle w, z\rangle$ be local coordinates on an open set U of a Riemann manifold with the property that the curves C_z given by $C_z(w) = (w, z)$ are geodesics parametrized according to arc length. Thus $z = $ const is a geodesic and $\|\partial/\partial w\| = 1$. Let

$$\left(\frac{\partial}{\partial w}, \frac{\partial}{\partial z}\right) = a.$$

Show that $\partial a/\partial w = 0$. [*Hint:* Show that by orthonormalizing $(\partial/\partial w, \partial/\partial z)$, we obtain a map ψ whose associated forms θ_1 and θ_2 are given by $\theta_1 = dw + a\, dz$, $\theta_2 = b\, dz$, where $b = \|\partial/\partial z - a(\partial/\partial w)\|$. Then compute l_1 and l_2 and use the fact that $z = 0$ is a geodesic.]

II.10 Construct *geodesic polar coordinates.* That is, for a fixed $p \in M$ let θ be an angular coordinate on the unit circle in $T_p(M)$. For each θ let $C_\theta(\cdot)$ be the unique geodesic, parametrized by arc length, such that $C(0) = p$ and $C'(0)$ corresponds to angle θ in $T_p(M)$. Show that the map $(r, \theta) \mapsto C_\theta(r)$ gives "coordinates" on $U - \{p\}$, where U is some neighborhood of p. By taking $\langle r, \theta\rangle$ to be the $\langle w, z\rangle$ of Exercise II.9 and passing to the limit $r = 0$, conclude that

$$\left(\frac{\partial}{\partial r}, \frac{\partial}{\partial \theta}\right) \equiv 0.$$

Thus in terms of the "coordinates" $\langle r, \theta\rangle$ on $U - \{p\}$, the Riemann metric takes the form

$$ds^2 = dr^2 + A\, d\theta^2.$$

These coordinates are the analogue, for a general two-dimensional Riemann manifold, of the polar coordinates introduced on the plane and on the sphere at the end of Chapter 9. The argument given there applies generally to give another proof of the fact that geodesics locally minimize arc length.

We now continue to study the consequences of the equation

$$d\omega_{12} = -\pi^*(K\Omega).$$

Let D be a domain with regular boundary on M, and let ψ be a map of some neighborhood of $D \to \mathfrak{F}$ satisfying $\pi \circ \psi = $ identity. Then, by Stokes' theorem,

$$\int_{\partial D} \psi^*(\omega_{12}) = -\int_D \psi^*\pi^*K\Omega = -\int_D K\Omega, \tag{II.12}$$

where Ω is the area form on M.

Let us apply this to the map ψ constructed as follows: Let Y be a vector field on a compact manifold M which vanishes at only a finite number of points $p_1, \ldots, p_r$. At any $x \in M$ where $Y_x \neq 0$, put $f_1(x) = Y_x/\|Y_x\|$, so $\psi(x) = \langle x, f_1(x), f_2(x)\rangle$. About each point p_i choose a chart (U_i, h_i) such that U_i contains no other point p_j and $h_i(p_i) = (0, 0)$ in the plane. Let $\gamma_{i,r}$ be $h_i^{-1}(C_r)$, where C_r is the circle of radius r in the plane. [Here r is chosen so small that C_r lies in $h_i(U_i)$.] Now let $D = M - U_i$ (interiors of $\gamma_{i,r}$). We must compute $\int_{\gamma_{i,r}} \psi^*(\omega_{12})$, where $\gamma_{i,r}$ is oriented clockwise rather than counterclockwise. For this purpose, we introduce about p_i the orthonormal frames coming from the coordinates x^1, x^2. Thus we have coordinates x^1, x^2, τ, where $\tau(x, f_1, f_2)$ measures the angle that $f_1(x)$ makes with

$$\left.\frac{\partial}{\partial x^1}\right|_x.$$

Thus (taking $\gamma_{i,r,\tau}$ clockwise)

$$-\int_{\gamma_{i,r}} d\tau(f_1(s)) = 2\pi$$

(index of Y at p_i). But $\psi^*\omega_{12} = \psi^*d\tau + \psi^*(k_1\alpha^1 + k_2\alpha^2)$, so

$$-\int_{\gamma_{i,r}} \psi^*\omega_{12} = 2\pi \text{ (index of } Y \text{ at } p_i) + \int_{\gamma_{i,r}} k_1\theta^1 + k_2\theta^2.$$

Now as $r \to 0$, the second term on the right vanishes and $\int_D \to \int_M$ on the right of (II.12). Thus we have proved the following:

Let Y be a vector field which vanishes at a finite number of points $p_1, \ldots, p_n$. Then

$$\sum_i \operatorname{index}_{p_i}(Y) = \frac{1}{2\pi}\int_M K\,dA. \tag{II.13}$$

In particular, the sum of the indices of a vector field is independent of the vector field, and the total integral of the curvature is independent of the Riemann metric.

Thus, for instance, if M is S^2, the vector field tangent to the meridian circles has zeros only at the north and south poles, and the index at each of these points is $+1$. Thus (II.13) says that the sum of indices of any vector field on S^2 must equal 2. (In particular, it is impossible to construct a vector field on S^2 which does not vanish anywhere.) Furthermore, (II.13) says that no matter what Riemann metric we put on S^2, we have

$$\int_{S^2} K\,dA = 4\pi.$$

Similarly, if M is the torus, we can put it on a vector field which does not vanish anywhere, so the integer in (II.13) is 0.

EXERCISES

Let M denote the torus with angular coordinates φ_1, φ_2 (defined up to 2π). The map F will denote the immersion of $M \to$ Euclidean three-space given by

$$\begin{aligned} x_1 &= (2 + \cos\varphi_1)\cos\varphi_2, \\ x_2 &= (2 + \cos\varphi_1)\sin\varphi_2, \\ x_3 &= \sin\varphi_1. \end{aligned}$$

II.11 What is the Riemann metric on M induced by F? That is, compute $(\partial/\partial\varphi_1, \partial/\partial\varphi_1)_p$, $(\partial/\partial\varphi_1, \partial/\partial\varphi_2)_p$ and $(\partial/\partial\varphi_2, \partial/\partial\varphi_2)_p$ for each $p \in M$. (These can be given as functions of φ_1, φ_2.) Let f_1, f_2 denote the vector fields obtained by orthonormalizing $\partial/\partial\varphi_1$, $\partial/\partial\varphi_2$. What is the explicit expression for f_1, f_2?

II.12 What is the total area of the torus relative to this Riemann metric?

II.13 In terms of the vector fields f_1, f_2 given in Exercise II.11 we can introduce (angular) coordinates φ_1, φ_2, τ on $\mathfrak{F}$. In terms of these coordinates, what are the explicit expressions for the vector-valued functions X, f^1, f^2, f^3, on $\mathfrak{F}$? (Each of these vector-valued functions should be given in terms of the fixed standard basis of Euclidean three-space, i.e., a triplet of functions of φ_1, φ_2, τ.) Thus

$$\mathrm{X}(\varphi_1, \varphi_2, \tau) = \langle (2 + \cos\varphi_1)\cos\varphi_2,\ (2 + \cos\varphi_1)\sin\varphi_2,\ \sin\varphi_1 \rangle.$$

What are the corresponding expressions for f_1, f_2, and f_3?

II.14 What are the explicit expressions for the forms ω_1, ω_2, and ω_{12} on $\mathfrak{F}$ given in terms of $d\varphi_1$, $d\varphi_2$, and $d\tau$?

II.15 What is the Gaussian curvature K of M? (Again, K can be given as a function of φ_1, φ_2.) For which points of M is the Gaussian curvature positive, and for which points is it negative?

II.16 On any Riemann manifold there is an obvious way of identifying a linear differential form with a vector field [via the identification of $T_x^*(M)$ with $T_x(M)$ given by the scalar product in $T_x(M)$]. If g is a function, the vector field associated to the form dg is denoted by grad g. Let M be the torus with Riemann metric as above, and let x_1 be the function given by $x_1(\varphi_1, \varphi_2) = (2 + \cos\varphi_1)\cos\varphi_2$, as before. Show that the vector field $X = \operatorname{grad} x_1$ vanishes at exactly four points, and compute its index at each of these points.

Let x^1, x^2 be a local system of coordinates on $U \subset M$, and let $\psi\colon U \to \mathfrak{F}$ be given by orthonormalizing $\langle \partial/\partial x^1, \partial/\partial x^2 \rangle$. Thus coordinates on $\pi^{-1}(U)$ are $x^1 \circ \pi$, $x^2 \circ \pi$, τ, where $\tau(z)$ is the angle that f_1 makes with $\partial/\partial x^1$ if $z = \langle f_1, f_2 \rangle$. Then

$$\omega_{12} = d\tau + k_1\alpha_1 + k_2\alpha_2 = d\tau + \pi^*\psi^*\omega_{12}$$

on $\pi^{-1}(U)$. Let C be a curve in U with $C'(t) > 0$, and let $\tilde{C}$ be the curve in $\tilde{\pi}(U)$ assigning to each t the unit tangent vector $C'(t)/\|C'(t)\|$. Then

$$\int_{\tilde{C}} \omega_{12} = \int \tilde{C}^*\omega_{12} = \int_C \psi^*\omega_{12} + \int_C d\tau,$$

where τ is the angle that $C'(t)$ makes with the x-axis.

From this formula we can deduce a number of interesting consequences which we state in exercise form.

II.17 Show that the sum of the exterior angles of a geodesic triangle D is given by $2\pi - \int_D K\Omega$. (A geodesic triangle is a domain whose boundary consists of three geodesics.)

II.18 Suppose that D is a domain which in the local coordinates is given by a simple polygon. Thus $\partial D = C_1 \cup \cdots \cup C_k$, where each C_i is a curve which is a straight line segment in the local coordinate system. Let $\alpha_1, \ldots, \alpha_k$ be the interior angles. (See Fig. 11.18.) Show that

$$k\pi + \int_D K\Omega = 2\pi - \sum \int_{\tilde{C}} \omega_{12}.$$

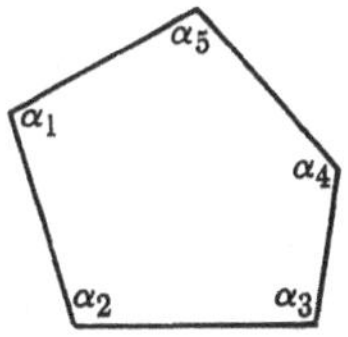

Fig. 11.18

II.19 This gives us another way of computing the integer in (II.13) if M is a compact surface.

Definition. A *cellulation* of a compact differentiable two-dimensional manifold M is a finite collection of closed subsets (called the cells) $F_1, \ldots, F_m$ such that

$$M = \bigcup_{i=1}^{m} F_i$$

satisfy the following:

1) For each F_i there is a one-to-one bidifferentiable map f_i of a neighborhood of F_i onto a polygon with n_i edges ($n_i \geq 3$).
2) For $i \neq j$, either $F_i \cap F_j$ is empty or $f_i(F_i \cap F_j)$ is a fixed edge or vertex of the corresponding polygon.

Let f be the number of faces, e the number of edges, and v the number of vertices in the cellulation of a two-dimensional Riemann manifold. Show that

$$f - e + v = \int_M K\Omega.$$

Thus $f - e + v$ does not depend on the cellulation. We have thus given three distinct ways of computing an integer attached to the manifold. This integer is called the *Euler characteristic* and is denoted by $\chi(M)$. Thus

$$\chi(M) = \int_M K\Omega = f - e + v = \sum \operatorname*{index}_{p_i} Y.$$

II.20 By a regular cellulation we mean a cellulation such that each face has the same number of edges and each vertex is the union of the same number of edges. Show that there are at most five possibilities for the number of faces in a regular cellulation of the sphere. Conclude that there are at most five "regular solids".

CHAPTER 12

POTENTIAL THEORY IN $\mathbb{E}^n$

1. SOLID ANGLE

In what follows $x^1, \ldots, x^n$ will denote Euclidean coordinates on $\mathbb{E}^n$. Let $r^2 = (x^1)^2 + \cdots + (x^n)^2$; then $dr^2 = 2r\,dr$, so that we have

$$\begin{aligned} r\,dr &= \sum x^i\,dx^i, \\ *r\,dr &= \sum (-1)^{i-1} x^i\,dx^1 \wedge \cdots \widehat{dx^i} \cdots \wedge dx^n, \\ d * r\,dr &= n\,dx^1 \wedge \cdots \wedge dx^n. \end{aligned}$$

Let i denote the injection of $S^{n-1} \to \mathbb{E}^n$ as the unit sphere. Let V_n denote the volume of the unit ball and A_{n-1} the volume of the unit $(n-1)$-sphere. The volume of the sphere of radius r is thus $r^{n-1}A_{n-1}$, and so

$$V_n = \int_0^1 r^{n-1}A_{n-1}\,dr = \frac{1}{n}A_{n-1}.$$

Since $i^*(*r\,dr)$ is an $(n-1)$-form invariant under rotations, it is some multiple of the volume form on S^{n-1}. By Stokes' theorem,

$$\int_{S^{n-1}} i^*(*r\,dr) = \int_{B^n} d*r\,dr = n\int_{B^n} dx^1 \wedge \cdots \wedge dx^n.$$

Comparing this with the above, we conclude that $i^*(*r\,dr)$ is the volume form on the unit sphere.

Let ρ denote the projection of $\mathbb{E}^n - \{0\}$ onto the unit sphere. Set

$$\tau = \rho^* i^*(*r\,dr).$$

Then τ is called the *element of solid angle*. Integrated over any $(n-1)$-surface, it gives the volume (counting sign of orientation and multiplicity) of its projection on the unit $(n-1)$-sphere.

We have

$$d\tau = 0,$$

since

$$\begin{aligned} d\tau &= d\rho^* i^*(*r\,dr) \\ &= \rho^*(di^*(*r\,dr)) = 0, \end{aligned}$$

because $i^*(*r\,dr)$ is an $(n-1)$-form on the $(n-1)$-dimensional manifold S^{n-1}, and d of any $(n-1)$-form on an $(n-1)$-dimensional manifold must vanish.

Let i_R denote the injection of the S^{n-1} into $\mathbb{E}^n$ as the sphere of radius R (so that $i_1 = i$). It then follows directly from the definitions that

$$i_R^*\tau = \frac{dS_R}{R^{n-1}},$$

where dS_R is the induced volume form. (See Fig. 12.1.)

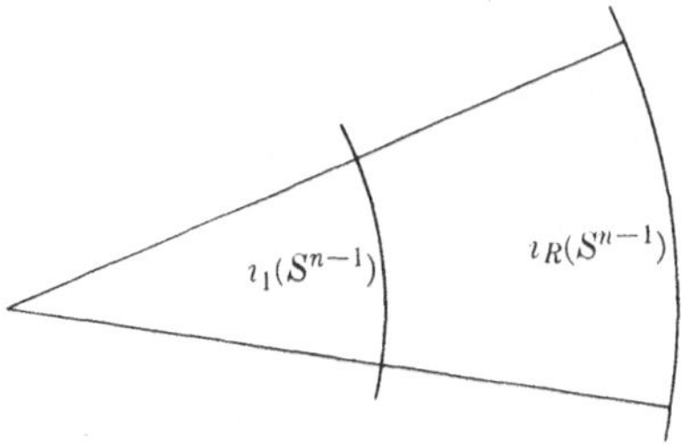

Fig. 12.1

Now the volume of the ball of radius R is R^nV_n and the surface volume (area) of the sphere of radius R is $R^{n-1}A_{n-1} = R^{-1}(nR^nV_n)$. Now $i_R^*(*dr)$ is an $(n-1)$-form and thus $i_R^*(*dr) = f\,dS_R$ for some function f. Since $*dr$ and dS_R are invariant under rotation, we conclude that f is a constant. But

$$\int_{S^{n-1}} i_R^*(*r\,dr) = n\int_{B^n(R)} dx^1 \wedge \cdots \wedge dx^n = nR^nV_n,$$

form which we conclude that

$$dS_R = i_R^*\left(\frac{*r\,dr}{r}\right) = i_R^*(*dr).$$

Thus

$$i_R^*\left(\tau - \frac{*dr}{r^{n-1}}\right) = 0.$$

Now let x be any point in $\mathbb{E}^n - \{0\}$, and let $\xi_1, \ldots, \xi_{n-1}$ be tangent vectors at x. Then

$$\left(\tau - \frac{*dr}{r^{n-1}}\right)(\xi_1, \ldots, \xi_{n-1})$$

will vanish if all the ξ's are tangent to the sphere centered at the origin and passing through x, by the above equation. If one of the ξ's, say ξ_1, is a multiple of $(\partial/\partial r)_x$, then again the expression will vanish, because $\tau(\xi_1, \ldots, \xi_{n-1}) = i^*(*r\,dr)(\rho_*\xi_1, \ldots, \rho_*\xi_{n-1})$ and $\rho_*\xi_1 = 0$, and because $*dr(\xi_1, \ldots, \cdot) \equiv 0$. By multilinearity, we conclude that the expression vanishes identically and therefore that

$$\tau = \frac{*dr}{r^{n-1}} = \frac{*r\,dr}{r^n} = \frac{\sum (-1)^{i-1}x^i\,dx^1 \wedge \cdots \wedge \widehat{dx^i} \wedge \cdots \wedge dx^n}{r^n},$$

where the $\widehat{\ }$ indicates that the corresponding term is omitted.

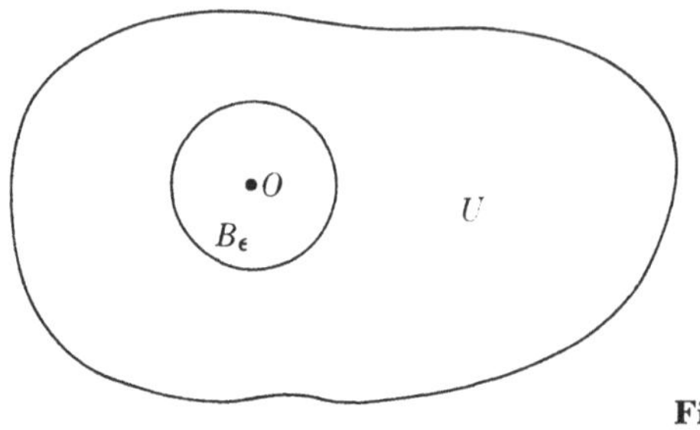

Fig. 12.2

2. GREEN'S FORMULAS

Let u and v be smooth functions on $\mathbb{E}^n$. Then

$$du = \sum \frac{\partial u}{\partial x^i}\, dx^i,$$

so

$$*du = \sum (-1)^{i-1} \frac{\partial u}{\partial x^i}\, dx^1 \wedge \cdots \widehat{dx^i} \cdots \wedge dx^n,$$

and therefore

$$d * du = \left[\sum \frac{\partial^2 u}{(\partial x^i)^2}\right] dx^1 \wedge \cdots \wedge dx^n = \Delta u\, dx^1 \wedge \cdots \wedge dx^n,$$

where the operator

$$\Delta = \frac{\partial^2}{(\partial x^1)^2} + \cdots + \frac{\partial^2}{(\partial x^n)^2}$$

is called the *Laplacian*.

Let U be an open subset with compact closure and almost regular boundary. (See Fig. 12.2.) Set

$$D_U[u, v] = \int_U du \wedge *dv = \int_U dv \wedge *du = \int \left(\sum_i \frac{\partial u}{\partial x^i}\frac{\partial v}{\partial x^i}\right) dx^1 \wedge \cdots \wedge dx^n.$$

Now $\int_{\partial U} u * dv = \int_U d(u * dv)$. But $d(u * dv) = du \wedge *dv + u \wedge d * dv$, so

$$\int_{\partial U} u * dv = D_U[u, v] + \int_U (u\Delta v)\, dx^1 \wedge \cdots \wedge dx^n. \tag{2.1}$$

Since D_U is symmetric in u and v, we have

$$\int_{\partial U} u * dv - v * du = \int_U (u\Delta v - v\Delta u)\, dx^1 \wedge \cdots \wedge dx^n. \tag{2.2}$$

Suppose U contains the origin, and let $U_\epsilon = U - \overline{B}_\epsilon$, where B_ϵ is the ball of radius ϵ centered at the origin. Let $v = r^{2-n}$.† Then $dv = (2 - n)r^{1-n}\, dr$ and $*dv = -(n-2)\tau$, so that $d * dv = 0$. Substituting into (2.2) and taking ϵ

---

†For $n = 2$, set $v = \log r$.

small enough so that $\partial U_\epsilon = \partial U - \partial B_\epsilon$, we get

$$-(n-2)\int_{\partial U}\left(u\tau + \frac{r^{2-n} * du}{n-2}\right) + (n-2)\int_{\partial B_\epsilon} u\tau + \epsilon^{n-2}\int_{\partial B_\epsilon} *du$$
$$= -\int_{U_\epsilon} r^{2-n}\Delta u\, dx^1 \wedge \cdots \wedge dx^n.$$

Let us examine what happens as $\epsilon \to 0$. The first term on the left does not depend on ϵ. For the second term, let us write $u = u(0) + \mathcal{O}(\epsilon)$. Then the second term becomes $(n-2)A_{n-1}u(0) + \mathcal{O}(\epsilon)$. For the third term on the left, we write

$$\int_{\partial B_\epsilon} *du = \int_{B_\epsilon} d * du = \int_{B_\epsilon} \Delta u\, dx^1 \wedge \cdots \wedge dx^n.$$

Thus the third term on the left vanishes to order ϵ^{n-2}. Since r^{2-n} is an integrable function on $\mathbb{E}^n$, the right-hand side tends to $-\int_U r^{2-n}\, \Delta u\, dx^1 \wedge \cdots \wedge dx^n$. Therefore, if we let $\epsilon \to 0$, we get

$$A_{n-1}u(0) = \int_{\partial U}\left(u\tau + \frac{r^{2-n} * du}{n-2}\right) - \frac{1}{n-2}\int_U r^{2-n}\, \Delta u\, dx^1 \wedge \cdots \wedge dx^n. \tag{2.3}$$

In particular, if u is a *harmonic* function, that is, $\Delta u = 0$, then

$$u(0) = \frac{1}{A_{n-1}}\int_{\partial U} u\tau + \frac{1}{A_{n-1}(n-2)}\int_{\partial U}\frac{*du}{r^{n-2}}. \tag{2.4}$$

In the special case $U = B_a$, the second term on the right-hand side of (2.4) becomes

$$\frac{1}{(n-2)A_{n-1}a^{n-2}}\int_{\partial B_a} *du = \frac{1}{(n-2)A_{n-1}a^{n-a}}\int_{B_a} d * du = 0.$$

Thus (by the definition of A_{n-1} and τ)

$$u(0) = \frac{\int_{S_a} u\, dS}{\int_{S_a} dS}, \tag{2.5}$$

where S_a is the sphere of radius ϵ and dS is its volume element. In other words, if u is a function that is harmonic in some domain, then the value of u at any point is equal to its average value on any sphere centered at that point whose interior is completely contained in the domain.

3. THE MAXIMUM PRINCIPLE

Equation (2.5) has a number of startling consequences which we shall now develop. Let u be a function that is harmonic in a domain U. Let x_0 be some point of U, and suppose that there is a neighborhood W of x_0 such that $W \subset U$ and

$$u(x) \le u(x_0) \qquad \text{for all} \quad x \in W.$$

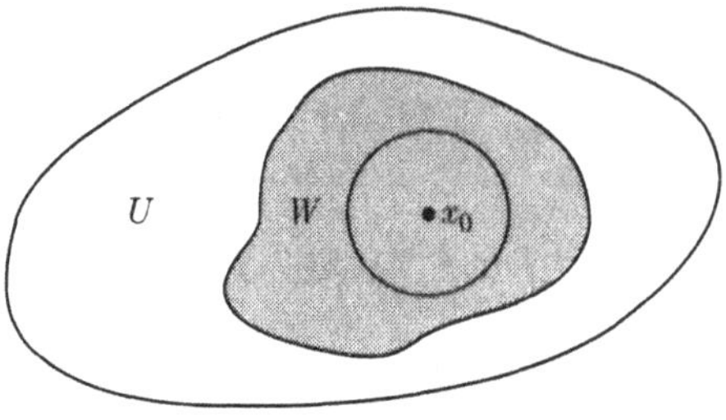

Fig. 12.3

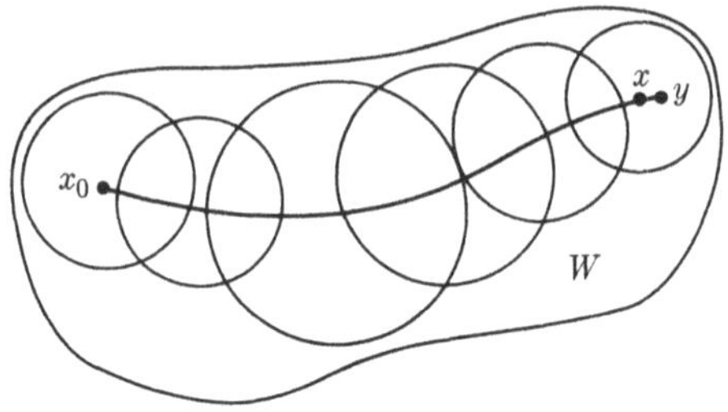

Fig. 12.4

(See Fig. 12.3.) Let S_a be a sphere of radius a centered at x_0, where a is so small that $\overline{B}_a \subset W$. Then $u(x) \leq u(x_0)$ for all $x \in S_a$. Therefore,

$$\int_{S_a} u \, dS \leq u(x_0) \int_{S_a} dS.$$

If there were some point $x \in S_a$ with $u(x) < u(x_0)$, then $u(y)$ would be less than $u(x_0)$ for all y sufficiently close to x, since u is continuous. But then the above inequality would be a strict inequality, i.e.,

$$\int_{S_a} u \, dS < u(x_0) \int_{S_a} dS,$$

which contradicts (2.5). We must therefore have

$$u(x) = u(x_0) \qquad \text{for all} \quad x \in S_a.$$

Now suppose that W is an open set that is *connected;* that is, suppose that any two points of W can be joined by a continuous curve lying entirely in W. Let y be any point of W, and let C be a curve joining x_0 to y. About each x on the curve we can find a sufficiently small ball with center x lying entirely in W. By the compactness of C we can choose a finite number of these balls which cover C. We can therefore formulate the following: There are a finite number of spheres $S_{a_1}, \ldots, S_{a_k}$ such that each sphere and its interior lie entirely in W, S_{a_1} has center x_0, the center x_i of $S_{a_{i+1}}$ lies on S_{a_i}, and $y \in S_{a_k}$. (See Fig. 12.4.) But this implies that $u(x_0) = u(x_1) = \cdots = u(x_k) = u(y)$. In other words, we have established:

Proposition 3.1. Let u be harmonic in a connected open set W, and suppose that u achieves its maximum value at some $x_0 \in W$. Then u is constant on W.

An immediate corollary of this result is:

Proposition 3.2. Let U be a connected open set with $\overline{U}$ compact. Then if u is a function that is continuous in $\overline{U}$ and harmonic in U,

$$u(x) < \max_{y \in \partial U} u(y) \tag{3.1}$$

unless u is a constant.

Proof. In fact, since $\overline{U}$ is compact and u is continuous, u must achieve its maximum at some point x_0 of $\overline{U}$. If we could actually choose $x_0 \in U$, then u would have to be a constant by Proposition 3.1. If u is not a constant, then $x_0 \in \partial U$, and we have (3.1). □

From this we deduce:

Proposition 3.3. Let U be a connected open set with $\overline{U}$ compact. Let u and v be functions that are continuous in $\overline{U}$ and harmonic in U. Suppose that

$$u(y) = v(y) \qquad \text{for all} \quad y \in \partial U.$$

Then

$$u(x) = v(x) \qquad \text{for all} \quad x \in U.$$

Proof. In fact, $u - v$ satisfies the hypotheses of Proposition 3.2 and vanishes on ∂U. Thus $u(x) - v(x) \leq 0$ for $x \in U$. Similarly $v(x) - u(x) \leq 0$, which implies the proposition. □

An alternative way of formulating Proposition 3.3 is to say that on a domain U a harmonic function is completely determined by its boundary values. This is a uniqueness theorem: there is at most one harmonic function with given boundary values. It suggests the problem of deciding whether the corresponding existence theorem is true. This problem is known as Dirichlet's problem.

Dirichlet's problem. Given a continuous function f defined on ∂U, does there exist a function u that is continuous in $\overline{U}$ and harmonic in U and such that $u(y) = f(y)$ for all $y \in \partial U$?

We shall show in Section 9 that we can always solve Dirichlet's problem for domains with almost regular boundary.

4. GREEN'S FUNCTIONS

Suppose that U is a domain for which the Dirichlet problem can be solved. We shall show in this section that the solution can be given "explicitly" in terms of a certain integral over the boundary.

We first introduce some notation. For each $x \in \mathbb{E}^n$ set†

$$K_n(x, y) = \frac{1}{(n-2)A_{n-1}\|x - y\|^{n-2}} \qquad \text{for} \quad y \in \mathbb{E}^n - \{x\} \qquad (n \geq 3),$$

so that (for fixed x)

$$*dK_n(x, \cdot) = \frac{1}{A_{n-1}}\tau_x \qquad \text{on} \quad \mathbb{E}^n - \{x\},$$

where τ_x denotes the solid angle about the point x. Then for $x \in U$, Eq. (2.3)

†This is for $n \geq 3$. For $n = 2$, set $K_2(x, y) = (1/2\pi) \ln |x - y|$.

can be rewritten as

$$u(x) = \int_{\partial U} \left[u * dK_n(x, \cdot) + K_n(x, \cdot) * du \right] - \int_U K_n(x, \cdot)\, \Delta u \, dV. \qquad (4.1)$$

Now for fixed $x \in U$ the function $K_n(x, y)$ is a differentiable function of y as y varies on ∂U. By assumption, we can therefore find a function $h(x, \cdot)$ that is harmonic in U and continuous in $\overline{U}$ and such that

$$h(x, y) = -K_n(x, y) \qquad \text{for all} \quad y \in \partial U. \qquad (4.2)$$

Furthermore, for fixed y, $h(x, y)$ is a continuous function of x. In fact, by the maximum principle,

$$|h(x_1, y) - h(x_2, y)| \leq \max_{z \in \partial U} |K_n(x_2, z) - K_n(x_1, z)|$$

and $K_n(x, z)$ is clearly uniformly continuous in x for all $z \in \partial U$ so long as x stays a fixed distance away from ∂U. We have thus constructed a function h_U such that

i) $h_U(x, y)$ is a continuous function of x and y for $x, y \in \overline{U}$, and is differentiable in y for $y \in U$;

ii) for each fixed x, $\Delta_y h_U = 0$; i.e., $\Delta h_U(x, \cdot) = 0$;

iii) $G_U(x, y) = K_n(x, y) + h_U(x, y) = 0$ for $y \in \partial U$.

The function G_U is called *the Green function* of the domain U. Let us suppose for a moment that G_U exists, and let us derive some of its properties. We write $G = G_U$ when there is no possibility of confusion. We first show that for $x \in U$ and $y \in U$ we have

$$G(x, y) = G(y, x). \qquad (4.3)$$

Let $B_{x,\epsilon}$ and $B_{y,\epsilon}$ be small balls about x and y. Let $u = G(x, \cdot)$ and $v = G(y, \cdot)$ in (2.2), where U is replaced by $U - B_{x,\epsilon} - B_{y,\epsilon}$. Since both functions are harmonic in the domain and vanish on ∂U, we obtain

$$\int_{\partial B_{x,\epsilon}} G(x, \cdot) * dG(y, \cdot) + \int_{\partial B_{y,\epsilon}} G(x, \cdot) * dG(y, \cdot)$$
$$= \int_{\partial B_{x,\epsilon}} G(y, \cdot) * dG(x, \cdot) + \int_{\partial B_{y,\epsilon}} G(y, \cdot) * dG(x, \cdot).$$

We will show that the left-hand side approaches $G(x, y)$ and the right-hand side approaches $G(y, x)$ as $\epsilon \to 0$. By symmetry, it suffices to look at the left-hand side.

Now

$$\int_{\partial B_{y,\epsilon}} G(x, \cdot) * dG(y, \cdot) = \int_{\partial B_{y,\epsilon}} G(x, \cdot) * dK_n(y, \cdot) + \int_{\partial B_{y,\epsilon}} G(x, \cdot) * dh.$$

The first term on the right-hand side approaches $G(x, y)$ as in the proof of (2.3), since $A_{n-1} * dK$ is the solid angle about y. The second term tends to zero, since $G(x, \cdot)$ and h are smooth functions on $B_{y,\epsilon}$. On the other hand,

$$\int_{\partial B_{x,\epsilon}} G(x, \cdot) * dG(y, \cdot) = \int_{\partial B_{x,\epsilon}} K(x, \cdot) * dG(y, \cdot) + \int_{\partial B_{x,\epsilon}} h(x, \cdot) * dG(y, \cdot).$$

The second term tends to zero, as above. The first term can be written (for $n \geq 3$)

$$\frac{1}{A_{n-1}(n-2)\epsilon^{n-2}} \int_{\partial B_{x,\epsilon}} *dG(y, \cdot) = \frac{1}{A_{n-1}(n-2)\epsilon^{n-2}} \int_{B_{x,\epsilon}} d * dG(y, \cdot) = 0,$$

since $G(y, \cdot)$ is harmonic in $B_{x,\epsilon}$. A similar argument works for $n = 2$ with ϵ^{n-2} replaced by $\log \epsilon$. This proves (4.3).

Let u be any smooth function on U. Apply (2.2) to u and $v = G(x, \cdot)$ on $U - B_{x,\epsilon}$. We get (since $G(x, \cdot) = 0$ on ∂U)

$$\int_{\partial U} u * dG(x, \cdot) - \int_{\partial B_{x,\epsilon}} u * dG(x, \cdot) + \int_{\partial B_{x,\epsilon}} G(x, \cdot) * du$$
$$= \int_U G(x, \cdot)\, \Delta u\, dx^1 \wedge \cdots \wedge dx^n.$$

The third integral on the left-hand side can be written as

$$\int_{\partial B_{x,\epsilon}} K_n(x, \cdot) * du + \int h(x, \cdot) * du$$
$$= \frac{1}{A_n(n-2)\epsilon^{n-2}} \int_{\partial B_{x,\epsilon}} * du + \int_{\partial B_{x,\epsilon}} h(x, \cdot)$$
$$= \frac{1}{A(n-2)\epsilon^{n-2}}\, \mathcal{O}(\epsilon^{n-1}) + \mathcal{O}(\epsilon^{n-1})$$

and so tends to zero. (The usual modification works for $n = 2$.) We get

$$u(x) = \int_U G(x, \cdot)\, \Delta u\, dx^1 \wedge \cdots \wedge dx^n + \int_{\partial U} u * dG(x, \cdot). \tag{4.4}$$

We observe that (4.4) shows that *if we know that there exists* a solution to $\Delta F = f$ with the boundary conditions $F = 0$, then it is given by

$$F(x) = \int G(x, y) f(y)\, dy.$$

Similarly, if we know that there exists a smooth solution to the problem

$$\Delta u = 0, \qquad u(x) = f(x) \qquad \text{for} \quad x \in \partial U, \tag{4.5}$$

then it is given by

$$u(x) = \int_{\partial U} u * dG(x, \cdot). \tag{4.6}$$

It is important to observe that these formulas are consequences of the existence of Green's function for U. Thus they are valid whenever we can find the function h such that properties (ii) and (iii) hold.

5. THE POISSON INTEGRAL FORMULA

In this section we shall explicitly construct the Green function for a ball of radius R. Let B_R be the ball of radius R centered at the origin.

For any $x \in \mathbb{E}^n - \{0\}$ let x' be the image of x under inversion with respect to the sphere of radius R:

$$x' = \frac{R^2}{\|x\|^2}\, x.$$

Define the function G_R by

$$A_{n-1}(n-2)G_R(x, y) = \begin{cases} \dfrac{1}{\|y-x\|^{n-2}} - \dfrac{R^{n-2}}{\|x\|^{n-2}\|y-x'\|^{n-2}} & \text{if } x \neq 0, \\ \dfrac{1}{\|y\|^{n-2}} - \dfrac{1}{R^{n-2}} & \text{if } x = 0.† \end{cases} \tag{5.1}$$

If $x \in B_R$, then $x' \notin B_R$, and so the second terms on the right-hand side of (5.1) are continuous and harmonic on $\overline{B_R}$. We must merely check that property (iii) holds. Now for $\|y\| = R$ we have, by similar triangles (or direct computation),

$$\frac{R}{\|x\|} = \frac{\|y - x'\|}{\|y - x\|}, \tag{5.2}$$

so that

$$G_R(x, y) = 0 \qquad \text{for} \quad \|y\| = R.$$

(See Fig. 12.5.) This is (iii), so we have verified that G_R is the Green function for the ball of radius R.

To apply (4.4) we must compute $*dG_R$ on the sphere of radius R. Now by (5.1) we have (for $x \neq 0$)

$$\begin{aligned} A_{n-1} * dG_R(x, \cdot) &= \tau_x - \frac{R^{n-2}}{\|x\|^{n-2}}\,\tau_{x'} \\ &= \sum \left\{\frac{y^i - x^i}{\|y - x\|^n} - \frac{R^{n-2}}{\|x\|^{n-2}}\,\frac{y^i - x'^i}{\|y - x'\|^n}\right\} * dy^i. \end{aligned}$$

But, by (5.2),

$$\frac{R^{n-2}}{\|x\|^{n-2}\|y - x'\|^n} = \frac{\|x\|^2}{R^2\|y - x\|^n} \qquad \text{if} \quad \|y\| = R.$$

---

†If $n = 2$, set

$$\begin{aligned} G_R(x, y) &= \log \|y - x\| - \log \frac{\|x\|}{R}\,\|y - x'\| && \text{if } x \neq 0, \\ &= \log \|y\| - \log R && \text{if } x = 0. \end{aligned}$$

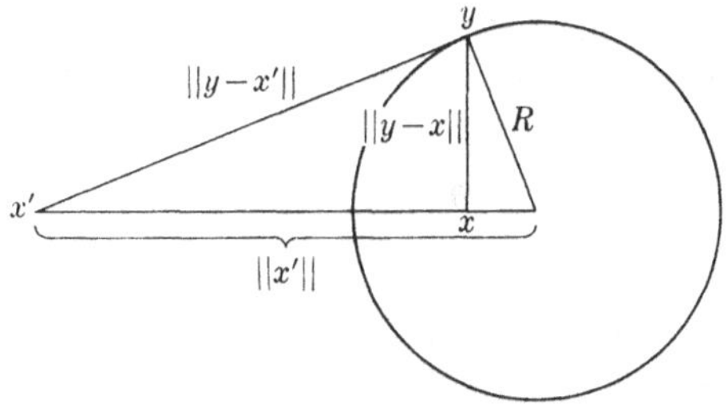

Fig. 12.5

We thus see that for $\|y\| = R$,

$$i_R^*(A_{n-1} * dG_R) = i_R^* \left\{ \frac{1}{\|y - x\|^n} \sum \left[y^i - x^i - \frac{\|x\|^2}{R^2} \left(y^i - \frac{R^2}{\|x\|^2} x^i \right) \right] * dy^i \right\}$$

$$= i_R^* \left\{ \frac{R^2 - \|x\|^2}{\|y - x\|^n R^2} \sum y^i * dy^i \right\}$$

$$= i_R^* \left\{ \frac{R^2 - \|x\|^2}{R^2 \|y - x\|^n} * r\, dr \right\}.$$

But $i_R^*(*r\, dr) = R\, dS_R$, where dS_R is the volume element on the sphere of radius R. If we substitute into (4.6), we obtain

$$u(x) = \frac{R^2 - \|x\|^2}{RA_{n-1}} \int_{S_R} \frac{u(y)}{\|y - x\|^n} dS_R \quad \text{(the Poisson integral formula).} \quad (5.3)$$

In the proof of (5.3) we used the assumption that the function u is differentiable in some neighborhood of the ball $\overline{B}_R$ and is harmonic for $\|x\| < R$. Actually, all that we need to assume is that u is differentiable and harmonic for $\|x\| < R$ and continuous on the closed ball $\|x\| \leq R$. In fact, for any $\|x\| < R$, Eq. (5.3) will be valid with R replaced by R_a, where $\|x\| < R_a < R$. If we then let R_a approach R, we recover (5.3) by virtue of the assumed continuity of u.

Equation (5.3) gives the solution of Dirichlet's problem for a ball *provided we know that the solution exists.* That is, if u is any function that is harmonic on the open ball and continuous on the closed ball, it satisfies (5.3). Now let us show that (5.3) is actually a solution of Dirichlet's problem for prescribed boundary values. Thus suppose we are given a continuous function u defined on the sphere S_R. Then we are given $u(y)$ for all $y \in S_R$. Define $u(x)$ for $\|x\| < R$ by (5.3). We must show that

a) u is harmonic for $\|x\| < R$, and

b) $u(x) \to u(y_0)$ if $x \to y_0$ and $\|y_0\| = R$.

To prove (a) we observe that $G_R(x, y)$ is a differentiable function of x and y in the range $\|x\| < R_1 < R$, $R_1 < \|y\| < R^2/R_1$, and is, by construction, a harmonic function of y. For $\|x\| < R$ and $\|y\| < R$, we know, by (4.4), that

$$G_R(x, y) = G_R(y, x).$$

Thus for fixed y with $\|y\| < R$, $G_R(\cdot, y)$ is a harmonic function on $B_R - \{y\}$. Letting $\|y\| \to R$, we see that $G_R(x, y)$ is a harmonic function of x for $\|x\| < R_1 < R$ for each fixed $y \in S_R$. Thus

$$\frac{\partial G_R}{\partial y^i}(x, y)$$

is a harmonic function of x for each $y \in S_R$. In other words, all the coefficients of $*dG_R(\cdot, y)$ are harmonic functions of x for each $y \in S_R$, and therefore so is each coefficient of $u(y) * dG_R(\cdot, y)$. It follows that the function $u(x) = \int_{S_R} u * dG_R(x, \cdot)$ is a harmonic function of x, since the integral converges uniformly (as do the integrals of the various derivatives with respect to x) for $\|x\| < R_1 < R$. This proves (a).

To prove (b) we first remark that the constant one is a harmonic function everywhere, so Eq. (5.3) applies to it. We thus have

$$\frac{R^2 - \|x\|^2}{A_{n-1}R} \int_{S_R} \frac{dS_R}{\|y - x\|^n} = 1 \qquad \text{for any} \quad \|x\| < R. \tag{5.4}$$

Now let y_0 be some point of S_R, and let u be a continuous function on S_R. For any $\epsilon > 0$ we can find a $\delta > 0$ such that

$$|u(y) - u(y_0)| < \epsilon \qquad \text{for} \quad \|y - y_0\| \le 2\delta \qquad (y \in S_R).$$

Let $Z_1 = \{y \in S_R : \|y - y_0\| > 2\delta\}$ and $Z_2 = \{y \in S_R : \|y - y_0\| \le 2\delta\}$. Then by (5.3) and (5.4) we have, for $\|x\| < R$,

$$u(x) - u(y_0) = \frac{R^2 - \|x\|^2}{(A_{n-1})R} \int_{S_R} \frac{u(y) - u(y_0)}{\|y - x\|^n}\, dS_R,$$

so

$$|u(x) - u(y_0)| \le I_1 + I_2,$$

where

$$I_1 = \frac{R^2 - \|x\|^2}{A_{n-1}R} \int_{Z_1} \frac{|u(y) - u(y_0)|}{\|y - x\|^n}\, dS_R$$

and

$$I_2 = \frac{R^2 - \|x\|^2}{A_{n-1}R} \int_{Z_2} \frac{|u(y) - u(y_0)|}{\|y - x\|^n}\, dS_R.$$

Now if $\|y_0 - x\| < \delta$, then for all $y \in Z_1$ we have $\|y - x\| > \|y - y_0\| - \|x - y_0\|$, so that $\|y - x\| > \delta$. Thus for all x such that $\|x - y_0\| < \delta$ the integral occurring in I_1 is uniformly bounded. Since $\|x\| \to R$ as $x \to y_0$, we conclude that $I_1 \to 0$ as $x \to y_0$.

With respect to I_2, we know that $|u(y) - u(y_0)| < \epsilon$ for all $y \in Z_2$, so that

$$I_2 < \frac{R^2 - \|x\|^2}{A_{n-1}R} \int_{Z_2} \frac{\epsilon\, dS_R}{\|y - x\|^n} < \epsilon\left(\frac{R^2 - \|x\|^2}{A_{n-1}R} \int_{S_R} \frac{dS_R}{\|y - x\|}\right) = \epsilon$$

by (5.4). This proves (b).

We have thus proved:

Theorem 5.1. Let u be a continuous function defined on the sphere S_R. There is a unique continuous function defined for $\|x\| \leq R$ which coincides with the given function on the sphere S_R and is harmonic for $\|x\| < R$. This function is given by (5.3) for all $\|x\| < R$.

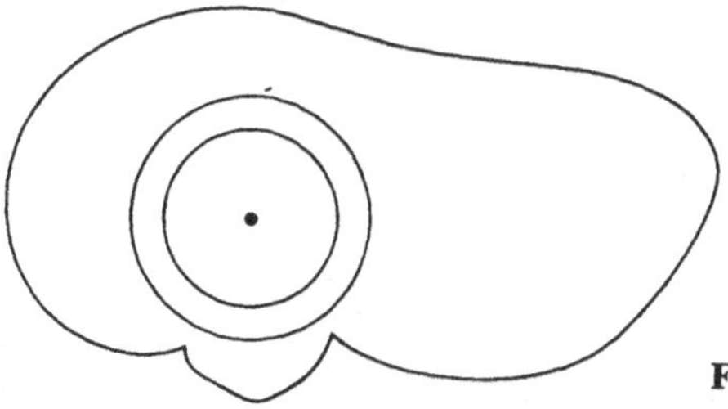

Fig. 12.6

6. CONSEQUENCES OF THE POISSON INTEGRAL FORMULA

In the proof of Theorem 5.1 we assumed that u is continuous in the closed ball, possesses two continuous derivatives in the open ball, and is harmonic in the open ball. However, it is clear from (5.3) that u possesses derivatives of all orders when $\|x\| < R$. In fact, if $\|x\| < R_1 < R$, we can differentiate (5.3) under the integral sign any number of times, since all the integrals we obtain will be uniformly integrable in $\|x\|$. Now if u is harmonic in some open set U and $x \in U$, we can choose R sufficiently small so that the ball of radius R centered at x will be contained in U. (See Fig. 12.6.) We can then apply (5.3). We thus conclude:

Proposition 6.1. Let u be a function defined on an open set U, possessing two continuous derivatives, and satisfying $\Delta u \equiv 0$. Then u has continuous derivatives of all orders.

We can improve Proposition 6.1 by examining (5.3) a little more closely. Let $\|y\| = R$, and let $R_1 < R$. Therefore, all $\|x\| \leq R_1$. The multinomial theorem allows us to expand

$$\frac{1}{\|y - x\|^n} = \sum A_{i_1,\ldots,i_n}(y) x_1^{i_1} \cdots x_n^{i_n}, \tag{6.1}$$

where the coefficients depend on y but the series converges uniformly in x for all $\|y\| = R$. Furthermore, if D is any operator of partial differentiation with respect to x, then we obtain an analogous expansion

$$D\{\|y - x\|^{-n}\} = \sum B_{\alpha_1,\ldots,\alpha_n}(y) x_1^{\alpha_1} \cdots x_n^{\alpha_n},$$

where this series is obtained from (5.4) by term-by-term differentiation and converges uniformly in the same region. If we now substitute (6.1) into (5.3), we can integrate the series term by term because of the uniform convergence in

y, and we conclude that

$$u(x) = \sum C_{\alpha_1,\ldots,\alpha_n} x_1^{\alpha_1} \cdots x_n^{\alpha_n},$$

where the series converges uniformly for $\|x\| < R_1$. Furthermore, we can differentiate the series term by term. Doing this and evaluating at $x = 0$, we see that

$$\alpha! C_\alpha = D^\alpha u(0),$$

where $\alpha = \langle \alpha_1, \ldots, \alpha_n \rangle$.

We thus have proved:

Proposition 6.2. Let u be harmonic in the ball $\|x\| < R$. Then

$$u(x) = \sum \frac{1}{\alpha!} D^\alpha u(0) x^\alpha, \tag{6.2}$$

where the Taylor series (6.2) converges for all $\|x\| < R$ and converges uniformly for all $\|x\| \leq R_1 < R$. In particular, u is determined throughout the ball by its value and the value of all its derivatives at the origin.

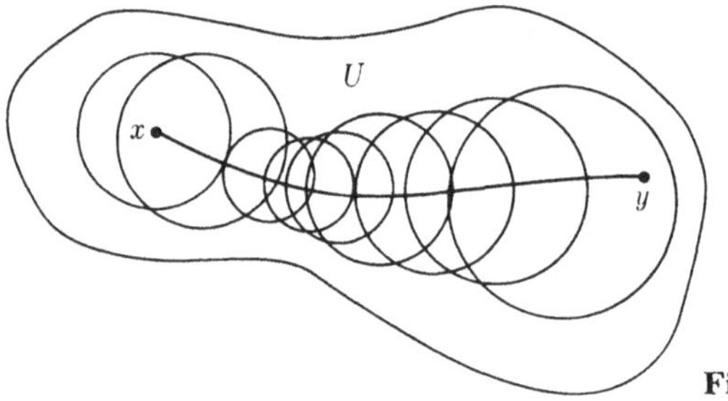

Fig. 12.7

Let U be some connected open subset of $\mathbb{E}^n$, and let u and v be two harmonic functions on U. Suppose that u and v have the same value and the same derivatives of all orders at some point $x \in U$. Let y be some other point of U. We can then connect x to y by a series of balls lying in U, where the center of each ball lies in the interior of the preceding ball, and such that x is the center of the first ball and y lies in the last ball. (See Fig. 12.7.) We thus conclude that

$$u(y) = v(y).$$

Thus we have:

Proposition 6.3. Let u and v be harmonic functions defined on an open connected set U. Suppose that u and v, together with all their derivatives, coincide at some point of U. Then $u \equiv v$ on U. In particular, if $u(x) = v(x)$ for all $x \in W$, where W is some open subset of U, then $u(x) = v(x)$ for all $x \in U$.

We continue to examine the consequences of (5.3). Let u be given by (5.3). Let D denote any operator of partial differentiation with respect to x. Thus

$$D^{\alpha}u = \frac{\partial^{\alpha_1+\cdots+\alpha_n}}{(\partial x^1)^{\alpha_1}\cdots(\partial x^n)^{\alpha_n}}.$$

Then $D^{\alpha}u(x)$ is given by an integral over S_R of a function involving, at worst, inverse powers of $\|y - x\|$ for y on S_R and the function u on S_R. In particular, if $\|x\| < R_1 < R$, then we can estimate the maximum absolute value of $D^{\alpha}u$ in terms of the values of u on S_R and the difference $R - R_1$. In short,

$$|D^{\alpha}u(x)| \leq c(D^{\alpha}, R, R_1) \max_{\|y\|=R} |u(y)| \qquad \text{for} \quad \|x\| < R_1, \tag{6.3}$$

where c depends only on α, R, and R_1. Now suppose that u is harmonic in some open set U, and let K_1 and K_2 be compact subsets of U with

$$K_1 \subset \operatorname{int} K_2 \subset K_2 \subset U.$$

About each $x \in K_1$ we can draw an open ball B_{R_x} such that $B_{R_x} \subset \operatorname{int} K_2$, so that

$$S_{R_x} \subset K_2.$$

We can also draw a ball $B_{R_{1x}}$ of slightly smaller radius about x. Now the open balls $B_{R_{1x}}$ cover K_1. Since K_1 is compact, we can select a finite number of these balls which cover K_1. By applying (6.3) to each of these balls, replacing $|u(y)|$ by the larger $\max_{z \in K_2} |u(z)|$, and using the smallest of the finitely many constants c, we conclude that

$$\max_{x \in K_1} |D^{\alpha}u(x)| < c(\alpha, K_1, K_2) \max_{z \in K_2} |u(z)|. \tag{6.4}$$

In particular, we have:

Proposition 6.4. Let $\{u_k\}$ be a sequence of harmonic functions defined on an open set U that converges uniformly on any compact subset of U. Then the sequence of partial derivatives $\{D^{\alpha}u\}$ also converges uniformly on any compact subset. In particular, the limit of the sequence is again a harmonic function.

In fact, for any compact subset K_1 we can always find a compact subset K_2 such that $K_1 \subset \operatorname{int} K_2$. Applying (6.4) to the harmonic functions $u_i - u_j$ establishes the uniform convergence of $\{D^{\alpha}u_k\}$ on K_1. Since the sequence of partial derivatives converges uniformly, $\Delta u = \lim \Delta u_k = 0$.

7. HARNACK'S THEOREM

We continue to reap the consequences of Eq. (5.3). In addition to what we assumed in the preceding section, let us suppose that $u(y) \geq 0$ for all $y \in S_R$.

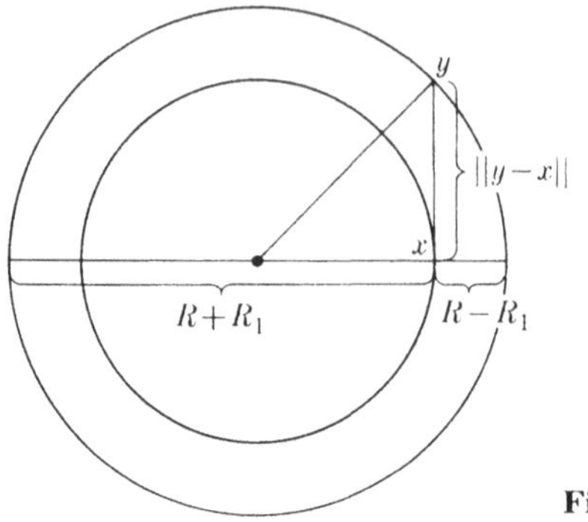

Fig. 12.8

Now for $\|x\| = R_1 < R$ we have (see Fig. 12.8)

$$R - R_1 \leq \|y - x\| \leq R + R_1 \qquad \text{for all} \quad \|y\| = R.$$

Then, by (5.3),

$$\frac{R^2 - R_1^2}{A_{n-1}R(R + R_1)^n} \int_{S_R} u \, dS_R \leq u(x) \leq \frac{R^2 - R_1^2}{A_{n-1}R(R - R_1)^n} \int_{S_R} u \, dS_R.$$

Now according to (2.5), the integrals occurring on the right and left of this inequality are equal to $A_{n-1}R^{n-1}u(0)$. Thus

$$\frac{(R^2 - R_1^2)R^{n-2}}{(R + R_1)^n} u(0) \leq u(x) \leq \frac{(R^2 - R_1^2)R^{n-2}}{(R - R_1)^n} u(0). \tag{7.1}$$

Inequality (7.1) is known as *Harnack's inequality*. It has the following consequence. Suppose that $\{u_k\}$ is a sequence of functions satisfying (5.3) and such that

$$u_1(y) \leq u_2(y) \leq \cdots \leq u_k(y) \leq \cdots \qquad \text{for all} \quad y \in S_R.$$

If we apply (7.1) to the functions $u_j - u_i$ $(j \geq i)$, we conclude that

$$|u_j(x) - u_i(x)| \leq d(R, R_1)|u_j(0) - u_i(0)| \qquad \text{for all} \quad \|x\| \leq R_1,$$

where $d(R, R_1)$ depends only on R and R_1.

In particular, if the sequence converges at the origin, it converges uniformly for $\|x\| \leq R_1 < R$. By applying our usual device of joining two points by a sequence of balls, we deduce:

Proposition 7.1 (*Harnack's theorem*). Let $\{u_k\}$ be a sequence of harmonic functions defined on a connected open set U and such that

$$u_1(x) \leq u_2(x) \leq \cdots \qquad \text{for all} \quad x \in U. \tag{7.2}$$

Suppose that the sequence $\{u_k(p)\}$ converges for some $p \in U$. Then the sequence of functions $\{u_k\}$ converges uniformly on any compact subset of U and, by Proposition 6.4, the limit function is again harmonic.

A useful variation of Harnack's theorem is:

Proposition 7.2. Let $\{u_k\}$ be a sequence of harmonic functions defined on an open set U and satisfying (7.2). Suppose that there is a constant M such that $u_k(x) < M$ for all k and all $x \in U$. Then the sequence of functions $\{u_k\}$ converges uniformly on any compact subset of U to a harmonic function.

To prove Proposition 7.2 we remark that this time the convergence of $\{u_k(x)\}$ at each $x \in U$ is automatic because it is a monotone (nondecreasing) sequence of bounded real numbers. Proposition 7.1 guarantees that the convergence is uniform on compact subsets to a harmonic limit function.

8. SUBHARMONIC FUNCTIONS

In this section and the next we shall show that Dirichlet's problem can be solved for bounded open sets U whose boundaries satisfy certain regularity conditions. The proof that we shall present (there are many others) is due to Perron and makes essential use of the concept of a subharmonic function. Let u be a function defined on the open set U. We say that u is *subharmonic* if

a) u is continuous; and

b) for any connected open subset W of U and any harmonic function, v defined on W, the function $u - v$ satisfies the maximum principle on W. In other words, for any such W and v, if there is some $x_0 \in W$ with

$$u(x) - v(x) \leq u(x_0) - v(x_0) \qquad \text{for all} \quad x \in W,$$

then $u(x) - v(x) \equiv u(x_0) - v(x_0)$ in W.

In order to understand condition (b) a little better, we study some of its consequences. First of all, we can let v be the harmonic function that is identically zero. Then (b) says that u satisfies the maximum principle on every open subset of U.

Next we let B_R be some ball with center z whose closure is contained in U. In particular, its boundary, S_R, is contained in U, and the function u is continuous on S_R. We can therefore find a harmonic function v defined on the open ball and taking on the values $u(y)$ for $y \in S_R$. We take W to be the open ball and v to be the function we have just constructed. Then $u - v$ vanishes on S_R, so (b) implies that $u(x) \leq v(x)$ for all $x \in W$. In particular, $u(z) \leq v(z)$. But

$$v(z) = \frac{1}{A_{n-1}R^{n-1}} \int_{S_R} u \, dS_R.$$

We have thus shown that if u is a subharmonic function defined on U and S_R is a sphere with center z lying in U, then

$$u(z) \leq \frac{1}{A_{n-1}R^{n-1}} \int_{S_R} u \, dS_R.$$

In other words, a subharmonic function is always less than or equal to its average value over a sphere about a point. This property is frequently taken as the definition of a function's being subharmonic, and the hypothesis of continuity is somewhat relaxed. However, the definition we gave above is more suitable for our present purposes.

Let w_1 and w_2 be two subharmonic functions defined on an open set U. Define the function $w_1 \vee w_2$ by setting

$$(w_1 \vee w_2)(x) = \max [w_1(x), w_2(x)].$$

The function $w_1 \vee w_2$ is again subharmonic. The fact that $w_1 \vee w_2$ is continuous is established as follows: Let x be any point of U. Then $w_1 \vee w_2(x)$ is either $w_1(x)$ or $w_2(x)$, say $w_1(x)$. Since w_1 is continuous, for any $\epsilon > 0$, we can find a $\delta > 0$ such that $|w_1(y) - w_1(x)| < \epsilon$ for $|y - x| < \delta$. Similarly, we can arrange to have $w_2(y) < w_2(x) + \epsilon$ for that same range of y, so that $w_2(y) < w_1(x) + \epsilon$, since $w_1(x) = \max [w_1(x), w_2(x)]$. For these values of y we thus have

$$w_1(x) - \epsilon < w_1(y) \leq w_1 \vee w_2(y) < w_1(x) + \epsilon,$$

so that $w_1 \vee w_2$ is continuous. Now let W be a connected open subset of U and let v be a harmonic function on W. Suppose that $w_1 \vee w_2 - v$ takes on its maximum value at some point x_0 of W. Suppose that $w_1 \vee w_2(x_0) = w_1(x_0)$. Then for all $x \in W$

$$w_1(x) - v(x) \leq w_1 \vee w_2(x) - v(x) \leq w_1(x_0) - v(x_0).$$

Since w_1 is subharmonic, the right and left sides of this inequality must be equal. Thus $w_1 \vee w_2 - v$ satisfies the maximum principle on W, which shows that $w_1 \vee w_2$ is subharmonic.

Let w be a subharmonic function defined on an open set U, and let B be a ball whose closure is contained in U. Let u be the solution of Dirichlet's problem for the ball B with boundary values w. Thus u is the unique function continuous in the closed ball, harmonic in the open ball, and coinciding with w on the boundary S of B. As we have already observed, $w(x) \leq u(x)$ for $x \in B$. Now define the function w_B by setting

$$w_B(x) = \begin{cases} w(x) & \text{for} \quad x \in U - B, \\ u(x) & \text{for} \quad x \in B. \end{cases}$$

We claim that the function w_B is again subharmonic. The proof is just as before. The continuity is obvious, as before. If W is subset of U and v is a harmonic function on W, then $w_B - v$ cannot achieve a strict maximum at any interior point: suppose $w_B(x) - v(x) \leq w_B(x_0) - v(x_0)$ for all $x \in W$. (See Fig. 12.9.) If $x_0 \in U - B$, then we have $w(x) - v(x) \leq w_B(x) - v(x) \leq w_B(x_0) - v(x_0) = w(x_0) - v(x_0)$. Since w is subharmonic, all these inequalities must be equalities. On the other hand, if $x_0 \in B$, then since w_B is harmonic in the open ball, we must have $w_B(x) - v(x) = w_B(x_0) - v(x_0)$ for all $x \in W$. By continuity, this implies

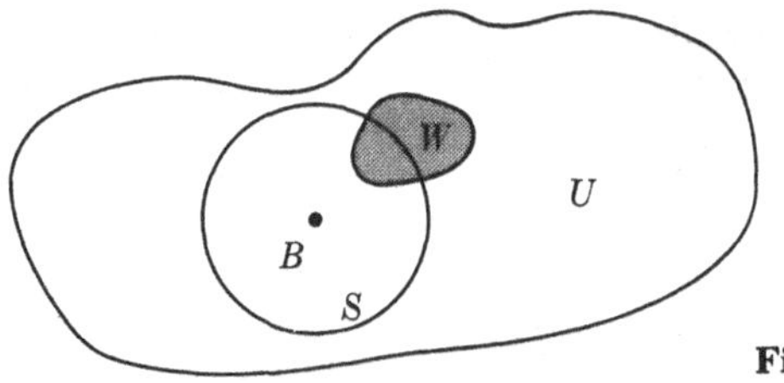

Fig. 12.9

that $w_B(y) - v(y) = w_B(x_0) - v(x_0)$ for $y \in W \cap S$. If $W \cap S \neq \emptyset$, we can take y as a new x_0 and the problem is reduced to the previous case.

In short, to every subharmonic function w defined on U we have attached another subharmonic function w_B such that w_B is actually harmonic in the interior of B and coincides with w in $U - B$, and such that $w \leq w_B$ throughout U. It is clear from the method of construction that if w_1 and w_2 are two subharmonic functions defined on U with $w_1 \leq w_2$, then

$$w_{1B} \leq w_{2B}.$$

9. DIRICHLET'S PROBLEM

Let U be an open subset of $\mathbb{E}^n$ with $\overline{U}$ compact. We say that U has a touchable boundary if for every $p \in \partial U$ there is a ball B such that $\overline{B} \cap \overline{U} = \{p\}$. Thus Fig. 12.10(a) represents a domain with a touchable boundary, while Fig. 12.10(b) and 12.10(c) represent domains that have untouchable points on their boundaries.

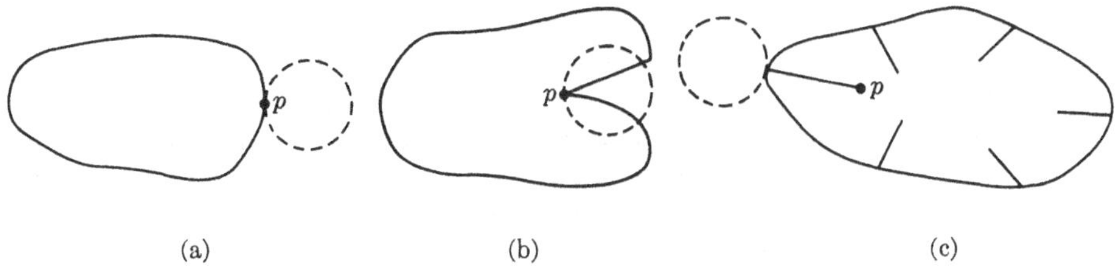

Fig. 12.10

Let U be an open subset of $\mathbb{E}^n$ with $\overline{U}$ compact and with touchable boundary. Let f be a bounded function defined on ∂U, and suppose that $|f(p)| < M$ for all $p \in \partial U$. Let W_f denote the class of all functions defined on U which satisfy the following two conditions:

a) each $w \in W_f$ is subharmonic; and

b) for each $p \in \partial U$,

$$\limsup_{\substack{x \in U \\ x \to p}} w(x) \leq f(p).$$

We can rephrase condition (b) as follows: For each $p \in \partial U$ and each $\epsilon > 0$ there is a $\delta > 0$ (which depends on p, w, and ϵ) such that

$$w(x) \leq f(p) + \epsilon \qquad \text{for} \quad \|x - p\| < \delta.$$

Note that the family of functions W_f is nonempty, since the constant function which is identically $-M$ clearly belongs to W_f. Also note that condition (b) implies that $\limsup w(x) < M$ as $x \to \partial U$. Since $w \in W_f$ is subharmonic, we conclude from the maximum principle that $|w| < M$ for all $w \in W_f$.

Now define the function u_f by setting

$$u_f(x) = \operatorname*{lub}_{w \in W_f} w(x).$$

In view of the preceding remarks, the function u_f is well defined and, in fact, $|u(x)| < M$ for all $x \in U$. We shall show that if f is continuous on ∂U, the function u_f is the solution of the Dirichlet problem for U. We must thus show that u_f is harmonic and takes on the boundary values f.

We first show, without any continuity restrictions on f, that u_f is harmonic. To do this it is sufficient to show that u_f is harmonic in any open ball B, where $\overline{B} \subset U$. Let x be some point of B. Let $w_1, w_2, \ldots, w_k, \ldots$, be a sequence of functions belonging to W_f such that $\lim w_i(x) = u_f(x)$. (Such a sequence exists by the definition of u_f.) Now define

$$\begin{aligned} w'_1 &= w_1, \\ w'_2 &= w_1 \vee w_2, \\ &\vdots \\ w'_i &= w_1 \vee \cdots \vee w_i. \end{aligned}$$

Then $w'_1 \leq w'_2 \leq \cdots \leq w'_i \leq \cdots$ is a monotone increasing sequence of functions belonging to W_f with $\lim w'_i(x) = u_f(x)$. Now replace each w'_i by w'_{iB}. The function w_{iB} belongs to W_f, since it is subharmonic and coincides with w'_i near ∂U. Furthermore, since $w \leq w_B$ for any subharmonic function, we can conclude that $\lim w'_{iB}(x) = u_f(x)$. Inside the ball B, each of the functions w'_{iB} is harmonic, so that the sequence $\{w'_{iB}\}$ is a monotone nondecreasing sequence of harmonic functions in B. Since all the functions of W_f are bounded by M, we conclude from Proposition 7.2 that the sequence $\{w'_{iB}\}$ converges uniformly on any compact subset of B to a function u which is harmonic in B, and we have $u_f(x) = u(x)$. Furthermore, by definition, we have $u(y) \leq u_f(y)$ for any $y \in B$. We would like to show that we actually have $u(y) = u_f(y)$ for all such y. To this effect, we follow the same procedure at y, namely, we select a sequence $\{v'_{iB}\}$ where each v'_{iB} is harmonic in B, belongs to W_f, and is such that the sequence is monotone nondecreasing and $\lim v'_{iB}(y) = u_f(y)$. In this way we obtain another function v which is harmonic in B and satisfies $v \leq u_f$ throughout B, and $v(y) = u_f(y)$. Finally, let the functions s_i be defined by

$$s_i = w'_{iB} \vee v'_{iB},$$

and consider the functions s_{iB}. On the one hand, they all belong to W_f and are harmonic in B. Furthermore, we have $w'_{iB} \leq s_{iB}$ and $v_{iB} \leq s_{iB}$. Finally, the sequence $\{s_{iB}\}$ is nondecreasing and therefore converges to a harmonic function s in B. By construction, we have

$$u \leq s \quad \text{and} \quad v \leq s$$

throughout B, while $u(x) = s(x) = u_f(x)$ and $u(y) = s(y) = u_f(y)$. Applying the maximum principle to the harmonic functions $u - s$ and $v - s$, we conclude (since the maximum value 0 is achieved at x and y, respectively) that $s = u = v$ in all of B. We thus see that $u_f(y) = u(y)$ for any $y \in B$, which shows that u_f is harmonic in U. Note that in proving that u_f is harmonic we did not make use of any properties of the boundary of U or any continuity assumptions of f.

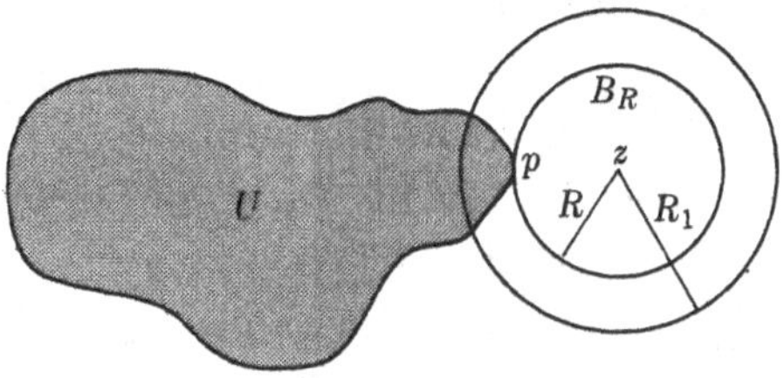

Fig. 12.11

Now let $p \in \partial U$ be a touchable point of the boundary, and assume that f is continuous at p. We shall show that $u_f(x) \to f(p)$ as $x \to p$. Let the ball B_R of radius R with center z touch $\bar{U}$ at p, that is, $B_R \cap \bar{U} = \{p\}$. (See Fig. 12.11.) Since f is continuous at p, for any $\epsilon > 0$ we can find an $R_1 > R$ such that

$$|f(q) - f(p)| < \epsilon \qquad \text{for all} \quad q \in \partial U \cap B_{R_1}. \tag{9.1}$$

Define the function b by setting $b(x) = \|x - z\|^{2-n} - R^{2-n}$.† Note that b is defined and harmonic on $\mathbb{E}^n - \{z\}$ and is negative for $\|x - z\| > R$. Furthermore, $b(x) \leq K < 0$ if $\|x - z\| \geq R_1$. In particular, for all $q \in \partial U$ we have

$$f(q) < \frac{2M}{K} b(q) + \epsilon + f(p). \tag{9.2}$$

In fact, this is just (9.1) if $\|q - z\| < R_1$, and it follows from $|f(q)| < M$ if $\|q - z\| \geq R_1$. By the maximum principle for subharmonic functions we conclude that $w(x) < (2M/K)b(x) + \epsilon + f(p)$ for any $w \in W_f$ and any $x \in U$. We thus have

$$u_f(x) < \frac{2M}{K} b(x) + f(p) + \epsilon \qquad \text{for any} \quad x \in U.$$

On the other hand, we have, for the same reason as before,

$$f(p) - \epsilon - \frac{2M}{K} b(q) < f(q) \qquad \text{for any} \quad q \in \partial U,$$

† $b(x) = \ln (R/\|x - z\|)$ if $n = 2$.

so that the harmonic function $f(p) - \epsilon - (2M/K)b$ actually belongs to W_f. In particular, we have

$$f(p) - \epsilon - \frac{2M}{K} b(x) \leq u_f(x) \qquad \text{for any} \quad x \in U.$$

Putting the two inequalities together, we conclude that

$$|u_f(x) - f(p)| \leq \frac{2M}{K} b(x) + \epsilon \qquad \text{for all} \quad x \in U. \tag{9.3}$$

Now the function b is continuous and vanishes at p. Thus by choosing x sufficiently close to p we can arrange that the right-hand side of (9.3) is less than 2ϵ. Thus $u_f(x) \to f(p)$ as $x \to p$.

In particular, if all points of ∂U are touchable, and if f is continuous at all points of the boundary, we have proved:

Theorem 9.1. Let U be an open subset of $\mathbb{E}^n$ with compact closure and touchable boundary. Let f be a continuous function defined on ∂U. Then there exists a unique function u_f which is continuous on $\overline{U}$ and harmonic in U, and which coincides with f on ∂U.

Some remarks concerning Theorem 9.1 and its proof are in order.

a) The only time we used the assumption that ∂U is touchable was when we constructed a function b which was subharmonic in $\overline{U}$, vanished at p, and took on values less than some negative constant on all points of ∂U outside some neighborhood of p. Now a more careful analysis will show that we can always construct such a function so long as we can touch p from the outside by a cone. Thus we can solve the Dirichlet problem even at points like the p in Fig. 12.12.

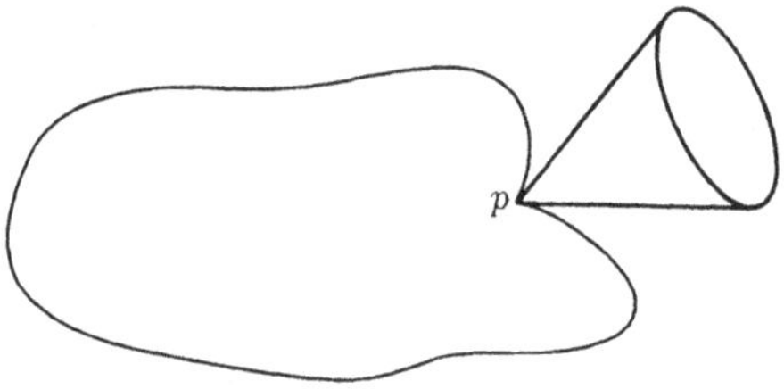

Fig. 12.12

b) On the other hand, *some* condition on the boundary is necessary. For instance, if ∂U contains an isolated point p, then we cannot assign the value at p arbitrarily. In fact, if u is continuous in a neighborhood W of p and harmonic in $W - \{p\}$, then the Poisson integral formula implies that the various derivatives of u will also be bounded in $W - \{p\}$. Therefore, the proof of the mean-value theorem applies to the point p itself, and so $u(p)$ is determined and cannot be assigned arbitrarily. More generally, a more delicate argument shows that the Dirichlet problem cannot be solved for domains whose boundaries contain spikes pointing inward or (in dim ≥ 3) sufficiently sharp cusps (as in

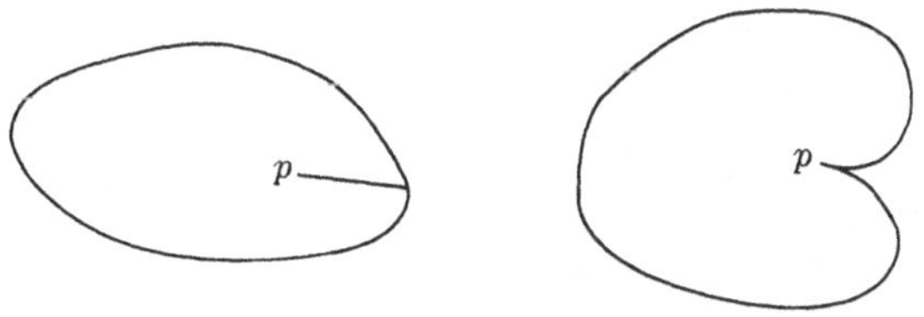

Fig. 12.13

Fig. 12.13). This is the mathematical analogue of the physical fact that a very sharp conductor cannot hold a charge, but will spark. (The relation with electrostatics will be discussed in Section 12.)

c) For the purpose of applying the results of Section 4, i.e., the construction of the Green function of the domain and the various identities, Theorem 9.1 is still not quite enough. We need to know not only that the Green function exists, but also that it has continuous derivatives up to the boundary. This fact requires further argument and additional assumptions about the nature of the boundary; it will be discussed in the next section.

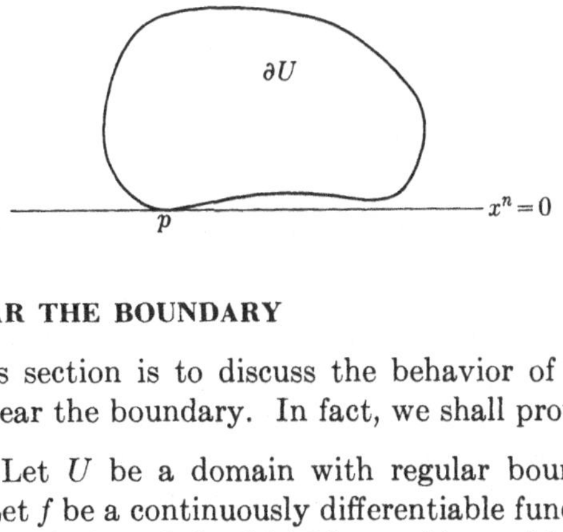

Fig. 12.14

10. BEHAVIOR NEAR THE BOUNDARY

The purpose of this section is to discuss the behavior of the solution of the Dirichlet problem near the boundary. In fact, we shall prove:

Theorem 10.1. Let U be a domain with regular boundary and compact closure in $\mathbb{E}^n$. Let f be a continuously differentiable function defined on ∂U, and let u be the solution of the Dirichlet problem with boundary values f. Then the partial derivatives $\partial u/\partial x^i$ can be extended to continuous functions on $\overline{U}$. Furthermore, if $p \in \partial U$ and $\xi = \langle \xi^1, \ldots, \xi^n \rangle$ is tangent to the boundary at p, then

$$\langle \xi, df \rangle = \langle \xi, du \rangle = \sum (\partial u/\partial x^i)\xi^i.$$

The following proof was suggested to us by Professor Ahlfors and is reproduced here with his kind permission.

Proof. Let p be some point of ∂U, and let us arrange, by a Euclidean motion, that the tangent plane to ∂U at p is the plane $x^n = 0$. (See Fig. 12.14.) Then near p the points of ∂U are all points of the form $(x^1, \ldots, x^{n-1}, \varphi(x^1, \ldots, x^{n-1}))$, where φ is a function defined near the origin of $\mathbb{R}^{n-1}$ and vanishing at the origin, together with all its first derivatives. In particular, there is a constant $C > 1$ such that

$$|\varphi(x^1, \ldots, x^{n-1})| \leq C\{(x^1)^2 + \cdots + (x^{n-1})^2\}$$

in some neighborhood of the origin of $\mathbb{R}^{n-1}$. We can therefore choose a sufficiently small $R < 1/2C$ so that the (open) ball of radius R with center $(0, \ldots, 0, R)$ lies entirely within U and the ball of radius R with center $(0, \ldots, 0, -R)$ lies inside the complement of $\overline{U}$. (See Fig. 12.15.)

Let $z = (0, \ldots, 0, -R)$. Then for all $q \in \partial U$ sufficiently close to p we have

$$\begin{aligned} \|q - z\|^2 &= (x^1)^2 + \cdots + (x^{n-1})^2 + R^2 - 2R\varphi(q) + \varphi^2(q) \\ &\geq (1 - 2RC)\{(x^1)^2 + \cdots + (x^{n-1})^2\} + R^2 + \varphi^2(q), \end{aligned}$$

so that

$$\|q - z\| - R \geq k\{(x^1)^2 + \cdots + (x^{n-1})^2\},$$

where k is some positive constant. Let $\varphi(r) = r^{n-2}$ ($=\ln r$ if $n = 2$). Then $\varphi'(R) \neq 0$, and therefore $|\varphi(\|q - z\|) - \varphi(R)| > C\big|\|q - z\| - R\big|$ for a suitable $C > 0$. In particular, we have the inequality

$$\left|\frac{1}{\|q - z\|^{n-2}} - \frac{1}{R^{n-2}}\right| \geq K\{(x^1)^2 + \cdots + (x^{n-1})^2\} \tag{10.1}$$

or

$$\left|\ln \frac{\|q - z\|}{R}\right| \geq K(x^1)^2 \qquad \text{if} \quad n = 2$$

for some positive constant K.

Now let us replace the function f by $f - [f(p) + \sum_1^{n-1} (\partial f/\partial x^i)(p)x^i]$. Then $u - [f(p) + \sum_1^{n-1} (\partial f/\partial x^i)(p)x^i]$ is the solution of the corresponding Dirichlet problem. In this way we may assume (changing our notation accordingly) that f, together with its first derivatives, vanishes at p. Since f is assumed to have continuous first derivatives, we may apply Taylor's theorem to conclude that there exists a $c > 0$ such that for all q on ∂U sufficiently close to p we have

$$|f(q)| \leq c\{(x^1)^2 + \cdots + (x^{n-1})^2\}. \tag{10.2}$$

As in the last section, let

$$\begin{aligned} b(x) &= \|x - z\|^{2-n} - R^{2-n} \\ &= \ln \frac{R}{\|q - z\|} \qquad \text{if} \quad n = 2. \end{aligned}$$

If we compare (10.2) with (10.1), we see that for all $q \in \partial U$ sufficiently close to p we have

$$|f(q)| \leq \frac{c}{K}\,|b(q)|.$$

On the other hand, since b is strictly negative outside some neighborhood of p

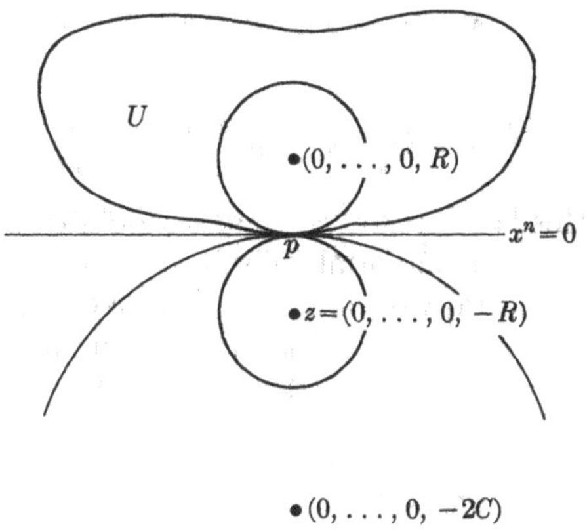

Fig. 12.15

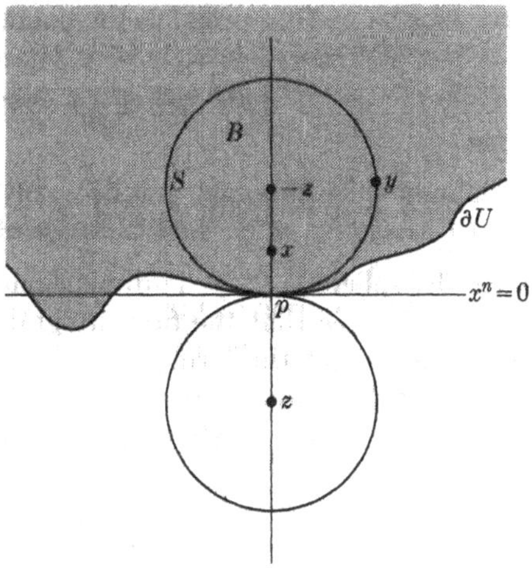

Fig. 12.16

(on ∂U), and since f is bounded on ∂U, we can (using the above inequality for all q near p) find a constant A such that

$$|f(q)| \leq A|b(q)| \qquad \text{for all} \quad q \in \partial U. \tag{10.3}$$

Since the function b is harmonic in U, the maximum principle applied to $u - Ab$ and $Ab - u$ allows us to conclude that

$$|u(x)| \leq A|b(x)| \qquad \text{for all} \quad x \in \overline{U}. \tag{10.4}$$

Now the function b is a differentiable function of the distance from z which vanishes when this distance is R. In particular, there is a constant $a > 0$ such that

$$|b(x)| \leq a\, d(x),$$

where $d(x)$ denotes the distance from x to the sphere of radius R with center z. Now let B be the ball of radius R with center $-z = (0, \ldots, 0, R)$, so that B lies in U, and let $S = \partial B$. Note that S is tangent to the sphere about z at the point p. (See Fig. 12.16.) Thus for points y on S, $d(y) \leq c_1\|y - p\|^2$. If we substitute this into (10.4) using the previous inequality, we conclude that

$$|u(y)| \leq L\|y - p\|^2 \qquad \text{for all} \quad y \in S, \tag{10.5}$$

for some constant L. We apply the Poisson integral formula to the ball B and the function u and differentiate with respect to x^i to obtain

$$\frac{\partial u}{\partial x^i} = \frac{-2(x^i + z^i)}{A_{n-1}R} \int \frac{u(y)}{\|x - y\|^n}\, dS + \frac{R^2 - \|x + z\|^2}{A_{n-1}R} \int \frac{u(y)(x^i - y^i)}{\|x - y\|^{n+2}}\, dS. \tag{10.6}$$

Now let x be on the normal to the boundary through p, so that $x = (0, \ldots, 0, x^n)$. If $|x^n| < R$, then

$$\|y - p\| \leq \|y - x\| + |x^n| \leq 2\|y - x\|.$$

Thus

$$\left|\int \frac{u(y)}{\|x - y\|^n}\, dS\right| \leq 2^n \int \frac{|u(y)|}{\|p - y\|^n}\, dS \leq 2^n D \int \|y - p\|^{2-n}\, dS.$$

Since the sphere is $(n - 1)$-dimensional, this last integral converges absolutely, and we thus see that the first integral in (10.6) is uniformly absolutely convergent. (Note that the term containing this integral vanishes if $i < n$.) We now show that the second term in (10.6) tends to zero as x^n tends to zero. In fact, $R^2 - \|x + z\|^2 = x^n(2R - x^n)$, so we can write the second term of (10.6) as

$$\frac{(2R - x^n)(x^n)^{1/2}}{A_{n-1}R} \int \frac{(x^n)^{1/2} u(y)(x^i - y^i)}{\|x - y\|^{n+2}}\, dS.$$

This time, since $x^n \leq \|x - y\|$ for all $y \in S$, we can assert that the integrand is smaller in absolute value than

$$\frac{|u(y)|\,|x^i - y^i|}{\|y - x\|^{n+3/2}} \leq \frac{|u(y)|}{\|y - x\|^{n+1/2}} \leq \frac{|u(y)| 2^{n+1/2}}{\|y - p\|^{n+1/2}} \leq D 2^{n+1/2} \|y - p\|^{3/2-n},$$

which is again absolutely integrable. Therefore, the integral occurring in the last expression is uniformly bounded for all values of $x^n \geq 0$. Since $(x^n)^{1/2} \to 0$, we conclude that the second term of (10.6) approaches zero.

If we now rephrase the result we have obtained independently of the special choice of coordinates, we see that we have proved the following: Let p be any point of ∂U, and let x be a point on the normal line to ∂U through p. If ξ is any vector parallel to a tangent vector at p, then $\langle \xi, du \rangle \to \langle \xi, df \rangle$ as $x \to p$ along the normal line. If $\partial/\partial n$ denotes the unit vector in the normal direction (pointing into U), then

$$\left\langle \frac{\partial}{\partial n}, du \right\rangle \to \frac{-2}{A_{n-1}} \int \frac{u(y) - u(p)}{\|y - p\|^n}\, dS, \tag{10.7}$$

where the integral is taken over the sphere of radius R tangent to ∂U at p and lying inside $\overline{U}$.

So far, we have proved convergence only if $x \to p$ along the normal direction. If we go back and examine the argument, we see that the radius R and the constant D in (10.5) can be chosen uniformly for all $p \in \partial U$. (In fact, since ∂U is compact, we can cover ∂U by a finite number of neighborhoods, of type (iii), such that a ball of radius R about each $p \in \partial U$ lies in one of these neighborhoods, where R is sufficiently small. In such a situation it is clear that the choice of R (still smaller, if necessary) depends only on the second derivatives of the change of charts from Euclidean coordinates to the adjusted charts. Also the choice of D depends only on the constants involved in the Taylor expansion of f and on R. Thus the choice of R and D can be made uniformly.) Since, by assumption, the partial derivatives of f are continuous and the right-hand side of (10.7) is continuous in p, we conclude that the limits of the partial derivatives of u exist as we approach the boundary in any direction, and that in fact we have proved Theorem 10.1. ☐

We can now construct a Green function for any domain with regular boundary by the solution of Dirichlet's problem, as in Section 4. Theorem 10.1 tells us that it is differentiable in the closed set $\bar{U}$ in the sense that the partial derivatives are continuous up to the boundary. If we want to derive the various formulas of Section 4, we can do so by applying a limit argument: We simply replace U by a smaller domain U^a such that $U^a \to U$ and $\partial U^a \to \partial U$ as $a \to 0$.

(Actually, a more careful and delicate argument will show that if f has two continuous derivatives, then the $\partial u/\partial x^i$ we obtain as limits on the boundary are in fact continuously differentiable. Since $\partial u/\partial x^i$ is the solution of the Dirichlet problem with these boundary values, we conclude that the second derivatives of u can be extended so as to be continuous on $\bar{U}$. In this way one can prove that if f is C^∞, all the derivatives of u can be extended so as to be continuous on $\bar{U}$. In particular, all the derivatives of the Green function $G(x, \cdot)$ can be extended to continuous functions at the boundary for any $x \in U$.)

11. DIRICHLET'S PRINCIPLE

The solution of Dirichlet's problem has an interesting and useful characterization as the solution of the problem of minimizing the Dirichlet integral $D[u, u] = \int \sum (\partial u/\partial x^i)^2$ with given boundary values. More precisely,

Theorem 11.1 (*Dirichlet's principle*). Let U be a domain with compact closure and regular boundary, and let f be a continuously differentiable function on ∂U. Let u be the solution of the Dirichlet problem with boundary values f, and let w be any function which is differentiable in U and takes on the values of f on the boundary, and whose derivatives can be extended to be continuous in $\bar{U}$. Then

$$D[u, u] \leq D[w, w], \tag{11.1}$$

and equality holds if and only if $u = w$.

Proof. Let us write $w = u + v$, where now v is continuously differentiable on U, continuous on $\bar{U}$, and vanishing on ∂U. Then

$$D[w, w] = D[u + v, u + v] = D[u, u] + D[v, v] + 2D[u, v], \tag{11.2}$$

since D is bilinear and symmetric in its arguments.

Now suppose for the moment that u possesses two continuous derivatives and v possesses one continuous derivative in a neighborhood of $\bar{U}$. Then

$$D[u, v] = D_U[u, v] = \int_U dv * du = \int_{\partial U} v * du - \int_U v\Delta u = 0. \tag{11.3}$$

Now $D[v, v] = \int \sum (\partial v/\partial x^i)^2 \geq 0$ and vanishes if and only if all the derivatives of v vanish, so that v is a constant on each component of U. Since v vanishes on ∂U, we conclude that $D[v, v] = 0$ if and only if $v = 0$.

In case u and v are not necessarily differentiable in a neighborhood of ∂U, we establish that $D[u, v] = 0$ by writing

$$D_U[u, v] = \lim D_{U^a}[u, v],$$

where U^a is a sequence of domains which approach U as $a \to 0$ and $\partial U^a \to \partial U$. Applying the previous argument to each of the domains U^a, we conclude that $D[u, v] = 0$ and $D[v, v] = 0$ if and only if $v = 0$. This proves the theorem. □

12. PHYSICAL APPLICATIONS

The study of Laplace's equation and its solutions plays an important role in many theories of classical physics. This is essentially due to the intimate connection between the Laplace operator and Euclidean geometry. Since this is not the place to elaborate on the physical applications, we refer the reader to any physics text for the details. (See, for instance, Feynman, Leighton, and Sands, *The Feynman Lectures on Physics*, Addison-Wesley, Reading, Mass., 1964, vol. II, especially Chapter 12.) We shall briefly mention some of the relevant physics in this section.

In *electrostatics* it is assumed that the electric field E satisfies the equations

$$d * E = \rho, \qquad dE = 0,$$

where ρ is the density of charge and we identify the vector field E with a linear differential form via the Euclidean metric on $\mathbb{E}^3$. Here ρ is assumed, for the moment, to be a smooth density. (It is also convenient to consider the limiting cases of "surface distributions" and "point distributions".) By the second of these equations, we know that E can be written as

$$E = -d\varphi,$$

where φ is a smooth function known as the potential of the electric field. The first equation then implies that

$$\Delta\varphi = d * d\varphi = -\rho.$$

If ρ is given, then φ is determined by the formula

$$\varphi(y) = \frac{1}{4\pi} \int \frac{\rho(x)}{\|y - x\|} \, dx$$

(since $A_2 = 4\pi$). As limiting cases, we obtain, for charge distributed along a surface S with density $\sigma\, dA$, the formulas

$$\varphi(y) = \frac{1}{4\pi} \int_S \frac{\sigma(x)}{\|y - x\|} \, dA,$$

where dA is the area density of the surface, and

$$\varphi(y) = \frac{1}{4\pi}\int_C \frac{l(x)}{\|y - x\|}\,ds,$$

where $l\,ds$ is a linear distribution along a curve C. For a point charge located at x, we obtain

$$\varphi(y) = \frac{e}{4\pi\|y - x\|},$$

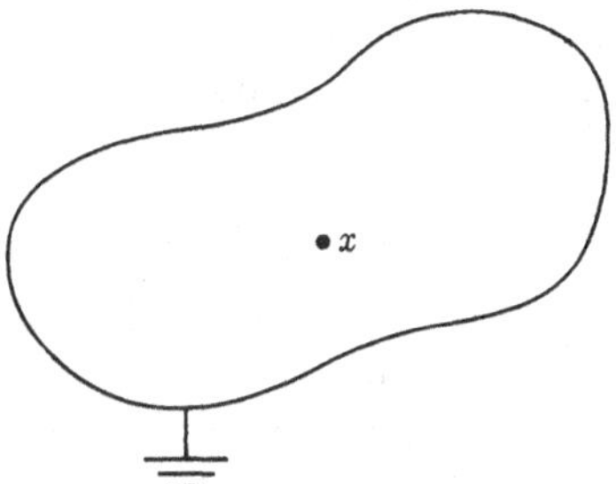

Fig. 12.17

where e is the magnitude of the charge.

There can be no electric field along a conductor (since this would result in a motion of the charge, which contradicts the assumption of a static field). Thus

$$d\varphi = 0 \qquad \text{and} \qquad \varphi = \text{const.}$$

In most problems that arise the distribution of charge is not known in advance, but must be determined. For example, suppose that a unit charge is placed at a point x inside a cavity whose boundary is a conductor which is grounded, i.e., kept at zero potential. (See Fig. 12.17.) We want to find the electric field inside the cavity. There will be a charge distribution induced on the boundary surface that we also wish to determine. Since there is no charge distribution anywhere inside the cavity, except at x, the potential φ must be harmonic everywhere except at x, and, in fact, differs from

$$\frac{1}{4\pi\|\cdot - x\|}$$

by a harmonic function. Since we want $\varphi = 0$ on the boundary, we see that the desired potential φ is exactly the Green function for the domain. The surface density can then be determined from $d\varphi$ along the boundary. (This is another reason why it is important to know that the solution of Dirichlet's problem is differentiable at the boundary.)

More frequently, we are interested in the electric field *outside* conducting surfaces. Here, strictly speaking, the theory we have developed of Dirichlet's problem does not apply, since we considered only bounded domains. We can handle this problem in one of two ways. First, we can reduce to the Dirichlet problem by considering everything contained inside a very large conducting sphere (at potential zero). Then we let the radius go to infinity to get the desired potential as a limit. Second, we can modify our arguments in Sections 3 through 10 to include the case of unbounded domains, but restrict all our functions to vanishing as c/r at infinity. The details are left to the reader.

In the theory of *diffusion* one assumes that material of some nature is flowing according to a vector field X. If the density of the material is $\rho\,dV$, then the amount of material flowing per unit time through an oriented piece of surface S is given by

$$\int_S \rho(X, \mathbf{n})\, dA = \int_S *\rho X,$$

where $\mathbf{n}$ is the unit normal vector to the surface, dA is the element of area, and, we are regarding ρX on the right-hand side as a linear differential form via the Euclidean metric. Thus the net amount of material flowing out of any region is $d * \rho X = \operatorname{div} \langle X, \rho \rangle$. Now "material" may be produced in the region at a rate $s\, dV$ (where s is the function describing the rate of productivity of the sources in the region). Thus the total change of density is given by

$$\frac{\partial \rho}{\partial t}\, dV = \operatorname{div} \langle X, \rho \rangle - s\, dV.$$

If, as we shall assume, the situation is stationary, i.e., the density does not change, we obtain the equations

$$\operatorname{div} \langle X, \rho \rangle = s\, dV.$$

On the other hand, it frequently happens that the flow is given by the gradient of some function, i.e.,

$$\rho_{\mathrm{id}} X = dN$$

for some function N. Combining these equations, we obtain

$$\Delta N = s.$$

In the theory of heat N is the temperature, $\rho_{\mathrm{id}} X$ is the rate of flow of heat, and s is the density of the sources of heat. The equation $\rho_{\mathrm{id}} X = dN$ says that the rate of flow of heat is proportional to the gradient of temperature. In the theory of diffusion of particles it is assumed that the rate of flow of the material is proportional to the change (i.e., gradient) of the density of the particles. (This says that the particles tend to flow from a region of higher density to a region of lower density.) Then N represents the density of the particles and s represents the rate of production of particles.

Suppose, for example, that the boundary of a region D is kept at some fixed temperature f, where f is a given function on ∂D and there are no sources of heat inside D. Then the distribution of temperature in D is given by the function N which satisfies the differential equation $\Delta N = 0$ and the boundary condition $N = f$ on ∂D. In other words, the temperature is determined by the solution of Dirichlet's problem.

In the theory of *steady, incompressible, irrotational flow* of a fluid it is assumed that a vector field X is given which represents the flow of the liquid. By the incompressibility it follows that $d * X = 0$. It is assumed that there is no circulation, that is, $dX = 0$. Thus $X = d\varphi$ for some function φ which is a solution of Laplace's equations. The natural boundary-value problems that arise in this case are different from those we have been discussing, so we simply refer the reader to standard works on fluid mechanics.

SUPPLEMENTARY EXERCISES

1. Let V_1 and V_2 be vector spaces with scalar products. A linear map $l\colon V_1 \to V_2$ is said to be conformal (or a similarity) if

$$(lv, lw) = c(v, w),$$

where c is a positive constant. (Note that if $c = 1$, then l is an isometry.)

Let M_1 and M_2 be manifolds each with a Riemann metric. A differentiable (C^∞) map φ is said to be conformal if φ_{*x} is conformal for each $x \in M_1$. Thus, to say that φ is conformal means that

$$(\varphi_{*x}\xi_1, \varphi_{*x}\xi_2) = c(x)(\xi_1, \xi_2) \qquad \text{if} \quad \xi_1, \xi_2 \in T_x(M_1)$$

for any $x \in M_1$. (Note that this time the c can depend on x.)

a) Let $\varphi\colon U \to \mathbb{R}^2$ be a conformal map, where $U \subset \mathbb{R}^2$ and we use the Euclidean metric on U and on $\mathbb{R}^2$. Suppose φ is given by

$$\varphi(x, y) = (u, v),$$

where $u = u(x, y)$ and $v = v(x, y)$, and where (x, y) and (u, v) are rectangular coordinates in the plane. Show that either the equations

$$\frac{\partial u}{\partial x} = \frac{\partial v}{\partial y}, \qquad \frac{\partial u}{\partial y} = -\frac{\partial v}{\partial x}$$

hold or the equations

$$\frac{\partial u}{\partial x} = -\frac{\partial v}{\partial y}, \qquad \frac{\partial u}{\partial y} = \frac{\partial v}{\partial x}$$

hold.

b) Conclude that if u and v are as in (a), then they are both harmonic functions.

2. a) Let u be a harmonic function defined on all of $\mathbb{R}^n$. Show that if u is bounded then it must be a constant.

b) Using (a) and Exercise 1, show that there is no conformal map of $\mathbb{R}^2$ onto a bounded subset $U \subset \mathbb{R}^2$.

13. PROBLEM SET: THE CALCULUS OF RESIDUES

On a manifold M we can consider complex-valued smooth functions, differential forms, etc. For instance, we say that the function

$$f = u + iv$$

is a complex-valued C^∞-function if each of the real-valued functions u and v are real-valued C^∞-functions. Similarly, we consider the complex-valued linear differential form

$$\omega = \sigma + i\tau,$$

where $\sigma_\tau \in \bigwedge^1(M)$ are linear (real-valued) differential forms. We can then perform the usual operators, remembering that $i^2 = -1$. Thus

$$f\omega = (u\sigma - v\tau) + i(v\sigma + \mu\tau).$$

Similarly, if

$$\omega_1 = \sigma_1 + i\tau_1 \qquad \text{and} \qquad \omega_2 = \sigma_2 + i\tau_2,$$

then

$$\omega_1 \wedge \omega_2 = (\sigma_1 \wedge \sigma_2 - \tau_1 \wedge \tau_2) + i(\sigma_1 \wedge \tau_2 + \tau_1 \wedge \sigma_2)$$

is a complex-valued exterior two-form.

We similarly have the operator d given by

$$df = du + i\,dv, \qquad d\omega = d\sigma + i\,d\tau, \quad \text{etc.}$$

If X_1 and X_2 are vector fields, we define the "complex vector field" $X_1 + iX_2$ as that differential operator on complex-valued functions which is given by

$$\begin{aligned}(X_1 + iX_2)f &= X_1 f + iX_2 f\\ &= (X_1 u - X_2 v) + i(X_1 v + X_2 u).\end{aligned}$$

From now on, let M be an open subset of $\mathbb{R}^2$ with rectangular coordinates (x, y). We set $z = x + iy$, so

$$dz = dx + i\,dy.$$

We let

$$\bar{z} = x - iy, \qquad d\bar{z} = dx - i\,dy,$$

and then define the complex vector fields $\partial/\partial z$ and $\partial/\partial \bar{z}$ by

$$\frac{\partial}{\partial z} = \frac{1}{2}\left(\frac{\partial}{\partial x} - i\frac{\partial}{\partial y}\right) \qquad \text{and} \qquad \frac{\partial}{\partial \bar{z}} = \frac{1}{2}\left(\frac{\partial}{\partial x} + i\frac{\partial}{\partial y}\right)\cdot$$

We then set

$$\frac{\partial f}{\partial z} = \left(\frac{\partial}{\partial z}\right) f \qquad \text{and} \qquad \frac{\partial f}{\partial \bar{z}} = \left(\frac{\partial}{\partial \bar{z}}\right) f$$

for any complex function f.

EXERCISES

13.1 Show that for any complex-valued smooth function f we have

$$df = \frac{\partial f}{\partial z}\,dz + \frac{\partial f}{\partial \bar{z}}\,d\bar{z}.$$

13.2 Show that Leibnitz's rule holds for $\partial/\partial z$ and $\partial/\partial \bar{z}$; that is,

$$\frac{\partial(fg)}{\partial z} = f\left(\frac{\partial g}{\partial z}\right) + g\left(\frac{\partial f}{\partial z}\right) \qquad \text{and} \qquad \frac{\partial(fg)}{\partial \bar{z}} = f\left(\frac{\partial g}{\partial \bar{z}}\right) + g\left(\frac{\partial f}{\partial \bar{z}}\right)\cdot$$

A function f is called *holomorphic* (or complex analytic) if $\partial f/\partial \bar{z} = 0$. For a holomorphic function f we write

$$f' = \frac{\partial f}{\partial z}\cdot$$

13.3 Show that dz and $d\bar{z}$ are independent in the sense that

$$\text{if} \quad a\,dz + b\,d\bar{z} = 0, \qquad \text{then} \quad a = 0 \text{ and } b = 0.$$

13.4 Conclude that f is holomorphic if and only if $df = h\,dz$ for some h.

13.5 Show that

$$d(z^n) = nz^{n-1}\,dz$$

(on $\mathbb{R}^2 - \{0\}$ if $n < 0$) for all integers n, so that z^n is holomorphic.

13.6 Conclude that every polynomial p given by $p(z) = \sum_0^n a_k z^k$ is holomorphic and that

$$p'(z) = \sum_1^n k a_k z^{k-1}.$$

13.7 Define the function e^z by setting $e^z = e^x(\cos y + i \sin y)$. Show that

$$de^z = e^z\,dz.$$

13.8 Show that the product of two holomorphic functions is holomorphic.

13.9 Let $f = u + iv$ be a holomorphic function, and consider the map sending $(x, y) \to (u, v)$ as a map of $M \subset \mathbb{R}^2$ into $\mathbb{R}^2$. Show that this map is conformal and that its Jacobian determinant at the point (x, y) is $|f'(z)|^2$ if $z = x + iy$.

13.10 Let g be a holomorphic function defined on the image of M under the map of $M \subset \mathbb{R}^2$ corresponding to f. Define $g \circ f$ by

$$g \circ f(x, y) = g\big(u(x, y), v(x, y)\big)$$

[which we can write, for short, as $g \circ f(z) = g\big(f(z)\big)$]. Show that $g \circ f$ is holomorphic and that

$$(g \circ f)'(z) = g'\big(f(z)\big)f'(z).$$

13.11 Let U be a domain with almost regular boundary in the plane. By Stokes' theorem we have

$$\int_{\partial U} \nu = \int_U d\nu$$

for any complex-valued linear differential form ν. (This is to be interpreted, as usual, as the equality of the real and imaginary parts of both sides.) Conclude that

$$\int_{\partial U} g\,dz = \int_U dg \wedge dz = \int_U \frac{\partial g}{\partial \bar{z}}\,d\bar{z} \wedge dz.$$

13.12 Show that $d\bar{z} \wedge dz = 2i\,dx \wedge dy$, so that

$$\int_{\partial U} g\,dz = 2i \int_U \frac{\partial g}{\partial \bar{z}}\,dx \wedge dy = 2i \int_U \frac{\partial g}{\partial \bar{z}}.$$

13.13 Conclude that if f is a smooth function defined in a neighborhood of $\overline{U}$, and if f is holomorphic in U, then

$$\int_{\partial U} f\,dz = 0.$$

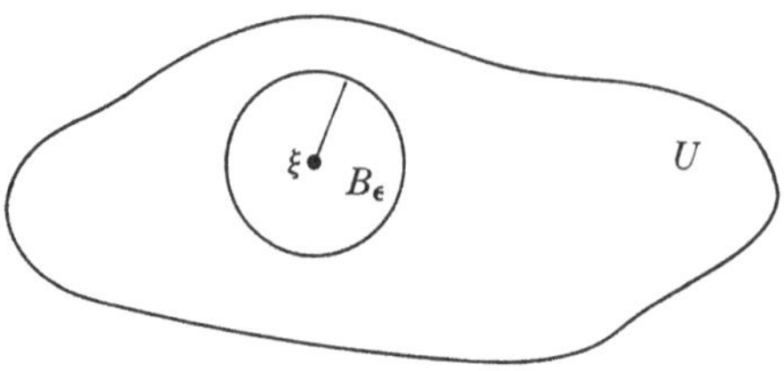

Fig. 12.18

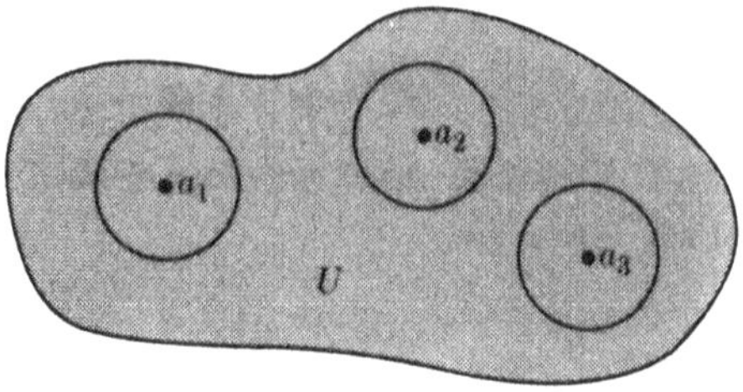

Fig. 12.19

13.14 Let $\xi = \zeta + i\eta$, where $(\zeta, \eta) \in U$ and g is a smooth function defined in a neighborhood of $\overline{U}$. (See Fig. 12.18.) Show that

$$g(\xi) = \frac{1}{2\pi i}\left[\int_{\partial U} \frac{U(z)}{z-\xi}\,dz + \int_U \frac{\partial U/\partial \bar{z}}{z-\xi}\,dz \wedge d\bar{z}\right].$$

[*Hint:* Apply Exercise 13.11 to the function $U(z)/(z-\xi)$ and the domain $U - B_\epsilon$, where B_ϵ is a disk of radius ϵ about ξ. Then let $\epsilon \to 0$.]

13.15 Conclude that if f is holomorphic in U, then for any $\xi \in U$,

$$f(\xi) = \frac{1}{2\pi i}\int_{\partial U} \frac{f(z)}{z-\xi}\,dz.$$

13.16 Let f be holomorphic in $U - \{a_1\} \cup \cdots \cup \{a_k\}$ (and let f be smooth in

$$W - \{a_1\} \cup \cdots \cup \{a_k\},$$

where $W \supset \overline{U}$). Suppose that in a neighborhood of each a_i

$$f(z) = B_{h_i,i}(z-a_i)^{-h_i} + \cdots + B_{1,i}(z-a)^{-1} + \varphi_i(z),$$

where φ_i is holomorphic in the neighborhood. (See Fig. 12.19.) Show that

$$\frac{1}{2\pi_i}\int_{\partial U} f(z)\,dz = B_{11} + B_{12} + \cdots + B_{1k}.$$

[*Hint:* Apply Exercise 13.13 to $U - \bigcup D_{i,\epsilon}$ where $D_{i,\epsilon}$ are small disks around the a_i, and let $\epsilon \to 0$.]

The result of Exercise 13.16 can frequently be used to evaluate definite integrals. For example, suppose we wish to evaluate

$$\int_0^{2\pi} R(\cos\theta, \sin\theta)\,d\theta,$$

where R is some rational function of two variables. If we set $z = e^{i\theta}$, then this integral becomes

$$-i\int_{\partial D_1} R\left[\frac{1}{2}(z+z^{-1}), \frac{1}{2i}(z-z^{-1})\right]\frac{dz}{z},$$

where D_1 is the unit disk. To apply Exercise 13.15, we must merely find the points a_j and the coefficients $B_{1,j}$ to evaluate the integral. For instance, if $a > 0$,

$$\int_0^{\pi} \frac{d\theta}{a + \cos\theta} = \frac{1}{2}\int_0^{2\pi} \frac{d\theta}{a + \cos\theta} = -i\Sigma - i\int_{\partial D_1} \frac{dz}{z^2 + 2az + 1}.$$

Now $z^2 + 2az + 1 = (z - a_1)(z - a_2)$, where $a_1 = -a + (a^2 - 1)^{1/2}$ and $a_2 = -a - (a^2 - 1)^{1/2}$. Thus only $a_1 \in D_1$, and

$$\frac{1}{z^2 + 2az + 1} = \frac{-1}{a_1 - a_2}\left(\frac{1}{z - a_1} - \frac{1}{z - a_2}\right),$$

so the integral is evaluated as

$$\int_0^{\pi} \frac{d\theta}{a + \cos\theta} = \frac{\pi}{\sqrt{a^2 - 1}}.$$

13.17 Evaluate

$$\int_0^{\pi/2} \frac{d\theta}{a + \sin^2\theta} \qquad (a > 0).$$

13.18 Evaluate

$$\int_0^{2\pi} \frac{\sin^2\theta}{a + b\cos\theta}\, d\theta.$$

Let P and Q be polynomials such that $Q(x) \neq 0$ for any real x. Suppose that

$$P(z) = a_{n-2}z^{n-2} + a_{n-3}z^{n-3} + \cdots + a_1 z + a_0$$

and

$$Q(z) = b_n z^n + b_{n-1}z^{n-1} + \cdots + b_1 z + b_0,$$

where $b_n \neq 0$. In other words, the degree of Q is at least two more than the degree of P. Then P/Q is absolutely integrable and

$$\int_{-\infty}^{\infty} \frac{P(x)}{Q(x)}\, dx = \lim_{R\to\infty} \int_{-R}^{R} \frac{P(x)}{Q(x)}.$$

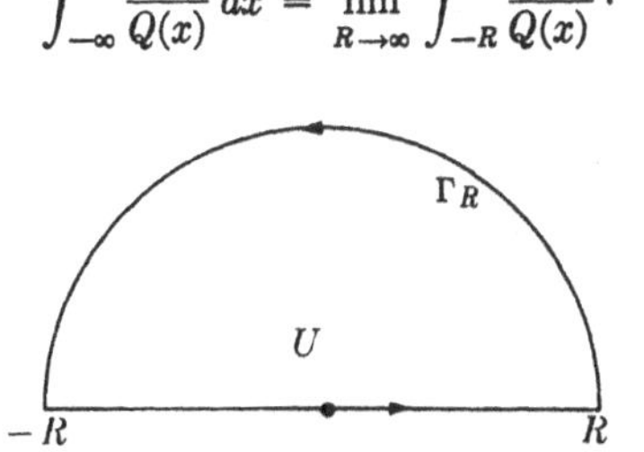

Fig. 12.20

Now consider $\int_{\partial U} [P(z)/Q(z)]\, dz$, where U is a semidisk of radius R. (See Fig. 12.20.) The integral over ∂U splits into two parts:

$$\int_{-R}^{R} \frac{P(x)}{Q(x)}\, dx + \int_{\Gamma_R} \frac{P(z)}{Q(z)}\, dz,$$

where Γ_R is half the circle of radius R. For large values of R we have, for $\|z\| = R$,

$$\left|\frac{P(z)}{Q(z)}\right| \le \frac{1}{R^2}\,\frac{|a_n| + |a_{n-1}|\cdot\frac{1}{R} + \cdots + |a_0|\cdot\frac{1}{R^{n-2}}}{\left(|b_n| - |b_{n-1}|\cdot\frac{1}{R} - \cdots - |b_0|\cdot\frac{1}{R^n}\right)} < \frac{C}{R^2},$$

so

$$\left|\int_{\Gamma_R} \frac{P(z)}{Q(z)}\,dz\right| = \left|iR\int_0^{\pi} \frac{P(Re^{i\theta})}{Q(Re^{i\theta})}\,d\theta\right| \le \frac{c\pi}{R}$$

and thus $\lim_{R\to\infty} \int_{\Gamma_R} = 0$. We can then apply Exercise 13.16, since there will be only a finite number of zeros of Q in the upper half-plane.

13.19 Evaluate

$$\int_{-\infty}^{\infty} \frac{dx}{(x^2 + a^2)^2} \qquad (a > 0).$$

13.20 Show that

$$\int_{-\infty}^{\infty} \frac{dx}{(x^2 + 1)^n} = \frac{\pi}{2^{2(n-1)}}\,\frac{(2n-2)!}{[(n-1)!]^2}.$$

CHAPTER 13

CLASSICAL MECHANICS

In this chapter we shall present a brief introduction to the study of classical theoretical mechanics. We emphasize that what we are studying is a branch of mathematics—idealized from the mathematical considerations common to many problems arising in the physics of mechanical systems. We will thus formulate, on an axiomatic basis, a mathematical model that describes features that arise in the study of the equations of mechanics. The model will apply to most situations arising in the mechanical applications. In order to simplify the mathematics, we have sacrificed a certain amount of generality, and therefore there will be some situations in classical mechanics where our model is inadequate.

Mechanics is devoted to the study of how a physical system evolves in time. The first fundamental assumption is that the system can be described (locally) by a collection of continuous parameters. More precisely, we assume that the set of all possible "positions" or configurations of the physical system is a differentiable manifold M which is called the *configuration space.*

For example, if the physical system consists of three particles free to move in space, we can describe the "position" of the system at any given instant by giving the positions of each of the three particles. Thus, in this case, the configuration space is

$$\mathbb{E}^3 \times \mathbb{E}^3 \times \mathbb{E}^3.$$

If we insist that no two particles be able to occupy the same position of space at the same time, then the configuration space M is given by

$$M = \mathbb{E}^3 \times \mathbb{E}^3 \times \mathbb{E}^3 - S,$$

where

$$S = \{ \langle v_1, v_2, v_3 \rangle : v_i \in \mathbb{E}^3 \text{ and } v_1 = v_2 \text{ or } v_1 = v_3 \text{ or } v_2 = v_3 \}.$$

If the physical system is a rigid body (say a top) spinning about some point held fixed in space, then the position of the system is completely described by giving the position in space of three orthonormal vectors on the body (drawn through the fixed point, say). Since these vectors can be placed in any position but must remain orthonormal and cannot change orientation (from right- to left-handedness), we see that M is the set of all oriented orthonormal bases of $\mathbb{E}^3$. In this case M is a three-dimensional manifold. If we fix some arbitrary initial orthonormal basis $\langle e_1, e_2, e_3 \rangle$, then any possible position of the system is given by $\langle v_1, v_2, v_3 \rangle$, where $v_i = Ae_i$ and A is a rotation, that is, A is an orthogonal linear transformation with positive determinant. Thus M is diffeomorphic to $0^+(3)$, the space of all orthogonal three-by-three matrices with positive determinant.

The basic problem of mechanics is to describe how the configuration of the system changes in time. As the system evolves, the configuration at any instant t will be given by some point $C(t) \in M$. Our problem is to give a reasonably simple description of the possible curves C which can arise as actual changes in the configuration of the mechanical system.

The second fundamental assumption of classical mechanics is that the curve $C(t)$ can be determined from a knowledge of the "state" of the system at any given time. That is, we may (and, in general, will) have to know more about the system at a given time than its configuration in order to be able to predict its future configuration. However, if we do have enough such instantaneous information, we can determine the curve $C(t)$. The total amount of relevant information is called the *state* of the system. It is assumed that the set of all states is itself a differentiable manifold, $\mathcal{S}$. Since the state of a system contains more information than the configuration, we can assign to every state s the configuration $\pi(s) \in M$ of the system in the state s. In other words, we have a map $\pi: \mathcal{S} \to M$. We assume that this map π is differentiable.

It is assumed that if we know the state s of the system at time t_0, we can predict its state at any future time t. Thus we are given a map

$$\varphi_{t,t_0}: \mathcal{S} \to \mathcal{S}$$

such that if the system is in the state s at time t_0, it will be in the state $\varphi_{t,t_0}(s)$ at time t. Now there is nothing special about the time t_0. If $t > t_1 > t_0$ and s is the state at time t_0, then $\varphi_{t_1,t_0}(s)$ is the state at time t_1, and therefore

$$\varphi_{t,t_1}(\varphi_{t_1,t_0}(s))$$

is the state at time t.

In other words,

$$\varphi_{t,t_1} \circ \varphi_{t_1,t_0} = \varphi_{t,t_0}.$$

Now it turns out that in most (basic) mechanical systems, if the time is suitably parametrized, the function φ_{t,t_0} really depends only on $t - t_0$. (This fails to hold in the so-called nonconservative systems. A typical nonconservative system is one involving friction. This is usually a consequence of not studying a sufficiently complete system. In the case of friction, for example, the heat loss must be taken into account.) Let us assume that φ_{t,t_0} depends only on $t - t_0$. Then, if we write

$$s_1 = t_1 - t_0 \qquad \text{and} \qquad s_2 = t - t_1,$$

then the previous equation can be written as

$$\varphi_{s_2} \circ \varphi_{s_1} = \varphi_{s_1+s_2}.$$

This looks like the defining relation for a one-parameter group except that so far we have been restricting ourselves to nonnegative values of s. In point of fact, it is assumed that this restriction is unnecessary and, in fact, we are given a flow φ on $\mathcal{S}$.

To repeat, we are given a differentiable manifold M representing the set of all possible configurations of our system, a differentiable manifold $\mathcal{S}$ representing all possible states of the system, a differentiable map $\pi: \mathcal{S} \to M$, and a flow φ on $\mathcal{S}$. Then the curves $C(t)$ are all assumed to be of the form

$$C(t) = \pi\varphi_t(x) \qquad \text{for some} \quad x \in \mathcal{S}.$$

We must therefore describe $\mathcal{S}$, π, and φ for any given configuration space M. Classical mechanics makes a very definite assumption about the nature of the space $\mathcal{S}$ (and the map π). It asserts that the state of a system is completely determined by its (configuration and its) "momentum". What is the momentum of our abstract setup? As usual, the momentum should be something that resists change in velocity. It turns out that an appropriate object representing "infinitesimal resistance to change in istantaneous velocity" is a cotangent vector. A heuristic motivation for this, which the reader may choose to ignore, is the following: At any given configuration $x \in M$, the set of all possible velocity vectors is just $T_x(M)$. At any given $v \in T_x(M)$, a "resistance to change in v" would be some function defined near v and vanishing at v. To first order we could replace such a function by its differential. Thus "infinitesimal resistance" is a linear function on $T_v(T_x(M))$. Since $T_x(M)$ is a vector space, we may identify $T_v(T_x(M))$ with $T_x(M)$ for each v. Thus all possible "momenta" become identified with all elements of $T_x^*(M)$.

The set $\mathcal{S}$ is thus taken to be the set of all momenta, i.e., the set of all cotangent vectors at all points of M. We must first show how to make this space into a differentiable manifold.

1. THE TANGENT AND COTANGENT BUNDLES

Let M be a differentiable manifold, and let us consider the set $T(M)$ of all tangent vectors to all points of M. Thus

$$T(M) = \bigcup_{x \in M} T_x(M).$$

Let π denote the map of $T(M)$ onto M which assigns to each tangent vector the point of M at which it is defined; that is, if $\xi \in T_x(M)$, then $\pi(\xi) = x$. We claim that $T(M)$ can be made into a differentiable manifold in a natural way, so that π is a differentiable map. In fact, let $\mathfrak{A}$ be an atlas on M. For any $(U, \alpha) \in \mathfrak{A}$, define $T(\alpha): \pi^{-1}(U) \to V \oplus V$ by setting

$$T(\alpha)\xi = \langle \alpha(x), \xi_\alpha \rangle \qquad \text{if} \quad \xi \in T_x(M), x \in U. \tag{1.1}$$

We claim that the collection $(\pi^{-1}U, T(\alpha))$ is an atlas on $T(M)$. To check that

$T(\alpha)$ is one-to-one, we observe that if $\xi \in T_x(M)$ and $\eta \in T_y(M)$ with $x \neq y$, then $\alpha(x) \neq \alpha(y)$; while if $\xi \neq \eta$ and both lie in $T_x(M)$, we have $\xi_\alpha \neq \eta_\alpha$. To check that the transition law is satisfied, we note that Eq. (4.3), Chapter 9 implies that if (U, α) and (W, β) are two charts of $\mathfrak{A}$,

$$T(\beta) \circ T(\alpha)^{-1}(v, \xi_\alpha) = \langle \beta \circ \alpha^{-1}(v), J_{\beta \circ \alpha^{-1}}(v)\xi_\alpha \rangle \tag{1.2}$$

for $\langle v, \xi_\alpha \rangle \in T(\alpha)\pi^{-1}(U \cap W)$, that is, $v \in \alpha(U \cap W)$ and ξ_α arbitrary in V.

The fact that the structure on $T(M)$ does not depend on the choice of $\mathfrak{A}$ is obvious. That π is differentiable is also clear; in fact, $(\pi^{-1}(U), T(\alpha))$ and (U, α) are compatible charts in terms of which $\alpha \circ \pi \circ T(\alpha)^{-1}(v, \xi_\alpha) = v$. We call $T(M)$, together with its structure as a differentiable manifold, the *tangent bundle* of M.

If M is finite-dimensional, let (U, α) be a chart of M with coordinates $x^1, \ldots, x^n$, so that

$$\alpha(x) = \langle x^1(x), \ldots, x^n(x) \rangle \in \mathbb{R}^n.$$

We will denote the coordinates associated to the chart $T(\alpha)$ on $\pi^{-1}(U)$ by $\langle q^1, \ldots, q^n, \dot{q}^1, \ldots, \dot{q}^n \rangle$. Hence if $\xi \in T_x(M)$, where $x \in U$, we have

$$\begin{aligned} T(\alpha)(\xi) &= \langle \alpha(x), \xi_\alpha \rangle \\ &= \langle q^1(\xi), \ldots, q^n(\xi), \dot{q}^1(\xi), \ldots, \dot{q}^n(\xi) \rangle, \end{aligned}$$

so that

$$q^i = x^i \circ \pi \qquad \text{and} \qquad \xi = \sum \dot{q}^i(\xi) \left(\frac{\partial}{\partial x^i}\right)_x. \tag{1.3}$$

In other words, the q^i are just the coordinates x^i regarded as functions on $T(M)$ via π, and the $\dot{q}^i(\xi)$ are the components of ξ relative to the basis

$$\left\{\left(\frac{\partial}{\partial x^i}\right)_x, \ldots, \left(\frac{\partial}{\partial x^n}\right)_x\right\}.$$

We can follow a similar procedure for

$$T^*(M) = \bigcup_{x \in M} T_x^*(M),$$

which we call the *cotangent bundle*. If (U, α) is a chart of an atlas $\mathfrak{A}$ of M, define the map

$$T^*(\alpha)\colon \pi^{-1}(U) \to V \oplus V^*$$

by setting

$$T^*(\alpha)(l) = \langle \alpha(x), l_\alpha \rangle \tag{1.4}$$

for $l \in T_x^*(M)$ and $x \in U$. As before, this defines an atlas on $T^*(M)$ which, in turn, defines a differentiable structure on $T^*(M)$ which is independent of the choice of atlas of M. (Note that we have used the same letter, π, to denote the two projections: that of $T(M) \to M$ and that of $T^*(M) \to M$. Whenever there is any confusion, we shall denote these maps by π_T and π_{T^*}.)

If M is finite-dimensional and (U, α) is a chart on the coordinates $x^1, \ldots, x^n$, we shall denote the coordinates of $(\pi^{-1}(U), T^*(\alpha))$ by

$$\langle q^1, \ldots, q^n, p^1, \ldots, p^n \rangle.$$

Thus, if $l \in T_x^*(M)$, where $x \in U$, we have

$$\begin{aligned} T^*(\alpha)l &= \langle \alpha(x), l_\alpha \rangle \\ &= \langle q^1(l), \ldots, q^n(l), p^1(l), \ldots, p^n(l) \rangle, \end{aligned}$$

so that

$$q^i = q^i \circ \pi \qquad \text{and} \qquad l = \sum p^i(l)(dx^i)_x. \tag{1.5}$$

2. EQUATIONS OF VARIATION

Let M_1 and M_2 be two differentiable manifolds, and let $\varphi: M_1 \to M_2$ be a differentiable map. Then φ induces a map $T(\varphi)$ of $T(M_1) \to T(M_2)$ when we set

$$T(\varphi)\xi = \varphi_{*x}(\xi) \qquad \text{if} \quad \xi \in T_x(M_1). \tag{2.1}$$

To check that $T(\varphi)$ is differentiable, let us choose compatible charts (U, α) on M_1 and (W, β) on M_2. Then it follows from the definition of $T(\varphi)$ that

$$(T(U), T(\alpha)) \qquad \text{and} \qquad (T(W), T(\beta))$$

are compatible charts. Furthermore,

$$T(\beta) \circ T(\varphi) \circ T(\alpha)^{-1}(v_1, v_2) = \langle (\beta \circ \varphi \circ \alpha^{-1})v_1, J_{\beta \circ \varphi \circ \alpha^{-1}}(v_1)v_2 \rangle. \tag{2.2}$$

This establishes that $T(\varphi)$ is differentiable. Observe that if $\psi: M_2 \to M_3$ is a second differentiable map, then it follows from Eq. (4.4) of Chapter 9 that

$$T(\psi \circ \varphi) = T(\psi) \circ T(\varphi). \tag{2.3}$$

Furthermore, it is clear from the definition that

$$\pi \circ T(\varphi) = \varphi \circ \pi. \tag{2.4}$$

In other words, the diagram

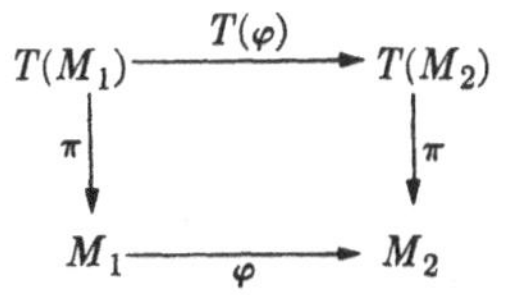

commutes in the sense that it doesn't matter which path one takes to get from $T(M_1)$ to M_2.

In particular, if $\{\varphi_t\}$ is a one-parameter group of diffeomorphisms of M, we get a one-parameter group $T(\varphi_t)$ on $T(M)$, where, by (2.4),

$$\pi \circ T(\varphi_t)\xi = \varphi_t(\pi(\xi)). \tag{2.5}$$

If X is the infinitesimal generator of $\{\varphi_t\}$, let us denote by $T(X)$ the infinitesimal generator of $\{T(\varphi_t)\}$. It follows from (2.5) that if $\xi \in T_x(M)$, then

$$\pi_{*\xi}(T(X)_\xi) = X_x. \tag{2.6}$$

Let us obtain the expression for $T(X)$ in terms of a chart: Let (U, α) be a chart, and choose an open set W such that $\overline{W} \subset U$ and $\epsilon > 0$ such that $\varphi_t(W) \subset U$ for $|t| < \epsilon$. Then

$$T(\alpha) \circ T(\varphi_t) \circ T(\alpha)^{-1} \langle v, w \rangle = \langle (\alpha \circ \varphi_t \circ \alpha^{-1})v, J_{(\alpha\circ\varphi_t\circ\alpha^{-1})}(v)w \rangle,$$

so that, differentiating with respect to t at $t = 0$, we see by Eq. (4.9) of Chapter 9 that

$$T(X)_{T(\alpha)} \langle v, w \rangle = \langle X_\alpha(v), dX_{\alpha(v)}(w) \rangle. \tag{2.7}$$

If M is finite-dimensional, $x^1, \ldots, x^n$ are the coordinates of (U, α), and

$$X_\alpha = \langle X^1, \ldots, X^n \rangle,$$

then we can rewrite (2.7) as

$$T(X)_{T(\alpha)} \langle v, w \rangle = \left\langle X^1(v), \ldots, X^n(v), \sum \frac{\partial X^1}{\partial x^j} w^j, \ldots, \sum \frac{\partial X^n}{\partial x^j} w^j \right\rangle, \tag{2.7'}$$

where $w = \langle w^1, \ldots, w^n \rangle$. In other words, in terms of the local coordinates $\langle q^1, \ldots, q^n, \dot{q}^1, \ldots, \dot{q}^n \rangle$ the differential equation corresponding to $T(X)$ take the form

$$\frac{dq^i}{dt} = X^i(q^1, \ldots, q^n) \tag{2.8}$$

and

$$\frac{d\dot{q}^i}{dt} = \frac{\partial X^i}{\partial x^1} \dot{q}^1 + \cdots + \frac{\partial X^i}{\partial x^n} \dot{q}^n. \tag{2.9}$$

Note that Eqs. (2.8) are just the local equations corresponding to the vector field X. Since $q^i = x^i \circ \pi$, this is simply another way of writing (2.6). Suppose that

$$\varphi_t(x) = \langle x^1(t), \ldots, x^n(t) \rangle$$

is a solution curve of X lying in U. Then we can regard the coefficients

$$\frac{\partial X^i}{\partial x^j} = \frac{\partial X^i}{\partial x^j}(x^1(t), \ldots, x^n(t))$$

as functions of t alone. Thus (2.9) takes the form

$$\frac{dw}{dt} = A(t)w$$

of a linear differential equation for w. This linear differential equation is called the *equation of variation* of X along the curve $\varphi(x)$. Roughly speaking, it represents, to linear approximation, how solution curves of X near $\varphi(x)$ are deviating from $\varphi(x)$.

We now go through a similar construction for the cotangent bundle. If $\varphi: M_1 \to M_2$ is a differentiable map, then $\varphi_x^*: T^*_{\varphi(x)}(M_2) \to T^*_x(M_1)$ is going in the wrong direction. We therefore restrict our attention to maps which are locally diffeomorphisms, and we define $T^*(\varphi)$ by setting

$$T^*(\varphi)l = (\varphi^{-1})^*l = (\varphi^*_{\varphi(x)})^{-1}l \qquad \text{if} \quad l \in T^*_x(M_1). \tag{2.10}$$

Since φ is a diffeomorphism,

$$(\varphi^*)^{-1}: T^*_x(M_1) \to T^*_{\varphi(x)}(M_2)$$

is well defined, and we have

$$\pi \circ T^*(\varphi) = \varphi \circ \pi. \tag{2.11}$$

If (U, α) and (W, β) are compatible charts on M_1 and M_2, then

$$T^*(\beta) \circ T^*(\varphi) \circ T^*(\alpha)^{-1} \langle v_1, v_2 \rangle = \langle (\beta \circ \varphi \circ \alpha^{-1})v_1, [J_{\beta\circ\varphi\circ\alpha^{-1}}(v_1)]^{*-1}v_2 \rangle, \tag{2.12}$$

and so $T^*(\varphi)$ is differentiable.

If $\varphi: M_1 \to M_2$ and $\psi: M_2 \to M_3$ are diffeomorphisms, then

$$T^*(\psi \circ \varphi)l = (\psi \circ \varphi)^{*-1}l = (\varphi^* \circ \psi^*)^{-1}l = \psi^{*-1} \circ \varphi^{*-1}l,$$

so that

$$T^*(\psi \circ \varphi) = T^*(\psi) \circ T^*(\varphi). \tag{2.13}$$

In particular, if $\{\varphi_t\}$ is a flow on a manifold M, we obtain a flow $T^*(\varphi_t)$ on $T^*(M)$. It satisfies

$$\pi \circ T^*(\varphi_t)l = \varphi_t(\pi(l)). \tag{2.14}$$

If X is the infinitesimal generator of $\{\varphi_t\}$, we denote by $T^*(X)$ the infinitesimal generator of $\{T^*(\varphi_t)\}$. It satisfies

$$\pi_* T^*(X)_l = X_x \qquad \text{for} \quad l \in T^*_x(M). \tag{2.15}$$

3. THE FUNDAMENTAL LINEAR DIFFERENTIAL FORM ON $T^*(M)$

Before returning to our study of mechanics, we study in some detail the geometry of the cotangent bundle. Let M be a differentiable manifold, and let z be a point of $T^*(M)$. Let ξ be a tangent vector to $T^*(M)$ at the point z, so that

$$\xi \in T_z(T^*(M)).$$

Then $\pi_*\xi$ is an element of $T_{\pi(z)}(M)$. Since $z \in T^*_{\pi(z)}(M)$ is a linear function on $T_{\pi(z)}(M)$, we can consider the expression

$$\langle \pi_*\xi, z \rangle$$

which depends linearly on ξ. We denote this linear function of ξ by θ_z. We have thus defined a linear differential form θ on $T^*(M)$ by setting

$$\langle \xi, \theta_z \rangle = \langle \pi_{*z}\xi, z \rangle \qquad \text{for} \quad \xi \in T_z(T^*(M)). \tag{3.1}$$

The form θ is called the *fundamental linear form* of $T^*(M)$. Let us obtain the expression for θ in terms of a chart $(\pi^{-1}(U), T^*(\alpha))$. Since $T^*(\alpha)$ maps $\pi^{-1}(U)$ onto an open set, O of $V \oplus V^*$, the expression $\theta_{T^*(\alpha)}$ should be a function from O to $(V \oplus V^*)^*$ which can be identified with $V^* \oplus V$. Let us evaluate this function. In terms of the chart (U, α) on M and $(\pi^{-1}(U), T^*(\alpha))$ on $T^*(M)$, we have

$$\xi_{T^*(\alpha)} = \langle v, w^* \rangle \in V \oplus V^* \qquad \text{and} \qquad (\pi_*\xi)_\alpha = v \in V,$$

so that

$$\langle \pi_*\xi, z \rangle = \langle v, z_\alpha \rangle.$$

Thus $\langle \langle v, w^* \rangle, (\theta_z)_{T^*(\alpha)} \rangle = \langle v, z_\alpha \rangle$; that is,

$$(\theta_z)_{T^*(\alpha)} = \langle z_\alpha, 0 \rangle \in V^* \oplus V = (V \oplus V^*)^*.$$

In other words, the local expression θ_α for the differential form θ in terms of the chart $T^*(\alpha)$ is given by

$$\theta_\alpha \langle u_1, u_2^* \rangle = \langle u_2^*, 0 \rangle. \tag{3.2}$$

If M is finite-dimensional and we use the local coordinates $\langle q^1, \ldots, q^n, p^1, \ldots, p^n \rangle$, then

$$\pi_* \left(\frac{\partial}{\partial q^i} \right)_z = \frac{\partial}{\partial x^i},$$

while

$$\pi_* \left(\frac{\partial}{\partial p^i} \right)_z = 0 \qquad \text{and} \qquad z = \sum p^j(z)\, dx_x^j.$$

Thus

$$\left\langle \left(\frac{\partial}{\partial q^i} \right)_z, \theta_z \right\rangle = \left\langle \frac{\partial}{\partial x^i}, \sum p^j(z)\, dx_x^j \right\rangle = p^i(z),$$

while

$$\left\langle \left(\frac{\partial}{\partial p^i} \right)_z, \theta_z \right\rangle = 0,$$

so that

$$\theta_z = \sum p^i(z)\, dq_z^i$$

or

$$\theta = \sum p^i\, dq^i. \tag{3.3}$$

Of course, (3.3) is just a way of writing (3.2) in terms of a basis of $V = \mathbb{R}^n$.

Let $\varphi: M_1 \to M_2$ be a diffeomorphism, let θ_1 be the fundamental linear form on $T^*(M_1)$, and let θ_2 be the fundamental linear form on $T^*(M_2)$. Let

$$\xi \in T_z(T^*(M_1)) \qquad \text{and} \qquad \pi(z) = x \in M_1.$$

Then

$$\pi_* T^*(\varphi)_* \xi = \varphi_* \pi_* \xi,$$

and so

$$\begin{aligned}\langle T^*(\varphi)_* \xi, \theta_{2T^*(\varphi)z} \rangle &= \langle \varphi_{*x} \pi_* \xi, T^*(\varphi) z \rangle \\ &= \langle \varphi_{*x} \pi_* \xi, (\varphi^*)^{-1}_{\varphi(x)} z \rangle \\ &= \langle \pi_* \xi, z \rangle \\ &= \langle \xi, \theta_{1z} \rangle.\end{aligned}$$

In other words,

$$\big(T^*(\varphi)\big)^* \theta_2 = \theta_1. \tag{3.4}$$

In particular, if $\{\varphi_t\}$ is a flow on M, then

$$\big(T^*(\varphi_t)\big)^* \theta = \theta.$$

If X is a vector field on M, then the infinitesimal version of the last equation is

$$D_{T^*(X)} \theta = 0. \tag{3.5}$$

Note that any vector field X on M defines a function f_X on $T^*(M)$ by

$$f_X(z) = \langle X_x, z \rangle \qquad \text{if} \quad \pi(z) = x. \tag{3.6}$$

We also have

$$f_X = \langle T^*(X), \theta \rangle, \tag{3.7}$$

since $\langle T^*(X)_z, \theta_z \rangle = \langle \pi_* T^*(X)_z, z \rangle = \langle X_x, z \rangle$ by (2.15).

Finally, in view of (3.5) and Eq. (6.14), Chapter 11, we have

$$0 = D_{T^*(X)} \theta = d\langle T^*(X), \theta \rangle + T^*(X) \lrcorner\, d\theta,$$

so that

$$df_X = -T^*(X) \lrcorner\, d\theta. \tag{3.8}$$

For reasons which will become clear later on, the function f_X is sometimes called the momentum function associated to the vector field X.

4. THE FUNDAMENTAL EXTERIOR TWO-FORM ON $T^*(M)$

It turns out that the exterior two-form

$$\Omega = d\theta \tag{4.1}$$

plays a fundamental role in mechanics, and we therefore study some of its properties. First of all, since $d^2 = 0$, we have

$$d\Omega = 0. \tag{4.2}$$

We claim that Ω_z is a nonsingular bilinear form on $T_z(T^*(M))$ for each $z \in T^*(M)$. That is,

$$\text{if} \quad \xi \in T_z(T^*(M)) \quad \text{is such that} \quad \xi \lrcorner\, \Omega_z = 0, \qquad \text{then} \quad \xi = 0. \tag{4.3}$$

For this purpose we compute the local expression for Ω in terms of a chart $(\pi^{-1}(U), T^*(\alpha))$ [where $z \in \pi^{-1}(U)$]. The map $T^*(\alpha)$ gives a diffeomorphism of $\pi^{-1}(U)$ with a subset of $V \oplus V^*$, and by (3.4) the form θ on M carries over to the corresponding form θ_α on $V \oplus V^* = T^*(V)$. Also, θ_α can be regarded as a $(V^* \oplus V)$-valued function which is given by (3.2). Let us denote $\xi_{T^*(\alpha)}$ by ξ_α, and let

$$\xi_\alpha = \langle X_1, X_2 \rangle \in V \oplus V^*$$

be considered a constant vector field on $V \oplus V^*$. Then

$$\langle \xi_\alpha, \theta_\alpha \rangle_{\langle u_1, u_2^* \rangle} = \langle X_1, u_2^* \rangle,$$

and so $d\langle \xi_\alpha, \theta_\alpha \rangle$ is the linear differential form given by

$$\langle \eta_\alpha, d\langle \xi_\alpha, \theta_\alpha \rangle \rangle = \langle X_1, Y_2 \rangle \qquad \text{if} \quad \eta_\alpha = \langle Y_1, Y_2 \rangle \in V \oplus V^*.$$

On the other hand, since ξ_α is a constant vector field, the Lie derivative $D_{\xi_\alpha}\theta_\alpha$ reduces to the ordinary derivative of the linear function $\langle u_2, u_2^* \rangle \mapsto \langle u_2^*, 0 \rangle$. This derivative is just the constant $\langle X_2, 0 \rangle$. Thus the Lie derivative $D_{\xi_\alpha}\theta_\alpha$ is given by the constant linear form

$$D_{\xi_\alpha}\theta_\alpha = \langle X_2, 0 \rangle \in V^* \oplus V;$$

that is,

$$\langle \eta_\alpha, D_{\xi_\alpha}\theta_\alpha \rangle = \langle Y_1, X_2 \rangle.$$

Now $\xi_\alpha \lrcorner \Omega_\alpha = \xi_\alpha \lrcorner d\theta_\alpha = D_{\xi_\alpha}\theta_\alpha - d\langle \xi_\alpha, \theta_\alpha \rangle$, so we see that $\xi_\alpha \lrcorner \Omega_\alpha$ is the constant form

$$\xi_\alpha \lrcorner d\theta = \langle X_2, -X_1 \rangle \in V^* \oplus V.$$

To recapitulate, if $\xi \in T_z(T^*(M))$ is such that $\xi_{T^*(\alpha)}$ is given by $\xi_{T^*(\alpha)} = \langle X_1, X_2 \rangle \in V \oplus V^*$, then

$$(\xi \lrcorner \Omega_z)_{T^*(\alpha)} = \langle X_2, -X_1 \rangle \in V^* \oplus V. \tag{4.4}$$

Equation (4.3) clearly follows from this.

Since (4.3) is of fundamental importance, we present an alternative derivation for the case of a finite-dimensional manifold. If we use the coordinates $\langle q^i, \ldots, q^n, p^i, \ldots, p^n \rangle$, then

$$\theta = \sum p^i \, dq^i,$$

so that

$$\Omega = \sum dp^i \wedge dq^i. \tag{4.5}$$

If

$$X = \sum \left(A^i \frac{\partial}{\partial q^i} + B^i \frac{\partial}{\partial p^i} \right),$$

then

$$X \lrcorner \Omega = \sum (B^i \, dq^i - A^i \, dp^i). \tag{4.6}$$

Thus $(X \lrcorner \Omega)_z = 0$ if and only if $X_z = 0$. This shows that on $T^*(M)$ we have a one-to-one correspondence, $X \mapsto X \lrcorner \Omega$, between vector fields and linear differential forms. Let us denote by ω_X the form corresponding to X, so that

$$\omega_X = X \lrcorner \Omega;$$

and let us denote by X_ω the vector field corresponding to the linear differential form ω. Thus

$$\omega = X_\omega \lrcorner \Omega = \omega_{X_\omega}.$$

Observe that

$$d\omega = 0 \qquad \text{if and only if} \qquad D_{X_\omega}\Omega = 0. \tag{4.7}$$

In fact, by (4.2),

$$D_{X_\omega}\Omega = d(X_\omega \lrcorner \Omega) = d\omega.$$

In particular, there is a distinguished class of vector fields on $T^*(M)$—those corresponding to functions, i.e., the vector fields of the form X_{dF}, where F is a function on $T^*(M)$. These vector fields are called *Hamiltonian* vector fields.

*In view of (4.6), if $D_X\Omega = 0$, then *locally*, at least, X is of the form X_{dF}, since any form ω with $d\omega = 0$ can be written locally as $\omega = dF$. If we make the topological assumption that $T^*(M)$ has vanishing first cohomology group, then $D_X\Omega = 0$ is equivalent to X being Hamiltonian. This assumption is really a restriction on the nature of the configuration space M. Since we do not wish to restrict M in this manner, we will not take $D_X\Omega = 0$ as the definition of a Hamiltonian vector field.*

Note that if X and Y are Hamiltonian vector fields, so is $aX + bY$, where a and b are constants. In fact, if $X = X_{dF}$ and $Y = X_{dG}$, then

$$aX + bY = X_{d(aF+bG)}.$$

Furthermore, $[X, Y]$ is also a Hamiltonian vector field. In fact,

$$\begin{aligned} D_X(Y \lrcorner \Omega) &= D_XY \lrcorner \Omega + Y \lrcorner D_X\Omega \\ &= D_XY \lrcorner \Omega, \end{aligned}$$

since $D_X\Omega = 0$. Since $D_XY = [X, Y]$, we see that

$$[X, Y] \lrcorner \Omega = D_X\, dG = dD_XG.$$

In other words, we have

$$[X_{dF}, X_{dG}] = X_{d(X_{dF}G)}. \tag{4.8}$$

We thus see that we have a binary operation on functions corresponding to the Lie bracket on Hamiltonian vector fields. It is called the *Poisson bracket* and is denoted by $\{F, G\}$. In other words, we define

$$\{F, G\} = X_{dF}G, \tag{4.9}$$

so that we can rewrite (4.8) as

$$[X_{dF}, X_{dG}] = X_{d\{F,G\}}. \tag{4.8$'$}$$

Note that

$$X_{dF}G = \langle X_{dF}, dG\rangle = \langle X_{dF}, X_{dG} \lrcorner \Omega\rangle = \langle X_{dG} \wedge X_{dF}, \Omega\rangle, \tag{4.10}$$

so that, in particular,

$$\{F, G\} = -\{G, F\}. \tag{4.11}$$

In terms of local coordinates $\langle q^1, \ldots, q^n, p^1, \ldots, p^n\rangle$, we have

$$dF = \sum \left(\frac{\partial F}{\partial q^i} dq^i + \frac{\partial F}{\partial p^i} dp^i\right),$$

so that by (4.4),

$$X_{dF} = \sum \left(\frac{\partial F}{\partial q^i}\frac{\partial}{\partial p^i} - \frac{\partial F}{\partial p^i}\frac{\partial}{\partial q^i}\right), \tag{4.12}$$

and therefore

$$\{F, G\} = \sum \left(\frac{\partial F}{\partial q^i}\frac{\partial G}{\partial p^i} - \frac{\partial F}{\partial p^i}\frac{\partial G}{\partial q^i}\right), \tag{4.13}$$

from which the antisymmetry (4.11) is apparent. A consequence of (4.11) is the following:

Proposition 4.1. If F and G are functions on $T^*(M)$ such that

$$X_{dF}G = 0,$$

then

$$X_{dG}F = 0.$$

In other words, if G is constant along the solution curves of X_{dF}, then F is constant along the solution curves of X_{dG}.

In fact,

$$X_{dF}G = \{F, G\} = -\{G, F\} = -X_{dG}F.$$

It will turn out that Proposition 4.1 is the prototype of all the "conservation laws" of mechanics.

We close this section with the following observation. Let Y be a vector field on M. Then the momentum function of Y is a function f_Y on $T^*(M)$. Equation (3.8) asserts that

$$-T^*(Y) = X_{df_Y}. \tag{4.14}$$

5. HAMILTONIAN MECHANICS

As we indicated in the introduction, the first fundamental assumption of mechanics is that the evolution of the system is determined by a flow on $T^*(M)$, where M is the configuration space of the system. The second fundamental assumption concerns the character of the flow. It is assumed that the infinitesimal generator of the flow is a Hamiltonian vector field. That is, it is assumed that there is a function H (called the energy) on $T^*(M)$ such that the vector field X_{-dH} is the infinitesimal generator of the flow on $T^*(M)$ describing the evolution of the system. (The minus sign is a consequence of certain standard conventions.) In order to see what this means, let us express the equations of motion in terms of q- and p-coordinates for a finite-dimensional system. Thus

$$X_{-dH} = \sum \left(\frac{\partial H}{\partial p^i} \frac{\partial}{\partial q^i} - \frac{\partial H}{\partial q^i} \frac{\partial}{\partial p^i} \right).$$

Thus, if $\langle q^1(\cdot), \ldots, q^n(\cdot), p^1(\cdot), \ldots, p^n(\cdot) \rangle$ is an integral curve of the flow, it must satisfy the differential equations

$$\frac{dq^i}{dt} = \frac{\partial H}{\partial p^i} \quad \text{and} \quad \frac{dp^i}{dt} = -\frac{\partial H}{\partial q^i}. \tag{5.1}$$

A trivial consequence of (4.11) is that

$$X_{-dH}H = 0.$$

In other words, the function H is constant along trajectories of the system. This principle is known as *the law of conservation of energy.*

More generally, we can formulate Proposition 4.1 as:

Proposition 5.1. Let X_{-dH} be the infinitesimal generator of a mechanical system with energy H. Let F be a function such that

$$X_{dF}H = 0.$$

Then F is constant on the trajectories, i.e., solution curves, of the flow generated by X_{-dH}.

Proposition 5.1 is the prototype of all the "momentum conservation" laws we shall derive later; see, for example, the discussion at the beginning of Section 6.

In order to specify the mechanical system, one must give the function H. It turns out (in many but not all cases) that the energy is the sum of two terms, $H = K + U$, where K is called the kinetic energy and U is called the potential energy. They each have a very special form which we now describe.

The kinetic energy is a function on $T^*(M)$ which is associated with a Riemann metric on M. Let $(\ ,\ )$ be a Riemann metric on M. It gives a scalar product $(\ ,\ )_x$ on each $T_x(M)$, and therefore induces an isomorphism of $T_x(M)$ with $T_x^*(M)$, and thus gives a scalar product on $T_x^*(M)$ which we will continue to denote by $(\ ,\ )$. The function K is then defined by

$$K(l) = \tfrac{1}{2}(l, l). \tag{5.2}$$

To understand the relevance of a Riemann metric to mechanics, let us consider the most elementary case – the study of a single particle of mass m in $\mathbb{E}^3$. The usual relation between velocity and momentum,

$$\mathbf{p} = m\dot{\mathbf{q}}$$

can be formulated as follows: Consider the Riemann metric on $\mathbb{E}^3$ which is $m \times$ the Euclidean metric. That is, if $\langle x, y, z \rangle$ are rectangular coordinates on

$\mathbb{E}^3$ and $\langle q_x, q_y, q_z, \dot{q}_x, \dot{q}_y, \dot{q}_z \rangle$ are the corresponding coordinates on $T(\mathbb{E}^3)$, then

$$\|(\dot{q}_x, \dot{q}_y, \dot{q}_z)\|^2 = m\dot{q}_x^2 + m\dot{q}_y^2 + m\dot{q}_z^2.$$

Then the map of $T_x(\mathbb{E}^3) \to T_x^*(\mathbb{E}^3)$ sends

$$\langle q_x, q_y, q_z, \dot{q}_x, \dot{q}_y, \dot{q}_z \rangle \mapsto \langle q_x, q_y, q_z, p_x, p_y, p_z \rangle,$$

where

$$p_x = m\dot{q}_x, \qquad p_y = m\dot{q}_y, \qquad p_z = m\dot{q}_z, \tag{5.3}$$

and

$$K(q_x, q_y, q_z, p_x, p_y, p_z) = \frac{1}{2m}(p_x^2 + p_y^2 + p_z^2). \tag{5.4}$$

Thus the passage from velocity to momentum depends on the choice of a Riemann metric, which determines a map from $T(M) \to T^*(M)$. This can be regarded as a generalized "choice of mass" for the configuration.

The function U is assumed to be of the form $U = \overline{U} \circ \pi$, where $\overline{U}$ is a function on M. The form $\mathbf{F} = -d\overline{U}$ is called the *force field* whose potential is $\overline{U}$. It can be regarded as a vector field on M in view of the Riemann metric on M:

$$(\xi, \mathbf{F}_x) = -\langle \xi, d\overline{U} \rangle \qquad \text{for any} \quad \xi \in T_x(M).$$

In the special case of a single particle in $\mathbb{E}^3$ with mass m, substituting (5.3) and (5.4) into Eq. (5.1) gives, when we write $H = K + U$,

$$\frac{dmq^i}{dt} = p^i \qquad \text{and} \qquad \frac{dp^i}{dt} = F^i \tag{5.5}$$

or, since m is constant,

$$m\frac{d^2q^i}{dt^2} = F^i, \tag{5.6}$$

which is the usual rule stating that force = mass $\times$ acceleration.

Returning to the general theory, we now formulate a useful corollary of Proposition 5.1.

Proposition 5.2. Let $H = K + U$ be the energy associated with the Riemann metric (,) and the function $\overline{U}$ on M. Let Y be a vector field on M which is an infinitesimal isometry of (,) and is such that $Y\overline{U} = 0$. Then the momentum function f_Y is constant under the flow generated by X_{-dH}.

Let $\{\varphi_t\}$ be the flow generated by Y. Since φ_t is a local isometry, $(\varphi_t^* l, \varphi_t^* l) = (l, l)$ wherever $\varphi_t^* l$ is defined, so that

$$K(T^*(\varphi_t^*)l) = K(l),$$

and thus

$$T^*(Y)K = 0.$$

Also,

$$\begin{aligned} T^*(Y)U &= \langle T^*(Y), dU\rangle \\ &= \langle T^*(Y), \pi^*\, d\overline{U}\rangle = \langle \pi_* T^*(Y), d\overline{U}\rangle = \langle Y, d\overline{U}\rangle = 0, \end{aligned}$$

so that $T^*(Y)H = 0$ and Proposition 5.1 applies [see Eq. (4.14)].

6. THE CENTRAL-FORCE PROBLEM

Before proceeding with the general theory, we illustrate the previous results in a simple but important case. We will study the motion of a particle in $\mathbb{E}^3$ acted on by a "force centered at the origin". That is, our configuration space M will be taken to be $\mathbb{E}^3 - \{0\}$, and the Riemann metric is $m \times$ the Euclidean metric, where m is the "mass" of the particle. We also assume that the function $\overline{U}$ depends only on the distance to the origin. Thus $\overline{U}(x, y, z) = P(r)$, where $r^2 = x^2 + y^2 + z^2$. Under these circumstances it is clear that any rotation about the origin is an isometry of the Riemann metric and preserves $\overline{U}$. We can therefore apply Proposition 5.2 to the infinitesimal rotations $x(\partial/\partial y) - y(\partial/\partial x)$ to conclude that the corresponding momentum function $xp_y - yp_x$ is constant. (This momentum function is known as the angular momentum about the z-axis.) Similarly, the functions $xp_z - zp_x$ and $yp_z - zp_y$ must be constant on any trajectory of the flow. If we write $\mathbf{x} = \langle x, y, z\rangle$ and $\mathbf{p} = \langle p_x, p_y, p_z\rangle$, then these three conservation laws can be combined to read

$$\mathbf{x} \wedge \mathbf{p} = \text{const}, \tag{6.1}$$

which is known as the *law of conservation of angular momentum*. Here $\mathbf{x}$ and $\mathbf{p}$ are considered vector-valued functions on $T^*(M)$. In order to study the implications of (6.1), let us first distinguish two cases: where the constant $\neq 0$ and where the constant vanishes. If the constant occurring in (6.1) does not vanish, then (6.1) implies that the plane spanned by $\mathbf{x}$ and $\mathbf{p}$ does not change. In particular, the motion is such that $\mathbf{x}$ lies in a fixed plane. If $\mathbf{x} \wedge \mathbf{p} = 0$, we argue as follows:

Since $\mathbf{p} = m\dot{\mathbf{x}}$, we have $\mathbf{x} \wedge \dot{\mathbf{x}} = 0$. Since $\mathbf{x} \neq 0$, this implies that

$$\dot{\mathbf{x}} = \lambda \mathbf{x}$$

for some function λ of time. Now $\|\mathbf{x}\| = (\mathbf{x}, \mathbf{x})^{1/2}$, so

$$\frac{d}{dt}\|\mathbf{x}\| = \frac{1}{\|\mathbf{x}\|}(\mathbf{x}, \dot{\mathbf{x}}) = \left(\frac{\mathbf{x}}{\|\mathbf{x}\|}, \dot{\mathbf{x}}\right) = \lambda\|\mathbf{x}\|, \tag{6.2}$$

and therefore

$$\frac{d}{dt}\left(\frac{\mathbf{x}}{\|\mathbf{x}\|}\right) = \frac{\lambda\mathbf{x}}{\|\mathbf{x}\|} - \frac{\lambda\|\mathbf{x}\|\mathbf{x}}{\|\mathbf{x}\|^2} = 0,$$

so that $\mathbf{x}/\|\mathbf{x}\| = \text{const}$; that is, $\mathbf{x}$ lies on a fixed ray through the origin.

If we differentiate (6.2) and make use of the fact that $\mathbf{x}/\|\mathbf{x}\|$ is constant, we get, by (5.6),

$$m\frac{d^2}{dt}\|\mathbf{x}\| = \left(\frac{\mathbf{x}}{\|\mathbf{x}\|}, m\frac{d}{dt}\dot{\mathbf{x}}\right) = -\left(\frac{\mathbf{x}}{\|\mathbf{x}\|}, P'(\|\mathbf{x}\|)\frac{\mathbf{x}}{\|\mathbf{x}\|}\right),$$

that is,

$$m\frac{d^2r}{dt^2} = -P'(r), \tag{6.3}$$

where $r = \|\mathbf{x}\|$.

In any event, in all cases the particle $\mathbf{x}$ moves in a plane. We may therefore restrict our attention to the plane. That is, we can consider a "new" mechanical system where $\overline{M} = \mathbb{E}^2 - \{0\}$ and its Riemann metric, the function $\overline{U} = P(r)$, etc. Let us introduce polar "coordinates" in the plane. Then $\partial/\partial\theta$ preserves the metric and the potential. If $\langle r, \theta, p_r, p_\theta\rangle$ are the corresponding coordinates on $T^*(\overline{M})$, then Proposition 5.2 implies that

$$p_\theta = \text{const.} \tag{6.4}$$

(Note that this is really just that part of (6.1) that we haven't yet used.) Now in terms of polar coordinates, the Euclidean metric has the form of Eq. (8.7) of Chapter 9, so that the Riemann metric associated to the mass m which is $m \times$ the Euclidean metric is given by

$$\|(\dot{r}, \dot{\theta})\|^2 = m\{\dot{r}^2 + r^2\dot{\theta}^2\}.$$

In particular, the associated map from $T(M)$ to $T^*(M)$ is given by

$$p_r = m\dot{r} \qquad \text{and} \qquad p_\theta = mr^2\dot{\theta}. \tag{6.5}$$

Thus (6.4) says that

$$r^2\dot{\theta} = \text{const.} \tag{6.4$'$}$$

To understand the significance of (6.4′), consider a curve $\mathbf{x}(\cdot)$ in the plane. Consider the region $\mathfrak{U}$ bounded by the portion of the ray from 0 to $\mathbf{x}(0)$, the portion of the ray from 0 to $\mathbf{x}(t)$, and the curve $\mathbf{x}(\cdot)$ from 0 to t. (See Fig. 13.1.) Then

$$\tfrac{1}{2}\int_{\partial\mathfrak{U}} r^2\,d\theta = \int_{\mathfrak{U}} r\,dr \wedge d\theta$$

is the area of $\mathfrak{U}$. On the other hand, since $d\theta = 0$ on the rays, we see that

$$\int_{\partial\mathfrak{U}} r^2\,d\theta = \int_0^t r^2\dot{\theta}\,dt.$$

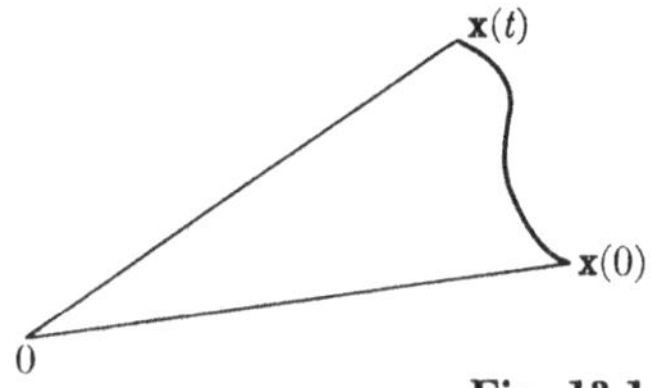

Fig. 13.1

Thus (6.4′) is exactly the content of Kepler's second law: *The particle sweeps out area at a constant rate.*

Thus we have seen that the spherical symmetry of the system implies that the motion is in a plane and that Kepler's second law holds.

We have not yet made use of the conservation of energy, which in the present context reads

$$\tfrac{1}{2}m\dot{r}^2 + \tfrac{1}{2}mr^2\dot{\theta}^2 + P(r) = \text{const} = \frac{1}{2m}p_r^2 + \frac{1}{2mr^2}p_\theta^2 + P(r). \tag{6.6}$$

Let us examine a particular solution curve for which $p_\theta = A = \text{const}$. Then differentiating (6.6) gives

$$\frac{dp_r}{dt} = m\frac{d^2r}{dt^2} = \frac{A^2}{mr^3} - P'(r). \tag{6.7}$$

We can interpret (6.7) as the equations of motion of a *one-dimensional* mechanical system. The kinetic energy of this system is $\tfrac{1}{2}m\dot{r}^2$, and the potential energy Q is given by

$$Q(r) = P(r) + \frac{A^2}{2mr^2}\cdot \tag{6.8}$$

The second term in (6.8) is known as the *centrifugal potential* and the corresponding term, A^2/mr^3, occurring in (6.7) is called the *centrifugal force*. Note that if $A = 0$, then (6.7) reduces to (6.3), which is what we would expect.

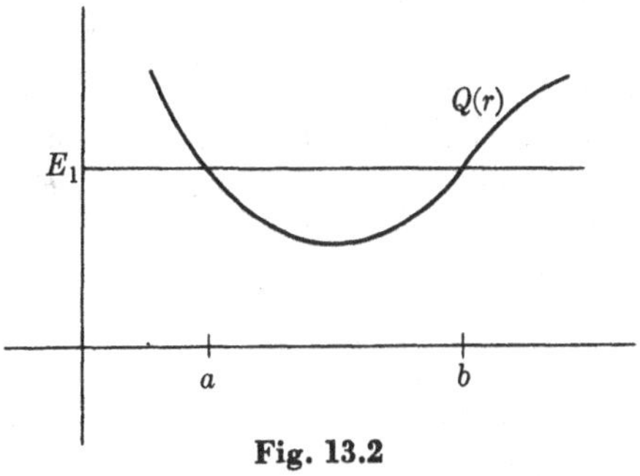

Fig. 13.2

We can now use the following procedure for solving the equations of motion: First, for a given angular momentum find the various solution curves $r(\cdot)$ of (6.7). For a solution $r(\cdot)$, determine $\theta(\cdot)$ by integrating $\dot{\theta} = A/mr^2$ to get

$$\theta(t) = \int_{t_0}^{t} \frac{A}{mr^2(s)}\,ds + \text{const}. \tag{6.9}$$

We can obtain a good bit of information about the nature of the solutions of (6.7) by using the law of conservation of energy. We draw a graph of the function, as shown in Fig. 13.2. Suppose that the trajectory $r(\cdot)$ has the constant energy $E_1 = \tfrac{1}{2}m\dot{r}^2 + Q(r)$. Then if the set $\{r : Q(r) \leq E_1\}$ is bounded, $r(t)$ must always be in this set, since $\tfrac{1}{2}m\dot{r}^2 \geq 0$. In fact, suppose that the interval $a \leq r \leq b$ is such that $Q(a) = Q(b) = E_1$ and $Q(b+\epsilon) > E_1$ and $Q(a-\epsilon) > E_1$ for small ϵ. Then if $a \leq r(t_0) \leq b$ for some value of t_0, it follows that $a \leq r(t) \leq b$ for all t. Furthermore, if $Q(r) < E_1$ for $a < r < b$, then for $a < r(t_0) < b$, $\dot{r}(t)$ cannot vanish. The particle is thus moving to one of the limits $r = a$ and $r = b$. If we use the law of conservation of energy, we see that

$$\tfrac{1}{2}m\dot{r}^2 + Q(r) = E_1,$$

so that

$$\frac{dr}{dt} = \dot{r} = \pm\sqrt{\frac{E_1 - Q(r)}{\frac{1}{2}m}}.$$

For $a < r < b$ we see that r is a monotone function of t, and we can solve to obtain t as a function of r by the formula

$$t(r) = \pm(\tfrac{1}{2}m)^{1/2}\int_{t_0}^{r}\frac{dx}{\sqrt{E_1 - Q(x)}} + t_0 \qquad \text{if} \quad r(t_0) = r_0. \tag{6.10}$$

In particular, if $Q'(a) \neq 0$ and $Q'(b) \neq 0$, the integral in (6.10) converges, so that it takes a finite amount of time for r to get to a or to b. Thus (since $d\dot{r}/dt = Q'(r) \neq 0$), the function r oscillates between a and b taking the time

$$T = (\tfrac{1}{2}m)^{1/2}\int_a^b \frac{dx}{\sqrt{E_1 - Q(x)}}$$

to get from one side to the other.

If $\{r : Q(r) \leq E_2\}$ is not bounded, then the motion with energy E_2 need not be bounded. For instance, suppose that $Q(r) < E_2$ for $r > a$. Then if $r(t_0) > a$ and $\dot{r}(t_0) < 0$, r will decrease until it hits a at which time it will turn around and then go off to infinity; if $\dot{r}(t_0) > 0$, then r will simply go to infinity. The trajectory in this case is that of r coming in from infinity, turning around at a, and then going back to infinity.

We recall that this is a descriptive analysis of the function $r(\cdot)$. The curve $\theta(\cdot)$ is then to be determined by (6.9).

Frequently we are interested in the tajectories as curves in the $r\theta$-plane without reference to the time dependence. Suppose that the energy is E and the angular momentum is A. For the range where $r \neq 0$ we can substitute

$$\frac{d\theta}{dt} = \frac{A}{mr^2}$$

into (6.10) to obtain [since $d\theta/dr = (d\theta/dt)(dt/dr)$]

$$\theta(r) = \pm\left(\frac{1}{2m}\right)^{1/2}\int_{r_0}^{r}\frac{A\,dx}{x^2\sqrt{E - Q(x)}} + \theta(r_0).$$

In the important case where $P(r) = -\alpha/r$ corresponding to the inverse-square law of gravitational attraction, we have

$$Q(x) = \frac{A^2}{2mx^2} - \frac{\alpha}{x},$$

so that the integrand is

$$\frac{1}{(2m)^{1/2}}\cdot\frac{A}{x^2\sqrt{E-\dfrac{A^2}{2mx^2}+\dfrac{\alpha}{x}}}=\frac{A}{x^2\left[2mE-\left(\dfrac{A}{x}-\dfrac{m\alpha}{A}\right)^2+\dfrac{m^2\alpha^2}{A^2}\right]^{1/2}}$$

$$=\frac{d}{dx}\left|\arccos\frac{\dfrac{A}{x}-\dfrac{m\alpha}{A}}{\sqrt{2mE+\dfrac{m^2\alpha^2}{A^2}}}\right|.$$

Thus

$$\theta(r)-\theta(r_0)=\arccos\frac{\dfrac{A}{r}-\dfrac{m\alpha}{A}}{\sqrt{2mE+\dfrac{m^2\alpha^2}{A^2}}}.$$

Let us choose r_0 to be the minimum value of r, and let us choose $\theta(r_0)=0$. Let us set

$$p=\frac{A^2}{m\alpha}\qquad\text{and}\qquad e=\sqrt{1+\frac{2EA^2}{m\alpha^2}}.$$

We see that the equations for the orbit are

$$\frac{p}{r}=1+e\cos\theta. \tag{6.11}$$

This is the equation of a conic section (Kepler's first law). If $E<0$, then $e<1$ and (6.11) represents an ellipse, whose major and minor semiaxes are

$$a=\frac{p}{1-e^2}=\frac{\alpha}{2|E|}$$

and

$$b=\frac{p}{\sqrt{1-e^2}}=\frac{A}{\sqrt{2m|E|}}.$$

We leave to the readers the details of working out the hyperbolic and parabolic orbits.

For the elliptic orbit, Kepler's second law implies that the total area of the interior of the ellipse is swept out at the uniform rate $A/2m$. Thus the area, πab, of the ellipse is given by $(A/2m)T$. Thus

$$\frac{A}{2m}T=\pi ab=\pi a\frac{A}{\sqrt{2m|E|}},$$

and hence

$$T=2\pi a^{3/2}\sqrt{m/\alpha}.$$

In other words, the square of the period of motion is proportional to the cube of the linear dimension of the orbit (Kepler's third law).

7. THE TWO-BODY PROBLEM

Let us consider the mechanical system consisting of n particles each having mass m_i, so that the configuration space is $\mathbb{E}^3 \times \cdots \times \mathbb{E}^3$ (n times), and if $a = \langle a^1, \ldots, a^n \rangle \in M$, where $a^1 = \langle x^1, y^1, z^1 \rangle \in \mathbb{E}^3$, then

$$\begin{aligned}\tfrac{1}{2}\|\dot{a}\|^2 &= \tfrac{1}{2}m_1\|\dot{a}^1\|^2 + \cdots + \tfrac{1}{2}m_n\|\dot{a}^n\|^2\\ &= \tfrac{1}{2}m_1[(\dot{x}^1)^2 + (\dot{y}^1)^2 + (\dot{z}^1)^2] + \cdots + \tfrac{1}{2}m_n[(\dot{x}^n)^2 + (\dot{y}^n)^2 + (\dot{z}^n)^2]\end{aligned}$$

is the kinetic energy of the system. As we indicated in Section 1, we may wish to restrict the manifold M to be that subset of $\mathbb{E}^3 \times \cdots \times \mathbb{E}^3$ for which no $a_i = a_j$ for any $i \neq j$. Let us further assume that the potential energy depends only on the mutual distances between the particles. That is, let us assume that

$$\overline{U}(a^1, \ldots, a^n) = P(\|a_2 - a_1\|, \|a_j - a_1\|, \ldots, \|a_n - a_{n-1}\|).$$

Then if A is a Euclidean motion of $\mathbb{E}^3$ and we apply A simultaneously to all the particles, the kinetic and potential energy are conserved. That is, the transformation

$$\langle a^1, \ldots, a^n \rangle \mapsto \langle Aa^1, \ldots, Aa^n \rangle$$

is an isometry of the Riemann metric on M and preserves $\overline{U}$.

Let $\langle x^1, y^1, z^1, \ldots, x^n, y^n, z^n \rangle$ be the Cartesian coordinates on M, and let

$$\langle q_{x1}, q_{y1}, q_{z1}, q_{x2}, \ldots, q_{yn}, q_{zn}, p_{x1}, \ldots, p_{zn} \rangle$$

be the corresponding coordinates on $T^*(M)$. Now $\partial/\partial x$ is the vector field representing the infinitesimal translation in the x-direction in $\mathbb{E}^3$. Therefore,

$$\frac{\partial}{\partial x^1} + \frac{\partial}{\partial x^2} + \cdots + \frac{\partial}{\partial x^n}$$

is the infinitesimal generator of the one-parameter group

$$\begin{aligned}&\langle x^1, y^1, z^1, x^2, \ldots, x^n, y^n, z^n \rangle\\ &\qquad \mapsto \langle x^1 + t, y^1, z^1, x^2 + t, y^2, z^2, \ldots, x^n + t, y^n, z^n \rangle.\end{aligned}$$

We can therefore apply Proposition 5.2 to conclude that the function

$$p_{x1} + p_{x2} + \cdots + p_{xn}$$

is constant in any trajectory. This function is known as the total linear momentum in the x-direction. Similarly, the total linear momentum in the y-direction,

$$p_{y1} + \cdots + p_{yn},$$

and the total linear momentum in the z-direction,

$$p_{z1} + \cdots + p_{zn},$$

must be conserved. If we define $p^i \in \mathbb{E}^3$ by setting $p^i = \langle p_{xi}, p_{yi}, p_{zi} \rangle$, then we can say that the $\mathbb{E}^3$-valued function

$$p^1 + p^2 + \cdots + p^n$$

must be conserved. This is the law of conservation of total linear momentum. (For two particles the assertion that $p_1 + p_2$ is conserved is just Newton's law of "equality of action and reaction". In our setup we see that this law is a reflection of the invariance of the physical situation under translations.)

The vector field $x(\partial/\partial y) - y(\partial/\partial x)$ represents infinitesimal rotation about the z-axis in $\mathbb{E}^3$. Therefore,

$$x^1 \frac{\partial}{\partial y^1} - y^1 \frac{\partial}{\partial x^1} + x^2 \frac{\partial}{\partial y^2} - y^2 \frac{\partial}{\partial x^2} + \cdots + x^n \frac{\partial}{\partial y^n} - y^n \frac{\partial}{\partial x^n}$$

represents simultaneous infinitesimal rotation of all the particles about the z-axis. Therefore, the function

$$q_{x^1}p_{y^1} - q_{y^1}p_{x^1} + \cdots + q_{x^n}p_{y^n} - q_{y^n}p_{x^n}$$

is conserved. This is the law of conservation of total angular momentum about the z-axis. Similarly, we obtain the law of conservation of total angular momentum about the x- and y-axes. If we set

$$q^i = \langle q_{x^i}, q_{y^i}, q_{z^i} \rangle \in \mathbb{E}^3,$$

we can combine the three equations by saying that the function

$$q^1 \wedge p^1 + \cdots + q^n \wedge p^n$$

with values in $\bigwedge^2(\mathbb{E}^3)$ is constant on the trajectories of the motion.

Let us examine more closely the law of conservation of total linear momentum. In view of the fact that

$$p_{x^i} = m_i \frac{dx^i}{dt},$$

etc., we have $p^i = m_i(da^i/dt)$, and thus

$$\frac{d}{dt}(m_1a^1 + m_2a^2 + \cdots + m_na^n) = \text{const.}$$

In other words, the center of mass

$$C = \frac{m_1a^1 + \cdots + m_na^n}{m_1 + \cdots + m_n}$$

moves in a straight line with constant velocity.

Suppose there are only two particles. Then it is reasonable to introduce the center of mass

$$C = \frac{m_1a^1 + m_2a^2}{m_1 + m_2}$$

and the relative position vector

$$d = a^1 - a^2$$

as new coordinates. If we solve for a^1 and a^2, we get

$$a^1 = \frac{m_2 d}{m_1 + m_2} + C, \qquad a^2 = -\frac{m_1 d}{m_1 + m_2} + C,$$

and therefore the kinetic energy is given by

$$K(C, \dot{d}) = \frac{1}{2}\frac{m_1 m_2}{m_1 + m_2}\|\dot{d}\|^2 + \tfrac{1}{2}(m_1 + m_2)\|C\|^2,$$

while $\overline{U}(C, d) = P(\|d\|)$.

Thus the motions of the system have the following description: The center of mass has constant linear motion, and the relative position vector $d = a^1 - a^2$ satisfies the equations of motion of a single particle with mass

$$\frac{m_1 m_2}{m_1 + m_2}$$

in the central-force field with potential $U = P(\|d\|)$. We can thus apply the results of the preceding section to determine the motions of the particle. In particular, if $P(\|d\|) = \alpha/\|d\|$ is the inverse-square potential, the corresponding two-body problems can be completely solved.

Note that if m_2 is very large compared to m_1, then C is very close to a^2. In this case $d(\cdot)$ is a good approximation to the motion of the smaller particle relative to the larger one.

This is the situation that arises, for example, in the study of the motion of of the planets relative to the sun.

8. LAGRANGE'S EQUATIONS

Our discussion of mechanics has led us to a certain kind of vector field X on $T^*(M)$, where M is the configuration manifold. In the case where $H = K + U$, the Riemann metric giving K induces an isomorphism $\mathcal{L}$ [that is, a diffeomorphism which is linear on each $T_x(M)$] from $T(M)$ to $T^*(M)$. We therefore obtain a vector field Y on $T(M)$ such that $\mathcal{L}_* Y_v = X_{\mathcal{L}(v)}$ at any $v \in T(M)$. We can therefore inquire about the form of this vector field Y in terms of local coordinates $\langle q^1, \ldots, q^n, \dot{q}^1, \ldots, \dot{q}^n \rangle$ on $T(M)$, associated with coordinates $\langle x^1, \ldots, x^n \rangle$ on M. Suppose that the Riemann metric is given by

$$\|(\dot{q}^1, \ldots, \dot{q}^n)\|^2 = \sum g_{ij}(q^1, \ldots, q^n)\dot{q}^i\dot{q}^j.$$

If $\langle q^1, \ldots, q^n, p^1, \ldots, p^n \rangle$ are corresponding local coordinates on $T^*(M)$, then the diffeomorphism $\mathcal{L}$ is given by

$$q^i = q^i, \qquad p^i = \sum g_{ij}(q^1, \ldots, q^n)\dot{q}^j. \tag{8.1}$$

We could proceed to use (8.1) to obtain the local expression for $\mathcal{L}$ and thereby the local expression for Y. However, it is more convenient to argue a little differently. Let L be the function on $T(M)$ defined by

$$L(q^1, \ldots, q^n, \dot{q}^1, \ldots, \dot{q}^n) = \tfrac{1}{2}\sum g_{ij}\dot{q}^i\dot{q}^j - \overline{U}(q^1, \ldots, q^n), \tag{8.2}$$

that is,

$$L = K - U,$$

where $K(v) = \frac{1}{2}\|v\|^2$ and $U(v) = \overline{U}(\pi(v))$ are the kinetic and potential energy expressed as functions on $T(M)$. We can rewrite (8.1) as

$$q^i = q^i, \qquad p^i = \frac{\partial L}{\partial \dot{q}^i}. \tag{8.3}$$

Then for any $l \in T^*(M)$, we have

$$H(l) = \langle \mathcal{L}^{-1}(l), l\rangle - L(\mathcal{L}^{-1}(l)). \tag{8.4}$$

In fact, by definition,

$$\langle \mathcal{L}^{-1}(l), l\rangle = \|l\|^2 = \|\mathcal{L}^{-1}(l)\|^2,$$

so that

$$\begin{aligned}\langle \mathcal{L}^{-1}(l), l\rangle - L(\mathcal{L}^{-1}(l)) &= \|l\|^2 - \tfrac{1}{2}\|l\|^2 + \overline{U}(\pi(l))\\ &= \tfrac{1}{2}\|l\|^2 + U(l)\\ &= H(l).\end{aligned}$$

In terms of local coordinates we can write (8.4) as

$$H(q^1, \ldots, q^n, p^1, \ldots, p^n) = \sum p^i\dot{q}^j - L(q^1, \ldots, q^n, \dot{q}^1, \ldots, \dot{q}^n), \tag{8.5}$$

where the $\dot{q}^i$ are regarded as functions of the q's and p's via $\mathcal{L}^{-1}$. We can write the map $\mathcal{L}^{-1}$ as

$$q^i = q^i, \qquad \dot{q}^i = \frac{\partial H}{\partial p^i}, \tag{8.6}$$

since $H(l) = \frac{1}{2}\|l\|^2 + \overline{U}(\pi(l))$. Furthermore, by (8.5),

$$\begin{aligned}\frac{\partial H}{\partial q^i} &= \frac{\partial}{\partial q^i}[\textstyle\sum \dot{q}^i p^i - L(q^1, \ldots, q^n, \dot{q}^1, \ldots, \dot{q}^n)]\\ &= \sum \frac{\partial(\dot{q}^i \circ \mathcal{L}^{-1})}{\partial q^i} p^i - \frac{\partial(L \circ \mathcal{L}^{-1})}{\partial q^i} - \frac{\partial L}{\partial q^i}\frac{\partial \dot{q}^i \circ \mathcal{L}^{-1}}{\partial q^i}\\ &= -\frac{\partial L \circ \mathcal{L}^{-1}}{\partial q^i}\end{aligned}$$

by (8.4).

Now let $v(\cdot) = \langle q^1(\cdot), \ldots, q^n(\cdot), \dot{q}^1(\cdot), \ldots, \dot{q}^n(\cdot)\rangle$ be a solution curve to the vector field Y, so that $\mathcal{L} \circ v(\cdot)$ is a solution curve of X on $T^*(M)$. Then by (5.1),

$$\frac{dq^i}{dt} = \frac{\partial H}{\partial p^i} = \dot{q}^i$$

and

$$\frac{dp^i}{dt} = \frac{d(\partial L/\partial \dot{q}^i)}{dt} = -\frac{\partial H}{\partial q^i} = \frac{\partial L}{\partial q^i}.$$

In other words, Eqs. (5.1) are equivalent to the equations

$$\frac{dq^i}{dt} = \dot{q}^i, \qquad \frac{d(\partial L/\partial \dot{q}^i)}{dt} - \frac{\partial L}{\partial q^i} = 0 \tag{8.7}$$

on $T(M)$. The first of Eqs. (8.7) says that we are dealing with a system of second-order differential equations which is given implicitly by the second of Eqs. (8.7). Equations (8.7) are known as Lagrange's equations.

For certain purposes Lagrange's equations are very convenient. We illustrate by establishing the "principle of mechanical similarity". Note that Eqs. (8.7) are unchanged if we replace L by cL, where c is any nonzero constant. Suppose that M is an open subset of a vector space with linear coordinates $x^1, \ldots, x^n$ and that the Riemann metric is given by $\sum g_{ij}\dot{q}^i\dot{q}^j$, where the g_{ij} are constant. Let m_α denote the linear map consisting of multiplication by $\alpha > 0$, that is, $m_\alpha(x^1, \ldots, x^n) = \langle \alpha x^1, \ldots, \alpha x^n \rangle$. Suppose that $\overline{U}$ is a homogeneous function of degree p, so that

$$\overline{U}(\alpha x^1, \ldots, \alpha x^n) = \alpha^p \overline{U}(x^1, \ldots, x^n).$$

Now let us change our time scale by a factor β, replacing t by $s = \beta t$. Then

$$\frac{d\alpha q^i}{ds} = \frac{\alpha}{\beta}\frac{dq^i}{dt},$$

and

$$\frac{1}{2}\sum g_{ij}\frac{d\alpha q^i}{ds}\frac{d\alpha q^j}{ds} = \frac{\alpha^2}{\beta^2}\frac{1}{2}\sum g_{ij}\frac{dq^i}{dt}\frac{dq^j}{dt}.$$

Let us choose β so that $\alpha^2/\beta^2 = \alpha^p$, that is,

$$\beta = \alpha^{1-(1/2)p}.$$

Then

$$L\left(\alpha q^1, \ldots, \alpha q^n, \frac{\alpha\, dq^1}{ds}, \ldots, \frac{\alpha\, dq}{ds}\right) = \alpha^p L\left(q^1, \ldots, q^n, \frac{dq^1}{dt}, \ldots, \frac{dq^n}{dt}\right).$$

In other words, replacing q^i by αq^i and t by βt carries solutions into solution: if we change the linear scale by α and change the time scale by $\alpha^{1-(1/2)p}$, we obtain an isomorphic situation. For instance, if U is homogeneous of degree -1 (as in the case of an inverse-square law of attraction), then $\beta = \alpha^{3/2}$. In particular, the period of any periodic orbit is proportional to the $\frac{3}{2}$-power of its linear dimension, which is just Kepler's third law. We thus see that Kepler's third law is an exact consequence of the inverse-square law of attraction.

Returning to the study of the general Lagrangian system, we observe again that it is a system of second-order differential equations. We can therefore apply the fundamental existence and uniqueness theorem to it to conclude that *for every $x \in M$ and for every $\xi \in T_x(M)$ there is a unique curve $C(\cdot)$ which is a trajectory of the system and for which $C(0) = x$ and $C'(0) = \xi$.*

9. VARIATIONAL PRINCIPLES

The function L plays a crucial role in the study of variational principles of mechanics. Consider the following problem: Let p and q be two points of M, and let $t_1 < t_2$ be two real numbers. For any differentiable curve C defined on the interval $[t_1, t_2]$ we set

$$I[C] = \int_{t_1}^{t_2} L(C'(t))\, dt. \tag{9.1}$$

Note that $C'(t) \in T(M)$ and L is a function on $T(M)$, so the integrand makes sense. Among all curves joining p to q, find that curve for which $I[C]$ takes on its minimum value. We shall see that a necessary condition for $I[C]$ to be a minimum is that it be a trajectory of a mechanical system. (In fact, if a suitable notion of neighborhood is introduced on the space of curves, it is also a necessary condition for C to be even a local minimum.)

Before establishing this result, it is convenient to have another expression for $I[C]$. As before, let $\mathcal{L}$ be the map of $T(M) \to T^*(M)$ given by the Riemann metric. Let $\overline{C}$ be the curve in $T^*(M)$ given by

$$\overline{C}(t) = \mathcal{L}(C'(t)).$$

Then

$$I[C] = \int_{\overline{C}} \theta - \int_{t_1}^{t_2} H(\overline{C}(t))\, dt, \tag{9.2}$$

where θ is the fundamental linear form on $T^*(M)$.

In fact, in terms of local coordinates,

$$\int_{\overline{C}} \theta = \sum \int_{\overline{C}} p^i\, dq^i = \sum \int_{C'} \frac{\partial L}{\partial \dot{q}^i}\, dq^i = \sum \int \frac{\partial L}{\partial \dot{q}^i} \dot{q}^i(t)\, dt,$$

since the curve C' by definition is such that

$$\dot{q}^i(t) = \frac{dq^i}{dt}(t).$$

But, by (8.5),

$$\sum \int \frac{\partial L}{\partial \dot{q}^i} \dot{q}^i(t)\, dt = \sum \int (p^i \circ \mathcal{L})(t)\dot{q}^i(t)\, dt = \int [H \circ \mathcal{L}(C'(t)) + L(C'(t))]\, dt,$$

so (9.2) holds.

Let Z be a vector field on M which generates a flow φ. For all sufficiently small s the curve $\varphi_s \circ C(\cdot)$ will be defined and

$$(\varphi_s \circ C)'(t) = \varphi_{s*}C'(t) = T[\varphi_s](C'(t)),$$

so that

$$\begin{aligned} I[\varphi_s \circ C] &= \int_{t_1}^{t_2} L(T[\varphi_s]C'(t))\, dt \\ &= \int_{\mathcal{L} \circ \varphi_s \circ C'} \theta - \int_{t_1}^{t_2} H(\mathcal{L} \circ T(\varphi_s)(C'(t)))\, dt. \end{aligned} \tag{9.3}$$

Since $I[C]$ is to be a minimum, we must have

$$\frac{dI[\varphi_s \circ C]}{ds} = 0.$$

We will now compute this derivative so as to derive the consequences of the above equation.

Now $\mathcal{L} \circ T[\varphi_s] \circ \mathcal{L}^{-1}$ is a flow on $T^*(M)$ which satisfies

$$\pi \circ \mathcal{L} \circ T(\varphi_s) \circ \mathcal{L}^{-1} = \varphi_s.$$

Let $\overline{Z}$ be the infinitesimal generator of this flow, so that at all points of $T^*(M)$ we have

$$\pi_*\overline{Z} = Z. \tag{9.4}$$

If we differentiate (9.3) with respect to s at $s = 0$, we obtain

$$\frac{d}{ds} I[\varphi_s \circ C] = \int_{\overline{C}} D_{\overline{Z}}\theta - \int_{t_1}^{t_2} D_{\overline{Z}}H(\overline{C}(t))\, dt.$$

Now

$$D_{\overline{Z}}\theta = d\langle \overline{Z}, \theta\rangle + \overline{Z} \lrcorner\, d\theta \qquad \text{and} \qquad \langle \overline{Z}, \theta\rangle_l = \langle \pi_*\overline{Z}, l\rangle = \langle Z, l\rangle = f_Z(l),$$

so we get

$$\frac{d}{ds} I[\varphi_s \circ C] = \int_{\overline{C}} \overline{Z} \lrcorner\, d\theta - \int_{t_1}^{t_2} D_{\overline{Z}}H(\overline{C}(t))\, dt + f_Z(\overline{C}(t_2)) - f_Z(\overline{C}(t_1)). \tag{9.5}$$

Now suppose that the vector field Z vanishes at p and q. Then all the curves $\varphi_s \circ C$ join p to q. If C is to minimize the integral I, then the derivative $d(I[\varphi_s \circ C])/ds$ must vanish at $s = 0$. Note that in this case the two last terms of (9.5) vanish and we must have

$$\int_{\overline{C}} \overline{Z} \lrcorner\, d\theta - \int_{t_1}^{t_2} D_Z H(\overline{C}(t)) = 0 \tag{9.6}$$

for all vector fields vanishing at p and q. In particular, let us take

$$\begin{aligned} Z &= \psi \frac{\partial}{\partial x^i} \qquad &&\text{for} \quad x \in U, \\ &= 0 &&\text{for} \quad x \notin U, \end{aligned}$$

where ψ is a C^∞-function whose support lies in some coordinate neighborhood U of M with coordinates $\langle x^1, \ldots, x^n \rangle$. Suppose further that $\psi(p) = \psi(q) = 0$ if $p \in U$ or $q \in U$. Then by (2.7′)

$$T(Z) = \psi \frac{\partial}{\partial q^i} + \sum \frac{\partial \psi}{\partial q^j} \frac{\partial}{\partial \dot{q}^j},$$

so that

$$\overline{Z} = \mathcal{L}_* T(Z) = \psi \frac{\partial}{\partial q^i} + \sum B^j \frac{\partial}{\partial p^j},$$

where B^j are some functions on $T^*(M)$ which depend linearly on Z. [This, of course, is just a restatement of (9.4).] Then

$$\overline{Z} \lrcorner\, d\theta = \overline{Z} \lrcorner \sum dp^i \wedge dq^i = \sum_j B^j\, dq^j - \psi\, dp^i,$$

while

$$D_{\bar{Z}}H = \psi \frac{\partial H}{\partial q^i} + \sum_j B^j \frac{\partial H}{\partial p^j}.$$

Thus if the curve $\bar{C}$ is given by $\bar{C}(t) = \langle q^1(t), \ldots, q^n(t), p^1(t), \ldots, p^n(t) \rangle$, Eq. (9.6) becomes

$$\int \left[\sum B^j \left(\frac{dq^j}{dt} - \frac{\partial H}{\partial p^j} \right) - \psi \left(\frac{dp^i}{dt} + \frac{\partial H}{\partial q^i} \right) \right] dt = 0.$$

Now by construction, $\bar{C}(t) = \mathcal{L} \circ C'(t)$, so on $\bar{C}$ we have

$$\frac{dq^j}{dt} = \dot{q}^j = \frac{\partial H}{\partial p^j} \tag{9.7}$$

by (8.6). Thus the first sum occurring in the above integral vanishes and we must have

$$\int \psi \left(\frac{dp^i}{dt} + \frac{\partial H}{\partial q^i} \right) dt = 0.$$

This must hold for all functions ψ whose support lies in a coordinate neighborhood and vanishes at p and q. Clearly, this can happen only if

$$\frac{dp^i}{dt} = -\frac{\partial H}{\partial q^i}. \tag{9.8}$$

This must hold for all i. Since (9.6) and (9.8) are exactly (5.2), we can assert:

Proposition 9.1. A necessary condition for C to minimize the functional $I[C]$ is that C is a trajectory of the corresponding dynamical system.

The question of when this necessary condition is also sufficient is a more complicated one. We shall not discuss it in any generality here, but refer the reader to any standard source on the calculus of variations.

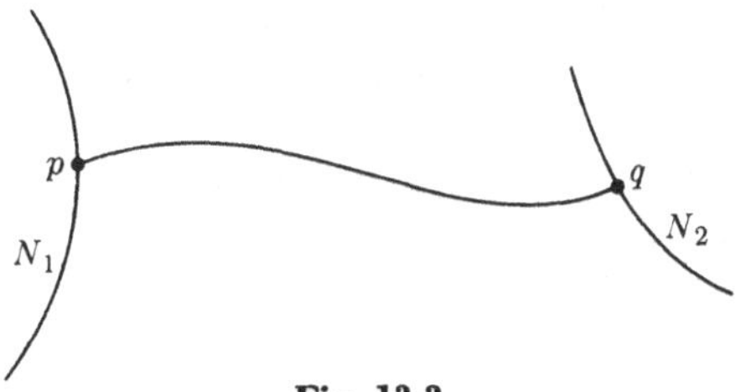

Fig. 13.3

By the way, we can derive a bonus from (9.5). Suppose that we consider the following problem: Let N_1 and N_2 be two submanifolds (Fig. 13.3) of M, and suppose that we require that C minimize I among all curves joining N_1 to N_2 and not merely among those joining p to q. In this case (9.5) will have to vanish for all vector fields Z which are tangent to N_1 at p and to N_2 at q. Now observe that if C is a solution to this minimum problem, it certainly is a solution to the problem of minimizing I among all curves joining p to q. In particular, C must be a trajectory of the mechanical system. As the reader can easily check, this implies that

$$\int_{\overline{C}} \overline{Z} \lrcorner \theta - \int (D_{\overline{Z}}H)(\overline{C}(t))\, dt = 0$$

for all vector fields Z. Thus, if C solves the more difficult minimization problem, we must have

$$f_Z(\overline{C}(t_1)) - f_Z(\overline{C}(t_2)) = 0.$$

Now

$$f_Z(\overline{C}(t_1)) = \langle Z_p, \overline{C}(t_1)\rangle \qquad \text{and} \qquad f_Z(\overline{C}(t_2)) = \langle Z_q, \overline{C}(t_2)\rangle.$$

Since if $p \neq q$ we can choose Z_p arbitrarily in $T_p(N_1)$ and Z_q arbitrarily in $T_q(N_2)$, we conclude in this case that C must also satisfy

$$\begin{aligned} \langle \xi, \overline{C}(t_1)\rangle &= 0 \qquad \text{for all} \quad \xi \in T_p(N_1), \\ \langle \eta, \overline{C}(t_2)\rangle &= 0 \qquad \text{for all} \quad \eta \in T_q(N_2). \end{aligned} \tag{9.9}$$

Since $\overline{C}(t_1) = \mathcal{L}C'(t_1)$, we have

$$\langle \xi, \overline{C}(t_1)\rangle = \langle \xi, \mathcal{L}C'(t_1)\rangle = (\xi, C'(t_1)),$$

so we can write (9.9) as

$$\begin{aligned} (\xi, C'(t_1)) &= 0 \qquad \text{for all} \quad \xi \in T_p(N_1), \\ (\eta, C'(t_2)) &= 0 \qquad \text{for all} \quad \eta \in T_q(N_2). \end{aligned} \tag{9.10}$$

In other words, the curve C must be orthogonal to the submanifolds N_1 and N_2.

Although our statement of Proposition 9.1 was couched in the framework of dynamical systems, it actually can be formulated in a more general context. Let L be any function on $T(M)$—not necessarily of the type $K - U$. We can then define the integral I as in (9.1) and again pose the minimization problem. This is the typical problem in the calculus of variations. We have already discussed this problem from a different point of view in Section 3.15. We leave to the reader the task of showing that our arguments carry over in this more general case if the matrix

$$\left(\frac{\partial L}{\partial \dot{x}^i\, \partial \dot{x}^j}\right)$$

is nowhere singular. (Here the map $\mathcal{L}$ of $T(M) \to T^*(M)$ is given by

$$\langle x^1, \ldots, x^n, \dot{x}^1, \ldots, \dot{x}^n\rangle \mapsto \left\langle x^1, \ldots, x^n, \frac{\partial L}{\partial \dot{x}^1}, \ldots, \frac{\partial L}{\partial \dot{x}^n}\right\rangle,$$

and the nonsingularity guarantees that this map is locally a diffeomorphism.)

10. GEODESIC COORDINATES

In this section we depart momentarily from the study of mechanics in order to exhibit some applications of the results of the preceding paragraphs to the study of Riemann manifolds. Note that a Riemann manifold always defines a mechanical system by its kinetic energy if we set the potential energy equal to zero. It is this special kind of mechanical system that we wish to study in this section.

Let M be a finite-dimensional manifold with a Riemann metric and, as above, define L on $T(M)$ by setting

$$L(v) = \tfrac{1}{2}\|v\|^2.$$

This then determines a vector field Y on $T(M)$ which corresponds to the system of differential equations (8.7) in terms of local coordinates. Let p be a point of M. For every $\xi \in T_p(M)$ there is a unique trajectory $C_\xi(\cdot)$ such that

$$C_\xi(0) = p, \qquad C'_\xi(0) = \xi.$$

In terms of local coordinates, if $\xi = \langle \xi^1, \ldots, \xi^n \rangle$, then

$$C_\xi(t) = \langle q^1_\xi(t), \ldots, q^n_\xi(t) \rangle,$$

where q^i_ξ are the unique solutions of the differential equations

$$\frac{dq^i_\xi}{dt} = \dot{q}^i_\xi, \qquad \frac{d(\partial L/\partial \dot{q}^i)}{dt} - \frac{\partial L}{\partial q^i} = 0$$

with the initial conditions

$$q^i_\xi(0) = x^i(p), \qquad \dot{q}^i_\xi(0) = \xi^i.$$

By the fundamental existence and uniqueness theorem the functions q^i_ξ are defined for sufficiently small t. We can regard $C_\xi(t)$ as dependent on both ξ and t. In other words, we have a map $C.(\cdot)$ assigning to each $\xi \in T_p(M)$ and each sufficiently small t a point of M. The map $C.(\cdot)$ is, in fact, defined in some neighborhood of $T_p(M) \times \{0\} \subset T_p(M) \times \mathbb{R}$.

The above is true for any Lagrangian function. In the case of no potential energy we can say a lot more. Let s be a real number. Consider the curve $C_\xi(s\,\cdot)$, that is,

$$C_\xi(st) = \langle q^1_\xi(st), \ldots, q^n_\xi(st) \rangle.$$

Then (suppressing the subscript ξ which is to be understood in the following computations)

$$\frac{dq^i(st)}{dt} = s\dot{q}^i(st),$$

$$\frac{\partial L}{\partial q^i}\left(q^1(st), \ldots, q^n(st), s\dot{q}^1(st), \ldots, s\dot{q}^n(st)\right) = \frac{1}{2}\, s^2 \sum \frac{\partial g_{kl}}{\partial q^i}\, \dot{q}^n(st)\dot{q}^l(st),$$

and

$$\frac{d}{dt}\left[\frac{\partial L}{\partial \dot{q}^i}\left(q^1(st), \ldots, q^n(st), s\dot{q}^1(st), \ldots, s\dot{q}^n(st)\right)\right]$$

$$= \frac{d}{dt} s \sum_k g_{ik}(q^1(st), \ldots, q^n(st))\dot{q}^k(st)$$

$$= s^2 \frac{d}{dt}\frac{\partial L}{\partial \dot{q}^i}\left(q^1(\cdot), \ldots, q^n(\cdot), \dot{q}^1(\cdot), \ldots, \dot{q}^n(\cdot)\right)\Big|_{st}.$$

In other words,

$$\frac{d}{dt}\frac{\partial L}{\partial \dot{q}^i}\left(q^1(s\,\cdot), \ldots, \dot{q}^n(s\,\cdot)\right) - \frac{\partial L}{\partial q^i}\left(q^1(s\,\cdot), \ldots, \ldots, s\dot{q}^n(s\,\cdot)\right)\Big|_t$$

$$= s^2\left[\frac{d}{dt}\frac{\partial L}{\partial \dot{q}^i}\left(q^1(\cdot), \ldots, \dot{q}^n(\cdot)\right) - \frac{\partial L}{\partial q^i}\left(q^1(\cdot), \ldots, \dot{q}^n(\cdot)\right)\right]\Big|_{st} = 0.$$

Thus the curve $C_\xi(s\,\cdot)$ is again a trajectory of the system, and we clearly have

$$C'_\xi(s\,\cdot)|_0 = sC'_\xi(\cdot)|_0 = s\xi.$$

By the uniqueness theorem for differential equations we thus must have

$$C_\xi(st) = C_{s\xi}(t). \tag{10.1}$$

We are therefore led to define a map, which we shall call exp, from $T_p(M) \to M$ by setting

$$\exp(\xi) = C_\xi(1). \tag{10.2}$$

Note that the map exp is defined and differentiable in some neighborhood of the origin in $T_p(M)$. In fact, by (10.1),

$$\exp(\xi) = C_{\xi/\|\xi\|}(\|\xi\|),$$

where now $\xi/\|\xi\|$ lies on the unit sphere in $T_p(M)$ [the unit sphere with respect to the Euclidean metric given by the scalar product on $T_p(M)$]. Since the unit sphere is compact, there is some $\epsilon > 0$ such that $C_\eta(t)$ is defined for all η on the unit sphere and all $|t| < \epsilon$. Thus exp will be defined for all $\|\xi\| < \epsilon$.

The map exp is a differentiable map from some neighborhood of the origin in the vector space $T_p(M)$ into the manifold. Let us compute

$$\exp_{*0}\colon T_0[T_p(M)] \to T_p(M).$$

Let $\xi \in T_p(M)$, and let us consider the straight line through 0, l_ξ, in $T_p(M)$ given by $l_\xi(t) = t\xi$. Since $T_p(M)$ is a vector space, we can identify $T_0[T_p(M)]$ with $T_p(M)$ via the identity chart, in which case we identify

$$l'_\xi(0) \quad \text{with} \quad \xi.$$

But

$$\exp[l_\xi(t)] = \exp(t\xi) = C_{t\xi}(1) = C_\xi(t), \tag{10.3}$$

and the tangent to this curve at 0 is just ξ. In other words,

$$\exp_{*0}[l'_\xi(0)] = \xi.$$

Thus, if we identify $T_0[T_p(M)]$ with $T_p(M)$, we can assert that

$$\exp_{*0}: T_0[T_p(M)] \to T_p(M)$$

is given by

$$\exp_{*0}(\xi) = \xi. \tag{10.4}$$

In particular, the map $\exp_{*0}$ is nonsingular, so that by the implicit-function theorem the map exp is a diffeomorphism in some neighborhood of the origin.

We have thus constructed a diffeomorphism exp from some neighborhood of the origin in $T_p(M)$ into M, which by (10.3) carries straight lines through the origin into trajectories through p. In the case of no potential energy the trajectories are called *geodesics* for reasons which will soon become apparent. If we identify $T_p(M)$ with V by some chart (U, α), we can then use exp to introduce a new chart (U', n_α) by setting

$$n_\alpha^{-1}(\xi_\alpha) = \exp(\xi).$$

The chart n_α has the property that $n_\alpha(p) = 0$ and n_α carries geodesics through p into straight lines through the origin. The chart (U', n_α) is called a *geodesic normal chart*, and corresponding coordinates are called *geodesic normal coordinates on M*.

Let us consider the curve $C_\xi(\cdot) = \exp(\cdot\ \xi)$ which is defined for $0 \le t \le 1$ so long as $\|\xi\| < \epsilon$. We have

$$I[C_\xi(\cdot)] = \int_0^1 L(C'_\xi(t))\,dt = \tfrac{1}{2}\int_0^1 \|C'_\xi(t)\|^2\,dt.$$

But, by the conservation of energy, for any trajectory of our system we have $H \circ \mathcal{L}(C'(t)) = \text{const}$. In this case, since $U = 0$, $H \circ \mathcal{L}(C'(t)) = L(C'(t)) = \frac{1}{2}\|C'_\xi(t)\|^2 = \text{const}$. Since $C'_\xi(0) = \xi$, we have $\|C'_\xi(t)\| = \|\xi\|$, and therefore

$$I[C_\xi(\cdot)] = \tfrac{1}{2}\int_0^1 \|\xi\|^2\,dt = \frac{\|\xi\|^2}{2}. \tag{10.5}$$

Now let $\{\beta_s\}$ be a one-parameter group of rotations in $T_p(M)$. Then

$$\varphi_s = \exp \circ \beta_s \circ \exp^{-1}$$

defines a one-parameter group on the open set $U = \exp\{v : \|v\| < \epsilon\} \subset M$. If $\|\xi\| < \epsilon$, we have by (10.5)

$$I[\varphi_s \circ C_\xi(\cdot)] = \frac{\|\beta_s\xi\|^2}{2} = \frac{\|\xi\|^2}{2},$$

so, by (9.10), we get

$$\frac{d}{ds} I[\varphi_s \circ C_\xi(\cdot)] = 0 = f_{Z_p}(\overline{C}(1)) = \langle C'(1), Z_{C(1)}\rangle,$$

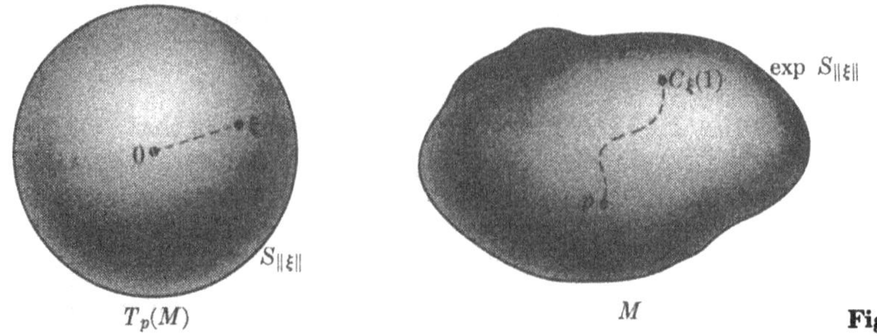

Fig. 13.4

where Z is the infinitesimal generator of φ. But $Z_{C(1)} = \exp_{*0} Y_\xi$, where Y is the vector field on $T_p(M)$ which is the infinitesimal generator of the one-parameter group of rotations. Now we can choose β arbitrarily. Therefore, Y_ξ can be *any* vector tangent to the sphere of radius $\|\xi\|$ in $T_p(M)$. We thus conclude that $(\eta, C'_\xi(1)) = 0$ for any η which is tangent to $\exp S_{\|\xi\|}$, where $S_{\|\xi\|}$ is the sphere of radius $\|\xi\|$ in $T_p(M)$. (See Fig. 13.4.) In other words, not only are the rays through the origin orthogonal to the spheres in the Euclidean metric of $T_p(M)$, but also the image of a ray through the origin under exp is orthogonal to the image of a sphere about p in the Riemann metric of M.

In particular, we can transfer "polar coordinates" on $T_p(M)$ to M so as to get "geodesic polar coordinates" on M. This has the following effect: Let r be the "radial coordinate", that is, r is the function defined in U by

$$r(x) = \|\exp^{-1} x\|.$$

Then for any $x \in U$, $x \neq p$, and any $\zeta \in T_x(M)$ we have

$$\|\zeta\| \geq |\langle \zeta, dr \rangle|, \tag{10.6}$$

with equality holding only if ζ is tangent to a geodesic through p. In fact, suppose that $\zeta \in T_x(M)$, where $x = \exp \xi$ for some $\xi \in T_p(M)$. Then we can write $\zeta = \zeta_1 + \zeta_2$, where ζ_1 is some multiple of $C'_\xi(1)$ and ζ_2 is tangent to $\exp S_{\|\xi\|}$. By the above result, $(\zeta_1, \zeta_2) = 0$, so

$$\|\zeta\|^2 = \|\zeta_1\|^2 + \|\zeta_2\|^2$$

and we obtain (10.6), with equality holding only if $\|\zeta_2\| = 0$.

We are now in the same fortunate position we were in Section 9 of Chapter 9. Let D be any curve joining p to x, where $x = \exp \xi$. Let t_1 be the first time that $D(t) \in \exp S_{\|\xi\|}$. Then the length of

$$\begin{aligned} D = \int_0^1 \|D'(t)\|\, dt &\geq \int_0^{t_1} \|D'(t)\|\, dt \geq \int_0^{t_1} |\langle D'(t), dr\rangle| \\ &\geq \int_0^{t_1} \langle D'(t), dr\rangle = r[D(t_1)] = \|\xi\|. \end{aligned}$$

Furthermore, equality holds only if $D'(t)$ is a nonnegative multiple of a tangent to a fixed geodesic through p. Then D must be the geodesic $C_\xi(\cdot)$. In short, we have proved:

Theorem 10.1. Let $\epsilon > 0$ be so small that the map exp is a diffeomorphism on $B_\epsilon = \{\xi \in T_p(M) : \|\xi\| < \epsilon\}$. Let $x = \exp \xi$ be a point of $U = \exp B_\epsilon$. Then the geodesic C_ξ joining p to x has length $\|\xi\|$, and any other curve D joining p to x is strictly longer unless D differs from C_ξ only in a (monotone) change of parameter. In other words, loosely speaking, geodesics are locally the shortest curves joining two points.

Since we have come this far, let us show in addition that geodesics also locally minimize the energy. Let D be any curve from $[0, 1]$ to M. Then by Schwarz's inequality we have

$$\left(\int_0^1 \|D'(t)\|\,dt\right)^2 \le \int_0^1 \|D'(t)\|^2\,dt \int_0^1 1\,dt$$
$$= 2I[D],$$

with equality holding only if $\|D'(t)\|$ is constant. If $C_\xi(\cdot)$ is the geodesic joining p to $x = \exp \xi$, we thus have

$$I[D] \ge \tfrac{1}{2}\left(\int_0^1 \|D'(t)\|\,dt\right)^2 \ge \tfrac{1}{2}\left(\int_0^1 \|C'(t)\|\,dt\right)^2 = \tfrac{1}{2}\|\xi\|^2.$$

Now equality holds in the second inequality only if $D'(t)$ is proportional to $C'(t)$, while equality holds in the first only if $\|D'(t)\| = $ const. We thus conclude that $\|D'(t)\| = \|\xi\|$, that is, $D'(t) = C'(t)$. In short, we have proved:

Proposition 10.1. Under the hypotheses of Theorem 10.1 the curve $C_\xi(\cdot)$ is a strict absolute minimum for $I[C]$ among all curves $C\colon [0, 1] \to M$ such that $C(0) = p$ and $C(1) = x$.

11. EULER'S EQUATIONS

Under certain circumstances, which we shall presently describe, the equations of motion take a particularly elegant form. The first special assumption that we shall make is that there is an isomorphism of $T(M)$ with $M \times V$. More precisely, we assume that there is a diffeomorphism, ι, of $T(M)$ with $M \times V$ such that $\iota(\xi) = \langle m, v \rangle$, where $m = \pi(\xi)$, and for each $x \in M$ the map $\xi \mapsto v$ of $T_x(M) \to V$ is a linear isomorphism of vector spaces. For example, if M is an open subset of V, then the identity chart defines such an isomorphism

$$\iota(\xi) = \langle \pi(\xi), \xi_{\mathrm{id}} \rangle.$$

A slightly less trivial example is furnished by the n-dimensional torus $T^n = S^1 \times \cdots \times S^1$. Then we can introduce "angular variables" $\theta^1, \ldots, \theta^n$, where θ^i is the angular variable on the ith circle. We thus obtain n vector fields $\partial/\partial\theta^1, \ldots, \partial/\partial\theta^n$ which are linearly independent at each point of M. We have a basis of $T_x(M)$ and therefore an isomorphism of $T_x(M)$ with $\mathbb{R}^n = V$ which defines the desired map ι. We shall encounter a more complicated example in the next section. We should point out that only very special kinds of manifolds admit such an isomorphism ι of $T(M)$ with $M \times V$.

For the rest of this section we shall identify $T(M)$ with $M \times V$ and $T^*(M)$ with $M \times V^*$ via the corresponding (adjoint) isomorphism.

The rule which identifies each $T_x(M)$ with V can be regarded as a V-valued linear differential form on M. Let us denote this form by ω. In other words, the identification ι is given by $\iota(\xi) = \langle \xi, \omega_x \rangle$ if $\xi \in T_x(M)$. We can therefore study the V-valued exterior two-form $d\omega$. For each pair of tangent vectors $\xi, \eta \in T_x(M)$ we obtain $\langle \eta, \xi \lrcorner\, d\omega \rangle$ as an element of V. Now we can identify ξ and η with vectors of V. We thus obtain a V-valued antisymmetric bilinear form on V; if we call it $\mathcal{A}_x$, we have

$$\mathcal{A}_x(v, w) = \langle \eta, \xi \lrcorner\, d\omega \rangle,$$

where $\langle \xi, \omega_x \rangle = v$, $\langle \eta, \omega_x \rangle = w$, and $\xi, \eta \in T_x(M)$. Note that, in general, the bilinear form $\mathcal{A}_x$ depends on x. Our second fundamental assumption about the identification ι is that $\mathcal{A}_x$ is independent of x. That is, we assume that there is a V-valued bilinear form $\mathcal{A}$ such that

$$\langle Y, X \lrcorner\, d\omega \rangle = \mathcal{A}(\langle X, \omega \rangle, \langle Y, \omega \rangle) \tag{11.1}$$

for all vector fields X and Y on M.

In the examples given above (the open subset of V on the torus) $d\omega = 0$, so that (11.1) holds trivially. In the next section we shall come across a case where $d\omega \neq 0$.

To understand (11.1) a little better, let us introduce the following notation: For any $v \in V$ let $\hat{v}$ be the vector field on M given by $\langle \hat{v}, \omega \rangle_x = v$ for all $x \in M$. Then for any $v, w \in V$, we have

$$\langle \hat{w}, \hat{v} \lrcorner\, d\omega \rangle = D_{\hat{v}} \langle \hat{w}, \omega \rangle - D_{\hat{w}} \langle \hat{v}, \omega \rangle - \langle [\hat{v}, \hat{w}], \omega \rangle.$$

Now $\langle \hat{w}, \omega \rangle = w$ and $\langle \hat{v}, \omega \rangle = v$ are constants, so the first two terms on the right disappear. Thus we can rewrite (11.1) as

$$\mathcal{A}\langle v, w \rangle = -\langle [\hat{v}, \hat{w}], \omega \rangle. \tag{11.1$'$}$$

For the kinetic energy of a mechanical system we need a Riemann metric on M. A Riemann metric on M gives a scalar product on each $T_x(M)$. This means giving a scalar product $(\ ,\ )_x$ on V for each $x \in M$. *Our third special assumption is that* $(\ ,\ )_x$ *does not depend on* x. Thus we are given a scalar product on V which gives the Riemann metric on M via the identification of $T(M)$ with $M \times V$.

We wish to describe the vector field on $T^*(M)$ as X_{-dH}, where $H = K + U$, K being the kinetic energy of the Riemann metric and U being some potential energy. Since $T^*(M) = M \times V^*$, a vector field X on $T^*(M)$ can be uniquely written as $X = X_1 + X_2$, where X_1 is tangent to M and X_2 is tangent to V^*. Furthermore, we can regard X_1 as a V-valued function and X_2 as a V^*-valued function (identifying the tangent space to vector space V^* with V^*). Then

$$\langle X, \theta\rangle_{<x,l>} = \langle X_1, l\rangle$$

at any $<x, l> \in M \times V^*$. We should really write this as

$$\langle \cdot, \theta\rangle = \langle\langle \cdot, \pi^*\omega\rangle, p\rangle,$$

where ω is the form defined above and the V^*-valued function $p: M \times V^* \to V^*$ is the projection onto the second factor, $p(m, l) = l$. Then

$$\langle \cdot, X \lrcorner\, d\theta\rangle = \langle\langle \cdot, \pi^*\omega\rangle, \langle X, dp\rangle\rangle - \langle\langle X, \pi^*\omega\rangle, \langle \cdot, dp\rangle\rangle - \langle\langle \cdot, X\pi^* \lrcorner\, d\omega\rangle, p\rangle.$$

Substituting $Y = Y_1 + Y_2$ and using (11.1), we obtain

$$\langle Y, X \lrcorner\, d\theta\rangle = \langle Y_1, X_2\rangle - \langle X_1, Y_2\rangle - \langle \mathfrak{a}(X_1, Y_1), l\rangle.$$

Now $H = K + U$, where $K(l) = \frac{1}{2}(l, l)$ and $U(x, l) = \overline{U}(x)$. Thus

$$\langle Y, dH\rangle = (Y_2, l) + Y_1\overline{U},$$

and the equation

$$\langle Y, X \lrcorner\, d\theta\rangle = -\langle Y, dH\rangle$$

becomes

$$\langle Y_1, X_2\rangle - \langle X_1, Y_2\rangle - \langle \mathfrak{a}(X_1, Y_1), l\rangle = -(Y_2, l) - \langle Y_1, d\overline{U}\rangle, \tag{11.2}$$

which must hold for all choices of Y.

Setting $Y_1 = 0$, we get

$$\langle X_1, \cdot\rangle = (\cdot, l). \tag{11.3}$$

In fact, the scalar product occurring in the last equation is on V^*. Transferring it to an equation on V, we get

$$(X_1, \cdot) = \langle \cdot, l\rangle. \tag{11.4}$$

Let $t \mapsto <C(t), v(t)>$ be a solution curve of our system transferred to $T(M) = M \times V$ by the Riemann metric. Then this last equation says $C'(t) = v(t)$.

If we set $Y_2 = 0$ in (11.2) and use (11.4), we get

$$\langle \cdot, X_2\rangle - (\mathfrak{a}(X_1, \cdot), X_1) = -\langle \cdot, d\overline{U}\rangle.$$

Now for any solution of (C, v) we have

$$\langle \cdot, X_2\rangle = \left(\cdot, \frac{dv}{dt}\right),$$

so that these last two equations can be rewritten as

$$\begin{gathered} C'(t) = v(t), \\ \left(\cdot, \frac{dv}{dt}\right) = (\mathfrak{a}(v, \cdot), v) - \langle v, d\overline{U}\rangle, \end{gathered} \tag{11.5}$$

which are known as Euler's equations.

12. RIGID-BODY MOTION

We shall apply the results of the previous section to the study of the motion of a rigid body. For simplicity, we shall confine ourselves to the study of the motion of a rigid body with one point fixed. The more general case where the body as a whole is allowed to move can be handled by similar methods. (Frequently, by considering first the motion of the center of gravity, the more general case can be split into two parts: the motion of the center of gravity and the motion of the body relative to the center of gravity. This then reduces the problem to the one we are studying.)

In order to exhibit the generality of our method, we shall consider the equations of motion of a rigid body in $\mathbb{E}^n$. Only at the end will we make use of the fact that $n = 3$. Let us fix some positive orthonormal system "drawn through the fixed point of the body". In other words, we fix some initial position $x_0 = \langle b^1, \ldots, b^n \rangle$ of the body. Any other position, x_1, of the body can be obtained from x_0 by a rotation: $x_1 = R_1 x_0$. Let $R(t) = e^{At}$ be a one-parameter group of rotations. Then

$$R_1 R(t) x_0 = R_1 R(t) R_1^{-1} x_1$$

is a curve of possible positions of the body. The tangent to this curve at X_1 will be denoted by $\hat{A}_{x_1}$. Thus $\hat{A}_{x_1} \in T_{x_1}(M)$. If $x_2 = R_2 x_0 = R_2 R_1^{-1} R_1 x_0 = R_2 R_1^{-1} x_1$, then $\hat{A}_{x_2}$ is the tangent to the curve $R_2 R(t) x_0 = R_2 R_1^{-1} R_1 R(t) x_0$, so that $\hat{A}_{x_2} = (R_2 R_1^{-1})_* \hat{A}_{x_1}$. It is clear from its definition that $\hat{A}_{x_1} = 0$ if and only if $A = 0$.

We can regard the map $A \mapsto \hat{A}_{x_1}$ as a map from the space of skew adjoint linear transformations to $T_{x_1}(M)$. Let V denote the space of skew adjoint linear transformations. Then since

$$\dim V = \dim M = \frac{n(n-1)}{2}$$

and the map $A \mapsto \hat{A}_{x_1}$ is an injection, we conclude that it is an isomorphism. We thus have a trivialization $T(M) \mapsto M \times V$. Consequently, we get a V-valued linear differential form ω, as in Section 11.

Let us describe once more the meaning of $\omega_x \colon T_x(M) \to V$. If $\xi \in T_x(M)$ represents an infinitesimal motion of the body, then since the body is rigid, we can regard ξ as an infinitesimal rotation of the body relative to an observer situated outside the body (fixed in space). Thus $\xi = B$ for some $B \in V$, say. Then $\langle \xi, \omega \rangle$ is the corresponding infinitesimal rotation expressed in terms of the basis attached to the body; that is, $\langle \xi, \omega \rangle = R_1^{-1} B R_1$ if $x = R_1 x_0$.

We denote by $\hat{A}$ the vector field $x \mapsto \hat{A}_x$ corresponding to $A \in V$. Let φ_t be the one-parameter group generated by $\hat{B}$ for some $B \in V$. Note that at any $x_2 = R_2 x_0$ we have $\varphi + x_2 = R_2 e^{Bt} x_0$. Then at any $x_1 = R_1 x_0$ we have

$$\begin{aligned} \varphi_t R_1 e^{As} R_1^{-1} x_1 &= \varphi_t R_1 e^{As} x_0 = R_1 e^{As} e^{Bt} x_0 \\ &= R_1 e^{Bt} (e^{-Bt} e^{As} e^{Bt}) x_0, \end{aligned}$$

so that

$$\varphi_{t*}\hat{A}_{x_1} = \widehat{(e^{-Bt}Ae^{Bt})}_{\varphi_t x_1}.$$

If we differentiate this equation with respect to t at $t = 0$, we conclude that

$$[\hat{A}, \hat{B}] = (AB - BA) = -[A, B],$$

so that, according to (11.1), we have

$$\mathfrak{a}(A, B) = [A, B] = BA - AB. \tag{12.1}$$

We now show how a mass distribution on the body determines a Riemann metric on M. Let p be a particle on the body with mass m. We will assume that the particle p has coordinates $\langle p^1, \ldots, p^n \rangle$ relative to the axes drawn on the body. Suppose that the body is at position $x = \langle b_1, \ldots, b_n \rangle$. Then the particle p will be situated at the point $p^1 b_1 + \cdots + p^n b_n \in \mathbb{E}^n$. If $R(t) = e^{At}$ is a one-parameter group of rotations, then when the body is at $R_1 R(t)x$ the particle p will be situated at

$$R_1[p^1 R(t) b_1 + \cdots + p^n R(t) b_n] = R_1 R(t) p.$$

Thus the velocity of the particle p as the body undergoes the motion generated by $\hat{A}$ is $R_1 A p$, and the kinetic energy of the particle p is $\frac{1}{2}m\|R_1 A p\|^2 = \frac{1}{2}m\|Ap\|^2$, since R_1 is an orthogonal linear transformation. We define the kinetic energy of $\hat{A}_x$ to be the total kinetic energy of all the particles of the body. Thus

$$\tfrac{1}{2}\|\hat{A}_x\|^2 = \tfrac{1}{2}\int_{\text{body}} m\|Ap\|^2. \tag{12.2}$$

Note that $\|\hat{A}_x\|$ depends only on A, so that our third requirement of the last section is satisfied, provided (12.2) does indeed define a norm on V. (*Note:* That the mass distribution could be such that $Ap = 0$ for all p in $\{p : m(p) > 0\}$ does not imply that $A = 0$. For example, suppose that all the mass were concentrated along a line l. If A represents infinitesimal rotation about l, so that $Ap = 0$ for $p \in l$, then $\|A\| = 0$. However, it is clear that if the set of p for which $m > 0$ spans $\mathbb{E}^n$, then (12.2) defines a Riemann metric. In fact,

$$\|A\| = 0 \Rightarrow Ap = 0$$

for all p belonging to a spanning set, and thus $A = 0$.)

Let us examine the scalar product given in (12.2) a little more closely. Let f denote the linear function on $\mathbb{E}^n \otimes \mathbb{E}^n$ corresponding to the scalar product on $\mathbb{E}^n$ (which is a bilinear form); in other words,

$$f(a_1 \otimes b_1 + \cdots + a_k \otimes b_k) = (a_1, b_1) + \cdots + (a_k, b_k).$$

Let s be any element of $V \otimes V$. Then s defines a bilinear form on $\mathrm{Hom}(\mathbb{E}^n)$ given by

$$s(A, B) = f((A \otimes B)s).$$

Note that if the tensor s is symmetric, so is the corresponding bilinear form. In our present context the scalar product (12.2) comes from the tensor $I \in V \otimes V$, where

$$I = \int_{\text{body}} mp \otimes p$$

is called the *inertia tensor of the body.** Thus (12.2) can be written as

$$\|\hat{A}\|^2 = I(A, A).$$

Euler's equations in this case become

$$C'(t) = A(t) \qquad \text{and} \qquad \left(\cdot, \frac{dA}{dt}\right) = I([\cdot, A], A) - \langle \hat{A}_{C(t)}, d\overline{U}\rangle. \tag{12.3}$$

Now the tensor I is symmetric. We can therefore find an orthonormal basis $e^1, \ldots, e^n$ of $\mathbb{E}^n$ which diagonalizes I, so that

$$I = I_1 e^1 \otimes e^1 + I_2 e^2 \otimes e^2 + \cdots + I_n e^n \otimes e^n.$$

Let E_{ij} $(i < j)$ be the antisymmetric matrix defined by

$$E_{ij}e^i = e^j, \qquad E_{ij}e^j = -e^i, \qquad E_{ij}e^l = 0 \qquad (l \neq i, j).$$

Then

$$I(E_{ij}, E_{kl}) = 0 \qquad \text{if} \quad i, j \neq kl$$

and

$$I(E_{ij}, E_{ij}) = I_i + I_j.$$

Now let us see what Eqs. (12.3) say for the case $n = 3$, where we have

$$[E_{12}, -E_{13}] = E_{23}, \qquad [E_{12}, E_{23}] = -E_{13}, \qquad [E_{13}, E_{23}] = E_{12}.$$

Suppose $A(t) = a_1(t)E_{23} - a_2(t)E_{13} + a_3(t)E_{12}$, and let

$$B = b_1E_{23} - b_2E_{13} + b_3E_{12} = \text{const.}$$

Then substituting B into (12.3) and comparing the coefficients of b_1, b_2, and b_3, we get

$$\begin{aligned}
(I_2 + I_3)\frac{da_1}{dt} &= (I_3 - I_2)a_2a_3 + a_1\langle E_{12}, d\overline{U}\rangle, \\
(I_1 + I_3)\frac{da_2}{dt} &= (I_1 - I_3)a_1a_3 + a_2\langle -E_{12}, d\overline{U}\rangle, \\
(I_1 + I_2)\frac{da_3}{dt} &= (I_2 - I_1)a_1a_2 + a_3\langle E_{13}, d\overline{U}\rangle.
\end{aligned} \tag{12.4}$$

A simple case of these equations arises when there is no potential, that is, $U = 0$. First of all, suppose that the body has a spherically symmetric distri-

* This differs slightly from that which is usually called the inertia tensor in physics texts. In terms of the coefficients I_i introduced below, what is usually called I_{ij} is $I_i + I_j$ in our setup.

bution of mass. Then $I_1 = I_2 = I_3$ and Eqs. (12.4) become

$$\frac{da_1}{dt} = \frac{da_2}{dt} = \frac{da_3}{dt} = 0.$$

In other words, $A =$ const. The motion of the body consists of a steady rotation about some axis fixed on the body. Of course, this means that for an observer in space also, the body undergoes a steady rotation about a fixed axis.

Next, let us consider the case of an axially symmetric rigid body moving freely; that is, $I_1 = I_2$ and $U = 0$. The equations of motion then become

$$\frac{da_1}{dt} = Ka_2a_3,$$

$$\frac{da_2}{dt} = -Ka_1a_3,$$

$$\frac{da_3}{dt} = 0,$$

where

$$K = \frac{I_3 - I_2}{I_3 + I_2}.$$

The solutions of these equations can be written down immediately: $a_3 = s =$ const and

$$a_1 = c_1 \cos Kst + c_2 \sin Kst,$$
$$a_2 = -c_1 \sin Kst + c_2 \cos Kst.$$

Thus for an observer situated on the body the instantaneous axis of rotation describes a circle around the axis of symmetry of the body. This motion is known as *regular precession*. (This motion should *not* be confused with the astronomical precession of the earth's axis, which is due to the gravitational action of the sun and moon.)

If no two of I_1, I_2, and I_3 are equal, then the equations of motion can still be solved in terms of integrals, although the expression is rather complicated. We refer the reader to any standard book on mechanics for the details.

So far we have been considering the motion with no potential. If $\overline{U} \neq 0$, then Eqs. (12.4) are usually not so easy to solve. Let us treat the case of a symmetrical top; that is, $I_1 = I_2$ and U is given by the gravitational potential. In order to solve this problem it is convenient to use the Euler angles of astronomy as the local coordinates on M. In order to avoid confusion, we reproduce the definitions here: We let δ_x, δ_y, δ_z be the basis vectors of $\mathbb{E}^3$ corresponding to the rectangular coordinates x, y, z, where we take the origin of $\mathbb{E}^3$ to be the fixed point of the body. We may assume that the center of mass of the body is distinct from the fixed point—otherwise there will be no gravitational force acting on the body. It is easy to see that the center of mass c lies on the axis of symmetry. We shall assume that the vectors drawn on the body are such that b_3 points in the direction from the fixed point to the center of mass, i.e., from 0 to c. Then

$0 < \theta < \pi$ is defined to be the angle between b_3 and δ_z:

$$\cos\theta = (b_3, \delta_z).$$

The line of nodes is the intersection of the plane spanned by b_1 and b_2 with the xy-plane. (Thus, in order for it to be defined, we must restrict to the open set $b \neq \pm\delta_z$.) We define the unit vector n along the line of nodes to be the one that makes δ_z, b_3, n a positive (right-handed) basis of $\mathbb{E}^3$.

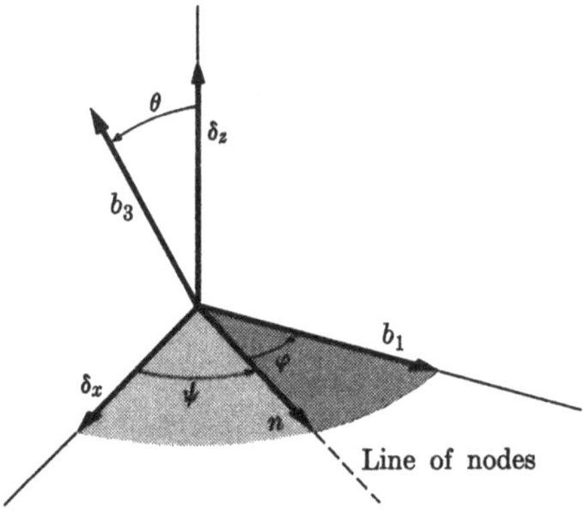

Fig. 13.5

The angle $0 < \psi < 2\pi$ is defined to be the angle that n makes with δ_x and $0 < \varphi < 2\pi$ is defined to be the angle that n makes with b_1. We now wish to find the transformation relating the basis of $T_x(M)$ given by $\partial/\partial\theta, \partial/\partial\varphi, \partial/\partial\psi$ with the orthogonal basis $E_{12}, -E_{13}, E_{23}$ introduced earlier. Suppose that x has the coordinate $\langle\theta, \psi, \varphi\rangle$. (See Fig. 13.5.) Now $(\partial/\partial\theta)_x$ represents an infinitesimal rotation about the line of nodes and $n = (\cos\varphi)b_1 - (\sin\varphi)b_2$, so that

$$\left(\frac{\partial}{\partial\theta}\right)_x = (\cos\varphi)E_{23} - (\sin\varphi)(-E_{13}).$$

The vector $(\partial/\partial\varphi)_x$ represents infinitesimal rotation about b_3, so that

$$\left(\frac{\partial}{\partial\varphi}\right)_x = E_{12}.$$

Finally, $(\partial/\partial\psi)_x$ represents infinitesimal rotation about δ_z. Now

$$(\delta_z, b_3) = \cos\theta.$$

Furthermore, since $(\delta_z, n) = 0$, the projection of δ_z onto the plane spanned by δ_1 and δ_2 must still be orthogonal to n, since n lies in that plane. It is therefore easy to check that this projection is given by $(\sin\varphi)b_1 + (\cos\varphi)b_2$. Thus

$$\delta_z = \sin\theta\,[(\sin\varphi)b_1 + (\cos\varphi)b_2] + (\cos\theta)b_3,$$

and therefore

$$\left(\frac{\partial}{\partial\psi}\right)_x = (\sin\theta\sin\varphi)E_{23} + (\sin\theta\cos\varphi)(-E_{13}) + (\cos\theta)E_{12}.$$

If $\xi \in T_x(M)$ is given by

$$\xi = \dot\theta\left(\frac{\partial}{\partial\theta}\right)_x + \dot\psi\left(\frac{\partial}{\partial\psi}\right)_x + \dot\varphi\left(\frac{\partial}{\partial\varphi}\right)_x,$$

we have

$$\begin{aligned} 2K(\xi) &= \|\xi\|^2 \\ &= \dot\theta^2(\cos\theta)^2(I_2+I_3) + (\sin\varphi)^2(I_1+I_3) + \dot\varphi^2(I_1+I_2) \\ &\quad + 2\dot\varphi\dot\psi(I_1+I_2)\cos\theta \\ &\quad + \dot\psi^2\{(\sin^2\theta)[(I_2+I_3)\sin^2\varphi + (I_1+I_3)\cos^2\varphi] + (I_1+I_2)\cos^2\theta\}. \end{aligned}$$

Now, by assumption $I_1 = I_2$. Let us set

$$M_1 = I_1 + I_3 = I_2 + I_3 \qquad \text{and} \qquad M_2 = I_1 + I_2.$$

Then the above expression simplifies to

$$K(\theta, \psi, \varphi; \dot\theta, \dot\psi, \dot\varphi) = \tfrac{1}{2}M_1(\dot\theta^2 + \dot\psi^2\sin^2\theta) + \tfrac{1}{2}M_2(\dot\varphi + \dot\psi\cos\theta)^2. \tag{12.5}$$

Let us now obtain the expression for the potential energy. It is proportional to the height of the center of gravity if we assume (as we shall) a uniformly vertical gravitational field. Thus

$$\overline{U}(\theta, \psi, \varphi) = k\cos\theta, \tag{12.6}$$

where $k = mg\|c\|$ and m is the total mass of the body, g is the force of the gravitational field, and $\|c\|$ is the distance of the center of gravity from the fixed point. Thus the Lagrangian is given by

$$L(\theta, \psi, \varphi; \dot\theta, \dot\psi, \dot\varphi) = \tfrac{1}{2}M_1(\dot\theta^2 + \dot\psi^2\sin^2\theta) + \tfrac{1}{2}M_2(\dot\varphi + \dot\psi\cos\theta)^2 - k\cos\theta. \tag{12.7}$$

Note that $\partial/\partial\varphi$ and $\partial/\partial\psi$ are both isometries of the Riemann metric which leave $\overline{U}$ invariant. Thus the corresponding momenta are constants of the motion. In other words, for any motion of the system we have

$$P_{\partial/\partial\psi} = C_1 \qquad \text{and} \qquad P_{\partial/\partial\varphi} = C_2,$$

where C_1 and C_2 are constants. But

$$P_{\partial/\partial\psi} = \frac{\partial L}{\partial\dot\psi} = M_1\dot\psi\sin^2\theta + M_2\cos\theta(\dot\varphi + \dot\psi\cos\theta) = C_1$$

and

$$P_{\partial/\partial\varphi} = M_2(\dot\varphi + \dot\psi\cos\theta) = C_2.$$

Solving these equations gives

$$\dot\varphi + \dot\psi\cos\theta = \frac{C_2}{M_2}, \qquad M_1\dot\psi\sin^2\theta = C_1 - C_2\cos\theta. \tag{12.8}$$

Let us substitute (12.8) into the expression for the energy,

$$E = K + \overline{U} = \tfrac{1}{2}M_1(\dot{\theta}^2 + \dot{\psi}^2 \sin^2 \theta) + \tfrac{1}{2}M_2(\dot{\varphi} + \dot{\psi} \cos \theta) + k \cos \theta,$$

to get, for a given value of C_1 and C_2, the expression

$$\frac{C_2^2}{2M_2} + \frac{1}{2} M_1\dot{\theta}^2 + \frac{(C_1 - C_2 \cos \theta)^2}{2M_1 \sin^2 \theta} + k \cos \theta = \text{const.} \tag{12.9}$$

Thus, just as in our treatment of the central-force problem, for a fixed value of $P_{\partial/\partial\varphi}$ and $P_{\partial/\partial\psi}$, the motion of θ is determined by a one-dimensional mechanical system whose energy is given by (12.9). After solving this mechanical system for θ, we can then obtain $\psi(\cdot)$ and $\varphi(\cdot)$ by integrations from (12.8). In order to obtain qualitative information about the behavior of the solutions $\theta(\cdot)$, we can apply the method of Section 5 to our one-dimensional mechanical system. Note that if $C_1 \neq C_2$ (as would be the case if the body were "spinning fast"), the kinetic energy tends to infinity as $\theta \to 0$ and as $\theta \to \pi$. We therefore conclude that θ oscillates between two values $0 < \theta_1 \leq \theta \leq \theta_2 < \pi$. In other words, the axis of symmetry of the body executes a periodic up and down motion (called *nutation*). As θ oscillates, $\dot{\psi}$ satisfies (12.8). Let us graph the curve that b_3 traces out on the unit sphere. If $C_1 > C_2 \cos \theta > 0$ for $\theta_1 \leq \theta \leq \theta_2$, then $\dot{\psi} > 0$ although it oscillates in magnitude. The motion is thus as given in Fig. 13.6. Another possibility is that

$$C_1 - C_2 \cos \theta_1 < 0 \qquad \text{and} \qquad C_1 - \cos \theta_2 > 0,$$

so that $\dot{\psi}$ is negative near $\theta = \theta_1$ and positive max $\theta = \theta_2$. In this case the average of $\dot{\psi}$ over a period is still positive, so that the motion is as shown in Fig. 13.7.

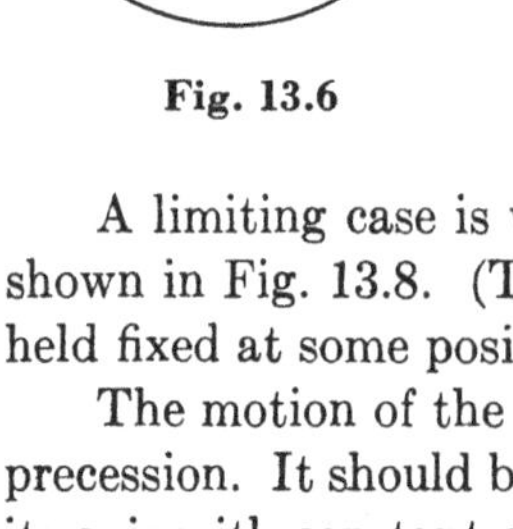

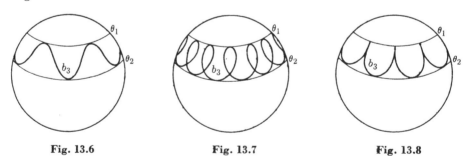

Fig. 13.6

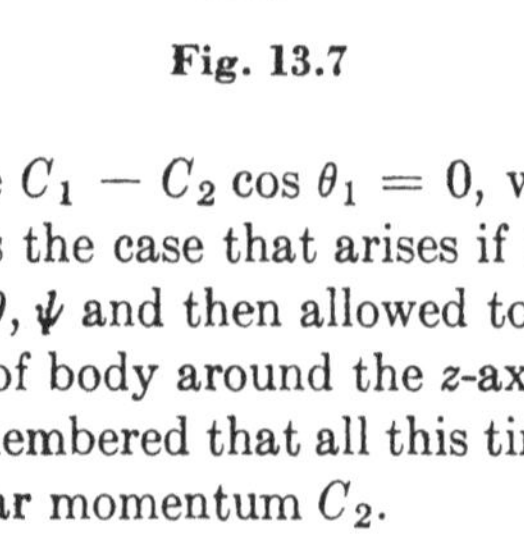

Fig. 13.7

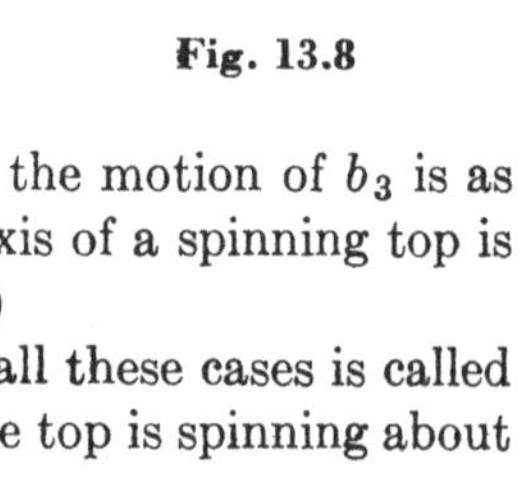

Fig. 13.8

A limiting case is where $C_1 - C_2 \cos \theta_1 = 0$, where the motion of b_3 is as shown in Fig. 13.8. (This is the case that arises if the axis of a spinning top is held fixed at some position θ, ψ and then allowed to fall.)

The motion of the axis of body around the z-axis in all these cases is called precession. It should be remembered that all this time the top is spinning about its axis with constant angular momentum C_2.

13. SMALL OSCILLATIONS

Suppose we are given a mechanical system on a manifold M with energy $H = K + U$. Suppose that the "force field" $d\overline{U}$ vanishes at some $x_0 \in M$. Then the constant curve $C(t) \equiv x_0$ is a trajectory of the system. In fact, let us choose a chart (W, α) with coordinates $x^1, \ldots, x^n$ such that

$$\alpha(x_0) = \langle 0, \ldots, 0 \rangle.$$

Then in the corresponding coordinates $\langle q^1, \ldots, q^n \rangle$ on $\pi^{-1}(W)$ we have at the point $\langle 0, \ldots, 0, \ldots, 0 \rangle$

$$\frac{\partial H}{\partial q^i} = \frac{\partial \overline{U}}{\partial x^i} = 0 \quad \text{and} \quad \frac{\partial H}{\partial p^i} = \frac{\partial K}{\partial p^i} = 0.$$

It is therefore natural to expect that for small initial values of q and p and for small intervals of time, the solutions of the system should be well approximated by the following linear system:

Replace the potential energy,

$$\overline{U}(x^1, \ldots, x^n) = \sum a_{ij}x^ix^j + U_3,$$

[where $U_3 = \mathcal{O}(\|x\|^3)$; that is, U_3 vanishes to third order at $x = 0$] by the quadratic potential energy,

$$\overline{U}_2(x) = \tfrac{1}{2}\sum a_{ij}x^ix^j,$$

and replace the kinetic energy corresponding to the given Riemann metric,

$$K(q, \dot{q}) = \tfrac{1}{2}\sum g_{ij}(q)\dot{q}^i\dot{q}^j,$$

by the one corresponding to the Euclidean metric at x_0,

$$K_2(q, \dot{q}) = \tfrac{1}{2}\sum g_{ij}(0, \ldots, 0)\dot{q}^i\dot{q}^j.$$

We thus obtain a mechanical system H_2 whose corresponding equations (5.1) are actually linear. [The reader should check, as an exercise, that these equations are exactly the equations of variation (introduced in Section 2) of the vector field X_{-dH} along the curve $\overline{C}(t) = \langle 0, \ldots, \ldots 0 \rangle$ in $T^*(M)$.]

Of course, as time increases, the values of q^i and p^i might become quite large and the linear approximations useless. However, under certain circumstances we can guarantee that q^i and p^i remain small for all time. In fact, suppose that the quadratic form $\sum a_{ij}x^ix^j$ is positive definite. Then $\overline{U}$ has a strict minimum at x_0—say $\overline{U}(x_0) = 0$. In particular, if we start at x_0 with a kinetic energy $K = E$, where E is sufficiently small, then x will be restricted to the neighborhood of x_0 defined by $\overline{U}(x) \leq E$, and the momenta will be restricted by the condition $K \leq E$, since, by the conservation of energy, $K + \overline{U} \equiv E$. (See Fig. 13.9.) Thus the q^i and the p^i will remain small. This does not mean that the solutions to the original mechanical system with Hamiltonian H will remain close to one fixed solution curve of the linearized system. It does mean

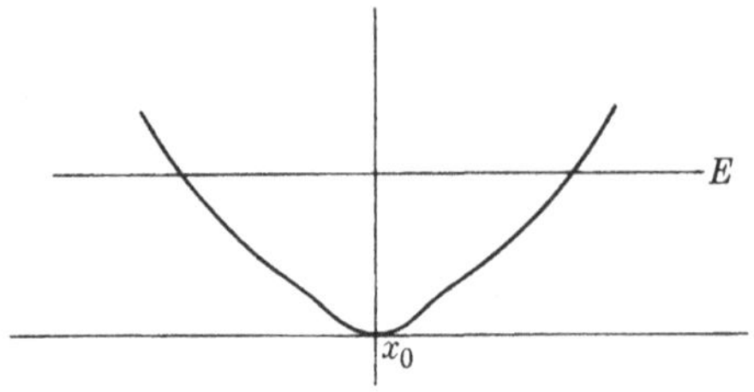

Fig. 13.9

that for any short interval of time (that is, a short interval near any time in the future), the trajectory will be close to some trajectory of the linearized system. It is therefore important to study the behavior of such mechanical systems.

We are thus interested in the following kind of mechanical system: The configuration space M is a vector space. The Riemann metric is given by a Euclidean metric on the manifold. The potential energy is a positive definite quadratic form on this vector space. Let us choose rectangular coordinates $x^1, \ldots, x^n$ with respect to the Euclidean metric. Thus

$$\|(\dot{q}^1, \ldots, \dot{q}^n)\|^2 = (\dot{q}^1)^2 + \cdots + (\dot{q}^n)^2,$$

and the map $\mathcal{L}$ is given by

$$p^i = \dot{q}^i.$$

Thus

$$\begin{aligned} L(q, \dot{q}) &= K(\dot{q}) - \overline{U}(q) \\ &= \tfrac{1}{2}(\textstyle\sum (\dot{q}^i)^2 - \sum a_{ij}q^i q^j) \end{aligned}$$

and Lagrange's equations become

$$\frac{dq^i}{dt} = \dot{q}^i, \qquad \frac{d\dot{q}^i}{dt} = -\sum a_{ij}q^j. \tag{13.1}$$

(Of course, these are just the Euler equations (11.15) for the case at hand where $\mathfrak{A} \equiv 0$.)

Equations (13.1) can be written more suggestively as follows: Let A be the linear transformation whose matrix is (a_{ij}). Stated invariantly, A is the unique self-adjoint linear transformation such that

$$\overline{U}(x) = (Ax, x). \tag{13.2}$$

Then the trajectories $v(\cdot)$ of the system are the solutions of the second-order differential equations

$$\frac{d^2v}{dt^2} = -Av. \tag{13.3}$$

To find the actual solutions of (13.1) and (13.3) we apply Theorem 3.1 of Chapter 5. According to that theorem, if M is finite-dimensional, we can find an orthonormal basis $e^1, \ldots, e^n$ of eigenvalues of A. In other words, if z^i are the rectangular coordinates corresponding to this basis,

$$\overline{U}(z^1, \ldots, z^n) = \lambda_1(z^1)^2 + \cdots + \lambda_n(z^n)^2$$

and Eqs. (13.1) and (13.3) become

$$\frac{d^2z^i}{dt^2} = -\lambda_i z^i.$$

Thus the general solution of (13.3) is given by

$$v(t) = (a_1 \cos \lambda_1 t + b_1 \sin \lambda_1 t)e^1 + \cdots + (a_n \cos \lambda_n t + b_n \sin \lambda_n t)e^n, \tag{13.4}$$

where the constants a^1 and b^1 are determined by

$$v(0) = a_1 e^1 + \cdots + a_n e^n$$

and

$$\frac{dv}{dt}(0) = \lambda_1 b_1 e^1 + \cdots + \lambda_n b_n e^n.$$

Thus the general motion is a superposition of independent oscillations, the frequency of each oscillation being determined by the eigenvalues $\{\lambda_i\}$. That is why the mechanical system (13.1) is called a system of "small oscillations".

14. SMALL OSCILLATIONS (Continued)

So far, we have been considering the linearized equations (13.1) as an approximation to a finite-dimensional mechanical system. The philosophy has been that the solutions to the actual mechanical system exist, but are hard to find. We use (13.3) as good approximating equations.

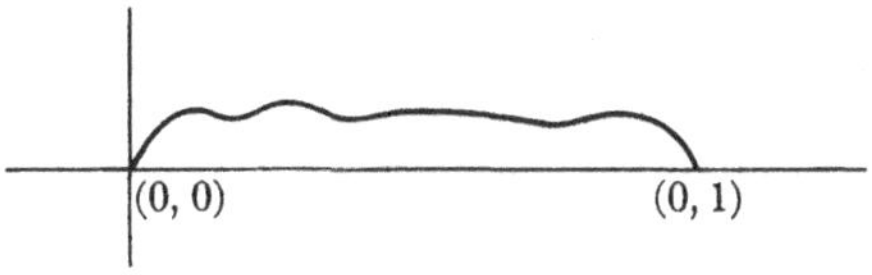

Fig. 13.10

It turns out that the method has more extensive applications, even to the case of infinite-dimensional systems where the very existence of solutions to the "actual" mechanical systems may be difficult to establish. Let us illustrate by the mechanical system consisting of a stretched string which is held fixed at two endpoints. For simplicity in illustration, we shall assume that the string is restricted to move in the xy-plane, although this is in no way essential to the argument. Let us assume that the string is homogeneous and that the two fixed points are (0, 0) and (0, 1). Then the configuration space should be all possible smooth curves joining (0, 0) to (0, 1). Thus the curve shown in Fig. 13.10 would be a possible element of our configuration space. In some sense, the configuration is an "infinite-dimensional manifold" and with suitable work this idea can be made precise. However, what is of interest to us is behavior near the "equilibrium" curve $C(t) = (t, 0)$. For such curves we will have $dx/dt > 0$, and so the curve can be described by using x as independent variable, i.e., by giving a function $u(x)$. In other words, we are replacing the big configuration space by the approximating vector space V of all functions u of one variable, with $u(0) = u(1) = 0$. Thus V is regarded as the "tangent space" to our system, in the sense that u is the "tangent vector" to the curve $C_s(\cdot)$, where $C_s(\tau) = (\tau, su(\tau))$. (Remember that our configuration space is a collection of curves, so that a curve in our configuration space is a one-parameter family of curves.) Now we expect the "kinetic energy" of u to be the total of the kinetic energy of all the particles on the string. The particle at τ has velocity $u(\tau)$ and therefore kinetic energy $\frac{1}{2}mu(\tau)^2$. If we assume that the mass density is constant, we thus get

$$K_2(u) = \tfrac{1}{2}m\int_0^1 u^2\,dx$$

as the expression for the kinetic energy. This makes V into a pre-Hilbert space as in Section 6 of Chapter 6.

We expect that the potential energy depends on the stretching of the string, i.e., that it is some function of the length:

$$\overline{U}(C) = F\left(\int_0^1 \|C'(t)\|\,dt\right) = F(L),$$

where L is the length of the curve and F is some smooth function with $F(1) = 0$. For curves parametrized by x, the length is given by

$$\int_0^1 \sqrt{1 + \left(\frac{du}{dx}\right)^2}\,dx.$$

Now

$$\sqrt{1 + a^2} = 1 + \tfrac{1}{2}a^2 + \text{higher-order terms.}$$

Thus using a Taylor expansion for F,

$$F(L) = F'(1)(L - 1) + \text{quadratic terms in } (L - 1)$$

and

$$L - 1 = \frac{1}{2}\int_0^1 \left(\frac{du}{dx}\right)^2 dx + \int \text{higher-order terms in } \frac{du}{dx},$$

we see that U_2, the quadratic approximation to $\overline{U}$, is given by

$$U_2(u) = \frac{C}{2}\int_0^1 \left(\frac{du}{dx}\right)^2 dx.$$

By analogy with (13.3) we expect that the "small oscillations" of the string are solutions $u_t(\cdot)$ of the equations

$$\frac{d^2u_t}{dt^2} = -Au_t,$$

where A is the self-adjoint linear operator such that

$$(Au, u) = \frac{C}{2}\int_0^1 \left(\frac{du}{dx}\right)^2 dx.$$

Now

$$(Au, u) = \frac{m}{2}\int_0^1 u(Au)\, dx.$$

Since $u(0) = u(1) = 0$, we have, by integration by parts,

$$\int_0^1 \frac{du}{dx}\cdot\frac{du}{dx} = -\int_0^1 u\left(\frac{d^2u}{dx^2}\right),$$

so that

$$Au = -\frac{C}{m}\frac{d^2u}{dt^2}.$$

Equation (13.3) thus becomes

$$\left(\frac{d}{dt}\right)^2 u_t = \frac{C}{m}\frac{d^2u}{dx^2}, \qquad u_t(0) = u_t(1) = 0. \tag{14.1}$$

Note that we have derived (14.1) by reasoning by analogy. We have not formulated the actual (nonlinear) infinite-dimensional mechanical system, nor have we any guarantee that there is one that can be solved. Furthermore, we don't actually know the function F. We only need to know its form and the value of $F'(1)$. Nevertheless, Eq. (14.1) gives a good explanation of observed physical phenomena.

The solution of (14.1) proceeds just as in the finite-dimensional case due to the fact that the operator A with the boundary conditions $u(0) = u(1) = 0$ is a Sturm-Liouville system, so the results of Sections 6 and 7 of Chapter 6 apply. In fact, we can choose the functions

$$u_n(x) = \sin(n\pi x)$$

as an orthogonal basis of eigenvectors of A, where u_n has the eigenvalue $n^2(c/m)\pi^2$. Thus the general solution is given by

$$u_t(x) = (a_1 \cos \alpha t + b_1 \sin \alpha t) \sin(\pi x) + (a_2 \cos 2\alpha t + b_2 \sin 2\alpha t) \sin(2\pi x) + \cdots,$$

where $\alpha = (c/m)\pi^2$. In other words, the general solution is a superposition (linear combination) of the "harmonics"

$$\sin n\alpha t \sin n\pi x, \qquad \cos n\alpha t \sin n\pi x.$$

These can be regarded as "standing waves", as, for example, in Fig. 13.11.

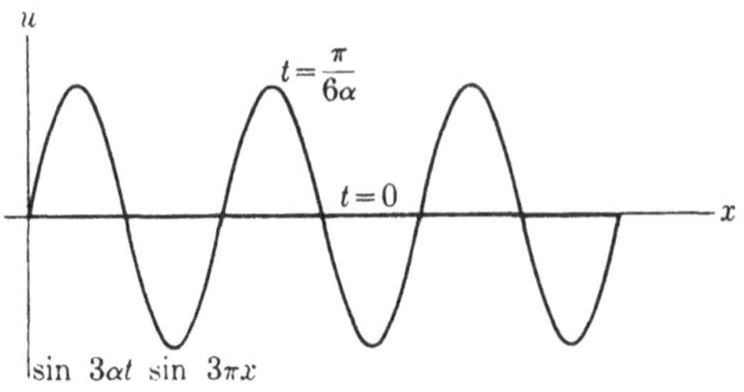

Fig. 13.11

As another illustration of this method, let us consider the "vibrating membrane" in n-dimensions. Here we are given a domain D with almost regular boundary in $\mathbb{E}^n$. We consider a stretched membrane in $\mathbb{E}^{n+1} = \mathbb{E}^n \times \mathbb{E}^1$ which is fastened along ∂D in $\mathbb{E}^n \times \{0\}$. Again, as our linear approximation V to the configuration manifold, we take the space of all functions u on D which vanish on ∂D. To be precise, we let V be the space of functions which we defined, are of class C^2 in some neighborhood $\overline{D}$, and vanish on ∂D. Thus the membrane is the surface in $\mathbb{E}^{n+1}$ whose points are of the form

$$\langle x^1, \ldots, x^n, u(x^1, \ldots, x^n)\rangle,$$

where $\langle x^1, \ldots, x^n\rangle \in D$. As before, we define the kinetic energy as

$$K(u) = \tfrac{1}{2}\int_D u^2,$$

while the potential energy is to be some function of the total volume (area) which vanishes at $\mu(D)$. Now the total volume (area) of the hypersurface is given by (see Exercise 4.3, Chapter 10)

$$\int_D \sqrt{1 + \sum \left(\frac{\partial u}{\partial x^i}\right)^2}.$$

Thus, as before,

$$\overline{U}_2(u) = \frac{c}{2}\int \sum \left(\frac{\partial u}{\partial x^i}\right)^2 = \frac{c}{2} D[u, u],$$

where D is the Dirichlet integral introduced in Section 11 of Chapter 12. By Green's formula we have (since $u = 0$ on ∂D)

$$D[u, u] = -\int_D u\Delta u.$$

Thus the operator A of (13.3) is given by $-(c/m)\Delta$, and Eq. (13.3) becomes

$$\frac{d^2}{dt^2} u_i = \frac{c}{m}\Delta u. \tag{14.2}$$

Note that this is exactly (14.1) if $n = 1$. In order to solve (14.2), we must find a complete set of eigenvalues for (14.2). If $\{u_n\}$ is such a basis, where λ_i is the eigenvalue associated to u_i, then the general solution of (14.2) would be given by

$$u_t(x) = \sum (a_n \cos \lambda_n t + b_n \sin \lambda_n t) u_n(x).$$

The problem of showing that $-\Delta$ has a complete set of eigenvectors is somewhat more difficult for $n > 1$, and will be discussed in the exercises.

EXERCISES

In order to study the eigenvalue problem it is convenient to replace the space of C^1-functions vanishing on ∂D by the space H_1^D (where we refer back to page 361 for the definition of the spaces H_s^D.) To show that this replacement is legitimate, we have:

14.1 Show that if φ is a function which is C^1 in a neighborhood of $\overline{D}$, then $\varphi \in H^D$ if and only if φ vanishes on ∂D.

14.2 Let the operator K be defined as $Kf = (1 - \Delta)f$ for smooth f and extended as in Section 14 of Chapter 8 to a map of $H_s \to H_{s-2}$. Let $g \in H_0$. Since

$$|(g, v)| \leq \|g\|_0 \|v\|_0 \leq \|g\|_0 \|v\|_1 \qquad \text{for any} \quad v \in H_1^D,$$

conclude that there is a bounded linear map L from H_0^D to H_1^D satisfying

$$(Lg, v)_1 = (g, v)_0 \qquad \text{for all} \quad v \in H_1^D.$$

14.3 Let f and g be locally integrable functions on D. By this we mean that $f\varphi$ and $g\varphi$ are integrable for every test function $\varphi \in C_0^\infty(D)$. We say that the differential equation $Kf = g$ holds *weakly* on D if $(f, K\varphi) = (g, \varphi)$ for all test functions $\varphi \in C_0^\infty(D)$. Show that this generalizes the notion of being a solution by proving the following lemma.

> **Lemma.** If the functions f and g are respectively in the classes $C^2(D)$ and $C^0(D)$, then the equation $Kf = g$ holds in the weak sense if and only if it holds in the classical sense.

In proving this lemma you may assume that if h is a continuous function on D such that $\int_D h\varphi = 0$ for every test function φ, then $h = 0$.

14.4 Prove that the operator L of Exercise 14.2 is a right inverse of K in the weak sense. That is, show that if $g \in H_0^D$ and $f = Lg$, then $Kf = -g$ weakly on D.

14.5 We now want to show that f is actually in $C^2(D)$ if g is suitably smooth. Roughly speaking, what we want to do is to "round off" f and g near the boundary of C in such a way that we can consider the adjusted functions to be defined on the whole of $\mathbb{R}^n$, and can then apply Exercises 14.25 and 14.30 of Chapter 8.

Our rounding off process will simply be multiplication by an arbitrary but fixed function ψ in $C_0^\infty(D)$. Prove, to begin with, that multiplication by such a ψ is a bounded linear mapping of H_s into itself for any s.

14.6 We know that $D_j = \partial/\partial x_j$ is a bounded linear mapping from H_s to H_{s-1} for every s. Combine this fact with the result of the above exercise to show that if $Kf = g$

weakly on D, and if ψ is any fixed element of $C_0^\infty(D)$, then there is a differential operator R of order 1 defined on the whole of $\mathbb{R}^n$ such that

1) $K(\psi f) = \psi g + Rg$ weakly on D;
2) $h \mapsto Rh$ is a bounded linear mapping from H_s to H_{s-1} for every s.

In order to consider R to be defined on $\mathbb{R}^n$ we have to extend ψ to $\mathbb{R}^n$ in an obvious way. The proof is essentially an integration by parts.

14.7 We say that a function h defined on D is *locally in* H_s if $\varphi h \in H_s$ (when extended to $\mathbb{R}^n$) for every $\varphi \in C_0^\infty(D)$. Use the above exercise to prove the following lemma:

> **Lemma.** Suppose that $Kf = g$ weakly in D, that $f \in H_j$ locally in D, and that $g \in H_m$ locally in D. Then $f \in H_{\min(m+2,j+1)}$ locally in D.

[*Hint:* In order to prove this crucial lemma, show first that the weak differential equation of Exercise 14.6 holds for all test functions φ on $\mathbb{R}^n$. The crux of the matter is that there is a function $\chi \in C_0^\infty(D)$ such that $\chi = 1$ on the support of ψ. We extend χ to $\mathbb{R}^n$ as above, and then for each test function φ on $\mathbb{R}^n$ we write

$$\varphi = \chi\varphi + (1 - \chi)\varphi,$$

where $\chi\varphi \in C_0^\infty(D)$. Now use the fact that the test functions φ on $\mathbb{R}^n$ are dense in H_s for every s, that $\psi f \in H_j$, and that $\psi g \in H_m$.]

14.8 Suppose now that $g \in H_0^D$, that $f = Lg \in H_1^D$ (Exercise 14.2), and that $kf = g$ weakly on D (Exercise 14.4). Apply the above lemma repeatedly to show that if $g \in H_m$ locally in D, then $f \in H_{m+2}$ locally in D. Conclude from Sobolev's lemma that if $m > n/2 + j$ and $g \in H_m$ locally in D, then $f = Lg \in C^{j+2}(D)$.

14.9 Show that $\|Lg\|_1 \leq \|g\|_0$ and conclude from Exercise 14.31 of Chapter 8 that if we regard L as an operator from H_0^D to H_0^D, it is compact, and all of its eigenvectors belong to H_1^D. Use Exercise 14.8 to show that every eigenvector belongs to $C^\infty(D)$.

15. CANONICAL TRANSFORMATIONS

In Sections 1 through 5 we formulated the notion of a mechanical system as a flow of a certain type on the cotangent bundle of the configuration space. The defining equations for the vector field X generating this flow were $X \lrcorner \Omega = -dH$. Thus the basic property of the cotangent bundle used in singling out the class of flows is the existence of the two-form Ω. It turns out that in studying the equations of mechanics it is sometimes convenient to forget that the flow is on the cotangent bundle and to concentrate on the form Ω. For instance, we may be able to introduce charts that don't arise from the configuration manifold but in terms of which the vector field X takes a particulary simple form. We shall therefore want to consider a manifold N which carries a two-form Ω, subject to certain restrictions which we shall describe below. On such manifolds we shall study vector fields satisfying $X \lrcorner \Omega = -dH$. It will be convenient to allow H to depend on the time t, as well as being a function on N, so that X will be a time-dependent vector field. The reason for this is twofold. First of all, it allows the consideration of "nonconservative" mechanical systems. Secondly, even in the study of the systems we have introduced so far, it is sometimes convenient to make a time-dependent change of coordinates to simplify the equations. This has the effect of changing a time-independent vector field into a time-dependent one. Now to the definitions:

Definition. A manifold N is said to possess a *Hamiltonian structure* (or to be a *Hamiltonian manifold*) if there is an exterior two-form Ω defined on N such that

i) $d\Omega = 0$, and

ii) Ω is of maximal rank in the sense that (4.3) holds.

Remarks

a) As we have seen, if $N = T^*(M)$, then N is a Hamiltonian manifold where Ω is given by (4.1).

b) If N is finite-dimensional, then it must be even-dimensional. In fact, condition (ii) says that Ω restricted to each tangent space is an antisymmetric bilinear form which is nonsingular. This can happen on a vector only if it is even-dimensional.

c) It can be proved that if N is a finite-dimensional Hamiltonian manifold, then one can always find local coordinates $q_1, \ldots, q_n, p_1, \ldots, p_n$ such that

$$\Omega = \sum dp_i \wedge dq_i.$$

[We know this to be the case if $N = T^*(M)$.] The point of this result (which we shall not prove here) is that locally all Hamiltonian manifolds of the same finite dimension look alike.

We shall now single out a class of vector fields on $N \times \mathbb{R}$.

Definition. The vector field $\overline{X}$ is a *Hamiltonian vector field* if there is a function $H = H_X$ on $N \times \mathbb{R}$ such that

$$\overline{X}t = \langle \overline{X}, dt \rangle \equiv 1, \tag{15.1}$$

where t is the standard coordinate on $\mathbb{R}$, regarded as a function on $N \times \mathbb{R}$, and

$$\overline{X} \lrcorner (\pi^*\Omega - dH \wedge dt) = 0, \tag{15.2}$$

where π is the projection of $N \times \mathbb{R}$ onto N; $\pi(x, t) = x$. Note that H is determined up to a function of t alone.

Let us set

$$\omega = \pi^*\Omega,$$

so that (15.2) can be written as

$$\overline{X} \lrcorner (\omega - dH \wedge dt) = 0. \tag{15.2'}$$

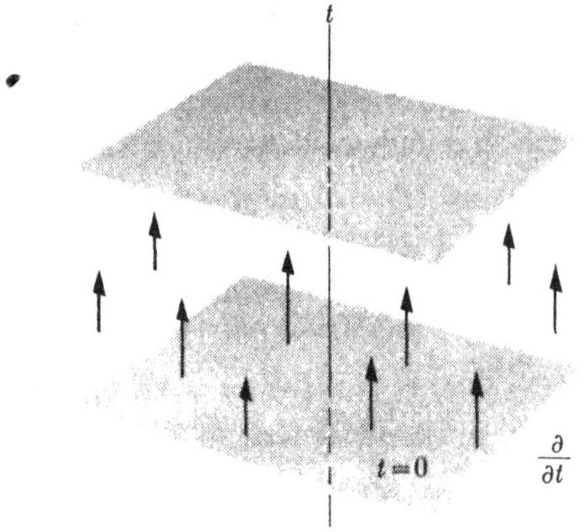

Fig. 13.12

If we consider the direct sum decomposition of the tangent space of $N \times \mathbb{R}$, then condition (15.1) says that we can write

$$\overline{X} = \left(X, \frac{\partial}{\partial t}\right),$$

where X is a time-dependent vector field on N; that is, X is a rule which assigns a tangent vector $X(x, t) \in T_x(N)$ to each x and t. Since ω does not involve dt, we can write

$$\overline{X} \lrcorner\, \omega = X(\cdot, t) \lrcorner\, \Omega \qquad \text{at any time } t. \tag{15.3}$$

[Strictly speaking, this equality should be written as follows: Let $i_t \colon N \to N \times \mathbb{R}$ be the map defined by $i_t(x) = (x, t)$. Then

$$i_t^*(\overline{X} \lrcorner\, \omega) = X(\cdot, t) \lrcorner\, \Omega.] \tag{15.3'}$$

Also,

$$\langle \overline{X}, dH \rangle = \langle X(\cdot, t), d\dot{H}(\cdot, t) \rangle + \frac{\partial H}{\partial t} \qquad \text{at any time } t.$$

Thus (15.2) can be split up into two equations. When we compare the terms not involving dt, we obtain

$$X(\cdot, t) \lrcorner\, \Omega = -dH(\cdot, t) \qquad \text{for every fixed } t. \tag{15.2a}$$

Thus

$$\overline{X} \lrcorner\, \omega = -dH + \frac{\partial H}{dt}\, dt. \tag{15.2b}$$

Note that (15.2a) is just the condition stated at the beginning of Section 5 with the novelty that H (and therefore X) can now depend on time.

Definition. A diffeomorphism, $\overline{\varphi}$, of $N \times \mathbb{R} \to N \times \mathbb{R}$ is called a *canonical transformation* if

i) $\overline{\varphi}^*(\omega) = \omega - dW \wedge dt$, where $W = W_{\overline{\varphi}}$ is some function depending on $\overline{\varphi}$; and

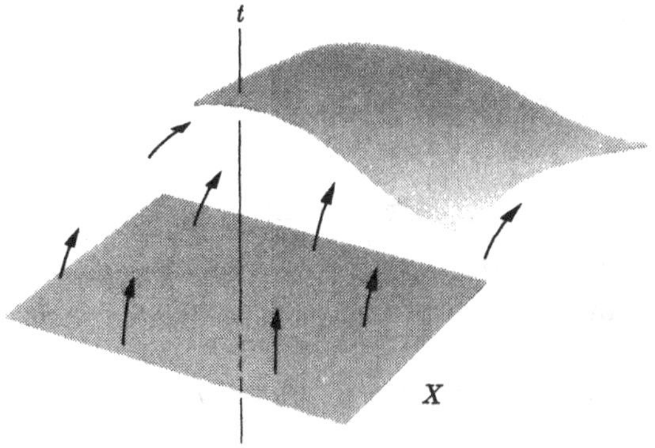

Fig. 13.13

ii) $\bar{\varphi}$ is time-preserving, i.e., $\bar{\varphi}$ has the form $\bar{\varphi}(x, t) = (\varphi(x, t), t)$, where $\varphi(\cdot, t)$ is a diffeomorphism of N for each t.

Observe that if $\bar{\varphi}$ is a canonical transformation, then so is $\bar{\varphi}^{-1}$ and

$$W_{\bar{\varphi}^{-1}} = -(\bar{\varphi}')^* W_{\bar{\varphi}}. \tag{15.4}$$

Also observe that if $\bar{\varphi}$ and $\bar{\psi}$ are canonical transformations, then so is $\bar{\psi} \circ \bar{\varphi}$, where

$$W_{\bar{\psi}\circ\bar{\varphi}} = \bar{\psi}^* W_{\bar{\varphi}} + W_{\bar{\psi}}. \tag{15.5}$$

These facts follow directly from the definitions and will be left as exercises for the reader.

We note next that if $\bar{\varphi}$ is a canonical transformation and $\bar{X}$ is a Hamiltonian vector field, then $\bar{\varphi}^*(X)$ is also a Hamiltonian vector field. In fact,

$$\begin{aligned}\bar{\varphi}^*\bar{X} \lrcorner [\omega - d(W_{\bar{\varphi}} + \bar{\varphi}^* H) \wedge dt] &= \bar{\varphi}^*\bar{X} \lrcorner [\bar{\varphi}^*\omega - \bar{\varphi}^*(dH \wedge dt)] \\ &= \bar{\varphi}^*[X \lrcorner (\omega - dH \wedge dt)] = 0.\end{aligned}$$

Thus we may take $H_{\varphi^* X}$ as

$$H_{\bar{\varphi}^* X} = W_{\bar{\varphi}} + \bar{\varphi}^* H. \tag{15.6}$$

Let $\bar{X}$ be a Hamiltonian vector field, and let $\bar{\varphi}$ be the map of $N \times \mathbb{R} \to N \times \mathbb{R}$ obtained by letting the system evolve from time $t = 0$ according to the flow generated by $\bar{X}$. That is, let the map $\varphi(\cdot, t)$ be defined so that the curve $t \mapsto \varphi(x, t)$ is a solution curve to the (time-dependent) vector field X which passes through x at time $t = 0$. To put it another way, the curve $t \mapsto (\varphi(x, t), t)$ is the solution curve to the vector field $\bar{X}$ which passes through $(x, 0)$ at time zero. (See Figs. 13.12 and 13.13.)

Note that it follows from the definition of $\bar{\varphi}$ that

$$\bar{\varphi}_* \left(\frac{\partial}{\partial t}\right) = X_{\varphi(x,t)}.$$

We claim that $\bar{\varphi}$ is a canonical transformation. In fact,

$$i_t^*(\bar{\varphi}^*\omega) = (\bar{\varphi} \circ i_t)^* \pi^* \Omega = \varphi(\cdot, t)^* \Omega,$$

since $\bar{\varphi} \circ i_t(x, t) = (\varphi(x, t), t)$. But

$$\frac{d}{ds}(\varphi(\cdot, s)^*\Omega) = \varphi(\cdot, s)^* D_{X(\cdot, t)}\Omega = 0$$

by (15.2a). Thus

$$i_t^* \bar{\varphi}^* \omega = \Omega,$$

or, in other words, $\varphi^*\omega$ is of the form $\omega + \theta \wedge dt$. To determine θ it suffices to take the interior product with $\partial/\partial t$, since ω doesn't involve dt. But

$$\left(\frac{\partial}{\partial t} \lrcorner \bar{\varphi}^* \pi^* \Omega\right) = \bar{\varphi}^* \left[\bar{\varphi}_* \left(\frac{\partial}{\partial t}\right) \lrcorner \pi^* \Omega\right] = \bar{\varphi}^*(\bar{X} \lrcorner \pi^* \Omega) = \bar{\varphi}^* \left(-dH + \frac{\partial H}{\partial t} dt\right),$$

so $\theta = \bar{\varphi}^* dH$. Thus $\bar{\varphi}$ is a canonical transformation and

$$W_{\bar{\varphi}} = -\bar{\varphi}^* H_X. \tag{15.7}$$

Note that (15.7) is just what we would expect from (15.6). In fact, $\varphi^*X = \partial/\partial t$, and we may take $H_{\partial/\partial t} = 0$.

Equations (15.6) and (15.7) are used in conjunction in the following way: Suppose that $H = H_0 + H_1$, where we know how to solve the differential equations corresponding to H_0. In other words, we can find the map $\bar{\varphi}$ corresponding to the vector field X_0, where $X_0 \lrcorner (\omega - dH_0 \wedge dt) = 0$. If

$$X \lrcorner (\omega - dH \wedge dt) = 0,$$

then φ^*X is a vector field whose corresponding Hamiltonian function is $\bar{\varphi}^*H_1$ by (15.6)

This method was first introduced by Lagrange in the study of the n-body problem. We can let H_0 be the Hamiltonian obtained by ignoring the terms in the potential energy coming from the interaction of the planets, and let H_1 be the rest of the Hamiltonian H. The solution for H_0 is then given by having the planets move about the sun according to Kepler's laws. For simplicity in discussion, let us restrict our attention to that portion of phase space where the motion is elliptical. Then the motion of the planets is specified by giving the various parameters of each ellipse (such as the plane of the ellipse, its major axis, its eccentricity, etc.) and telling the position of the planet on its ellipse at time $t = 0$. This corresponds to the use of the map φ. One then regards the equations of motion of the whole system as differential equations for the parameters of each ellipse. This corresponds to studying the vector field φ^*X. This idea of introducing the parameters of the ellipses as "generalized" coordinates was one of the key steps leading to the notion of the invariant calculus on manifolds.

We have seen that solving the differential equations corresponding to a Hamiltonian H is the same as looking for a map $\bar{\varphi}$ satisfying (15.7). Under certain circumstances this can be reduced to looking for the solution to a certain partial differential equation. Suppose that we have local coordinates $q_1, \ldots, q_n$,

$p_1, \ldots, p_n$ such that $\Omega = \sum dp_i \wedge dq_i$. Let $V = V(q_1, \ldots, q_n, p_1, \ldots, p_n, t)$ be a function such that the maps φ_1 and φ_2 are diffeomorphisms, where

$$\varphi_1(q_1, \ldots, q_n, p_1, \ldots, p_n, t) = \left\langle q_1, \ldots, q_n, \frac{\partial V}{\partial q_1}, \ldots, \frac{\partial V}{\partial q_n}, t \right\rangle$$

and

$$\varphi_2(q_1, \ldots, q_n, p_1, \ldots, p_n, t) = \left\langle \frac{\partial V}{\partial p_1}, \ldots, \frac{\partial V}{\partial p_n}, p_1, \ldots, p_n, t \right\rangle$$

We claim that $\bar{\varphi} = \varphi_1 \circ \varphi_2^{-1}$ is a canonical transformation and that

$$W_{\bar{\varphi}} = \varphi_2^{-1*} \frac{\partial V}{\partial t}. \tag{15.8}$$

Note that $\omega = d(\sum p_i\, dq_i) = -d(\sum q_i\, dp_i)$. Thus

$$\begin{aligned}
\varphi^*\omega - \omega &= d(\bar{\varphi}^* \textstyle\sum p_i\, dq_i + \sum q_i\, dp_i) \\
&= d\varphi_2^{-1*}(\varphi_1^* \textstyle\sum p_i\, dq_i + \varphi_2^* \sum q_i\, dp_i) \\
&= d\varphi_2^{-1*}\left(\sum \frac{\partial V}{\partial q^i} dq_i + \sum \frac{\partial V}{\partial p_i} dp^i\right) \\
&= d\varphi_2^{-1*}\left(dV - \frac{\partial V}{\partial t} dt\right) \\
&= \varphi_2^{-1*} d\, dV - d\varphi_2^{-1*} \frac{\partial V}{\partial t} \wedge dt \\
&= -d\left(\varphi_2^{-1*} \frac{\partial V}{\partial t}\right) \wedge dt.
\end{aligned}$$

If we substitute into (15.7) we see that φ solves our differential equations if and only if

$$\varphi_2^{-1*} \frac{\partial V}{\partial t} + \bar{\varphi}^* H = 0.$$

But $\bar{\varphi}^* H = \varphi_2^{-1*} \varphi_1^* H$, so we can write this equation as

$$\frac{\partial V}{\partial t} + \varphi_1^* H = 0 \tag{15.9}$$

or

$$\frac{\partial V}{\partial t} + H\left(q_1, \ldots, q_n, \frac{\partial V}{\partial q_1}, \ldots, \frac{\partial V}{\partial q_n}, t\right) = 0. \tag{15.9'}$$

Equation (15.9) is known as the *Hamilton-Jacobi equation.*

We therefore have a prescription for (locally) solving the equations of motion: Find a solution V of (15.9) which has the property that φ_1 and φ_2 are diffeomorphisms. Under certain circumstances, a proper choice of coordinates allows us to solve (15.9) by the method of "separation of variables". We illustrate this method in the following examples.

Example 1. *Central-force motion again.* Here

$$H = \frac{1}{2m}\left(p_r^2 + \frac{p_\theta^2}{r^2}\right) + U(r),$$

when we use polar coordinates in the plane. Equation (15.9) becomes

$$\frac{\partial V}{\partial t} + \frac{1}{2m}\left[\left(\frac{\partial V}{\partial r}\right)^2 + \frac{1}{r^2}\left(\frac{\partial V}{\partial \theta}\right)^2\right] + U(r) = 0.$$

Since the variables t and θ do not occur explicitly in this equation, we seek a solution of the form

$$V = V_1(t) + V_2(\theta) + V_3(r)$$

and conclude that both $V_1'(t)$ and $V_2'(\theta)$ depend only on r, and so must be constants. We may thus write

$$V_1'(t) = -E, \qquad V_2'(\theta) = A_\theta,$$

where E and A_θ are constants. (This just reflects the conservation of energy and angular momentum.) We then get the equation

$$\frac{\partial V}{\partial r} = V_3'(r) = \sqrt{2m(E - U(r)) - \frac{A_\theta^2}{r^2}}.$$

Thus

$$V = A_\theta \theta + \int_0^r \left[2m(E - U(s)) - \frac{A_\theta^2}{s^2}\right]^{1/2} ds - Et$$

is a solution of (15.9). Here we consider V a function of the variables r, θ, E, A_θ. Then the map φ_2 is given by

$$\varphi_2(r, \theta; E, A_\theta) = \left\{ -t + \int_0^r \frac{m\, ds}{\left[2m(E - U(s)) - \frac{A_\theta^2}{s^2}\right]^{1/2}}, \right.$$

$$\left. \theta - \int_0^r \frac{A_\theta\, ds}{s^2\left[2m(E - U(s)) - \frac{A_\theta^2}{s^2}\right]^{1/2}}, E, A_\theta \right\}.$$

Now the map $\varphi_1 \circ \varphi_2^{-1}$ takes the flow into the constant flow, so that we must have

$$t - t_0 = \int_0^r \frac{m\, ds}{\left[2m(E - U(s)) - \frac{A_\theta^2}{s^2}\right]^{1/2}}$$

and

$$\theta - \theta_0 = \int_0^r \frac{A_\theta\, ds}{s^2\left[2m(E - U(s)) - \frac{A_\theta^2}{s^2}\right]^{1/2}},$$

where t_0 and θ_0 are constants. Note that the second of these equations gives the orbit explicitly, which can then be solved to give r as a function of t.

Example 2. *The simple harmonic oscillator.* Here

$$H = \frac{p^2}{2m} + \frac{kq^2}{2},$$

so Eq. (15.9) becomes

$$\frac{1}{2m}\left(\frac{\partial V}{\partial q}\right)^2 + \frac{kq^2}{2} + \frac{\partial V}{\partial t} = 0.$$

Again, since time doesn't enter explicitly, we can write

$$V = -Et + W,$$

where W is a function of q alone which must satisfy

$$\frac{1}{2m}(W^1)^2 + \frac{kq^2}{2} = E$$

or

$$W = \sqrt{mk}\int_0^q \left(\frac{2E}{k} - s^2\right)^{1/2} ds.$$

Then

$$\frac{\partial V}{\partial E} = \left(\frac{m}{k}\right)^{1/2}\int_0^q \left(\frac{2E}{k} - s^2\right)^{-1/2} ds - t.$$

Thus

$$t - t_0 = -\left(\frac{m}{k}\right)^{1/2}\int_0^q \frac{ds}{\left(\frac{2E}{k} - s^2\right)^{1/2}}$$

$$= -\left(\frac{m}{k}\right)^{1/2} \arccos\left(\frac{k}{2E}\right)^{1/2} q.$$

Solving for q in terms of t gives

$$q = \sqrt{\frac{2E}{k}}\cos\left(\frac{k}{m}\right)^{1/2}(t - t_0).$$

Example 3. *The motion of a particle attracted by two fixed point masses.* Here

$$H = \frac{1}{2m}(P_x^2 + P_y^2 + P_z^2) + \frac{A}{r_1} + \frac{B}{r_2},$$

where r_1 and r_2 are the distances to the two points and A and B are constants (determined by the masses of these two points).

For the purpose of solving this problem, it is convenient to introduce so-called elliptical coordinates. Let us assume that the two fixed points lie on the x-axis, each at a distance c from the origin. We may take $c = 1$ for simplicity.

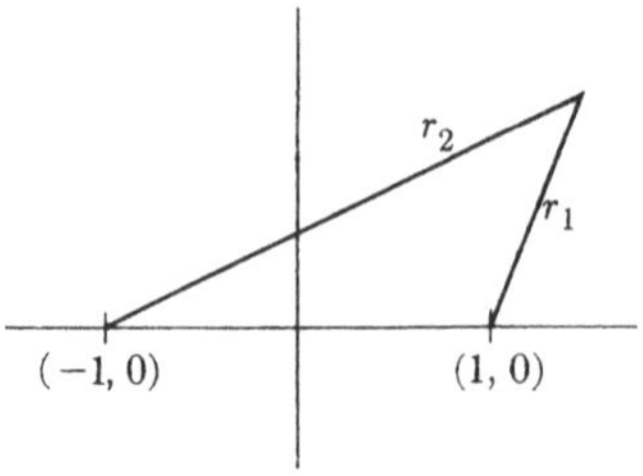

Fig. 13.14

(See Fig. 13.14.) In the xy-plane define the local coordinates ξ and η by setting

$$\xi = \tfrac{1}{2}(r_1 + r_2), \qquad \eta = \tfrac{1}{2}(r_1 - r_2).$$

Thus the curves $\xi =$ const represent ellipses with semimajor axis ξ and foci at the two fixed points, while the curves $\eta =$ const are hyperbolas with semimajor axis η and the same foci. Note that $0 < |\eta| \le 1 \le \xi < \infty$. The equations of these two curves are

$$\frac{x^2}{\xi^2} + \frac{y^2}{\xi^2 - 1} = 1 \qquad \text{and} \qquad \frac{x^2}{\eta^2} - \frac{y^2}{1 - \eta^2} = 1,$$

so that

$$x^2 = \xi^2\eta^2 \qquad \text{and} \qquad y^2 = (\xi^2 - 1)(1 - \eta^2).$$

Thus

$$\frac{dx}{x} = \frac{d\xi}{\xi} + \frac{d\eta}{\eta}, \qquad \frac{dy}{y} = \frac{\xi\, d\xi}{\xi^2 - 1} - \frac{\eta\, d\eta}{1 - \eta^2},$$

and therefore

$$dx^2 + dy^2 = (\xi^2 - \eta^2)\left(\frac{d\xi^2}{\xi^2 - 1} + \frac{d\eta^2}{1 - \eta^2}\right).$$

If we now rotate about the x-axis to get the analogue of cylindrical coordinates in space, we have the coordinates

$$\langle x, \rho, \theta \rangle \qquad \text{and} \qquad \langle \xi, \eta, \theta \rangle$$

in space, and the Euclidean metric is given by

$$\begin{aligned} dx^2 + dy^2 + dz^2 &= dx^2 + d\rho^2 + \rho^2\, d\theta^2 \\ &= (\xi^2 - \eta^2)\left(\frac{d\xi^2}{\xi^2 - 1} + \frac{d\eta^2}{1 - \eta^2}\right) + (\xi^2 - 1)(1 - \eta^2)\, d\theta^2. \end{aligned}$$

Also,

$$\begin{aligned} U &= \frac{A}{r_1} + \frac{B}{r_2} = \frac{Ar_2 + Br_1}{r_1 r_2} \\ &= \frac{1}{\xi^2 - \eta^2}(\alpha\xi + \beta\eta), \end{aligned}$$

where $\alpha = A + B$ and $\beta = A - B$. Then H takes the form

$$H(\xi, \eta, \theta, P_\xi, P_\eta, P_\theta, t) = \frac{1}{2m(\xi^2 - \eta^2)} \times \left[(\xi^2 - 1)P_\xi^2 + (\eta^2 - 1)P_\eta^2 + \left(\frac{1}{\xi^2 - 1} + \frac{1}{1 - \eta^2}\right)P_\theta^2 + \alpha\xi + \beta\eta\right].$$

Since t and θ do not occur explicitly in (15.9), we may write

$$V = -Et + A_\theta \theta + W,$$

where now W must satisfy

$$(\xi^2 - 1)\left(\frac{\partial W}{\partial \xi}\right)^2 + (\eta^2 - 1)\left(\frac{\partial W}{\partial \eta^2}\right) + \left(\frac{1}{\xi^2 - 1} + \frac{1}{1 - \eta^2}\right)A_\theta^2 + \alpha\xi + \beta\eta = 2m(\xi^2 - \eta^2)E.$$

Note that if we set $W = W_1(\xi) + W_2(\eta)$, this equation separates into two, and we can explicitly solve each of them by quadratures. This gives the solution to the original equations of motion. We leave the details to the reader.

SELECTED REFERENCES

Chapters 1, 2, and 7

A more extensive treatment of linear algebra, over arbitrary coefficient fields, can be found in

BIRKHOFF, G., and S. MACLANE, *A Survey of Modern Algebra*, 3rd ed., Macmillan, New York, 1965.

HOFFMAN, K., and R. KUNZE, *Linear Algebra*, Prentice-Hall, Englewood Cliffs, N. J., 1961.

For linear algebra over commutative rings that are not fields, see

LANG, S., *Algebra*, Addison-Wesley, Reading, Mass., 1965.

ZARISKI, O., and P. SAMUEL, *Commutative Algebra*, 2 vols., Van Nostrand, Princeton, 1959, 1960.

Chapter 3

Difficult but rewarding is

DIEUDONNÉ, J., *Foundations of Modern Analysis*, Academic Press, New York, 1960. The differential notation in this book is $Df(\alpha)$ (instead of dF_α) and $Df(\alpha) \cdot \xi$ (instead of $dF_\alpha(\xi)$).

Chapters 4 and 5

Good books on general (point set) topology are

KELLEY, J., *General Topology*, Van Nostrand, Princeton, 1961.

SIMMONS, G., *Topology and Modern Analysis*, McGraw-Hill, New York, 1963.

The remaining standard examples of Banach spaces and Hilbert space require the Lebesgue integral and are therefore beyond our scope. However, the interested reader can pursue the abstract theory of Banach and Hilbert spaces in Simmons and in such books as

HILLE, E., and R. S. PHILLIPS, *Functional Analysis and Semigroups*, American Mathematical Society, Providence, R. I., 1957.

MURRAY, F. J., *An Introduction to Linear Transformations in Hilbert Space*, Princeton University Press, Princeton, 1941.

RIESZ, F., and B. SZ-NAGY, *Functional Analysis*, Ungar, New York, 1955.

YOSIDA, K., *Functional Analysis*, Springer, Berlin, 1965.

The books by Murray and Yosida are more advanced and harder.

Chapter 6

Standard books on ordinary differential equations are

BIRKHOFF, G., and G. C. ROTA, *Ordinary Differential Equations*, Ginn, Boston, 1962.

HUREWICZ, W., *Lectures on Ordinary Differential Equations*, M.I.T. Press, Cambridge, Mass., 1958.

Advanced treatises are

CODDINGTON, E., and N. LEVINSON, *Theory of Ordinary Differential Equations*, McGraw-Hill, New York, 1955.

HARTMAN, P., *Ordinary Differential Equations*, Wiley, New York, 1964.

Chapter 8

This chapter has been devoted to the theory of content. For modern mathematics the more powerful theory of Lebesgue measure and integration is needed. For one-dimensional theory the reader can consult

RUDIN, W., *Principles of Mathematical Analysis*, 2nd ed., McGraw-Hill, New York, 1964.

For the general theory the reader can consult

HALMOS, P., *Measure Theory*, Van Nostrand, Princeton, 1961.

Another way of computing certain definite integrals, involving the "residue calculus", is discussed in a set of exercises at the end of Chapter 12, and can be read independently after Chapters 8 and 11.

For the relationship of integration to "generalized functions" see

GEL'FAND, I. M., and G. E. SHILOV, *Generalized Functions*, vol. 1, Academic Press, New York, 1964. A little complex variable theory is necessary to read this book.

Chapters 9, 10, and 11

For a more abstract treatment see

LANG, S., *Introduction to Differentiable Manifolds*, Interscience, New York, 1962.

For a less abstract treatment see

FLANDERS, H., *Differential Forms with Applications to Physical Sciences*, Academic Press, New York, 1963.

FLEMING, W., *Functions of Several Variables*, Addison-Wesley, Reading, Mass., 1965.

SPIVAK, M., *Calculus on Manifolds*, Benjamin, New York, 1965.

A more extensive treatment of differential geometry will be found in

WILLMORE, T., *An Introduction to Differential Geometry*, Oxford University Press, London, 1959.

Chapter 12

The classical book on potential theory is

KELLOGG, O., *Foundations of Potential Theory*, Springer, Berlin, 1929.

The relationship between harmonic functions on the plane and analytic functions is studied in standard texts on the latter subject, such as

AHLFORS, L., *Complex Analysis*, 2nd ed., McGraw-Hill, New York, 1966.

HILLE, E., *Analytic Function Theory*, Ginn, Boston, 1959.

Chapter 13

A standard modern book on mechanics is

GOLDSTEIN, H., *Classical Mechanics*, Addison-Wesley, Reading, Mass., 1952.

For the classical astronomy-oriented aspect of mechanics see

POINCARÉ, H., *Leçons de mécanique céleste*, Gauthier-Villars, Paris, 1905–1910.

WHITTAKER, E. T., *Analytical Dynamics*, 4th ed., Cambridge University Press, Cambridge, England, 1937.

A leisurely geometrical study of classical mechanics will be found in

LANCZOS, C., *The Variational Principles of Mechanics*, University of Toronto Press, Toronto, 1960.

For a book which treats classical mechanics in the spirit of this chapter, and which also studies statistical mechanics and quantum mechanics, see

MACKEY, G., *Mathematical Foundations of Quantum Mechanics*, Benjamin, New York, 1965.

An elegant treatment of the geometrical applications of the calculus of variations will be found in

MILNOR, J., *Morse Theory*, Princeton University Press, Princeton, 1965.

NOTATION INDEX

General Conventions

□ end of proof

$\mathbb{R}$ the real number system

$\mathbb{Z}$ the integers

$\mathbb{C}$ the complex number system

$\mathbb{E}^n$ Euclidean n-space

boldface letters for n-tuplets: $\boldsymbol{\alpha} = \langle \alpha_1, \ldots, \alpha_n \rangle$

Special Symbols

INDEX

Printed in the United States
By Bookmasters